14th International Conference on Strangeness in Quark Matter

(SQM 2013)

Journal of Physics: Conference Series Volume 509

Birmingham, United Kingdom
22 – 27 July 2013

Editors:

D. Evans
S. Hands
R. Lietava
R. Romita
O. Baillie

ISBN: 978-1-63439-289-1
ISSN: 1742-6588

Printed from e-media with permission by:

Curran Associates, Inc.
57 Morehouse Lane
Red Hook, NY 12571

Some format issues inherent in the e-media version may also appear in this print version.

Printed by Curran Associates, Inc. (2014)

For permission requests, please contact the Institute of Physics
at the address below.

Institute of Physics
Dirac House, Temple Back
Bristol BS1 6BE UK

Phone: 44 1 17 929 7481
Fax: 44 1 17 920 0979

techtracking@iop.org

Additional copies of this publication are available from:

Curran Associates, Inc.
57 Morehouse Lane
Red Hook, NY 12571 USA
Phone: 845-758-0400
Fax: 845-758-2634
Email: curran@proceedings.com
Web: www.proceedings.com

TABLE OF CONTENTS

FLOW

LOW ENERGY / HEAVY FLAVOUR

STRANGENESS 1

HEAVY FLAVOUR 1

PHASE TRANSITION

POLARISATION OBSERVABLES AND EA

POSTERS

BEAM ENERGY SCAN

HARD PROBES

HEAVY FLAVOUR 2

FUTURE

RESONANCES

STRANGENESS 2

QUARKONIA AND HEAVY FLAVOUR

THERMAL MODELS AND HYDRO

PA COLLISIONS

HYPERNUCLEI

QUARKONIA AND HEAVY FLAVOUR

THERMAL MODELS

Preface

The 2013 Strangeness in Quark Matter conference took place at the University of Birmingham in July 2013, in a period of remarkably good weather that gave a very welcome boost to the event. 158 physicists from 25 countries gathered for a week of discussions on the production of strangeness and heavy flavour in heavy ion collisions. The dates for the meeting had been set two years earlier in Cracow, so as to allow it to offer the first major examination of proton-lead collisions from the Large Hadron Collider. It had originally been thought that these collisions would be recorded by the end of 2012, but in the event it turned out that the running period was postponed until January and February of 2013, giving the LHC collaborations – all four major collaborations since LHCb also took part in pPb data taking – very little time to prepare their results. Nevertheless, new results were provided, and their presentation and interpretation formed one of the highlights of the conference. In addition, there was a more detailed assessment of the RHIC beam energy scan, many new heavy flavour results from the RHIC and LHC heavy ion runs and detailed discussions of the future FAIR and NICA programmes.

The conference also hosted a good cross-section of current topics in theoretical talks. In the last few years there has been much interest in thermalization and in the use of hydrodynamics to describe the fluctuations visible in higher-order flow coefficients. Discussions of both of these features were well represented, indicating the maturing of this field.

We gratefully acknowledge support from The University of Birmingham, CERN, The Extreme Matter Institute (EMMI), the Institute for Particle Physics Phenomenology (IPPP), Institute of Physics Publishing and from the UK Science and Technology Facilities Council (STFC). This assistance allowed us to support over twenty young physicists, and allowed us to ensure that we had the breadth of programme needed for a successful conference. We thank the International Organizing Committee for their help and advice in planning the conference, and we are grateful to the University of Birmingham Conference Service and to the Birmingham Botanical Gardens for the excellent way in which the catering and room provision was provided.

David Evans *School of Physics and Astronomy, The University of Birmingham*
Simon Hands *Department of Physics, Swansea University*
Roman Lietava *School of Physics and Astronomy, The University of Birmingham*
Rosa Romita *Oliver Lodge Laboratory, The University of Liverpool*
Orlando Villalobos Baillie *School of Physics and Astronomy, The University of Birmingham*
Editors

Journal of Physics: Conference Series **509** (2014) 011001 doi:10.1088/1742-6596/509/1/011001

Conference Photo

Journal of Physics: Conference Series **509** (2014) 011002 doi:10.1088/1742-6596/509/1/011002

Committees

International Advisory Committee

J Aichelin	*Nantes*	J Nagle	*Colorado*
F Antinori	*Padova*	G Odyniec	*Berkeley*
F Becattini	*Florence*	J Rafelski	*Arizona*
T S Biro	*Budapest*	H G Ritter	*Berkeley*
P Braun-Munzinger	*GSI*	G Roland	*MIT*
J Cleymans	*Cape Town*	K Šafařík	*CERN*
L Csernai	*Bergen*	L Šándor	*Kosice*
A Dainese	*Padova*	P Steinberg	*BNL*
W Florkowski	*Kraków*	G S F Stephans	*MIT*
P Giubellino	*Torino*	H Stöcker	*Frankfurt*
H Z Huang	*UCLA*	J Takahashi	*Campinas*
Y G Ma	*Shanghai*	F Wang	*Purdue*
A Mócsy	*Pratt*	N Xu	*Berkeley*
B Mohanty	*NISER*	P F Zhuang	*Tsinghua*
B Müller	*BNL*		

Local Organizing Committee

L S Barnby	*Birmingham*	R Lemmon	*Daresbury*
D Evans	*Birmingham*	R Romita	*Liverpool*
S Hands	*Swansea*	A Starinets	*Oxford*
P G Jones	*Birmingham*	O Villalobos Baillie	*Birmingham*
R Lietava	*Birmingham*		

Peer review statement

All papers published in this volume of *Journal of Physics: Conference Series* have been peer reviewed through processes administered by the proceedings Editors. Reviews were conducted by expert referees to the professional and scientific standards expected of a proceedings journal published by IOP Publishing.

A Snapshot of our Experimental Knowledge Circa Winter 2012-13

Helen Caines

Physics Dept., Yale University, New Haven, CT 06511, U.S.A

E-mail: helen.caines@yale.edu

Abstract. I present a snapshot of some experimental results from the LHC and RHIC experiments as of Winter 2012-13. This sets the scene for the new data that will be presented at this conference.

Introduction

By presenting edited highlights of the recent experimental results from RHIC and the LHC, I aim to describe where our field is standing prior to the new data that will be presented at this conference. This includes results from proton-proton, proton-Pb, deuteron-Au and heavy-ion collisions over a wide range in collision energies. The RHIC collaborations have been focused on the new Beam Energy Scan (BES) data, collected to search for evidence of a "turn-off" of the QGP, a Critical Point in the QCD phase diagram or evidence of a first order phase transition. Meanwhile at the LHC the Pb-Pb results indicate that the medium created at $\sqrt{s_{NN}}$= 2.76 TeV is hotter, denser, and longer lived than that created via collisions at RHIC. There are also remarkable first results emerging from the p-A run at the LHC, which was designed as a control dataset with which to explore cold nuclear matter effects and search for possible evidence for the Color Glass Condensate (CGC).

Pb-Pb collisions - a more extreme QGP

The temperature formed in the collision can be probed using direct photons. In the range $p_T \leq 4$ GeV/c thermal photons emitted during the early stages of the collision are believed to dominate the measured photon spectrum. Once the contributions from NLO pQCD processes have been subtracted the resulting distribution can be fit with an exponential and an estimate of the temperature extracted. Preliminary analysis by ALICE of the 0-40% Pb-Pb data indicate that $T_{init} = 304 \pm 51$ MeV more than twice the critical temperature, T_c, required to form a QGP (Fig. 1) [1] and significantly above that reported by PHENIX of $T_{init} = 221 \pm 19 \pm 19$ MeV for Au-Au collisions at $\sqrt{s_{NN}}$=200 GeV [3]. Interpretation of this result is however complex since the photon spectrum, while expected to be dominated by early times, is actually an integration over the whole lifetime of the medium.

An alternative approach, first suggested nearly three decades ago [4], is to access the initial temperature via measurements of quarkonia suppression (J/Ψ, $\Psi^{'}$, Υ states etc). These heavy quark-anti-quark states are loosely bound and therefore expected to disassociate in the QGP due

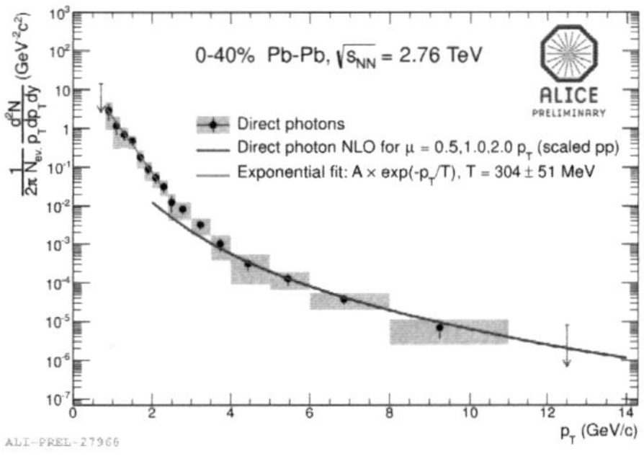

Figure 1. Direct photon p_T distribution for 0-40% Pb-Pb events at $\sqrt{s_{NN}}$=2.76 TeV. Blue curve: scaled pp NLO calculation. Red curve: exponential fit to the data. Figure from [1].

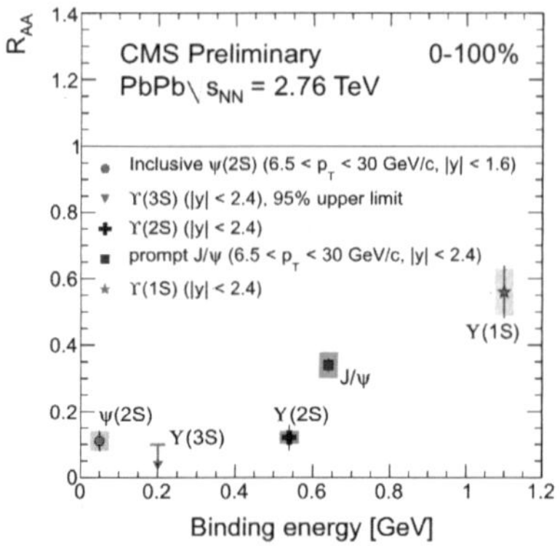

Figure 2. Minimum R_{AA} as a function of binding energy for quarkonia in Pb-Pb collisions at $\sqrt{s_{NN}}$=2.76 TeV. Figure from [2].

to color screening. The temperature at which each quarkonium "melts" depends on its binding energy. Therefore measurements of various quarkonia yields in A-A collisions relative to the binary collision scaled rates in pp should act as a QGP thermometer. Figure 2 shows one such set of measurements by CMS in $\sqrt{s_{NN}}$=2.76 TeV Pb-Pb collisions [2]. The expected dependence of the suppression on binding energy is observed. Although it should be noted that the p_T and rapidity ranges are not the same for all the measurements shown, these results suggest that T_{init} is well above T_c, in agreement with the temperature deduced from direct photons [5].

Statistical model fits to the central Pb-Pb particle ratios reported ALICE show some tension when trying to fit the central data. While the measured K/π ratio is in agreement with calculations from thermal models, the p/π ratio is a factor of 1.5 lower [6]. Discussions of interpretations of this result are still underway as this conference begins. However, one possible explanation is that hadronic re-scatterings after chemical freeze-out drive the system out of chemical equilibrium. If the protons are not included in the fits the extracted temperature is $T_{ch} \sim 164$ MeV with a baryon chemical potential of $\mu_b = 1$ MeV. Spectral analyses of identified protons, kaons, and pions show that the medium is expanding more explosively than at RHIC, $\langle\beta_T\rangle = 0.65c$, but that the temperature at kinetic freeze-out stays approximately the same at $T_{kin} \sim$ 80-95 MeV [6]. Femtoscopic radii have been utilized to extract information about the volume and lifetime of the medium at kinetic freeze-out [7]. Results from ALICE show that in central events with a volume of ~5000 fm^3 and a lifetime of ~11 fm/c at freeze-out the medium made at the LHC has twice the volume of that made at RHIC and lives 40% longer.

Measurements of photons, W bosons, and Z bosons by CMS exhibit binary scaling of the central Pb-Pb data, as expected for colorless objects, Fig. 3. However, the nuclear modification factor for charged particles and b quarks, inferred from displaced J/Ψ measurements, are highly suppressed in the same centrality class [8]. These hadrons have similar R_{AA} values to those of fully reconstructed jets with a resolution parameter of R=0.2 [8,10,11]. Both ATLAS and CMS report that jets continue to be highly suppressed out to $p_T \sim 300$ GeV/c in central events, even for R=0.4-0.5. These results indicate that (heavy) quarks and gluons remain strongly coupled to the medium at the LHC, and that the lost energy appears at large angles with respect to the jet axis.

Journal of Physics: Conference Series **509** (2014) 012001 doi:10.1088/1742-6596/509/1/012001

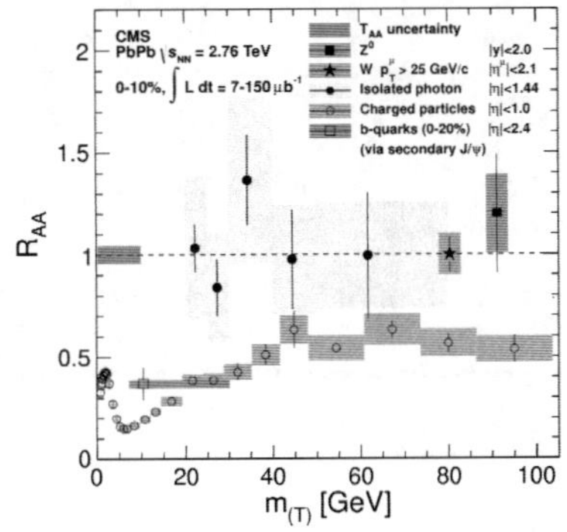

Figure 3. W, Z, b quarks, photon, and charged particle Pb-Pb R_{AA} for $\sqrt{s_{NN}}$=2.76 TeV. Figure from [8].

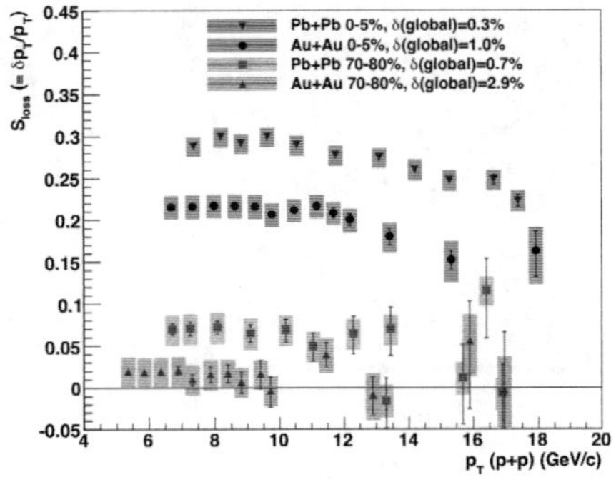

Figure 4. $\delta p_T/p_T$ as a function of pp p_T for central and peripheral A-A collisions at the LHC and RHIC energies. Figure from [9].

In a new analysis from PHENIX the fractional shift in p_T required to match the yield in a given A-A bin to that in pp, $\delta p_T/p_T$, is calculated instead of R_{AA} [9]. Figure 4 shows this energy loss estimate for central and peripheral data at both RHIC and the LHC. These results indicate that $\delta p_T/p_T$ is $\sim$ 1.3 times larger for a given centrality and p_T bin at the LHC than at RHIC, but that at both facilities $\delta p_T/p_T$ decreases at high p_T. It should be noted that at the same time dN/dy is $\sim$ 2.2 times that at RHIC, suggesting that the coupling to the QGP is smaller at the LHC. Much of this lost energy re-emerges as numerous low momentum particles at large angles to, but still correlated with, the jet axis. This has been demonstrated both at RHIC using jet-hadron [12] and gamma-hadron correlations [13] and at the LHC by measuring, for example, fragmentation functions down to low p_T [14]. Such a measurement from ATLAS is illustrated in Fig. 5. The central to peripheral ratio of the unfolded fragmentation functions for R=0.3 jets with $p_T > 92$ GeV in Pb-Pb events is shown, and a clear enhancement of low p_T hadrons is observed. highly suppressed R_{AA}s can be reconciled with the notion of enhanced soft particle production during the jet's fragmentation by noting that the jet R_{AA} results are for R=0.4. While there is increased low p_T production within this radius, the majority of the soft particles are emitted at larger angles. This causes significantly less of the initial parton's energy to be recovered within a cone radius of R=0.4 in Pb-Pb collisions than for the same R in the pp data; which in turn results in a jet and hadron R_{AA} that is less than unity.

The precision vertexing capabilities at all the LHC experiments has produced a wealth of detailed heavy flavor measurements. In the charm sector ALICE reports that the D mesons have similar R_{AA} to that of light quark mesons [15] and that v_2 remains significant for Ds at high p_T; both these results indicate significant energy loss of the c quark that is comparable to that of light quarks and gluons [16]. However the non-prompt J/Ψ results, Fig. 3, suggest that b quarks loose less energy at an equivalent initial p_T.

The J/Ψ results from ALICE at high p_T also indicate a suppression when compared to scaled pp data [18]. However, the value of R_{AA} is closer to unity than that reported by PHENIX [19], this suggests that there is significant regeneration of the J/Ψ at $\sqrt{s_{NN}}$= 2.76 TeV, which is possible due to the copious production of charm quarks at these energies. The fact that the J/Ψ also has a significant v_2 [20] supports the regeneration hypotheses, as the v_2 could then be due to the "bulk" thermalized charm rather than intrinsic flow of the original quarkonium.

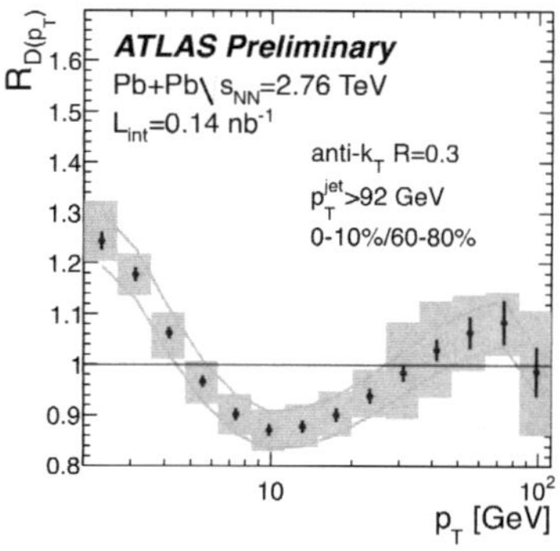

Figure 5. Ratio of 0-10% to 60-80% Pb-Pb unfolded fragmentation functions. Figure from [14].

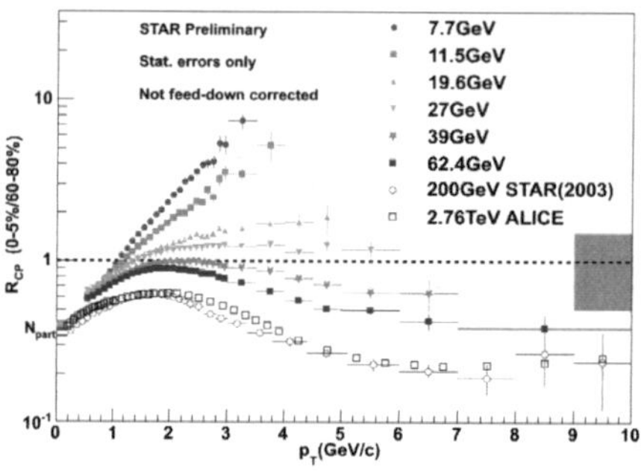

Figure 6. R_{CP} for charged hadrons for $\sqrt{s_{NN}}$=7.7-2760 GeV. Figure from [17].

Beam Energy Scan - exploring the phase diagram

The chemical potential and freeze-out temperature as extracted from particle ratios by STAR [21] show the expected trends as a function of beam energy; falling along the $\langle E\rangle/N \sim 1$ boundary predicted by Cleymans et al. [22]. However, a recent paper from Beccattini et al. [23] suggests that the quality of such fits can be improved at the lower collision energies by taking into account hadronic re-scattering. These scatterings cause the system to fall out of equilibrium and reduce the apparent T_{ch}. Once this hadronic phase is taken into account the hadronization temperature is at, or close to, that of the phase transition boundary predicted by lattice QCD calculations. It will be interesting to see such an analysis applied to the recent RHIC BES data. Additional indications that a transition to the QGP possibly occurs down to the lowest RHIC collision energies comes from estimates of the Bjorken energy density which remain above critical values for central events even at $\sqrt{s_{NN}}$= 7.7 GeV if the formation time is less than ~2 fm/c [24].

As discussed above, the suppression of high p_T particle production when compared to that in scaled pp events is used to probe partonic energy loss to the QGP. Measurements of the nuclear modification factor, R_{CP}, where peripheral events are used as a proxy for pp data have been made by STAR in the BES [17]. The results, including data from ALICE are shown in Fig. 6. A clear transition from suppression to enhancement is revealed with decreasing beam energy, with the ratio crossing unity around $\sqrt{s_{NN}}$=27 GeV. However, care must be taken when interpreting these results since the absence of suppression does not necessarily imply an absence of jet quenching. This is because a low beam energies a phenomenon known as the "Cronin Effect", which results in enhanced particle production at intermediate p_T, has been previously observed in asymmetric collisions [25]. While the sizes of cold nuclear matter (CNM) effects such as this are known to vary with beam energy and particle species, the exact causes and magnitudes are currently not understood. It is therefore not yet possible to state at what beam energy, if any recorded so far, jet quenching "turns off" and further studies are clearly required.

Azimuthal HBT and directed flow, v_1, have both been predicted as sensitive to a first order phase transition or a softening of the Equation of State (EoS) [28]. The slope of the directed flow, dv_1/dy', of net-protons near mid-rapidity as a function of beam energy is shown in Fig. 7 [26] along with calculations from URQMD for 10-40% central Au-Au events. A prominent dip in the v_1 slope is observed between $\sqrt{s_{NN}}$=10-20 GeV. Such a feature is not observed in the model, which predicts a monotonic trend with beam energy. While this dip shape is consistent with the general expectations for a changing EoS, further theoretical studies are needed before conclusive

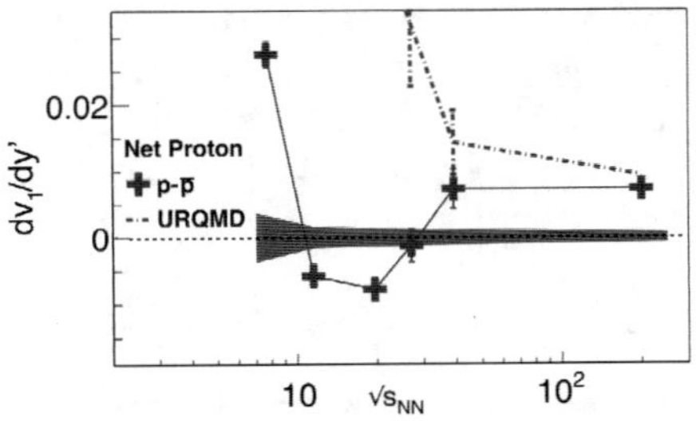

Figure 7. Net-proton directed flow as a function of beam energy compared to URQMD. Figure from [26].

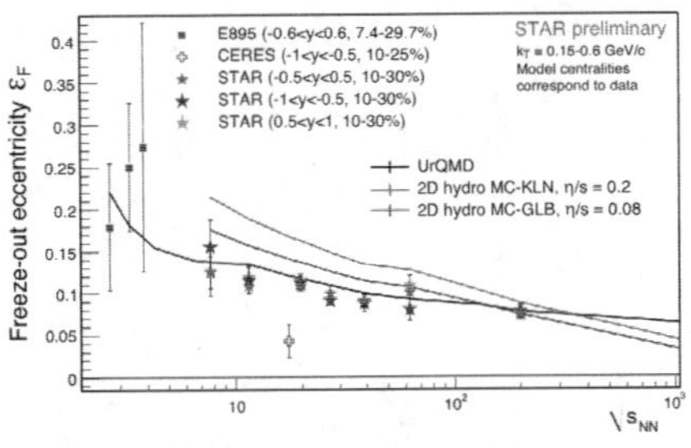

Figure 8. The kinetic freeze-out eccentricity as a function of beam energy. Figure adapted from [27].

statements can be made. The freeze-out eccentricities extracted from azimuthal HBT analyses, Fig. 8, exhibit no anomalous features in this region of $\sqrt{s_{NN}}$, instead they smoothly decrease with increasing beam energy [27]. These apparently contradictory results may be due to the fact that HBT is more sensitive to late stage hadronic interactions which wash out such early time signals. Interestingly, these HBT results demonstrate a different dependence to the modeling of the initial state, such as CGC or Glauber, and the viscosity to entropy ratio than does elliptic flow. Combining the results of spatial and momentum anisotropy analyses may therefore help us resolve ambiguities in the modeling of the initial conditions of heavy-ion collisions and the viscosity of the medium.

p-A collisions - the control dataset

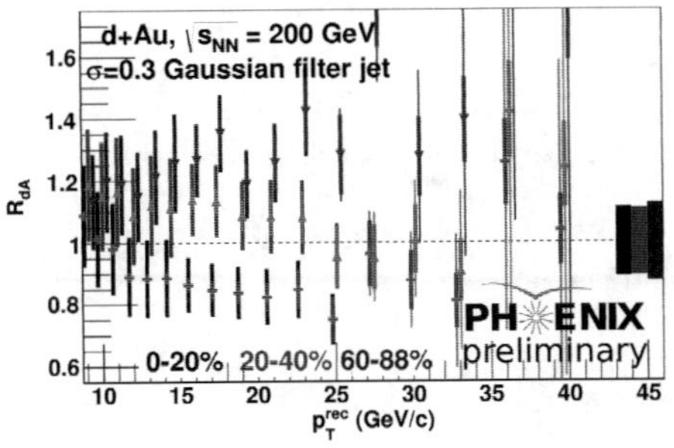

Figure 9. Jet R_{dA} for various centralities using the Gaussian filter at $\sqrt{s_{NN}}$=200 GeV. Figure from [29].

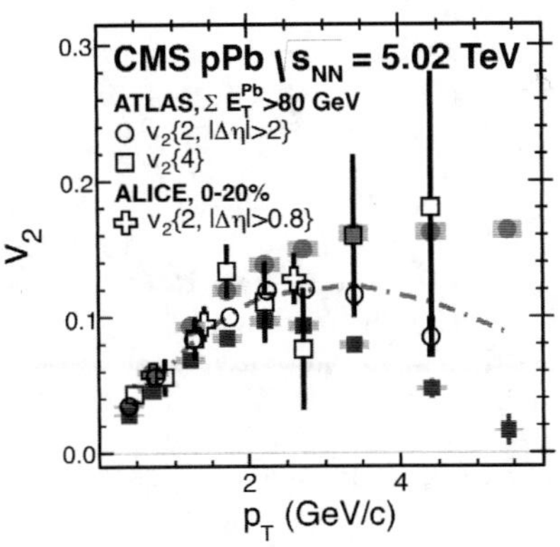

Figure 10. p-Pb $v_2\{4\}$ (filled squares) and $v_2\{2\}$ (filled circles) for $120<N_{trk}<150$ compared to similar events classes in ALICE and ATLAS. Figure from [30].

In minimum-bias p-A and d-A datasets the measured nuclear modification factors are all at or slightly above unity [31–33], suggesting that cold nuclear matter effects are small for light flavor hadrons. In p-A collisions at 5.02 TeV the rapidity distribution of the data appears to favor saturation models [34]. However, at both RHIC and the LHC things get interesting when efforts are made to analyze the data in centrality or multiplicity bins. For example PHENIX report that the R_{dA} of jets for peripheral events is more enhanced than that in central d-Au events [29] (Fig. 9) and all the LHC experiments now report that long range $\Delta\eta$ correlations are seen in high multiplicity p-A collisions [35–37] with a double peak structure consistent with the events having strong v_n terms, (see for example Fig. 10) demanding the question "do p-A collisions also exhibit collective motion?"

Summary

In summary, the field is still highly active with a large number of surprising results emerging from both the LHC and RHIC in the past year. There is a wealth of experimental data now available from $\sqrt{s_{NN}}$= 7.7 to 2760 GeV and I expect to see a significant number of new results presented at this conference, along with novel theoretical explanations and predictions with which to challenge the data.

Acknowledgments
Spectral thanks to the SQM2013 organizers for giving me the opportunity to give this overview talk and a reason to revisit my Alma Mater.

References

[1] Wilde M (ALICE Collaboration) 2013 *Nucl.Phys.A904-905* **2013** 573c–576c (*Preprint* 1210.5958)
[2] Roland G (CMS Collaboration) 2013 *Nucl.Phys.* **A904-905** 43c–50c
[3] Adare A *et al.* (PHENIX Collaboration) 2010 *Phys.Rev.Lett.* **104** 132301 (*Preprint* 0804.4168)
[4] Matsui T and Satz H 1986 *Phys.Lett.* **B178** 416
[5] Calderon M (for the CMS Collaboration) 2013 *J.Phys.Conf.Ser.* **458** 012011 (*Preprint* 1305.5560)
[6] Abelev B *et al.* (ALICE Collaboration) 2012 *Phys.Rev.Lett.* **109** 252301 (*Preprint* 1208.1974)
[7] Aamodt K *et al.* (ALICE Collaboration) 2011 *Phys.Lett.* **B696** 328–337 (*Preprint* 1012.4035)
[8] Veres G I (CMS Collaboration) 2013 *Nucl.Phys.* **A904-905** 146c–153c
[9] Adare A *et al.* (PHENIX Collaboration) 2012 (*Preprint* 1208.2254)
[10] Reed R (ALICE) 2013 *J.Phys.Conf.Ser.* **446** 012006 (*Preprint* 1304.5945)
[11] Aad G *et al.* (ATLAS Collaboration) 2013 *Phys.Lett.* **B719** 220–241 (*Preprint* 1208.1967)
[12] Adamczyk L *et al.* (STAR Collaboration) 2013 (*Preprint* 1302.6184)
[13] Adare A *et al.* (PHENIX Collaboration) 2012 (*Preprint* 1212.3323)
[14] Spousta M (ATLAS Collaboration) 2013 *Nucl.Phys.* **A904-905** 130c–137c (*Preprint* 1211.3334)
[15] Abelev B *et al.* (ALICE Collaboration) 2012 *JHEP* **1209** 112 (*Preprint* 1203.2160)
[16] Abelev B *et al.* (ALICE Collaboration) 2013 *Phys.Rev.Lett.* **111** 102301 (*Preprint* 1305.2707)
[17] Horvat S (STAR) 2013 *PoS* **CPOD2013** 002
[18] Abelev B *et al.* (ALICE Collaboration) 2012 *Phys.Rev.Lett.* **109** 072301 (*Preprint* 1202.1383)
[19] Adare A *et al.* (PHENIX Collaboration) 2011 *Phys. Rev. C* **84**(5) 054912
[20] Abbas E *et al.* (ALICE Collaboration) 2013 (*Preprint* 1303.5880)
[21] Kumar L (STAR collaboration) 2012 *Central Eur.J.Phys.* **10** 1274–1277 (*Preprint* 1201.4203)
[22] Cleymans J and Redlich K 1998 *Phys.Rev.Lett.* **81** 5284–5286 (*Preprint* nucl-th/9808030)
[23] Becattini F *et al.* 2013 *Phys.Rev.Lett.* **111** 082302 (*Preprint* 1212.2431)
[24] Mitchell J (PHENIX) 2013 *PoS* **CPOD2013** 003
[25] Cronin J, Frisch H J, Shochet M, Boymond J, Mermod R *et al.* 1975 *Phys.Rev.* **D11** 3105
[26] Pandit Y (STAR Collaboration) 2013 *Nucl.Phys.A904-905* **2013** 357c–360c (*Preprint* 1210.5315)
[27] Anson C (STAR Collaboration) 2011 *J.Phys.* **G38** 124148 (*Preprint* 1107.1527)
[28] Stoecker H 2005 *Nucl.Phys.* **A750** 121–147 (*Preprint* nucl-th/0406018)
[29] Wysocki M G (PHENIX Collaboration) 2013 *Nucl.Phys.* **A904-905** 67c–74c
[30] Chatrchyan S *et al.* (CMS Collaboration) 2013 *Phys.Lett.* **B724** 213–240 (*Preprint* 1305.0609)
[31] Abelev B *et al.* (ALICE Collaboration) 2013 *Phys.Rev.Lett.* **110** 082302 (*Preprint* 1210.4520)
[32] Adler S *et al.* (PHENIX Collaboration) 2003 *Phys.Rev.Lett.* **91** 072303 (*Preprint* nucl-ex/0306021)
[33] Adams J *et al.* (STAR Collaboration) 2003 *Phys.Rev.Lett.* **91** 072304 (*Preprint* nucl-ex/0306024)
[34] Abelev B *et al.* (ALICE Collaboration) 2013 *Phys.Rev.Lett.* **110** 032301 (*Preprint* 1210.3615)
[35] Chatrchyan S *et al.* (CMS Collaboration) 2013 *Phys.Lett.* **B718** 795–814 (*Preprint* 1210.5482)
[36] Abelev B *et al.* (ALICE Collaboration) 2013 *Phys.Lett.* **B719** 29–41 (*Preprint* 1212.2001)
[37] Aad G *et al.* (ATLAS Collaboration) 2013 *Phys.Rev.Lett.* **110** 182302 (*Preprint* 1212.5198)

Strangeness in Quark Matter: Opening Talk

J. Steinheimer[1], T. Lang[1], H. van Hees[1], A.S. Botvina[2,3], K.K. Gudima[4], I.N. Mishustin[1,5], H. Stöcker[1,6,7] and M. Bleicher[1,7]

E-mail: steinheimer@fias.uni-frankfurt.de
[1] Frankfurt Institute for Advanced Studies, Johann Wolfgang Goethe Universität, Ruth-Moufang-Str. 1, 60438 Frankfurt am Main, Germany
[2]Institute for Nuclear Research, Russian Academy of Sciences, 117312 Moscow, Russia
[3]Helmholtz-Institut Mainz, J.Gutenberg-Universität, 55099 Mainz, Germany
[4]Institute of Applied Physics, Academy of Sciences of Moldova, MD-2028 Kishinev, Moldova
[5]Kurchatov Institute, Russian Research Center, 123182 Moscow, Russia
[6]GSI Helmholtzzentrum für Schwerionenforschung GmbH, Planckstr. 1, 64291 Darmstadt, Germany
[7]Institut für Theoretische Physik, Goethe Universität, Max-von-Laue-Str. 1, 60438 Frankfurt am Main, Germany

Abstract. We discuss several new developments in the field of strange and heavy flavor physics in high energy heavy ion collisions. As shown by many recent theoretical works, heavy flavored particles give us a unique opportunity to study the properties of systems created in these collisions. Two in particular important aspects, the production of (multi) strange hypernuclei and the properties of heavy flavor mesons, are at the core of several future facilities and will be discussed in detail.

As strange quarks have to be newly produced during the hot and dense stage of a relativistic nuclear collision, they carry information, on the properties of the matter that was created [1, 2, 3, 4], to the observed final particle state. The enhancement of strange particle production is discussed [5, 6, 7, 8, 9] as a possible signal for the creation of a deconfined phase. Recently several observables, regarding strange and charm quarks have shown the importance of understanding the dynamics of strangeness and charm production in heavy ion collisions:

- Strange particle ratios and yields from the ALICE collaboration may indicate that there is either no unique chemical freeze out point for strange an non-strange particles [10, 11, 12, 13], or the light quark phase space is severely over-saturated [14].
- Lattice calculations on the stability of the H-dibaryon indicate it might be either very loosely bound or a resonant state [15, 16, 17].
- Viscous hydrodynamics, with fluctuating initial conditions [18] and finite but small viscous corrections, seems to describe strange hadron observables even at large baryon densities [19, 20].
- The hydrodynamic model calculations show sensitivity on the life time of the system and the applied equation of state [21].
- There are indications that systems created in high energy p+p and p+Pb collisions can thermalize/equilibrate to a certain degree and show signs of collectivity [22, 23, 24, 25].
- A polarization of Λ's due to the finite angular momentum of the fireball is expected [26].

Journal of Physics: Conference Series **509** (2014) 012002 doi:10.1088/1742-6596/509/1/012002

- Possible signals for the observation of quarkionic matter where proposed [27].
- There still is the unexplained puzzle of the strongly enhanced Ξ^- yield at the HADES experiment [28].

In addition to these interesting results we will focus on two topics which are at the core of present and upcoming experimental programs, FAIR with CBM [29] and NICA, namely the production of strange clusters and hypernuclei and the description of heavy quark transport in nuclear collisions.

1. Hypernuclei

Although abundantly produced, the interactions of strange hadrons are not very well understood but important for the description of the hadronic phase of a heavy ion collision and dense hadronic matter as can be found in compact stars. Hypernuclear physics offers a direct experimental way to study hyperon–nucleon (YN) and hyperon–hyperon (YY) interactions. Exotic forms of deeply bound objects with strangeness [30] and later the H di-baryon [31] was proposed by theory. Recent lattice QCD calculations suggest that the H-dibaryon is either a weakly bound system or a resonant state [32, 33, 16, 17], and there could be strange di-baryon systems including Ξ's that can be bound [34]. An experimental confirmation of such a state would therefore be an enormous advance in the understanding of the hyperon interaction.
Hypernuclei are known to be produced in heavy ion collisions already for a long time [35, 36, 37, 38], and the recent discoveries of the first anti-hypertriton [39] and anti-α [40] has fueled the interest in the field of hypernuclear physics. One can discriminate two distinct mechanisms for hypercluster formation in heavy ion collisions. First, the absorption of hyperons in the spectator fragments of non central collisions [41, 42, 43, 44]. The hyper-systems obtained here are rather large and moderately excited, decaying into hyper fragments later on [44, 45].

Alternatively, (hyper-)nuclear clusters can emerge from the hot and dense fireball region of the reaction. In this scenario the cluster is formed at, or shortly after, the (chemical-)freeze out of the system. A general assumption is, that these clusters are then formed through coalescence of different newly produced hadrons [46]. To estimate the production yield we compare two distinct approaches. First we use the hadronic transport model DCM [47] to provide us with the phase space information of all hadrons produced in a heavy ion collision. This information then serves as an input for a coalescence prescription [48]. On the other hand it has been shown that thermal models consistently describe the production yields of hadrons (and nuclei [49]) very well. We can therefore assume thermal production of clusters from a fluid dynamical description to heavy ion collisions [50, 51]. Figures 1 and 2 show the results for di-baryon and hypernuclei yields in the mid-rapidity region of central, $b < 3.4$ fm, heavy ion collisions at different beam energies. We compare results from the coalescence approach (symbols) with those from the fluid dynamical model (solid lines). It is clearly visible that large and multi-strange nuclear clusters exhibit a production maximum at a lower beam energy of $E_{lab} = 10A$ GeV. This means that experiments at the future facilities in Dubna and at FAIR will be well suited for the search of these new and exotic clusters.

Λ-hyperons are produced mainly in the hot and dense fireball, however, they have a broad rapidity distribution, so that a certain fraction of them can be found in the spectator kinematical region. Some of these Λ-hyperons are captured by nuclear spectator fragments produced in peripheral collisions [42]. The production of large excited spectator residues is well established in relativistic heavy-ion collisions[55]. We expect that the capture of hyperons by spectators leads to their exitation and break-up into fragments [45]. Using the DCM [47] and UrQMD [56, 57] models we simulate peripheral (b=8.5 fm) relativistic nuclear reactions to calculate the local nucleon density ρ at the positions of the hyperons created. This local density is then used to determine the effective potential $V_\Lambda(\rho) = -\alpha\frac{\rho}{\rho_0}\left[1-\beta(\frac{\rho}{\rho_0})^{2/3}\right]$, parameterized in Ref. [58].

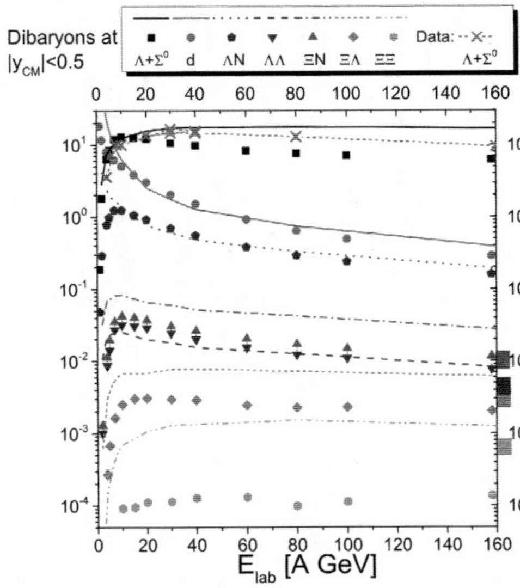

Figure 1. Yields of di-baryons in the mid rapidity region ($|y| < 0.5$) of most central collisions of Pb+Pb/Au+Au. Thermal production in the UrQMD hybrid model (lines) is compared to coalescence results with the DCM model (symbols). Black lines and symbols depict the production of Λ's from both models, compared to data (grey crosses) from [52, 53, 54].

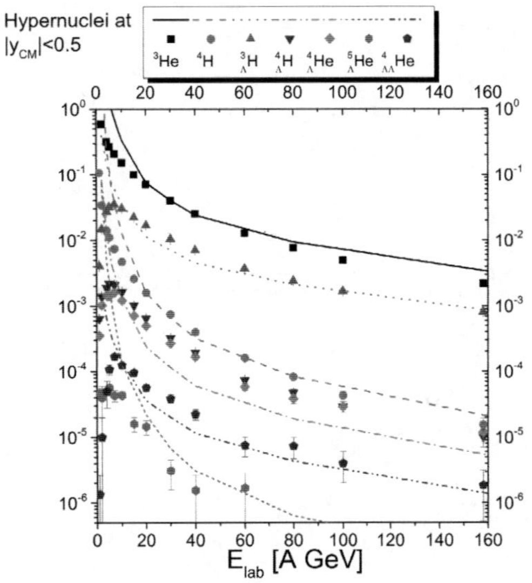

Figure 2. Yields per event of different (hyper-)nuclei in the mid rapidity region ($|y| < 0.5$) of most central collisions of Pb+Pb/Au+Au. Shown are the results from the thermal production in the UrQMD hybrid model (lines) as compared to coalescence results with the DCM model (symbols).

Here $\alpha = 57.5$ MeV, and $\beta = 0.522$. A Λ can be considered as absorbed when its kinetic energy, relative to the spectator, is smaller then the binding energy due to the nuclear density. Figure 3 shows the space coordinates of the absorption of Λ's, by the spectators, for different time intervals, calculated with the DCM transport model. One can clearly see how the Λ's are absorbed in the spectator region. In figure 4 we show the probabilities for the formation of a strange spectator residual for two different beam energies, calculated with the DCM and UrQMD transport model. Both models predict residuals with even 3 absorbed Λ's.

2. Charm Transport

Heavy quarks are an ideal probe for the QGP, mainly produced in the initial hard processes of a heavy ion collision. The most interesting observables are the nuclear modification factor, R_{AA}, and the elliptic flow, v_2. The measured large elliptic flow, v_2, of open heavy-flavor mesons indicate a high degree of thermalization of the heavy quarks with the bulk medium. A quantitative analysis of the degree of thermalization of heavy-quark degrees of freedom may lead to an understanding of the transport properties of QCD. We use a hybrid model, consisting of the UrQMD model [57, 56] and a full (3+1)-dimensional ideal hydrodynamical model [62, 63] to simulate the bulk medium created in an ultra relativistic nuclear collision (for alternative approaches see [64, 65, 66]). The heavy-quark propagation in the medium is described by a relativistic Langevin approach [67]. Within this framework we use different drag and diffusion coefficients of heavy flavors on the heavy-quark observables and compare the results with the experimental data from the Relativistic Heavy Ion Collider (RHIC) and the Large Hadron Collider (LHC). To make comparisons to experiment we coalesce a light quark with a heavy

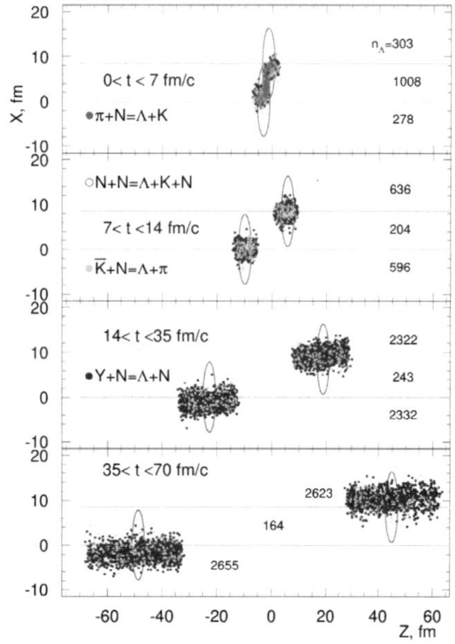

Figure 3. Coordinates of Λ absorption in the X–Z plane. The number of hyperons n_Λ (per $2{\cdot}10^5$ events) captured in the participant and spectator zones during these intervals is noted on the right side.

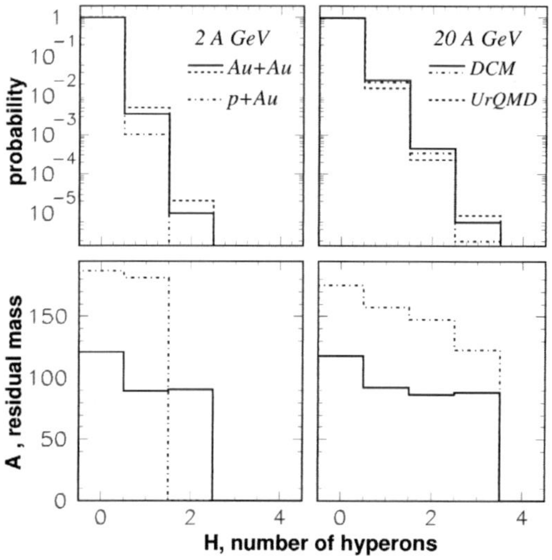

Figure 4. Formation probability of strange spectator residuals (top), and their mean mass numbers (bottom) versus the number of captured Λ hyperons (H), calculated with DCM and UrQMD model for p + Au and Au + Au collisions with energy of 2 GeV (left), and 20 GeV per nucleon (right).

quark [68]. Figures 5 through 8 show our results on the v_2 and R_{AA} of electrons from heavy quark decays for two different beam energies. One can see that our model is in rather good agreement with the data when a late decoupling criterion is chosen.

3. Summary

We have shown two important aspects of present and future experimental programs dedicated on the study of strange and heavy flavor physics. The study of hypernuclei and multi strange clusters can deepen our understanding of hyperon interactions as well as hadron formation in heavy ion collisions. Charm quarks on the other hand can serve as an important probe for the hot and dense QGP phase of the collision and are now used to extract information on the transport properties of the QCD medium.

4. Acknowledgements

This work has been supported by GSI and Hessian initiative for excellence (LOEWE) through the Helmholtz International Center for FAIR (HIC for FAIR). T. Lang gratefully acknowledges support from the Helmholtz Research School on Quark Matter Studies. I.M. acknowledges partial support from grant NS-215.2012.2 (Russia). The computational resources were provided by the LOEWE Frankfurt Center for Scientific Computing (LOEWE-CSC).

Journal of Physics: Conference Series **509** (2014) 012002 doi:10.1088/1742-6596/509/1/012002

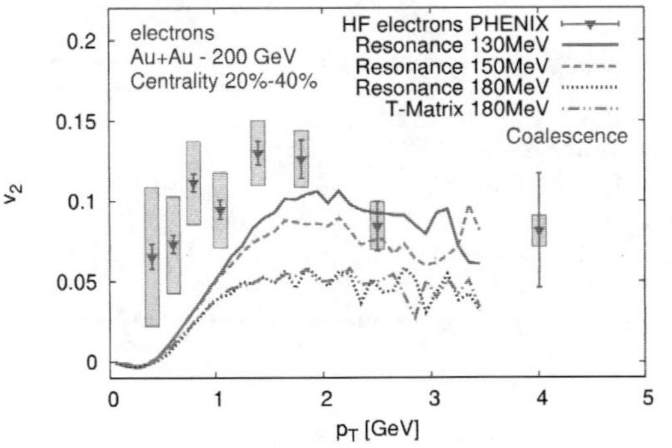

Figure 5. Elliptic flow v_2 of electrons from heavy quark decays in Au+Au collisions at $\sqrt{s_{NN}} = 200$ GeV using a coalescence mechanism, compared to data [59]. We use a rapidity cut of $|y| < 0.35$.

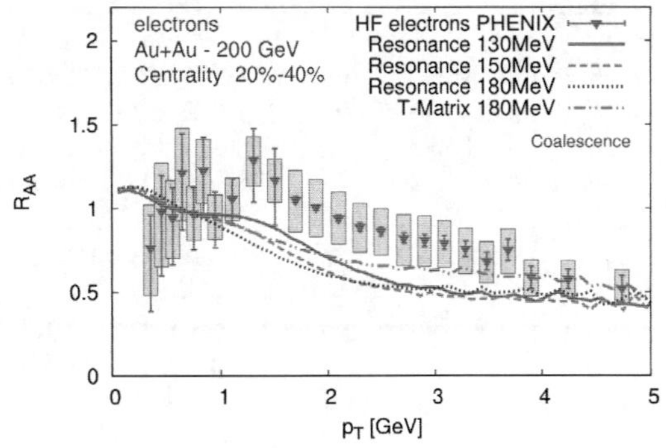

Figure 6. Nuclear modification factor R_{AA} of electrons from heavy quark decays in Au+Au collisions at $\sqrt{s_{NN}} = 200$ GeV using a coalescence mechanism, compared to data [59]. We use a rapidity cut of $|y| < 0.35$.

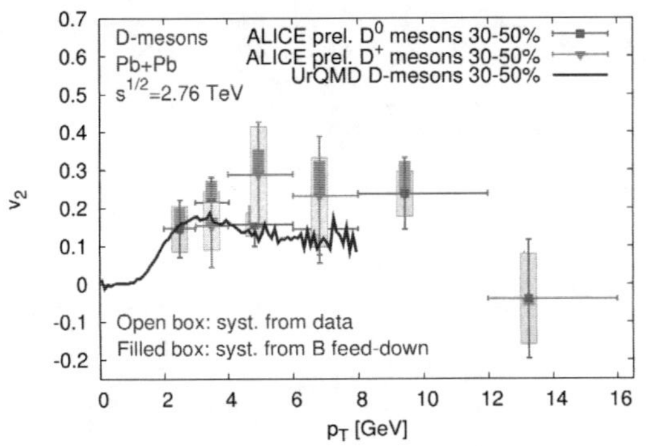

Figure 7. Flow v_2 of D-mesons in Pb+Pb collisions at $\sqrt{s_{NN}} = 2.76$ TeV compared to data from the ALICE experiment [60]. A rapidity cut of $|y| < 0.35$ is employed.

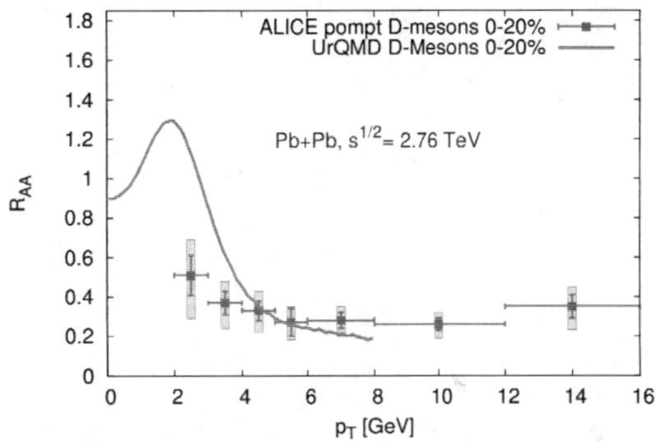

Figure 8. R_{AA} of D-mesons in Pb Pb collisions at $\sqrt{s_{NN}} = 2.76$ TeV compared to experimental data from ALICE [61]. A rapidity cut of $|y| < 0.35$ is employed.

References

[1] Koch P, Muller B and Rafelski J 1986 *Phys. Rept.* **142** 167
[2] Rafelski J 1982 *Phys. Rept.* **88** 331
[3] Rafelski J and Muller B 1982 *Phys. Rev. Lett.* **48** 1066 [Erratum-ibid. **56** 2334].
[4] Biro T and Zimanyi J 1982 Phys. Lett. B **113** 6
[5] Adams J *et al* [STAR Collaboration] 2005 *Nucl. Phys.* A **757** 102
[6] Back B B *et al* 2005 *Nucl. Phys.* A **757** 28
[7] Arsene I *et al* [BRAHMS Collaboration] 2005 *Nucl. Phys.* A **757** 1
[8] Gazdzicki M *et al* [NA49 Collaboration] 2004 *J. Phys.* G **30** S701
[9] Gazdzicki M and Gorenstein M I 1999 *Acta Phys. Polon.* B **30** 2705
[10] Bellwied R, Borsanyi S, Fodor Z, Katz S D and Ratti C 2013 *Phys. Rev. Lett.* **111** 202302
[11] Becattini F *et al* 2013 *Phys. Rev. Lett.* **111** 082302
[12] Milano L *et al* [ALICE Collaboration] 2013 *Nucl. Phys.* A904-905 **2013** 531c
[13] Bass S A and Dumitru A *Phys. Rev.* C **61** 064909
[14] Petran M, Letessier J, Petracek V and Rafelski J 2013 *Phys. Rev.* C **88** 034907

[15] Beane S R *et al.* 2011 *Mod. Phys. Lett.* A **26** 2587
[16] Inoue T *et al* [HAL QCD Collaboration] 2011 *Phys. Rev. Lett.* **106** 162002
[17] Buchoff M I, Luu T C and Wasem J 2012 *Phys. Rev.* D **85** 094511
[18] Bleicher M *et al* 1998 *Nucl. Phys.* A **638** 391
[19] Karpenko I A, Bleicher M, Huovinen P and Petersen H 2013 arXiv:1310.0702 [nucl-th]
[20] Petersen H, Li Q, Zhu X and Bleicher M 2006 *Phys. Rev.* C **74** 064908
[21] Li Q F, Steinheimer J, Petersen H, Bleicher M and Stöcker H 2009 *Phys. Lett.* B **674** 111
[22] Vogel S, Gossiaux P B, Werner K and Aichelin J 2011 *Phys. Rev. Lett.* **107** 032302
[23] Lang T and Bleicher M 2013 *Phys. Rev.* C **87** 024907
[24] Werner K, Bleicher M, Guiot B, Karpenko I and Pierog T 2013 arXiv:1307.4379 [nucl-th]
[25] Bozek P, Broniowski W and Torrieri G 2013 *Phys. Rev. Lett.* **111** 172303
[26] Becattini F, Csernai L and Wang D J 2013 *Phys. Rev.* C **88** 034905
[27] Torrieri G, Vogel S and Bäuchle B 2013 *Phys. Rev. Lett.* **111** 012301
[28] Agakishiev G *et al.* [HADES Collaboration] 2011 *Eur. Phys. J.* A **47** 21
[29] Senger P *et al Lect. Notes Phys.* **814** 681
[30] Bodmer A R 1971 *Phys. Rev.* D **4** 1601
[31] Jaffe R L 1977 *Phys. Rev. Lett.* **38** 195
[32] Beane S R *et al* [NPLQCD Collaboration] 2011 *Phys. Rev. Lett.* **106** 162001
[33] Beane S R *et al* 2011 *Mod. Phys. Lett.* A **26** 2587
[34] Beane S R *et al* [NPLQCD Collaboration] 2012 *Phys. Rev.* D **85** 054511
[35] Braun-Munzinger P and Stachel J 1995 *J. Phys.* G **21** L17
[36] Ahn J K *et al.* 2001 *Phys. Rev. Lett.* **87** 132504
[37] Takahashi H *et al* 2001 *Phys. Rev. Lett.* **87** 212502
[38] Andronic A, Braun-Munzinger P, Stachel J and Stöcker H 2011 *Phys. Lett.* B **697** 203
[39] STAR Collaboration 2010 *Science* **328** 58
[40] STAR Collaboration 2011 *Nature* **473** 353
[41] Ko C M 1985 *Phys. Rev.* C **32** 326
[42] Gaitanos T, Lenske H and Mosel U 2008 *Phys. Lett.* B **663** 197
[43] Gaitanos T, Lenske H and Mosel U 2009 *Phys. Lett.* B **675** 297
[44] Botvina A S *et al* 2011 *Phys. Rev.* C **84** 064904
[45] Botvina A S and Pochodzalla J 2007 *Phys. Rev.* C **76** 024909
[46] Scheibl R and Heinz U W 1999 *Phys. Rev.* C **59** 1585
[47] Toneev V D and Gudima K K 1983 *Nucl. Phys.* A **400** 173C
[48] Steinheimer J *et al* 2012 *Phys. Lett.* B **714** 85
[49] Andronic A, Braun-Munzinger P and Stachel J 2009 *Phys. Lett.* B **673** 142
[50] Steinheimer J *et al* 2009 *Phys. Lett.* B **676** 126
[51] Petersen H, Steinheimer J, Burau G, Bleicher M and Stöcker H 2008 *Phys. Rev.* C **78** 044901
[52] Ahmad S *et al* 1996 *Phys. Lett.* B **382** 35
[53] Mischke A *et al* [NA49 Collaboration] 2002 *J. Phys.* G **28** 1761
[54] Alt C *et al* [NA49 Collaboration] 2008 *Phys. Rev.* C **78** 034918
[55] Botvina A S *et al* 1995 *Nucl. Phys.* A **584** 737
[56] Bleicher M *et al* 1999 *J. Phys.* G **25** 1859
[57] Bass S A *et al* 1989 *Prog. Part. Nucl. Phys.* **41** 255 [Prog. Part. Nucl. Phys. **41** 225]
[58] Ahmad I, Mian M and Rahman Khan M Z 1985 *Phys. Rev.* C **31** 1590
[59] Adare A *et al* [PHENIX Collaboration] 2011 *Phys. Rev.* C **84** 044905
[60] Abelev B *et al* [ALICE Collaboration] 2013 *Phys. Rev. Lett.* **111** 102301
[61] Abelev B *et al* [ALICE Collaboration] 2012 *JHEP* **1209** 112
[62] Rischke D H, Bernard S and Maruhn J A 1995 *Nucl. Phys.* A **595** 346
[63] Rischke D H, Pursun Y and Maruhn J A 1995 *Nucl. Phys.* A **595** 383
[64] Uphoff J, Fochler O, Xu Z and Greiner C 2010 *Phys. Rev.* C **82** 044906
[65] Gossiaux P B, Aichelin J, Gousset T and Guiho V 2010 *J. Phys.* G **37** 094019
[66] Linnyk O *et al* 2013 *Phys. Rev.* C **87** 014905
[67] Rapp R and van Hees H 2010 *Quark Gluon Plasma 4, World Scientific* 111
[68] Lang T, van Hees H, Steinheimer J and Bleicher M 2012 arXiv:1211.6912 [hep-ph]

Journal of Physics: Conference Series **509** (2014) 012003 doi:10.1088/1742-6596/509/1/012003

Overview of results from ALICE

Mateusz Płoskoń, for the ALICE Collaboration

Lawrence Berkeley National Laboratory, 1 Cyclotron Road, 94720 Berkeley, CA, USA

E-mail: MPloskon@lbl.gov

Abstract. ALICE is a dedicated experiment for measurements of heavy-ion collisions at the Large Hadron Collider (LHC). A wealth of experimental data recorded in 2010, 2011 and 2012 suggests that a strongly interacting de-confined medium is created in collisions of lead ions at a centre-of-mass energy $\sqrt{s_{\mathrm{NN}}} = 2.76$ TeV. In order to quantify the properties of this hot and dense matter, measurements were performed in smaller systems, such as proton-proton and proton-lead, where effects related to the medium are expected to be negligible. We present an overview of recent measurements of particle production and particle correlations in proton-proton, Pb–Pb and p–Pb collisions at the LHC by ALICE Collaboration.

1. Identified particle production

Recent ALICE results on the inclusive particle production from different collision systems can be found in [1, 2, 3] for pp collisions, in [4] for p–Pb collisions and in [5, 6, 7] (and ref. therein) for Pb–Pb collisions. Figure 1 shows the integrated yields of various hadrons per unit rapidity in most central Pb–Pb collisions. The particle yields are compared to a simultaneous fit (to all particles except K*) of a thermal equilibrium (statistical hadronization) model defined with three parameters: the volume of the fireball V, baryochemical potential μ_b and temperature T. The fit to ALICE data with the fixed V for Pb–Pb collisions and $\mu_b = 1$ results in T of 156 MeV. We note the tension in the fit results between protons and multi-strange baryons as well as much heavier, deuterons. There are several potential explanations for such a tension: additional late stage baryon-antibaryon annihilation (specifically p–$\bar{\mathrm{p}}$), sequential freeze-out of different quark flavors, non-equilibrium freeze-out conditions, or incomplete knowledge of the baryon resonance spectrum. For a more in-depth discussion on the particle production see [8] and references therein.

Modifications of particle production in heavy-ion collisions can be established by studying the nuclear modification factor R_{AA}. R_{AA} is defined as the ratio of yields in heavy-ion collisions to yields measured in pp collisions scaled with appropriate number of binary nucleon-nucleon collisions. The R_{AA} in the absence of nuclear or medium effects is by construction unity at high transverse momenta where hard processes dominate. The particular shape differences of R_{AA} shown in Fig. 2 between baryons and mesons are a consequence of the interplay between the shapes of the particle p_{T} distributions, the collective flow (both depending on particle mass), jet quenching phenomena (sensitive to parton type), and possible novel hadronization phenomena such as parton recombination. One easily identifies the similarities of the curves for two groups of particles: baryons and mesons. The two special cases are the Ω baryon for which the large R_{AA} values can be related to the suppressed strangeness production in pp collisions, and the ϕ meson for which R_{AA} at p_{T} below 2 GeV/c resembles the suppression of baryons, whereas for

Journal of Physics: Conference Series **509** (2014) 012003 doi:10.1088/1742-6596/509/1/012003

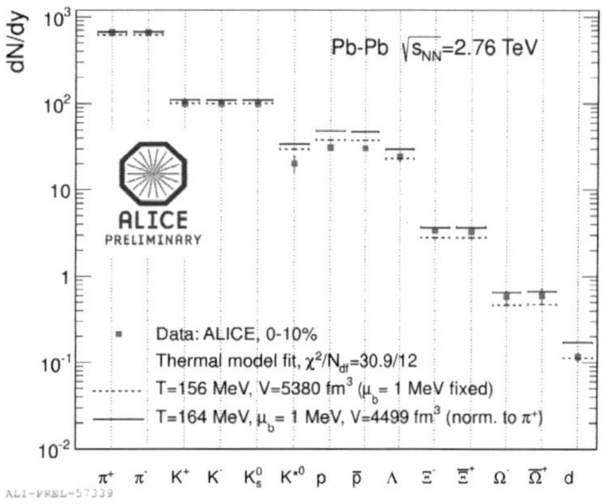

Figure 1. Identified particle abundances in most central heavy-ion collisions compared to a thermal model fits.

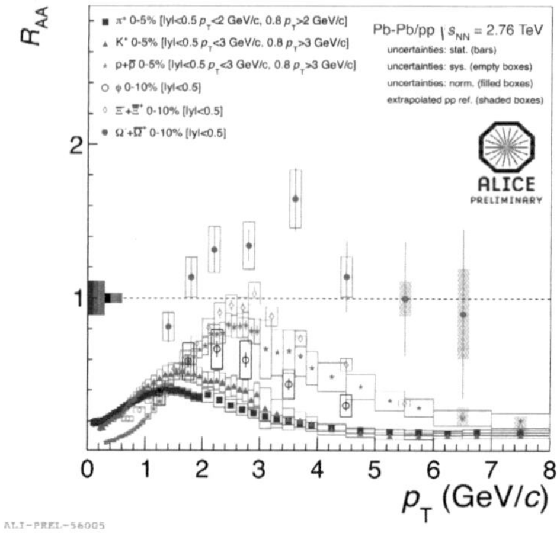

Figure 2. Nuclear modification factor for identified particles in most central Pb–Pb collisions.

$p_{\rm T} > 2$ GeV/c deviates from both, mesons and baryons. For an overview of measurements of short lived resonances see [9, 10].

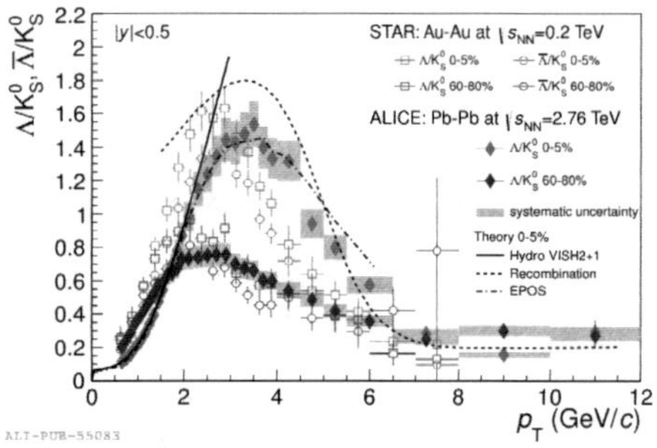

Figure 3. Ratio of Λ baryons to K^0_s mesons as a function of the transverse momentum in most central and peripheral heavy-ion collisions at RHIC and at the LHC [6].

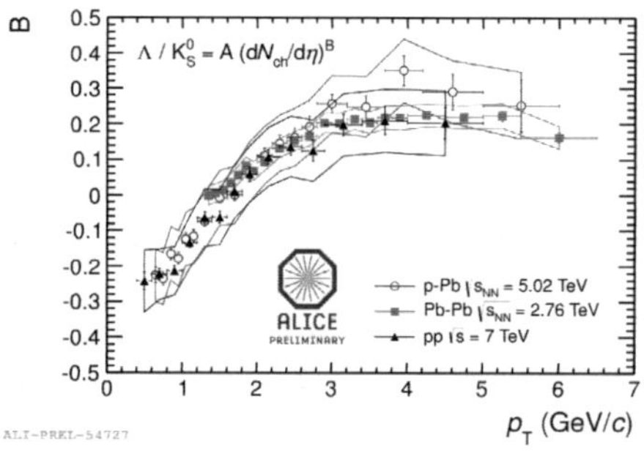

Figure 4. Exponent B calculated from the $\Lambda/\mathrm{K^0_S}$ratio as a function of transverse momentum for three collision systems: pp p–Pb and Pb–Pbat the LHC [4].

Concurrently, the $p_{\rm T}$ dependence of the baryon to meson ratio in heavy-ion collisions is likewise sensitive to the various dynamical phenomena of particle production (collective expansion of the system, parton recombination into hadrons, and jet fragmentation and hadronization). Figure 3 shows the measurements performed at RHIC and at the LHC of the $\Lambda/\mathrm{K^0_S}$ ratio in most-central and peripheral collisions. For the most central collisions at the LHC the ratio reaches its maximum at larger momenta than at RHIC. This observation is consistent with an expectation of stronger radial flow at higher energies. The ratio declines in the 3-6 GeV/c interval, where its shape, in some models is ascribed to parton recombination effects, reaching values of about 0.2 measured in peripheral collisions and driven by jet fragmentation

Journal of Physics: Conference Series **509** (2014) 012003 doi:10.1088/1742-6596/509/1/012003

processes. Moreover, the baryon to meson ratio for a given momentum bin shows a power law behavior with respect to the particle multiplicity density, such that a simple functional form can be fitted with $\Lambda/\mathrm{K}^0_s = A(\mathrm{d}N_{ch}/\mathrm{d}\eta)^B$, where A and B are free parameters. Figure 4 shows remarkable similarity of the extracted exponent between pp, p–Pb and Pb–Pb systems. The importance of this finding, hinting at a possibly common mechanism driving the baryon to meson ratio, remains to be fully understood [4].

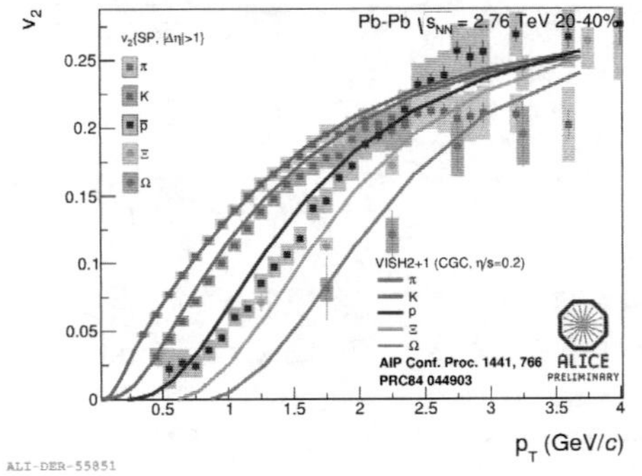

Figure 5. Elliptic flow coefficient (v_2) for various particle species as a function of transverse momentum for semi-central heavy-ion collisions at the LHC. The measurements are compared to hydrodynamical predictions (see [11] and references therein).

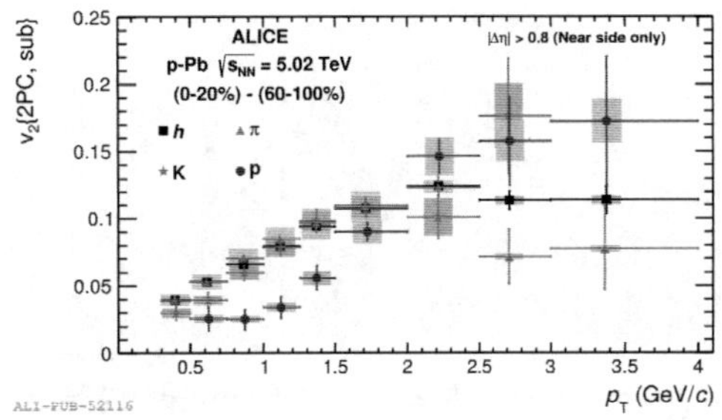

Figure 6. The v_2 coefficient extracted from two-particle correlations in most violent (highest multiplicity) p–Pb collisions for unidentified hadrons, pions, kaons and protons [13].

The collective behavior of the created matter within high-energy heavy-ion collisions is quantified by the second Fourier coefficient (v_2) of the azimuthal angular distribution of measured particles. The initial spatial anisotropy of the system of two nuclei colliding with non-zero impact parameter is transformed into a final state momentum anisotropy by interactions between the particles of the hot medium. Figure 5 shows good agreement between the measurements of v_2 for pions, kaons, protons and hydrodynamical modeling of the plasma. Similar agreement is seen also for heavier particles, such as Ξ and Ω, obeying the particular hydrodynamical prediction of mass ordering in the reported dependence of v_2 on transverse momenta of particles. More results on v_2 and higher harmonics measured with ALICE are reported in [11]. Furthermore, a striking observation has been made by analyzing the highest multiplicity p–Pb collisions and decomposing the two-particle azimuthal correlations into the single particle Fourier coefficients. For unidentified particles a positive v_2 coefficient has been found [12] for the event class of $0 - 20\%$ of the multiplicity distribution selected by the VZERO–A multiplicity detector ($2.8 < \eta_{\mathrm{det}}^{\mathrm{VZERO-A}} < 5.1$). The recent analysis of these two-particle correlations with emphasis on the particle identification (see Fig. 6) has revealed that a similar mass ordering as seen in Pb–Pb collisions also holds for the highest multiplicity p–Pb collisions [13].

2. Hard probes

Hard scatterings occurring early within high energy heavy-ion collisions are utilized to study the properties of the hot and dense medium. Partons, including heavy-quarks, are expected to loose energy within the medium depending on the traversed path length and the medium density. The parton energy loss is manifested by suppression of particle and jet yields measured in heavy-ion

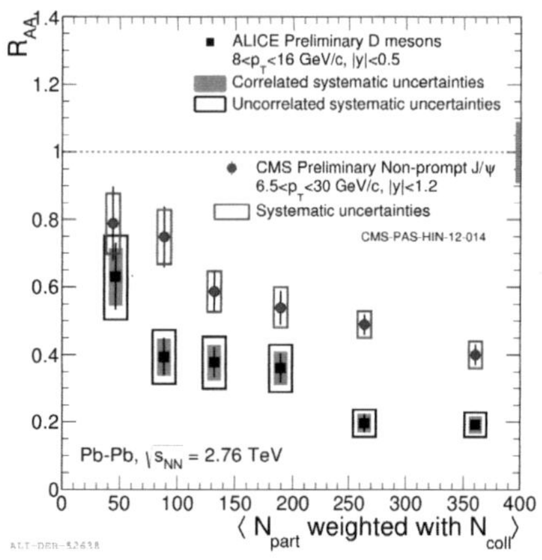

Figure 7. Nuclear modification factor for selected $p_{\rm T}$ ranges of D-meson and J/ψ from B-meson decays as a function of the average number of participating nucleons (see text for details) from ALICE and CMS at the LHC [17].

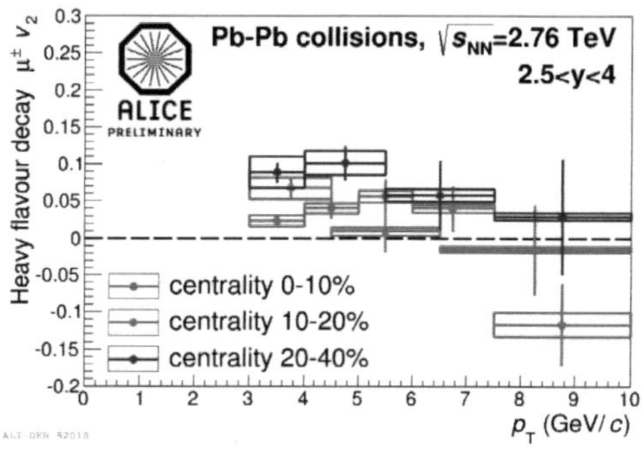

Figure 8. Elliptic flow coefficient (v_2) measured for heavy-flavor decay muons at forward rapidities in three centrality bins of Pb–Pb collisions at the LHC [18].

collisions as compared to pp collisions [14]. ALICE measurements on D-meson production in pp collisions are reported in [15], and the strong suppression of D-meson production in heavy-ion collisions is reported in [16]. Figure 7 shows the $R_{\rm AA}$ of D-mesons measured with ALICE together with non-prompt J/ψ preliminary results measured with CMS. The $p_{\rm T}$ range of the J/ψ has been chosen such that the $p_{\rm T}$ range of B-meson (parent in case of J/ψ) match the D-meson $p_{\rm T}$ range (see [17] for details). The observation that $R_{\rm AA}^{\rm D} < R_{\rm AA}^{\rm non-prompt\ J/\psi}$ is consistent with the predicted mass ordering in the energy-loss (larger loss for lighter quarks) within the hot plasma created in heavy-ion collisions. Furthermore, the measurement of elliptic flow (v_2) of heavy-flavor decay muons at large rapidities shown in Fig. 8 [18] is consistent with the substantial interaction of heavy quarks within the medium. Similar observations of positive heavy-flavor v_2 have been previously reported by ALICE for D-mesons in [19], and [20] for heavy-flavor decay electrons at mid-rapidity. The measurements of the $R_{\rm AA}$ together with the flow measurements are an essential input to modeling of the heavy-quark in-medium energy-loss.

In order to assess the importance of *cold nuclear matter effects* for the strong suppression of the hard probes observed in heavy-ion collisions, up to now, ALICE measured the nuclear modification factor in minimum-bias p–Pb collisions for charged particles [21], D-mesons (see Fig. 9 and [22]), heavy-flavor decay electrons [23], and jets reconstructed with charged particles [24] (see Fig. 10). In all cases $R_{\rm pPb}$ is consistent with unity within the uncertainties present in the current measurements strongly suggesting that the suppression seen in heavy-ion collisions must be related to final state interaction of the particles and jets with the hot and dense plasma. For more on p–Pb results and on the feasibility for reporting multiplicity differential nuclear modification factors see [25].

3. Quarkonia

ALICE has measured the production of quarkonia in pp, p–Pb and Pb–Pb collisions at the LHC. The suppression of J/ψ production at forward rapidities ($2.5 < y < 4.0$) in heavy-ion collisions

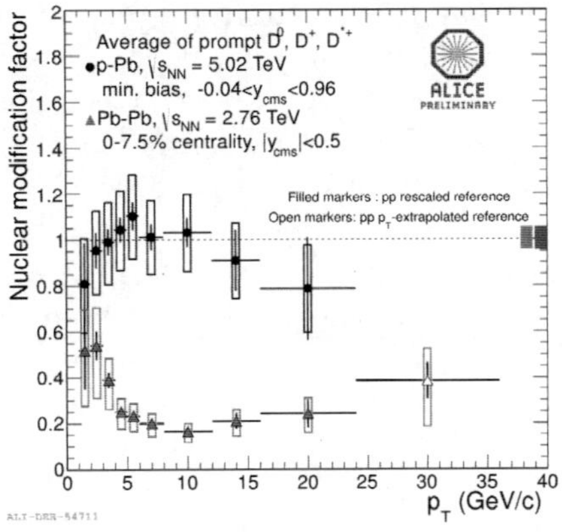

Figure 9. Comparison of the nuclear modification factor for D-mesons in heavy-ion and proton-lead collisions at the LHC.

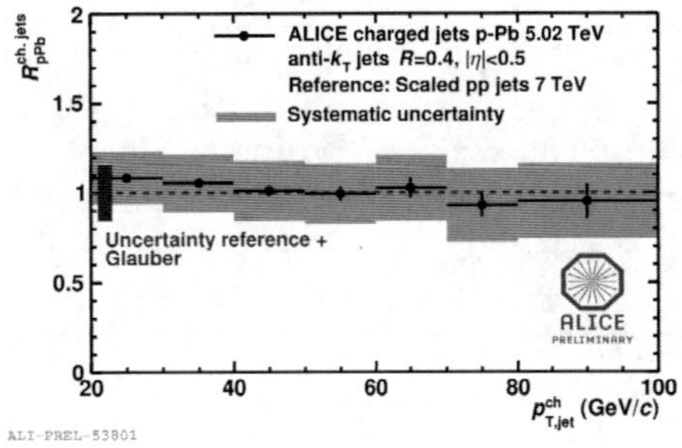

Figure 10. Nuclear modification factor for anti-k_T jets with $R = 0.4$ reconstructed from charged particles in minimum-bias proton-lead collisions at the LHC.

as a function of average number of participating nucleons has been reported by ALICE in [26] and is shown in Fig. 11 together with the R_{AA} of Υ(1S) [27]. The suppression pattern as a function of collision centrality reaches values of about 0.5 and remains constant over a wide range of centralities up to the most central events. Together with observation of the positive v_2 in the semi-central collisions this result suggests that the (re-)combination of $c\bar{c}$ pairs in the plasma or at the phase boundary plays a sizable role in the J/ψ production. The R_{AA} of both J/ψ and $\Upsilon(1S)$ have been measured for down to zero transverse momentum. Within the uncertainties a similar suppression of $\Upsilon(1S)$ as for inclusive J/ψ is observed. The suppression is stronger for central than semi-peripheral collisions. Together with the measurement from CMS performed for $|y| < 2.4$, no rapidity dependence of the Υ(1S) suppression is observed within the large rapidity range probed by the two experiments (for details see [28] and ref. therein).

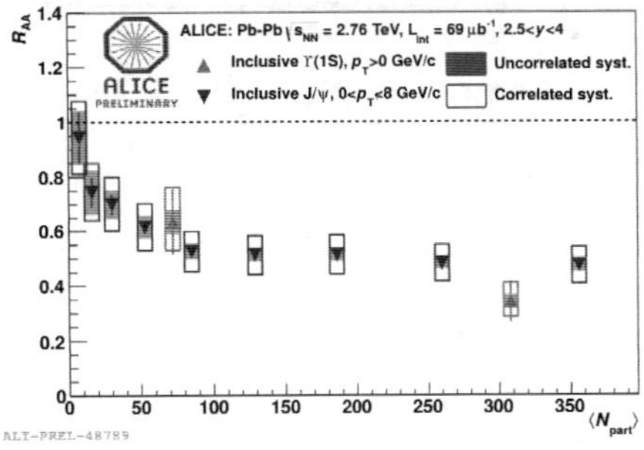

Figure 11. Nuclear modification factor for J/ψ and Υ as a function of the average number of participating nucleons in Pb–Pb collisions at the LHC.

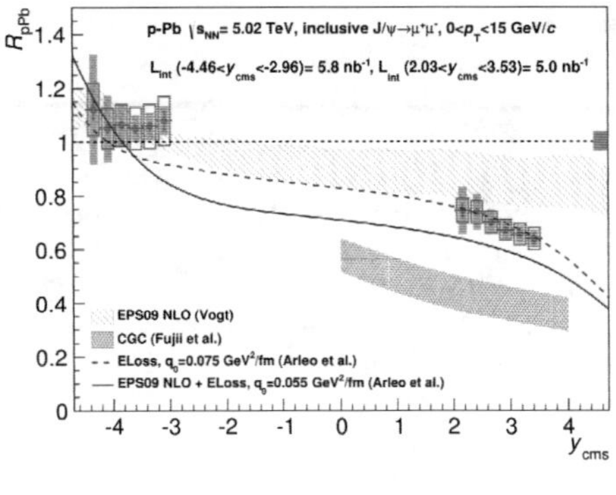

Figure 12. Nuclear modification factor for J/ψ as a function of center-of-mass rapidity in minimum bias proton-lead collisions at the LHC. The R_{pPb} is compared a theoretical models (see [29] for details).

To investigate *cold nuclear matter effects* relevant for quarkonium production (parton shadowing, gluon saturation, parton energy loss or final state break-up), ALICE studied proton-nucleus interactions. The nuclear modification factor R_{pPb} for the J/ψ [29] as a function of the quarkonium rapidity y_{cms} is shown in Fig. 12. The results are compared with theory predictions, based on a pure nuclear shadowing scenario, as well as partonic energy loss, either in addition to shadowing or as the only nuclear effect. Within the uncertainties, the model based on nuclear shadowing is in fair agreement with the data. Models including a contribution from parton energy loss are likewise in fair agreement with the data, while the prediction based on Color Glass Condensate model underestimate the results. A more complete overview of the quarkonia results has been given in [27]. For more results and discussions on quarkonia (including Υ(1S)) production in pp, p–Pb and Pb–Pb collisions see [28, 30, 31, 32] and references therein.

4. More ALICE highlights

This writeup presents a selected list of measurements performed with ALICE. The reader is encouraged to get familiar with ALICE measurements of low-mass di-leptons [33], on the search for exotic hyper-matter and nuclei [34], production of hyper nuclei in heavy-ion collisions [35], and particle production in ultra-peripheral collisions [36].

References

[1] Aamodt K *et al.* [ALICE Collaboration] 2011 "Strange particle production in proton-proton collisions at $\sqrt{s}$ = 0.9 TeV with ALICE at the LHC" *Eur. Phys. J.* C **71** 1594

[2] Aamodt K *et al.* [ALICE Collaboration] 2011 "Production of pions, kaons and protons in *pp* collisions at $\sqrt{s}$= 900 GeV with ALICE at the LHC" *Eur. Phys. J.* C **71** 1655

[3] Abelev B *et al.* [ALICE Collaboration] 2012 "Multi-strange baryon production in pp collisions at $(s)^{1/2} = 7$ TeV with ALICE" *Phys. Lett.* B **712** 309

[4] Abelev B B *et al.* [ALICE Collaboration] 2013 "Multiplicity Dependence of Pion, Kaon, Proton and Lambda Production in p–Pb Collisions at $\sqrt{s_{NN}} = 5.02TeV$" Preprint nucl-ex/1307.6796

[5] Abelev B B *et al.* [ALICE Collaboration] 2013 "Centrality dependence of π, K, p production in Pb-Pb collisions at $\sqrt{s_{NN}} = 2.76$ TeV" *Phys. Rev.* C **88** 044910

[6] Abelev B B *et al.* [ALICE Collaboration] 2013 "K^0_S and Λ production in Pb-Pb collisions at $\sqrt{s_{NN}} = 2.76$ TeV" Preprint nucl-ex/1307.5530

[7] Abelev B B *et al.* [ALICE Collaboration] 2013 "Multi-strange baryon production at mid-rapidity in Pb-Pb collisions at $\sqrt{s_{NN}} = 2.76$ TeV" Preprint nucl-ex/1307.5543

[8] Barnby L [for the ALICE Collaboration] 2013 "Strangeness in ALICE" (SQM 2013, this conference)

[9] Knospe A G [for the ALICE Collaboration] 2013 "Hadronic resonances in heavy-ion collisions at ALICE (Strangeness in Quark Matter 2013)" Preprint nucl-ex/1309.3322 (these proceedings)

[10] Lee G R [for the ALICE Collaboration] 2013 "Resonance Production in $\sqrt{s} = 7$ TeV *pp* collisions with ALICE" (SQM 2013, this conference)

[11] Zhou Y [for the ALICE Collaboration] 2013 "Anisotropic flow of identified particles in Pb–Pb collisions at $\sqrt{s_{\mathrm{NN}}} = 2.76$ TeV with the ALICE detector" Preprint nucl-ex/1309.3237 (these proceedings)

[12] Abelev B *et al.* [ALICE Collaboration] 2013 "Long-range angular correlations on the near and away side in p–Pb collisions at $\sqrt{s_{NN}} = 5.02$ TeV" *Phys. Lett.* B **719** 29

[13] Abelev B B *et al.* [ALICE Collaboration] 2013 "Long-range angular correlations of pi, K and p in p–Pb collisions at $\sqrt{s_{NN}} = 5.02$ TeV" *Phys. Lett.* B **726** 164

[14] Abelev B B *et al.* [ALICE Collaboration] 2013 "Measurement of charged jet suppression n Pb-Pb collisions at $\sqrt{s_{NN}} = 2.76$TeV" Preprint nucl-ex/1311.0633

[15] Abelev B *et al.* [ALICE Collaboration] 2012 "Measurement of charm production at central rapidity in proton-proton collisions at $\sqrt{s} = 7$ TeV" *JHEP* **1201** 128 (Preprint hep-ex/1111.1553)

[16] Abelev B *et al.* [ALICE Collaboration] 2012 "Suppression of high transverse momentum D mesons in central Pb-Pb collisions at $\sqrt{s_{NN}}$=2.76 TeV" *JHEP* **1209** 112 (Preprint nucl-ex/1203.2160)

[17] Bruna E [for the ALICE Collaboration] 2013 "D meson nuclear modification factor and v_2 in Pb–Pb collisions" (SQM 2013, this conference)

[18] Zhang X [for the ALICE Collaboration] 2013 "Nuclear Modification Factor and Elliptic Flow of Muons from Heavy-Flavour Hadron Decays in Pb–Pb Collisions at $\sqrt{s_{\mathrm{NN}}} = 2.76$ TeV with ALICE" (SQM 2013, this conference)

[19] Abelev B *et al.* [ALICE Collaboration] 2013 "D meson elliptic flow in non-central Pb-Pb collisions at $\sqrt{s_{NN}}$ = 2.76 TeV" *Phys. Rev. Lett.* **111** 102301 (Preprint nucl-ex/1305.2707)

[20] Thomas D [for the ALICE Collaboration] 2013 "Measurement of electrons from heavy-flavour decays in Pb–Pb collisions at $\sqrt{s_{\rm NN}}$ = 2.76 TeV with ALICE" (SQM 2013, this conference)

[21] Abelev B *et al.* [ALICE Collaboration] 2013 "Transverse Momentum Distribution and Nuclear Modification Factor of Charged Particles in *p*-Pb Collisions at $\sqrt{s_{NN}} = 5.02$ TeV" *Phys. Rev. Lett.* **110** 082302 (Preprint nucl-ex/1210.4520)

[22] Luparello G [for the ALICE Collaboration] 2013 "Measurement of D meson production in p–Pb collisions with the ALICE detector" (SQM 2013, this conference)

[23] Heide M A [for the ALICE Collaboration] 2013 "Measurement of electrons from heavy-flavour hadron decays in pp and p–Pb collisions with ALICE at the LHC" (SQM 2013, this conference)

[24] Verweji M [for the ALICE Collaboration] 2013 "Jet production and structure in pp, p-Pb and Pb-Pb collisions measured by ALICE" (SQM 2013, this conference)

[25] Morsch A [for the ALICE Collaboration] 2013 "p–Pb Results from ALICE with an Emphasis on Centrality Determination" Preprint nucl-ex/1309.5525 (these proceedings)

[26] Abelev B B *et al.* [ALICE Collaboration] 2013 "Centrality, rapidity and transverse momentum dependence of J/ψ suppression in Pb-Pb collisions at $\sqrt(s_{NN}) = 2.76$TeV" Preprint nucl-ex/1311.0214

[27] Bruno G E [for the ALICE Collaboration] 2013 "ALICE quarkonia" (SQM 2013, this conference)

[28] Khan P [for the ALICE Collaboration] "Upsilon Production in Pb–Pb Collisions at Forward Rapidity with ALICE at the LHC" (SQM 2013, this conference)

[29] Abelev B B *et al.* [ALICE Collaboration] 2013 "J/ψ production and nuclear effects in p-Pb collisions at $\sqrt{s_{NN}}$=5.02 TeV" Preprint nucl-ex/1308.6726

[30] Lakomov I [for the ALICE Collaboration] 2013 "J/ψ production in p–Pb collisions with ALICE at the LHC" (SQM 2013, this proceedings)

[31] Fionda F [for the ALICE Collaboration] 2013 "Measurements of J/ψ to e^+e^- with ALICE at the LHC" (SQM 2013, this conference)

[32] Valencia Palomo L [for the ALICE Collaboration] 2013 "Charmonium production measurements in Pb–Pb collisions with ALICE at the LHC" (SQM 2013, this conference)

[33] De Falco A [for the ALICE Collaboration] 2013 "Low mass vector meson production in pp, p–Pb and Pb–Pb collisions measured with ALICE at the LHC" (SQM 2013, this conference)

[34] Doenigus B [for the ALICE Collaboration] 2013 "Search for exotic hyper-matter and measurement of (anti-)nuclei yields with ALICE at the LHC" (SQM 2013, this conference)

[35] Lea R [for the ALICE Collaboration] 2013 "Production of hypernuclei in Pb–Pb collisions at $\sqrt{s_{\rm NN}} = 2.76$ TeV with ALICE at the LHC" (SQM 2013, this conference)

[36] Tapia Takaki D. [for the ALICE Collaboration] 2013 "ALICE results on ultra-peripheral collisions" (SQM 2013, this conference)

Journal of Physics: Conference Series **509** (2014) 012004 doi:10.1088/1742-6596/509/1/012004

Recent results on strangeness production at STAR

Xianglei Zhu (for the STAR Collaboration)
Department of Engineering Physics, Tsinghua University, Beijing 100084, China
E-mail: zhux@tsinghua.edu.cn

Abstract.
We report on the measurements of strange hadrons ($K^{\pm}$, K^0_S, Λ, Ξ, Ω and ϕ) production at mid-rapidity in Au+Au collisions at $\sqrt{s_{NN}}$ = 7.7 – 39 GeV from the STAR beam energy scan (BES) program. The collision energy dependence of strange hadron yields are presented. The strange hadron related particle ratios are compared with calculations from statistical and hadronic cascade models. The nuclear modification factors and baryon to meson ratios ($\overline{\Lambda}/K^0_S$ and Ω/ϕ) are presented. The physics implications for collision dynamics are discussed.

1. Introduction

The main motivation of STAR beam energy scan program is to study the QCD phase diagram [1]. Systematic study of Au+Au collisions from $\sqrt{s_{NN}}$ = 39 GeV down to 7.7 GeV in the RHIC BES program may help to achieve the following goals: 1) searching for the QCD critical point where the first order phase transition at finite μ_B ends and identifying the phase boundary of the first order phase transition; 2) locating the particular collision energy where the deconfinement starts to happen, i.e. the onset of deconfinement.

Strange hadrons are excellent probes for identifying the phase boundary and onset of deconfinement. Strangeness enhancement in A+A with respect to p+p collisions has long been suggested as a signature of Quark-Gluon Plasma (QGP) formation in these collisions [2]. Until now, strangeness has been extensively measured in many experiments at different accelerator facilities [3–19]. Generally, the yields of strange hadrons in nuclear collisions are close to those expected from statistical models [20–23]. The precise measurement of strange hadrons yield in heavy ion collisions in BES will certainly lead to a better understanding of strangeness production mechanism in nuclear collisions and a more reliable extraction of the possible phase boundary.

On the other hand, it has been observed at RHIC that, at high p_T, the nuclear modification factor R_{CP} of various particles is much less than unity for Au+Au collisions at $\sqrt{s_{NN}}$ = 200 GeV [24], indicating the significant energy loss of the scattered partons in the dense matter ("jet quenching"). By measuring the R_{CP} of strange hadrons in BES, one can potentially pin down the beam energy at which interactions with the medium begin to affect hard partons [1]. At intermediate p_T, it was found at RHIC that, the p/π and Λ/K^0_S ratios are larger than unity in more central events. These ratios are much higher than that observed in elementary collisions [24–26]. These phenomena hint toward different hadronization mechanisms in this p_T range. There are recombination/coalescence models which allow soft partons to coalesce into hadrons, or soft and hard partons to recombine into hadrons [27]. They naturally reproduce an enhanced baryon to meson ratio when the parton p_T distribution is exponential. Such models require constituent quarks to coalesce or recombine, and hence the observation of coalescence or recombination behavior is one of the corner-stone pieces of evidence for the formation of the strongly interacting Quark-Gluon Plasma. It is also interesting to investigate at which collision energies these phenomena are prevalent [1], in order to locate the energy point at which the onset of the deconfinement happens.

In this paper, we present the strangeness data obtained from the high statistics Au+Au collisions at $\sqrt{s_{NN}} = 7.7$, 11.5, 19.6, 27 and 39 GeV, collected during the first phase of the STAR Beam Energy Scan (BES) program in 2010 and 2011.

2. Results

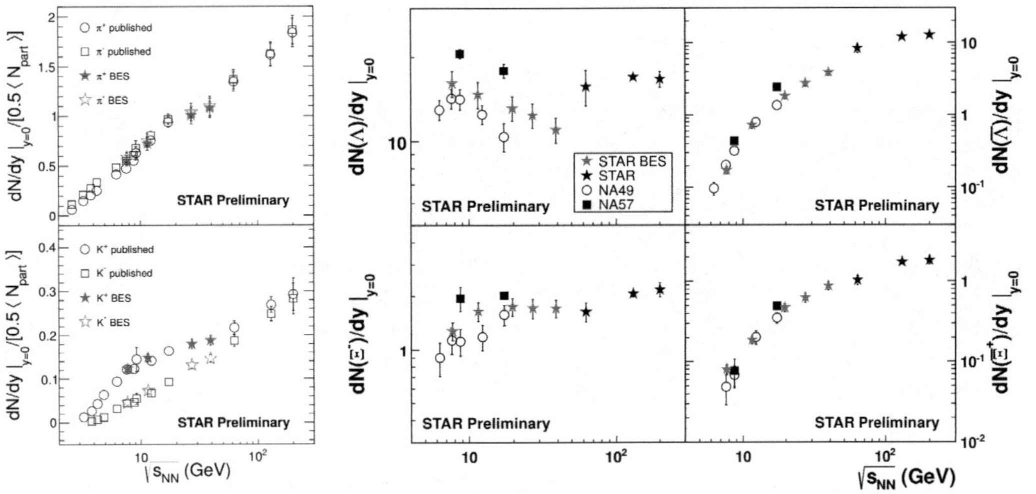

Figure 1. Collision energy dependence of strange hadron yields at mid-rapidity in most central (0-5%) Au+Au collisions. Also shown are the results from most central heavy ion collisions at STAR [13, 15–17, 28], NA57 [8], NA49 [9, 29] and AGS [30]. Errors are the quadratic sum of statistical and systematic errors.

Figure 1 shows the collision energy dependence of the particle yield (dN/dy) at mid-rapidity for $\pi^{\pm}$, $K^{\pm}$, Λ, $\overline{\Lambda}$, Ξ^- and $\overline{\Xi}^+$ from the most central (0-5%) Au+Au collisions, compared to corresponding data from AGS, NA49, NA57, as well as the STAR data at higher energies. $\pi^{\pm}$, $K^{\pm}$ yields have been divided by the number of participants in the most central collisions. Figure 1 shows that STAR BES data lie on a trend established by the corresponding data from NA49, NA57 and previous STAR data. There seems to be a non-monotonic $\sqrt{s_{NN}}$ dependence in the Λ dN/dy, which may originate from the interplay of slow rise of Λ total multiplicity and the change of shape in rapidity distribution in the same energy range [9].

Figure 2 shows the $\sqrt{s_{NN}}$ dependence of the ratios of $K^{\pm}$, Λ, $\overline{\Lambda}$, Ξ^- and $\overline{\Xi}^+$ at mid-rapidity to that of pions in STAR BES, as well as the existing data from various experiments at different energies and the calculations from hadronic transport models (UrQMD v1.3 and HSD) [35] and a statistical hadron gas model (SHM) [22]. The BES data are in good agreement with the trend of the published results. Though the hadronic models seem to reproduce the Λ/π data, the default UrQMD (v1.3) fails in reproducing the Ξ^-/π and $\overline{\Xi}^+/\pi$ ratios due to a smaller Ξ yield in the model. On the other hand, the SHM model predictions agree well with data across the whole energy range from AGS to top RHIC energies. The K^+/π^+, Λ/π and Ξ^-/π ratios all show a maximum at $\sqrt{s_{NN}} \sim 8$ GeV, which seems to be consistent with the picture of maximum net-baryon density at freeze-out at this collision energy [36].

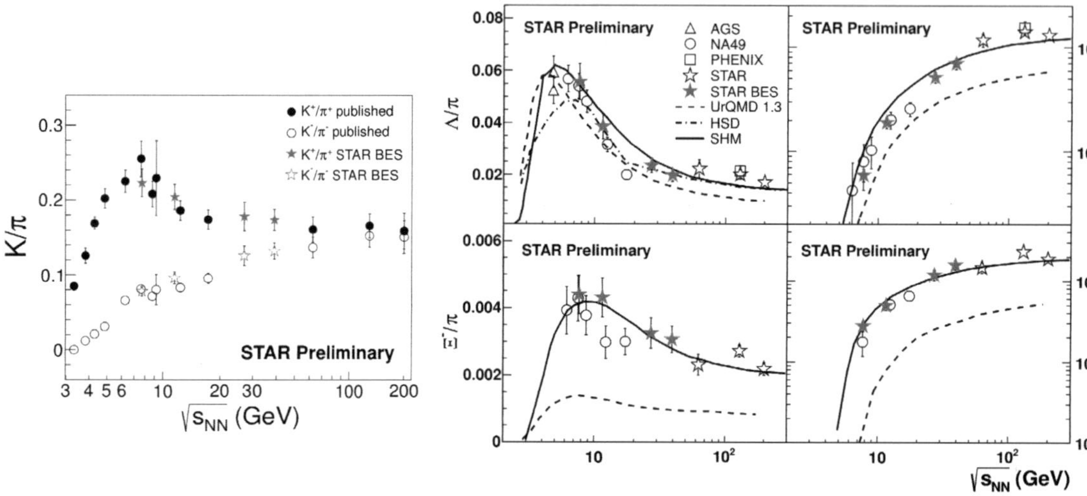

Figure 2. (left) Energy dependence of $K^{\pm}/\pi^{\pm}$ ratio for central collisions at mid-rapidity. Also shown are previous published data from RHIC [28], AGS [30], SPS [29]. (right) Energy dependence of Λ, $\overline{\Lambda}$, Ξ^- and $\overline{\Xi}^+$to pions ($1.5(\pi^+ + \pi^-)$) ratios at mid-rapidity in central Au+Au and Pb+Pb collisions. The pion yield in STAR BES is taken from [31]. Also shown are AGS [3,4,6,7], NA49 [9], PHENIX [18,32] and STAR [13,15–17,33,34] data, as well as calculations from two hadronic transport models: HSD and UrQMD (v1.3) [35] and a statistical hadron gas model (SHM) [22]. Errors are the quadratic sum of statistical and systematic errors.

Figure 3 (left) shows the R_{CP} of K^0_S, Λ, Ξ and Ω at mid-rapidity from STAR BES. At $\sqrt{s_{NN}} \geq 19.6$ GeV, the K^0_S $R_{CP} \leq 1$ for $p_T > 1.5$ GeV/c and much less than those of baryons. This K^0_S high-p_T suppression and baryon/meson separation is qualitatively consistent with results from higher RHIC energies [24]. However, for $\sqrt{s_{NN}} \leq 11.5$ GeV, although the maximum accessible p_T is smaller at these two energies, the R_{CP} data seem qualitatively different from that at $\sqrt{s_{NN}} \geq 19.6$ GeV: there is no suppression for K^0_S at $p_T > 1.5$ GeV/c; and the baryon/meson separation becomes less significant at intermediate p_T. Figure 3 (right) shows the $\overline{\Lambda}/K^0_S$ ratios as a function of p_T in different centralities from Au+Au collisions in STAR BES. At $\sqrt{s_{NN}} \geq 19.6$ GeV, the $\overline{\Lambda}/K^0_S$ can reach a maximum value of unity at $p_T \sim 2.5$ GeV/c in the most central collisions, while in the peripheral collisions, the maximum value is only about 0.3 – 0.5. This shows that there is baryon enhancement at intermediate p_T for $\sqrt{s_{NN}} \geq 19.6$ GeV, which is similar to that observed at higher energies. However, for Au+Au collisions at $\sqrt{s_{NN}} \leq 11.5$ GeV, the maximum values of $\overline{\Lambda}/K^0_S$ in the measured p_T range is much smaller than unity, and the difference between 0-5% and 40-60% is also less significant. This indicates much less baryon to meson enhancement at intermediate p_T in Au+Au collisions at $\sqrt{s_{NN}} \leq 11.5$ GeV.

Figure 4 shows the baryon-to-meson ratio, $N(\Omega^- + \Omega^+)/2N(\phi)$, as a function of p_T in Au+Au central collisions at $\sqrt{s_{NN}} = 11.5 - 200$ GeV. The 200 GeV data are from previous STAR measurements [37]. In central Au+Au collisions at $\sqrt{s_{NN}} = 200$ GeV, the intermediate-p_T (2 – 4 GeV/c) Ω to ϕ ratio is explained by mainly thermal strange quark recombination in the deconfined matter [38]. The $N(\Omega^- + \Omega^+)/2N(\phi)$ ratios in $\sqrt{s_{NN}} = 19.6$, 27, and 39 GeV are close to that in 200 GeV. However, the ratios at 11.5 GeV seem to deviate from the trend with a χ^2/ndf of 8.3/2 for $p_T > 2.4$ GeV/c. The difference in the ratios between 11.5 GeV data and those above 19.6 GeV may indicate a significant change in the underlying p_T distributions for strange quarks which recombine/coalesce to form the final Ω and ϕ particles in our measurement,

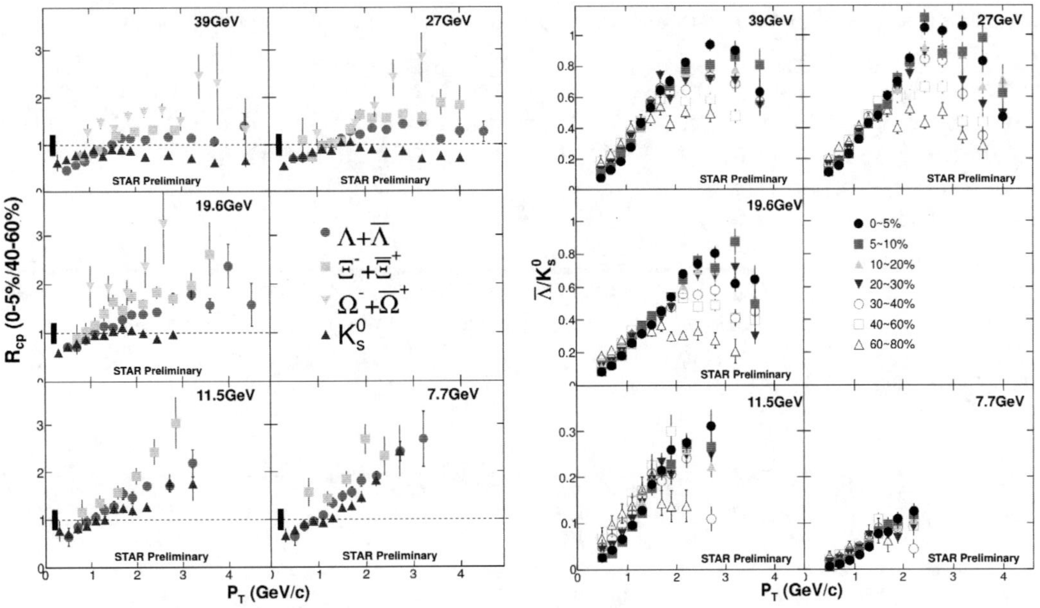

Figure 3. (left) K^0_S, Λ, Ξ and Ω R_{CP}(0-5%)/(40-60%) at mid-rapidity ($|y| < 0.5$) in Au+Au collisions at $\sqrt{s_{NN}} = 39 - 7.7$ GeV. Errors are statistical. The left black band is the normalization error from N_{bin}. (right) $\bar{\Lambda}/K^0_S$ ratio as a function of p_T at mid-rapidity ($|y| < 0.5$) in different centralities from Au+Au collisions at $\sqrt{s_{NN}} = 39 - 7.7$ GeV. Errors are statistical.

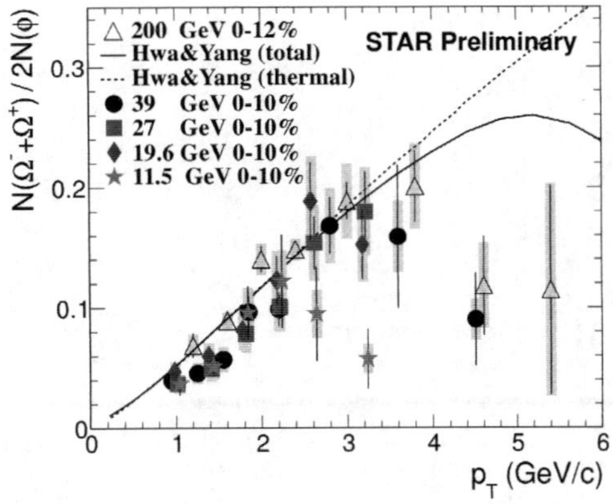

Figure 4. The baryon-to-meson ratio, $N(\Omega^-+\Omega^+)/2N(\phi)$ as a function of p_T at mid-rapidity ($|y| < 0.5$) from central Au+Au collisions at $\sqrt{s_{NN}} = 11.5 - 200$ GeV. Green bands denote systematical errors. The solid and dashed lines represent recombination model calculations for central collisions at $\sqrt{s_{NN}} = 200$ GeV [38] with total and thermal strange quark contributions, respectively.

or indeed that there is no substantial deconfined phase in 11.5 GeV collisions.

3. Summary

We presented recent measurements on strange hadrons production in the STAR beam energy scan. The strange hadron yields from STAR BES seem to follow a trend with published data from RHIC, SPS and AGS. The ratios of Λ, $\bar{\Lambda}$, Ξ^- and $\bar{\Xi}^+$ to pions at mid-rapidity agree well with the calculations of the statistical hadron gas model. For $\sqrt{s_{NN}} \leq 11.5$ GeV, we also observe: K^0_S R_{CP} are larger than unity for $p_T > 1.5$ GeV/c; $\overline{\Lambda}/K^0_S$ ratio in the most central collisions is much less than unity at intermediate p_T and the difference between central and peripheral collisions gets reduced; Ω to ϕ ratio seems to turn down much earlier. These measurements point to a beam energy region between 11.5 and 19.6 GeV as a favored range for further investigation of the deconfinement phase transition.

Acknowledgments

X. Zhu thanks the support by the National Natural Science Foundation of China (Grant Nos. 11035009, 11335005), the Major State Basic Research Development Program in China (No. 2014CB845400) and the Foundation for the Authors of National Excellent Doctoral Dissertation of P.R. China (FANEDD) (No. 201021).

References

[1] Aggarwal M M *et al* (STAR Collaboration) 2010 arXiv:1007.2613
[2] Rafelski J and Müller B 1982 *Phys. Rev. Lett.* **48** 1066
[3] Albergo S *et al* 2002 *Phys. Rev. Lett.* **88** 062301
[4] Ahle L *et al* 1998 *Phys. Rev. C* **57** 466
[5] Chung P *et al* 2003 *Phys. Rev. Lett.* **91** 202301
[6] Back B B *et al* 2001 *Phys. Rev. Lett.* **87** 242301
[7] Ahmad S *et al* 1996 *Phys. Lett. B* **382** 35; *ibid* 1998 *Nucl. Phys. A* **636** 507
[8] Antinori F *et al* 2004 *Phys. Lett. B* **595** 68; *ibid* 2006 *J. Phys. G* **32** 427; http://wa97.web.cern.ch/WA97/yields.html
[9] Alt C *et al* 2008 *Phys. Rev. C* **78** 034918
[10] Anticic T *et al* 2009 *Phys. Rev. C* **80** 034906
[11] Milosevic J 2006 *J. Phys. G* **32** S97
[12] Abelev B I *et al* 2008 *Phys. Rev. C* **77** 044908
[13] Adams J *et al* 2007 *Phys. Rev. Lett.* **98** 062301
[14] Abelev B I *et al* 2009 *Phys. Lett. B* **673** 183
[15] Adler C *et al* 2002 *Phys. Rev. Lett.* **89** 092301
[16] Adams J *et al* 2004 *Phys. Rev. Lett.* **92** 182301
[17] Aggarwal M M *et al* 2011 *Phys. Rev. C* **83** 024901
[18] Adcox K *et al* 2002 *Phys. Rev. Lett.* **89** 092302
[19] Abelev B B *et al* (ALICE Collaboration) 2013 arXiv:1307.5530; arXiv:1307.5543
[20] Becattini F *et al* 2001 *Phys. Rev. C* **64** 024901
[21] Braun-Munzinger P *et al* 2002 *Nucl. Phys. A* **697** 902
[22] Andronic A *et al* 2006 *Nucl. Phys. A* **772** 167
[23] Redlich K and Tounsi A 2002 *Eur. Phys. J. C* **24** 589
[24] Lamont M A C (for the STAR Collaboration) 2006 *J. Phys. Conf. Ser.* **50** 192
[25] Adcox K *et al* (PHENIX Collaboration) 2004 *Phys. Rev. C* **69** 024904
[26] Abelev B I *et al* (STAR Collaboration) 2007 *Phys. Lett. B* **655** 104; *ibid* 2006 *Phys. Rev. Lett.* **97** 152301
[27] Hwa R C and Yang C B 2002 *Phys. Rev. C* **66** 025205
[28] Abelev B I *et al* (STAR Collaboration) 2009 *Phys. Rev. C* **79** 034909; *ibid.* 2010 **81** 024911
[29] Afanasiev S V *et al* (NA49 Collaboration) 2002 *Phys. Rev. C* **66** 054902; Alt C *et al* (NA49 Collaboration) 2008 *Phys. Rev. C* **77** 024903; *ibid.* 2006 **73** 044910
[30] Ahle L *et al* (E866 and E917 Collaborations) 2000 *Phys. Lett. B* **490** 53; *ibid.* 2000 **476** 1; Klay J L *et al* (E895 Collaboration) 2002 *Phys. Rev. Lett.* **88** 102301
[31] Kumar L (for the STAR Collaboration) 2011 *J. Phys. G* **38** 124145; Das S (for the STAR Collaboration) these proceedings
[32] Adcox K *et al* 2002 *Phys. Rev. Lett.* **88** 242301
[33] Adler C *et al* 2004 *Phys. Lett. B* **595** 143
[34] Adams J *et al* 2004 *Phys. Rev. Lett.* **92** 112301
[35] Bratkovskaya E L *et al* 2004 *Phys. Rev. C* **69** 054907; taken from [9]
[36] Randrup J and Cleymans J 2006 *Phys. Rev. C* **74** 047901
[37] Abelev B I *et al* (STAR Collaboration) 2009 *Phys. Rev. C* **79** 064903
[38] Hwa R C and Yang C B 2007 *Phys. Rev. C* **75** 054904

Identified charge hadron spectra and ratios in Au+Au and d+Au collisions at $\sqrt{s_{NN}} = 200$ GeV

Ron Belmont
University of Michigan, Ann Arbor, MI 48108, USA
E-mail: rbelmont@umich.edu

Abstract. PHENIX has recently reported [1] measurements of identified charged hadron spectra and ratios in Au+Au and d+Au collisions at $\sqrt{s_{NN}} = 200$ GeV. Identified hadrons are an important probe of both hot and cold nuclear matter. The intermediate p_T region, 2–5 GeV/c, is of particular interest. In Au+Au collisions, the production of mesons is suppressed in this p_T region relative to that in p+p collisions, while the production of baryons is nearly unmodified. On the other hand, in d+Au collisions, the meson production exhibits a slight enhancement in this p_T region while the baryon production exhibits a much stronger enhancement. In these proceedings, the p_T spectra and ratios of identified charged hadrons $\pi^{\pm}$, $K^{\pm}$, p, and $\bar{p}$ in 5 different centrality classes for each collision species will be discussed.

1. Introduction

The content presented this talk and in these proceedings is from a recent PHENIX paper on identified charged hadrons [1]. The particles are identified using the time-of-flight detector in the west arm (TOFW) of the PHENIX central spectrometer. The TOFW provides excellent PID capabilities, with 4σ π/K separation up to p_T of 2.5 GeV/c, and 4σ K/p separation up to p_T of 4 GeV/c. The p_T reach can be extended greatly by exclusion cuts, where in addition to requiring that the m^2 of the particle is within 2σ of the mean, a further requirement that the m^2 of the particle is beyond 2σ of the mean of the other particles is also employed. In this analysis, this allows identification of pions and protons up to 6 GeV/c (5 GeV/c) in p_T and kaons up to 4 GeV/c (3.5 GeV/c) in p_T in Au+Au (d+Au) collisions.

2. Spectra

The p_T spectra for all species and both collision systems are presented in Ref. [1]. In principle all the information reported therein and in these proceedings is encoded in the spectra. Particle ratios, however, often help illustrate differences between particle species, centrality classes, collisions systems, etc. There has been tremendous interest at this conference and elsewhere in hydrodynamical flow in small systems like d+Au and p+Pb, which is best evinced in the spectra themselves. The author anticipates and encourages theoretical studies of these spectra. The spectra are not shown in these proceedings in the interest of not exceeding the page limit.

Journal of Physics: Conference Series **509** (2014) 012005 doi:10.1088/1742-6596/509/1/012005

3. Ratios

3.1. p/π and K/π ratios

Presented in Figures 1 and 2 are the K^+/π^+ (upper panels) and K^-/π^- (lower panels) as a function of p_T for Au+Au and d+Au collisions, respectively. PHENIX data from p+p collisions [2] are shown as a reference. The K/π ratio in Au+Au shows a weak centrality dependence, with the ratio rising slightly as the collisions become more central. This is indicative of the well-known strangeness enhancement in the medium. The enhancement is also p_T dependent, with the ratios being roughly equal at low p_T and then the enhancement steadily increasing up to $p_T \approx 2$ GeV/c, above which the enhancement levels off and is flat. This detailed measurement of the p_T dependence of this ratio can hopefully shed light on the strangeness production mechanism(s) in the medium. In d+Au collisions this effect is notably absent, as no centrality dependence is seen whatsoever, with the caveat that the span of N_{part} values is small. Finally, note that for the most peripheral Au+Au and all d+Au centralities the K/π ratios are consistent with those in p+p collisions in the overlapping p_T region.

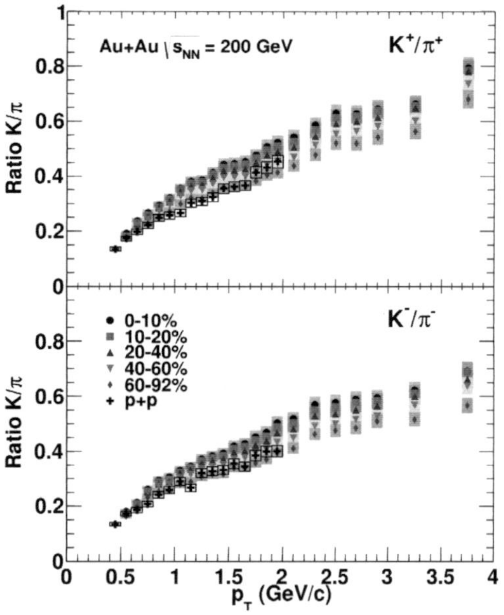

Figure 1. K^+/π^+ (upper panel) and K^-/π^- (lower panel) as a function of p_T in Au+Au collisions.

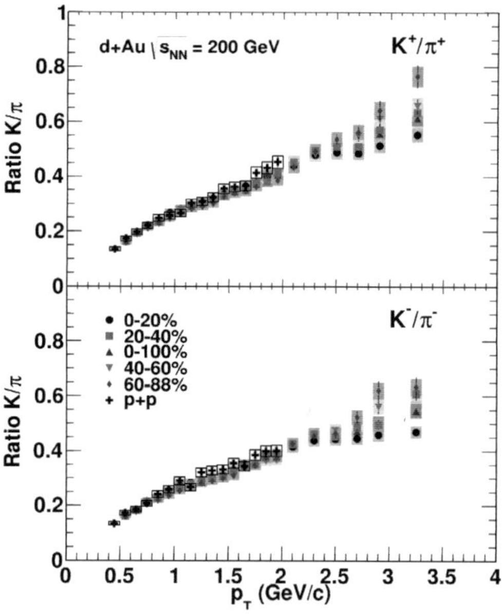

Figure 2. K^+/π^+ (upper panel) and K^-/π^- (lower panel) as a function of p_T in Au+Au collisions.

Presented in Figures 3 and 4 are the p/π^+ (upper panels) and $\bar{p}/\pi^-$ (lower panels) as a function of p_T for Au+Au and d+Au collisions, respectively. PHENIX data from p+p collisions [2] are shown as a reference. The p/π ratio in Au+Au shows a strong centrality dependence. In d+Au the centrality dependence at first appears weak, but in fact should be regarded as quite strong given the small range of N_{part} in the d+Au centrality selections. In both collision systems, the ratio rises quickly, reaches a maximum between 2–2.5 GeV/c in p_T, and then falls off slowly. This maximum point is closer to 2 GeV/c in d+Au. There appears to be a centrality dependence to this maximum in Au+Au, as for example the maximum for peripheral Au+Au appears to be closer to 2 GeV/c (like the d+Au) but closer to 2.5 GeV/c for the most central.

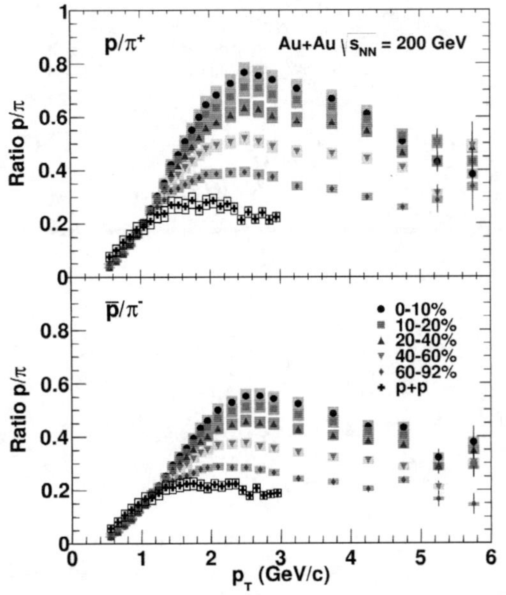

Figure 3. p/π^+ (upper panel) and $\bar{p}/\pi^-$ (lower panel) as a function of p_T in Au+Au collisions.

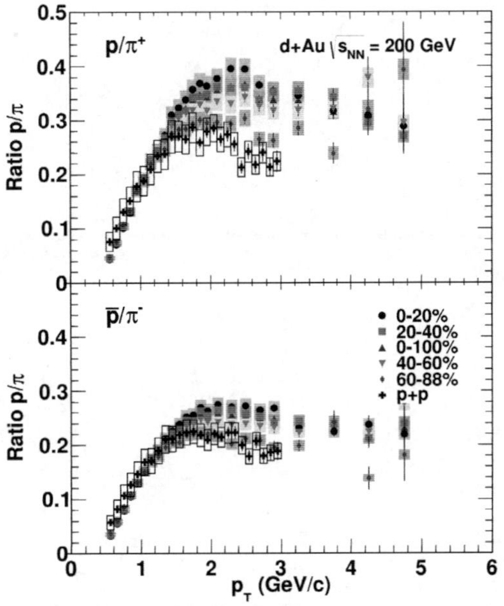

Figure 4. p/π^+ (upper panel) and $\bar{p}/\pi^-$ (lower panel) as a function of p_T in d+Au collisions.

3.2. Nuclear modification factors

Shown in Figures 5 and 6 are R_{AA} and R_{dA}, respectively, as a function of p_T in the various centrality classes. The p+p reference data for $\pi/K/p$ are from Ref. [2]. Also shown are π^0 data taken from Refs. [3, 4] and ϕ data taken from Ref. [5].

As exhibited in the R_{AA}, charged and neutral pions show fairly significant suppression. The charged kaons exhibit a very similar p_T shape as the charged pions, although an overall higher level. The ϕ meson is at a yet higher level in turn. All three mesons π, K, and ϕ are suppressed but with decreasing suppression in that order. This ordering cannot be attributed to mass alone since the ϕ is heavier than the proton but has a significantly lower R_{AA}. On the other hand, this ordering also coincides with increasing strangeness content, meaning it could be attributable to strangeness enhancement. In the R_{dA} all mesons exhibit essentially identical behavior, showing little or no modification. This suggests that any differences in particle production between d+Au and p+p collisions are not attributable to particle mass or strangeness content.

In both the R_{AA} and R_{dA} the protons exhibit a strong enhancement relative the mesons, which is also seen in the p/π ratios. The behavior is clearly consistent with different behaviors for baryons and mesons. The physical origin of differences between baryons and mesons has been frequently attributed to hadronization by parton recombination, which has some success in describing the experimental data [6]. However, an alternative explanation, such as that involving baryon junctions, should not be forgotten, and the author invites further theoretical investigation of the present work.

3.3. Comparison between peripheral Au+Au and central d+Au

As alluded to in previous sections, there are remarkable similarities between peripheral Au+Au and central d+Au. A direct comparison between the two is justified by this alone. However, this can be further motivated by the very similar Glauber parameters for the two systems. The N_{coll} values are 14.8±3.0 and 15.1±1.0 for peripheral Au+Au and central d+Au, respectively; the

Journal of Physics: Conference Series **509** (2014) 012005 doi:10.1088/1742-6596/509/1/012005

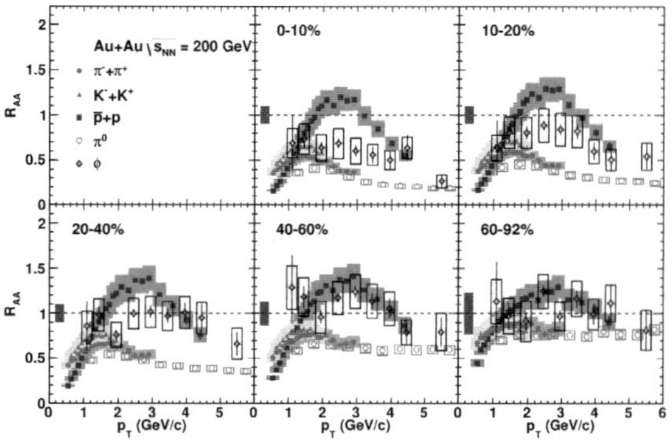

Figure 5. R_{AA} as a function of p_T. A dashed line is drawn at unity indicating non-modification as a visual aid.

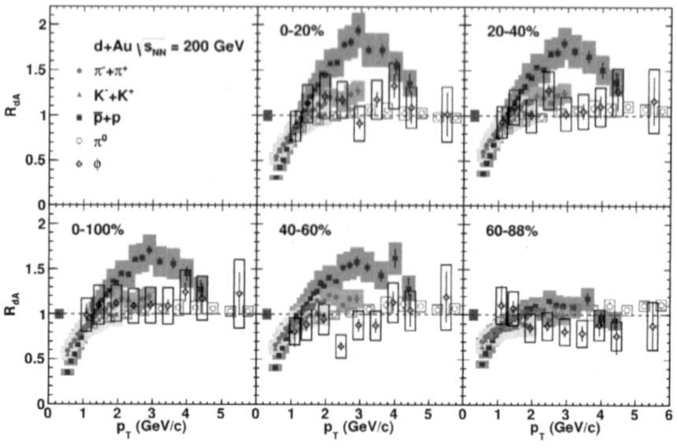

Figure 6. R_{dA} as a function of p_T. A dashed line is drawn at unity indicating non-modification as a visual aid.

N_{part} values are 14.7±2.9 and 15.3±0.8 for peripheral Au+Au and central d+Au, respectively.

Shown in Figures 7 and 8 are the K/π and p/π ratios, respectively, as a function of p_T in both peripheral Au+Au and central d+Au. As can be seen, the ratios are essentially identical in the two systems.

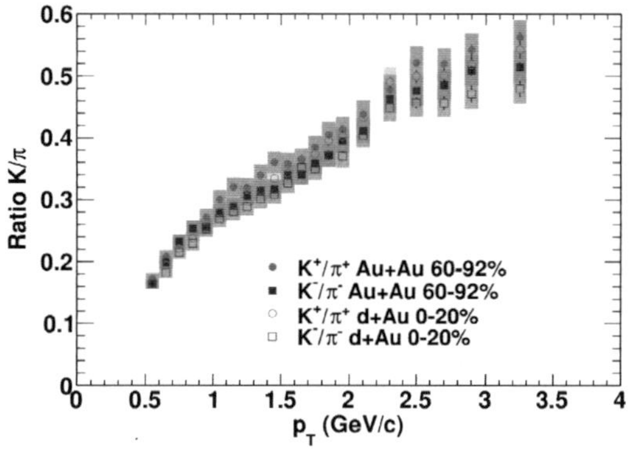

Figure 7. K^+/π^+ and K^-/π^- as a function of p_T in peripheral Au+Au collisions and central d+Au collisions.

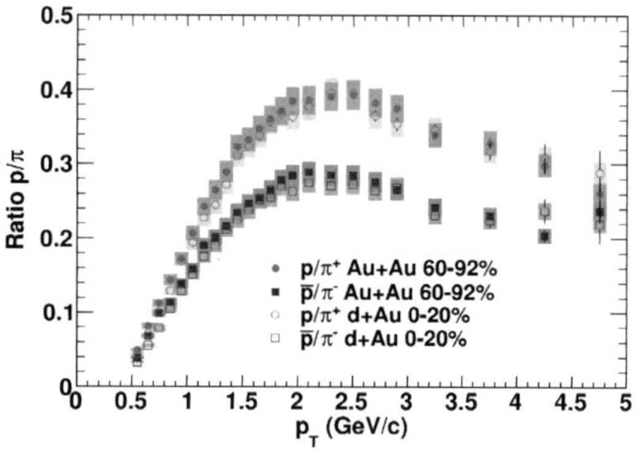

Figure 8. p/π^+ and $\bar{p}/\pi^-$ as a function of p_T in peripheral Au+Au collisions and central d+Au collisions.

A direct ratio of the spectra in the two systems is presented in Figure 9. No scaling is applied. However, as noted above, the respective N_{coll} and N_{part} are very similar, so a scaling by either of these values would result in a negligible change.

This ratio is very striking. For p_T values of 2.5 GeV/c and higher, the ratio is flat and independent of p_T, and is further identical for each particle species. This indicates that the baryon enhancement in each system is quantitatively the same and that the baryon enhancement is likely driven by the same mechanism. This in turn provides significant theoretical constraints.

At lower p_T the ratio trends upwards. A mass and/or particle type separation may be evident, although the three species are consistent within systematic uncertainties. There are many potential physical effects at play. For example, there is a rapidity shift in soft particle production in d+Au collisions due to the participant asymmetry [7]. This naturally creates a

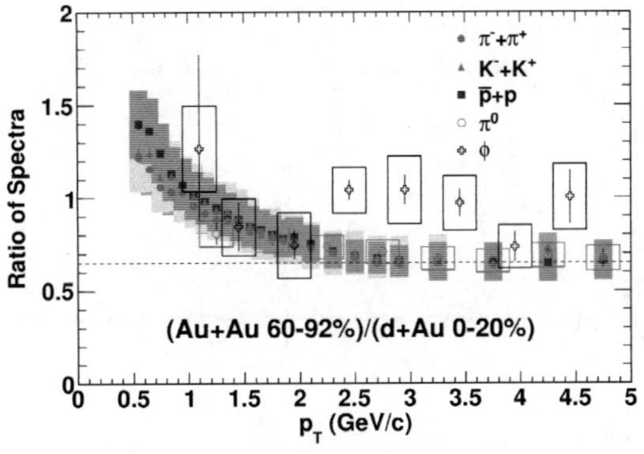

Figure 9. Ratio of spectra in peripheral Au+Au to central d+Au. No scaling is applied. A dashed line is drawn at 0.65 as a visual aid.

deficit of particle production at low p_T and could explain the observed upward trend in this ratio. Additionally, if there is radial flow in central d+Au collisions and it is stronger than that in peripheral Au+Au, that could also explain this trend, with additional benefit of explaining the mass ordering that may be present.

References

[1] Adare A *et al* 2013 *Phys. Rev.* C **88** 024906
[2] Adare A *et al* 2011 *Phys. Rev.* C **83** 064903
[3] Adler S S *et al* 2007 *Phys. Rev. Lett.* **98** 172302
[4] Adare A *et al* 2008 *Phys. Rev. Lett.* **101** 232301
[5] Adare A *et al* 2011 *Phys. Rev.* C **83** 024909
[6] Zhu L and Hwa R C 2013 *Preprint* 1307.3328
[7] Back B B *et al* 2004 *Phys. Rev. Lett.* **93** 082301

Journal of Physics: Conference Series **509** (2014) 012006 doi:10.1088/1742-6596/509/1/012006

Heavy flavours and quarkonia

Raphael Granier de Cassagnac
Laboratoire Leprince-Ringuet, CNRS/IN2P3, École polytechnique, France
E-mail: raphael@in2p3.fr

Abstract.
This contribution aims at reviewing the experimental measurements of heavy flavours and quarkonia, prior to the new results presented at the Strange Quark Matter 2013 conference.

Heavy quarks (charm and beauty) are created in the earliest phase of ultra-relativistic heavy-ion collisions. They thus probe the entire evolution of the produced medium often referred to as the quark-gluon plasma. The large fraction of heavy quarks that are separated from their initial antiquark partner, the so-called *open flavour*, are susceptible to interact with the surrounding medium resulting in elliptic flow and/or energy loss, with possibly different magnitudes for different quarks. The small fraction of heavy quark pairs that usually bound themselves into *quarkonia* possiblly melt in the plasma and could potentially be formed again when the medium freezes out, from uncorrelated pairs. In addition to these *hot* effects, normal *cold* nuclear effects (modifications of the parton distribution functions, absorption by initial nucleons, etc.) can also affect heavy flavours. The variety of influences the medium can have on heavy flavours makes them an interesting but challenging ground to study its properties.

The current results on open charm and beauty are reviewed in sections 1 and 2, followed by the results on charmonia and bottomonia in sections 3 and 4. For each, the RHIC measurements and conclusions are given first, before presenting the new results produced at LHC.

1. Open charm

Two ways are typically used to measure open heavy flavour production. One consists in separating so-called *non-photonic electrons*, coming from a mix of charm and beauty leptonic decays, from the copious background coming from lighter quarks and photon conversions. The advantage of this method lays in the fact that electrons are easy to trigger upon, especially if of high transverse momentum (p_T). The other way consists in measuring hadronic decays of (hence identified) D or B mesons. These topologies (such as $D \to K\pi$) are more difficult to trigger on and usually sit on a large background, but provide an unambiguous access to the meson flavour and kinematics. Obviously, the performances of both techniques are dramatically improved if the position where the decay occurs is measured, which is doable with Silicon vertex detectors since D and B mesons have typical proper decay length of the order of hundreds of micrometers.

In 2004, the PHENIX collaboration released an article [1] showing that the total charm production is essentially scaling with the number of elementary nucleon-nucleon collisions. This was based on non-photonic electrons detected at mid-rapidity and of p_T greater than 300 MeV/c. A more modern version of this result was published in [2] and is shown on Fig. 1, as blue circles. The uncertainly is typically of the order of 25%, mostly due to the lack of a vertex detector.

The STAR experiment carried out similar studies [3] but is particularly well suited to measure hadronic decays. Thanks to the addition of a Time-of-Flight detector and a massive minimum-bias sample, the scaling of open charm with elementary nucleon-nucleon collisions is now also observed with D mesons, through a STAR preliminary result (see, e.g. Ref. [4]).

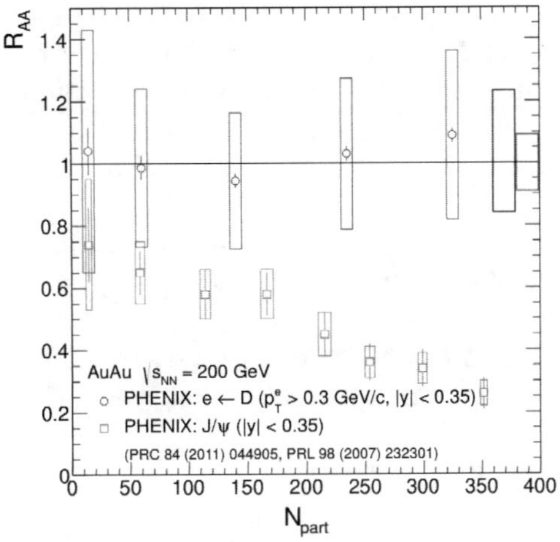

Figure 1. At RHIC, nuclear modification factor of low p_T non-photonic electrons [2] (blue circles) and inclusive J/ψ [5] (red squares) as a function of centrality depicted as the number of participating nucleons.

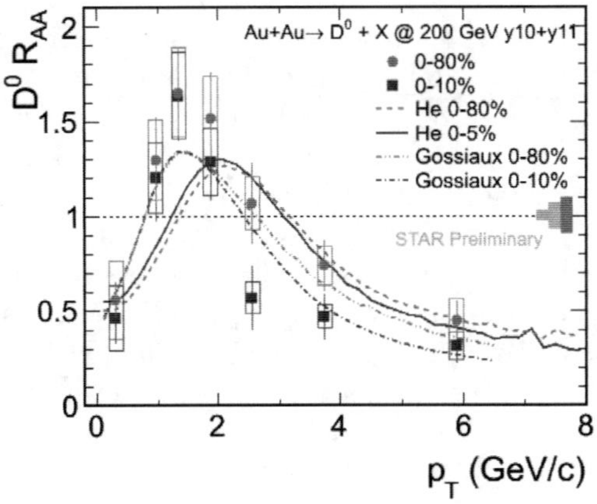

Figure 2. At RHIC, (preliminary) nuclear modification factor of D mesons from the STAR experiment [4], as a function of their transverse momentum and for two centrality selections.

When binned in transverse momentum, the above results show that most of the charm yield is concentrated at low p_T, typically below 2 GeV/c. Concerning D mesons, a preliminary result shows a remarkable feature with a suppression at low p_T and an enhancement in the moderate p_T region (1–2 GeV/c), as shown on Fig. 2 and in Ref. [4], a behaviour that could arise from the radial flow of the light quark contained in the D mesons [6, 7]. For non-photonic electrons, this effect dilutes itself in the leptonic decay and the nuclear modification factor is indeed compatible with unity at low p_T [8]. At higher p_T, all measurements related to open heavy flavours (such as the blue circles of Fig. 3) exhibit a strong suppression which is attributed to heavy quark energy loss. This is further confirmed when comparing AuAu collisions to dAu, in which no suppression, and even an enhancement [9], is observed.

The last remarkable behaviour of charm at RHIC is its strong elliptic flow, observed in non-photonic electrons [8] and revealing a strong coupling of the heavy quarks to the medium. At high p_T, STAR shows that the charm v_2 parameter grows above hydrodynamic expectations [4], a behaviour that can probably be attributed to the path-length dependence of energy loss.

Overall, modeling both the quenching and the flow of charm quark observed at RHIC has become a challenge to theorists.

At LHC, the ALICE experiment has released extensive results concerning open charm, using different techniques at mid-rapidity, in particular D meson measurements via hadronic decay [10], but also from muons at forward rapidity [11]. Other complementary preliminary results can be found in Ref. [12]. A strong suppression is observed over the whole probed p_T range, in particular at 1–3 GeV/c and contrary to RHIC. A possible scaling with the number of elementary nucleon-nucleon collisions is thus not observed but could still lay in the lowest p_T (below 1 GeV/c) that are not yet accessible. It is to be noted that a violation of this scaling could arise from the modification of parton distribution functions (shadowing).

The ALICE experiment also showed that D mesons significantly flow at the LHC [13], and the situation is now the same than at RHIC: Theorists are trying to model the energy loss and the collective behaviour of charm quarks in the medium.

2. Open beauty

Open beauty was marginally accessible at RHIC, though non-photonic electrons at high p_T must come from a mix of charm and beauty decays. At LHC, the CMS experiment provided the first unambiguous b-quark measurements in heavy-ion collisions. The (published) *non-prompt* component of J/ψ mesons [14] and (preliminary) topologically-identified jets from b quarks [15] provide information on the fate of beauty at relatively low and high p_T, respectively. Both show a significant suppression (of the same order of the inclusive jet in the later case), pointing to an important energy loss of b quarks in the medium.

A comparison of the measurements from the CMS and ALICE experiments indicate that D mesons are more suppressed than B mesons, as shown on Fig. 4. This is often quoted as reflecting a stronger energy loss of the c quarks, though a quantitative analysis is yet to be done, with a proper treatment of the quark fragmentation, production spectrum and decay kinematics.

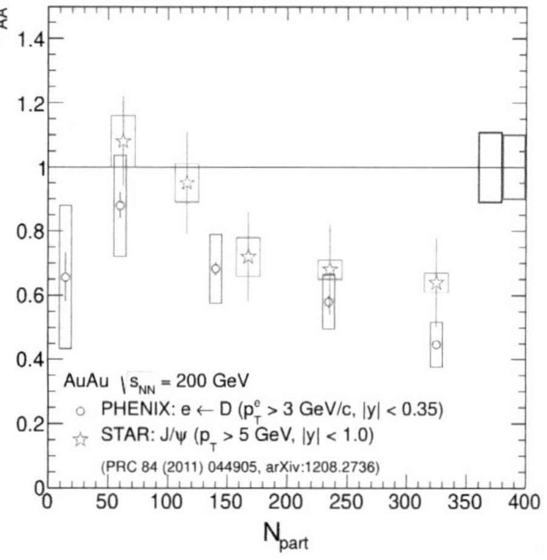

Figure 3. At RHIC, nuclear modification factor of high p_T non-photonic electrons [2] (blue circles) and high p_T J/ψ [16] (green stars) as a function of centrality depicted as the number of participating nucleons.

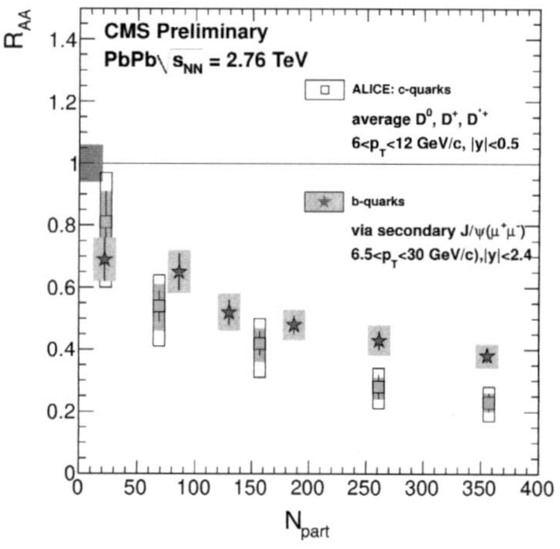

Figure 4. At LHC, nuclear modification factor of D mesons [12] (stars) and non-prompt J/ψ [17] (squares) as a function of centrality depicted as the number of participating nucleons.

3. Charmonia

In 2006, the first J/ψ significant measurements at RHIC brought up two surprises [5] that were later confirmed with more statistics [18] and are illustrated by Fig. 5. First, J/ψ are as suppressed at RHIC (red circles) than at SPS (black cross) [19], at mid rapidity. Second, the suppression is stronger at forward rapidity (blue squares), where the J/ψ should rather see lower densities. Two popular answers were proposed to solve this puzzle. On one hand, J/ψ mesons could be regenerated, especially at mid-rapidity, from initially-uncorrelated c and $\bar{c}$ quarks streaming through the medium and meeting at hadronization. On the other hand, shadowing or saturation could further suppress the forward J/ψ. In this second case, it is possible that directly produced J/ψ survive at both SPS and RHIC, the observed suppression being due to the melting of the higher states decaying into J/ψ (feed-down).

Journal of Physics: Conference Series **509** (2014) 012006 doi:10.1088/1742-6596/509/1/012006

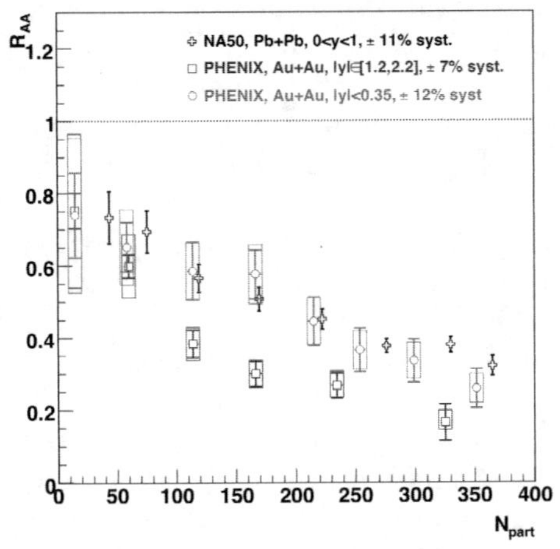

Figure 5. Nuclear modification factor of J/ψ from the first PHENIX [5] and last NA50 [19] publications, as a function of centrality depicted as the number of participating nucleons.

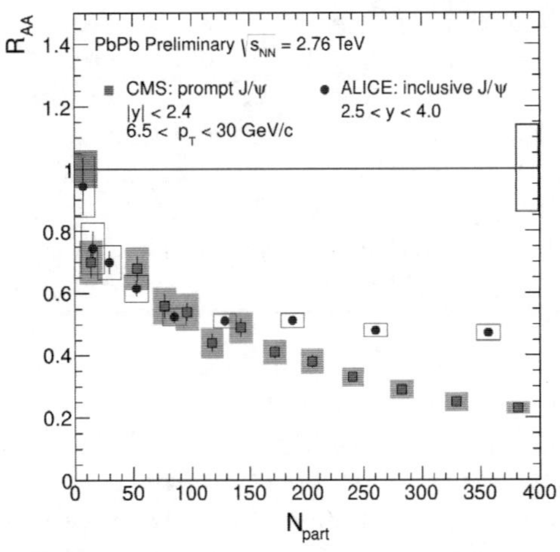

Figure 6. Nuclear modification factor of J/ψ from the ALICE [20] (all p_T) and CMS [17] (prompt, high p_T) preliminary results, as a function of centrality depicted as the number of participating nucleons.

The dAu collisions indeed revealed a similar effect, with more suppression in the forward region [21] but a detailed understanding of the observed yields is still missing. In particular, a simple combination of nuclear shadowing and nuclear absorption does not reproduce the modification factors observed in different rapidity ranges [22]. This makes a quantitative extrapolation of the nuclear effects observed in pA to AA collisions difficult.

An important finding at RHIC is the fact that J/ψ are less suppressed at high p_T, both in AuAu [16] and dAu [23] collisions. For AuAu collisions, this can be seen by comparing PHENIX low p_T J/ψ from Fig. 1 and STAR high p_T J/ψ from Fig. 3. The latter shows that the D and J/ψ R_{AA} somewhat meet for selected p_T thresholds, but this has to happen for an increasing and decreasing suppression as a function of p_T.

There was a hope that the situation will be clearer at LHC, with a centrality-growing modification factor in case of regeneration, and a step-like behaviour in case of sequential suppression, as frequently advertised by a popular but naive cartoon. The suppression of low p_T (inclusive) J/ψ was reported by the ALICE collaboration at forward rapidity and shows a remarkably saturating nuclear modification factor $R_{AA} \simeq 0.5$ ([24] even more remarkable in the new preliminary result with finer centrality binning shown on Fig. 6). At mid-rapidity, statistically-limited results show a similar feature, with a higher R_{AA} [20]. Though the observed plateau reminds the topology of a long-awaited sequential melting, the current interpretation is the opposite: regeneration is at play at LHC! This twist is justified by three related observations: 1) These low p_T J/ψ are *less* suppressed than at RHIC, as can be seen by comparing Figs.5 and 6, while nuclear effects (shadowing) should be larger, 2) High p_T J/ψ are *more* suppressed than at RHIC, as seen for instance by the CMS experiment [14] and displayed also on Fig. 6, 3) A hint of J/ψ elliptic flow is observed [25] which is supposed to arise from inheriting the individual charm quark flow at hadronization time.

In pPb collisions, the adventure is just beginning at LHC, with results from the ALICE [26] and LHCb [27] collaborations showing a J/ψ backward/forward asymmetry, which is compatible with various predictions of cold nuclear matter effects, within (large) uncertainties.

Journal of Physics: Conference Series **509** (2014) 012006 doi:10.1088/1742-6596/509/1/012006

The study of pp collisions at LHC also revealed interesting features. The average yield of J/ψ was shown by the ALICE collaboration to grow with the average multiplicity of the events [28]. In other words, events with more multiplicity also produce more J/ψ mesons on average, which is interpreted as a consequence of multi-parton interaction. A preliminary result also showed this behaviour for D mesons [12] and a comprehensive study of all heavy and hard probes versus pp event multiplicity will be interesting.

Both at RHIC and LHC, the excited ψ' state brought up surprises. At RHIC, the PHENIX collaboration has published its nuclear modification factor in dAu collisions [29], showing a dramatic dependence on event centrality, with a rather strong suppression in the most central collisions. At LHC, the CMS collaboration released a preliminary results showing that, in PbPb collisions, ψ' are more suppressed than J/ψ at high p_T (more than 6.5 GeV/c) but might be less suppressed at lower p_T (down to 3 GeV/c) [30]. This is not observed on a preliminary result by the ALICE collaboration [20], though the two results are not incompatible. These ψ' results call for more statistics. The CMS result in particular suffered from the low statistics of the first pp sample at 2.76 TeV and should be updated with the larger sample.

4. Bottomonia

The available bottomonium statistics were marginal at RHIC, and the three $\Upsilon(1S)$, $\Upsilon(2S)$ and $\Upsilon(3S)$ states unresolved. However a few results exist, such as the nuclear modification factor of the (grouped) three states in dAu collisions from PHENIX [31] and a corresponding preliminary result on AuAu collisions from STAR [4]. They show, within large uncertainties, no sizeable effect in dAu collisions and a significant suppression in AuAu collisions ($R_{AA} \simeq 0.3 - 0.6$).

Thanks to the enhanced production cross section and high luminosity, bottomonia are copiously produced at LHC. A first publication by the CMS collaboration indicated that the excited states ($2S$ and $3S$) were more suppressed than the ground ($1S$) state [32]. It was further confirmed in Ref. [33], which shows a clear sequential pattern in the nuclear modification factors, with $0.56 \pm 0.08 \pm 0.07$, $0.12 \pm 0.04 \pm 0.02$ and less than 0.10 with a 95% confidence level, for the $\Upsilon(1S)$, $\Upsilon(2S)$ and $\Upsilon(3S)$, respectively, as shown on the right sub panel of Fig. 7. Also on this figure is the centrality dependence of the $\Upsilon(1S)$ and $\Upsilon(2S)$, together with a preliminary $\Upsilon(1S)$ result from the ALICE collaboration [20]. With possibly half of the $\Upsilon(1S)$ coming from the decay of higher-mass bottomonia (in particular from χ_b), this result leaves open the possibility that the $\Upsilon(1S)$ ground state survives in the plasma created at LHC.

Five quarkonia are thus now accessible, as shown for instance on Fig. 8 which summarizes the nuclear modification factors observed with the CMS experiment (hence for high p_T ψ and inclusive Υ). Taken as such, the quarkonium suppression shows an ordering as a function of their binding energy, hence materilazing the old dream suggested in the seminal quarkonium suppression paper [34]. If one considers that selecting high p_T charmonia is a way of going away from the regeneration regime, it might indeed be relevant, but it is far too premature to come to such a conclusion and works need to be carried out to disentangle all cold and hot effects, on the five quarkonia.

Conclusions

Heavy flavours have proven themselves to be an interesting and surprising playground at RHIC energies. With the arrival of LHC and the wide opening of the b-quark sector, both for open heavy flavour and bottomonia, a rich phenomenology should arise in the near future.

Acknowledgements

The author received funding from the European Research Council under the FP7 Grant Agreement no. 259612. He wishes to thanks Torsten Dahms, Daniel Kikola and Camelia Mironov for providing figures.

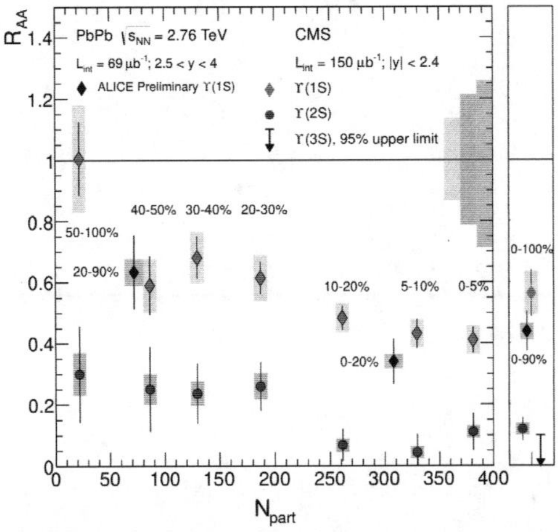

Figure 7. Nuclear modification factor of $\Upsilon(1S)$ (diamonds, CMS in green, ALICE in black [20]) and $\Upsilon(2S)$ (blue circles, CMS [33]) versus centrality depicted as the number of participating nucleons.

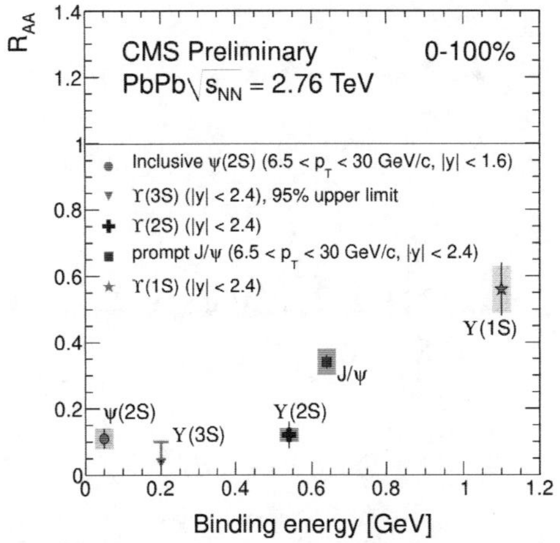

Figure 8. Nuclear modification factors of 5 quarkonia from the CMS experiment [14, 30, 33], in minimum bias collisions and versus their binding energy. Bottomonia are inclusive but charmonia lay at high p_T.

References

[1] Adler S S *et al.* (PHENIX) 2005 *Phys. Rev. Lett.* **94** 082301 (*Preprint* nucl-ex/0409028)
[2] Adare A *et al.* (PHENIX) 2011 *Phys.Rev.* **C84** 044905 (*Preprint* 1005.1627)
[3] Abelev B *et al.* (STAR) 2008 (*Preprint* 0805.0364)
[4] Kikola D (STAR) *These proceedings*
[5] Adare A *et al.* (PHENIX) 2007 *Phys. Rev. Lett.* **98** 232301 (*Preprint* hep-ex/0611020)
[6] He M, Fries R J and Rapp R 2013 *Phys.Rev.Lett.* **110** 112301 (*Preprint* 1204.4442)
[7] Gossiaux P B *et al.* 2012 *PoS* **QNP2012** 160 (*Preprint* 1207.5445)
[8] Adare A *et al.* (PHENIX) 2007 *Phys. Rev. Lett.* **98** 172301 (*Preprint* nucl-ex/0611018)
[9] Adare A *et al.* (PHENIX) 2012 *Phys.Rev.Lett.* **109** 242301 (*Preprint* 1208.1293)
[10] Abelev B *et al.* (ALICE) 2012 *JHEP* **1209** 112 (*Preprint* 1203.2160)
[11] Abelev B *et al.* (ALICE) 2012 *Phys.Rev.Lett.* **109** 112301 (*Preprint* 1205.6443)
[12] Stocco D (ALICE) *These proceedings*
[13] Abelev B *et al.* (ALICE) 2013 *Phys.Rev.Lett.* **111** 102301 (*Preprint* 1305.2707)
[14] Chatrchyan S *et al.* (CMS) 2012 *JHEP* **1205** 063 (*Preprint* 1201.5069)
[15] CMS 2012 *CMS-PAS-HIN-12-003*
[16] Adamczyk L *et al.* (STAR) 2013 *Phys.Lett.* **B722** 55–62 (*Preprint* 1208.2736)
[17] CMS 2012 *CMS-PAS-HIN-12-014*
[18] Adare A *et al.* (PHENIX) 2011 *Phys.Rev.* **C84** 054912 (*Preprint* 1103.6269)
[19] Alessandro B *et al.* (NA50) 2005 *Eur.Phys.J.* **C39** 335–345 (*Preprint* hep-ex/0412036)
[20] Bruno G (ALICE) *These proceedings*
[21] Adare A *et al.* (PHENIX) 2011 *Phys.Rev.Lett.* **107** 142301 (*Preprint* 1010.1246)
[22] Adare A *et al.* (PHENIX) 2008 *Phys.Rev.* **C77** 024912 (*Preprint* 0711.3917)
[23] Adare A *et al.* (PHENIX) 2012 (*Preprint* 1204.0777)
[24] Abelev B *et al.* (ALICE) 2012 *Phys.Rev.Lett.* **109** 072301 (*Preprint* 1202.1383)
[25] Abbas E *et al.* (ALICE) 2013 (*Preprint* 1303.5880)
[26] Abelev B B *et al.* (ALICE) 2013 (*Preprint* 1308.6726)
[27] Aaij R *et al.* (LHCb) 2013 (*Preprint* 1308.6729)
[28] Abelev B *et al.* (ALICE) 2012 *Phys.Lett.* **B712** 165–175 (*Preprint* 1202.2816)
[29] Adare A *et al.* (PHENIX) 2013 (*Preprint* 1305.5516)
[30] CMS 2012 *CMS-PAS-HIN-12-007*
[31] Adare A *et al.* (PHENIX) 2013 *Phys.Rev.* **C87** 044909 (*Preprint* 1211.4017)
[32] Chatrchyan S *et al.* (CMS) 2011 *Phys.Rev.Lett.* **107** 052302 (*Preprint* 1105.4894)
[33] Chatrchyan S *et al.* (CMS) 2012 *Phys.Rev.Lett.* **109** 222301 (*Preprint* 1208.2826)
[34] Matsui T and Satz H 1986 *Phys.Lett.* **B178** 416

Journal of Physics: Conference Series **509** (2014) 012007 doi:10.1088/1742-6596/509/1/012007

Open and Hidden Heavy Flavor Production in pp, pA and AA Collisions

Ramona Vogt
Physics Division, Lawrence Livermore National Laboratory, Livermore, CA 94551, USA
Physics Department, University of California, Davis, CA 95616, USA
E-mail: vogt@physics.ucdavis.edu

Abstract. In these proceedings we present a brief overview of recent open heavy flavor and quarkonium production theory in proton-proton (pp), proton/deuteron-nucleus (pA, dA) and nucleus-nucleus collisions.

1. pp Interactions

1.1. *Open heavy flavor production*

There are currently two approaches to heavy flavor production at colliders: those that employ collinear factorization and the low x k_T-factorization approach. We briefly describe each one.

Complete collinearly-factorized calculations are only available up to next-to-leading (NLO) order in the perturbative expansion. Because of the finite heavy quark mass, it is possible to calculate the total cross section, not possible for processes involving light partons in the final state. However, the 'light' charm quark mass results in large cross section uncertainties [1]. While these can be tamed by fitting the factorization and renormalization scales to data [2], they remain non-negligible. The perturbative expansion works better for the heavier bottom quarks. There are a number of approaches available to calculate distributions. They should all give equivalent total cross sections when integrated over all phase space.

The Fixed-Order Next-to-Leading Logarithm (FONLL) [3] calculation, which includes next-to-leading log resummation at high p_T, only addresses single inclusive production. All other kinematic variables are integrated over. It has been employed successfully to study heavy flavor production at RHIC via both reconstructed D mesons and semileptonic decays of D and B mesons [4]. Semileptonic decays are employed to study single lepton spectra. The decay $B \to J/\psi X$ is also available. Since the heavy quark is treated as light in the FONLL approach, it uses *e.g.* 4 active flavors for charm production instead of 3 so that total cross sections calculated with FONLL, even at fixed order, are typically smaller than in the other approaches.

The other NLO codes calculate exclusive $Q\overline{Q}$ pair production which allows for studies of correlated observables. The first, `HVQMNR` [5], also includes single Q and total $Q\overline{Q}$ production, as well as fragmentation and decays to leptons. `HVQMNR` uses negative weight events used to cancel divergences numerically and which often tend to give a negative value for the cross section in the lowest pair p_T bin. This negative value can be removed by adding smearing the parton momentum through the introduction of intrinsic transverse momenta, k_T. In addition, it is not an event generator and also does not include any resummation. More recent codes employing

NLO matrix elements include complete events. `MC@NLO` [6] is a more general event generator employing NLO matrix elements for a number of hard processes (including $b\bar{b}$ production) while `POWHEG − hvq` [7] is specific to heavy flavor production. `POWHEG − hvq` is a positive weight generator and also includes leading-log resummation. In both codes, the entire event is available since `PYTHIA` [8] and `HERWIG` [9] are employed to produce the complete event.

In all cases, a theoretical uncertainty band is obtained by varying the quark mass with fixed scales and varying the scales in some prescribed way for fixed mass and adding the differences in quadrature [4, 2, 10]. The single particle observables calculated using collinear factorization are all in good agreement with the RHIC and LHC data and with each other.

In the k_T-factorization approach, off-shell leading order matrix elements for $g^*g^* \to c\bar{c}$ together with unintegrated gluon densities that depend on the transverse momentum of the gluon, k_T, as well as the usual dependence on x and μ_F. The unintegrated gluon distributions are normalized so that $g(x,\mu_F^2) = \int_0^{\mu_F^2} dk_T^2 f_g(x,k_T^2,\mu_F^2)$. There are a variety of unintegrated gluon densities on the market but only a few give shapes consistent with the D^0 distributions at $\sqrt{s} = 7$ TeV. There is only a small enhancement in the charm quark p_T distribution at low p_T over the FONLL result. This approach has also been used to calculate correlated $c\bar{c}$ production at forward rapidity, as measured by LHCb and is in generally rather good agreement with the data [11]. The results with collinear factorization were not included in the comparison.

1.2. Quarkonium Production

The simplest approach to quarkonium production is the color evaporation model (CEM) which treats heavy flavor and quarkonium production on an equal footing. In the CEM, the quarkonium production cross section is some fraction of all $Q\overline{Q}$ pairs below the $H\overline{H}$ threshold where H is the lowest mass heavy-flavor hadron. Thus the CEM cross section is simply the $Q\overline{Q}$ production cross section with a cut on the pair mass but without any constraints on the color or spin of the final state. The color of the octet $Q\overline{Q}$ state is evaporated through an unspecified process which does not change the momentum. The additional energy needed to produce heavy-flavored hadrons when the partonic center of mass energy, $\sqrt{\hat{s}}$, is less than $2m_H$, the $H\overline{H}$ threshold energy, is nonperturbatively obtained from the color field in the interaction region. The quarkonium yield is only a fraction of the total $Q\overline{Q}$ cross section below $2m_H$. The actual fraction depends on the heavy quark mass, the scale parameters, the parton densities and the order of the calculation [12]. For the most recent results of the CEM on J/ψ production, see Ref. [2]. CEM calculations do not currently address quarkonium polarization.

Two often used models of quarkonium production depend on the color of the $Q\overline{Q}$ state at production. The color singlet model (CSM) assumes perturbative production of an on-shell Q and $\overline{Q}$ at the final state scale. The color and spin of the pair is unchanged by binding. The LO CSM calculations predicted that χ_c production should dominate high p_T J/ψ production. Further refinements with higher order contributions, up to a partial NNLO result, give better agreement at high p_T. The non-relativistic QCD (NRQCD) approach is based on a systematic expansion of the quarkonium wavefunction in the strong coupling constant and the relative velocity of the Q and $\overline{Q}$. The velocity expansion is assumed to work better for Υ production. It assumes factorization of the perturbative contribution and non-perturbative hadronization and includes both singlet and octet matrix elements. The octet matrix elements are fit parameters determined from comparison to quarkonium p_T distributions. For a recent review of quarkonium production, see Ref. [13].

Most recently, several groups have attempted to fix the octet matrix elements from a global fit to inclusive J/ψ data from RHIC, the Tevatron, the LHC and, in some cases, HERA [14, 16, 17]. By fitting to a global data set, they attempt to obtain a set of universal matrix elements. The NLO results are obtained by fixing the charm quark mass and scale; fitting the matrix elements with these parameters; and determining the uncertainty on the fit by varying the scale parameters around the central value while keeping the mass and matrix elements fixed. There are some caveats to these analyses: they are limited to prompt J/ψ only; feed down from ψ' and χ_c is either neglected or subtracted assuming that the p_T shapes are the same for all quarkonium states (like the CEM); fixed-target results are not included; and the dependence of the matrix elements on mass and scale is neglected [14]. That said, the results show that octet components are necessary to obtain agreement with the data.

The crucial test of the quarkonium production models is the polarization. In NRQCD, the dominant production mechanism is gluon fragmentation into a color octet. Since this gluon is nearly on mass shell, it is transversely polarized. This polarization should be reflected in the final-state quarkonium distribution. At NLO the CSM result is longitudinally polarized. Neither approach is in agreement with the J/ψ polarization data [13, 15]. Most recently, the CMS Υ data [18] shows no significant polarization, in contradiction to both results. Before the issue of quarkonium production in pp collisions can be settled, the polarization results must be reconciled.

2. p/d $+A$ Interactions

There are several important cold nuclear matter (CNM) effects that need to be taken into account when determining the strength of deconfinement effects on a particular final state. We first discuss mechanisms common to both open and hidden charm and then discuss final-state absorption of quarkonium.

The most general CNM effect, affecting all production processes, is the modification of the parton distributions in nuclei, often referred to as shadowing. Since nuclear deep inelastic scattering experiments probe only charged parton densities, the nature and magnitude of the effect on the nuclear gluon density is not well known. One possible experimental means of probing the nuclear gluon density at low x is through ultra-peripheral collisions (UPCs) at the LHC [19]. In these collisions, the nuclei do not touch and only interact electromagnetically so that J/ψ photoproduction involves the low x gluon density in a single nucleus. The ALICE UPC J/ψ data shows that this method can eliminate certain shadowing parameterizations [20, 21].

The effects of shadowing in nuclei are parameterized by various groups using global fitting methods similar to those used to evaluate the parton densities in the proton. Currently, LO and NLO sets are available with up to 31 error sets, such as EPS09 [22], evolving quarks, antiquarks and gluons separately with Q^2. The uncertainty on the gluon density in the nucleus remains large without general agreement on the best fit shape.

Another significant unknown relating to nuclear shadowing is its dependence on impact parameter or collision centrality. The impact parameter dependence was neglected in most previous parameterizations. Instead, assumptions based on either a linear dependence on path length through the nucleus or the nuclear density were introduced [23]. The PHENIX d+Au J/ψ data suggested a stronger than linear dependence [24]. These results prove challenging for the recent EPS09s spatially-dependent modifications which retain up to quartic powers in the expansion of the centrality dependence as a function of path length for A-independent coefficients [25]. Instead these data suggest that shadowing is concentrated in the core of the nucleus with radius of $R \sim 2.4$ fm with a relatively sharp surface, a width of $d \sim 0.12$ fm [26]. These studies need to be backed up with more data over more final states.

A second cold matter effect is energy loss in medium. Initial-state energy loss has been studied in Drell-Yan production and found to be small. However, there is an inherent ambiguity when applying initial-state energy loss to Drell-Yan production since most groups parameterizing the nuclear parton densities include these same Drell-Yan data to extract the strength of shadowing on the antiquark densities [22]. Also, by forcing the loss to be large enough to explain the high x_F behavior of J/ψ production in fixed-target interactions [27] violates the upper bound on energy loss established by small angle forward gluon emission [28]. More recently, it has been proposed that cold matter energy loss should be treated as a final-state effect [29]. The final-state J/ψ energy loss in pA collisions is currently implemented as a probability distribution dependent on the energy loss parameter. The effect modifies the x_F and p_T distributions in a rather crude fashion since the quarkonium distribution in pp collisions is parameterized as a convolution of factorized power laws, $\propto (1-x)^n (p_0^2/(p_0^2+p_T^2))^m$, rather than using a quarkonium production model [29, 30]. It has yet to be implemented for other processes.

Initial-state energy loss in the medium can be connected to transverse momentum broadening in nuclei relative to pp collisions, as in the Cronin effect [31].

Final-state nuclear absorption, which affects only quarkonium states, involves CNM breakup of the (proto)quarkonium state. Absorption is related to the size and production mechanism of the interacting state and can be described by a survival probability, $S_A^{\rm abs} = \exp\{-\int_z^\infty dz' \rho_A(b,z') \sigma_{\rm abs}^C(z-z')\}$ where z' is the production point and z is the dissociation point; $\rho_A(b,z')$ is the nuclear matter density; and $\sigma_{\rm abs}^C$ is the effective absorption cross section for quarkonium state C [32]. The J/ψ has been most studied. Larger effects, at least at midrapidity, have been seen for the ψ' [27, 33]. Such effects on Υ production may also be expected with stronger nuclear effects on the $\Upsilon(2S)$ and $\Upsilon(3S)$ relative to the $\Upsilon(1S)$ [34] although the overall absorption should be reduced.

Previous studies have shown the absorption cross section to depend on rapidity (or x_F) as well as the nucleon-nucleon center of mass energy, $\sqrt{s_{NN}}$ with stronger absorption at lower energies [35]. Increased effective absorption at backward rapidity may be due to interaction or conversion inside the target while increased effective absorption at forward rapidity may be due to energy loss.

All these cold matter effects should be reduced for bottom quark and Υ production due to the larger masses and scales involved.

3. AA Interactions

3.1. *Open Heavy Flavor*

Most open heavy flavor AA results are focused on energy loss in the medium and typically rely on thermalization and hydrodynamic evolution. The differences are in the details of the calculation. For example, both He *et al* [36] and Alberico *et al* [10] use a Langevin approach. However, the former relates the drag coefficient to the heavy quark relaxation rate, calculated using in-medium heavy-light quark resonant rescattering in a T-matrix model and includes recombinant production of D mesons while in the latter the transport coefficients includes both soft (obtained either in the hard thermal loop approximation or from lattice QCD) and hard (calculated perturbatively) contributions. Both are compared to the nuclear modification factor R_{AA} and to the elliptic flow coefficient v_2. While both can obtain general agreement with R_{AA} at high p_T, without recombination, not enough flow is generated to reach agreement with v_2 [36, 10]. A number of other talks at this conference covered related calculations employing energy loss in the medium, see Refs. [37, 38, 39, 40, 41, 42, 43, 44, 45].

A recent paper has addressed b-jet quenching in Pb+Pb collisions [46]. The b-quark jets are generated in PYTHIA via three mechanisms: standard LO $b\bar{b}$ production diagrams ($gg \to b\bar{b}$, $q\bar{q} \to b\bar{b}$); gluon splitting into $b\bar{b}$ pairs in the final state (*e.g.* $gg \to gg$, $g \to b\bar{b}$); and other standard LO jet production processes *e.g.* $qq \to qq$ followed by fragmentation. The gluon splitting contribution dominates b-jet production at high p_T unless the b quark is the leading particle in the jet. The results depend on the cone size (larger cone radii reduces suppression), inclusion of collisional dissipation (increases suppression), in-medium coupling (greater suppression with larger coupling), and mass of propagating parton (increase uncertainty in R_{AA} at low p_T). Cold nuclear matter effects are small. The results are in agreement with very preliminary CMS data, see Ref. [46] for details.

Journal of Physics: Conference Series **509** (2014) 012007 doi:10.1088/1742-6596/509/1/012007

3.2. Quarkonium

3.2.1. Lattice-based results The lattice results presented here were summarized in a recent review [47]. Please refer to that work and Allton's presentation [48] for more details.

A comparison of singlet and octet free energies shows that the temperature dependence of the total free energy (octet and singlet) is much stronger than for the singlet alone. This is consistent with recent calculations of quarkonia spectral functions and correlators which show that no bound states persist in the medium. It is also consistent with effective field theory calculations with separation of scales that depend on the relative size of the binding energy of the quarkonium state (BE) and the temperature of the system (T). If BE $> T$, the heavy quark potential is unmodified by the medium. However, the bound state acquires a finite thermal width. If BE $< T$, the singlet and octet potentials become temperature dependent and acquire an imaginary part since gluons exchanged in octet-singlet transitions scatter from thermal excitations in the medium. With increasing temperature, BE $\rightarrow 0$ and medium effects are incorporated into the potential. Now the separation of scales fails and lattice results are required to constrain the potential. See Ref. [47] for a full description and more references.

One way to define the quarkonium dissociation temperature is the point at which the real and imaginary parts of the binding energy are equal. However, since the state undergoes in-medium decays below the defined dissociation temperature, its actual value is not all that important. This in-medium decay rate is directly proportional to the imaginary part of the potential [47].

Another important recent development is the use of viscous hydrodynamics and anisotropic systems to describe quarkonium propagation in the medium. Typical hydrodynamic calculations assume that the system is close to thermal equilibrium and is thus also isotropic in momentum space. However, large initial momentum anisotropies can persist throughout the lifetime of the medium which can increase the dissociation temperature since the areas of anisotropy can lead to reduced Debye screening. In calculations, the anisotropy is accounted for by introducing an anisotropy parameter related to the ellipticity of the momentum distribution. In such calculations the dissociation temperature, defined as the point at which the real and imaginary parts of the binding energy are equal, can increase as much as 25% [47]. In recent calculations of Υ suppression, the screening mass was taken to be dependent on the anisotropy of the medium. The value of $4\pi\eta/S$ was tuned to achieve good agreement with the CMS Υ data as a function of the number of participants, $N_{\rm part}$. The value obtained, $4\pi\eta/S = 3$ [49], is consistent with IP-Sat light particle flow results [50].

3.2.2. Normalizing quarkonium suppression Finally, we briefly address the question of how to define quarkonium suppression in AA collisions. Recently it was argued that the ratio of hidden to open heavy flavor is a better determination of true quarkonium suppression than the value of R_{AA}, the ratio of *e.g.* hidden charm in AA to pp collisions [51]. This is because similar suppression patterns are seen for D mesons/semileptonic decays of heavy flavor mesons and J/ψ at intermediate and high p_T at the LHC and at high p_T at RHIC. The only apparent difference in the suppression patterns is at low p_T at RHIC where R_{AA} for J/ψ is lower than that for non-photonic electrons, assumed to be predominantly from D decays in this p_T range [51]. At intermediate and high p_T, one may assume that the p_T range is outside the region where Debye screening dominates J/ψ suppression and a similar energy loss mechanism is responsible for the suppression pattern for both J/ψ and open charm. Full p_T phase space comparisons are needed.

So far the LHC data are in somewhat different rapidity regions and there is no low p_T point of comparison, as at RHIC. The data shown in Ref. [51] are all as a function of $N_{\rm part}$. Comparisons of D meson and J/ψ suppression as a function of p_T in different centrality bins would be useful. A similar study of B meson and Υ family suppression would also be enlightening because here ATLAS and CMS can reconstruct $p_T = 0$ Υ states so a comparison at low p_T may be possible, as long as the B decays can be identified.

This work was performed under the auspices of the U.S. Department of Energy by Lawrence Livermore National Laboratory under Contract DE-AC52-07NA27344, and supported in part by the JET collaboration.

References

[1] Vogt R 2008 *Eur. Phys. J. ST* **155** 213
[2] Nelson R E, Vogt R and Frawley A D 2013 *Phys. Rev.* C **87** 014908
[3] Cacciari M, Greco M and Nason P 1998 *JHEP* **9805** 007 Cacciari M, Frixione S and Nason P 2001 *JHEP* **0103** (2001) 006
[4] Cacciari M, Nason P and Vogt R 2005 *Phys. Rev. Lett.* **95** 122001
[5] Mangano M L, Nason P, and Ridolfi G 1992 *Nucl. Phys.* B **373** 295
[6] Frixione S and Webber B R 2002 *JHEP* **0206** 029; Frixione S, Nason P and Webber B R 2003 *JHEP* **0308** 007
[7] Frixione S, Nason P, and Ridolfi G 2007 *JHEP* **0709** 126; arXiv:0707.3081 [hep-ph].
[8] Sjostrand T *et al.* 2001 *Comput. Phys. Commun.* **135** 238; Sjostrand T *et al.* 2003 arXiv:hep-ph/0308153.
[9] Corcella G *et al.* 2001 *JHEP* **0101** 010; arXiv:hep-ph/0210213.
[10] Alberico W M *et al.* 2013 *Eur. Phys. J.* C **73** 2481
[11] Maciula R and Szczurek A 2013 *Phys. Rev.* D **87** 094022
[12] Gavai R *et al.* 1995 *Int. J. Mod. Phys.* A **10** 3043
[13] Brambilla N *et al.* 2011 *Eur. Phys. J.* C **71** 1534
[14] Butenschoen M and Kniehl B A 2011 *Phys. Rev.* D **84** 051501
[15] Butenschoen M and Kniehl B A 2012 *Phys. Rev. Lett.* **108** 172002
[16] Ma Y-Q, Wang K and Chao K-T 2011 *Phys. Rev. Lett.* **106** 042002
[17] Sharma R and Vitev I 2013 *Phys. Rev.* C **87** 044905
[18] Chatrchyan S *et al.* [CMS Collaboration] 2013 *Phys. Rev. Lett.* **110** 081802
[19] Baltz A J *et al.* 2008 *Phys. Rept.* **458** 1
[20] Abelev B *et al.* [ALICE Collaboration] 2013 *Phys. Lett.* B **718** 1273
[21] Abbas E *et al.* [ALICE Collaboration] 2013 arXiv:1305.1467 [nucl-ex].
[22] Eskola K J, Paukkunen H and Salgado C A 2009 *JHEP* **0904** 065
[23] Emel'yanov V, Khodinov A, Klein S R and Vogt R 1998 *Phys. Rev. Lett.* **81** 1801
[24] Adare A *et al.* [PHENIX Collaboration] 2011 *Phys. Rev. Lett.* **107** 142301
[25] Helenius I, Eskola K J, Honkanen H and Salgado C A 2012 *JHEP* **1207** 073
[26] McGlinchey D C, Frawley A D and Vogt R 2013 *Phys. Rev.* C **87** 054910
[27] Leitch M J *et al.* [E866/NuSea Collaboration] 2000 *Phys. Rev. Lett.* **84** 3256
[28] Brodsky S J and Hoyer P 1993 *Phys. Lett.* B **298** 165
[29] Arleo F and Peigne S 2013 *JHEP* **1303** 122
[30] Arleo F, Kolevatov R, Peigne S and Rustamova M 2013 *JHEP* **1305** 155
[31] Kluberg L *et al.* 1977 *Phys. Rev. Lett.* **38** 670
[32] Vogt R 2002 *Nucl. Phys.* A **700** 539
[33] Adare A *et al.* [PHENIX Collaboration] 2013 arXiv:1305.5516 [nucl-ex]
[34] Alde D M *et al.* 1991 *Phys. Rev. Lett.* **66** 2285
[35] Lourenco C, Vogt R and Woehri H K 2009 *JHEP* **0902** 014
[36] He M, Fries R J and Rapp R 2012 *Phys. Rev.* C **86** 014903
[37] Gousset T 2013 these proceedings
[38] Horowitz W A 2013 these proceedings
[39] Nahrgang M 2013 these proceedings
[40] Das S K 2013 these proceedings
[41] Berrehrah 2013 these proceedings
[42] Uphoff J 2013 these proceedings
[43] Renk T 2013 these proceedings
[44] Gossiaux P B 2013 these proceedings
[45] Djordjevic M 2013 these proceedings
[46] Huang J, Kang Z-B and Vitev I 2013 arXiv:1306.0909 [hep-ph]
[47] Mocsy A, Petreczky P and Strickland M 2013 *Int. J. Mod. Phys.* A **28** 1340012
[48] Allton C 2013 these proceedings
[49] Strickland M 2013 *AIP Conf. Proc.* **1520** 179
[50] Gale C *et al.* 2013 *Phys. Rev. Lett.* **110** 012302
[51] Satz S 2013 *Adv. High Energy Phys.* **2013** 242918

Journal of Physics: Conference Series **509** (2014) 012008 doi:10.1088/1742-6596/509/1/012008

ALICE results on quarkonium production in pp, p-Pb and Pb-Pb collisions

Giuseppe Eugenio Bruno for the ALICE Collaboration
Dipartimento Interateneo di Fisica "M. Merlin" and Sezione INFN, Bari, Italy
E-mail: Giuseppe.Bruno@ba.infn.it

Abstract. The study of quarkonia, bound states of heavy (charm or bottom) quark-antiquark pairs such as the J/ψ or the Υ, provides insight into the earliest and hottest stages of high-energy nucleus-nucleus collisions where the formation of a Quark-Gluon Plasma is expected. High-precision data from proton-proton collisions represent an essential baseline for the measurement of nuclear modifications in nucleus-nucleus collisions and serve also as a crucial test for models of quarkonium hadroproduction. Another fundamental tool to understand the quarkonium production in nucleus-nucleus collisions is the the study of proton-nucleus interactions, which allows one to investigate cold nuclear matter effects, such as parton shadowing or gluon saturation. The ALICE detector provides excellent capabilities to study quarkonium production at the Large Hadron Collider at both central and forward rapidity. An overview on ALICE results on quarkonium production in pp, p–Pb and Pb–Pb collisions is presented. Results are compared to theoretical model predictions.

1. Introduction

The ALICE experiment [1] studies proton-proton (pp), proton-nucleus (p–A) and nucleus-nucleus (A–A) collisions at the Large Hadron Collider (LHC), with the main goal of investigating the properties of the high-density, colour-deconfined state of strongly-interacting matter (the Quark-Gluon Plasma, QGP) that is expected to be formed in high-energy nuclear collisions. The suppression of quarkonium states was proposed long ago [2] as a signature of the formation of the QGP: in a deconfined medium, the binding potential between the constituents of a quarkonium state, a heavy quark (Q) and its antiquark ($\bar{\mathrm{Q}}$), is screened by the colour charges of the surrounding light quarks and gluons. In this scenario, quarkonium suppression occurs sequentially, according to the binding energy of each meson: strongly bound states, as the J/ψ and $\Upsilon(1S)$, should melt at higher temperatures with respect to the more loosely bound ones, as the $\psi(2S)$ and χ_c, for the charmonium family, or the $\Upsilon(2S)$ and $\Upsilon(3S)$ for the bottomonium one. The in-medium modification of quarkonium production is usually quantified by the nuclear modification factor: $R_{AA} = \frac{d^2N_{AA}/dp_T dy}{\langle N_{coll}\rangle\, d^2N_{pp}/dp_T dy}$, where $d^2N/dp_T dy$ denotes the transverse momentum (p_T) and rapidity (y) differential yield of a given particle measured in A–A or pp collisions and $\langle N_{coll}\rangle$ is the average number of nucleon-nucleon collisions over the given centrality interval of A–A collisions; $\langle N_{coll}\rangle$ is calculated using the Glauber model [3]. However, there are further possible effects on the quarkonium production in heavy-ion collisions. On the one hand, the large number of heavy quarks produced in heavy-ion collisions, in particular in the charm sector at the energies accessible by the LHC, may lead to an increased production of charmonia via (re-)combination in the QGP [4, 5] or at the phase boundary [6, 7]. On the other hand, modifications to the parton distribution functions inside the nucleus (shadowing) and other cold nuclear matter effects can change the production of quarkonia without the presence of a QGP [6, 8]. Hence, for the interpretation of A–A results, data from p–A collisions are crucial because they allow us to disentangle these cold nuclear matter effects from those related to the formation of a hot QCD medium. Finally, besides being the natural reference for A–A studies, quarkonium hadro-production in elementary pp collisions are important *per se*, representing a challenging testing ground for models based on QCD. ALICE results on quarkonium production in pp are therefore also summarized in the following.

2. Experiment, data taking conditions and analysis

The ALICE experiment consists of a central barrel embedded in a solenoidal magnet and a forward muon spectrometer. Details on the experimental set-up can be found in [1]. The measurement of quarkonium production is carried out in the central barrel ($|y| < 0.9$) through their e^+e^- decay, while at forward rapidity ($2.5 < y < 4$) the $\mu^+\mu^-$ decay is studied in the muon spectrometer.

For the measurement in the central barrel described hereafter, events are triggered by a minimum bias (MB) condition, which corresponds roughly to the presence of at least one charged particle in about 8 units of pseudo-rapidity. This condition is defined using the information from the two innermost pixel layers of the silicon Inner Tracking System (ITS) ($|\eta| < 2$ and $|\eta| < 1.4$) and from two arrays of scintillators (VZERO) placed at forward rapidities ($-3.7 < \eta < -1.7$ and $2.8 < \eta < 5.1$). In Pb-Pb, several centrality classes can be further defined at the trigger level by means of thresholds on the total signal amplitude in the VZERO. For the measurement with the forward muon spectrometer one can require, in addition to the MB condition, one or two candidate muons to be detected by the muon triggering system.

For Pb–Pb collisions at central rapidity, the analysis is based on a combination of 2010 data, taken with a MB trigger, and 2011 data where, in addition, centrality selections corresponding to central events (0–10%) and semi-central events (10–40%) were performed. The integrated luminosity is $L_{\rm int} = 15\ \mu b^{-1}$. The Pb–Pb forward rapidity results refer to the 2011 data ($L_{\rm int} = 70\ \mu b^{-1}$), and were collected by requiring a dimuon trigger, which can select both opposite- and like-sign pairs. In order to define $R_{\rm AA}$, one needs a sample of pp data collected at the same energy ($\sqrt{s} = 2.76$ TeV) and in the same kinematic domain of Pb-Pb data. Such data were taken in 2011, collecting $L_{\rm int} = 1.1\ {\rm nb}^{-1}$ with a MB trigger for the central rapidity analysis and $L_{\rm int} = 20\ {\rm nb}^{-1}$ with a single-muon trigger for the forward rapidity analysis. Moreover, pp data at $\sqrt{s} = 7$ TeV, the top energy of the 2010 and 2011 LHC data taking, were also collected and analysed, the results discussed here referring to integrated luminosities up to 100 ${\rm nb}^{-1}$.

Data were collected in 2013 for p-Pb collisions using, for the quarkonium studies with the muon spectrometer, the dimuon trigger in coincidence with the MB trigger. Due to the energy asymmetry of the LHC beams ($E_{\rm p} = 4$ TeV, $E_{\rm Pb} = 1.58\,$A TeV) the nucleon-nucleon center-of-mass system of the collisions does not coincide with the laboratory system, but is shifted by $\Delta y = 0.465$ in the direction of the proton beam. Data have been taken with two beam configurations, by inverting the direction of the orbits of the two particle species. In this way in the dimuon channel the regions $2.03 < y_{\rm cms} < 3.53$ and $-4.46 < y_{\rm cms} < -2.96$ have been studied, where positive rapidities refer to the situation where the proton beam is travelling towards the muon spectrometer (in the following these configurations are referred to as p-Pb and Pb-p, respectively). The integrated luminosities of the analysed data for the two configurations are about 5 ${\rm nb}^{-1}$ (p-Pb) and 6 ${\rm nb}^{-1}$ (Pb-p). For the central barrel studies, about 130 M MB events (i.e. an integrated luminosity of about 50 μb^{-1}) have been collected. In addition, other rarer trigger data sample, of interest for charmonia studies, e.g. that requiring an electron in the Transition Radiation Detector (TRD), corresponding to $L_{\rm int} \approx 1.5\ {\rm nb}^{-1}$ are also available.

These data samples are being analysed at the time of the conference.

In the central barrel the analyses are mainly based on the charged track reconstruciton performed with the ITS and Time Projection Chamber (TPC) detectors, on the electron identification via the measurement of the specific energy loss in the TPC and on the auxilary information from the Time Of Flight (TOF) detector and the TRD for the rejection of hadrons. Thanks to the excellent spatial resolution of the ITS, the non-prompt J/ψ contribution can be separated, thus allowing a measurement of beauty production down to very low $p_{\rm T}$ ($p_{\rm T}$(J/ψ) $\approx$ 1.5 GeV/c). Details of the analyses are discussed in [9]. In the muon channel, opposite sign (OS) muons, which have been filtered by the hadron absorber and reconstructed in the muon spectrometer, are paired to obtain invariant mass distributions with clean mass peaks for the quarkonium states. Quarkonium signals are extracted by means of fit to the OS mass distribution, using a phenomenological shape for the background, and Crystal Ball shape for the signals. Further details can be found in [10, 11, 12]. Examples of the dileptons invariant mass distributions obtained in Pb–Pb and p–Pb collisions are shown in Fig. 1.

3. Results

3.1. pp

ALICE has measured the production cross section of inclusive J/ψ at $\sqrt{s} = 7$ and 2.76 TeV [13, 14]. At $\sqrt{s} = 2.76$ TeV a $p_{\rm T}$ differential study was possible only in the forward region, due to the rather small MB integrated luminosity. At central rapidity, the fraction of J/ψ from beauty hadron decay could be separated at low $p_{\rm T}$ [15], which also led to an estimation of the total $b\bar{b}$ cross section at $\sqrt{s} = 7$ TeV. The first LHC study of the inclusive J/ψ polarization has been performed at forward rapidity, showing that the J/ψ production is essentially unpolarized up to $p_{\rm T} = 8$ GeV/c [16]. The comparisons of these results with NLO NRQCD calculations that include color octet processes show good agreement. We have also performed a unique measurement of the J/ψ yield as a function of the charged particle multiplicity [17, 18], which shows an almost linear increase of the yield with the multiplicity. The latter result may either indicate that J/ψ production in pp is connected with strong hadronic activity, or that multiparton interactions could also affect the harder momentum scales relevant for quarkonium production.

Preliminary results on Υ production in pp collisions at $\sqrt{s} = 7$ TeV have also been delivered [11], which are in good agreement with the published LHCb results [19].

Reference pp data at the same energies of the p–Pb ($\sqrt{s_{\rm NN}} = 5.02$ TeV) and Pb–Pb ($\sqrt{s_{\rm NN}} = 2.76$ TeV) interactions are, respectively, not available and of limited statistics (a few days of data taking was devoted to the study of pp collisions at $\sqrt{s} = 2.76$ TeV). Therefore, interpolation procedures are often introduced (except for the forward rapidity J/ψ results at $\sqrt{s} = 2.76$ TeV) to estimate the production cross section of quarkonium states at these energies, which lead to large uncertainties.

3.2. Pb–Pb

The inclusive J/ψ nuclear modification factor as a function of centrality, $p_{\rm T}$ and y in Pb-Pb collisions at $\sqrt{s_{\rm NN}} = 2.76$ TeV, has been measured down to zero $p_{\rm T}$ [9]. At forward rapidity, $R_{\rm AA}$ shows a clear suppression of the J/ψ yield, with no significant dependence on centrality for $\langle N_{\rm part}\rangle$ larger than 70 (left panel of Fig. 2). At mid rapidity the J/ψ $R_{\rm AA}$ is compatible with a constant value (middle panel of Fig. 2). At forward rapidity the J/ψ $R_{\rm AA}$ exhibits a strong $p_{\rm T}$ dependence and decreases by a factor of 2 from low $p_{\rm T}$ to high $p_{\rm T}$. This behavior shows a striking difference with the one observed by the PHENIX experiment at RHIC energy [21], which measured a larger suppression at low $p_{\rm T}$ than that measured by ALICE at the LHC. The measurement of the inclusive J/ψ elliptic flow at forward rapidity [22] has provided an indication of non-zero v_2 in semi-central Pb–Pb collisions. Both results are in agreement with the global

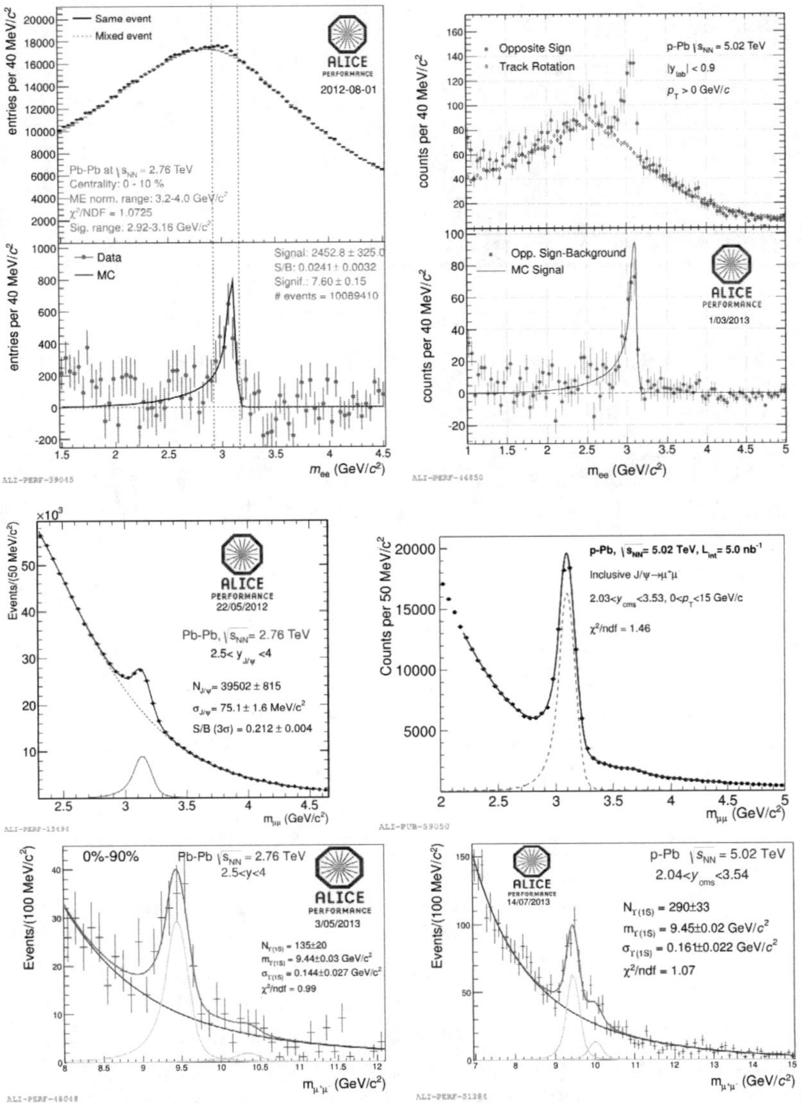

Figure 1. Invariant mass distributions of opposite sign electron (top panels) or muon (middle and bottom panels) pairs in Pb–Pb (left panels) and p–Pb (right panels) collisions. In the dielectron channel the combinatorial background is described with the event mixing or the track rotation methods and subtracted from the same-event opposite-sign distributions.

picture in which a significant fraction of the observed J/ψ is produced from (re-)combination of charm quarks in the QGP phase or at the phase boundary.

The fraction of non-prompt J/ψ ($f_{\rm B}$) has been measured at central rapidity in the range $2 < p_{\rm T} < 10$ GeV/c [9]. No significant dependence of $f_{\rm B}$ on centrality could be determined. Considering the ALICE and CMS results [23] together, an indication for a similar trend of $f_{\rm B}$ as a function of $p_{\rm T}$ in pp and Pb–Pb collisions is observed. The preliminary results on the ψ(2S) are discussed in [20]. The nuclear modification factor of the Υ(1S) has been measured at forward rapidity ($2.5 < y < 4.0$) for $p_{\rm T} > 0$ [11]. Within the uncertainty a similar suppression as that measured for inclusive J/ψ is observed. The suppression is stronger for central than

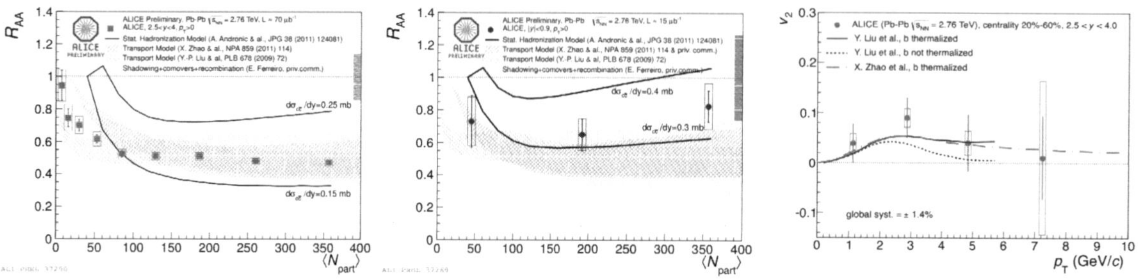

Figure 2. Inclusive J/ψ $R_{\rm AA}$ measured in Pb–Pb collisions at $\sqrt{s_{\rm NN}} = 2.76$ TeV at forward (left panel) and central (middle pannel) rapidity and inclusive J/ψ $v_2(p_{\rm T})$ for semi-central Pb-Pb collisions (right panel) compared to theoretical models.

semi-peripheral collisions. Combining this measurement with that of CMS [24] performed for $|y| < 2.4$, no rapidity dependence of the Υ(1S) suppression is observed within the large rapidity range probed by the two experiments.

3.3. p–Pb

The nuclear modification factor in p-Pb collisions, $R_{\rm pPb}$, has been measured in the muon spectrometer for the J/ψ and Υ(1S) states [11, 12]. Since no pp collisions at $\sqrt{s} = 5.02$ TeV have been delivered, the pp references are obtained interpolating the results collected at higher and lower energies, and these pp references introduce an important contribution to the total uncertainty on $R_{\rm pPb}$, especially for the J/ψ. The variable $R_{\rm FB}$, which is defined as the ratio of the J/ψ (and similarly for the Υ) forward to backward yields measured in the common J/ψ rapidity range $2.96 < |y_{\rm cms}| < 3.53$, has been introduced to avoid the need of the pp reference, and its relative large uncertainty. The drawback of this observable, which is measured in a restricted rapidity interval, is a minor sensitivity to discriminate among models that can describe the cold nuclear matter effects, with respect to $R_{\rm pPb}$.

The nuclear modification factors for the J/ψ and Υ(1S) as a function of the quarkonium rapidity $y_{\rm cms}$ are shown in the left panel of Fig. 3. The $R_{\rm FB}$ of J/ψ as a function of $p_{\rm T}$ is shown in the right hand panel of Fig. 3. The results are compared with theory predictions, based on a pure nuclear shadowing scenario, as well as partonic energy loss, either in addition to shadowing or as the only nuclear effect (see [11, 12] for a detailed discussion, and references therein). Within the uncertainties, both the models based on shadowing only and the coherent energy loss approach are in fair agreement with the data. The present theoretical and experimental uncertainties prevent from drawing more precise conclusions on the role of the different contributions. The prediction based on the Color Glass Condensate model underestimates our results (not shown in Fig. 3, see [12] and references therein).

4. Conclusions

ALICE has measured the production of quarkonia in pp, p–Pb and Pb–Pb collisions at the energies of the LHC. Recent NLO NRQCD calculations that include color octet processes describe well the production in pp collisions. In Pb–Pb interactions, the results suggest that the (re-)combination of $c\bar{c}$ pairs in the QGP or at the phase boundary plays a sizeable role in the J/ψ production. The Υ(1S) has been found to be suppressed at forward rapidity to the same extent as measured at midrapidity by the CMS Collaboration. First p–Pb results are in fair agreement with predictions of models based on nuclear shadowing and coherent energy loss.

Journal of Physics: Conference Series **509** (2014) 012008 doi:10.1088/1742-6596/509/1/012008

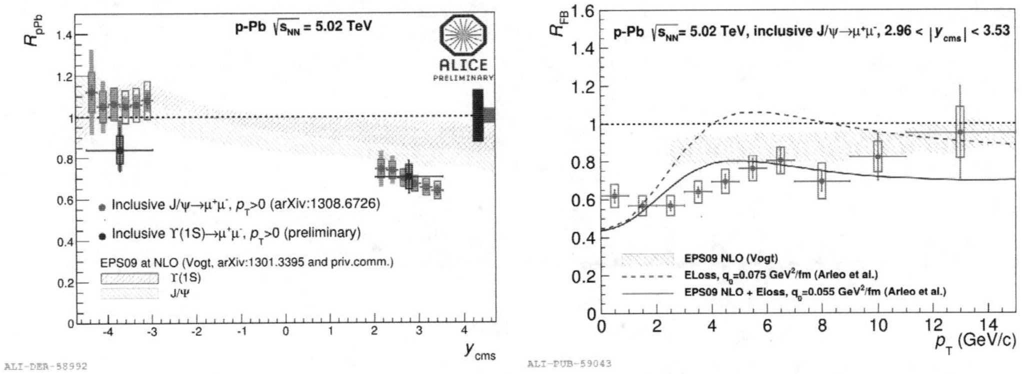

Figure 3. Left panel: the nuclear modification factor for $\Upsilon(1S)$ and inclusive J/ψ as a function of $y_{\rm cms}$ in p–Pb collisions at $\sqrt{sNN}$ = 5.02 TeV. Error bars correspond to statistical uncertainties, open boxes represent uncorrelated systematic uncertainties, shaded boxes show partially correlated systematic uncertainties and the boxes around $R_{\rm pPb} = 1$ shows the fully correlated uncertainties. Right panel: the forward to backward ratio for inclusive J/ψ production as a function of $p_{\rm T}$. Statistical uncertainties are shown as bars, uncorrelated systematic uncertainties correspond to open boxes. Theoretical model predictions are superimposed, see text for details.

References

[1] Aamodt K et al. (ALICE Collaboration) 2008 *JINST* **3** S08002
[2] Matsui T and Satz H 1986 *Phys. Lett.* B **178** 416
[3] Miller M L, Reygers K, Sanders S J and Steinberg P 2007 *Ann. Rev. Nucl. Part. Sci.* **57** 205-243
[4] Braun-Munzinger P and Stachel J 2000 *Phys. Lett.* B **490** 196
[5] Andronic A, Braun-Munzinger P, Redlich K. and Stachel J 2007 *Nucl. Phys.* A **789** 334
[6] Zhao X and Rapp R 2011 *Nucl. Phys.* A **859** 114
[7] Liu Y-P, Qu Z, Xu N and Zhuang P 2009 *Phys. Lett.* B **678** 72
[8] Vogt R 2010 *Phys. Rev.* C **81** 044903
[9] Fiorella Fionda et al. (ALICE Collaboration), *these proceedings*
[10] Lizardo Valencia Palomo et al. (ALICE Collaboration), *these proceeding*
[11] Palash Khan et al. (ALICE Collaboration), *these proceedings*
[12] Igor Lakomov et al. (ALICE Collaboration), *these proceedings*
[13] Aamodt K et al. (ALICE Collaboration) 2011 *Phys. Lett. B* **704** 442
[14] Abelev B et al. (ALICE Collaboration) 2012 *Phys. Lett. B* **718** 295
[15] Abelev B et al. (ALICE Collaboration) 2012 *JHEP* **11** 065
[16] Abelev B et al. (ALICE Collaboration) 2012 *Phys. Rev. Lett.* **108** 082001
[17] Abelev B et al. (ALICE Collaboration) 2012 *Phys. Lett. B* **712** 165
[18] Bala R et al. (ALICE Collaboration), *these proceedings*
[19] Aaij R et al. (LHCb Collaboration) 2012 *Eur. Phys. J. C* **72** 2025
[20] Scomparin E et al. (ALICE Collaboration 2013 Quark Matter proceedings)
[21] Adare A et al. (PHENIX COllaboration) 2011 *Phys.Rev. C* **84** 054912
[22] Abbas E et al. (ALICE Collaboration) 2013 *Phys. Rev. Lett.* **111** 162301
[23] Chatrchyan S et al. (CMS Collaboration) 2012 *JHEP* **05** 063
[24] Chatrchyan S et al. (CMS Collaboration) 2012 *Phys. Rev. Lett.* **109** 222301

Heavy flavor production in STAR

Daniel Kikoła[1] for the STAR Collaboration
Purdue University, West Lafayette, IN, US
E-mail: kikola@if.pw.edu.pl

Abstract.
We report on the recent STAR results on heavy flavor production in Au+Au collisions at $\sqrt{s_{NN}}$ = 200, 62.4 and 39 GeV. We present the nuclear modification factor and elliptic flow of open charm mesons and electrons from semileptonic decays of heavy flavor hadrons. We also report on new measurements of energy dependence of J/ψ production. STAR data are compared to theoretical model calculations and physics implications are discussed.

1. Introduction

Heavy quarks, charm and bottom, are unique probes of the Quark Gluon Plasma (QGP) properties at RHIC because they are produced in the initial interactions with large momentum transfer, before the QGP phase, and they are expected to interact with the QGP differently from light quarks. Experimentally, we study heavy quark production via decay products of charmed and beauty hadrons. At the moment, there are two feasible techniques for such studies at RHIC: using electrons from semi-leptonic decay of open heavy flavor mesons (so called non-photonic electrons, NPE), or through hadronic decays of charmed mesons. In the first case, the yields are larger and a specialized trigger can be used, although the information about the parent meson kinematics is incomplete and such an electron sample is a mixture of electrons from charmed and beauty hadron decays. In the latter, we identify charmed mesons via hadronic decays (e.g. $D^0 \rightarrow K^-\pi^+$) which give access to the kinematic of parent meson, but it suffers from large combinatorial background if a vertex detector is not available.

Elliptic flow of heavy quarks is another important tool for studying the QGP properties. It can shed new light on possible collective behavior and degree of thermalization of nuclear matter created in heavy ion collisions. Simultaneous measurements of the c and b quark production and elliptic flow are crucial for understanding the nature of interactions of heavy quarks with the surrounding partonic medium, and the parton energy loss mechanism in general.

Measurements of the production of various quarkonia states can provide insight into thermodynamic properties of the QGP. Different states have different binding energy and therefore disassociate, due to the Debye screening of the quark-antiquark potential, at different temperatures. However, there are other effects that may affect the observed production, such as cold nuclear matter effects, final state nuclear absorption and statistical coalescence of quark-antiquark pairs. The study of quarkonia production for different colliding systems, collision energies and centralities can help to disentangle the interplay of these mechanisms and to understand the medium properties.

[1] Present address: Faculty of Physics, Warsaw University of Technology, Warsaw, Poland

Besides studies of the QGP properties at top RHIC energy, we measure heavy quark production as a function of colliding energy in RHIC Beam Energy Scan program. Such measurements, complementary to the results for light quarks, provide us with additional means to map out the phase diagram of nuclear matter.

2. Data analysis

STAR [1] is a large acceptance ($|\eta|$ <1, full coverage in ϕ) multipurpose detector composed of several subsystems, which is well suited for the measurement of heavy quark production. In the analyses reported here, STAR Time Projection Chamber (TPC), Barrel Electromagnetic Calorimeter (BEMC) and Time-Of-Flight (TOF) detectors are used. We reconstruct quarkonia (J/ψ, Υ) via the di-electron decay channel. We identify charmed mesons via their hadronic decays (e.g. $D^0 \to K^-\pi^+$). The TPC is used for charged track reconstruction and particle identification via specific ionization energy loss (dE/dx). BEMC is used at high p_T for electron selection and also for a high-p_T (p_T > 2 GeV) electron trigger, while TOF provides electron identification at low p_T. For direct reconstruction of open charm mesons, TOF information together with the dE/dx are combined to select daughter pion and kaon candidates. In this article, we present results obtained with Au+Au data collected in 2010 and 2011 at $\sqrt{s_{\rm NN}} = 200$, 62 and 39 GeV and $p+p$ $\sqrt{s} = 200$ GeV data taken in 2009. More details about NPE and charmed meson measurements at STAR can be found in Refs. [2, 3, 4].

3. Results

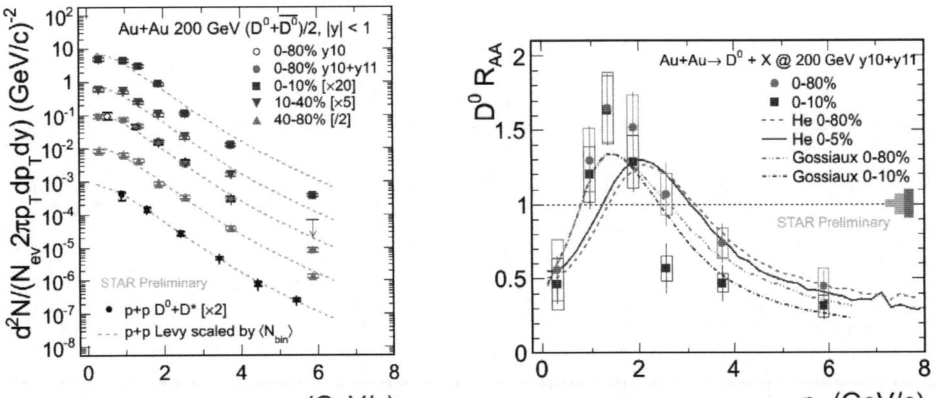

Figure 1. Left panel: D^0 invariant yield as a function of p_T in $p+p$ and Au+Au collisions at $\sqrt{s_{\rm NN}} = 200$ GeV. The dashed lines represent a Levi fit to $p+p$ data scaled by the number of binary collisions in a given centrality bin. Right panel: D^0 nuclear modification factor as a function of p_T in central (0-10%) and min-bias (0-80%) Au+Au collisions. The lines represent model calculations by He et al. [5] and Gossiaux et al. [6].

The left panel of Fig. 1 shows D^0 p_T spectrum in $p+p$ and Au+Au collisions at $\sqrt{s_{\rm NN}} = 200$ GeV. The dashed lines represent a Levi fit to $p+p$ data scaled by the number of binary collisions ($N_{\rm bin}$) in a given centrality bin. The Levi fit describes the $p+p$ data well; however, there are clear discrepancies when the fit is compared to Au+Au data. To quantify a modification of the D^0 production in Au+Au collisions, we calculate the nuclear modification factor R_{AA}: $R_{AA} = \frac{{\rm d}^2N_{Au+Au}/dydp_T}{N_{\rm bin}\times {\rm d}^2N_{p+p}/dydp_T}$, where ${\rm d}^2N_{Au+Au}/dydp_T$ and ${\rm d}^2N_{p+p}/dydp_T$ are invariant yields in Au+Au and $p+p$ collisions, respectively, and N_{bin} is the number of binary collisions in a given Au+Au centrality class. The right panel of Fig. 1 shows the D^0 R_{AA} for 0-10% most central and minimum bias Au+Au collusions. At high p_T, D^0 production is significantly suppressed. Moreover, D^0 production is enhanced at intermediate p_T (a "bump" at $p_T \sim 1.5$ GeV/c). The STAR data are compared to model predictions by He et al. [5] and Gossiaux et al. [6]. Both models assume that heavy quarks are strongly coupled with the surrounding partonic medium and predict a large suppression at high p_T. The models qualitatively reproduce the shape of R_{AA} as a function of p_T. The enhancement at $p_T \sim 1.5$ GeV/c in those calculations is due to the charm quark and medium interaction as well as the radial flow of light quarks which coalesce with charm quarks to produce D mesons. However, a similar effect for other particles (protons, pions) is observed at intermediate p_T in d+Au collisions [7] (the Cronin effect).

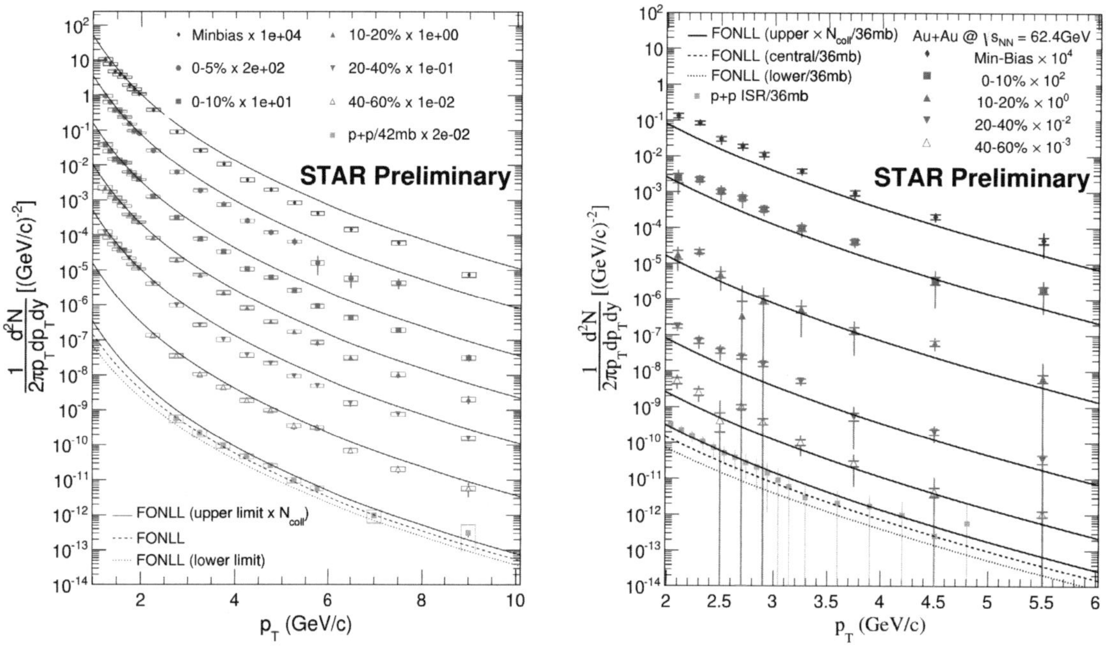

Figure 2. Invariant yield of non-photonic electrons as a function of p_T in Au+Au collisions at $\sqrt{s_{\text{NN}}} = 200$ GeV (left panel) and at $\sqrt{s_{\text{NN}}} = 62$ GeV (right panel). Solid lines represent an upper limit on FONLL predictions for NPE production scaled by number of binary collisions (N_{bin}) in a given centrality. For 62 GeV, ISR results for NPE p_T spectrum [8] are also presented.

Figure 2 shows the NPE p_T spectra for Au+Au collisions at $\sqrt{s_{\text{NN}}} = 200$ and 62.4 GeV. The data are compared with FONLL calculations scaled by N_{bin} corresponding to a given centrality bin. At 200 GeV, NPE production in central and mid-central collisions is suppressed compared to FONLL; however, the suppression is not observed at 62.4 GeV. Note, that results for 62.4 GeV are not corrected for $J/\psi \to e^+e^-$ contribution, which is as large as $\sim 20\%$ at high p_T at 200 GeV. Figure 3 shows the R_{AA} and elliptic flow of non-photonic electrons. We observe a strong NPE suppression at high p_T in central collisions, similar to D^0 and light hadrons ($\pi^{\pm}$). Figure 3 (right panel) shows v_2 results obtained using two- and four-particle correlations ($v_2\{2\}$ and $v_2\{4\}$, respectively). $v_2\{2\}$ and $v_2\{4\}$ have different sensitivity to the flow fluctuations and so called nonflow (correlations not related to the reaction plane) and thus provide an upper and lower limit on the average elliptic flow [9]. We compare our data to a few model calculations.

The DGLV model with radiative energy loss only [10] underestimates the observed suppression; however, it gives a reasonable description of the data when collisional energy loss is added. The collisional dissociation model [11] and AdS/CFT correspondence calculations [12] can also describe the measured R_{AA} reasonably well. In a partonic transport model, BAMPS [13, 14], heavy quarks lose energy due to elastic collisions with the rest of the medium. The dashed-dotted green line shows predictions for v_2 from the implementation of radiative and collisional energy loss from Gossiaux et al. [15, 6]. He et al. employ the TMatrix interactions model [5] which is a non-perturbative approach to heavy quark energy loss. Each of these models predicts a non-zero v_2 of charm quarks, thus observed finite NPE v_2 suggest a finite elliptic flow of charm quarks which is acquired due to interactions with quarks and gluons in the QGP phase. At 200 GeV, v_2 increases with p_T for $p_T > 3$ GeV which is likely due to decay or jet-like correlations.

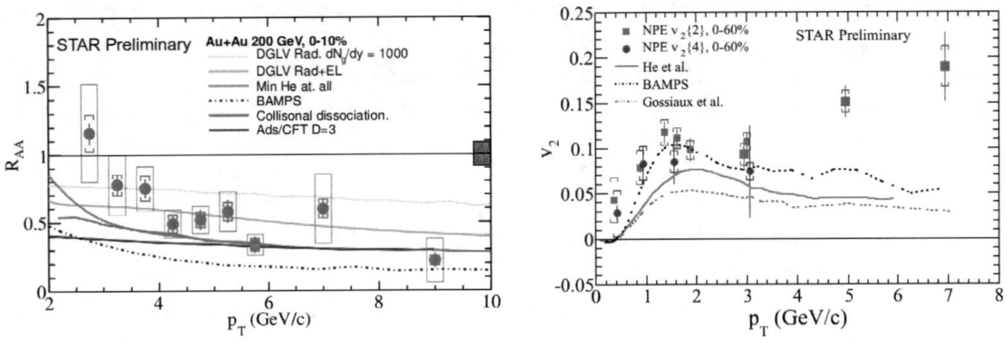

Figure 3. Non-photonic electron nuclear modification factor (left panel) and elliptic flow (v_2) (right panel) at $\sqrt{s_{\mathrm{NN}}} = 200$ GeV.

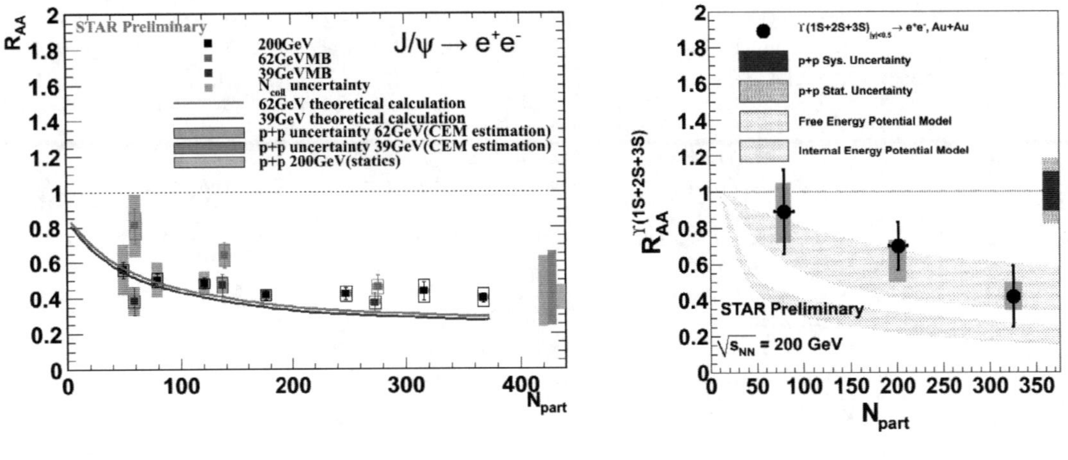

Figure 4. Left panel: nuclear modification factor R_{AA} for low$-p_T$ ($p_T < 5$) GeV/c J/ψ at $\sqrt{s_{\mathrm{NN}}} = 200$, 62.4 and 39 GeV. Right panel: R_{AA} for Υ at top RHIC energy.

The left panel of Fig. 4 shows the nuclear modification factor as a function of centrality (represented by number of participants, N_{part}) for low$-p_T$ ($p_T < 5$) GeV/c J/ψ in Au+Au

Journal of Physics: Conference Series **509** (2014) 012009 doi:10.1088/1742-6596/509/1/012009

collisions at $\sqrt{s_{\rm NN}} = 200$, 62.4 and 39 GeV. In the case of 39 and 62 GeV, predictions of the Color Evaporation Model [16] are used as a baseline, with uncertainties indicated by boxes on the right hand side of the plot. We observe a significant suppression at 39 and 62 GeV, at the similar level to $\sqrt{s_{\rm NN}} = 200$ GeV. STAR data are compared to predictions from model by Zhao and Rapp [17] which assumes two main components: suppression of directly produced J/ψ due to color screening and secondary production via recombination of $c, \bar{c}$ quarks. The model predicts very similar suppression for both 39 and 62 GeV, despite the significant difference in energy and thus different initial energy densities. In this model, larger direct suppression at higher energy is compensated by an enhanced regeneration rate due to the larger cross section for charm quarks. These calculations agree reasonably well with our data within errors.

The right panel of Fig. 4 shows the nuclear modification factor as a function of centrality for Υ in Au+Au collisions at $\sqrt{s_{\rm NN}} = 200$ GeV. The suppression increases with centrality and a significant suppression is observed in the most central collisions. Data are compared to a model by Strickland et al. [18], which assumes a thermal dissociation of Υ in a thermalized, expanding partonic medium, simulated with relativistic hydrodynamics. These calculations predict almost complete suppression of $\Upsilon(2S)$ and $\Upsilon(3S)$ states in the most central collisions. The model reproduces the STAR results well and the data favors the internal energy potential scenario.

4. Summary

D^0 and non-photonic electron production are suppressed at high p_T in Au+Au collision at $\sqrt{s_{\rm NN}} = 200$ GeV. On the other hand, NPE spectra at $\sqrt{s_{\rm NN}} = 62.4$ GeV is not suppressed compared to the pQCD calculations. We also observe a finite elliptic flow of non-photonic electrons $\sqrt{s_{\rm NN}} = 200$ GeV and comparison with models suggests a finite elliptic flow of charm quarks. J/ψ suppression at lower energies (39 and 62 GeV) is similar to top RHIC energy. The Υ suppression at $\sqrt{s_{\rm NN}} = 200$ GeV increases with centrality and it is consistent with a model that assume complete suppression of $\Upsilon(2S)$ and $\Upsilon(3S)$ states.

References

[1] Anderson M *et al.* 2003 *Nucl. Instrum. Meth. A* **499** 659–678 (*Preprint* nucl-ex/0301015)
[2] Agakishiev H *et al.* (STAR Collaboration) 2011 *Phys. Rev. D* **83** 052006 (*Preprint* 1102.2611)
[3] Tlusty D (STAR collaboration) 2013 *Nucl. Phys. A 904-905* **2013** 639c–642c (*Preprint* 1211.5995)
[4] Mustafa M (STAR Collaboration) 2013 *Nucl. Phys. A 904-905* **2013** 665c–668c (*Preprint* 1210.5199)
[5] He M, Fries R J and Rapp R 2012 *Phys. Rev. C* **86** 014903 (*Preprint* 1106.6006)
[6] Gossiaux P, Aichelin J, Gousset T and Guiho V 2010 *J. Phys. G* **37** 094019 (*Preprint* 1001.4166)
[7] Adare A *et al.* (PHENIX Collaboration) 2013 (*Preprint* 1304.3410)
[8] Basile M *et al.* 1981 *Il Nuovo Cimento A* **65** 421–456 ISSN 0369-3546
[9] Voloshin S A, Poskanzer A M, Tang A and Wang G 2008 *Phys. Lett. B* **659** 537–541 (*Preprint* 0708.0800)
[10] Djordjevic M, Gyulassy M, Vogt R and Wicks S 2006 *Phys. Lett. B* **632** 81 – 86
[11] Sharma R, Vitev I and Zhang B W 2009 *Phys. Rev. C* **80**(5) 054902
[12] Horowitz W 2008 *J. Phys. G* **35** 044025 (*Preprint* 0710.0703)
[13] Uphoff J, Fochler O, Xu Z and Greiner C 2011 *Phys. Rev. C* **84** 024908 (*Preprint* 1104.2295)
[14] Uphoff J, Fochler O, Xu Z and Greiner C 2012 *Phys. Lett. B* **717** 430–435 (*Preprint* 1205.4945)
[15] Gossiaux P and Aichelin J 2008 *Phys. Rev. C* **78** 014904 (*Preprint* 0802.2525)
[16] Nelson R E, Vogt R and Frawley A D 2013 *Phys. Rev. C* **87**(1) 014908 (and private communication)
[17] Zhao X and Rapp R 2010 *Phys. Rev. C* **82** 064905 (and private communication) (*Preprint* 1008.5328)
[18] Strickland M 2011 *Phys. Rev. Lett.* **107** 132301 (*Preprint* 1106.2571)

Journal of Physics: Conference Series **509** (2014) 012010 doi:10.1088/1742-6596/509/1/012010

Quarkonia and heavy flavors latest PHENIX results

Denis Jouan for the PHENIX collaboration

Institut de Physique Nucléaire, 91406 Orsay, France,

Université Paris Sud et IN2P3/CNRS

jouan@ipno.in2p3.fr

Abstract Heavy quarkonia are direct probes of deconfinement in the quark gluon plasma formed in heavy ion collisions. Other effects also occur, such as modification of the parton distributions in cold nuclear matter, and a multidimensional description is necessary to disentangle the various contributions. Characterization of the production processes in p+p and p or d induced collisions with ions is critical in the extraction of the additional effects in ion-ion collisions. The Relativistic Heavy Ion Collider has delivered a wide range of systems and energies, from p+p to Au+Au through d+Au and Cu+Au, from 7.6 to 200 GeV (N-N collision energy), allowing the PHENIX experiment to start exploring these experimentally available dimensions for open and hidden charm and beauty as a function of rapidity and transverse momentum. Recent PHENIX results on heavy flavor production measured through lepton pairs or singles, including J/ψ, ψ', χ_c, Υ and open charm and beauty, are presented.

1. Introduction

Heavy quarks and quarkonia are produced in the first steps of a collision, and they are sensitive to the whole evolution of the system, and in particular to the initial stage, which makes heavy quarkonia an excellent probe of the quark gluon plasma formed in heavy ion collisions [1]. As a bound pair of rare quarks, charmonia and bottomia should be direct probes of the deconfined nature of the medium, even possibly a QGP thermometer.

As the energy of the collision increases, the expected energy density and the lifetime of the expected deconfined phase increase. But other effects make the quarkonium picture more complex. Breaking of the bound quarkonium pair, or modifications of their kinematic characteristics, could come from collisions with nucleons or particles produced during the collision. Also, even though they were discovered several decades ago, quarkonia production is not yet fully understood (see for instance [2]). In the last 20 years Tevatron results and, recently, polarization measurements at LHC [3] have raised questions about our understanding of the production process.

The situation is complex, with a range of possible physics processes and strengths. But the recent start of LHC and the increasingly precise and detailed results coming from RHIC are putting more constraints on models of heavy flavor modification in nuclear collisions.

2. Expérimental set up

The PHENIX experimental set up is described in detail in [4].
Open and hidden heavy flavor yields are deduced from measurements of electrons at mid-rapidity ($-0.35<y<0.35$), and muons at forward and backward rapidity ($1.2<|y|<2.2$).
As a function of the centrality of the collision between two nuclei A and B, the modification of the yield is characterized by R_{AB}, the ratio of invariant yields in A+B collisions to those in p+p collisions, scaled by the number of nucleon nucleon collisions estimated in a geometrical Glauber model.

3. Quarkonia production

Quarkonia are measured using electron and muon pairs. Early PHENIX Au-Au results already suggested a J/ψ suppression at RHIC energies ($\sqrt{s}$=200 GeV), increasing with the centrality of the collision. These results, compatible with the formation of a quark gluon plasma, were comparable with the Pb-Pb ones at SPS ($\sqrt{s}$=17.3 GeV). This was confirmed by higher statistics PHENIX measurements, showing also that suppression was stronger at forward rapidity. Recent observations [5] at $\sqrt{s}$=39 GeV and 62.4 GeV showed similar suppression. This apparent stability from SPS to RHIC could hide a more complex phenomenon: the increase of energy density leading to more suppression, but also to more coalescence or recombination due to higher yields of the underlying heavy quark population [6]. The reduced suppression observed at higher collision energy [7] supports this interpretation.

For the smaller colliding system Cu+Cu, results [8] have been found to be very consistent with Au-Au for the same numbers of participating nucleons. For the d-Au system a decrease in the direction of the lighter partner (forward rapidity) is systematically observed. This is a general trend, observed for the bulk of the production [10]. Recent results from d+Au collisions with high statistical precision [11] bring a new sensitivity in the p_T and y dimensions. Calculations based on the NLO EPS09 nuclear PDF (nPDF) are found to not reproduce the p_T and y distributions simultaneously [11]. On the other hand a multiple scattering and energy loss model has an impressive ability [12] to reproduce the changes of distributions over wide p_T ranges at mid and forward rapidity, at all collision energies.

In Figure 1 the R_{dAu} for the J/ψ , measured in the central rapidity region, is shown to depend weakly on the centrality, while that for the ψ ' [13] displays unexpectedly strong suppression for the most central collisions. Stronger suppression of the ψ' than the J/ψ has been observed at lower energies, where the difference seems to be consistent with breakup by collisions with nucleons. But at RHIC breakup by nucleons is unable to explain the large suppression because the two quarkonia are expected [14] to cross the target nucleus too early in their development to have a different sensitivity to breakup effects. Beside their interest as a reference for heavy ion collisions, p(d)+A quarkonia results across a wide range of rapidities provide sensitivity to the time spent in the target nucleus by the quarkonia precursor. This allowed [15], a separation of the J/ψ data into regimes dominated by breakup and energy loss - respectively associated with the forward rapidity domains of the heavy or light colliding partners - in addition to the nuclear effects on the PDF.

The recent measurement of χ_c in d–Au [13] is an additional step towards the study of the relationship between the characteristics of the quarkonia, in particular their binding energy, and the CNM suppression, at these energies.

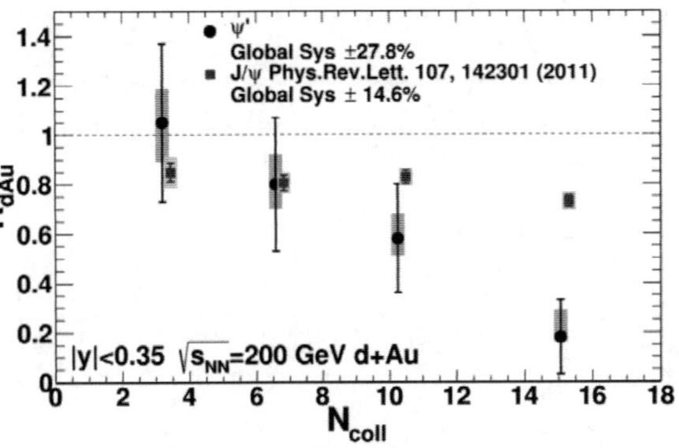

Figure 1 *J/ψ and ψ' R_{dAu} with respect to centrality. ψ' displays an unexpected additional suppression.*

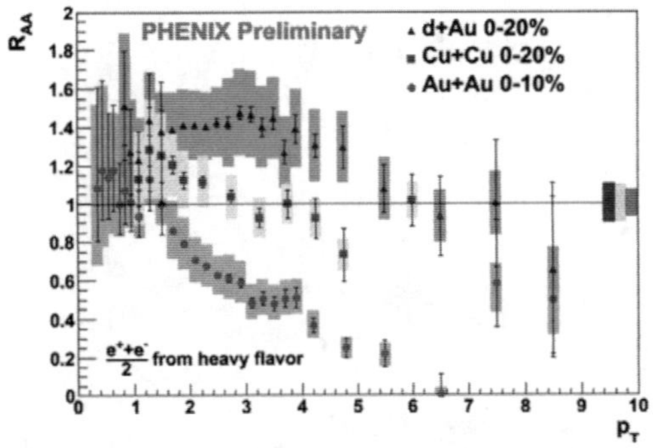

Figure 3 *evolution from d-Au to Au+Au of R_{AA} for open heavy flavor production at central rapidity*

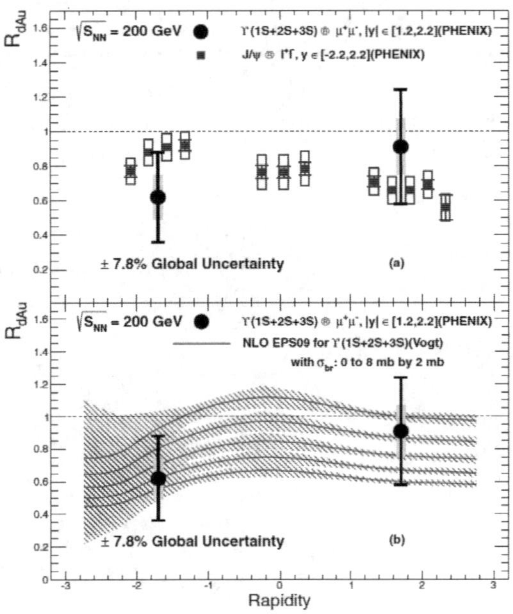

Figure 2 (from [16])

R_{dAu} for ϒ, with respect to rapidity, and compared to
a) R_{dAu} for J/ψ
b) calculation based on NLO EPS09 and breakup cross sections

Figure 2 [16] displays R_{dAu} versus rapidity, for the ϒ, compared to J/ψ (top) and to a model calculation (bottom). In contrast to the J/ψ, the ϒ suppression appears stronger in the backward region. The large error bars preclude physics conclusions at this stage, but it is noteworthy that the calculation [17] suggests a stronger decrease in the backward region. This could be linked to the breakup effect but also to the EMC effect in the gluon structure functions [18]. This EMC gluon effect can be accessed mainly through ϒ measurements at these energies.

4. Open heavy flavour production

Open heavy flavor (HF) originate from the same production mechanism as heavy quarkonia (for instance [19] reproduces quarkonia measurements with a calculation tuned on open flavor). The produced b and c quarks combine with light quarks leading to B and D excited mesons, mostly decaying to ground states, which then can decay in the semileptonic channels. This HF lepton source has to be separated from the leptons coming from lower mass mesons, and, in the case of electrons, from photon decays. In PHENIX it is done [20][21] using a Monte Carlo simulation of these backgrounds, and for the muon arms by using the variation of their decay rate with distance from the production point to the hadron absorber in front of the muon arm. For electrons in the central arm, the measurement of background at low p_T using foils as a photon converter is also employed.

Figure 3 shows the p_T evolution of R_{AA} for electrons in central collisions for three systems. High p_T Au-Au displays a strong suppression. That was unexpected for heavy quarks due to the dead cone effect in heavy particle energy loss. The d+Au collision data allow estimation of CNM effects. In contrast to the Au+Au data, they mostly show enhancement at mid p_T. Cu+Cu collisions are intermediate between d+Au and Au+Au. When integrated in p_T, a continuous evolution from enhancement to suppression is observed from d-Au peripheral to Au-Au central.

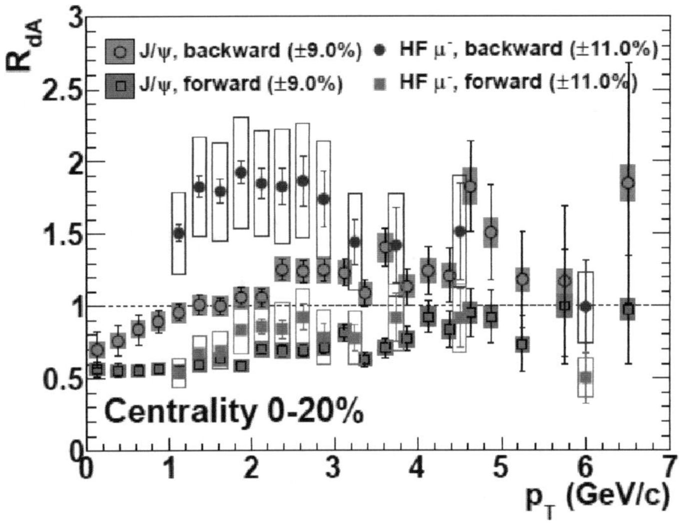

Figure 4 *For the most central collisions, Heavy flavor $R_{dAu}(p_T)$, displays a backward enhancement. Whereas the J/ψ $R_{dAu}(p_T)$ [11] decreases at low p_T, and displays a smaller forward/backward ratio.*

The modification of open HF for d-Au at mid-rapidity lies in between that observed in the backward and forward regions. As shown in Figure 4, for the most central collisions, $R_{dAu}(p_T)$ displays [22] an increased enhancement in the backward region, and a suppression in the forward one. This behavior is different from the one observed for quarkonia [11] shown for comparison in Figure 4. The J/ψ at low p_T is always suppressed, and the difference between backward/forward rapidities is smaller. The differences might be due to different dominant effects at forward and backward rapidities [15]. In the forward region the similar R_{dAu} of J/ψ and HF could reflect the same shadowing and/or energy loss effect with no nuclear breakup at the very short target crossing time, whereas in the backward region the suppression of the J/ψ relative to HF could be due to breakup of expanding quarkonia.

5. Summary

Thanks to the exploration in various colliding systems and energies made possible by RHIC, a more detailed landscape is emerging in ultrarelativistic collisions with heavy ions. Even dimensions like the time evolution of the system seem to become experimentally accessible, thanks to the diversity of probes provided by the quarkonia families and their various formation times, seen from the target nucleus.

Open HF is a reference for quarkonia production, as well as an extremely interesting probe of the energy loss of heavy quarks in the QGP through the observed Au+Au high p_T suppression. The suppression pattern of J/ψ with the centrality of the nucleus nucleus collisions remains similar from 200 to 20 GeV collision energy, at the current precision. More p+p and h+A measurements are required to go further. But the d+Au at 200 GeV results have already provided some perspective. As a function of centrality in d+Au collisions, quarkonia and open HF production are modified, with strong suppression in the forward (light partner) rapidity domain. At backward rapidity however there is notably different behavior: the decrease in R_{dAu} of the Υ, and the increase for open HF. The evolution with p_T also puts constraints on models. The ψ' is more suppressed than the J/ψ, possibly due to its different binding energy. But this difference is not expected at these energies given the small nuclear crossing time scale of the charmonia precursor, which would lead to a similar breakup cross section due to nucleon collisions. The explanation is not clear yet, but may involve the small binding energy of the ψ' making it very sensitive to the small hot regions produced in d+Au collisions.

The increase of precision and kinematic range has led to stronger constraints on models but also revealed more complexity, and raises new questions. With the vertex detector recently added, allowing experimental separation of open bottom and charm contributions at all rapidities in PHENIX, the near future addition of a new forward calorimeter, and the proposal to measure data from a range of p+A collisions as well as Au+Au with these new detectors, the next generation of results will bring additional information to constrain models and advance our understanding of the mechanisms that influence heavy flavor production in p+A and Au+Au collisions.

References

[1] Satz H 2007 *Nuclear Physics* A **783** 249
[2] Vogt R 2011 *Nuclear Physics* B **214** 147
[3] Kartvelishvili V 2013 *EPJ Web of Conferences* **49** 13008
[4] Adcox K et al. 2003 *Nucl. Instrum. Meth.* A **499** 469
[5] Adare A et al. 2013 *Physical Review* C **86** 064901
[6] Zhao X, Rapp R 2010 *Physical Review* C **82** 064905
[7] Bruno G E 2013 *this conference*
[8] Adare A et al. 2008 *Physical Review Letters* **101** 122301
[9] Hollis R S et al. 2013 *J. Phys.: Conf. Ser.* 458 012007
[10] Back B B et al. 2005 *Physical Review* C **72** 031901(R)
[11] Adare A et al. 2013 *Physical Review* C **87** 034904
[12] Arleo F, Kolevatov R, Peigné S, Rustamova M 2013, *J. High Energy Phys.* **1305** 155
[13] Adare A et al. 2013 *Physical Review Letters* **111** 202301
[14] Arleo F, Gossiaux P, Gousset T, Aichelin J 2000, *Physical Review* C **61** 054906
[15] McGlinchey D, Frawley A D, Vogt R *Physical Review* C **87** 054910
[16] Adare A et al. 2013 *Physical Review* C **87** 044909
[17] Vogt R 2010 *Physical Review* C **81** 044903
[18] Ferreiro E G, Fleuret F, Lansberg J P, Matagne N, Rakotozafindrabe A 2013 *The European Physical Journal C* **73** 2427
[19] Nelson R, Vogt R, Frawley A D, 2013 *Physical Review* C **87** 014908
[20] Adare A et al. 2012 *Physical Review Letters* **109** 242301
[21] Adare A et al. 2012 *Physical Review* C **86** 024909
[22] Adare A et al. 2013 *preprint* arXiv:1310.1005 [nucl-ex]

Open heavy-flavour results from ALICE

Diego Stocco for the ALICE Collaboration
SUBATECH (Ecole des Mines, CNRS-IN2P3, Université de Nantes), Nantes, France
E-mail: diego.stocco@subatech.in2p3.fr

Abstract. The ALICE Collaboration has measured heavy-flavour production in a wide rapidity range and in several decay channels in pp, Pb–Pb and, recently, in p–Pb collisions. An overview of some recent results is presented in this paper, with a particular emphasis on the first measurements in p–Pb collisions at $\sqrt{s_{\rm NN}} = 5.02$ TeV. A selection of the new results in Pb–Pb collisions at $\sqrt{s_{\rm NN}} = 2.76$ TeV is discussed as well.

1. Introduction

The ALICE experiment [1] studies the properties of the strongly interacting matter at the extreme energy densities reached in ultra-relativistic heavy-ion collisions at the LHC. The measurement of open charm and open beauty production plays an important role in the investigation, shedding light on the mechanisms of parton energy loss and hadronization in the hot and dense medium formed in high-energy nucleus-nucleus collisions. In particular, it allows one to test theoretical models predicting that the energy loss increases with the colour charge of the parton and decreases with its mass [2]. Heavy-flavour measurements in nucleon–nucleon and nucleon–nucleus collisions are mandatory for this study: the former sets the reference, while the latter allows to disentangle the initial-state effects (such as the modification of the Parton Distribution Functions in nuclei [3], the saturation of the gluon distribution at low Bjorken-x values and transverse momentum broadening of partons [4]), from the ones due to the interaction with the high energy density medium.

The excellent performance of the ALICE detector allows heavy-flavour measurements in several decay channels and in a wide rapidity range. At forward rapidity, open heavy-flavour production is studied inclusively in the semi-muonic decay channel. At mid-rapidity, D mesons are reconstructed via their hadronic decays and, in addition, electrons from semi-leptonic decays of charm and beauty hadrons are measured. After a description of the experimental apparatus and the strategy used for the measurements (Section 2), the most recent results in Pb–Pb collisions at $\sqrt{s_{\rm NN}} = 2.76$ TeV (Section 3) and the very first results in p–Pb collisions at $\sqrt{s_{\rm NN}} = 5.02$ TeV (Section 4) will be discussed. The pp results are not treated in this paper, but an interesting recent study on D meson production as a function of charged particle multiplicity is discussed elsewhere in these proceedings [5].

2. Heavy-flavour measurements with ALICE

ALICE [1] consists of a central barrel, a Muon Spectrometer and a set of detectors at large rapidities for global event characterisation and trigger. The central barrel ($|\eta| < 0.9$) is equipped with a solenoidal magnet providing a magnetic field up to 0.5 T, which allows the measurement

Journal of Physics: Conference Series **509** (2014) 012011 doi:10.1088/1742-6596/509/1/012011

of charged particles down to low $p_{\rm T}$. The tracking is performed by combining the information from the Time Projection Chamber (TPC) and the Inner Tracking System (ITS). The latter is made of six layers of silicon detectors providing a resolution on the distance between the track and the primary vertex (impact parameter) of better than 65 μm for $p_{\rm T} > 1$ GeV/c [6]. Charged particles are identified via their specific energy loss in the TPC and their time of flight, measured with the TOF detector. The electron identification can include, in addition, the information from the Transition Radiation Detector (TRD) and the Electromagnetic Calorimeter (EMCal).

Muons with momenta larger than 4 GeV/c are reconstructed and identified in the pseudo-rapidity interval $-4 < \eta < -2.5$ with the Muon Spectrometer. This consists, as seen from the interaction point, of a composite frontal absorber, 5 stations of tracking chambers, an iron wall and 2 stations of trigger chambers [1]. Charged particles are bent by a dipole magnet providing a 3 T$\times$m magnetic field and are identified as muons if they reach the trigger stations placed downstream the iron wall of about 7.2 interaction length thickness.

The event centrality is estimated event by event from the signal amplitudes in two arrays of scintillator hodoscopes (VZERO), covering the full azimuth in the pseudo-rapidity regions $2.8 < \eta < 5.1$ and $-3.1 < \eta < -1.7$.

Charmed mesons are measured through the exclusive reconstruction of the hadronic decay channels: $\rm D^0 \to K^-\pi^+$, $\rm D^+ \to K^-\pi^+\pi^+$, $\rm D^{*+} \to D^0\pi^+ \to K^-\pi^+\pi^+$ and $\rm D_s^+ \to \phi\pi^+ \to K^+K^-\pi^+$ and their charge conjugates. The large combinatorial background is reduced by selections based on the decay topology. At low momenta, the background is further suppressed by the kaon identification with the TPC and TOF detectors. D-meson raw yields are then extracted through the analysis of the invariant mass of their decay products. The results include both the prompt contribution and the feed down from the decay of hadrons containing beauty. The latter is estimated with perturbative QCD calculations in the framework of Fixed-Order Next-to-Leading Log (FONLL) [7] and subtracted (see [6] for further details).

The measurement of electrons from heavy-flavour decays at mid-rapidity ($|\eta| < 0.8$) requires the subtraction of several background contributions from the inclusive electron sample. The most important are photon conversions in the beam pipe and detector material, Dalitz decays of light neutral mesons, di-electron decays of vector mesons and direct radiation. Such contributions are described by a Monte Carlo cocktail based on the measured distributions of π^0, with the resonance decay contributions obtained through m_T scaling. Alternatively, the contribution of electrons from Dalitz decay and gamma-conversions was evaluated from the opposite-sign electron pairs with an invariant mass close to zero [8].

Muons from heavy-flavour decays are measured at forward rapidity. The main source of background in the inclusive transverse momentum distribution consists of muons from the decay of light hadrons. This contribution dominates the low-$p_{\rm T}$ region of the spectrum, and can therefore be rejected by a cut on the muon $p_{\rm T}$. The remaining contamination at larger transverse momenta is estimated with Monte Carlo simulations in pp collisions and with a data driven method (based on the the extrapolation at forward rapidity of the pion and kaon yields measured at mid-rapidity) in Pb–Pb collisions and it is subtracted [9].

3. Suppression and azimuthal anisotropy in Pb–Pb collisions at $\sqrt{s_{\rm NN}} = 2.76$ TeV

The study of the relative variation of particle production in heavy-ion collisions with respect to the pp reference is commonly performed through the nuclear modification factor ($\mathsf{R}_{\rm AA}$), which is defined as:

$$\mathsf{R}_{\rm AA}(x) = \frac{1}{\langle T_{\rm AA}\rangle}\frac{{\rm d}N_{\rm AA}/{\rm d}x}{{\rm d}\sigma_{\rm pp}/{\rm d}x} \quad (1)$$

where ${\rm d}N_{\rm AA}/{\rm d}x$ (${\rm d}\sigma_{\rm pp}/{\rm d}x$) is the yield (cross-section) of particles as a function of a variable x measured in heavy-ion (pp) collisions, and $\langle T_{\rm AA}\rangle$ is the nuclear overlap function calculated in the framework of the Glauber model. In the absence of initial-state or medium effects, $\mathsf{R}_{\rm AA}$ is expected to be one for heavy flavours, whose yields are proportional to the number of binary nucleon-nucleon collisions (binary-scaling).

The nuclear modification factors of D mesons [6] and electrons from heavy-flavour decays [10] at mid-rapidity and of muons from heavy-flavour decays at forward rapidity [9] have been measured with ALICE: all of the results show a strong reduction of the yields at large transverse momenta ($p_{\mathrm{T}} \gtrsim 5$ GeV/c) in the most central collisions. The R_{AA} of D mesons was also compared with the one of charged particles and pions [11, 12]: a similar suppression is observed, although the uncertainties do not allow yet to draw a strong conclusion on the colour charge dependence of the parton energy loss. The nuclear modification factor of D mesons was recently measured as a function of centrality (estimated as the average number of participant nucleons, scaled with the number of nucleon-nucleon collisions from the Glauber model) [11]. The results are shown in the left panel of Figure 1 (black squares): R_{AA} depends on centrality, decreasing from peripheral to central collisions. The blue filled circles in the same figure show the equivalent measurement for non-prompt J/ψ (i.e. coming from beauty hadron decays) from CMS [13]. A similar suppression pattern is observed, but with a value of R_{AA} which is about two times larger than for D mesons. The p_{T} range of the two results is chosen in such a way that the average p_{T} of the parent B mesons is compatible with the one of D mesons, thus allowing for a consistent comparison of charm and beauty hadrons. The result is therefore an indication of a smaller in-medium energy loss for beauty than for charm quarks.

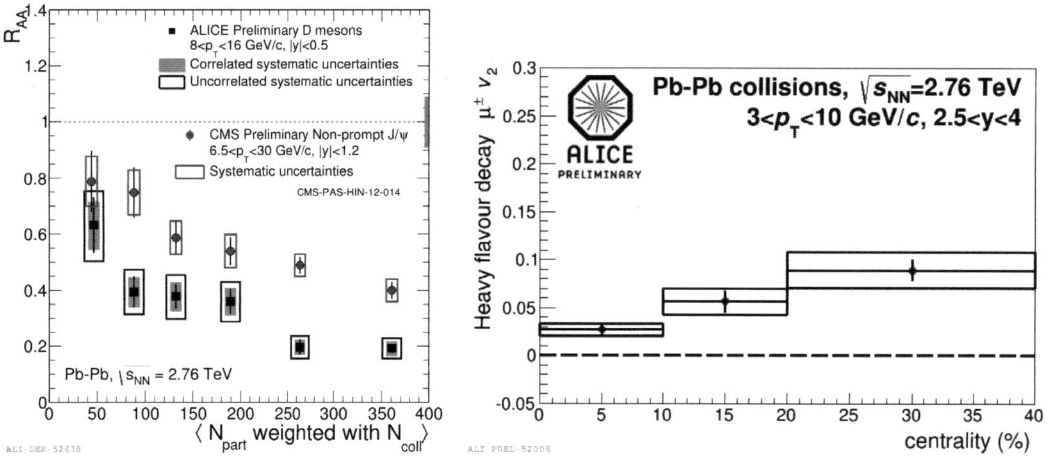

Figure 1. Left panel: nuclear modification factor as a function of centrality for D mesons measured with ALICE (black squares) and non-prompt J/ψ from CMS [13] (blue filled circles). Right panel: elliptic flow of muons from heavy-flavour decays as a function of centrality. The bars (open boxes) are the statistical (systematic) uncertainties.

In semi-central heavy-ion collisions, the initial geometrical azimuthal anisotropy can be converted into an anisotropy in momentum space in case of thermalisation or rescattering of particles with the medium. This anisotropy is quantified by the Fourier expansion of the azimuthal distributions of particles with respect to the plane defined by the impact parameter and the collision axis (Ψ_{RP}):

$$E\frac{\mathrm{d}^3N}{\mathrm{d}^3p} = \frac{1}{2\pi}\frac{\mathrm{d}^2N}{p_{\mathrm{T}}\mathrm{d}p_{\mathrm{T}}\mathrm{d}y}\left(1 + \sum_{n=1}^{\infty} 2v_n \cos\left(n(\phi - \Psi_{RP})\right)\right) \tag{2}$$

The coefficient of the second order harmonic of the expansion (v_2) is called elliptic flow.

Journal of Physics: Conference Series **509** (2014) 012011 doi:10.1088/1742-6596/509/1/012011

The ALICE Collaboration measured the elliptic flow of D mesons [14] and electrons from heavy-flavour decays [10] at mid-rapidity: a positive value of v_2 is observed at intermediate p_T in semi-central collisions. The measured v_2 of D mesons in the 30-50% centrality is comparable in magnitude to the one of charged hadrons, thus suggesting that the relaxation time of charm quarks in the medium is similar to that of light partons [14]. The v_2 of muons from heavy-flavour decays at forward rapidity was also measured for the first time with ALICE, in the transverse momentum range $3 < p_T < 10$ GeV/c [15]. The results are shown in the right panel of Figure 1 as a function of the collision centrality: the value of v_2 increases from central to peripheral collisions. A positive value of v_2 is observed, thus indicating that heavy quarks participate in the collective motion of the system. These measurements are a powerful test for theoretical models implementing different mechanisms for the energy loss of charm and beauty quarks: while many of them are able to reproduce the reduction of yields at high p_T (see e.g. [12] and references therein), the simultaneous description of R_{AA} and v_2 is challenging.

4. Nuclear modification factor in p–Pb collisions at $\sqrt{s_{NN}} = 5.02$ TeV

The ALICE Collaboration measured heavy-flavour production in proton-lead collisions at $\sqrt{s_{NN}} = 5.02$ TeV: the first results are shown in Figure 2. The left panel shows R_{pPb} of electrons

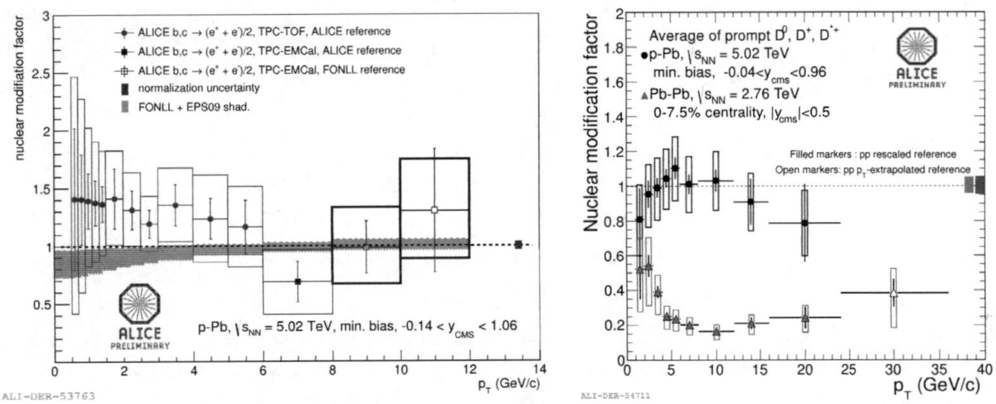

Figure 2. Left panel: nuclear modification factor of electrons from heavy-flavour decays in p–Pb collisions at $\sqrt{s_{NN}} = 5.02$ TeV. Measurements were performed using the TPC-TOF (red circles) or TPC-EMCal (blue squares). At high p_T, the open symbols denote that FONLL [7] calculations were used as a reference instead of rescaled data (see text). pQCD predictions [16] with EPS09 [3] implementation of the nuclear modification of PDFs are also shown (grey band). Right panel: nuclear modification factor of D mesons in p–Pb collisions at $\sqrt{s_{NN}} = 5.02$ TeV (black circles) compared to Pb–Pb collisions at $\sqrt{s_{NN}} = 2.76$ TeV (red triangles).

from heavy-flavour decays (see also [8]). At low transverse momenta, the analysis is performed by using the TPC-TOF information (red circles) while at larger momenta, the EMCal is also used for electron identification (blue squares). The reference consists of p_T-differential cross sections of heavy flavour decay electrons measured in pp collisions at $\sqrt{s} = 7$ TeV [8] and extrapolated to $\sqrt{s} = 5.02$ TeV with a scaling procedure based on the FONLL [7] predictions at both energies. At high transverse momenta, however, where the statistical uncertainties of the pp results are large, the FONLL predictions at $\sqrt{s} = 5.02$ TeV are used directly (open blue square). The results are consistent within uncertainties with pQCD calculations [16] with EPS09 [3] implementation of the nuclear modification of PDFs (grey band).

The right panel of Figure 2 shows the R_{pPb} of D mesons (black circles). Also in this case, the data are well described by pQCD calculations with shadowing (see [17] for further details).

The initial-state effects are indeed rather small, and cannot account for the small R_{AA} observed in heavy-ion collisions (red triangles), which is therefore a genuine effect of the interaction with the hot and dense medium.

5. Summary

The ALICE Collaboration measured heavy-flavour production in pp, p–Pb and Pb–Pb collisions at the LHC. In Pb–Pb collisions, new results on the nuclear modification factor of D mesons as a function of the collision centrality allow for a comparison with the non-prompt J/ψ measured with CMS in a consistent transverse momentum range of the D and B mesons. The results give an indication of a smaller energy loss for beauty than for charm. The first measurement of the elliptic flow of muons from heavy-flavour decays in $2.5 < y < 4$ has been presented: the value of v_2 is larger than zero, thus indicating that heavy quarks participate in the collective motion of the system. The simultaneous reproduction of R_{AA} and v_2 measurements in the wide rapidity range of ALICE is a challenge for theoretical models implementing different mechanisms of heavy quark energy loss in medium.

The nuclear modification factor of D mesons and electrons from heavy-flavour decays was also measured for the first time in p–Pb collisions at $\sqrt{s_{\mathrm{NN}}} = 5.02$ TeV. The $\mathsf{R}_{\mathsf{pPb}}$ is consistent within uncertainties with perturbative QCD calculations [16] with EPS09 [3] implementation of the nuclear modification of PDFs. The resulting effect is small with respect to the large reduction of heavy-flavour yields observed in Pb–Pb collisions, which is therefore an effect of the charm (beauty) quark in-medium energy loss.

References

[1] Aamodt K *et al.* (ALICE Collaboration) 2008 *JINST* **3** S08002
[2] Armesto N, Dainese A, Salgado C A and Wiedemann U A 2005 *Phys.Rev.* **D71** 054027
[3] Eskola K, Paukkunen H and Salgado C 2009 *JHEP* **0904** 065
[4] Blaizot J P, Gelis F and Venugopalan R 2004 *Nucl.Phys.* **A743** 13–56
[5] Bala R (ALICE Collaboration) Open-charm production as a function of charged particle multiplicity in pp collisions at $\sqrt{s} = 7$ TeV with ALICE *These proceedings*
[6] Abelev B *et al.* (ALICE Collaboration) 2012 *JHEP* **1209** 112
[7] Cacciari M, Frixione S, Houdeau N, Mangano M L, Nason P *et al.* 2012 *JHEP* **1210** 137
[8] Heide M (ALICE Collaboration) Measurement of electrons from heavy-flavour hadron decays in pp and p–Pb collisions with ALICE at the LHC *These proceedings*
[9] Abelev B *et al.* (ALICE Collaboration) 2012 *Phys.Rev.Lett.* **109** 112301
[10] Thomas D (ALICE Collaboration) Measurement of electrons from heavy flavour decays in Pb–Pb collisions at $\sqrt{s_{\mathrm{NN}}} = 2.76$ TeV with ALICE *These proceedings*
[11] Bruna E (ALICE Collaboration) D-meson nuclear modification factor and v_2 in Pb–Pb collisions *These proceedings*
[12] Conesa del Valle Z (ALICE Collaboration) 2013 *Nucl.Phys.* **A904-905** 178c–185c
[13] Chatrchyan S *et al.* (CMS Collaboration) 2012 *CMS-PAS-HIN-12-014*
[14] Abelev B *et al.* (ALICE Collaboration) 2013 *Phys.Rev.Lett.* **111** 102301
[15] Zhang X (ALICE Collaboration) Nuclear modification factor and elliptic flow of muons from heavy-flavour decays in Pb-Pb collisions at $\sqrt{s_{\mathrm{NN}}} = 2.76$ TeV with ALICE *These proceedings*
[16] Mangano M L, Nason P and Ridolfi G 1992 *Nucl.Phys.* **B373** 295–345
[17] Luparello G (ALICE Collaboration) Measurement of D-meson production in p–Pb collisions with the ALICE detector *These proceedings*

Open and closed heavy-flavour suppression in heavy-ion collisions with CMS

Mihee Jo for the CMS collaboration
Korea University, Seoul, 136-701, Republic of Korea
E-mail: mihee.jo@cern.ch

Abstract. The Compact Muon Solenoid (CMS) is fully equipped to measure quarkonia in the dimuon decay channel in the high multiplicity environment of nucleus-nucleus collisions. Quarkonia are especially relevant for studying the quark-gluon plasma since they are produced at early times and propagate through the medium, mapping its evolution. CMS has measured the nuclear modification factors of non-prompt J/ψ (from b-hadron decays), prompt J/ψ, inclusive ψ(2S), and the first three Υ states in PbPb collisions at $\sqrt{s_{NN}} = 2.76$ TeV. A suppression of non-prompt J/ψ, which is sensitive to the in-medium b-quark energy loss, has been measured at relatively high p_T ($6.5 < p_T < 30$ GeV/c) in PbPb collisions, compared to the yield in pp collisions scaled by the number of inelastic nucleon-nucleon collisions. For prompt J/ψ in the same kinematic range, a strong, centrality-dependent suppression is observed. Such strong suppression at high p_T has previously not been observed at RHIC. At mid-rapidity and high p_T, ψ(2S) show an even stronger suppression than J/ψ. Furthermore, CMS has measured the suppression of the three Υ states, separately, down to $p_T = 0$ GeV/c. A clear ordering of the suppression with binding energy is observed, as expected from sequential melting of quarkonium states.

1. Introduction

A deconfined state of quarks and gluons, the Quark-Gluon Plasma (QGP), is predicted to be formed at high parton density and medium temperature by lattice Quantum Chromodynamics (QCD) in high energy heavy-ion collisions. Heavy quarks are created at the early stage of collisions thereby they help to study the properties of QGP. The confining potential of quark and anti-quark is expected to be screened in the QGP and the suppression of heavy quarkonium production is predicted [1]. At higher medium temperature, quarkonia with stronger binding energy are expected to be dissociated, hence quarkonium states are melted sequentially.

Partons that are produced in the initial hard scattering process undergo energy loss as they traverse the medium. Gluons, light quarks and heavy quarks lose energy in the medium differently by inelastic and elastic interactions. Especially for heavy quarks, the radiative energy loss by gluon emission at small angle is expected to be reduced [2]. The independent measurement of b-quark is essential for better understanding of these phenomenon and this can be obtained via non-prompt J/ψ which is originated from b-hadron decays.

In these proceedings, the nuclear modification factors, R_{AA}, of Υ(1S), Υ(2S), Υ(3S) [3], prompt J/ψ and non-prompt J/ψ [4] are presented as well as the $\psi(2S)/J/\psi$ ratio [5] as measured with the Compact Muon Solenoid (CMS) [6] detector.

2. Experimental Methods

The measurement of J/ψ, ψ(2S) and Υ via $\mu^+\mu^-$ decays at $\sqrt{s_{NN}} = 2.76$ TeV are performed with PbPb and pp datasets corresponding to integrated luminosities of 150 μb^{-1} and 230 nb^{-1}, respectively. Muons are identified in the range $|\eta| < 2.4$ with the muon system and the p_T measurement is based on the information from tracker system. The forward calorimetry system covers the range $2.9 < |\eta| < 5.2$ and is used for event selection and centrality determination.

Reconstructed $\mu^+\mu^-$ pairs are selected in the range of $|y| < 1.6$ and $6.5 < p_T < 30$ GeV/c, $1.6 < |y| < 2.4$ and $3 < p_T < 30$ GeV/c for J/ψ and ψ(2S). Only muons with $p_T > 4$ GeV/c and dimuon pairs in $|y| < 2.4$ are used for Υ(nS).

Invariant mass of J/ψ, ψ(2S), Υ(nS) and lifetime distributions of J/ψ are presented in Fig.1. Invariant mass spectrum of J/ψ is fitted with a sum of Crystal Ball function and a Gaussian to describe the signal and an exponential for the background in the region of $2.6 < m_{\mu^+\mu^-} < 3.5$ GeV/c^2. Non-prompt J/ψ originated from b-hadron decays is identified via the measurement of secondary $\mu^+\mu^-$ vertex [7]. In order to obtain the fraction of non-prompt J/ψ, 2-dimensional simultaneous extended unbinned maximum likelihood fitting with the invariant mass and the pseudo-proper decay length is performed.

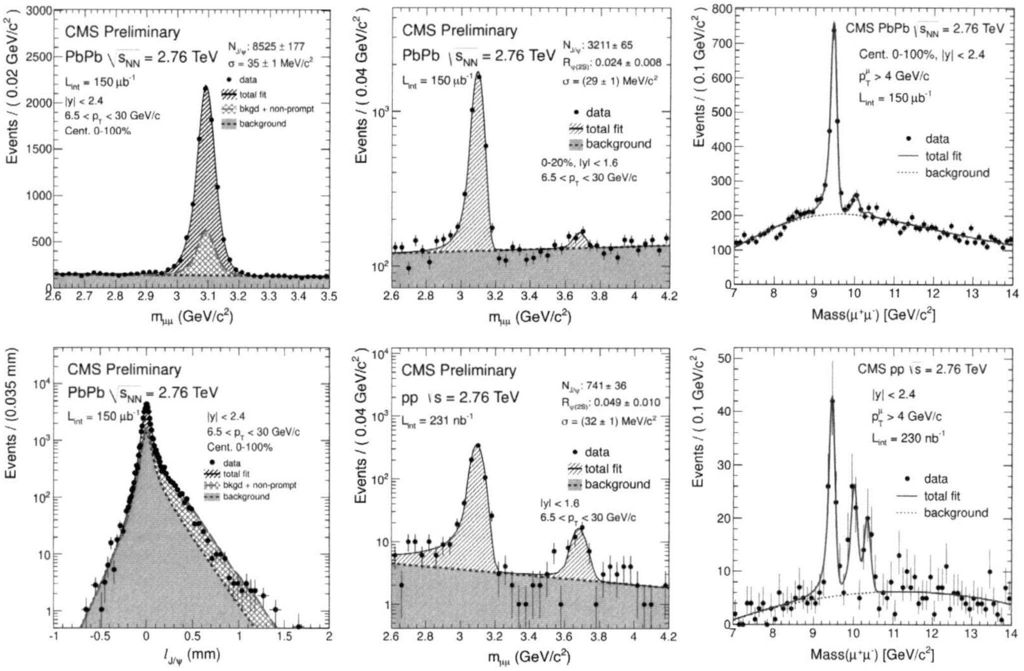

Figure 1. Invariant mass spectrum *(top left)* and pseudo-proper decay length distribution *(bottom left)* of J/ψ by PbPb collisions in $|y| < 2.4$, $6.5 < p_T < 30$ GeV/c, centrality 0-100%. Wider invariant mass spectrum of J/ψ and ψ(2S) in PbPb *(top center)* and in pp *(bottom center)* in $|y| < 1.6$, $6.5 < p_T < 30$ GeV/c, centrality 0-20% for PbPb. Invariant mass spectrum of Υ(nS) in PbPb *(top right)* and in pp *(bottom right)* in $|y| < 2.4$, $p_T^{\mu} > 4$ GeV/c, centrality 0-100% for PbPb.

The $\psi(2S)/J/\psi$ ratio is obtained by simultaneous unbinned maximum likelihood fitting in the range of $2.6 < m_{\mu^+\mu^-} < 4.2$ GeV/c^2 independently in PbPb and pp datasets. For J/ψ, a Crystal Ball function is used at mid-rapidity while additional a Gaussian function is used at forward rapidity. Same shapes were used for $\psi(2S)$. The two width parameters of the Crystal Ball and Gaussian functions are set to the values of J/ψ that are multiplied by Particle Data Group(PDG) mass ratio of $\psi(2S)$ and J/ψ. The mass ratio of $\psi(2S)$ and J/ψ is also fixed to the PDG value. An exponential function is used to describe the background.

The invariant mass spectrum of Υ(nS) is also described by an extended unbinned maximum likelihood fit. The signals consist of three Crystal Ball functions and the mass differences between Υ(nS) are fixed to their PDG values and the widths are scaled with the resonance mass. The shape parameters are common for PbPb and pp datasets via a simultaneous fit. The background model is a second-order polynomial for pp dataset and a exponential for PbPb dataset.

Efficiencies are estimated from Monte-Carlo simulations for all quarkonium states. Data-driven efficiency measurement, Tag and Probe, is also performed as a cross-check. The systematic uncertainties for efficiencies, 1-16% for the J/ψ R_{AA}, 4-8% for the Υ(nS) R_{AA} and 2-10% for the $\psi(2S)/J/\psi$ ratio, are estimated by comparing the efficiencies obtained from data and MC simulations and considering variations of simulated kinematic distributions.

The nuclear modification factor, R_{AA}, is defined in the form

$$R_{AA} = \frac{\mathcal{L}_{pp}}{T_{AA} N_{\mathrm{MB}}} \frac{N_{\mathrm{PbPb}}}{N_{pp}} \cdot \frac{\varepsilon_{pp}}{\varepsilon_{\mathrm{PbPb}}} \tag{1}$$

where $\mathcal{L}_{pp}$ is the luminosity of the *pp* dataset, T_{AA} is the nuclear overlap function, N_{MB} is the number of the sampled minimum bias PbPb events, N is the raw yield of quarkonium, and ε is the efficiency. If the R_{AA} is less(larger) than 1, that means the production of particle is relatively reduced(enhanced) in heavy-ion collisions with respect to proton-proton collisions, respectively.

3. Results

The R_{AA} for Υ(nS) states have been measured as well as the double ratio of Υ(1S) and Υ(2S) in Fig.2. For the minimum bias(0–100%) case, R_{AA} of Υ(nS) are

$$\begin{aligned} R_{AA}(\Upsilon(1S)) &= 0.56 \pm 0.08(stat) \pm 0.07(syst), \\ R_{AA}(\Upsilon(2S)) &= 0.12 \pm 0.04(stat) \pm 0.02(syst), \\ R_{AA}(\Upsilon(3S)) &= 0.03 \pm 0.04(stat) \pm 0.01(syst), < 0.10 \text{ (95\%CL)}. \end{aligned} \tag{2}$$

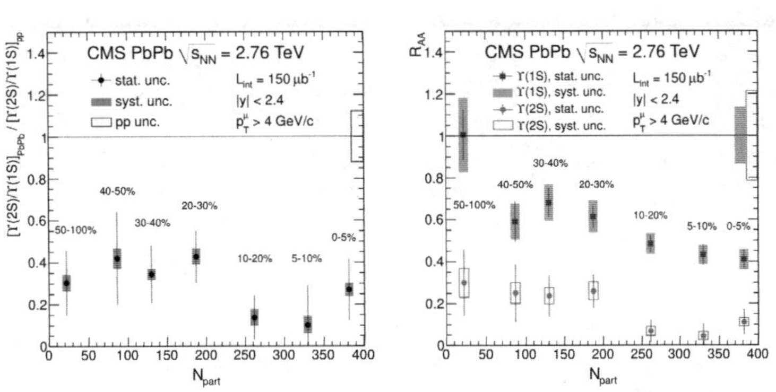

Figure 2. The double ratio *(left)* and the R_{AA} *(right)* of the Υ(2S) and Υ(1S) states are presented as a function of number of participants. The uncertainties that are independent from N_{part} are shown by box at unity.

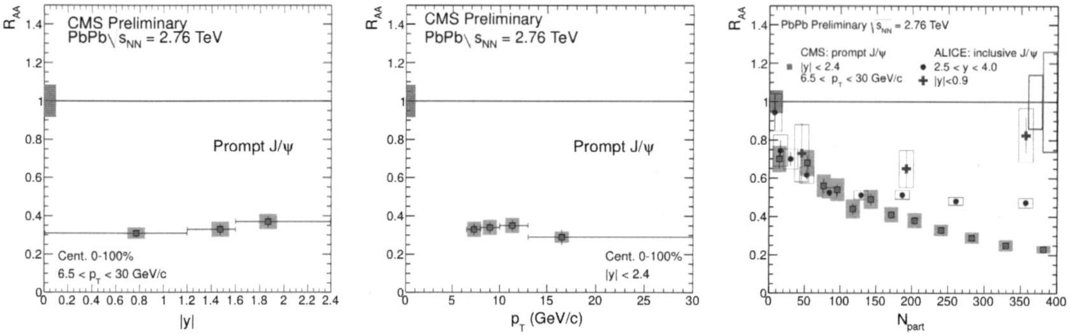

Figure 3. The R_{AA} of prompt J/ψ with respect to rapidity*(left)*, p_T*(center)* and centrality*(right)* are presented. Inclusive J/ψ measurement by ALICE*(blue, green markers)* are compared as a function of centrality.

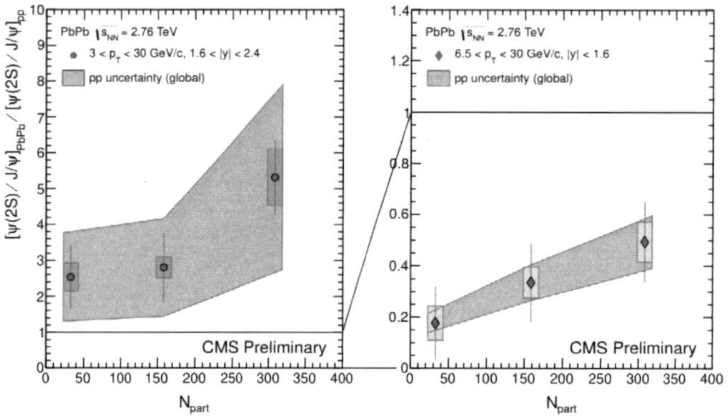

Figure 4. The double ratio of J/ψ and $\psi(2S)$ at lower p_T *(left)*, $3 < p_T < 30$ GeV/c, and higher p_T *(right)*, $6.5 < p_T < 30$ GeV/c, are shown.

A clear suppression of higher Υ states is observed. Different amount of Υ(nS) suppression can be explained by increased suppression of higher order resonances that are more loosely bounded.

The R_{AA} of prompt J/ψ are presented as a function of centrality, p_T and rapidity. Prompt J/ψ is suppressed over all measured p_T and rapidity regions. A clear centrality dependent suppression is observed. In Fig.3, inclusive J/ψ R_{AA} results from ALICE [8] is compared. Note that inclusive J/ψ measured by ALICE consist of particles with $p_T > 0$ GeV/c while CMS measurements start from higher p_T, 6.5 GeV/c.

The first measurement of the $\psi(2S)/J/\psi$ ratio in heavy-ion collisions is presented in Fig.4. At lower p_T and forward rapidity region, the production of $\psi(2S)$ is slightly larger than $\psi(2S)$ however the error driven by lack of pp collision statistics is too large to draw a conclusion. At higher p_T region, $\psi(2S)$ is more suppressed than J/ψ over all measured centrality regions.

The open bottom R_{AA} is measured in p_T, rapidity and centrality via non-prompt J/ψ and presented in Fig.5. Hints of rapidity and p_T dependent suppression are observed. A comparison of the open charm R_{AA} by ALICE [9] shows a slightly less suppression at the most central collisions. Note that non-prompt J/ψ is measured from 6.5 to 30 GeV/c by CMS while D meson measured by ALICE starts from 6 to 12 GeV/c at mid-rapidity, and p_T of open bottom is expected to be higher than that of non-prompt J/ψ.

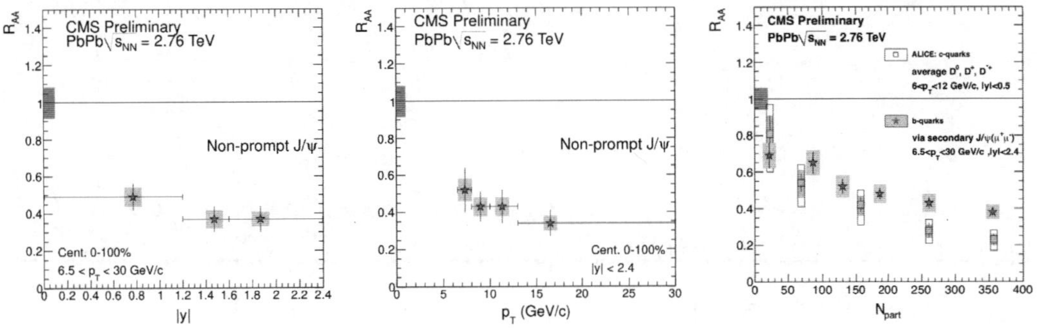

Figure 5. The non-prompt J/ψ R_{AA} with respect to rapidity*(left)*, p_T*(center)* and centrality*(right)* are presented. Centrality dependence of open charm R_{AA} via D meson measurement by ALICE is compared to non-prompt J/ψ.

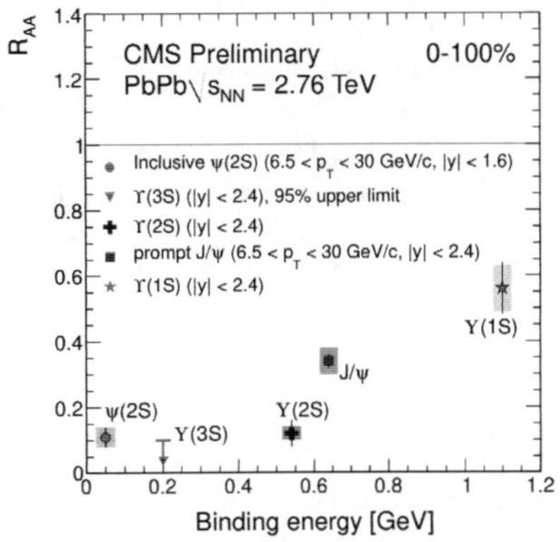

Figure 6. R_{AA} of quarkonia is presented with prompt J/ψ, $\psi(2S)$ and $\Upsilon(nS)$ as a function of their binding energy.

4. Summary

CMS has measured R_{AA} and double ratios of open and closed heavy-flavour with PbPb and pp collisions at $\sqrt{s_{NN}} = 2.76$ TeV. Quarkonium states with smaller binding energy are more suppressed than quarkonia with higher binding energy, such as $\Upsilon(1S)$. Quarkonium R_{AA} in various p_T and rapidity regions as a function of the binding energy is presented at Fig.6. The centrality, p_T and rapidity dependent suppression of non-prompt J/ψ from B meson decays has been observed. This reflects the in-medium energy loss of b-quark.

References

[1] Matsui T and Satz H 1986 *Phys. Lett.* *B*178 416.
[2] Dokshitzer Yuri L and Kharzeev D E 2001 *Phys. Lett.* *B*519 199-206
[3] CMS Collaboration 2012 *Phys. Rev. Lett* 109 222301
[4] CMS Collaboration 2012 *CMS PAS HIN-12-014* http://cdsweb.cern.ch/record/1472735
[5] CMS Collaboration 2012 *CMS PAS HIN-12-007* http://cdsweb.cern.ch/record/1455477
[6] CMS Collaboration 2008 *JINST* 3 S08004
[7] CMS Collaboration 2012 *JHEP* 05 063
[8] ALICE Collaboration 2013 *Nucl. Phys. A* 904-905 202c-209c
[9] ALICE Collaboration 2012 *JHEP* 1209 112

Journal of Physics: Conference Series **509** (2014) 012013 doi:10.1088/1742-6596/509/1/012013

Strange particle production in Monte Carlo generators in pp and pPb collisions at the LHC

K. Werner

SUBATECH, University of Nantes – IN2P3/CNRS– EMN, Nantes, France

E-mail: werner@subatech.in2p3.fr; Invited talk, SQM 2013

Abstract. Experimental data on strange particle production in p-Pb collisions and even in p-p scattering at several TeV show many similarities to the corresponding Pb-Pb results, the latter ones usually being interpreted in terms of hydrodynamic flow. We therefore analyse p-p and p-Pb data, comparing strange particle results to the simulations from various event generators: EPOS3, EPOS LHC, QGSJETII, SIBYLL, PHOJET, AMPT, PYTHIA6, PYTHIA8, HERWIG++, SHERPA. We show only selected results, a much more complete analysis can be found in [1].

Collective hydrodynamic flow seems to be well established in heavy ion (HI) collisions at energies between 200 and 2760 AGeV, whereas p-p and p-nucleus (p-A) collisions are often considered to be simple reference systems, showing "normal" behavior, such that deviations of HI results with respect to p-p or p-A reveal "new physics". Surprisingly, the first results from p-Pb at 5 TeV on the transverse momentum dependence of azimuthal anisotropies and particle yields are very similar to the observations in HI scattering [2, 3].

Do we see flow already in p-Pb or even in p-p collisions? In order to answer this question, we will analyse the predictions of different Monte Carlo generators, concerning identified (strange) particle production, a list of models is given in table 1. The EPOS3 approach [1] is the successor of EPOS2 [4, 5]. The main new ingredients: a more sophisticated treatment of nonlinear effects in the parton evolution by considering individual (per Pomeron) saturation scales [6], and a 3D+1 viscous hydrodynamical evolution. EPOS LHC [7] is a tune of EPOS1.99, containing flow put in by hand, parametrizing the collective flow at freeze-out. Both EPOS3 and EPOS LHC are based on initial conditions from Gribov-Regge multiple scattering [8]. The QGSJETII [9], SIBYLL [10], and PHOJET model [11] are also based on Gribov-Regge multiple scattering, but there is no fluid component. The main ingredients of the AMPT model [12] are a partonic cascade and a subsequent hadronic cascade, providing in this way some "collectivity" in nuclear collisions, but not in proton-proton. In addition, we will also show results from the so-called "general-purpose event generators for LHC physics" [13], as there are PYTHIA6 [14], PYTHIA8 [15], HERWIG++ [16], and SHERPA [17]. All these models are based on the factorization formula for inclusive cross sections, with a more or less sophisticated treatment of multiple scattering, whereas Gribov-Regge theory provides a multiple scattering scheme from the beginning.

There are few other studies of hydrodynamic expansion in proton-nucleus systems. In [18], fluctuating initial conditions based on the so-called Monte Carlo Glauber model (which is actually a wounded nucleon model) are employed, followed by a viscous hydrodynamical evolution. Also [19] uses fluctuating initial conditions, here based on both Glauber Monte Carlo

Journal of Physics: Conference Series **509** (2014) 012013 doi:10.1088/1742-6596/509/1/012013

Model	Theoretical concept	Flow	Ref.
EPOS3	GR	hydro	[1]
EPOS LHC	GR	parametrized	[7]
QGSJETII	GR	no	[9]
SIBYLL	GR	no	[10]
PHOJET	GR	no	[11]
AMPT	PHC	partly	[12]
PYTHIA6	Fact	no	[14]
PYTHIA8	Fact	no	[15]
HERWIG++	Fact	no	[16]
SHERPA	Fact	no	[17]

Table 1. List of models used to analyse identified particle production. "GR" stands for Gribov-Regge approach, "PHC" for partonic and hadronic cascade, "Fact" for factorization approach.

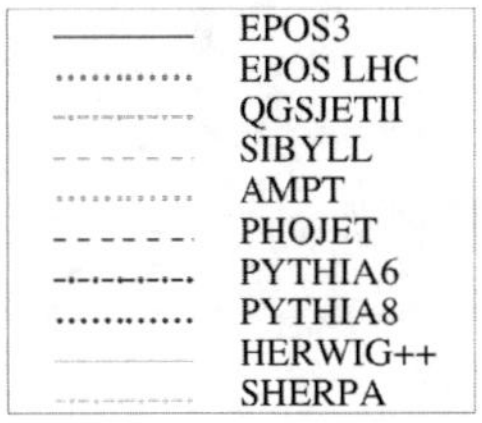

Table 2. (Color online) Line codes for the different models.

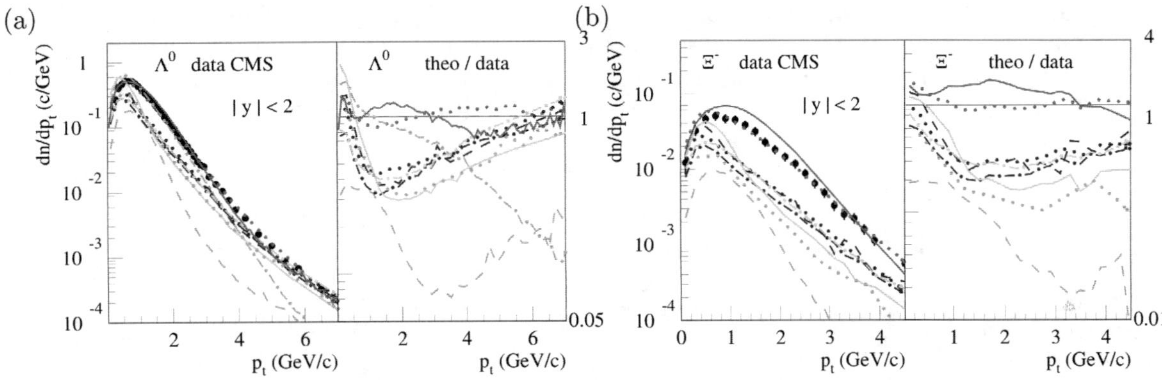

Figure 1. (Color online) **p-p scattering at 7 TeV:** Transverse momentum spectra and ratios "theory over data" of Λ baryons (a) and Ξ baryons (b). We show data from CMS [23, 24] (symbols) and simulations from the different models, using the line codes defined in tab. 2.

and Glasma initial conditions. Finally in [20], ideal hydrodynamical calculations are performed, starting from smooth Glauber model initial conditions.

We first study **p-p scattering at 7TeV**. We investigate the transverse momentum spectra of Λ baryons and Ξ baryons (see fig. 1), comparing simulations with data from CMS [23, 24], since for these particles the spectra from the different models are very different. One can distinguish three groups of models: (1) QGSJETII and SIBYLL are far off the data, they are simply not constructed to produce these kind of baryons. (2) The so-called QCD generators like PYTHIA, HERWIG, SHERPA etc show a profound "dip" in the region between 1 and 5 GeV/c, underpredicting the data by a factor of 4-5 for the Ξ baryons, and by a factor of around 3 for the Λ baryons. (3) The two EPOS versions are relatively close to the data. We recall that EPOS LHC contains collective flow (put in by hand) and EPOS3 hydrodynamic flow.

From the above study we conclude that flow seems to help to explain particle spectra. To understand better the flow contribution, we sketch in fig. 2(a) the core-corona procedure employed in EPOS to determine the flow contribution: String segments with high pt escape

Journal of Physics: Conference Series **509** (2014) 012013 doi:10.1088/1742-6596/509/1/012013

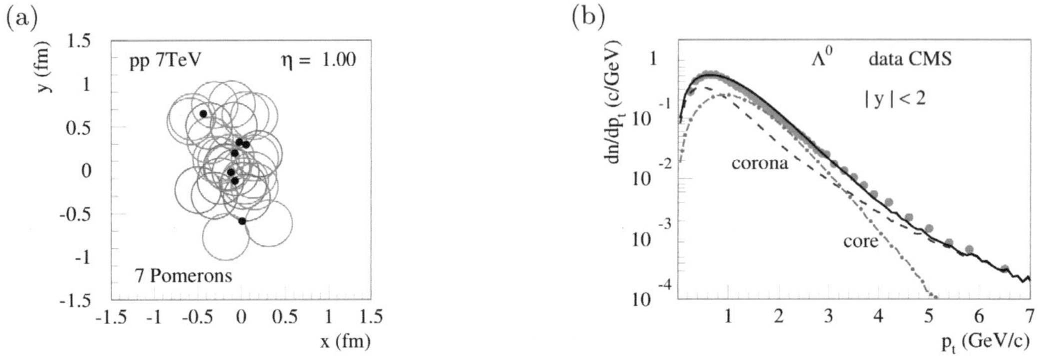

Figure 2. (Color online) **p-p scattering at 7 TeV:** (a) Core-corona separation. (b) Transverse momentum spectra of Λ baryons. We show data from CMS [23] (symbols) and simulations from EPOS3. The dashed line is the corona contributions, the dashed-dotted one the core contribution, the full line is the sums of all contributions.

(green circles, corona), the others form the "core" i.e. the initial condition for hydro (red circles). We plot in 2(b) again the transverse momentum spectra of Λ baryons, but this time only for EPOS3, also showing the corona and the core contribution. The core evolved hydrodynamically, and one can see clearly the intermediate p_t enhancement due to flow, as compared to "normal" production from (kinky) strings in the corona contributions. So we get huge flow effects for heavy particles like Λ (and Ξ) baryons.

We will now turn to **p-Pb scattering**. The CMS collaboration published a detailed study [2] of the multiplicity dependence of (normalized) transverse momentum spectra in p-Pb scattering at 5.02 TeV. The multiplicity (referred to as N_{track}) counts the number of charged particles in

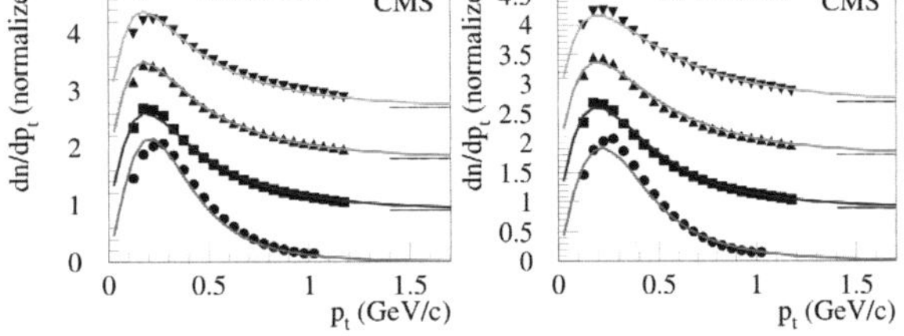

Figure 3. (Color online) Transverse momentum spectra of pions in p-Pb scattering at 5.02 TeV, for four different multiplicity classes with mean values (from bottom to top) of 8, 84, 160, and 235 charged tracks.

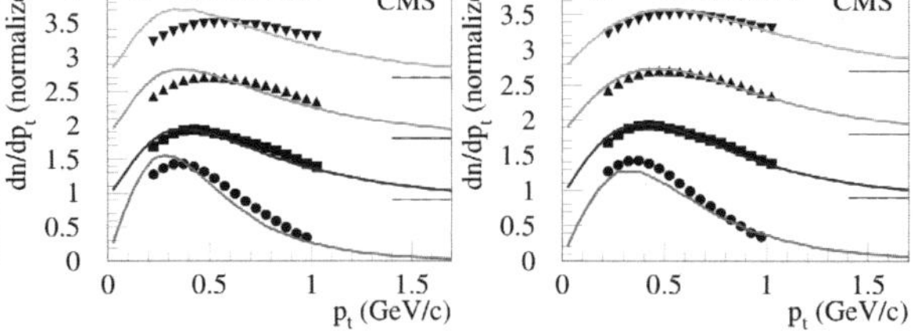

Figure 4. (Color online) Same as fig. 3, but for kaons. We show data from CMS [2] (symbols) and simulations from QGSJETII, AMPT, EPOS LHC, and EPOS3, as indicated in the figures.

Journal of Physics: Conference Series **509** (2014) 012013 doi:10.1088/1742-6596/509/1/012013

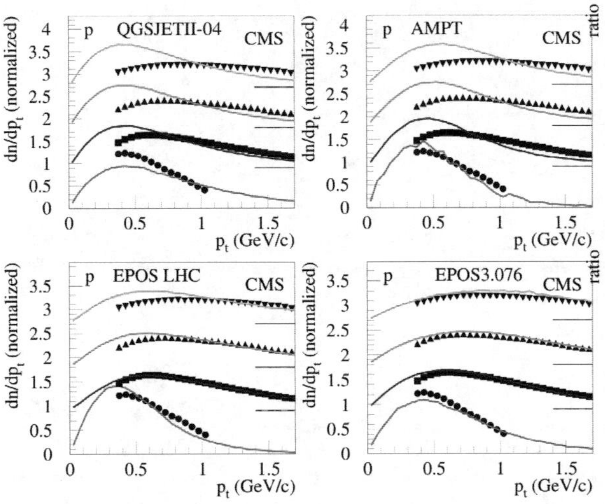

Figure 5. (Color online) Same as fig. 3, but for protons.

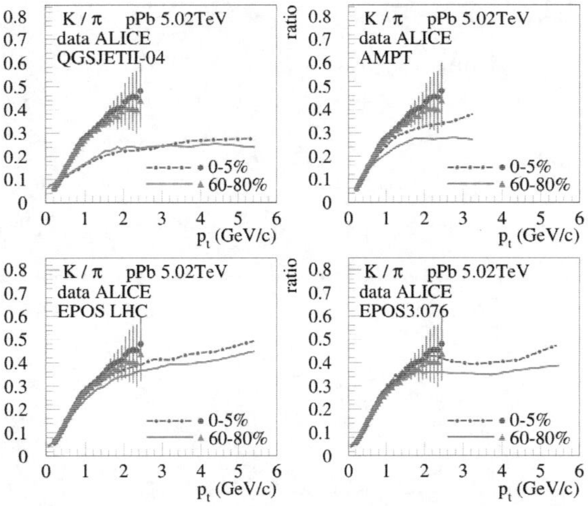

Figure 6. (Color online) Kaon over pion ratio as a function of transverse momentum in p-Pb scattering at 5.02 TeV, for the 0-5% highest multiplicity (red dashed-dotted lines, circles) and 60-80% (green solid lines, triangles).

the range $|\eta| < 2.4$. In fig. 3, we compare experimental data [2] for pions (black symbols) with the simulations from QGSJETII (upper left figure), AMPT (upper right), EPOS LHC (lower left), and EPOS3 (lower right). The different curves in each figure refer to different centralities, with mean values (from bottom to top) of 8, 84, 160, and 235 charged tracks. They are shifted relative to each other by a constant amount. Concerning the models, QGSJETII is the easiest to discuss, since here there are no flow features at all, and the curves for the different multiplicities are identical. The data, however, show a slight centrality dependence: the spectra get somewhat harder with increasing multiplicity. The other models, AMPT, EPOS LHC, and EPOS3 are close to the data.

In figs. 4, 5, we compare experimental data [2] for kaons and protons (black symbols) with the simulations. The experimental shapes of the p_t spectra change considerably, getting much harder with increasing multiplicity. In QGSJETII, having no flow, the curves for the different multiplicities are identical. The AMPT model shows some (but too little) change with multiplicity. EPOS LHC goes into the right direction, whereas EPOS3 gives a reasonable description of the data. **It seems that hydrodynamical flow helps considerably to reproduce these data.**

Also ALICE [3] has measured identified particle production for different multiplicities in p-Pb scattering at 5.02 TeV. Here, multiplicity counts the number of charged particles in the range $2.8 < \eta_{\rm lab} < 5.1$. It is useful to study the multiplicity dependence, best done by looking at ratios. In fig. 6, we show the pion over kaon (K/π) ratio as a function of transverse momentum in p-Pb scattering at 5.02 TeV, for high multiplicity (red dashed-dotted lines, circles) and low multiplicity events (green solid lines, triangles), comparing data from ALICE [3] (symbols) and simulations from QGSJETII, AMPT, EPOS LHC, and EPOS3 (lines). In all models, as in the data, there is little multiplicity dependence. However, the QGSJETII model is considerably below the data, for both high and low multiplicity events. AMPT is slightly below, whereas EPOS LHC and EPOS3 do a reasonable job. Concerning the proton over pion (p/π) ratio, fig. 7,

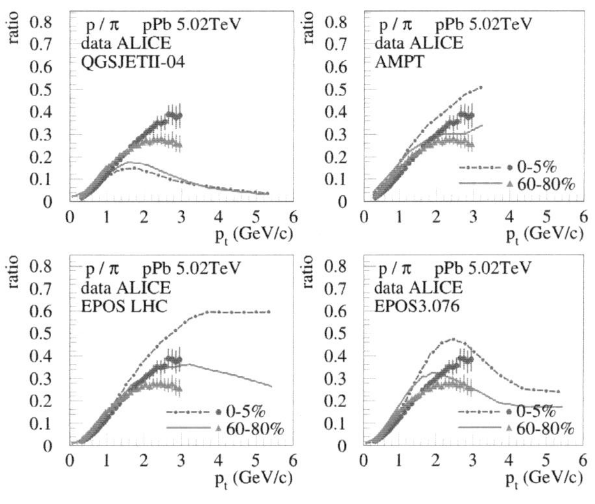

Figure 7. (Color online) Same as fig. 6, but proton over pion ratio.

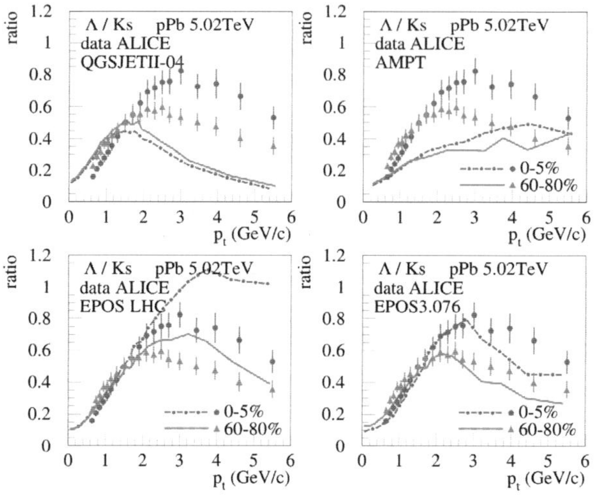

Figure 8. (Color online) Same as fig. 6, but Λ over K_s ratio.

again QGSJETII is way below the data, for both high and low multiplicity events, whereas the three other models show the trend correctly, but being slightly above the data. Most interesting are the lambdas over kaon (Λ/K_s) ratios, as shown in fig. 8, because here a wider transverse momentum range is considered, showing a clear peak structure with a maximum around 2-3 GeV/c and a slightly more pronounced peak for the higher multiplicities. QGSJETII and AMPT cannot (even qualitatively) reproduce this structure. EPOS LHC shows the right trend, but the peak is much too high for the high multiplicities. EPOS3 is close to the data.

To summarize these ratio plots (keeping in mind that the QGSJETII model has no flow, AMPT "some" flow, EPOS LHC a parametrized flow, and EPOS3 hydrodynamic flow): Flow seems to help considerably. However, from the Λ/K_s ratios, we conclude that EPOS LHC uses a too strong radial flow for high multiplicity events. The hydrodynamic flow employed in EPOS3 seems to get the experimental features reasonably well. Crucial is the core-corona procedure discussed earlier: there is more core (compared to corona) in more central collisions, but the centrality (or multiplicity) dependence is not so strong, and there is already an important core (=flow) contribution in peripheral events.

Finally, we sketch very briefly results on elliptical flow v_2 obtained from dihadron correlations, showing ALICE results [25, 26] and EPOS3 simulation, see ref. [27] for details. In fig. 9. we plot v_2 as a function of p_t. Clearly visible in data and in the simulations: a separation of the results for the three hadron species: in the p_t range of 1-1.5 GeV/c, the kaon v_2 is somewhat below the pion one, whereas the proton result is clearly below the two others. Within our fluid dynamical approach, the above results are nothing but a "mass splitting". The effect is based on an asymmetric (mainly elliptical) flow, which translates into the corresponding azimuthal asymmetry for particle spectra. Since a given velocity translates into momentum as $m_A\gamma v$, with m_A being the mass of hadron type A, flow effects show up at higher values of p_t for higher mass particles.

To summarize : Comparing experimental data on identified particle production to various Monte Carlo generators, we conclude that hydrodynamical flow seems to play an important role in p-Pb and even in p-p scattering.

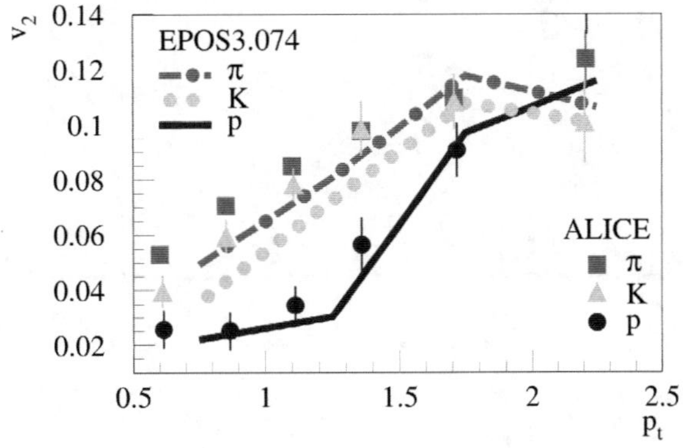

Figure 9. (Color online) Elliptical flow coefficients v_2 for pions, kaons, and protons. We show ALICE results (squares) and EPOS3 simulations (lines). Pions appear red, kaons green, protons blue.

[1] Werner K et al 2013 arXiv:1312.1233
[2] CMS collaboration 2013 arXiv:1307.3442
[3] ALICE collaboration 2013 arXiv:1307.6796
[4] Werner K, Karpenko Iu, Pierog T, Bleicher M, Mikhailov K 2010 Phys. Rev. C **82**, 044904
[5] Werner K, Karpenko Iu, Bleicher M, Pierog T, Porteboeuf-Houssais S 2012 Phys. Rev. C **85** 064907, arXiv:1203.5704
[6] McLerran L, Venugopalan R 1994 Phys. Rev. D **49** 2233; McLerran L, Venugopalan R 1994 Phys. Rev. D **49** 3352; McLerran L, Venugopalan R 1994 Phys. Rev. D **50** 2225
[7] Pierog T, Karpenko Iu, Katzy J M, Yatsenko E, Werner K 2013 arXiv:1306.5413
[8] Drescher H J, Hladik M, Ostapchenko S, Pierog T and Werner K 2001 Phys. Rept. **350**, 93
[9] Ostapchenko S 2006 Phys. Rev. D **74** 014026; Ostapchenko S 2011 Phys.Rev. D **83** 014018
[10] Ahn E J et al 2009 Phys. Rev. D **80** 094003.
[11] Engel R and Ranft J 1996 Phys. Rev. D **54** 4244.
[12] Lin Z W, Ko C M, Li B A, Zhang B and Pal S 2005 Phys. Rev. C **72** 064901
[13] Buckleya A et al. 2011 arXiv:1101.2599
[14] Sjostrand T, Mrenna S, Skands P 2006 JHEP **05** 026 hep-ph/0603175
[15] Sjostrand T, Mrenna S, Skands P 2008 Comput. Phys. Commun. **178** 852-867 arXiv:0710.3820
[16] Bahr M et al 2008 Eur. Phys. J. C **58** 639 arXiv:0803.0883
[17] Gleisberg T et al 2009 JHEP **02** 007 arXiv:0811.4622
[18] Bozek P, Broniowski W 2013 arXiv:1304.3044
[19] Bzdak A, Schenke B, Tribedy P, Venugopalan R 2013 arXiv:1304.3403
[20] Qin G Y, Mueller B 2013 arXiv:1306.3439
[21] Werner K, Karpenko Iu, Pierog T, Bleicher M, Mikhailov K 2010 Phys. Rev. C **83** 044915 arXiv:1010.0400
[22] Werner K, Karpenko Iu, Pierog T 2011 Phys. Rev. Lett. **106** 122004 arXiv:1011.0375
[23] CMS collaboration 2011 JHEP **1105** 064
[24] CMS collaboration 2011 JHEP **1105** 064 (via http://mcplots.cern.ch)
[25] ALICE collaboration 2013 Phys.Lett. B **719** 29-41 arXiv:1212.2001
[26] ALICE collaboration 2013 CERN-PH-EP-2013-115 arXiv:1307.4379
[27] Werner K, Bleicher M, Guiot B, Karpenko Iu, Pierog T 2013 arXiv:1307.4379 [nucl-th]

Traveling Through the Universe: Back in Time to the Quark-Gluon Plasma Era

Johann Rafelski[1], and Jeremiah Birrell[2]
[1]Department of Physics, and [2]Program in Applied Mathematics
The University of Arizona, Tucson, AZ 85721, USA

Abstract. We survey the early history of the discovery of quark gluon plasma and the early history of the Universe, beginning with the present day and reaching deep into QGP and almost beyond. We introduce cosmological Universe dynamics and connect the different Universe epochs with one another. We describe some of the many remaining open questions that emerge.

1. A long, long time ago in our Universe

Matter emerged from a primordial state we call quark-gluon plasma (QGP). We describe how relativistic heavy ion collision laboratory experiments shape how we think about the origin of matter, and we study the quark-hadron Universe applying these results. The discovery of QGP can be dated back to the CERN press event of February 10, 2000, when the evidence for "a new state of matter discovered at CERN" was presented. These experimental results originated in the relatively low SPS energy scale.

'QGP' is not a new particle but a paradigm-shift of how we understand matter in extreme conditions. Thus there is no clear criterion by which one can claim an experimental discovery. For this reason QGP was rediscovered again with new experimental results obtained at the order of magnitude higher RHIC collision energies. In addition to confirming the CERN results, RHIC produced new puzzling phenomena; some will be discussed below. The circumstance repeats for the third time today: LHC data confirms SPS and RHIC results, and is offering another rich field of new experimental results.

Since no one plans to announce the QGP discovery at the LHC, we conclude that QGP has gained considerable acceptance as a new form of matter. This then suggests using this new paradigm in the study of other phenomena. Here we will travel back in time to explore the properties of the Universe when it was much hotter and much denser, journeying beyond the era of hadrons into the quark Universe. We will briefly visit three stages in the Universe evolution, beginning from today's condition as we go back in time:

(i) The period reaching back to shortly after big-bang nucleosynthesis (BBN) completed is the directly 'observable' history of the Universe, in the sense that some aspects of this evolution are directly visible today and are being analyzed in depth. Our knowledge is based primarily on two observables. The cosmic microwave background (CMB), the red-shifted heat of the big-bang, reaches us without much interaction from an epoch which can be dated back to about 1/1000 of the age of the Universe. The cosmic abundance of light elements is interpreted as the consequence of nuclear burning in the early Universe that ended when the Universe was a fraction of an hour old.

(ii) The era connecting the period just after baryon antimatter disappeared to BBN passes through 7 orders of magnitude in time, from 1ms to 1000 s. The Universe emerging from the baryon antimatter annihilation phase still comprises muons and pions. When these disappear the Universe expands, pushed apart by effectively massless $e^+e^-\gamma$-QED plasma, and neutrinos. The three neutrino flavors decouple just before BBN and become a separate radiation fraction, while e^+e^--annihilation continues through the BBN process. There is good hope of identifying observables from this period which could still be connected to the present day Universe. Current interest is centered on improving the understanding of the energy content and momentum distribution of neutrino cosmic background radiation.

(iii) The earliest stage we can now begin to discuss in detail is the evolution of the Universe from about 10ps to the end of QGP hadronization at 20μs and the end of baryon antimatter annihilation at 1 ms, thus 8 orders of magnitudes in time scale. During this period we see the Universe emerge from the electro-weak symmetry restored stage into the quark stage, and evolve into a dense hadron stage, which in turn disappears when both antibaryons annihilate and mesons decay. It is in this period that our recent experimental heavy ion insights into the quark-hadron Universe become of use.

Because the QGP hadronization process is complex and could involve inhomogeneity of the matter distribution, there has always been the intuitive feeling that some macroscopic observables will ultimately emerge that relate to matter distribution in the Universe. Here we recall the mystery of the large scale structure of visible matter, which is localized to the surface or even to the edges of bubble-like boundaries with a scale difference of matter structure thickness size to the matter void size being as large as 50.

After presenting in section 2 a few motivational remarks about the process of discovery and importance of QGP, we offer in section 3 a short introduction to the essential elements of cosmology required to follow our trip back in time. In sections 4, 5, 6 we offer in turn discussions of these stages of the Universe evolution. We connect the three eras and describe some of the implications of our findings in section 7. We complete our discussion with a short listing of open questions in section 8. The results and discussion presented here complement our recent related considerations which were addressing specific questions about the QGP era in the Universe [1, 2].

2. How we got to QGP from the hadron side

The 20 year long march to the QGP announcement at CERN also started at CERN. In the early days this was the one and only location in the world with a considerable tradition of thermal particle physics. One of us (JR) arrived at CERN in 1977 when Hagedorn temperature $T_H = 160$ MeV was already a local trademark, introduced to characterize multiparticle production in what were at the time high energy *pp* collisions. In search of an explanation of the cause for this thermal behavior, a rather unexpected novel feature of particle physics, Rolf Hagedorn proposed that point hadrons could be made of other hadrons, within a 'bootstrap' concept. Once quarks as building blocks of hadrons were introduced, the bootstrap model in which hadrons are made of hadrons was justified in terms of small domains in space filled with quarks. However, the quark-hadron bootstrap was theoretically consistent only once hadrons were allowed to have a proper volume. This refined model allowed the study of how hadrons, occupying a finite volume and stressed in collision, dissolve into their constituents, creating in this way a hot blob of quarks we call QGP. At the 1984 Quark Matter meeting in Helsinki, Hagedorn described these developments in the last pages of the very accurate historical reminiscences "How we got to QCD matter from the hadron side by trial and error," [3].

With Hagedorn reporting the tale of the theoretical formulation of QGP formation in heavy ion collisions nearly 30 years ago, and strangeness as a signature of QGP already a well developed idea at that time [4], what exactly have we done since? Many ideas and developments were devoted to alternate diagnostic tools, and many related novel theoretical ideas emerged. The situation is presented succinctly within the abstract of a recent review of Jacak and Muller [5] "...this new state of matter is a nearly ideal, highly opaque liquid. A description based on string theory and black holes in five dimensions has made the quark-gluon plasma an archetypal strongly coupled quantum system."

These courageous words relate to the finding that QGP matter flows with minimal friction. This is somewhat surprising since we also observe both at RHIC and prominently at LHC, that in QGP high energy parton jets rapidly quench. These two observations in fact are a contradiction, since QCD naively predicts just the opposite, specifically that at the low energy scales the QCD interactions should be hindering flow, and at the high energy scale of jets, the parton interaction is weakening, removing quenching. Clearly we do have some remaining theoretical challenges in the study of the behavior and properties of QGP – we note that no one is able to precisely tell based on QCD when the expanding QGP 'hadronizes', breaking up into individual hadrons, and conversely, when as function of heavy ion collision energy we form QGP.

An important epistemological question is how the introduction of QGP into the physics vocabulary changed our global understanding of natural phenomena:
a) QGP exhibits quark deconfinement.
b) Conversely, the quark confinement effect makes hadrons massive, explaining the origin of 99% of the mass of matter [6].
c) Quark confinement is recognized as a transport property of empty space, the structured quantum vacuum, rather than a direct property of the quark-quark interaction.
d) The dominant interaction of quarks (and gluons) is with empty space, the quantum vacuum, a major departure from two body interactions being the dominant force in elementary processes.

This insight creates another paradigm shift; the vacuum state is strongly structured. With this, a further challenge arises relating to the understanding of quantum vacuum structures both in principle and at a technical level. The technical effort has been centered for many years on the ever improving lattice-QCD description of the deconfinement phenomenon [7]. However, we lack a true in principle understanding of how the quantum vacuum structure acts. What is the mechanism that we could present in the classroom to explain why the mass of a small quark bubble such as a proton is at the scale of 1 GeV and not 0.5 or 2 GeV? Where exactly does the scale come from?

3. Standard Cosmology

In order to travel back in time to the period when QGP dominated the Universe, we first need to elaborate on the relation between the expansion dynamics of the Universe and temperature. For this purpose we need some preparation in the standard cosmological (FLRW-Universe) model. We use the spacetime metric

$$ds^2 = c^2dt^2 - a^2(t)\left[\frac{dr^2}{1-kr^2} + r^2(d\theta^2 + \sin^2(\theta)d\phi^2)\right] \tag{1}$$

characterized by the scale parameter $a(t)$ of a homogeneous spatial Universe. The geometric parameter k identifies the geometry of the spacial hypersurfaces defined by comoving observers. Space is a flat-sheet for the observationally preferred value $k = 0$ [8]. The global Universe dynamics can be characterized by two quantities: the Hubble parameter H, a strongly time dependent quantity on cosmological time scales, and the deceleration parameter q:

$$\frac{\dot{a}}{a} \equiv H(t), \quad \frac{\ddot{a}}{a} = -qH^2, \quad q \equiv -\frac{a\ddot{a}}{\dot{a}^2}, \quad \dot{H} = -H^2(1+q). \tag{2}$$

The Einstein equations are:

$$G^{\mu\nu} = R^{\mu\nu} - \left(\frac{R}{2} + \Lambda\right) g^{\mu\nu} = 8\pi G_N T^{\mu\nu}, \quad R = g_{\mu\nu}R^{\mu\nu}, \quad T^\mu_\nu = \mathrm{diag}(\varepsilon, -P, -P, -P). \tag{3}$$

It is common to absorb the Einstein cosmological constant Λ into the energy and pressure

$$\varepsilon_\Lambda = \frac{\Lambda}{8\pi G_N}, \qquad P_\Lambda = -\frac{\Lambda}{8\pi G_N} \tag{4}$$

and we implicitly consider this done from now on.

One should note that the bag constant $\mathcal{B}$ of the quark-bag model has the same behavior in regard to energy and momentum as has the Einstein cosmological parameter $\mathcal{B} \leftrightarrow \Lambda/8\pi G_N$. $\mathcal{B}$ adds positively to the energy density but negatively to the pressure, counteracting the positive particle pressure. Contrary to the initial expectation based on the quark-bag model where quark pressure is in equilibrium with the bag constant, in the dynamical Universe the appearance of such a bag term will accelerate the expansion just like today where we see an acceleration due to dark energy. The parallel meaning of $\mathcal{B}$ and $\Lambda/8\pi G_N$ relies on both quantities acting within the volume of their respective 'Universe', in the sense that $\mathcal{B}$ is strictly and only present within the volume of quark blobs – hadrons, or QGP.

Two dynamically independent equations arise using the metric Eq. (1) in Eq. (3):

$$\frac{8\pi G_N}{3}\varepsilon = \frac{\dot{a}^2 + k}{a^2} = H^2\left(1 + \frac{k}{\dot{a}^2}\right), \qquad \frac{4\pi G_N}{3}(\varepsilon + 3P) = -\frac{\ddot{a}}{a} = qH^2. \tag{5}$$

We can eliminate the strength of the interaction, G_N, solving both these equations for $8\pi G_N/3$, and equating the result to find a relatively simple constraint for the deceleration parameter:

$$q = \frac{1}{2}\left(1 + 3\frac{P}{\varepsilon}\right)\left(1 + \frac{k}{\dot{a}^2}\right). \tag{6}$$

For $k = 0$ note that in a matter-dominated Universe where $P/\varepsilon << 1$ we have $q \simeq 1/2$; for a radiative Universe where $3P = \varepsilon$ we find $q \simeq 1$; and in a dark energy Universe in which $P = -\varepsilon$ we find $q = -1$. We will work here within the standard cosmology model with $k = 0$.

As must be the case for any solution of Einstein's equations, Eq. (5) implies that the energy momentum tensor of matter is divergence free:

$$T^{\mu\nu}||_\nu = 0 \Rightarrow -\frac{\dot{\varepsilon}}{\varepsilon + P} = 3\frac{\dot{a}}{a} = 3H. \tag{7}$$

The same relation also follows from conservation of entropy, $dE + PdV = TdS = 0$, $dE = d(\varepsilon V)$, $dV = d(a^3)$. A dynamical evolution equation for $\varepsilon(t)$ arises once we combine Eq. (7) with Eq. (5), eliminating H. Given an equation of state $P(\epsilon)$, solution of this equation describes the dynamical evolution of matter in the Universe. We can study the entropy conserving Universe evolution both forwards and backwards in time. However, there are entropy producing processes such as antibaryon annihilation which will need further attention in the future.

4. Back in time to BBN

Given equations of state connecting ε and P for the various components of the Universe matter content, Eq. (5) and Eq. (7) for $k = 0$ are immediately soluble and the outcome is widely discussed for the period of time spanning BBN to the present era, passing through photon freeze-out at the ion recombination era. In the following we use the mix of matter (31%) and dark energy (69%) with photon and neutrino backgrounds favored by the latest Planck results [8], where we gave two neutrino species mass of $m_\nu = 30$ meV and a third neutrino remains massless – other mass choices are possible within present day constraints and will impact to some degree where exactly matter dominance emerges from the radiative Universe. We presume that neutrino kinetic freeze-out completed before the onset of e^+e^--annihilation into photons. This means that

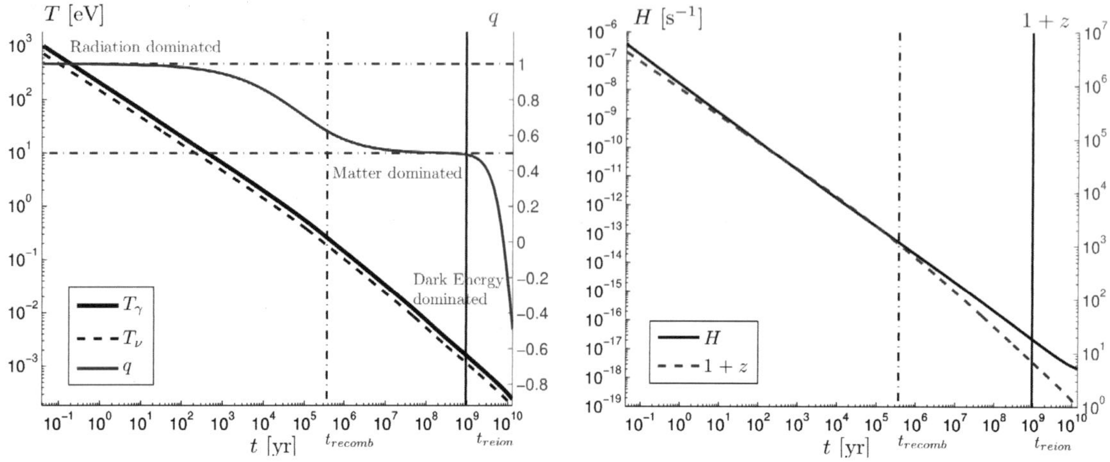

Figure 1. From the present day until near BBN: on left – evolution of temperature T and deceleration parameter q; on right – evolution of the Hubble parameter H and redshift z.

all entropy flows from e^+e^--annihilation into the photons and leads to the ambient temperature of neutrinos in the current epoch being lower by a factor $(4/11)^{1/3}$. This is a common simplifying assumption, but small refinements are possible. The currently accepted model of the neutrino freeze-out is given in [9].

Figure 1 shows in the left frame the temperature (left axis) and deceleration parameter (right axis) from shortly after the completion of BBN until today. This type of presentation will be repeated below for the two earlier eras; the connection from BBN to QGP and from QGP to the electroweak (EW) transition. The horizontal dot-dashed lines show the pure radiation-dominated value of $q = 1$ and the matter-dominated value of $q = 1/2$. The expansion in this era starts off as radiation-dominated, but transitions to matter-dominated starting around $T = \mathcal{O}(10\text{ eV})$ and begins to transition to a dark energy dominated era at $T = \mathcal{O}(1\text{ meV})$. We are still in the midst of this transition today. The dashed line shows the neutrino temperature, which maintains a constant ratio of $T_\nu/T_\gamma \approx (4/11)^{1/3}$ since neutrino decoupling and e^+e^- annihilation around $T = \mathcal{O}(1\text{ MeV})$.

The vertical dot-dashed lines show the time of recombination at $T \simeq 0.3$ eV and reionization at $T \simeq \mathcal{O}(1\text{ meV})$. On the right in figure 1 we show the Hubble parameter H and redshift $z + 1 \equiv a_0/a(t)$. We can see in figure 1 a visible deviation from power law behavior due to the transitions from radiation to matter dominated and from matter to dark energy dominated expansion. These transitions are accentuated and more easily visualized in the form of the deceleration parameter q. The time span covered by the figure 1 is in essence the entire lifespan of the Universe. All that happens before is a blip in comparison, but of course on a logarithmic time scale, the two eras we study next are of comparable duration.

5. From BBN passing through QED-neutrino plasma to the hadron Universe

The era separating the photon-neutrino-matter-dark energy Universe we just described from the QGP and hadron gas Universe is quite complex in its evolution. It begins when the number of e^+e^--pairs has decayed to the same abundance as the number of baryons in the Universe at the temperature $T = \mathcal{O}(10\text{ keV})$. It reaches back to $T = \mathcal{O}(30\text{ MeV})$ where there is no baryon antimatter as yet – no antiprotons, antineutrons, antilambdas. However there is significant abundance of muons, mesons, strangeness. Seen from the QGP side, this era begins after all baryon antimatter formed in QGP hadronization has annihilated. Pions and muons disappear

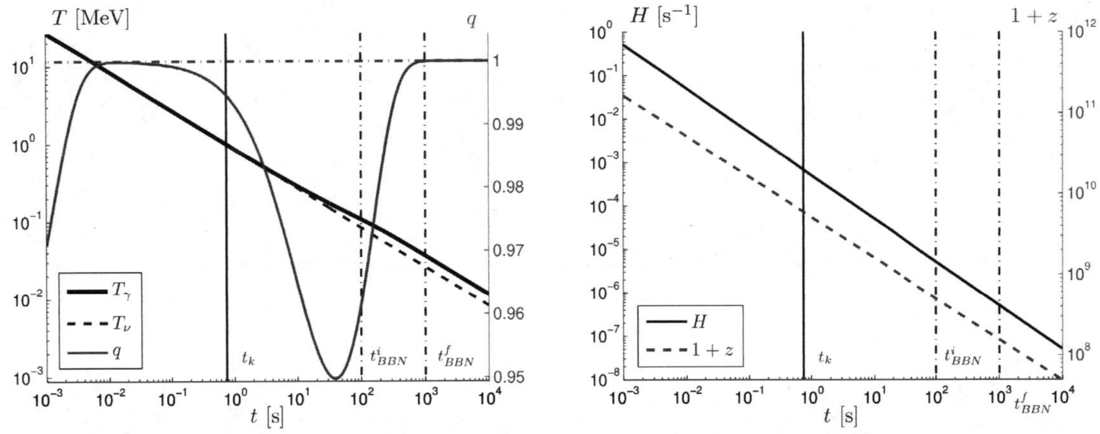

Figure 2. From the end of baryon antimatter annihilation through BBN, see figure 1.

in abundance as compared to residual nucleons at $T \simeq \mathcal{O}(5\text{ MeV})$, and we pass through neutrino kinetic freeze-out at $T \equiv T_k \simeq \mathcal{O}(1\text{ MeV})$ followed by $e^\pm$ annihilation and photon reheating at $T = \mathcal{O}(m_e)$, and big bang nucleosynthesis (BBN) at $T = \mathcal{O}(100\text{ keV})$ which actually is still immersed in a rich e^+e^--pair 'bath'.

In figure 2 the horizontal dot-dashed line for $q = 1$ shows the pure radiation dominated value with two exceptions. First, the presence of massive pions and muons reduce the value of q near to the maximal temperature shown. Second, when temperature is near the value of the electron mass the e^+e^--pairs are not yet fully depleted but already sufficiently non-relativistic to cause another dip in q. The dashed line shows the neutrino temperature, which decouples from the $e^\pm$ and photon temperature at $T = \mathcal{O}(1\text{ MeV})$ when neutrinos freeze-out and begin free streaming. In figure 2 the time unit is seconds, and the range spans the domain from fractions of a millisecond to a few hours. We left some time gap between this and the domain shown in figure 1 describing the current era – there is an uneventful evolution between the two domains.

6. Connecting with the Quark Universe

Going further back in time, the integration through the baryon antimatter requires a detailed study of the hadron chemistry using the methods originally developed for heavy ion collision physics as presented in Refs. [1, 10]. There are some structural uncertainties in doing this computation which were discussed in these references. One should further note that the Universe was for a short time filled with mesons, and most remarkably, a great number of diverse baryonic pairs. At a time of 10–15μs hadrons dissolve in the QGP phase. Freely propagating quarks or gluons filled the early Universe – quarks remain unbound for as long as the Universe is above the critical temperature $T_c \simeq 155 \pm 10\,\text{MeV}$. The Universe is pushed apart by the thermal pressure generated predominantly by relativistic particles and thanks to the presence of numerous gluons which have additional degeneracy 8 from color charge compared to photons, and essentially massless u, d, s quarks, Universe expansion was rather rapid.

Figure 3 shows the temperature (left axis) and deceleration parameter (right axis) from the end of the electroweak era at $T = \mathcal{O}(100\text{ GeV})$, when the electroweak symmetry was broken and particles became massive, through the quark gluon plasma (QGP) era and up to the QGP phase transition at $T = \mathcal{O}(150\text{ MeV})$. The horizontal dot-dashed line shows the pure radiation dominated value of $q = 1$. From the right hand scale, one can see that the expansion in this era is almost purely radiation dominated. There are two smaller deviations from $q = 1$ that correspond to T on the order of the top, Higgs, Z, and W masses in the first case, and the bottom,

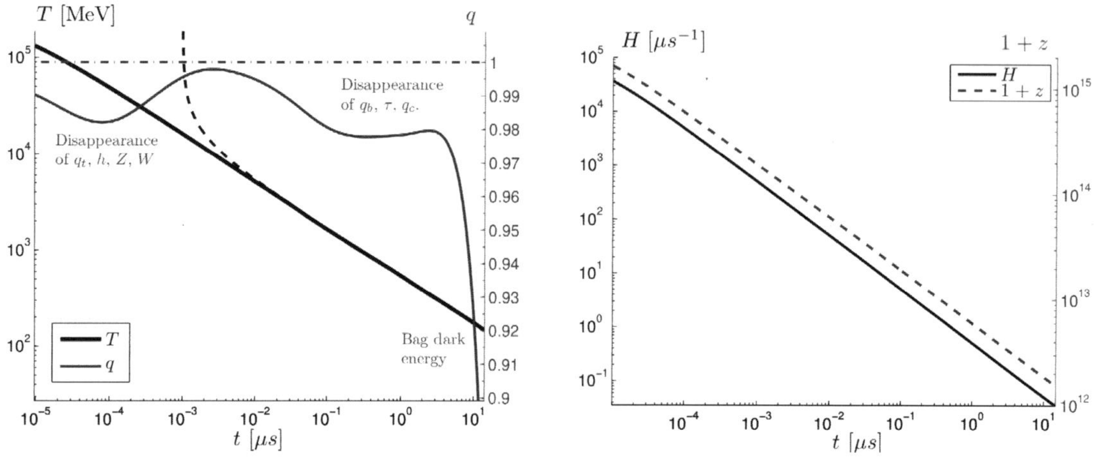

Figure 3. From before our particle world to end of quark-gluon plasma, see figure 1.

tau, and charm masses in the other. The comparatively larger drop in q as T approaches the QGP phase transition temperature arises because we injected the bag constant dark energy to demonstrate how a first order phase transition would accelerate the Universe dynamics. On the right in figure 3 we show the Hubble parameter and redshift through the QGP phase. Because the Universe is very nearly radiation dominated during this era, they obey simple power laws to high precision. All nuances that can be seen in the deceleration parameter are lost at the double-logarithmic resolution level.

The solution presented is the consequence of a forward integration of Einstein equations. A naive (i.e. ignoring any physics emerging at higher energy, including the details of the EW phase transition) backward integration establishes a pico-second timescale between the big bang and the end of the electroweak era. The precise timing is insignificant for the subsequent evolution and properties of the QGP era, as can be seen in the dashed line in figure 3, where the starting time was shifted by 1ns. On the other hand, this means that we can precisely show the time scale at which the presumed hadronization condition will be reached for a free quark-gluon gas. Inclusion of QCD interactions can be a future task.

Figure 4 zooms-in on the temperature range where the QGP phase transition occurs, with several values of the bag energy density $\mathcal{B}$. We see that the temperature dynamics are relatively insensitive to $\mathcal{B}$. On the other hand, $\mathcal{B}$ impacts strongly the critical temperature T_c where the particle pressure cannot continue to support the deconfined state. This determines the onset of the transition which will be determined entirely by the appropriate treatment of QCD degrees of freedom near the transition condition. Considering that quark Universe hadronization is expected at $T = 155 \pm 10$ MeV within the QCD-lattice study of QGP[7], we recognize in figure 4 the probable life span of 10–15μs of the quark Universe. The reason that we used a forward integration in the QGP era is that there are quite a few uncertainties about the dynamics and physics of the slow hadronization process and we cannot yet fully characterize the time scales at which hadronization is complete as is discussed in in Refs. [1, 10]. This also means that the redshift curve in figure 3 is only a (perhaps crude) approximation, as it will depend on the precise evolution of the scale factor $a(t)$ during hadronization.

7. Degrees of freedom and connection of all three eras of evolution

Before the QGP era, when the Higgs vacuum was not frozen, all particles were nearly massless, possibly retaining mass at the scale we observe today for neutrinos. Later by the Higgs

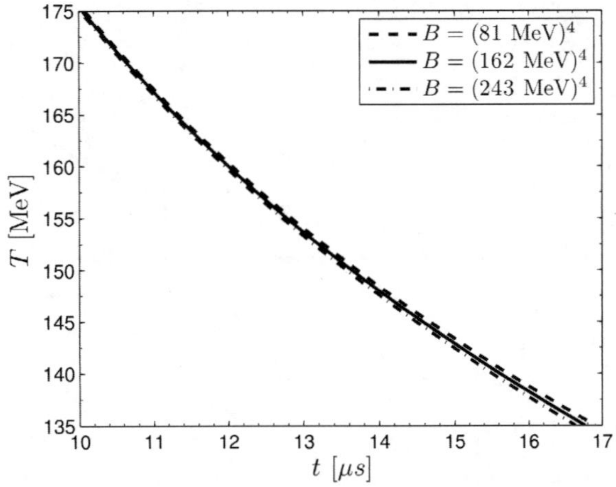

Figure 4. Evolution of temperature T near the QGP phase transition for several values of the bag energy density.

mechanism mass was given to many of the QGP era particles. Before the Higgs vacuum froze, the Universe was pushed apart by 28 bosonic and 90 fermionic degrees of freedom. Let us verify: the doublet of charged Higgs particles has $4 = 2 \times 2 = 1 + 3$ degrees of freedom – three will migrate to the longitudinal components of $W^{\pm}, Z$ when the electro-weak vacuum freezes and the EW symmetry breaking arises, while one is retained in the one single dynamical charge neutral Higgs component. In the massless stage, the SU(2)×U(1) theory has 4×2=8 gauge degrees of freedom where the first coefficient is the number of particles ($\gamma, Z, W^{\pm}$) and each massless gauge boson has two transverse polarizations. Adding in $8_c \times 2_s = 16$ gluonic degrees of freedom we obtain 4+8+16=28 bosonic degrees of freedom.

The count of fermionic degrees of freedom includes three f families, two spins s, another factor two for particle-antiparticle duality. We have in each family of flavors a doublet of $2 \times 3_c$ quarks, 1-lepton and 1/2 neutrinos (due left-handedness which was not implemented counting spin). Thus we find that a total $3_f \times 2_p \times 2_s \times (2 \times 3 + 1_l + 1/2 - \nu) = 90$ fermionic degrees of freedom. We further recall that massless fermions contribute 7/8 of that of bosons in both pressure and energy density. Thus the total number of massless non-interacting particles at a temperature above the top quark mass scale, referring by convention to bosonic degrees of freedom, is is $g_{\rm SM} = 28 + 90 \times 7/8 = 106.75$

At times where dimensional scales are irrelevant and the expansion is radiative, entropy conservation means that temperature scales inversely with the scale factor $a(t)$. This follows from Eq. (7) when $\varepsilon \simeq 3P \propto T^4$. However, as the temperature drops and at their respective $m \simeq T$ scales, successively less massive particles annihilate and disappear from the thermal Universe. Their entropy reheats the other degrees of freedom and thus in the process the entropy originating in a massive degree of freedom is shifted into the other, still present effectively massless degrees of freedom. This causes the $T \propto 1/a(t)$ scaling to break down; during each of these 'reorganization' periods the drop in temperature is slowed by the concentration of entropy in fewer degrees of freedom. To quantify this we study the ratio

$$r_U \equiv \left(\frac{1+z}{T/T_{\rm now}}\right)^3, \qquad 1+z \equiv \frac{a_{\rm now}}{a(t)}. \tag{8}$$

The third power is chosen to connect with the growth of the volume of the Universe.

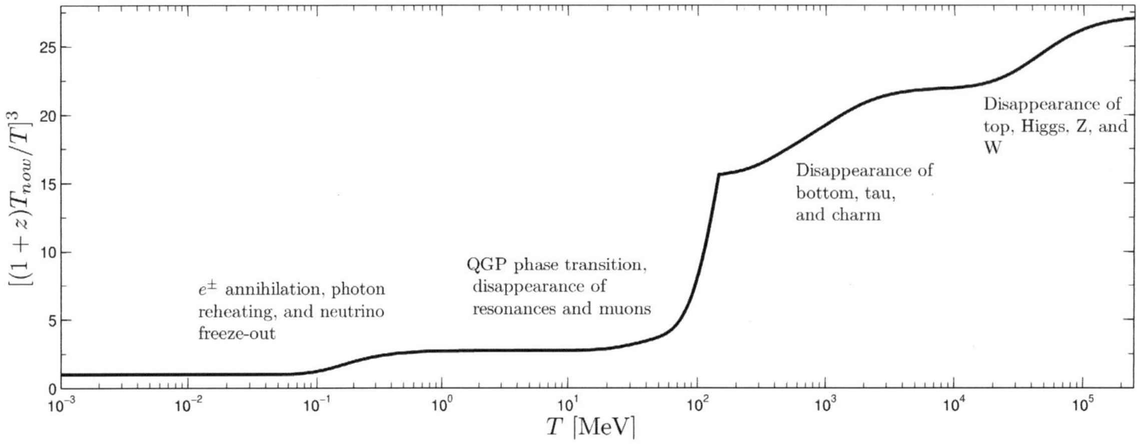

Figure 5. Disappearance of degrees of freedom through the evolution of the Universe in time and how this affects the fractional drop of temperature compared to red-shift.

In figure 5 we show ratio r_U Eq. (8) as a function of photon temperature T from the primordial high temperature T in early Universe on right to the present on the left, where r_U must be by definition unity. The periods of change seen in figure 5 come when the temperature crosses the mass of a particle species that is in equilibrium. One can see drops corresponding to the disappearance of particles as indicated. After e^+e^- annihilation on left, there are no significant degrees of freedom remaining to annihilate and feed entropy into photons, and so r_U remains constant until today. In producing this plot, we used a very rough model for the QGP-hadron phase transition.

As long as the dynamics are at least approximately entropy conserving, the total drop in r_U is entirely determined by entropy conservation. Namely, the magnitude of the drop in r_U figure 5 is a measure of the inflation factor of the Universe, beyond what would be estimated from the thermal redshift. We now show that it is also a measure of the number of degrees of freedom that have disappeared from the Universe. Consider two times t_1 and t_2 at which all particle species that have not yet annihilated are effectively massless. By conservation of comoving entropy and scaling $T = 1/a$ we have

$$1 = \frac{a_1^3 S_1}{a_2^3 S_2} = \frac{a_1^3 \sum_i g_i T_{1,i}^3}{a_2^3 \sum_j g_j T_{2,j}^3}, \qquad \frac{r_{U1}}{r_{U2}} = \frac{\sum_i g_i (T_{1,i}/T_{1,\gamma})^3}{\sum_j g_j (T_{2,j}/T_{2,\gamma})^3} \tag{9}$$

where the sums are over the total number of degrees of freedom present at the indicated time and the degeneracy factors g_i contain the 7/8 factor for fermions. In the second form we divided the numerator and denominator by $a_{now}T_{\gamma,now}$. We distinguish between the temperature of each particle species and our reference temperature, the photon temperature. This is important since today neutrinos are colder than photons, due to e^+e^--reheating occurring after neutrinos decoupled.

Using Eq. (9) we compute the total drop in r_U shown in figure 5: at $T = \mathcal{O}(100\text{ GeV})$ the number of active degrees of freedom is slightly below $g_{\rm SM} = 106.75$ due to the partial disappearance of top quarks, but this approximation will be good enough for our purposes. At this time, all the species are in thermal equilibrium with photons and so $T_{1,i}/T_{1,\gamma} = 1$ for all i. Today we have 2 photon and $7/8 \times 6$ neutrino degrees of freedom with a neutrino to photon reheating ratio of approximately $T_\nu/T_\gamma = (4/11)^{1/3}$. Therefore we have

$$\frac{r_{100GeV}}{r_{now}} = \frac{g_{SM}}{g_{\text{now}}} = \frac{106.75}{2 + \frac{7}{8} \times 6 \times \frac{4}{11}} = 27 \tag{10}$$

which is the fractional change we see in figure 5. The meaning of this factor is that the Universe inflated by a factor 27 above the thermal red shift scale as massive particles disappeared successively from the inventory.

8. Conclusions

Many open challenges of cosmic evolution deserving in depth investigation come to mind. These are related to QGP-hadronization and matter emergence from baryon-antimatter annihilation [1, 2], as well as neutrinos transiting to a free-streaming decoupled gas [13]. We list a few that we are either ready to address and/or we already mentioned in this report:

(i) The role of equilibrated flavor physics in u, d, s, c, b QGP and its hadronization;
(ii) Degree of matter inhomogeneity at QGP hadronization;
(iii) Strangeness is present in a significant abundance in early Universe down to $T = 10$ MeV: potential for production of strange nuclearites;
(iv) What happens during baryon antimatter annihilation, and how much entropy is produced;
(v) How did the process of neutrino kinetic decoupling occur and in particular when precisely;
(vi) Are there effects in BBN due to the presence of dense e^+e^--plasma.

We conclude by recalling that the baryon-matter asymmetry is an enigmatic riddle [14] of cosmology, of particle physics, and arguably, could be part of the physics of quark-gluon plasma and its hadronization transformation into matter. The study of QGP touches on other foundational challenges such as how transport properties of the structured quantum vacuum, depending on temperature, allow quarks to move freely in the Universe, while when this motion freezes, the mass of particles is created.

Acknowledgments

This work has been supported by a grant from the U.S. Department of Energy, grant DE-FG02-04ER41318. JB is supported by the Department of Defense (DoD) through the National Defense Science & Engineering Graduate Fellowship (NDSEG) Program.

References

[1] Fromerth M J, Kuznetsova I, Labun L, Letessier J, and Rafelski J 2012 *Acta Phys. Polon.* B **43** 2261-84
[2] Rafelski J 2013 Nucl. Phys. Proc. Suppl. **243-244** 155
[3] Hagedorn R 1985 *Lect. Notes Phys.* **221** 53-76
[4] Rafelski J 1982 *Phys. Rept.* **88** 331-47
[5] Jacak B V and Muller B 2012 *Science* **337** 310-14
[6] Kronfeld A S 2013 "Lattice Gauge Theory and the Origin of Mass" in *100 Years of Subatomic Physics* Henley E M, and Ellis S D eds.(World Scientific), pp. 493-518
[7] Philipsen O 2013 *Prog. Part. Nucl. Phys.* **70** 55-107
[8] Ade P A R *et al.* [Planck Collaboration] 2013 Planck 2013 results. XVI. Cosmological parameters *Preprint arXiv:1303.5076*
[9] Mangano G, Miele G, Pastor S, Pinto T, Pisanti O and Serpico P D 2005 *Nucl. Phys.* B **729** 221-34
[10] Fromerth M J and Rafelski J 2002 Hadronization of the Quark Universe *Preprint astro-ph/0211346*
[11] Beringer J *et al.* [Particle Data Group] 2012 *Phys. Rev. D* **86** 010001
[12] Letessier J and Rafelski J 2002 *Hadrons and Quark-Gluon Plasma, Camb. Monogr. Part. Phys. Nucl. Phys. Cosmol.* **18** (Cambridge, UK)
[13] Birrell J, Yang C T, Chen P and Rafelski J 2013 Mod. Phys. Lett. A **28** 1350188
[14] Canetti L, Drewes M and Shaposhnikov M 2012 *New J. Phys.* **14** 095012

Journal of Physics: Conference Series **509** (2014) 012015 doi:10.1088/1742-6596/509/1/012015

Quark-gluon plasma phenomenology from the lattice

Chris Allton[1,*], Gert Aarts[1], Alessandro Amato[1,2], Wynne Evans[1,3], Pietro Giudice[1,4], Simon Hands[1], Aoife Kelly[5], Seyong Kim[6], Maria-Paola Lombardo[7], Sinead Ryan[8], Jon-Ivar Skullerud[5], Tim Harris[8]

[1] Department of Physics, College of Science, Swansea University, Swansea SA2 8PP, U.K.
[2] Institute for Theoretical Physics, Universität Regensburg, D-93040 Regensburg, Germany
[3] Fakultät für Physik, Universität Bielefeld, D-33615 Bielefeld, Germany
[4] Institut für Theoretische Physik, Universität Münster, Germany
[5] Department of Mathematical Physics, NUIM, Maynooth, Co. Kildare, Ireland
[6] Department of Physics, Sejong University, Seoul 143-747, Korea
[7] INFN-Laboratori Nazionali di Frascati, I-00044, Frascati (RM) Italy
[8] School of Mathematics, Trinity College, Dublin 2, Ireland
[*] Speaker

E-mail: c.allton@swan.ac.uk

Abstract.
The FASTSUM Collaboration has calculated several quantities relevant for QCD studies at non-zero temperature using the lattice technique. We report here our results for the (i) interquark potential in charmonium; (ii) bottomonium spectral functions; and (iii) electrical conductivity. All results were obtained with 2+1 flavours of dynamical fermions on an anisotropic lattice which allows greater resolution in the temporal direction.

1. Introduction
The Particle Data Book [1] is a repository of particle physics knowledge, and yet it contains no entries on the deconfined phase of QCD. We present here some lattice calculations of phenomena in the quark-gluon plasma (QGP) phase with the ultimate aim of addressing this omission.

The lattice approach is based on the non-perturbative study of correlation functions of operators calculated in a background of glue and dynamical (sea) quarks using the imaginary time (i.e. Euclidean) formulation[1]. For instance, to study the η_c system, correlators, $C(\tau)$, of the operator $J = \overline{c}\gamma_5 c$ are determined (c is the charm quark field). Each state i which has the same quantum numbers as J (and therefore can be excited by it) contributes a term $\sim \mathrm{e}^{-M_i\tau}$ to $C(\tau)$, where M_i is the mass of state i. For large τ, only the ground state remains, and its properties can be determined.

While the lattice technique has had many successes in calculating phenomenologically relevant quantities at zero chemical potential, μ, particularly in the confined phase of QCD, it has a significant limitation for $\mu > 0$. This is the well-known "sign problem" which has haunted efforts to extend our knowledge into the entire (T, μ) plane[2].

[1] For a general introduction to lattice gauge theory see [2], and non-zero temperature reviews, see [3, 4].
[2] For reviews of the sign problem, see [5].

In this talk, I will discuss three lattice results obtained by the FASTSUM Collaboration at $\mu = 0$ and non-zero temperature, T, above and below the deconfining temperature, T_c, all produced on our 2+1 flavour, anisotropic lattices. These topics are a calculation of the potential in the charmonium system (Sec.2), bottomonium spectral functions (Sec.3), and a determination of the electrical conductivity of QCD as a function of T (Sec.4).

This work uses lattice simulations with the parameters in Table 1 and a carefully crafted action with reduced lattice artefacts [6]. Anisotropic rather than isotropic lattices provide a distinct advantage due to the finer temporal resolution and correspondingly larger sampling of the correlator, $C(\tau)$. This is particularly significant in the $T > 0$ case where the temporal extent is limited to $0 \le \tau \le 1/T$. Our T_c value is obtained from the Polyakov Loop.

N_s	32	24	24	24	32	24	32	24	32	24	24	32
N_τ	48	40	36	32	32	28	28	24	24	20	16	16
T(MeV)	117	141	156	176	176	201	201	235	235	281	352	352
T/T_c	0.63	0.76	0.84	0.95	0.95	1.09	1.09	1.27	1.27	1.52	1.90	1.90

Table 1. Lattice parameters used where $N_{(s)\tau}$ is the number of sites in the spatial (temporal) direction. Our spatial and temporal lattice spacings are $a_s = 0.1227(8)$ and $a_\tau = 0.0351(2)$ fm.

2. Charmonium potential

The interquark potential in charmonium is of great interest to phenomenologists modelling the QGP phase in heavy ion collisions. Knowledge of this potential aids our understanding of the J/ψ system, and the extent to which states become unbound with increasing temperature.

Lattice calculations of the finite temperature interquark potential have, until recently, been restricted to the static (i.e. infinitely heavy) quark limit [7]. On the other hand, the finite-mass interquark potential has been calculated at $T = 0$ [8] using the method developed by the HAL QCD collaboration for internucleon potentials [9]. This method first calculates the wavefunction of the two-particle system, $\psi(r)$, which is then used as input into the Schrödinger equation, yielding the potential, $V(r)$, as output.

In our case, we consider non-local operators of (charm) quark fields, $J(x, \mathbf{r}) = c(x)\Gamma U(x, x + \mathbf{r})\bar{c}(x + \mathbf{r})$, where Γ is an appropriately chosen Dirac (gamma) matrix, and $U(x, x + \mathbf{r})$ is the gauge connection between x and $x + \mathbf{r}$ [10]. The correlation function of these operators is

$$C(\mathbf{r}, \tau) = \sum_{\mathbf{x}} \langle J(x, \mathbf{r}) J^\dagger(x, \mathbf{0}) \rangle = \sum_i \frac{\psi_i(\mathbf{r})\psi_i^*(\mathbf{0})}{2M_i} \mathrm{e}^{-M_i \tau}, \tag{1}$$

where the second sum is over the states i. Following [11], $C(\mathbf{r}, \tau)$ satisfies the Schrödinger equation which can then be used to extract the potential.

As usual, the potential can be decomposed into the central, $V_C(r)$, and the spin dependent, $V_S(r)$, terms, $V_\Gamma(r) = V_C(r) + \mathbf{s}_1 \cdot \mathbf{s}_2\, V_S(r)$, where $\mathbf{s}_{1,2}$ are the spins of the charm quarks. In Fig.1 (Left), we show the charmonium central potential, $V_C(r)$ thus obtained at various temperatures above and below the deconfining temperature. We use $N_s = 24$ in this work. As can be seen, there is a clear temperature dependence in our calculated potential which becomes less confining as T increases. In Fig.1 (Right) $V_S(r)$, is plotted showing a repulsive core with a temperature variation at large distances.

3. Bottomonium spectral functions

In this section we discuss our studies of bottomonium at $T > 0$. Recent results from CMS show that the 1S and 2S/3S Υ states have different relative multiplicities in Pb-Pb compared

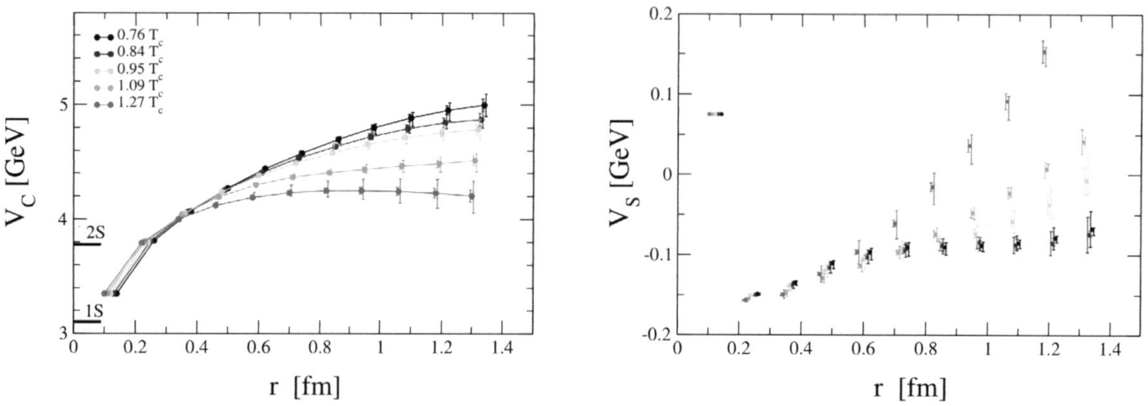

Figure 1. (Left) The charmonium central potential, $V_C(r)$, showing the 1S and 2S masses for the J/ψ system. (Right) The charmonium spin dependent potential, $V_S(r)$.

to p-p collisions at the LHC [12]. This result confirms the picture, originally proposed in the charmonium system, in which the higher mass states in quarkonium are the first to become unbound as the temperature increases beyond T_c [13].

We have performed lattice simulations of the bottomonium system using an $\mathcal{O}(v^4)$ NRQCD lattice action [14] to represent the b-quarks. This extends our earlier work [15]. NRQCD is an approximation obtained from QCD as a velocity expansion in v/c, and is thus applicable for b-quarks. The advantages of NRQCD over the (exact) relativistic quark formulation are two-fold. There is no periodicity in time which complicates the correlation function: in the relativistic case, backward movers effectively half the number of time points that carry independent information. The lack of these thermal boundary effects means that the NRQCD quarks should be viewed as test colour charges moving in a thermal bath of dynamical light quarks and gluons. Secondly, the solution of the NRQCD propagator is an initial value problem which is much easier to solve than the matrix inversion required in the relativistic case. Thus, for a given computer resource, much higher statistics can be achieved in the NRQCD case.

The spectral function, $\rho(\omega)$, can be defined from the two-point correlation function, $C(\tau)$, of bottomonium operators via

$$C(\tau) = \int \rho(\omega) K(\tau, \omega) d\omega, \tag{2}$$

where in the NRQCD case the kernel is $K(\tau, \omega) = \exp(-\omega\tau)$. In principle, the spectral function contains complete information on the states in the channel considered. For a stable particle of mass M, $\rho(\omega) \propto \delta(\omega - M)$. For a resonance, this δ-function broadens acquiring a non-zero width, and if the state becomes unbound, the spectral feature disappears.

We have used the Maximum Entropy Method (MEM), a Bayesian technique, to de-convolve eq.(2) to extract $\rho(\omega)$ [16]. The results for the S-wave (Υ) channel and P-wave (χ_{b1}) channels are shown in Fig.2, all obtained with $N_s = 24$. There is a distinct temperature dependence in both channels. While the Υ ground state (1S) peak is seen to decrease with T, it remains a distinct (i.e. bound) feature up to $T \approx 1.90T_c$. However, the excited state (2S) peak seems to disappear for $T \gtrsim T_c$. This agrees with the experimental result obtained by the CMS collaboration [12]. These results contrast with the χ_{b1} case where the ground state (1P) appears to melt at $T \approx T_c$, see Fig.2 (Right).

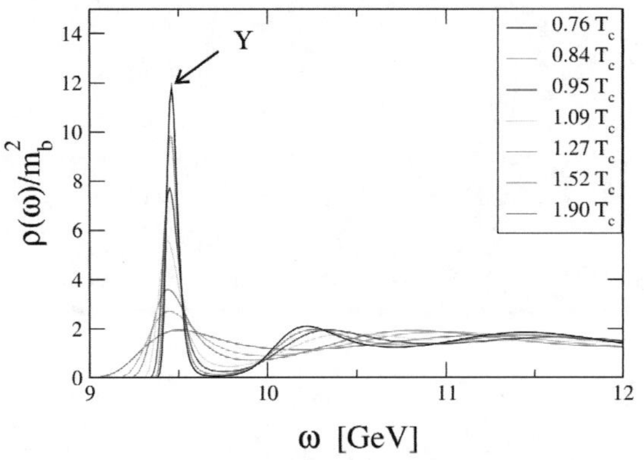

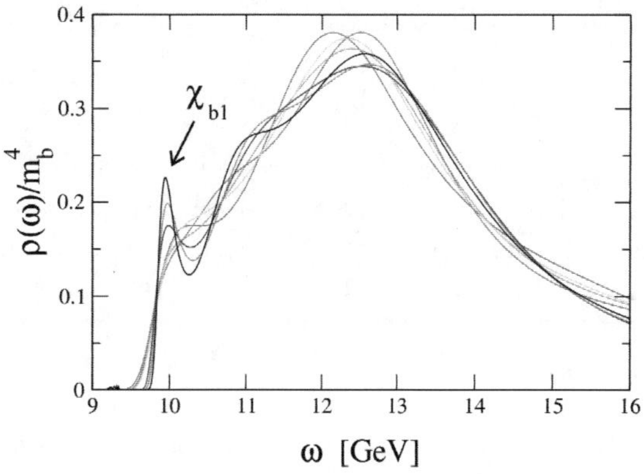

Figure 2. The NRQCD Bottomonium S-wave (Left) and P-wave (Right) spectral functions.

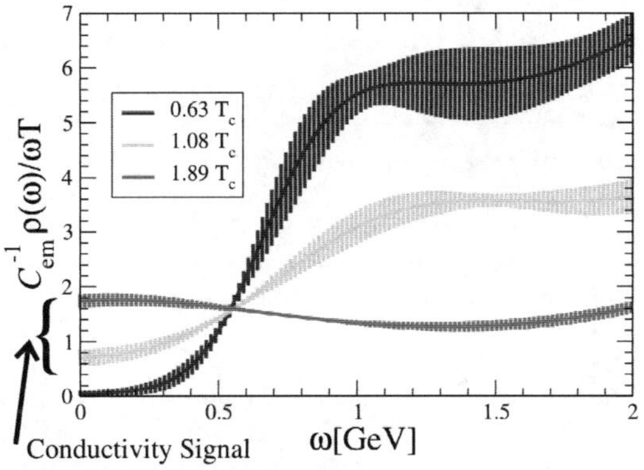

Figure 3. The (light-quark) electromagnetic spectral function, $\rho^{\text{em}}(\omega)$, showing the conductivity signal in the $\omega \to 0$ limit. $C_{\text{em}} = 5/9e^2$ for two light flavours.

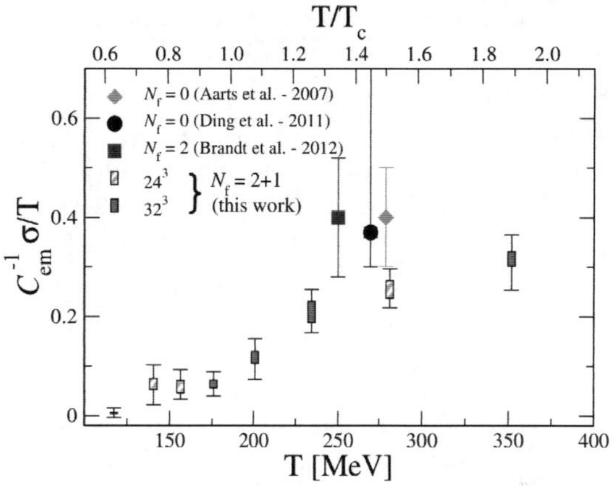

Figure 4. The conductivity, σ, as a function of temperature. As well as the results from our work, quenched ($N_f = 0$) [18, 19] and dynamical results [20] are also shown.

4. Electrical conductivity

We now discuss our calculation of the electrical conductivity, σ [17]. Transport coefficients such as σ are of special relevance for understanding the bulk properties of the fireball in relativistic heavy-ion collision experiments. Typically, they can be determined from the zero energy limit of the appropriate spectral function, i.e. $\lim_{\omega\to 0} \rho/\omega$. In the case of the conductivity, the correlator to consider is that of the electromagnetic current of quark flavour f,

$$C^{\text{em}}(\tau) = \sum_i \int d^3x \langle j_i^{\text{em}}(\tau, \mathbf{x}) j_i^{\text{em}}(0, \mathbf{x})^\dagger \rangle \quad \text{with} \quad j_i^{\text{em}}(x) = e \sum_f q_f j_i^f(x), \tag{3}$$

where e is the electron's charge, $q_f = \frac{2}{3}, -\frac{1}{3}$, is the fractional charge and $j_i^f \sim \overline{\psi}^f \gamma_i \psi^f$ the number density current of quark field ψ^f. In our case [17] we use the conserved lattice vector current (i.e. $\partial_\mu j_\mu^f = 0$) which means we do not have to renormalise our results. We again use MEM to invert the correlator to extract the corresponding $\rho^{\text{em}}(\omega)$, see Eq(2), but this time with the relativistic kernel $K(\tau, \omega) = \cosh[\omega(\tau - 1/2T)]/\sinh[\omega/2T]$.

The conductivity is obtained via the Kubo relation,

$$\frac{\sigma}{T} = \frac{1}{6T} \lim_{\omega \to 0} \frac{\rho^{\rm em}(\omega)}{\omega}. \tag{4}$$

Fig.3 shows $\rho^{\rm em}(\omega)$ around $\omega \approx 0$ for three temperatures spanning T_c. There is a clear non-zero intercept at $\omega = 0$ for $T \gtrsim T_c$, indicating a conductivity signal. In Fig.4, we present our σ values for all the T values we studied. The dimensionless ratio σ/T increases across the transition temperature. Results from three other published works are also shown [18, 19, 20].

5. Conclusion

In this talk, I have summarised our collaboration's lattice calculations showing that the charmonium potential becomes less binding with higher T, that the $\Upsilon(1S)$ state survives above T_c but the $\Upsilon(2S)$ and χ_{b1} don't, and that the conductivity increases with T across the transition.

References

[1] Beringer J *et al.* [Particle Data Group Collaboration], 2012 *Phys. Rev. D* **86** 010001.
[2] Montvay I, Münster G "Quantum Fields on a Lattice", Cambridge Monographs on Mathematical Physics.
[3] Levkova L, 2011 *PoS LATTICE 2011* 011 *(Preprint arXiv:1201.1516 [hep-lat]).*
[4] Lombardo M P, 2012 *PoS LATTICE 2012* 016 *(Preprint arXiv:1301.7324 [hep-lat]).*
[5] de Forcrand P, 2009 *PoS LAT 2009* 010 *(Preprint arXiv:1005.0539 [hep-lat]).* Aarts G, 2012 *PoS LATTICE 2012* 017 *(Preprint arXiv:1302.3028 [hep-lat]).*
[6] Edwards R G, Joo B and Lin H W, 2008 *Phys. Rev. D* **78** 054501 *(Preprint arXiv:0803.3960 [hep-lat]).*
[7] Kaczmarek O, Karsch F, Zantow F and Petreczky P, 2004 *Phys. Rev. D* **70** 074505 [*Erratum - 2005 Phys. Rev. D* **72** 059903], Maezawa Y *et al.* [WHOT-QCD Collaboration], 2007 *Phys. Rev. D* **75** 074501 *(Preprint [hep-lat/0702004]),* Mócsy A and Petreczky P, 2008 *Phys. Rev. D* **77** 014501 *(Preprint arXiv:0705.2559 [hep-ph]),* Fodor Z, Jakovác A, Katz S D and Szabo K K, 2007 *PoS LAT 2007* 196 *(Preprint arXiv:0710.4119 [hep-lat]),* Petreczky P, Miao C and Mócsy A, 2001 *Nucl. Phys. A 855* 125 *(Preprint arXiv:1012.4433 [hep-ph]).* Bazavov A and Petreczky P, 2013 *Nucl. Phys. A* **904-905** 599c *(Preprint arXiv:1210.6314 [hep-lat]).* Burnier Y and Rothkopf A, 2012 *Phys. Rev. D* **86** 051503 *(Preprint arXiv:1208.1899 [hep-ph]).*
[8] Ikeda Y and Iida H, 2010 *PoS LATTICE 2010* 143 *(Preprint arXiv:1011.2866 [hep-lat]),* Kawanai T and Sasaki S, 2011 *Phys. Rev. Lett.* **107** 091601 *(Preprint arXiv:1102.3246 [hep-lat]),* 2012 *Phys. Rev. D* **85** 091503 *(Preprint arXiv:1110.0888 [hep-lat]),* 2011 *PoS LATTICE 2011* 126 *(Preprint arXiv:1111.0256 [hep-lat]),* Iida H and Ikeda Y, 2011 *PoS LATTICE 2011* 195
[9] Ishii N, Aoki S and Hatsuda T, 2007 *Phys. Rev. Lett.* **99** 022001 *(Preprint [nucl-th/0611096]),* Aoki S, Hatsuda T and Ishii N, 2010 *Prog. Theor. Phys.* **123** 89 *(Preprint arXiv:0909.5585 [hep-lat]),* Aoki S *et al.* [HAL QCD Collaboration], 2012 *PTEP* **2012** 01A105 *(Preprint arXiv:1206.5088 [hep-lat]).* Ishii N [HAL QCD Collaboration], 2011 *PoS LATTICE 2011* 160, Ishii N *et al.* [HAL QCD Collaboration], 2012 *Phys. Lett. B* **712** 437 *(Preprint arXiv:1203.3642 [hep-lat]).*
[10] Allton C R, Evans P W M and Skullerud J-I, 2012 *PoS LATTICE 2012* 082 *(Preprint arXiv:1306.3140 [hep-lat]).* Evans P W M, Allton C R and Skullerud J-I, arXiv:1303.5331 [hep-lat]. Evans P W M, Allton C, Giudice P and Skullerud J-I, 2013 *PoS LATTICE 2013* 168 *(Preprint arXiv:1309.3415 [hep-lat]).*
[11] Ishii N [HAL QCD Collaboration], 2011 *PoS LATTICE 2011* 160, Ishii N *et al.* [HAL QCD Collaboration], 2012 *Phys. Lett. B* **712** 437 *(Preprint arXiv:1203.3642 [hep-lat]).*
[12] Chatrchyan S *et al.* [CMS Collaboration], 2011 *Phys. Rev. Lett.* **107** 052302.
[13] Matsui T and Satz H, 1986 *Phys. Lett. B* **178** 416.
[14] Bodwin G T, Braaten E and Lepage G P, 1995 *Phys. Rev. D* **51** 1125 [*Erratum - Phys. Rev. D* **55** 5853].
[15] Aarts G, Kim S, Lombardo M-P, Oktay M B, Ryan S M, Sinclair D K and Skullerud J-I, 2011 *Phys. Rev. Lett.* **106** 061602 *(Preprint arXiv:1010.3725 [hep-lat]),* Aarts G, Allton C, Kim S, Lombardo M-P, Oktay M B, Ryan S M, Sinclair D K and Skullerud J-I, 2011 *JHEP* **1111** 103 *(Preprint arXiv:1109.4496 [hep-lat]),* 2013 *JHEP* **1303** 084 *(Preprint arXiv:1210.2903 [hep-lat]).*
[16] Asakawa M, Hatsuda T and Nakahara Y, 2011 *Prog. Part. Nucl. Phys.* **46** 459 *(Preprint [hep-lat/0011040]).*
[17] Amato A, Aarts G, Allton C, Giudice P, Hands S and Skullerud J-I, 2013 *Phys. Rev. Lett.* **111** 172001 *(Preprint arXiv:1307.6763 [hep-lat]).*
[18] Aarts G, Allton C, Foley J, Hands S and Kim S, 2007 *Phys. Rev. Lett.* **99** 022002. *(Preprint [hep-lat/0703008]).*
[19] Ding H T *et al.*, 2011 *Phys. Rev. D* **83** 034504. *(Preprint arXiv:1012.4963 [hep-lat]).*
[20] Brandt B B, Francis A, Meyer H B and Wittig H, 2013 *JHEP* **1303** 100. *(Preprint arXiv:1212.4200 [hep-lat]).*

Coarse Graining in Hydrodynamics and Effects of Fluctuations

T. Kodama and T. Koide

Abstract. We address the physical meaning of coarse graining scale and the hidden statistical ensemble in event by event analysis in hydrodynamic approach to heavy-ion collisions. It is shown that these features are appropriately understood in the frame work of variational formulation and pointed out that the local thermal equilibrium does not necessarily play a critical role in the description of the collective flow patterns. We further discuss that the effect of viscosity is also formlated in the form of the variational method including fluctuations.

Instituto de Física, Universidade Federal do Rio de Janeiro, C.P. 68528, 21941-972, Rio de Janeiro, Brazil

E-mail: tkodama@if.ufrj.br, koide@if.ufrj.br

1. Introduction

Hydrodynamic approach on event-by-event (EbE) basis has been applied successfully to describe the global and collective features of the data from relativistic heavy-ion collisions, particularly to the behavior of the Fourier components of flow pattern $\{v_n\}$ as a function of centrality and transverse momenta data [1], leading to a general expectation that we may determine the equation of state (EoS) and transport coefficients in addition to the information on initial collision dynamics. On the other hand, these successes brought us several new interesting questions and mysteries. The most crucial one is why at all the hydrodynamic approaches work so well for such violent and almost microscopic collisional processes. It is commonly believed that the fundamental hypothesis for the validity of hydrodynamics is the local thermal equilibrium (LTE). In this vision, we would conclude that the thermalization time and correlation length should be extremely small. For AA collisions this could still be acceptable due to the large number of partons involved, but surprisingly, the recent ALICE experiment reports a similar collective flow pattern in the pA data, too [2]. This casts a very serious question for the proper physical meaning of hydrodynamic description in pA collisions or, even in AA collisions [3]. In this report, we address this question and point out that hydrodynamic behavior for a finite set of observables does not necessarily mean the validity of LTE on real EbE basis, but works effectively in statistical basis for the ensemble of events specified by these observables.

2. Relativistic Hydrodynamics and Role of Coarse Graining

Let us denote the conserved four-current density by $n^\mu(x)$, satisfying the continuity equation, $\partial_\mu n^\mu(x) = 0$. The energy-momentum tensor $T^{\mu\nu}(x)$ of the system also conserves, $\partial_\mu T^{\mu\nu}(x) = 0$. Although these 4 equations are not enough to determine the time evolution of these quantities, in some special physical situations they become sufficient since the total number of variables drastically reduces. Suppose that, in the Landau frame (the energy flow rest frame), the spatial part of $T^{\mu\nu}(x)$ becomes isotropic for any x. In addition, if there is no matter diffusion in this frame and there exists a functional relation among the local quantities, ε, P and n, as $P = P(\varepsilon, n)$, then we obtain a set of closed dynamical equations for $n^\mu(x)$ and $T^{\mu\nu}(x)$, constituting the ideal hydrodynamics. As was mentioned, these conditions can be realized simply by LTE so that usually the success of (almost) ideal hydrodynamic description of heavy-ion collisions means the realization of LTE. However, as we will discuss here, LTE is a sufficient but not a necessary condition for the hydrodynamic picture on EbE basis, especially for the description of a limited set of observables.

Although hydrodynamic equation has a form of local classical field theory, its variables such as n and ε are defined as averages in a certain finite volume which we refer to as *fluid element*. In this sense, the assumption that EoS $P = P(\varepsilon, n)$ should be satisfied strictly at a local point is physically meaningless in a finite-size system. For the hydrodynamic description of relativistic heavy-ion collisions, the locality means that size of the typical fluid element should be sufficiently small compared to that of the whole system (i.e., at least less than half fm), but at the same time it cannot be taken too much small, otherwise the number of degrees of freedom contained in the fluid element would not be enough to define thermodynamical quantities. Similarly, the time scale of the hydrodynamic motion should be much larger than the microscopic one so that LTE can be attained, but it should be much smaller than the time-scale of the expansion of the system, which is already very small due to the rapid expansion of the system. Therefore, for an allowed size of fluid element for the description of heavy-ion collisions, there exist large fluctuations and inhomogeneity in terms of the microscopic configurations, leading to a large deviation from the thermodynamic limit necessary to define the EoS.

As an example, let us consider a classical microscopic system which contains a large number of quickly moving point-like particles. Then, the density n_0^* is a sum of the Dirac delta functions. However, we usually do not require a very precise resolution both in space and in time to describe the collective flow behaviors. Thus we introduce an averaged smooth density distribution $\tilde{n}^*(\vec{x}, t)$ from $n_0^*(x)$ as $\tilde{n}^*(\vec{x}, t) = \int dt' \int d^3\vec{X}\ n_0^*\left(\vec{X}\right)\ U_\tau(t'-t)\, W_h\left(\left|\vec{x} - \vec{r}\left(t'; \vec{X}\right)\right|\right)$ using a 4 dimensional smoothing kernel [4] $W(x) = U_\tau(t) \times W_h(\vec{x})$. Typically U and W are given by the Gaussian distributions with, respectively, width τ and h, which characterize the scales of the time and space resolutions. Similarly, the smoothed spatial current vector $\tilde{\mathbf{j}}(\vec{x}, t)$ can be defined, satisfying the continuity equation, $\partial\tilde{n}^*/\partial t + \nabla_x \cdot \tilde{\mathbf{j}} = 0$. Using these current and density, the four-current, $\tilde{j}^\mu = \left(\tilde{n}^*, \tilde{\mathbf{j}}\right)$ and the proper density, $\tilde{n} = \sqrt{\tilde{j}^\nu \tilde{j}_\nu}$ can be composed. The smoothed four-velocity field is then defined as $\tilde{u}^\mu = \tilde{j}^\mu/n$.

On the other hand, the smoothed energy-momentum tensor $\tilde{T}^{\mu\nu}$ can be introduced in analogous way as the convolution of the original $T^{\mu\nu}$ using the same smoothing kernel. Such an energy-momentum tensor again satisfies the continuity equation, $\partial_\mu \tilde{T}^{\mu\nu} = 0$. From this smoothed energy-momentum tensor, we can calculate the smoothed proper energy density as $\tilde{\varepsilon} \equiv \tilde{u}_\mu \tilde{u}_\nu \tilde{T}^{\mu\nu}$. The energy density defined in this way is an average of the energy density observed in the rest frame of the *matter* flow. The average is taken over all contributions within the range of the coarse-graining scale in space-time. In the following, we take these smoothed quantities as dynamical variables to represent this system [4].

When the above procedure is applied to one collision event which is characterized by a microscopic state, then we can construct the corresponding hydrodynamic description. However, it is obvious that there exist many different microscopic configurations which give the same hydrodynamic response. Let us prepare the set of collision events described by microscopic configurations which gives a specified four-current $\tilde{j}^\mu$ at the initial time t_0, and call this set Ω. If we calculate $\tilde{\varepsilon}$ at a space-time point x for each event in Ω, the value of $\tilde{\varepsilon}$ is not the same in general.

However, if the coarse-graining size is increased, the number of independent microscopic configurations in Ω may become sufficiently large in a way that $\tilde{\varepsilon}$ and $\tilde{n}$ distribute sharply around their mean-values, $\tilde{\boldsymbol{\varepsilon}}$ and $\tilde{\mathbf{n}}$, respectively, as a consequence of the central limit theorem. If this happens, since $\tilde{\boldsymbol{\varepsilon}}$ and $\tilde{\mathbf{n}}$ are the averaged energy and matter densities belonging to the same fluid element, it is possible that they are strongly correlated so that $\tilde{\boldsymbol{\varepsilon}}$ can be expressed as a function of $\tilde{\mathbf{n}}$, $\tilde{\boldsymbol{\varepsilon}} = \tilde{\boldsymbol{\varepsilon}}(\tilde{\mathbf{n}})$ (barotropic fluid).

Suppose that the fluctuations in $\tilde{\varepsilon}$ and $\tilde{j}^\mu$ are not important in the way that the system is characterized basically by the densities $\tilde{\mathbf{n}}$ and $\tilde{\boldsymbol{\varepsilon}}$. In such a case, we expect that the most promising dynamics will be determined by the optimization of the model action,

$$I = -\int d^4x \, \tilde{\boldsymbol{\varepsilon}}\left(\frac{1}{\gamma}\tilde{\mathbf{n}}^*\right), \tag{1}$$

where $\tilde{\mathbf{n}}^* \equiv \gamma\tilde{\mathbf{n}}$ denotes the mean matter density observed in the reference frame and γ is the Lorentz factor. This variational procedure leads to the ideal hydrodynamics. In other words, the hydrodynamic model can be considered as the optimized dynamics of the coarse-grained system (of course only the largeness of coarse-graining is not a sufficient condition to realize such situations). See Ref. [5] for details.

This is a mere single example to show that the collective behavior does not necessarily mean LTE in an single event. In realistic situations, we further extend the ensemble Ω in a way that Ω is a whole set of events which have the same final state observables, such as collective flow parameters. Such Ω might be a huge statistical ensemble of many different events. That is, we have to have in mind that the final state observables are still far from exclusive even for the collective behaviors of the system to conclude that the realization of LTE is attained in a real single event. For the effect of the fluctuations in $\tilde{\varepsilon}$ and $\tilde{j}^\mu$, see Sec. 4.

3. Necessity of Real Event by Event Analysis

As shown above, the hydrodynamic description in heavy-ion collisions reduces to a coarse-grained dynamics obtained by the optimization of the model action (1) under the assumption of existence of an effective EoS, $\tilde{\boldsymbol{\varepsilon}} = \tilde{\boldsymbol{\varepsilon}}(\tilde{\mathbf{n}})$. Therefore, the success of the ideal hydrodynamic modeling of relativistic heavy-ion collisions depends on the consistent choice of this EoS and the model action. These two conditions will be satisfied for a broader range of microscopic configurations than those required by the real "local thermal equilibrium" for each event.

On the other hand, the size of Ω depends on the coarse-graining scale and final state observables. For larger Ω, the two conditions are more likely to be satisfied. We however loose the better resolution in the space-time recognition for larger Ω. In fact, we cannot observe inhomogeneities with smaller wavelength than the coarse-graining scale. This affects directly the class of observables that the model can describe. Even though some observables might be insensitive to inhomogeneities in each event. As an extreme example, we take the situation where the coarse-graining size is larger than the system size and the total time evolution. Then the ensemble Ω can be regarded as the statistical ensemble of the whole system itself, and the resultant system reduces simply to the so-called fire-ball. The thermal model for particle ratio can be considered in this category.

In general, the less the resolution is, the larger Ω becomes. Thus for some observables which do not require a precise space-time resolution, the effective hydrodynamic description for the statistical ensemble Ω will be sufficient for the understanding of the physics of these observables, without implying LTE on EbE basis. As a matter of fact, the experimental observables are usually averaged over collision events classified in terms of their initial configurations rather loosely defined, such as centrality, event plane, etc. In other words, the present collective flow data are still of inclusive nature. In order to claim that the true hydrodynamics with LTE is valid, we need to have observables that reflect the genuine hydrodynamic profile in EbE basis. For example, the remnant of a sharp shock wave propagation, if exists, would be a good evidence and it also tells the possible limit size of coarse-graining for the collective flow. The shock thickness should not be larger than the coarse-graining scale of the collective flow.

The key point is that when we apply the hydrodynamic modeling, we do not know a priori the suitable coarse-graining scale in the real scenario. This puts a certain limitation in extracting the meaningful information of the initial condition from the hydrodynamic analysis. For this purpose, it is essential to find out the set of observables which carry the information on the inhomogeneities of the initial conditions on EbE basis. The flow parameters $\{v_n\}$, often called "event-by-event" analysis, in the sense that correlations among different observables measured for each event in coincidence, but there still exists a huge statistical ensemble which gives the same observed correlation as we have discussed. For example, the cumulant method to determine the flow parameters eliminates the information of event plane. In the recent paper, it is pointed out that event plane may differ in low and high p_T domain [4], according to the coarse-graining scale. If it can be experimentally measured, it would furnish some information on coarse graining scale in heavy-ion collisions.

4. Fluctuation of Fluid Variables and Stochastic Variational Method

Within the vision that hydrodynamic evolution is an effective dynamics for coarse-grained variables of the energy-momentum tensor, each real collision event is an element of the statistical ensemble Ω and does not obey a unique time evolution equation due to the difference in the microscopic degrees of freedom to which our macroscopic hydrodynamic variables are blind. When the fluctuation of events in Ω is large, they should be taken into account in the determination of dynamics of coarse-grained hydrodynamic variables. As was discussed so far, the variational approach is an optimized method for the formulation of the coarse-grained dynamics. Then the variation procedure in Eq. (1) should be modified so as to include the effect of the fluctuation which was ignored in Sec. 3. The stochastic variational method (SVM) is known as an appropriate approach for such situations [6].

In order to formulate the variational approach involving stochastic processes, we have to introduce two stochastic differential equations (SDE), one for the forward direction in time (FSDE), and the other, backward in time (BSDE) which describes the time reversed process of FSDE. These two SDEs are necessary to accommodate the fixed initial and final boundary conditions in the variational procedure, but they cannot be completely independent. For further discussion on this point, see [6]. Such stochastic processes are known as Bernstein process. For this process, the two Fokker-Planck equations for FSDE and BSDE should be equivalent. This leads to the consistency condition, $\mathbf{u} = \tilde{\mathbf{u}} + 2\nu\nabla \ln \rho$,where $\mathbf{u}$ and $\tilde{\mathbf{u}}$ are the velocity fields for FSDE and BSDE, respectively. Here ρ is the particle density which is given by the solution of the Fokker-Planck equation and ν is a parameter representing the intensity of the noise.

The purpose of SVM is to determine these velocity fields $\mathbf{u}$ and $\tilde{\mathbf{u}}$ from an action through the variation principle introducing noises, starting from the classical (non-dissipative) Lagrangian. For a non-relativistic fluid, the most natural form of the stochastic Lagrangian density is given by [6] $\mathcal{L} = 1/2\ \rho^m \left[(1/2+\alpha_2)\left\{(1/2+\alpha_1)\,\mathbf{u}^2 + (1/2-\alpha_1)\,\tilde{\mathbf{u}}^2\right\} + (1/2-\alpha_2)\,\tilde{\mathbf{u}}\cdot\mathbf{u}\right] - \varepsilon$, here α_1 and α_2 is arbitrary constants and ρ^m is the mass density. The corresponding action is an average over the whole SDE solutions, and the variation is taken with respect to these stochastic trajectories to determine the unknown fields, $\mathbf{u}$ and $\tilde{\mathbf{u}}$. The dynamical equation obtained from the stochastic variation is found to be

$$\rho^m\left(\partial_t + \mathbf{u}_m\cdot\nabla\right)u_m^i - \sum_j \partial_j(\mu\rho_m e_{ij}^m) - 2\kappa\rho^m\sum_j \partial_i(\sqrt{\rho^m}^{-1}\partial_j^2\sqrt{\rho^m}) = -\nabla P,$$

where $\eta = \alpha_1(1+2\alpha_2)\nu\rho^m$, $\kappa = 2\alpha_2\nu^2$ and $e_{ij}^m = \partial_j u_m^i + \partial_i u_m^j$. The pressure P is defined by $(\rho_m)^2 d(\varepsilon/\rho_m)/d\rho_m$. One can see easily that the second term on the left hand side corresponds to the viscous term (containing the gradient of velocity field) and this equation is reduced to the Navier-Stokes-Fourier equation when we set $\alpha_2 = 0$. That is, the fluctuation effects which were ignored in Sec. 3 induces the effects of viscosity in accordance with the fluctuation-dissipation theorem (more precisely, the variation of entropy should be taken into account to get the second coefficient of viscosity). For other results of different values of $\alpha' s$, see Ref. [6].

5. Concluding remarks

In this work, we addressed the physical meaning of coarse-graining scale and the hidden statistical ensemble in EbE analysis in hydrodynamic approach to heavy-ion collisions. We introduced explicitly the coarse-graining procedure for the hydrodynamic modeling together with its variational formulation. In this picture, the collective flow patterns can be reproduced without requiring the LTE in a strict sense for event by event. That is, the hydrodynamic behavior observed in relativistic heavy-ion collisions does not necessarily imply the realization of LTE for a one collisional event. Furthermore, we call attention that the validity of EoS does not mean the thermal equilibrium as has already been pointed out before (see for example, Ref. [7] and also the recent work on the dynamical isotropization of pressure in the Color Glass Condensate scheme [8]). We further discussed possible signals for coarse-graining scale and genuine hydrodynamic behaviors on EbE basis. For example, the remnant of a sharp shock wave propagation would be a good observable which tells the possible coarse-graining size of the collective flow. Another example is to determine the event plane for different transverse momentum domain. Finally we showed that the coarse-graining is intimately related to the origin of viscosity and this effect can be formulated in the variational method extending dynamical variables to stochastic domain. In order to quantify the questions raised here, it will be useful to perform the analysis of coarse-graining described in this work for a certain microscopic model which gives complete dynamical evolution of the energy-momentum tensor, such as PHSD [9].

This work is supported by CNPq, FAPERJ and PRONEX of Brazil, and also a part of the work has been supported by EMMI-Helmholtz Institute, Germany.

References

[1] Paiva S et al 1997 Phys. Rev. C **55** 1455
Aguiar C E et al 2002 Nucl. Phys. A **698** 639
Andrade R et al 2006 Phys. Rev. Lett. **97** 202302
Andrade R et al 2007 Braz. J. Phys. **37** 717
Takahashi J et al 2009 Phys. Rev. Lett. **103** 242301
Wang D J et al 2012 arXiv:1205.4934
Qiu Z et al 2012 Phys. Lett. B **707** 151
Alver B and Roland G 2010 Phys. Rev. C **81** 054905
Teaney D and Yan Li 2011 Phys. Rev. C **83** 064904
Schenke B et al 2012 Phys. Rev. Lett. **108** 252301
Dumitru A and Nara Y 2012 Phys. Rev. C **85** 034907
[2] ALICE Collaboration, arXiv:1307.3237
[3] Gyulassy M talk in 9th International Workshop on Relativistic Aspects of Nuclear Physics, Rio de Janeiro, October, 2013
[4] Mota Ph et al 2012 Eur. Phys. J. A **48** 165
[5] Elze H-T et al 1999 J. Phys. G **25** 1935
[6] See for example Koide T and Kodama T 2012 J. Phys. A**45** 255204 and references therein
[7] Kodama T et al 2007 Int. J. Mod. Phys. E**16** 763.
[8] Epelbaum T and Gelis F 2013 Phys. Rev. Lett. **111** 232301
[9] See for example, Bratkovskaya E L et al 2011 Nucl.Phys. A**856** 162 and references therein

Hydrodynamic models of particle production - p-Pb collisions

Piotr Bożek[1,2], **Wojciech Broniowski**[2,3,4], **Giorgio Torrieri**[5,6]

[1] AGH University of Science and Technology, Faculty of Physics and Applied Computer Science, al. Mickiewicza 30, 30-059 Krakow, Poland
[2] Institute of Nuclear Physics PAN, PL-31342 Kraków, Poland
[3] Institute of Physics, Jan Kochanowski University, PL-25406 Kielce, Poland
[4] CNRS, URA2306, Institut de Physique Théorique de Saclay, F-91191 Gif-sur-Yvette, France
[5] FIAS, J.W. Goethe Universität, Frankfurt A.M., Germany
[6] Pupin Physics Laboratory, Columbia University, 538 West 120th Street NY 10027, USA

Abstract. Viscous hydrodynamics gives a satisfactory description of the transverse momentum spectra, of the elliptic and triangular flow, and of the femtoscopic correlations for particles produced in relativistic heavy-ion collisions. On general grounds, a similar collective behavior has been predicted for proton-lead (p-Pb) collisions at the LHC. We present results of the hydrodynamics calculation of the elliptic and triangular flow in p-Pb. We discuss the mass dependence of flow coefficients and of the average transverse momentum for identified particles.

1. Hydrodynamic model

The observation of the jet quenching and of the strong azimuthally asymmetric flow shows that in a high energy heavy-ion collision at ultrarelativistic energies a dense and hot droplet of matter is formed. Event-by-event hydrodynamic calculations [1, 2, 3, 4, 5, 6] are used to provide the response of the collective flow to the deformations of the fluctuating initial state. The eccentricity and triangularity of the initial density result in the formation of the elliptic and triangular flow in the spectra [3, 7]. Not only the average, but the whole distribution of the harmonic flow coefficients v_n [8] can be reproduced [9] for the IP-Glasma initial conditions. The event-by-event distribution of the hydrodynamic response shows that the distributions of v_2 and v_3 follow the initial distributions of ϵ_2 and ϵ_3.

Non-flow correlations between particles with soft momenta can be included in the simulation giving correlations between unlike-sign particles, mainly at small relative angle and relative pseudorapidity [10]. The hydrodynamic model gives predictions for the spectra of identified particles that are in satisfactory agreement with experiment at LHC energies [11, 12, 13, 14, 15]. Advanced hydrodynamic calculations are used to extract the properties of the quark-gluon plasma at different temperatures [16, 17, 18, 9, 19] and baryonic densities [20, 21]. The hard equation of state of the quark-gluon plasma leads to a rapid build up of the transverse flow, which implies a strong dependence of the femtoscopic radii on the pion pair momentum [22, 23].

The p-Pb collisions have been proposed as a laboratory to study the initial state effects and to obtain reference data for Pb-Pb collisions [24]. On the other hand, it has been predicted that the collective expansion of the fireball formed in central p-Pb collisions is significant and can lead to observable elliptic and triangular flow [25]. Experiments with p-Pb collisions at the

LHC measure the two-particle correlation function in relative pseudorapidity $\Delta\eta$ and relative azimuthal angle $\Delta\phi$ [26, 27, 28]. The two-dimensional correlation function presents two ridge-like structures, elongated in pseudorapidity, with pairs collimated in the same direction (same-side ridge, $\Delta\phi \simeq 0$) and in the opposite direction (away-side ridge, $\Delta\phi \simeq \pi$). The azimuthal structure in the two-dimensional correlation function can be understood as an effect of the azimuthally asymmetric collective flow and of the transverse momentum conservation [29], in a similar way as in heavy-ion collisions [30, 31]. The formation of the two ridges is predicted in the color glass condensate approach as well [32, 33, 34]. In Ref. [25] it has been proposed to look for collectivity in small systems using d-A collisions. In central d-A collisions the density profile in the transverse plane is determined by the deuteron wave-function, hence the eccentricity ϵ_2 is large $\simeq 0.5$, which gives a large elliptic flow [25]. The analysis of correlation functions in d-Au collisions at RHIC shows a strong v_2 component [35], but quantitative conclusions are difficult due to large non-flow contributions in the low multiplicity environment at the RHIC energies.

2. Transverse flow in p-Pb interactions

The multiplicity in central p-Pb collisions is comparable to peripheral Pb-Pb collisions. The source size can be estimated using the Glauber Monte Carlo model [36, 37]. The value of the root mean square radius, 1-2 fm, depends on the details of the energy deposition in the model, and grows for more central collisions. The expansion of the large energy density deposited in the

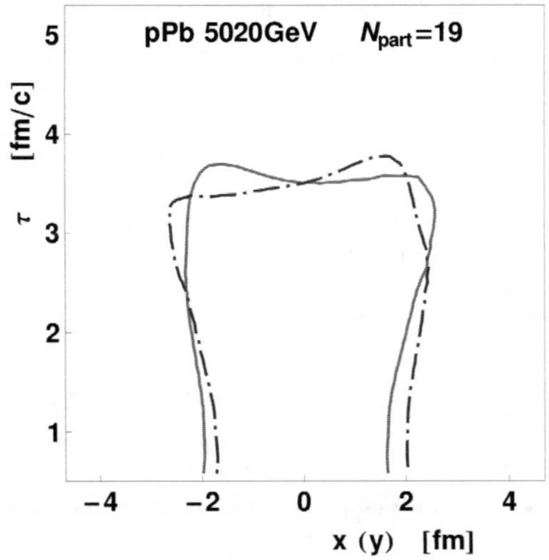

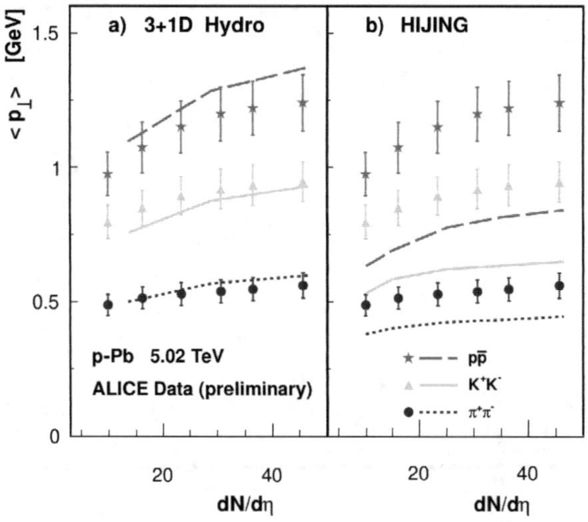

Figure 1. Left: The freeze-out isotherms in the $x-\tau$ plane (solid) and the $y-\tau$ plane (dashed) in a sample event. Right: Mean transverse momentum of identified particles as a function of charged particle density from hydrodynamics (panel (a), lines) and HIJING (panel (b), lines), ALICE Collaboration data [38] (symbols in panels a) and b)).

small volume generates large transverse collective flow [36, 37, 39, 40]. In the left panel of Fig. 1 is shown the freeze-out hypersurface for an event with 19 participant nucleons. The transverse size increases during the freeze-out and can be measured using interferometry radii. The values of the interferometry radii grow as a power of the multiplicity $N^{1/3}$ [41]. The slope of this dependence is different for A-A and for p-p collisions. It may suggest that the dynamics of typical p-p interactions is different than in peripheral A-A collisions. It would be very interesting to investigate experimentally this dependence also for p-Pb collisions to see, whether the space-time development of the p-Pb system is similar to p-p or to Pb-Pb interactions. The hydrodynamic

Journal of Physics: Conference Series **509** (2014) 012017 doi:10.1088/1742-6596/509/1/012017

model predicts for the p-Pb system the interferometry radii are close to those corresponding to A-A interactions [39].

The average transverse momentum of particles produced in the p-p, p-Pb and A-A interactions increases with the event multiplicity [38, 42]. For p-p interactions this increase can be explained as a color reconnection effect [43]. Modeling the particle production in p-Pb and A-A collisions as a superposition of nucleon-nucleon interactions one finds the average transverse momentum below the measured value [44, 38]. This effect is visible in the right panel of Fig. 1, where the HIJING model, based on the superposition of nucleon-nucleon collisions, underpredicts the average transverse momentum. This difference leaves room for additional collective transverse velocity to be generated during the expansion phase in p-Pb collisions. An important characteristic of the collective transverse flow is the mass hierarchy. The average transverse momentum from collective flow is larger for heavier particles. The data for π, K, and p transverse momenta can be reproduced naturally in the hydrodynamic framework [45] (Fig. 1 right panel). The consistency of the hydrodynamic calculations with the experimental data validates the collective flow interpretation, while we note that the mass hierarchy of the average transverse momentum can also be understood as coming from geometrical scaling [46]. The transverse momentum spectra in p-Pb collisions contain a soft component coming from the collective expansion with statistical emission at freeze-out [25, 40, 47]. The spectra can be reproduced in the region of intermediate $p_\perp$ in the EPOS LHC model [40, 47, 42].

3. Elliptic and triangular flow in p-Pb interactions

The most important evidence for the collective expansion in A-A collisions is the observation of elliptic and triangular flow. The elliptic flow coefficient v_2 has been measured using 2- and 4-particle cumulants in p-Pb collisions [48, 49]. The data is consistent with the predictions of the hydrodynamic model [25, 36, 50, 51] (Fig. 2 left panel). In p-Pb interaction the model

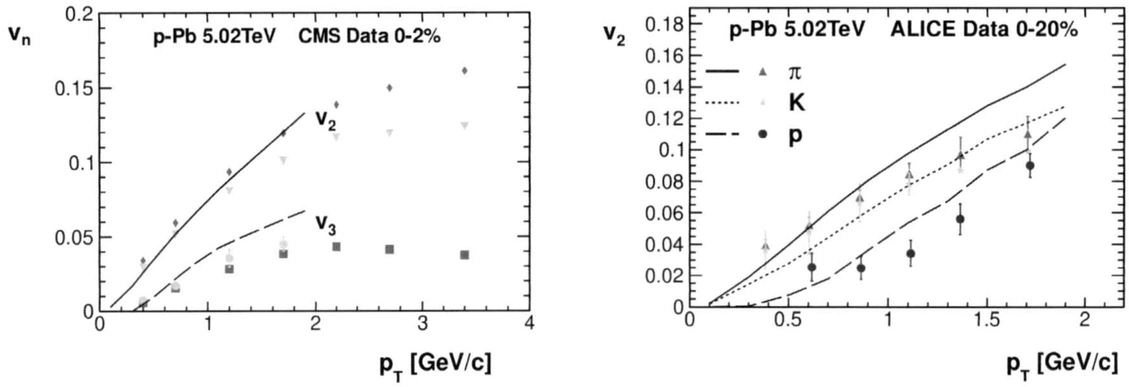

Figure 2. Left: v_2 and v_3 for charged particles from the hydrodynamic calculation, CMS Collaboration data [49]. Right: $v_2(p_\perp)$ for pions, kaons and protons from the hydrodynamic model, ALICE Collaboration data [52]).

uncertainty on the value of the initial eccentricity is a more severe problem than for the A-A system [36, 44]. The small size and short lifetime of the system make the final results depend significantly on other parameters of the model as well: the initial time of the expansion, the freeze-out temperature, the shear viscosity. The relativistic viscous hydrodynamics can be reliably applied if the velocity gradients in the system are moderate and if the mean free path is smaller than the size of the system. In small systems the contribution of non-thermal corona is relatively more important, which can reduce the value of the elliptic flow. These assumptions become more and more questionable when applying the model to more peripheral p-Pb collisions. An additional difficulty in comparing to the experimental data is related to the large contribution of non-flow correlations that are not implemented in the models. All of the above mentioned factors mean that the agreement of the calculation with the data can be at best semi-quantitative.

An important experimental result is the observation of mass splitting of the elliptic flow in p-Pb interactions [52]. The smaller value of the elliptic flow coefficient $v_2(p_\perp)$ for heavy particles is a characteristic of the collective elliptic flow. This feature is reproduced by the calculation [45, 51] (Fig. 2 right panel). The observation of triangular flow in p-Pb interactions is a strong argument for the collective expansion scenario. The predictions of the hydrodynamic model, with fluctuating initial conditions are consistent with measurements [25, 36, 50] (Fig. 2 right panel). This means that the model captures correctly the fluctuations of the initial state and describes realistically its collective expansion. The mass splitting for the triangular flow v_3 in p-Pb collisions is smaller than for v_2, moreover, it is distorted by resonance decays.

4. Collectivity in p-Pb

The formation of a fireball of strongly interacting fluid in ultrarelativistic heavy-ion collisions is well established. The observation of elliptic and triangular flow in agreement with predictions of nearly prefect fluid hydrodynamics, the quantitative description of transverse momentum spectra, and jet quenching are strong evidence of the creation of the quark-gluon plasma. Quantitatively the hydrodynamic model can be reliably applied in central A-A collisions. Many experimental efforts have been devoted to the study the properties of the matter in the fireball when changing the system size and the energy. The data indicate that a collectively expanding system is formed in peripheral A-A collisions and at different energies of the RHIC beam energy scan.

It has been generally expected that in p-Pb collisions at the LHC, the final state interactions would be negligible [24], but the hydrodynamic model applied to initial conditions extrapolated to the p-Pb system at TeV energies predicted a significant collective expansion [25] that could be observed as the elliptic and triangular flow. Many experimental data from p-Pb interactions are consistent with the collective expansion scenario, the double-ridge [29], the elliptic and triangular flow [25, 36, 50, 51], the mass hierarchy of average transverse momenta [45, 40], and of the elliptic flow [45, 51]. The observed appearance of the flow in low multiplicity p-Pb events [49] and general arguments suggest that the p-Pb system can be used as a testing ground for the onset of collectivity in small systems.

Acknowledgments

PB and WB acknowledge the support of the Polish National Science Centre, grant DEC-2012/06/A/ST2/00390 and PL-Grid infrastructure. GT acknowledges the financial support received from the Helmholtz International Centre for FAIR within the framework of the LOEWE program (Landesoffensive zur Entwicklung Wissenschaftlich-Ökonomischer Exzellenz) launched by the State of Hesse, and from DOE under Grant No. DE-FG02-93ER40764.

References

[1] Schenke B, Jeon S and Gale C 2011 *Phys. Rev. Lett.* **106** 042301
[2] Petersen H, Qin G Y, Bass S A and Muller B 2010 *Phys.Rev.* **C82** 041901
[3] Gardim F G, Grassi F, Luzum M and Ollitrault J Y 2012 *Phys. Rev.* **C85** 024908
[4] Bożek P and Broniowski W 2012 *Phys. Rev.* **C85** 044910
[5] Qiu Z, Shen C and Heinz U 2012 *Phys. Lett.* **B707** 151
[6] Pang L, Wang Q and Wang X N 2012 *Phys.Rev.* **C86** 024911
[7] Niemi H, Denicol G, Holopainen H and Huovinen P 2013 *Phys.Rev.* **C87** 054901
[8] Aad G *et al.* (ATLAS Collaboration) 2013 (*Preprint* 1305.2942)

[9] Gale C, Jeon S, Schenke B, Tribedy P and Venugopalan R 2013 *Phys.Rev.Lett.* **110** 012302
[10] Bożek P and Broniowski W 2012 *Phys.Rev.Lett.* **109** 062301
[11] Shen C, Heinz U, Huovinen P and Song H 2011 *Phys. Rev.* **C84** 044903
[12] Bożek P and Wyskiel-Piekarska I 2012 *Phys. Rev.* **C85** 064915
[13] Karpenko I, Sinyukov Y and Werner K 2013 *Phys.Rev.* **C87** 024914
[14] Werner K, Karpenko I, Bleicher M, Pierog T and Porteboeuf-Houssais S 2012 *Phys.Rev.* **C85** 064907
[15] Abelev B *et al.* (ALICE Collaboration) 2013 *Phys.Rev.* **C88** 044910
[16] Shen C, Bass S A, Hirano T, Huovinen P, Qiu Z *et al.* 2011 *J.Phys.* **G38** 124045
[17] Niemi H, Denicol G S, Huovinen P, Molnar E and Rischke D H 2011 *Phys. Rev. Lett.* **106** 212302
[18] Bożek P 2010 *Phys. Rev.* **C81** 034909
[19] Luzum M and Ollitrault J Y 2013 *Nucl.Phys.A904-905* **2013** 377c
[20] Steinheimer J, Koch V and Bleicher M 2012 *Phys.Rev.* **C86** 044903
[21] Karpenko I *et al.* 2013 *these proceedings*
[22] Broniowski W, Chojnacki M, Florkowski W and Kisiel A 2008 *Phys. Rev. Lett.* **101** 022301
[23] Pratt S 2009 *Phys. Rev. Lett.* **102** 232301
[24] Salgado C A 2011 *J.Phys.G* **G38** 124036
[25] Bożek P 2012 *Phys. Rev.* **C85** 014911
[26] Chatrchyan S *et al.* (CMS Collaboration) 2013 *Phys. Lett.* **B718** 795
[27] Abelev B *et al.* (ALICE Collaboration) 2013 *Phys.Lett.* **B719** 29
[28] Aad G *et al.* (ATLAS Collaboration) 2013 *Phys.Rev.Lett.* **110** 182302
[29] Bożek P and Broniowski W 2013 *Phys. Lett.* **B718** 1557
[30] Takahashi J, Tavares B, Qian W, Andrade R, Grassi F *et al.* 2009 *Phys. Rev. Lett.* **103** 242301
[31] Luzum M 2011 *Phys.Lett.* **B696** 499
[32] Dusling K and Venugopalan R 2013 *Phys. Rev. D* **87** 051502
[33] Dusling K and Venugopalan R 2013 *Phys. Rev.* **D87** 094034
[34] Dusling K and Venugopalan R 2013 *Phys. Rev. D* **87** 054014
[35] Adare A *et al.* (PHENIX Collaboration) 2013 (*Preprint* 1303.1794)
[36] Bożek P and Broniowski W 2013 *Phys. Rev.* **C88** 014903
[37] Bzdak A, Schenke B, Tribedy P and Venugopalan R 2013 *Phys. Rev.* **C87** 064906
[38] Abelev B B *et al.* (ALICE Collaboration) 2013 (*Preprint* 1307.1094)
[39] Bożek P and Broniowski W 2013 *Phys. Lett.* **B720** 250
[40] Pierog T, Karpenko I, Katzy J, Yatsenko E and Werner K 2013 (*Preprint* 1306.0121)
[41] Kisiel A (ALICE Collaboration) 2011 *J.Phys.* **G38** 124008
[42] Chatrchyan S *et al.* (CMS Collaboration) 2013
[43] Ortiz A, Christiansen P, Cuautle E, Maldonado I and Paic G 2013 *Phys.Rev.Lett.* **111** 042001
[44] Bzdak A and Skokov V 2013 *Phys.Lett.* **B726** 408
[45] Bożek P, Broniowski W and Torrieri G 2013 *Phys.Rev.Lett.* **111** 172303
[46] McLerran L, Praszalowicz M and Schenke B 2013 *Nucl.Phys.* **A916** 210
[47] Abelev B B *et al.* (ALICE Collaboration) 2013 (*Preprint* 1307.6796)
[48] Aad G *et al.* (ATLAS Collaboration) 2013 *Phys.Lett.* **B725** 60
[49] Chatrchyan S *et al.* (CMS Collaboration) 2013 *Phys.Lett.* **B724** 213
[50] Qin G Y and Müller B 2013 (*Preprint* 1306.3439)
[51] Werner K, Bleicher M, Guiot B, Karpenko I and Pierog T 2013 (*Preprint* 1307.4379)
[52] Abelev B B *et al.* (ALICE Collaboration) 2013 *Phys.Lett.* **B726** 164

Journal of Physics: Conference Series **509** (2014) 012018 doi:10.1088/1742-6596/509/1/012018

Interpretation of strange hadron production at LHC

M Petráň[1], J Letessier[2], V Petráček[3] and J Rafelski[1]

[1] Department of Physics, University of Arizona, Tucson, AZ 85719, USA
[2] Laboratoire de Physique Théorique et Hautes Energies, Université Paris 6, Paris 75005, France
[3] Czech Technical University in Prague, Brehova 7, 115 19 Praha 1, Czech Republic

Abstract. We extend the SHM analysis of hadron production results showing here consistency with the increased experimental data set, stability of the fit with regard to inclusion of finite resonance widths and 2-star hyperon resonances. We present new results on strangeness yield as a function of centrality and present their interpretation in terms of QGP inspired model of strangeness abundance in the hadronizing fireball.

1. Inclusion of new data

We interpret strange soft hadron multiplicity results obtained in Pb–Pb collisions at $\sqrt{s_{NN}} = 2.76\,\mathrm{TeV}$ at CERN Large Hadron Collider (LHC). This contribution extends the more detailed presentation of Ref. [1], whereas Ref. [2] provided the related analysis without data inter- or extrapolation. In the results presented here we can use final data for K^0_S, Λ, $\Xi^\pm$ and $\Omega^\pm$ [3, 4]. Our fit with these results converges to the same set of thermal parameters as in [1].

Considering limited space and in order to rely solely on directly measured and final experimental results we show here as an example the centrality 10%–20% data bin. In the top two result rows of Table 1, we compare the prior with the present results for statistical parameters. As before Λ-abundance is the dominant contributor to the error as is seen also in direct comparison of input and output results shown in Figure 1. Thus in Table 1 the χ^2_{tot} is found to be greater, yet relatively small given the 9 degrees of freedom.

As seen in the top two rows of results, the differences between these two fits are well within the error. Looking at μ_B, recall that we did constrain the value in Ref. [1] by its centrality dependence. Our present result is the outcome of an unconstrained fit for this centrality. This is

Table 1. Comparison of non-equilibrium SHM model parameter obtained in fit to 10–20% centrality results for Pb–Pb collisions at $\sqrt{s_{NN}} = 2.76\,\mathrm{TeV}$. See text for discussion.

$\Sigma(1560)$	widths	$dV/dy\,[\mathrm{fm}^3]$	$T\,[\mathrm{MeV}]$	$\mu_B\,[\mathrm{MeV}]$	γ_q	γ_s	χ^2_{tot}
reported in [1]		2003 ± 47	138.6	1.23 ± 0.06	1.63 ± 0.01	2.06 ± 0.13	3.943
NO	NO	2033 ± 105	138.6	1.36 ± 7.94	1.63 ± 0.01	2.02 ± 0.08	6.599
NO	YES	1978 ± 488	139.3	1.15 ± 0.97	1.62 ± 0.19	2.00 ± 0.31	5.169
YES	NO	2042 ± 409	138.5	0.49 ± 0.35	1.63 ± 0.15	2.03 ± 0.25	4.766
YES	YES	1976 ± 398	139.2	0.74 ± 0.09	1.62 ± 0.01	2.01 ± 0.14	3.472

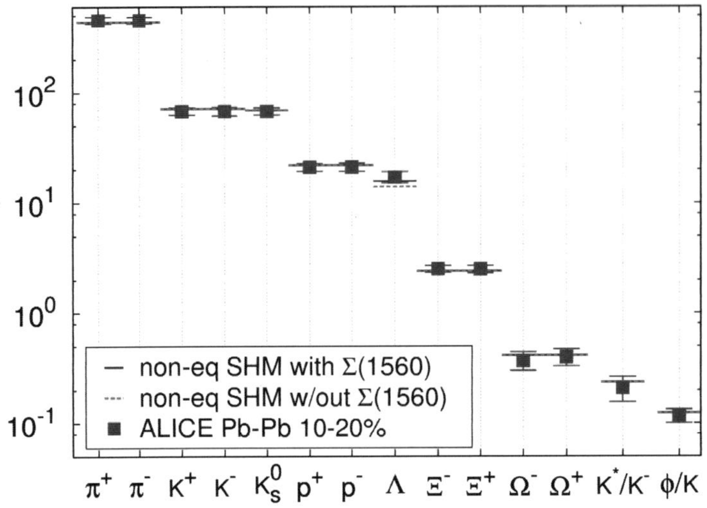

Figure 1. (color online) Comparison of non-equilibrium SHM fit with most recent data from Pb–Pb collisions at $\sqrt{s_{NN}} = 2.76\,\text{TeV}$ for the 10%–20% centrality bin measured by the ALICE collaboration (darkgreen squares), with the 2-star resonance $\Sigma(1560)$ feeding the yield of Λ (solid blue bars), and without the $\Sigma(1560)$ feed (red dashed bars).

possible since the added experimental data for $\Xi^\pm$ and $\Omega^\pm$ contain a small particle-antiparticle asymmetry even if this is well within the particle yield error bar. The value of μ_B we find is near to our earlier expectations; however, the error in μ_B is large. There is no error shown for T since the statistical fit error is much smaller than the precision of the result shown. The error that relates directly to the error of fitted particle yields is found to mainly impact dV/dy and γ_s. This shows that there are two types of error in the data: a common normalization error for all data points, and a further strange particle normalization error.

The bulk properties of the fireball at hadronization are shown in Table 2. These are presented without an error estimate: the impact of the error in dV/dy impacts directly the total strangeness yield $N_{s+\bar{s}}$ and entropy S. The other three properties that we show, pressure P, energy density ε, trace anomaly $(\varepsilon - 3P)/T^4$, are impacted by other statistical parameters but the error of γ_s is the most important. The two top rows are practically equal showing that the additional data, and changes in final data do not impact any of the results and discussion presented in Ref. [1].

Table 2. Comparison of models (see table 1 for all other details) in terms of physical bulk properties. From left to right, we show pressure P, energy density ε, trace anomaly $\varepsilon - 3P$ in units of T^4, total strangeness $N_{s+\bar{s}}$ and entropy S.

$\Sigma(1560)$	widths	$P\,[\text{MeV/fm}^3]$	$\varepsilon\,[\text{GeV/fm}^3]$	$(\varepsilon - 3P)/T^4$	$N_{s+\bar{s}}$	S
reported in [1]		79.1	0.467	4.77	384	6466
NO	NO	79.2	0.468	4.79	388	6589
NO	YES	82.0	0.483	4.85	385	6593
YES	NO	78.7	0.465	4.78	387	6574
YES	YES	81.7	0.481	4.84	384	6570

2. Improving the fit

2.1. Influence of hadron resonance width

The SHARE package [6, 7] was developed with the option of allowing the resonance widths presented by the Particle Data Group [5] (PDG) values. The computing time required to allow the resonance widths is significant with fits needing often more than 24h CPU time. The reason for this is that many numerical foldings need to be performed in order to evaluate the total particle yields; see [6] for details. In the past the small change in outcome did not justify the numerical effort. Thus this test has not been performed often.

The third result row in Table 1 shows the statistical parameters obtained allowing for hadron resonance widths. While $\chi^2_{\rm tot}$ of the fit slightly decreases and central points of statistical parameters aside of T agree within error, we note that the error is now more distributed among parameters and the progression is larger indicating slightly less consistency between fit and data.

The fact that there is an increase in T by $\delta T = 0.7$ MeV, which is well above statistical error, is not surprising considering the systematic effect that the resonance widths have on particle yields. Note that to compensate the increase in value of T the central value of dV/dy decreased. Another consequence of the slight increase of T is a slight (1%) increase of the intensive bulk properties P, ε, $(\varepsilon - 3P)/T^4$, see the third result row in Table 2. The extensive bulk properties directly related to the measured particle yields i.e. $N_{s+\bar{s}}$ and S remain unchanged.

2.2. $\Sigma(1560)$ *as a source of* Λ

The only data point that is not fitted within the 1 s.d. experimental error margin is the yield of Λ. This situation was also discussed in Ref. [1], where the preliminary Λ/π ratio was fitted and was systematically under–predicted for all centralities as the only data point standing out with model value just outside the one standard deviation error margin. The newly reported experimental value for 10–20% centrality, $\Lambda = 17 \pm 2$, is to be compared to the fitted value, $\Lambda = 14$. Inclusion of hadron resonances with their widths discussed above does not improve the fit to the Λ data point.

The question arises if perhaps a contribution to the Λ yield was inadvertently omitted. The SHARE implementation of SHM [6, 7] includes 3-star (***) and 4-star (****) hadron resonances from among all resonances reported by PDG [5]. There are many omitted 2-star (**) resonances. Generally these are so massive that their contribution to particle yields is insignificant. However, we found one exception to this rule, a $\Sigma(1560)$ which decays into Λ and which would thus systematically increase the Λ yield within the error across all centrality.

The charged states $\Sigma(1560)^{\pm}$ have been clearly observed (6σ signal) by two independent experiments [8]. On the other hand, no evidence of $\Sigma(1560)^0 \to \Lambda\pi^0$ has been found in a recent crystal ball experiment [9] and hence the resonance remains unconfirmed at (**) level. $\Sigma(1560)$ quantum numbers have not been measured. Inspired by the close in mass resonance $\Sigma(1580)\frac{3}{2}^-$ [10], we assign spin $\frac{3}{2}$ also to $\Sigma(1560)$.

Once we include $\Sigma(1560)$ in the list of hadron states, the new resonance decay feeds the Λ yield by $\Sigma(1560) \to \Lambda\pi$ (100%) decay. When we repeat the fit, the resulting statistical model parameters are all almost identical to those without $\Sigma(1560)$ as is seen in Table 1. This means that the other 13 data points alone constrain enough the fit parameters so that the effect of an omitted resonance simply reduces the χ^2 without changing other results. This result confirms 'missing' resonance hypothesis as the probable origin of the Λ yield underprediction.

In the final result row in Table 1, we consider the finite width of all resonances and the influence of the 2-star $\Sigma(1560)$ resonance. The overall $\chi^2_{\rm tot}$ is decreased by a factor of two, the fitted value of μ_B is showing a relatively small error. The main fit error is in overall normalization dV/dy. The physical bulk properties remain as already obtained and discussed without $\Sigma(1560)$.

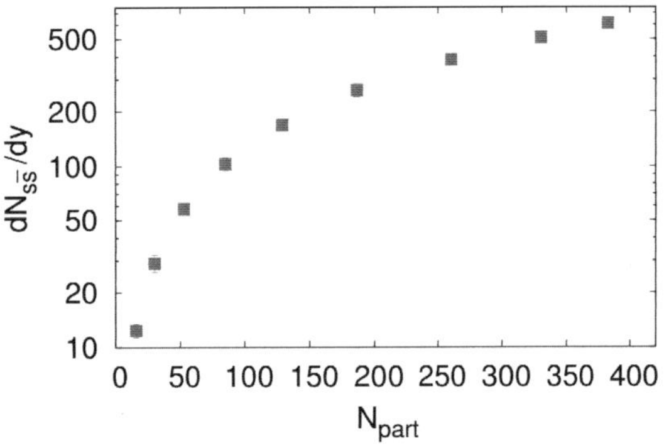

Figure 2. (color online) Total strangeness $dN_{s+\bar{s}}/dy$ produced in Pb–Pb collisions at $\sqrt{s_{NN}} = 2.76\,\text{TeV}$ as a function of centrality.

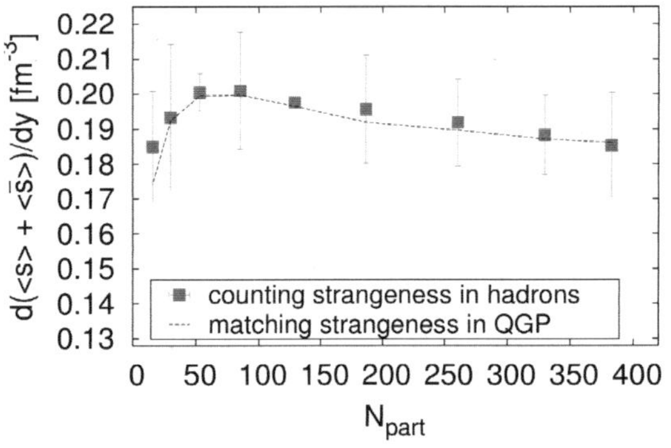

Figure 3. (color online) $s + \bar{s}$ strangeness density per unit rapidity in the hadron phase (red squares), fitted with strangeness in the QGP phase (blue dashed line). See text.

3. Strangeness conservation and strange quark mass at hadronization

In the central Pb–Pb collisions at the LHC, an unprecedented amount of strangeness is produced, a total of $dN_{s\bar{s}}/dy \simeq 600$ strange and anti-strange quarks per unit of rapidity for the most head-on 5% central collisions, see Figure 2. For the peripheral collisions the rise of the total strangeness yield is very rapid, as both the size of the volume and saturation of strangeness production combine. For the more head-on collisions we see a power (~ 1.17) law rise similar to particle interpolation in [1]. Normalization error bar contained in dV/dy is typically 10% is not shown, only the error from γ_s is shown in Figure 2 (often hidden in the symbol size).

The corresponding total observed strangeness density in the hadron phase is depicted in Figure 3. The uncertainty is evaluated using the relative uncertainty of the model parameter γ_s reported in [1]. Note that the variation in density shown is only about $\pm 3.5\%$. Such constancy of the strangeness yield, assuming a recombinant hadronization, should result in strange hadron yield ratios independent of centrality. In Figure 4 the flat line at the bottom is a ratio of two doubly strange particle yields, Ξ/ϕ. Its constancy provides a reference of precision of the argument as this result should be constant [12] even if the strangeness density varies.

The other ratios compare the yield of doubly strange particles with the yield of single strange particles and in one case doubly strange with pions. When strangeness is not saturated as a function of centrality in QGP, one expects an increase in these ratios which we clearly see at lower SPS and RHIC energies. However at LHC a different pattern emerges: there is a bit of increase looking at some of the most peripheral bins which is followed by a slow decrease. The strangeness density shown in Figure 3 mirrors this behavior.

It is of considerable interest to understand if the 'measured' strangeness density seen in Figure 3 can be interpreted in terms of sudden hadronization of a QGP fireball. Sudden hadronization implies the conservation of strangeness yield. In sudden hadronization model the volume does not change thus the density in the hadron phase equals that in QGP as well. The chemical freeze-out temperature T is also the hadronization temperature, i.e. the temperature of QGP breakup. We evaluate the QGP phase strangeness density for a given T given by integral of Fermi gas strangeness density in the QGP using [11]:

$$s(m_s, T; \gamma_s^{\text{QGP}}) = -\frac{g}{2\pi^2}\left(\frac{T}{\hbar c}\right)^3 \sum_{n=1}^{\infty} (-\gamma_s^{\text{QGP}})^n \frac{1}{n^3}\left(\frac{nm_s}{T}\right)^2 K_2\left(\frac{nm_s}{T}\right), \tag{1}$$

where m_s is the strange quark mass, γ_s^{QGP} is the phase space occupancy: here superscript QGP

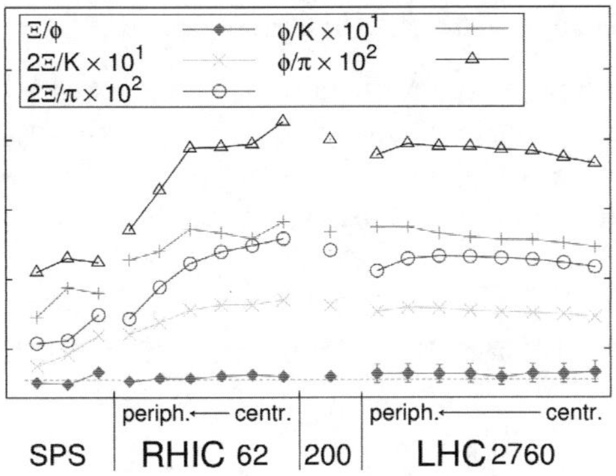

Figure 4. Measured particle ratios from relativistic heavy ion collisions across different collisional energies at SPS, RHIC and LHC, and centralities.

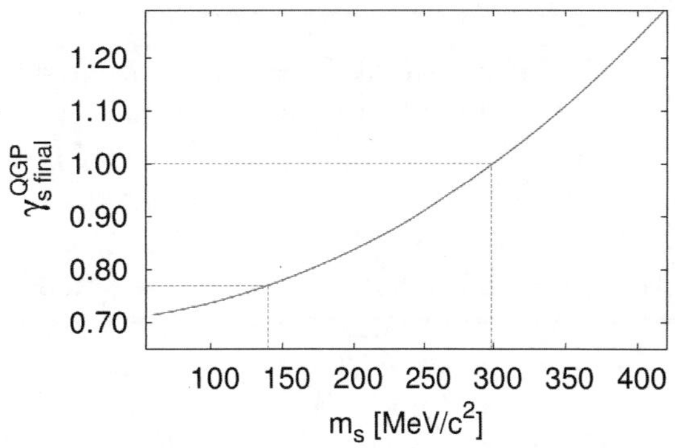

Figure 5. (color online) The strangeness phase space occupancy in the QGP phase as defined by Eq. (2) and correlated in the fit to strangeness density with the strange quark mass. See text.

helps to distinguish from that in the hadron phase where we used γ_s without a superscript. The degeneracy $g = 12 = 2_{\rm spin}3_{\rm color}2_{\rm p}$ where the last factor accounts for the presence of both quarks and antiquarks. We will discuss below the reduction in s due to color-interactions.

In central LHC collisions, the large volume (longer lifespan) suggests that strangeness approaches saturated yield in the QGP. This hypothesis is confirmed by almost constant particle ratios we discussed above, see Figure 4. However, in peripheral collisions, the short lifespan of the fireball may not be sufficient to reach chemical equilibrium. Therefore we introduce a centrality dependent strangeness phase space occupancy $\gamma_s^{\rm QGP}(N_{part})$ which is to be used in Eq.(1).

To model the centrality dependence of $\gamma_s^{\rm QGP}(N_{part})$, we recall that the lifespan τ of the fireball depends on the transverse surface of the fireball and hence the time is proportional to the transverse radius $\tau \propto r_\perp$. The transverse radius is related to number of participants as $r_\perp \propto N_{part}^{1/3}$. Thus, at mid-rapidity, the strangeness production will be assumed to be proportional to $r_\perp^2 \propto N_{part}^{2/3}$ and modeled by the usual saturating functional form:

$$\gamma_s^{\rm QGP}(N_{part}) = \gamma_{s\,\rm final}^{\rm QGP} \tanh\left[\left(\frac{N_{part}}{N_0}\right)^{2/3}\right]. \tag{2}$$

$\gamma_{s\,\rm final}^{\rm QGP}$ is the asymptotic saturation of strangeness phase space in the QGP for large systems and N_0 controls the scale of the fireball transverse size.

We now match the strangeness density measured in the hadron phase shown in Figure 3 with the centrality dependent strangeness density in the QGP, using Eq.(2) in Eq.(1). Our fit to the observed strangeness density is shown in Figure 3. We observe a very strong correlation between m_s and $\gamma_{s\,\rm final}^{\rm QGP}$, which we show as a function in Figure 5, while choosing the best N_0 which converges within a narrow interval of $N_0 \in (11, 14)$. This shows a quick saturation of $\gamma_s^{\rm QGP}(N_{part})$, which reaches $0.95\,\gamma_{s\,\rm final}^{\rm QGP}$ already for $N_{part} \simeq 2N_0 = 30$.

To choose a set of values of $(m_s,\ \gamma_{s\,\rm final}^{\rm QGP})$ we consider two cases shown in Figure 5:

(i) The strangeness in QGP is chemically equilibrated in central collisions, $\gamma_{s\,\rm final}^{\rm QGP} \simeq 1$. This requires the strange quark to have an effective mass of $m_s = 299\,{\rm MeV}/c^2$ at hadronization;

(ii) We assume the PDG value of strange quark mass [5] $m_s \simeq 140\,{\rm MeV}/c^2$ at a scale of $\mu \simeq 2\pi T \simeq 0.9\,{\rm GeV}$. This requires $\gamma_{s\,\rm final}^{\rm QGP} \simeq 0.77$.

We believe that both approaches can coincide in an improved theoretical description:

(i) We omitted in Eq. (1) the thermal QCD prefactor which reduces the expected QGP strangeness density due to thermal QCD many body interactions. The typical reduction of s is by a factor $(1 - c\alpha_s/\pi)$. With $\alpha_s \simeq 0.65$ and $c \simeq 1$ this effect reduces the QGP density by $\simeq 20\%$. The resultant $\gamma_{s\,\mathrm{final}}^{\mathrm{QGP}}$ can be 20% larger. This effect is present; we do not know its exact magnitude.

(ii) The measured value of s could be reduced from hadronization value by longitudinal dilution of strangeness during matter expansion after hadronization. This effect vanishes in the limit of Bjorken scaling as for every particle that moves out, another particle moves back into the central rapidity acceptance domain. The rapidity plateau has not been demonstrated experimentally for strange hadrons at LHC.

We note that longitudinal dilution by 15% restores strangeness abundance to prior expectations [14]. Both effects would allow $m_s \simeq 140\,\mathrm{MeV}/c^2$ to be consistent with $\gamma_{s\,\mathrm{final}}^{\mathrm{QGP}} \simeq 1$.

4. Conclusions

The most important result of this analysis is complete stability at 1%-level of results presented in Ref. [1]. This earlier analysis is fully compatible with the latest results [3, 4]. The 1–1.5 s.d. discrepancy of Λ yield systematically below the experimental result inspired us to explore potential Λ sources. We investigated the finite width of all resonances. We found that our fit is very stable and confidence level is slightly improved, however to explain Λ yield we needed the 2-star (**) resonance $\Sigma(1560)$, increasing the model yield of Λ to within $^1/_2$ s.d. of the experimental yield. $\Sigma(1560)$ causes no other change in the outcome of our analysis.

We than considered in detail how to interpret the strangeness yield present at hadronization in terms of a QGP inspired model. We considered strangeness conservation during hadronization and concluded that for a fully consistent description we must account for possible reduction of s by interactions. We than argued that the relatively low hadronization strangeness density could be a consequence of particle dilution in central rapidity region should Bjorken scaling not apply fully. Hadrons from jet quenching and charm decay are produced predominantly in central rapidity domain.

Acknowledgments

This work has been supported by a grant from the U.S. Department of Energy, grant DE-FG02-04ER41318, Laboratoire de Physique Théorique et Hautes Energies, LPTHE, at University Paris 6 is supported by CNRS as Unité Mixte de Recherche, UMR7589.

References

[1] Petran M, Letessier J, Petracek V and Rafelski J 2013 *Phys. Rev.* C **88**, 034907
[2] Petran M and Rafelski J 2013 *Phys. Rev.* C **88**, 021901
[3] Abelev B *et al.* [ALICE Collaboration] 2013 *Preprint* 1307.5530 [nucl-ex]
[4] Abelev B *et al.* [ALICE Collaboration] 2013 *Preprint* 1307.5543 [nucl-ex]
[5] Beringer J *et al.* [Particle Data Group Collaboration] 2012 *Phys. Rev.* D **86**, 010001
[6] Torrieri G, Steinke S, Broniowski W, Florkowski W, Letessier J and Rafelski J 2005 *Comput. Phys. Commun.* **167**, 229
[7] Torrieri G, Jeon S, Letessier J and Rafelski J 2006 *Comput. Phys. Commun.* **175**, 635
[8] Meadows B T 1980 *Proc Baryon 1980, Toronto* ed N Isgur (Toronto: Toronto U) p 283
[9] Olmsted J *et al.* [Crystal Ball Collaboration] 2004 *Phys. Lett.* B **588**, 29
[10] Litchfield P J 1974 Spin Parity of the Sigma (1580) *Phys. Lett.* B **51**, 509
[11] Letessier J and Rafelski J 2002 (Cambridge: Camb. Monogr. Part. Phys. Nucl. Phys. Cosmol. **18**).
[12] Petran M and Rafelski J 2010 *Phys. Rev.* C **82**, 011901
[13] Aamodt K *et al.* [ALICE Collaboration] 2011 *Phys. Lett.* B **696**, 328
[14] Rafelski J and Letessier J 2011 *Phys. Rev.* C **83**, 054909

Confronting LHC data with the statistical hadronization model

J Stachel[1], A Andronic[2], P Braun-Munzinger[2,3,4], K Redlich[5]
[1]Physikalisches Institut, Universität Heidelberg, Heidelberg, Germany
[2]Research Division and ExtreMe Matter Institute EMMI, GSI Helmholtzzentrum für Schwerionenforschung, Darmstadt, Germany
[3] Technische Universität Darmstadt, Darmstadt, Germany
[4] Frankfurt Institute for Advanced Studies, J.W. Goethe Universität, Frankfurt, Germany
[5] Institute of Theoretical Physics, University of Wroclaw, Wroclaw, Poland

E-mail: stachel@physi.uni-heidelberg.de

Abstract.
The most recent data from the CERN LHC are compared with calculations within the statistical hadronization model. The parameters temperature und baryon chemical potential are fitted to the data. The best fit yields a temperature of 156 MeV, slightly below the expectation from RHIC data. Proton yields are nearly three standard deviations below this fit and possible reasons are discussed.

1. Introduction

In ultra-relativistic nuclear collisions the phase diagram of strongly interacting matter is studied [1]. One of the interesting questions is about the location of the boundary between hadronic matter and the quark-gluon plasma (QGP). Over the past 20 years the understanding has arisen, that hadron yields can be very successfully compared to a simple statistical model. For central collisions of heavy nuclei (Au or Pb) and energies $E_{lab}/A \geq 5$ GeV, the grand canonical ensemble gives an appropriate representation. The resulting parameters, temperature T and baryon chemical potential μ_b, represent the so-called chemical freeze-out. It was further realized [2] that the very high densities close to the transition temperature, when a QGP hadronizes, and the rapid fall-off of density below the transition temperature lead to rapid equilibration and chemical freeze-out only a few MeV below the transition temperature. The current understanding of the connection between the QCD phase diagram and chemical freeze-out is summarized in [3].

The statistical hadronization model was first successfully applied to data from ultra-relativistic nuclear collisions from the AGS [4] and not much later to first data from the CERN SPS [5]. Both comparisons yielded excellent agreement. Later, a much more complete set of hadron yields was obtained from the SPS experiments, including data from several energies below the maximum energy down to $E_{lab}/A = 20$ GeV. And from the year 2000 on an increasing set of hadron yields from Au+Au collisions at RHIC became available and was compared to statistical model calculations [6]. All data sofar were found to be in very good agreement with the grand canonical statistical model. For a summary see [7].

Inspecting the extracted statistical model parameters as a function of center of mass energy, two characteristic features arose: (i) The baryon chemical potential drops monotonically with increasing energy and (ii) the temperature initially increases but appears to level off at center of mass energies per colliding nucleon pair of about 20 GeV at a temperature around 160 MeV. This led to a prediction for LHC [8] that chemical freeze-out would occur at $\mu_b = 0.8^{+1.2}_{-0.6}$ MeV and $T = 161\pm$ 4 MeV.

2. LHC data

First results of ALICE at the LHC were discussed in the framework of the statistical hadronization model in [9, 10]. Now a much extended set of published data is available as well as some preliminary data on the production yields of nuclei. A fit of the statistical hadronization model to the data currently available from ALICE at the LHC is shown in Fig. 1. The data point for K^{0*} is not included in the fit. The fit yields a baryon chemical potential of zero and a temperature of 156 MeV with a reduced χ^2 of 2.4. A fit to the preliminary data [10] was significantly worse with $T = 152$ MeV and a reduced $\chi^2 = 4.3$. In Fig. 2 the deviations between fit and data are shown. The proton and antiproton yields are under the model by 18.0 and 19.4 % which, due to the small experimental errors, amounts to a deviation of 2.7 and 2.9 sigma, respectively. The cascade yields, on the other hand, are above the model by about 2 sigma. Otherwise the agreement of data and fit is excellent. The deviation for the K^{0*} meson should be ignored; as a strongly decaying resonance it's yield can be significantly modified after chemical freeze-out.

To demonstrate the sensitivity of the model prediction to the temperature, we also show in Fig. 1 results for a statistical model calculation using $T = 164$ MeV and $\mu_b = 1$ MeV. This higher temperature increases the disagreement for the antiprotons to about 50 % and the yields of nuclei are much overpredicted. Entirely leaving protons out of the fit, the temperature would increase by 2 MeV to $T = 158$ MeV with an otherwise prefect fit with a reduced χ^2 less than one. In that sense one could talk about a proton anomaly, albeit not a very strong one (see above).

3. Comments on the proton anomaly

Already for the RHIC data there is some indication of a low proton yields as compared to the statistical model (see e.g. [10]). But due to bigger uncertainties in removing the contributions from feeding by weak decays - the published hadron yields were obtained without vertex detectors - there are deviations between experiments and no clear picture emerges.

In the following, a few arguments in connection with the close to 3 standard deviation difference of the experimental proton yields from the statistical hadronization model calculations will be made.

The hadron spectrum that enters in the statistical model incorporates all known hadronic states contained in the 2008 compilation by the particle data group [17]. It is clear that this spectrum is still incomplete, there are as of yet undiscovered states. A study was made of the effect of an incomplete hadron spectrum on the K/π ratio [18]. It was based on a Hagedorn resonance gas assumption for states above 3 GeV in mass. This was found to reduce the calculated K/π ratio by about 15 %. While a similar study for the p/π ratio does not yet exist, we would like to argue that it will also reduce the calculated ratio; high mass hadrons will contribute at most one (anti-)proton but multiple pions. In addition, results from lattice QCD predict [19] that numerous additional baryon resonances exist at low masses, partly with high spin and therefore degeneracy. Some of these states have already been found [20]. The effect of an incomplete hadronic spectrum in the statistical hadronization model is to produce relatively too many protons as compared to pions.

Alternatively, it has been argued, that annihilation in the hadronic phase would reduce the number of (anti-)protons [21, 22, 23]. Employing UrQMD to model a hadron gas after hadronization significant reductions in the yields of (anti-)protons and cascades are observed

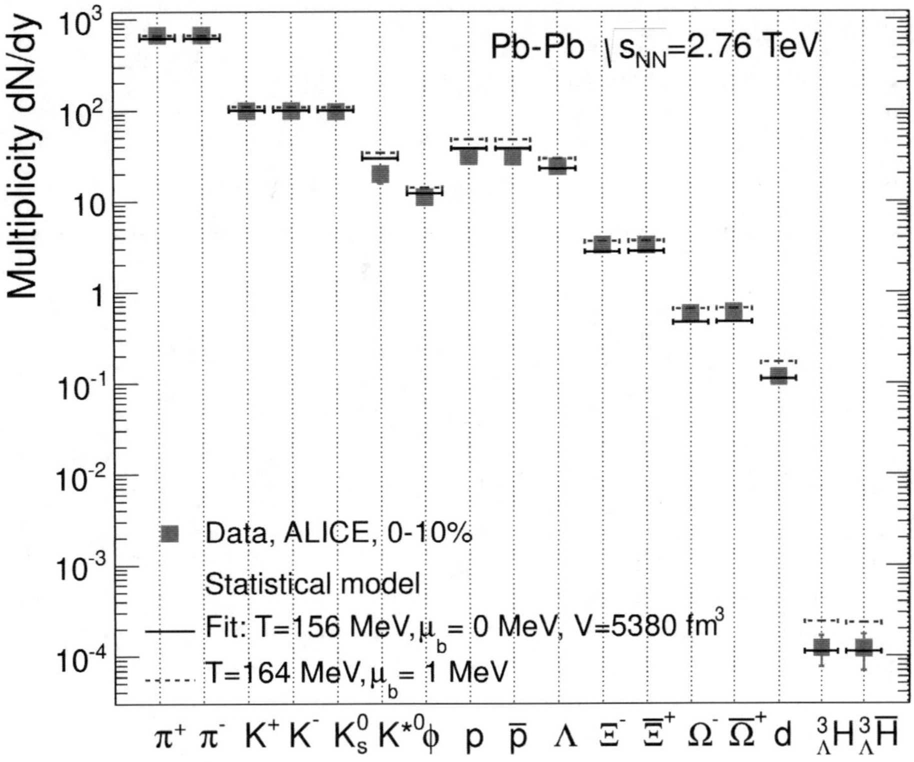

Figure 1. Hadron yields from ALICE at the LHC [11, 12, 13, 14, 15, 16] and fit with the statistical hadronization model. In addition to the fit, yielding T=156 MeV, also results of the model for $T = 164$ MeV are shown, normalized to the value for π^+. The data point for the K^{0*} is not included in the fit.

for the current LHC energy. This leads to a good description of the observed (anti-)proton yields. However, employing this mechanism, the discrepancy for (anti-)cascades is increased. Also, it has been noted by the authors themselves [22], that in UrQMD detailed balance is not implemented for some of the important annihilation reactions. Already in [24] it was argued, that implementing detailed balance would not lead to a depletion of the antiprotons. The effect of annihilation alone and of then in addition including the back reactions with full detailed balance was studied for full SPS energy [25] (and also AGS energy. There it was shown that the annihilation plus back reaction nearly fully compensate for central collisions reaching the equilibrium value for (anti-)proton yields. In a more recent study for collider energies it was shown [26] that properly taking into account the back reactions reduces the effect of annihilation in the hadronic phase to about one half. Here, (anti-)protons, lambdas, cascades and omegas are equally affected, making the agreement for the last 2 species worse. Another argument why one should not put too much trust in the quantitative changes of hadron yields in the hadronic phase within the UrQMD model is the lifetime of the fireball. From 2-pion Hanbury Brown-Twiss correlations an overall lifetime of the system including QGP phase and hadronization of 10 fm/c is deduced [27] for central PbPb collisions at the LHC. Coupling UrQMD to a hydrodynamics evolution the system, the integral time until thermal freeze-out is significantly longer.

Annihilation in the hadronic phase should affect nuclei as well and it can be seen from Fig. 2

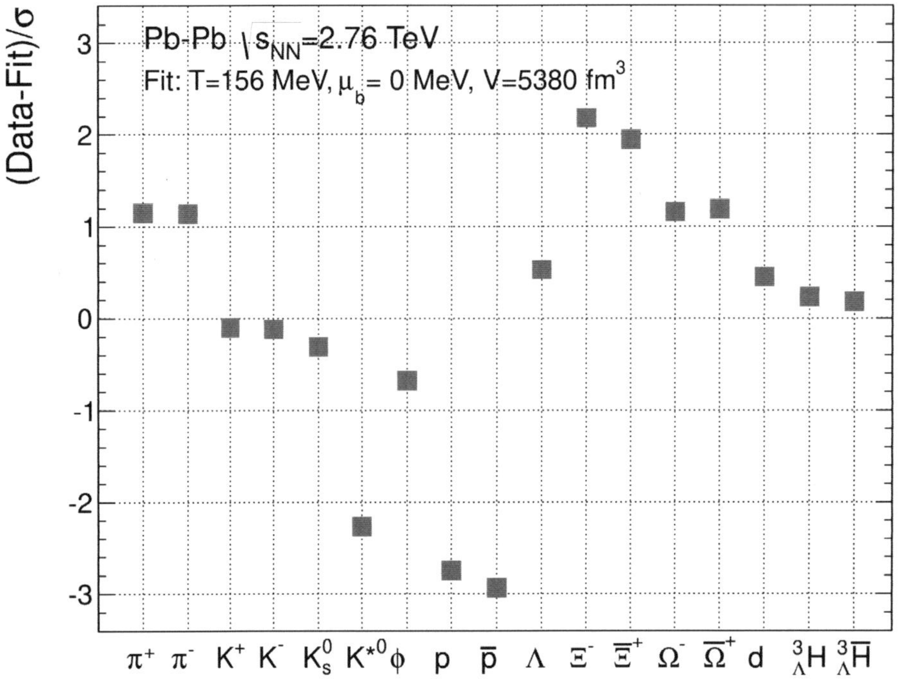

Figure 2. Deviations between thermal fit and data normalized to the error of the data points.

that they are perfectly reproduced without annihilation. One could ask, why (lightly) bound nuclei should also follow the statistical hadronization approach. This is plausible since hadronic reinteractions do not change the entropy per baryon. For much lower energies this argument was already made by Siemens and Kapusta [28] to deduce the entropy per baryon from the yields of light nuclei relative to protons (essentially the d/p ratio). It has been shown [29] that for a system in equilibrium the statistical hadronization and the coalescence approaches agree over many orders of magnitude. In fact, the statistical hadronization approach reproduces very well measured yields of nuclei for central collisions at all energies from AGS up to LHC.

A puzzle is currently the centrality dependence of the proton to pion ratio, which is increasing with centrality for the PHENIX data from RHIC and is decreasing for the ALICE data at the LHC [11]. This opposite trend does not support the annihilation picture of protons in the hadronic phase. If it is a real effect (the current significance is only at the 2 sigma level) it has currently no physics explanation.

A proposal has been made [30, 31] to extend the statistical hadronization model to include out-of-equilibrium features and according new parameters. It is therefore not surprizing that the proton yield can be brought into agreement with such a model calculation. A stringent test for this approach will be the yields of light (anti-)nuclei, since then no additional free parameter is available and they are sensitive to increasing powers of the quark chemical potential. Already the yields of (anti-)hypertriton presented here are in good agreement with the standard statistical hadronization picture employed in this contribution, while they are overpredicted by a factor of 6 in the out-of-equilibrium model of [31].

4. Conclusion

The ALICE data from the LHC pose with their small errors an unprecedented test to the statistical hadronization model. An increasing number of final data has become available over the past year. Excellent agreement with the statistical hadronization model has been achieved with exception of the (anti-)proton yields. Sofar no convincing explanation for this 2.8 sigma deviation is available. Beyond further phenomenological studies, we are looking forward to more data and in particular also to high precision data for the full LHC energy to become available in the next few years.

Acknowledgments

Support by the BMBF Verbundforschung and the Helmholtz Alliance EMMI is gratefully acknowledged.

References

[1] Braun-Munzinger P and Wambach J 2009 *Rev. Mod. Phys.* **81** 1031
[2] Braun-Munzinger P, Stachel J and Wetterich C 2004 *Phys. Lett.* B **596** 61
[3] Braun-Munzinger P and Stachel J 2011 *Festschrift in honor of G.E. Brown's 85th birthday* ed S Lee (Singapore: World Scientific) p 103 (*Preprint* arXiv:1101.3167 [nucl-th])
[4] Braun-Munzinger P, Stachel J, Wessels J P, Xu N 1999 *Phys. Lett.* B **344** 43
[5] Braun-Munzinger P, Stachel J, Wessels J P, Xu N 2000 *Phys. Lett.* B **365** 1
[6] Braun-Munzinger P, Magestro D, Redlich K and Stachel J 2001 *Phys. Lett.* B **518** 41
[7] Andronic A, Braun–Munzinger P and Stachel J 2006 *Nucl. Phys.* A **772** 167
[8] Andronic A, Braun-Munzinger P and Stachel J 2007, Workshop on Heavy Ion Collisions at the LHC: Last Call for Predictions, Geneva, Switzerland, May - June 2007, (*Preprint* arXiv:0707.4076 [nucl-th])
[9] Andronic A, Braun-Munzinger P, Redlich K and Stachel J 2011 *J. Phys.* G **38** 124081
[10] Andronic A, Braun-Munzinger P, Redlich K and Stachel J 2013 *Nucl. Phys.* A **904-905** 535c
[11] Abelev B *et al.* [ALICE Collaboration] 2013 *Phys. Rev.* C in print (*Preprint* arXiv:1303.0737 [hep-ex])
[12] Abelev B *et al.* [ALICE Collaboration] 2013 (*Preprint* arXiv:1307.5530 [nucl-ex])
[13] Abelev B *et al.* [ALICE Collaboration] 2013 (*Preprint* arXiv:1307.5543 [nucl-ex])
[14] Knospe A G [ALICE Collaboration] 2013 *these proceedings* (*Preprint* arXiv:1309.3322)
[15] Doenigus B [ALICE Collaboration] 2013 *these proceedings*
[16] Camerini P [ALICE Collaboration] 2013, presented at NUFRA2013 4rth Int. Conf. on Nucl. Frag., Kemer, Turkey
[17] Amsler C *et al.* [Particle Data Group] 2008 *Phys. Lett.* B **667** 1
[18] Andronic A, Braun-Munzinger P and Stachel J 2009 *Phys. Lett.* B **673** 142
[19] Edwards R *et al.* 2011 *Phys. Rev.* D **84** 074508
[20] Anisovich V *et al.* 2012 *Eur. Phys. J.* A **48** 15
[21] Bass S A *et al.* 1999 *Phys. Rev.* C **60** 021902
[22] Becattini F *et al.* 2012 *Phys. Rev.* C **85** 044921
[23] Becattini F *et al.* 2013 *Phys. Rev. Lett.* **111** 082302
[24] Rapp R and Shuryak E V 2001 *Phys. Rev. Lett.* **86** 2980
[25] Cassing W 2002 *Nucl. Phys.* A **700** 618
[26] Pan Y and Pratt S 2013 *Phys. Rev. Lett.* **110** 042501
[27] Aamodt K *et al.* [ALICE Collaboration] 2011 *Phys. Lett.* B **696** 328
[28] Siemens P J and Kapusta J I 1979 *Phys. Rev. Lett.* **43** 1486
[29] Braun-Munzinger P and Stachel J 1995 *J. Phys.* G **21** L17
[30] Letessier J and Rafelski J 1999 *Phys. Rev.* C **59** 947
[31] Petran M, Letessier J, Petracek V and Rafelski J 2013 *these proceedings* (*Preprint* arXiv:1309.2098)

Journal of Physics: Conference Series **509** (2014) 012020 doi:10.1088/1742-6596/509/1/012020

Latest LHCb results from the pA data

Liang Zhong
Tsinghua University, Beijing, China
E-mail: liang.zhong@cern.ch

Abstract. The J/ψ production cross-sections in proton-nucleus collisions at $\sqrt{s} = 5$ TeV are measured using the LHCb detector. A sizable suppression is observed in pA forward collisions, while only a slight suppression is seen in pA backward collisions. The nuclear attenuation factor $R_{p\mathrm{A}}$ and the forward-backward production ratio r_{FB} are determined.

1. Introduction

The suppression of heavy quarkonia production is a probe to explore quark-gluon plasma (QGP). However, the suppression can also be caused by cold nuclear matter effects (CNM). In proton-nucleus (pA) collisions QGP cannot be created, while CNM can still occur. Hence pA collision can separate the contribution of QGP and CNM. This provides important inputs to heavy ion collisions, where QGP and CNM are entangled.

The LHCb detector [1] is a single-arm forward spectrometer covering the pseudorapidity range $2 < \eta < 5$. The detector includes a high precision tracking system consisting of a silicon-strip vertex detector (VELO) surrounding the pp interaction region, a large-area silicon-strip detector located upstream of a dipole magnet with a bending power of about 4 Tm, and three stations of silicon-strip detectors and straw drift tubes placed downstream. The combined tracking system has a momentum resolution that varies from 0.4% at 5 GeV/c to 0.6% at 100 GeV/c, and an impact parameter resolution of 20 μm for tracks with high transverse momentum (p_{T}). Charged hadrons are identified using two ring-imaging Cherenkov detectors. Photon, electron and hadron candidates are identified by a calorimeter system consisting of scintillating-pad and preshower detectors, an electromagnetic calorimeter and a hadronic calorimeter. Muons are identified by a system composed of alternating layers of iron and multiwire proportional chambers. The trigger [2] consists of a hardware stage, based on information from the calorimeter and muon systems, followed by a software stage which applies a full event reconstruction.

The LHC carries out pA collisions by colliding a proton beam with an energy of 4 TeV and a lead beam with an energy of 1.58 TeV per nucleon. This corresponds to a proton-nucleon centre-of-mass energy of $\sqrt{s} = 5$ TeV. The instantaneous luminosity in pA collisions is $5 \times 10^{27}\mathrm{cm}^{-2}\mathrm{s}^{-1}$, which is five orders of magnitude below the nominal luminosity in pp collisions. The directions of the proton and lead beams are swapped between the data taking periods. The configuration with the direction of travel of the protons from the VELO to the muon system (the muon system to the VELO) is referred to as producing pA forward (backward) collisions. Positive rapidity (y) in the proton-nucleon centre-of-mass system (CM) is always defined to be the direction of the proton beam. The rapidity coverage in the CM is shifted due to the asymmetric beam

Journal of Physics: Conference Series **509** (2014) 012020 doi:10.1088/1742-6596/509/1/012020

energy, which results in a rapidity coverage of $1.5 < y < 4.0$ for pA forward collisions and $-5.0 < y < -2.5$ for pA backward collisions.

The dataset used in this analysis was collected during the proton-nucleus run in early 2013. The samples used for this analysis correspond to integrated luminosities of 0.75 nb^{-1} for forward collisions and 0.30 nb^{-1} for backward collisions. Simulated samples based on pp collisions are used to estimate acceptance and reconstruction efficiencies. In case of significant differences between simulated sample and data, the simulated samples are re-weighted to match the data.

2. J/ψ production measurement

The strategy of the J/ψ production cross-section measurement follows the previous measurements at LHCb [3, 4, 5]. There are three sources of J/ψ mesons in proton-nucleus collisions: direct production in the collisions, feed-down from higher charmonium states, and the decay of b hadrons. Those produced by the first two sources are referred to as prompt J/ψ mesons, and those from the third as J/ψ mesons from b hadrons (abbreviated as "J/ψ from b" in the following). The prompt J/ψ mesons and J/ψ from b carry different information of CNM, therefore their cross-sections should be measured separately.

The double differential cross-section is defined as

$$\frac{\mathrm{d}^2\sigma}{\mathrm{d}p_{\mathrm{T}}\mathrm{d}y} = \frac{N(J/\psi \to \mu^+\mu^-)}{\varepsilon L \mathcal{B}(J/\psi \to \mu^+\mu^-)\Delta p_{\mathrm{T}} \Delta y}, \tag{1}$$

where $N(J/\psi \to \mu^+\mu^-)$ is the observed number of J/ψ, ε the efficiency, L the integrated luminosity, $\mathcal{B}(J/\psi \to \mu^+\mu^-)$ the branching ratio of $J/\psi \to \mu^+\mu^-$ [6], Δp_{T} the p_{T} bin width and Δy the y bin width.

The J/ψ candidates are reconstructed from dimuon final states. Muons with high p_{T} in the final states are selected, and good fit qualities of tracks and vertices are required. An unbinned maximum likelihood fit is performed to extract the number of signal candidates. The signal peak is fitted with a Crystal Ball function [7], and the background with an exponential function.

The prompt J/ψ mesons decay at the primary vertex (PV), while the J/ψ from b hadrons decay at a sizable distance from the PV because of the large lifetimes of b hadrons. The pseudo-proper-time t_z is employed to distinguish prompt J/ψ mesons and J/ψ from b. It is defined as

$$t_z \equiv \frac{(z_{J/\psi} - z_{\mathrm{PV}})M_{J/\psi}}{p_z}, \tag{2}$$

where $z_{J/\psi}$ is the z position of the J/ψ decay vertex, z_{PV} that of the primary vertex refitted after removing the two muon tracks of the J/ψ candidate, p_z the momentum projection of the J/ψ candidate along the beam axis, and $M_{J/\psi}$ the invariant mass of the J/ψ candidate.

The t_z distribution of prompt J/ψ mesons is described by a δ-function convolved with a double-Gaussian function, and that of J/ψ from b by an exponential function convolved with a double Gaussian function. For the background, the t_z distribution is modelled by an empirical function based on the shape of the t_z distribution observed in the J/ψ mass sidebands.

The cross-sections of the prompt J/ψ mesons and J/ψ from b are obtained by a simultaneous fit to the invariant mass and the t_z distribution of the J/ψ candidates. The projections of the fit are shown in Fig. 1.

The measured double differential cross-sections for prompt J/ψ mesons and J/ψ from b in pA forward collisions as a function of p_{T} and y, assuming no J/ψ polarisation, are shown in Fig. 2. The single differential cross-sections for prompt J/ψ mesons and J/ψ from b in pA forward and pA backward collisions as a function of p_{T} and y are displayed in Fig. 3 and 4, respectively.

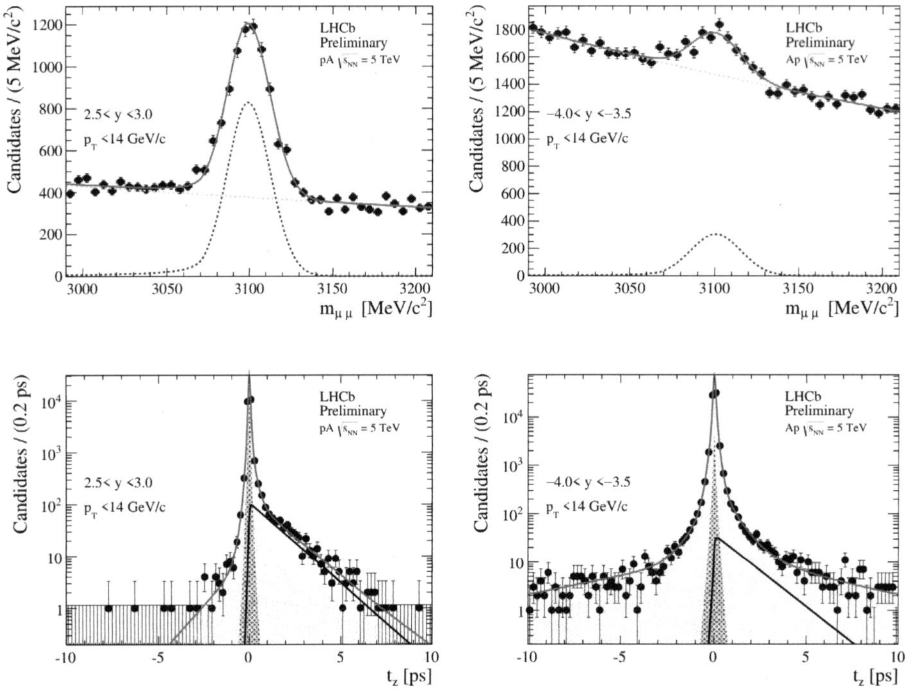

Figure 1. Fit to (top) the invariant mass and (bottom) pseudo-proper-time of the J/ψ candidates for (left) pA forward and (right) pA backward collisions.

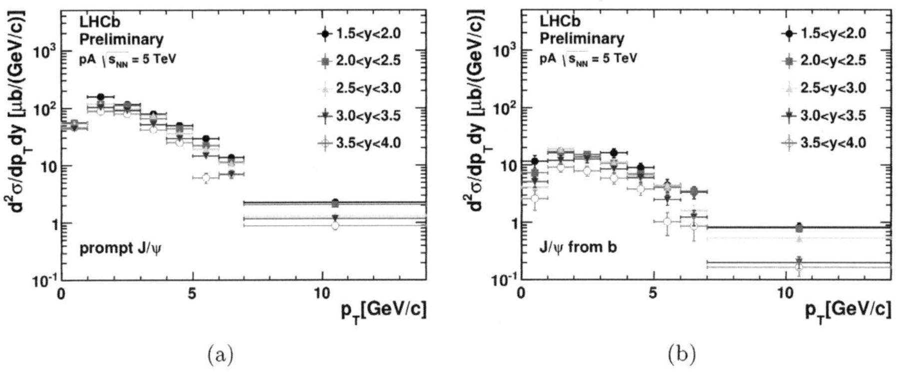

Figure 2. Double differential cross-section of (left) prompt J/ψ mesons and (right) J/ψ from b as functions of p_{T} in bins of y in pA forward collisions.

The integrated cross-sections for prompt J/ψ mesons in pA forward and backward collisions and in pp collisions are compared in the common rapidity region, $2.5 < |y| < 4.0$. The J/ψ cross-section in pp collisions at $\sqrt{s} = 5$ TeV is determined to be $4.78 \pm 0.23 \pm 0.15$ μb by interpolating the cross-sections at 2.76, 7, and 8 TeV [3, 4, 5]. After integration over the common rapidity region and the full p_{T} range, the cross-sections are scaled by a factor of $1/A$, where A is 208 for pA collisions, and 1 for pp collisions. The scaled results are shown in Fig. 5. Clear suppression is observed in forward collisions, while only a mild suppression is observed in backward collisions.

Quarkonia production cross-sections in pA collisions are expected to be sensitive to CNM.

Journal of Physics: Conference Series **509** (2014) 012020 doi:10.1088/1742-6596/509/1/012020

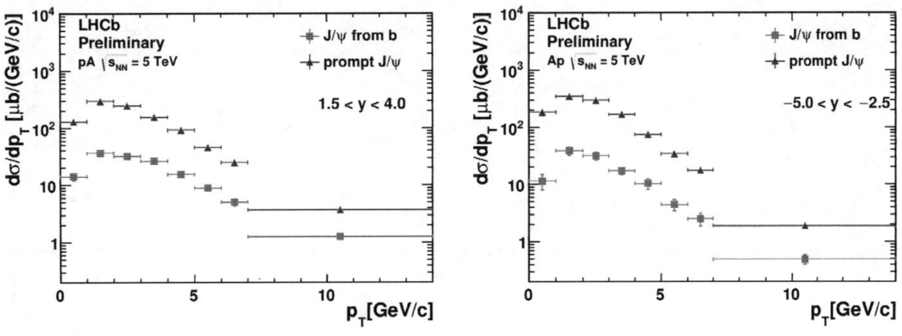

(a) (b)

Figure 3. Single differential cross-section of (red) prompt J/ψ mesons and (blue) J/ψ from b for (left) pA and (right) pA backward collisions as a function of p_{T}.

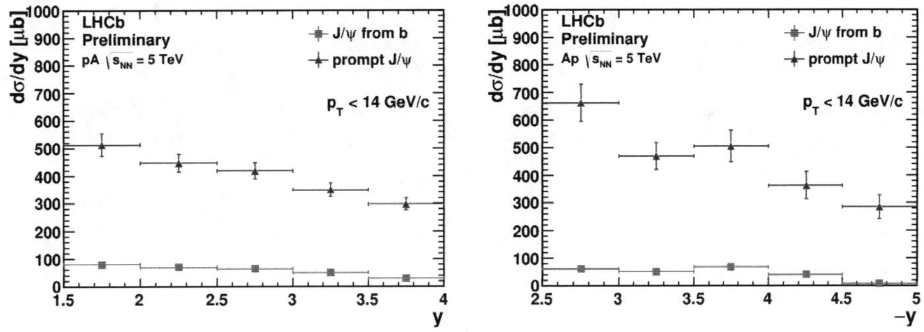

(a) (b)

Figure 4. Single differential cross-section of (red) prompt J/ψ mesons and (blue) J/ψ from b for (left) pA and (right) pA backward collisions as a function of y.

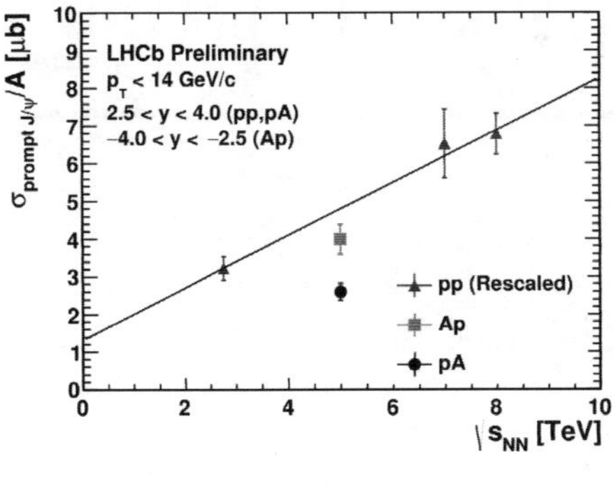

Figure 5. The cross-sections for prompt J/ψ mesons scaled by a factor of $1/A$ in the common rapidity range of $2.5 < |y| < 4.0$ for pp, pA forward, and pA backward collisions. The blue triangles are the cross-sections at 2.76, 7 and 8 TeV in the same kinematic region, the black line is a linear fit to the blue triangles, the red dot is the cross-section in pA backward collisions, and the black dot is the cross-section in pA forward collisions.

Comparisons of J/ψ production cross-sections in pA and pp collisions can shed light on these effects. The nuclear attenuation factor

$$R_{p\mathrm{A}}(|y|, \sqrt{s_{\mathrm{NN}}}) = \frac{1}{A} \frac{\frac{\mathrm{d}\sigma_{\mathrm{pA}}}{\mathrm{dy}}(|y|, \sqrt{s_{\mathrm{NN}}})}{\frac{\mathrm{d}\sigma_{\mathrm{pp}}}{\mathrm{dy}}(|y|, \sqrt{s_{\mathrm{NN}}})} \quad (3)$$

is evaluated as a function of y and compared to theoretical models [8] in Fig. 6(a). Within the

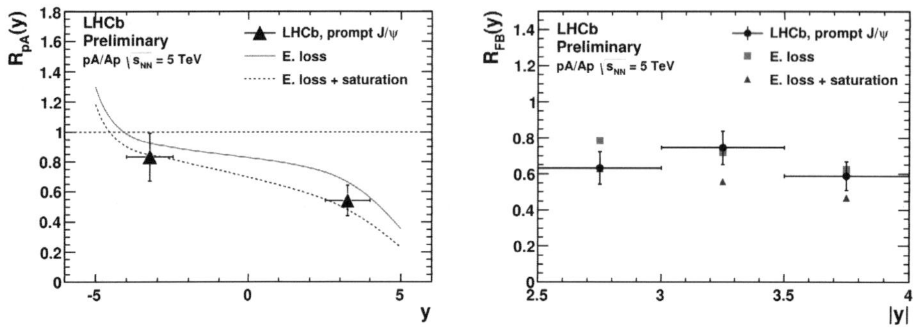

(a) (b)

Figure 6. Comparisons of (a) the attenuation factor R_{pA} and (b) the forward-backward production ratio r_{FB} between data and theory. The black triangles (diamonds) are the LHCb results, the red solid line (squares) are the theoretical prediction based on parton energy loss effects, and the blue dotted line (triangles) are a model with additional saturation effects in the left (right) plot.

current precision the data do not distinguish between the models with or without saturation effects.

The production of J/ψ mesons is found to have a forward-backward asymmetry [9, 10]. This is quantified by the forward-backward production ratio, $r_{FB}(y) = R_{pA}(y)/R_{Ap}(y)$. With the current precision, the data do not discriminate between the models from Ref. [8] with and without saturation effects.

3. Conclusion

The differential cross-sections of J/ψ mesons are measured with the LHCb detector using data collected in pA collisions at $\sqrt{s_{NN}} = 5$ TeV [11]. Clear suppression is observed in forward collisions, while only a mild suppression is seen in backward collisions. The nuclear attenuation factor R_{pA} and the forward-backward production ratio r_{FB} are determined, and the preliminary results are consistent with theoretical predictions and the results of ALICE [12].

References

[1] Alves Jr A A *et al.* 2008 *JINST* **3** S08005
[2] Aaij R *et al.* 2013 *JINST* **8** P04022 (*Preprint* `hep-ex/1211.3055`)
[3] Aaij R *et al.* (LHCb Collaboration) 2011 *Eur. Phys. J. C* **71** 1645 (*Preprint* `hep-ex/1103.0423`)
[4] Aaij R *et al.* (LHCb Collaboration) 2013 *J. High Energy Phys.* **1302** 041 (*Preprint* `hep-ex/1212.1045`)
[5] Aaij R *et al.* (LHCb Collaboration) 2013 *J. High Energy Phys.* **1306** 064 (*Preprint* `hep-ex/1304.6977`)
[6] Beringer J *et al.* 2012 *Phys. Rev.* D **86** 010001
[7] Skwarnicki T 1986 *A study of the radiative cascade transitions between the Upsilon-Prime and Upsilon resonances* Ph.D. thesis
[8] Arleo F and Peigne S 2013 *J. High Energy Phys.* **1303** 122 (*Preprint* `hep-ph/1212.0434`)
[9] Leitch M *et al.* (FNAL E866/NuSea Collaboration) 2000 *Phys. Rev. Lett.* **84** 3256–3260 (*Preprint* `nucl-ex/9909007`)
[10] Adare A *et al.* (PHENIX Collaboration) 2011 *Phys.Rev.Lett.* **107** 142301 (*Preprint* `nucl-ex/1010.1246`)
[11] Aaij R *et al.* (LHCb Collaboration) 2013 (*Preprint* `nucl-ex/1308.6729`)
[12] Abelev B B *et al.* (ALICE Collaboration) 2013 (*Preprint* `nucl-ex/1308.6726`)

14th International Conference on Strangeness in Quark Matter (SQM2013) IOP Publishing
Journal of Physics: Conference Series **509** (2014) 012021 doi:10.1088/1742-6596/509/1/012021

p–Pb Results from ALICE with an Emphasis on Centrality Determination

Andreas Morsch (for the ALICE Collaboration)
CERN PH Division, 1211 Geneva, Switzerland
E-mail: andreas.morsch@cern.ch

Abstract. New ALICE results concerning particle production at low and intermediate p_T in p–Pb collisions at $\sqrt{s_{NN}} = 5.02$ TeV are briefly discussed. Emphasis is given to the determination of centrality in p–Pb and their implications for binary scaling of hard processes.

The analysis of p–Pb collisions is essential for the study of initial and final state effects in cold nuclear matter and has the main goal to establish a baseline for the interpretation of the heavy-ion results [1]. In September 2012, an LHC pilot p–Pb run took place at $\sqrt{s_{NN}} = 5.02$ TeV followed by a long run in February 2013 delivering $\approx 30\,\mathrm{nb}^{-1}$ of p–Pb collisions for each experiment - precious reference data for the Pb–Pb studies, but also, as soon turned out, good for a few surprises. Several measurements of particle production in the low and intermediate p_T region clearly show that p–Pb collisions can not be explained by an incoherent superposition of p-p collisions and indicate the presence of collective effects. In the first part of this paper, we will briefly discuss recent ALICE results on this topic. The above mentioned studies have been performed as a function of particle multiplicity without making any use of the collision geometry e.g. the number of binary collisions N_{coll} or the number of participants $N_{part} = N_{coll}+1$. In order to study nuclear modifications of particle production at high p_T we have to compare particle yields measured in p–Pb to the pp reference spectrum scaled by N_{coll}. The determination of N_{coll} and biases on the binary scaling will be discussed in sections 2 and 3, respectively.

1. Particle production at low and intermediate transverse momentum

ALICE has measured the average transverse momentum, $\langle p_T \rangle$, as a function of the charged particle multiplicity, N_{ch}, in pp, p–Pb and Pb–Pb collisions at $\sqrt{s_{NN}}$ = 7 TeV, 5.02 TeV and 2.76 TeV, respectively (Fig. 1) [2]. From measurements of $\langle p_T \rangle$ in pp at several energies we expect that the collision energy dependence is weak and, hence, we assume the results for the three collision systems can be directly compared. With respect to Pb–Pb, in p–Pb, $\langle p_T \rangle$ shows a much stronger increase with multiplicity following the pp data up to $N_{ch} = 14$. Note that multiplicities around 14 correspond to typical p–Pb collision, whereas pp collisions at this multiplicity are already strongly biased ($N_{ch} > 14$ corresponds to 50% (10%) of the p–Pb (pp) cross-section).

In pp, the strong rise of $\langle p_T \rangle$ with multiplicity can be attributed to the presence of color reconnections (CR) between strings resulting from multiple parton scatterings [3] which can be interpreted as a collective final state effect. In other words, the data can not be described by an independent superposition of parton scatterings. In a similar way, attempts to describe

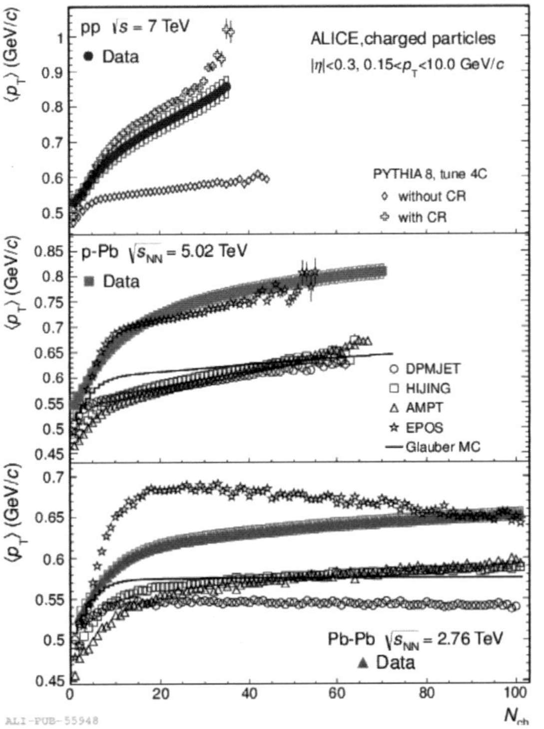

Figure 1. $\langle p_T \rangle$ as a function of charged-particle multiplicity N_{ch} measured in pp (upper panel), p–Pb (middle), and Pb–Pb (lower) collisions in comparison to model calculations [2].

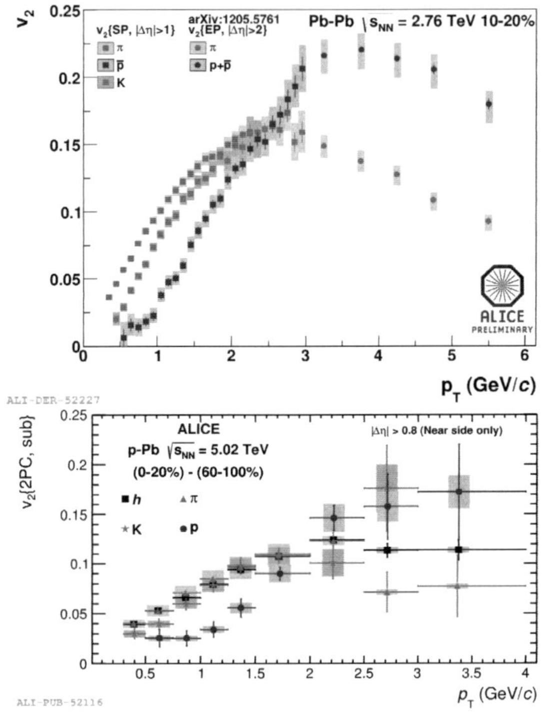

Figure 2. For hadrons, pions, kaons and protons: (upper panel) $v_2(p_T)$ for mid-central Pb–Pb collisions and (lower panel) $v_2\{2\text{PC}, \text{sub}\}(p_T)$ from 2-particle correlations in the 0–20% multiplicity class after subtraction of the correlation from the 60–100% multiplicity class [9].

the rise of $\langle p_T \rangle$ in p–Pb collisions by a superposition of parton scatterings from an incoherent superposition of pp collisions fail suggesting that also in this case coherent final state effects are at work. Indeed, the EPOS generator which includes such effects can reproduce the p-Pb data, however, it fails to describe peripheral Pb–Pb collisions. ALICE has also measured the multiplicity dependence of $\langle p_T \rangle$ for identified particles (π, K, p). Here, a clear mass ordering $\langle p_T \rangle_p > \langle p_T \rangle_K > \langle p_T \rangle_\pi$ is observed [5], which is an indication for collective expansion with a common velocity field.

Further evidence for collective effects in p–Pb results from the study of triggered 2-particle angular correlations in the azimuthal ($\Delta\varphi$) and pseudo-rapidity ($\Delta\eta$) differences. Already analyzing high-multiplicity p–Pb collisions from the pilot run CMS has reported the presence of a near-side ridge structure elongated in $\Delta\eta$ [6]. Using low-multiplicity events as a reference, ALICE and ATLAS found that the near-side ridge actually has an almost perfectly symmetrical counter-part, back-to-back in azimuth [7, 8] very similar to the momentum anisotropy observed in Pb–Pb, where the effect is attributed to collectivity (flow). In Pb–Pb, this interpretation is further corroborated by the mass dependence of the elliptic flow coefficient v_2. In mid-central Pb–Pb collision (10-20%) and $p_T < 2\,\text{GeV}$, a clear mass ordering $v_2^\pi > v_2^K > v_2^p$ is observed. Also the slopes are different and this leads to a crossing-point at around 2.5 GeV. The same behavior is observed in p–Pb collisions (Fig. 2) [9].

2. Centrality in p-Pb

ALICE has measured the nuclear modification factor R_{pPb} for minimum bias p-Pb collisions and it is found to be unity for p_{T} above $\approx 6\,\mathrm{GeV}$ [10]. For minimum bias collisions, N_{coll} is fixed by the total p–Pb and proton–nucleon (p–N) cross-sections: $N_{\mathrm{coll}}^{\mathrm{MB}} = 208\,\sigma_{\mathrm{pN}}/\sigma_{\mathrm{pPb}} = 6.9$. How can N_{coll} be determined for different centrality classes? In general, centrality is defined via centrality estimators that depend monotonically on the number of collisions, e.g. multiplicity and summed energy in a certain pseudo-rapidity range. In contrast to Pb–Pb collisions, for p–Pb the multiplicity fluctuations for a fixed N_{coll} are large with respect to the relatively small range of N_{coll} (typically between 1 and 16). The presence of large fluctuations can bias the p–N collisions themselves. Hence, for each centrality class we have to answer two independent questions: *What is the mean number of collisions?* and *How much are the p–N collisions biased?* Let us start with the answer to the first question. ALICE employs various sub-detector systems covering disjunct

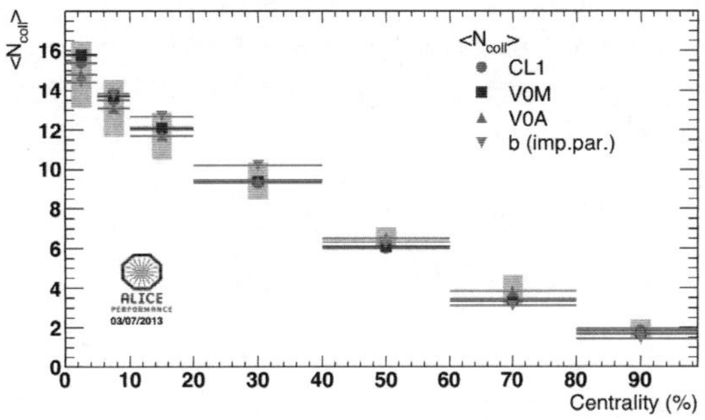

Figure 3. Average N_{coll} as a function of centrality for different estimators. Shaded areas represent the systematic uncertainties for the V0A estimator.

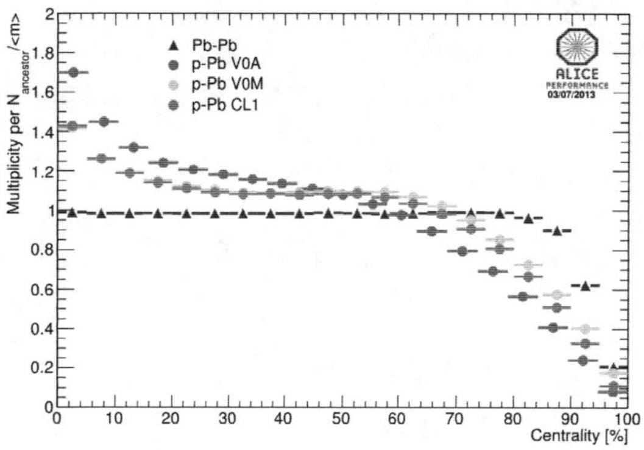

Figure 4. Ratio of multiplicity per ancestor to the NBD mean multiplicity.

pseudo-rapidity ranges to estimate the centrality. The number of tracks are reconstructed and counted using ITS and TPC covering $|\eta| < 0.9$. Clusters are counted in the innermost layers of the ITS, the Silicon Pixel Detector, in $|\eta| < 2.0$ and $|\eta| < 1.4$. A pair of scintillator hodoscopes, VZEROA and VZEROC, measure charged particle multiplicity in the ranges $2.8 < \eta < 5.1$ and $-3.7 < \eta < -1.7$, respectively. The zero degree calorimeters (ZDC) measure forward proton and neutron energies at beam rapidity. The centrality estimators discussed in this paper are: *CL1*: number of clusters in the 2nd pixel layer, *V0A*: the VZEROA multiplicity, *V0M*: the sum of VZEROA and VZEROC multiplicities, *ZNA*: the ZDC neutron calorimeter energy in direction of the Pb-beam.

Using these estimators we are sensitive to the reaction products of p-N collisions, the Pb fragmentation products that go mainly in the direction of the Pb beam (*V0A*) and the so called slow nucleons from evaporation and knock-out that are emitted into the very forward directions and are detected by the zero degree calorimeters. For centrality estimation, particle production in the central part is modeled by negative binomial distributions (NBD) and forward slow nucleon production by fragmentation models for evaporation and knock-out [11].

As for Pb–Pb collisions, in p–Pb, we define centrality classes as percentiles of the multiplicity distributions. In order to extract for each class $\langle N_{\mathrm{coll}} \rangle$, we use the Glauber fit approach with N_{part} as the number of particle sources (ancestors) [12]. Fig. 3 shows $\langle N_{\mathrm{coll}} \rangle$ as a function of centrality for the different estimators. The variation of $\langle N_{\mathrm{coll}} \rangle$ between different estimators is small and of similar magnitude as the systematic error obtained by varying the Glauber parameters and from a closure test using HIJING [13] simulations.

3. Bias on binary scaling

Let us now tackle the second question: *How biased are* p–N *collisions for a given centrality class?* Compared to Pb–Pb collisions, we find that in p–Pb collisions the correlation between the centrality estimator and N_{coll} is very loose. In other words the same value of N_{coll} can contribute to several adjacent centrality classes. So, what exactly distinguishes two centrality classes for the same N_{coll}? Is this relevant for other physics observables? The Glauber Monte Carlo itself can give us an indication on the strength of these biases. Fig. 4 shows the ratio of the generated multiplicity per ancestor to the mean multiplicity of the NBD. This ratio is constant in case all collisions are unbiased. What one sees is that in p–Pb the ratio is above unity for central collisions and below unity for peripheral collisions, whereas in Pb-Pb the bias is much smaller and restricted to very peripheral collisions. Also the mean p–N impact parameter, b_{NN}, can be extracted from the Glauber MC. For peripheral collisions, it is larger than the average. Note that in some models large b_{NN} correspond to softer than average interactions.

Multiplicity fluctuations described by the NBD have no immediate dynamical interpretation. However, models based on multi-parton interaction include intrinsically a fluctuating number of particles sources, the hard scatterings. For example, in HIJING [13], the mean number of hard scatterings, $\langle n_{\mathrm{h}}\rangle$, for a p–N collision is obtained from an impact parameter dependent p–N overlap function, T_{N}, and the hard cross-section, σ_{hard}, via $\langle n_{\mathrm{h}}\rangle = T_{\mathrm{N}}(b_{\mathrm{NN}})\sigma_{\mathrm{hard}}$. The number of scatterings, i, itself follows a Poissonian distribution $p_i = \langle n_{\mathrm{h}}\rangle^i / i! \exp(-\langle n_{\mathrm{h}}\rangle)\rangle$. Hence, in these models, there is a natural link between multiplicity fluctuations and the number of hard scatterings.

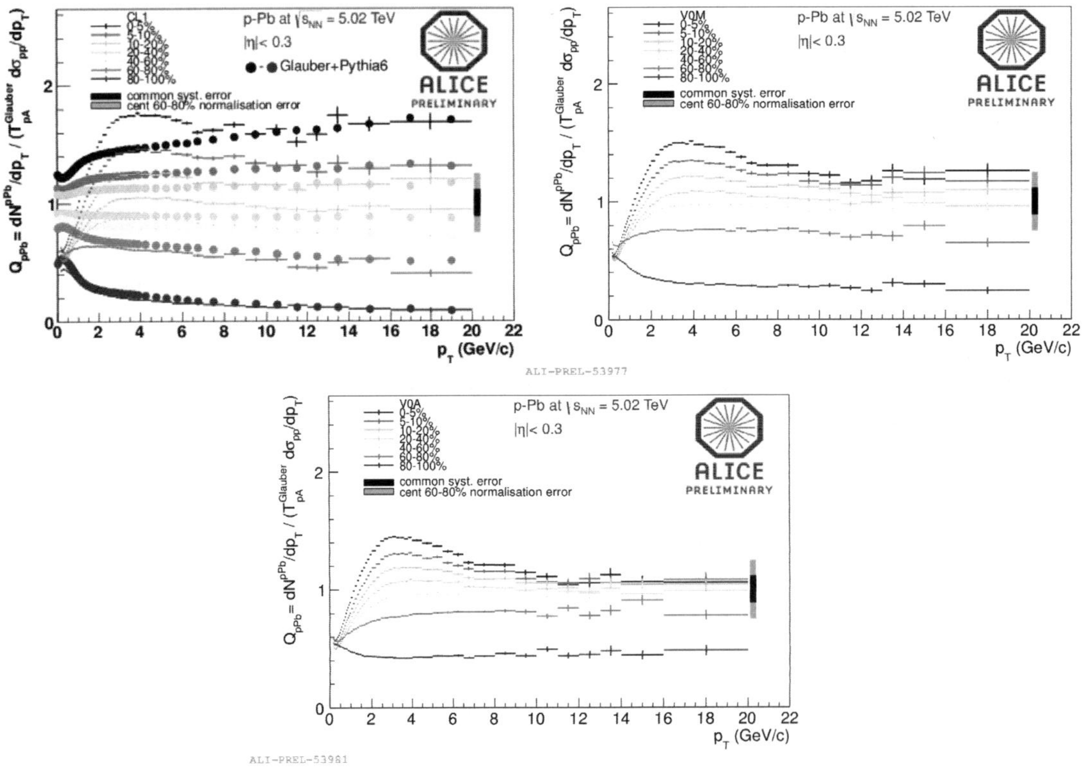

Figure 5. $Q_{\mathrm{pPb}}(p_{\mathrm{T}})$ for three centrality estimators CL1, V0M and V0A. The CL1 results are compared to a Pythia6 simulation (solid circles, see text)

The number of binary p–N collisions is used to scale the reference pp yields and obtain the nuclear modification factor. However, from the discussion above we have at least two new elements: For a given centrality class, hard processes scale with $\langle N_{\rm coll}\rangle\langle n_{\rm h}\rangle_{pN}/\langle n_{\rm h}\rangle_{pp}$. Further, for a given p–Pb impact parameter, b, $\langle n_{\rm h}\rangle$ depends on the average p–N impact parameter. This is mainly important for peripheral collisions. Here, the multiplicity estimator acts also as a veto on hard processes which contribute to the overall multiplicity (jet-veto). Being aware of this biases we define the observable $Q_{\rm pPb} = {\rm d}N^{\rm pPb}/{\rm d}p_{\rm T}/(\langle N_{\rm coll}\rangle {\rm d}N^{pp}/{\rm d}p_{\rm T}) = {\rm d}N^{\rm pPb}/{\rm d}p_{\rm T}/(\langle T_{\rm pPb}\rangle {\rm d}\sigma^{pp}/{\rm d}p_{\rm T})$, which is not equal to $R_{\rm pPb}$, since we do not take into account the bias on the mean number of hard scatterings. Fig. 5 shows $Q_{\rm pPb}(p_{\rm T})$ for the CL1, V0M and V0A centrality estimators. For all centrality classes $Q_{\rm pPb}$ strongly deviates from unity at high $p_{\rm T}$. However the spread between different centrality classes reduces when moving from CL1 to V0M to V0A, increasing the η-separation between the estimator and the $p_{\rm T}$ measurement. The smallest bias is expected for the ZNA-estimator, the analysis of which is still in progress. For the most peripheral collisions with CL1, there is a clear indication for a jet veto bias: $Q_{\rm pPb}$ has a significant negative slope due to the fact that the contribution of jets to the overall multiplicity increases with $p_{\rm T}$. The slope is reduced for V0M and absent for V0A. The CL1 results are compared to a Pythia6 [14] simulation, where $N_{\rm coll}$ pp-collisions have been superimposed with a probability proportional to the Glauber $N_{\rm coll}$ distribution. As for the data, the centrality has been obtained from the charged particle multiplicity in $|\eta| < 1.4$ and $\langle N_{\rm coll}\rangle$ is directly obtained from the Monte Carlo. Quite surprisingly this simple model can reproduce the large bias at high $p_{\rm T}$. It also agrees with the low-$p_{\rm T}$ region of the most peripheral collisions and reproduces the jet-veto bias. For all other centralities, it does not reproduce the Cronin-like enhancement at $\approx 4\,{\rm GeV}$ and the low-$p_{\rm T}$ region is overestimated. The latter is expected since in this region also the minimum bias $R_{\rm pPb}$ is below unity.

In summary, in pp and p-Pb a strong increase in $\langle p_{\rm T}\rangle$ with multiplicity is observed. In models for pp this is understood as a consequence of color reconnections between strings produced in multiple parton interactions raising the question whether similar effects are at work in p–Pb or whether it is the result of collective flow or initial state effects. The v_2 coefficients obtained from jet-subtracted identified 2-particle correlations show a mass ordering and crossing of v_2 of protons and pions. The pattern is reminiscent of Pb-Pb collisions, where this is effects is attributed to hydrodynamic flow. Centrality estimators based on multiplicity measurements in $|\eta| < 5$ induce a bias on the hardness of the p–N collisions that can be quantified by the number of hard scatterings per p–N collision. Low (high) multiplicity p–Pb corresponds to lower (higher) than average number of hard scatterings. The comparisons to incoherent superposition of p–N collisions have to be performed including this bias.

References

[1] Accardi A *et al* 2003 *Preprint* hep-ph/0308248
[2] Abelev B *et al* 2013 *Phys. Lett.* B **727** 371-380
[3] Skands P Z and Wicke D 2007 *Eur. Phys. J.* C **52** 133
[4] Pierog T *et al* 2013 *Preprint* hep-ph/13060121
[5] Abelev B *et al* 2014 *Phys. Lett.* B **728** 25-38
[6] Chatrchyan S *et al* 2013 *Phys. Lett.* B **718** 795
[7] Abelev B *et al* 2013 *Phys. Lett.* B **719** 29-41
[8] Aad G *et al* 2013 *Phys. Rev. Lett.* **110** 182302
[9] Abelev B *et al* 2013 *Phys. Lett.* B **726** 164-177
[10] Abelev B *et al* 2013 *Phys. Rev. Lett.* **110** 082302
[11] Sikler F 2003 *Preprint* hep-ph/0304065
[12] Abelev B *et al* 2013 *Phys. Rev.* C **88** 044909
[13] Gyulassy M and Wang X N 1994 *Comput. Phys. Commun.* **83** 307
[14] Sjostrand T, Mrenna S and Skands P Z 2006 *JHEP* **0605** 026
[15] Abelev B *et al* 2013 *Phys. Rev. Lett.* **110** 032301

Jet quenching and heavy quarks

Thorsten Renk

Department of Physics, P.O. Box 35, FI-40014 University of Jyväskylä, Finland

E-mail: `thorsten.i.renk@jyu.fi`

Abstract. Jet quenching and more generally physics at high transverse momentum P_T scales is a cornerstone of the heavy-ion physics program at both RHIC and LHC. In this work, the current understanding of jet quenching in terms of a QCD shower evolution being modified by the surrounding medium is reviewed along with the evidence for this picture from light parton high P_T observables. Conceptually, the same QCD shower description should also be relevant for heavy quarks, but with several important modifications introduced by the quark masses. Thus especially in the limit of small jet energy over quark mass E_{jet}/m_q, the relevant physics may be rather different from light quark jets, and several attempts to explain the observed phenomenology of heavy quarks at high P_T are discussed here.

1. The QCD shower in vacuum

Perturbative Quantum-Chromodynamics (pQCD) is used very successfully to understand particle production at high P_T in proton-proton (p-p) collisions by factorizing the perturbatively calculable hard process from the non-perturbative initial and final state which is parametrized by universal objects, the parton distributions (PDFs) and fragmentation functions (FFs) respectively. It has been realized early on that hard processes can hence be utilized in the context of heavy-ion collisions by providing a probe with a perturbatively calculable rate, which then allows medium tomography via the parton final state interaction with the bulk QCD medium created in such collisions [1].

The partons emerging from a hard process are highly virtual and hence utilize the phase space provided by their virtuality for intense QCD radiation, known as a parton shower, which after hadronization is experimentally observed as a hadron jet. Due to the hard virtuality scale, such showers are perturbatively tractable and simulated e.g. by Monte-Carlo (MC) techniques such as the PYSHOW algorithm [2], leading to the scale evolution of the FF. Let us for illustration briefly investigate the physics encoded by PYSHOW. The algorithm treats the QCD shower as an interated series of $1 \rightarrow 2$ splittings of a parent parton a into two daughters b, c where the energy is distributed as $E_a = zE_b + (1-z)E_c$ and the relevant virtuality scale (and hence the radiation phase space) decreases with every splitting as parametrized by $t = \log Q^2/\Lambda^2_{QCD}$ until it reaches a non-perturbative virtuality $Q_0 \sim 1$ GeV. The differential splitting probability at a scale t is then given by the splitting kernel $P_{a\rightarrow bc}(z)$ which can be computed in pQCD for the characteristic subprocess, integrated over the kinematically available range in z

$$I_{a\rightarrow bc}(t) = \int_{z_-(t)}^{z_+(t)} dz \frac{\alpha_s}{2\pi} P_{a\rightarrow bc}(z).$$

Journal of Physics: Conference Series **509** (2014) 012022 doi:10.1088/1742-6596/509/1/012022

where the kinematic limits available for z depend on a combination of parent and daughter virtualities Q_{abc} and masses m_{abc} via $M_{abc} = \sqrt{m_{abc}^2 + Q_{abc}^2}$ as

$$z_\pm = \frac{1}{2}\left(1 + \frac{M_b^2 - M_c^2}{M_a^2} \pm \frac{|\mathbf{p}_a|}{E_a}\frac{\sqrt{(M_a^2 - M_b^2 - M_c^2)^2 - 4M_b^2 M_c^2}}{M_a^2}\right).$$

It is this restriction of radiation phase space due to quark mass which makes the fragmentation in vacuum for heavy quarks much harder than for light quarks.

2. The QCD shower in the medium

The uncertainty relation suggests that the formation time of a virtual parton a is parametrically $\tau \sim E_a/Q_a^2$ in the c.m. frame. This allows one to argue that the medium can affect neither the hard process itself nor the initial hard branchings in the shower but only later subsequent branchings. Thus, conceptually including the medium corresponds to replacing the FF by a medium-modified FF (MMFF) (see e.g. [3] for a more comprehensive discussion). There are various proposals for how the medium might affect the shower, some assume a modification of the splitting kernel leading to fractional energy loss [4]; others add to this elastic scattering of partons with the medium [5]; so far the best tested idea is however to modify the available radiation phase space by introducing a flow of energy and momentum between medium and jet as exemplified by the MC code YaJEM [6, 7].

Such a perturbative picture of the medium-modified shower leads to an impressive description of multiple observables, in particular it describes correctly the pattern of hadronizing medium-induced radiation double differentially in P_T and angle with respect to the jet axis [8] as observed by the STAR collaboration in jet-h correlations [9], the evolution of the nuclear suppression factor from RHIC to LHC [3, 10] or the full range of energy dependence of the dijet imbalance [11] as measured by CMS [12]. No viable description of these observables has been suggested so far which is not based on medium perturbations of a perturbative shower picture, and pQCD based jet quenching works across a substantially larger range of observables than discussed here. This striking success of a pQCD picture for light quarks thus suggests that heavy-quark showers (where in addition a hard mass scale exists) chould also be tractable by the same techniques.

3. Heavy-quark showers

The interactions of QCD are flavour-blind, i.e. gluons couple only to the color charge of quarks, and hence there is no fundamental difference between light and heavy quark jets. Nevertheless, the presence of a substantial quark mass can change the kinematics. As mentioned before, a quark mass introduces a constraint on radiation phase space, thus generically suppressing the vacuum as well as the medium induced QCD radiation (this is known as 'dead cone effect' [13]) and led to the expectation that heavy quarks should lose less energy due to interactions with the medium than light quarks). Also the formation time of heavy quark showers can be estimated to $\tau \sim E_{jet}/\sqrt{Q_0^2 + m_q^2}$, i.e. a b-quark shower terminates much earlier than a light quark shower.

Another important aspect is that $m_q \gg T$ with T the medium temperature, i.e. there are no thermally excited heavy-quarks in the medium, which makes conversion reactions like $b\bar{b} \to gg$ impossible (note however that such conversions are a numerically small effect even for light quarks [14]). Furthermore, heavy quarks are tagged [15], i.e. once a hard gluon taking 90% of the quark's energy is radiated, that energy is lost from the quark, whereas the same emission from a light quark would be interpreted as a change in the identity of the leading light parton since gluon jets cannot easily be identified. Finally, note that for $E \sim$ few GeV light quarks are highly relativistic and can be assumed by move with a velocity $v \approx 1$, where such an assumption can not be made for b-quarks.

Since for $Q^2 \gg m_q^2$ the shower evolution is not influenced by the dead cone effect, one can use the uncertainty relation for the shower formation length $L \sim E_{jet}/m_q^2$ to establish conservative estimates for what jet energies E_{jet} a shower surely probes the whole medium ($L > 10$ GeV) and when the shower terminates before the medium forms ($L < 0.5$ GeV). Using these relations, with $E_c > 78$ GeV and $E_b > 900$ GeV, no differences between the medium modification of heavy and light quark jets are expected as they probe the medium for the same length in a kinematical regime where the quark mass does not matter whereas for $E_c < 3.9$ GeV or $E_b < 45$ GeV one finds that the shower is over before the medium is formed, i.e. in this situation an on-shell quark rather than an evolving shower is the relevant state which is perturbed by the medium. Between these scales is an intermediate region of complicated physics in which one can neither assume that a shower treatment alone nor an on-shell treatment alone would lead to the right physics, and comparison with experiment is needed to understand which approximation works best.

Despite claims in the literature that heavy quark energy loss would be a good benchmark test for models designed for light parton high P_T physics, this is hence not true in general but rather depends on the kinematic regime under consideration, as the relevant physics processes can be qualitatively different.

4. The heavy quark puzzle

While the dead cone effect discussed above would suggest that the medium is able to induce less radiation (and hence less suppression) for heavy quarks as compared with light quarks, the experimental observation of non-photonic single electron nuclear suppression factor (with the electrons predominantly originating from heavy quark decays) showed a suppression to the same level as for charged hadrons [16, 17]. This is known as the 'heavy quark puzzle'.

Since the dead cone effect works strongest for the b-quark, early attempts at an explanation argued that the ratio of c to b contribution to the electron spectrum isn't well understood, however template fits to the decay kinematics later extracted experimental ratios compatible with theory expectations. Another proposal was that elastic energy loss is quite strong for all partons, i.e. close to 50% of all energy loss is not via medium-induced radiation but through elastic recoil of medium constituents [18]. However, for the light quark sector, the contribution of such elastic reactions (which would be incoherent) is very well constrained by pathlength dependent observables such as the reaction plane dependence of R_{AA} or the strength of back-to-back dihadron correlations [3, 19, 20] to be about 10% of the total.

A more radical proposal argues that they physics of in-medium showers (especially for heavy quarks) might better be described by strong coupling techniques using the AdS/CFT duality [21, 22]. However, with the first data from LHC available, it could be established that strong coupling techniques fail to predict the evolution of the medium effect from 200 AGeV to 2.76 ATeV correctly, i.e. they overpredict the quenching both for heavy quarks [23] and light quarks [3] (cf. Fig. 1), and to date there is no viable strong coupling model which can account for even a fraction of the large body of high P_T data available.

The currently most promising idea to solve the puzzle involves a selective enhancement of the heavy quark elastic cross section with the medium via resonance formation — heavy quarks may form proto-hadrons even in a deconfined medium, and while these are unstable, they may still leave a pronounced effect in the scattering cross section [24, 25, 26]. Such an effect could work selectively for heavy quarks since light quarks have a long shower formation time and typically do not come on-shell throughout the medium evolution time.

5. Current heavy-quark phenomenology at high P_T

While modern calculations for the high P_T phenomenology of light partons are mainly concerned with QCD shower evolution and vacuum and medium-induced radiation, heavy quark phenomenology is usually computed with elastic reactions of on-shell quarks dominating

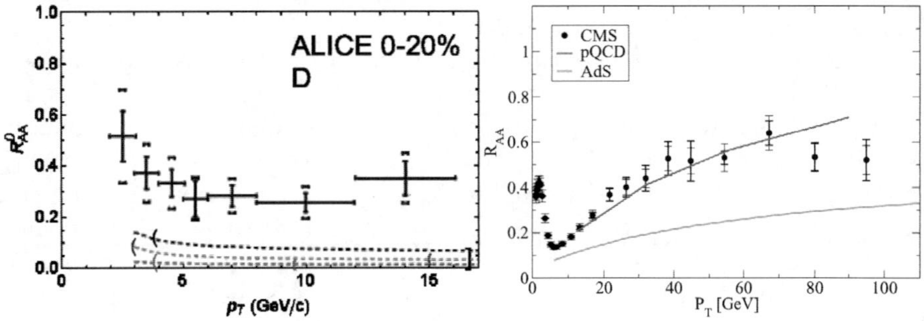

Figure 1. Failure of AdS/CFT strong coupling techniques to extrapolate R_{AA} from RHIC to LHC energies for heavy quarks (left)[23] and light quarks (right) [3]

energy loss. As discussed above, these distinctly different approaches are well motivated by the respective kinematical regimes of the observables and the dead cone effect.

Usually a K-factor, motivated by the resonance formation picture, is used to enhance the heavy quark elastic cross section in such computations which is then fit to data, aiming for a simultaneous description of the P_T dependence of both the mean electron nuclear suppression factor R_{AA} and its event plane angle dependence in terms of the second harmonic coefficient v_2. Such approaches may include only elastic reactions (e.g. [27]) or include medium-induced radiation, albeit suppressed by the dead cone effect [28] and are able to achieve a fair agreement with the available data. Of particular importance here is the use of realistic and constrained models of the medium evolution. In the light parton sector, it could be demonstrated that the choice of the medium evolution model has a large influence on the obtained value of v_2 [29], and a corresponding study of the systematical effect of the influence of the choice of the medium model on v_2 for heavy quark observables is crucial to judge models correctly against the data.

6. Outlook

In many computations of elastic energy loss, a fixed value of the QCD coupling α_s is used. However, in reality the running of α_s with the momentum scale is a well established phenomenon, leading to the problem that in the low momentum transfer scatterings expected with medium constituents, α_s may run into a non-perturbative region. One proposal for the determination of the coupling is to obtain it self-consistently with the scale of the Debye screening mass which limits small-angle scatterings in a thermal medium [30] via

$$m_D^2 = 4\pi(1 + \frac{1}{6}n_f)\alpha_s(m_D^2)T^2.$$

Numerically this results in an enhancement of the elastic scattering cross section as compared to fixed α_s computations. However, the same enhancement would also be relevant for the light quark sector, leading again to a large fraction of elastic energy transfer into the medium which is not observed (note that since a radiation vertex in a shower probes $\alpha_s(Q^2)$ for a high virtuality scale, the situation is different for radiative processes). Ultimately, this raises the question of what degrees of freedom in the medium light and heavy quarks are scattering off. Both the success of fluid dynamics in the description of the bulk medium evolution (which can not be obtained in perturbation theory) and the failure of models with a large component of elastic energy loss, assuming scattering off a gas of perturbative quarks and gluons (e.g. [20]) strongly suggest that the medium degrees of freedom are not light quasiparticles. A precise determination of the energy transfer by elastic reactions both in the light and in the heavy quark sector might provide an important clue as to their true nature.

The current physics picture of light quark phenomenology in terms of medium-modified shower evolution vs. heavy quark phenomenology in terms of largely elastic processes for on-shell quarks which are enhanced by resonant processes is rather compelling for the momentum regimes currently under investigation. The CMS collaboration has openend a new frontier using b-tagged jets, demonstrating that at 100 GeV b-tagged jets show an R_{AA} compatible with untagged jet R_{AA}, a possible indication that at this scale b-quarks might already effectively behave 'light'.

One crucial test whether the picture of the dominance of elastic energy loss is correct or not are back-to-back correlations which can be shown to be very sensitive to quantum coherence effects and the resulting pathlength-dependence of the medium modification [3, 19, 32]. If the observed suppression of non-photonic electrons is indeed the result of elastic scattering processes, then the away side conditional yield ratio I_{AA} of medium over vacuum e.g. D meson yield, given a triggered D meson should be much higher than in the case of, e.g. $\pi - \pi$ correlations. Measuring and successfully describing such correlations would thus go a long way in confirming our current consistent understanding of light and heavy quark phenomenology at high P_T and would open the path for a precision determination of medium properties in the future.

References

[1] Gyulassy M. and Wang X. N. 1994 *Nucl. Phys.* B **420** 583 ; Baier R., Dokshitzer Y. L., Mueller A. H., Peigne S. and Schiff D. 1997 *Nucl. Phys.* B **484** 265; Zakharov B. G. 1997 *JETP Lett.* **65** 615; Wiedemann U. A. 2000 *Nucl. Phys.* B **588** 303; Gyulassy M., Levai P. and Vitev I. 2001 *Nucl. Phys.* B **594** 371; Wang X. N. and Guo X. F. 2001 *Nucl. Phys.* A **696** 788.
[2] Bengtsson M. and Sjöstrand T. 1987 *Phys. Lett.* B **185** 435; Bengtsson M. and Sjöstrand T. 1987 *Nucl. Phys.* B **289** 810; Norrbin E. and Sjöstrand T. 2001 *Nucl. Phys.* B **603** 297.
[3] Renk T. 2012 *Phys. Rev.* C **85** 044903.
[4] Armesto N., Cunqueiro L. and Salgado C. A. 2009 *Eur. Phys. J.* C **63** 679.
[5] Zapp K., Ingelman G., Rathsman J., Stachel J. and Wiedemann U. A. 2009 *Eur. Phys. J.* C **60** 617.
[6] Renk T. 2008 *Phys. Rev.* C **78** 034908.
[7] Renk T. 2009 *Phys. Rev.* C **79** 054906.
[8] Renk T. 2013 *Phys. Rev.* C **87**, no. 2, 024905.
[9] Adamczyk L. *et al.* [STAR Collaboration] 2013 1302.6184 [nucl-ex].
[10] Renk T. 2013 *Phys. Rev.* C **88**, 014905.
[11] Renk T. 2012 *Phys. Rev.* C **86** 061901.
[12] Chatrchyan S. *et al.* [CMS Collaboration] 2012 *Phys. Lett.* B **712** 176.
[13] Dokshitzer Y. L. and Kharzeev D. E. 2001 *Phys. Lett.* B **519** 199.
[14] Auvinen J., Eskola K. J. and Renk T. 2010 *Phys. Rev.* C **82** 024906.
[15] Peigne S. and Peshier A. 2008 *Phys. Rev.* D **77** 114017.
[16] Adare A. *et al.* [PHENIX Collaboration] 2007 *Phys. Rev. Lett.* **98** 172301.
[17] Abelev B. I. *et al.* [STAR Collaboration] 2007 *Phys. Rev. Lett.* **98** 192301 [2011 Erratum-ibid. **106** 159902].
[18] Djordjevic M. 2006 *J. Phys.* G **32**, S333.
[19] Renk T. 2011 *Phys. Rev.* C **84** 067902.
[20] Auvinen J., Eskola K. J., Holopainen H. and Renk T. 2010 *Phys. Rev.* C **82**, 051901 .
[21] Gubser S. S. 2007 *Phys. Rev.* D **76** 126003.
[22] Horowitz W. A. and Gyulassy M. 2008 *Phys. Lett.* B **666** 320.
[23] Horowitz W. A. 2013 *Nucl. Phys.* A904-905 **2013** 186c.
[24] Adil A. and Vitev I. 2007 *Phys. Lett.* B **649** 139.
[25] Sharma R., Vitev I. and Zhang B. -W. 2009 *Phys. Rev.* C **80** 054902.
[26] He M., Fries R. J. and Rapp R. 2012 *Phys. Rev.* C **86** 014903.
[27] Uphoff J., Fochler O., Xu Z. and Greiner C. 2012 *Phys. Lett.* B **717**, 430.
[28] Nahrgang M., Aichelin J., Gossiaux P. B. and Werner K. 2013 1305.6544 [hep-ph].
[29] Renk T., Holopainen H., Heinz U. and Shen C. 2011 *Phys. Rev.* C **83**, 014910.
[30] Peshier A. 2008 *J. Phys.* G **35** 044028.
[31] Nguyen M. for the CMS collaboration, presentation Jet Workshop Paris 2013.
[32] Renk T. 2007 *Phys. Rev.* C **76** 064905.

14th International Conference on Strangeness in Quark Matter (SQM2013) IOP Publishing
Journal of Physics: Conference Series **509** (2014) 012023 doi:10.1088/1742-6596/509/1/012023

Jet results from PbPb and pPb collisions with CMS

George S. F. Stephans
For the CMS Collaboration

Massachusetts Institute of Technology, Cambridge, MA, USA

E-mail: gsfs@mit.edu

Abstract. The Compact Muon Solenoid (CMS) experiment at the Large Hadron Collider (LHC) has used a wide variety of observables to study the hot and dense medium created in heavy ion collisions. The extraction of fully reconstructed jets is one area in which the higher center of mass energy provided by the LHC is particularly helpful. This paper provides an overview of some of the jet and jet-related results that have been extracted from CMS data. The systems studied include PbPb and pp at $\sqrt{s_{NN}}$ = 2.76 TeV and more recently pPb at $\sqrt{s_{NN}}$ = 5.02 TeV.

The start of data taking at the Large Hadron Collider (LHC) at CERN expanded the accessible heavy ion center-of-mass energy range by more than an order of magnitude. These higher-energy collisions provide greatly enhanced production cross sections for a number of entities, most significantly heavy particles and high transverse momentum (p_T) jets. More specifically, these higher-p_T jets have sufficient energy to stand out very clearly from the background of soft particles (the so-called "underlying event") created in the heavy ion collision. This allows the extraction of a number of observables using fully reconstructed jets alone or correlations between jets and charged particles. Compared to a high-p_T particle, the properties of an observed jet are more closely related to those of the initially hard-scattered parton. As a result jet studies allow the extraction of more direct information about the interactions of those partons with the hot and dense medium created in heavy ion collisions.

The Compact Muon Solenoid (CMS) detector at the LHC is particularly well suited for these types of studies. Its high-precisions calorimetry and tracking allow very accurate reconstruction of both jets and charged particle tracks. The combination of a fast Level-1 (L1) trigger with a sophisticated High Level Trigger (HLT) capable of running full particle and jet algorithms allows efficient extraction of the most interesting events. Thus far, the systems studied include PbPb and pp at $\sqrt{s_{NN}}$ = 2.76 TeV and more recently pPb at $\sqrt{s_{NN}}$ = 5.02 TeV. Jets are reconstructed using the anti-k_T algorithm with a radius parameter of $R = 0.3$. A full description of the CMS detector can be found in Ref. [1]. More information on all of the CMS heavy ion physics results can be found on the "Public Physics Results" web page [2].

The studies of jets in heavy ion collisions using CMS began with the observation that the momentum imbalance between pairs of back-to-back jets was much larger in central PbPb collisions than in either pp or peripheral collisions [3]. Despite this evidence for significant parton energy loss, there was no sign of any deflection, the two jets still emerged essentially back to back. By studying charged particles in events with unbalanced jets, the missing energy was found to emerge predominantly in low momentum particles at large angles with respect to the jet axis [3]. In a different approach, the ratio of the transverse momenta of the two jets was studied as a function of centrality and the transverse momentum of the leading jet. This ratio rises smoothly with increasing leading-jet p_T in pp collisions. The same pattern was found in PbPb at all centralities, except that the ratio is systematically shifted down, i.e. the p_T of the sub-leading jet is smaller [4]. The size of this shift was found to increase monotonically with increasing centrality of the PbPb collision. Furthermore, at a fixed centrality, the shift was found to be independent of jet p_T up to the highest values studied. CMS has also measured the yield of both high-p_T jets [5] and charged particles [6] and found comparable levels of suppression.

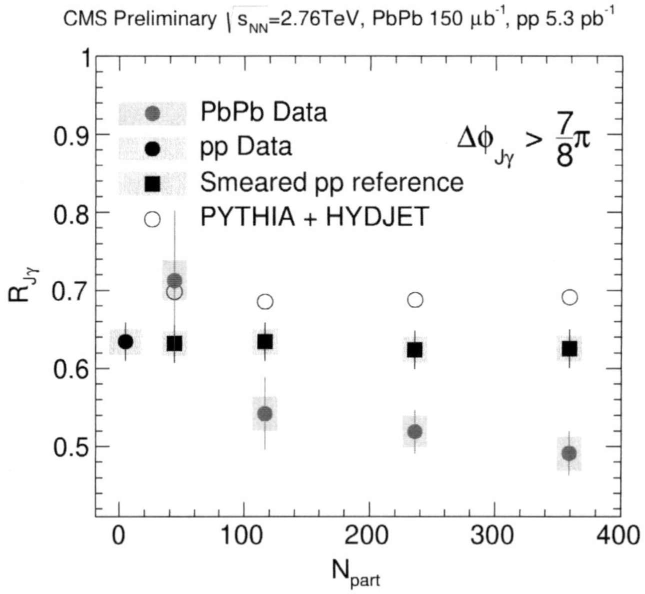

Figure 1. Associated jet fraction in γ-jet events as a function of centrality for PbPb collisions at $\sqrt{s_{NN}} = 2.76$ TeV are compared to a reference generated using pp data at the same energy. The fraction $R_{\gamma j}$ represents the fraction of events with a high-p_T photon in which an associated jet is found. The momenta of jets in pp were smeared by the relative jet energy resolution to be used as the reference of each centrality bin.

One disadvantage of using dijet events to study parton energy loss is that both partons are traversing the medium. Thus, only the *difference* in energy loss can be measured, Furthermore, the lower cutoff on jet p_T in the analysis leads to a surface bias in the event sample. Dijets originating deep inside the medium are more likely to have both jets fall below the limit. An alternative that avoids both problems is the study of γ-jet events. Data from several experiments, including CMS [7], have already established that isolated photons are not quenched by the medium. The p_T spectrum of γs, even in central PbPb events is identical within uncertainties to that found in pp at the same center of mass energy. Furthermore, the yield of high p_T photons matches the prediction based on pp data and the number of nucleon-nucleon collisions for a given PbPb centrality. The biggest disadvantage of γ-jet compared to dijets is the significantly smaller production rate. Nonetheless, the first measurement of parton energy loss using isolated photons to tag the initial parton momentum (both magnitude and direction) has been performed by CMS [8]. As for the dijets, the data showed evidence for increasing energy loss with increasing centrality but no evidence for any parton deflection. This conclusion has recently been strengthened by repeating the analysis using a reference spectrum extracted from CMS pp data as opposed to PYTHIA [9]. Figure 1 shows $R_{\gamma j}$, the fraction of events containing a high-p_T

photon in which a jet was also found as a function of centrality. The PbPb data are compared to pp data and also to a reference in which the pp data was smeared to account for the different jet energy resolution in PbPb events of different centralities.

Having established the basic parameters of parton energy loss, CMS data have also been used to investigate the detailed properties of the quenched jets, primarily through the use of jet-particle correlations. Two different approaches have been used, analyzing the distribution of both the transverse momentum of the associated particles and their angular distribution with respect to the jet axis. The former analysis, in the form of fragmentation functions, initially found essentially no change with respect to similar p_T jets in pp collisions, even for the most unbalanced events in very central PbPb collisions [10]. However, this first study was restricted to particles with p_T above 4 GeV/c. Further analysis did reveal a difference, namely an excess of low momentum particles which increased with increasing PbPb centrality [11]. The angular distribution of the jet fragmentation pattern has also been studied. Rather than simply counting particles, the distribution of the total jet p_T in rings of increasing radius around the jet axis was extracted [12]. This so-called "jet shape" revealed significant deviations from the pattern seen in pp, especially far from the jet axis in the most central PbPb collisions. Some evidence for a modest decrease slightly closer to the jet axis were also seen. Both analyses suggest that the core of the jet fragmentation pattern, consisting predominantly of high p_T particles is not significantly modified by the presence of the hot and dense medium.

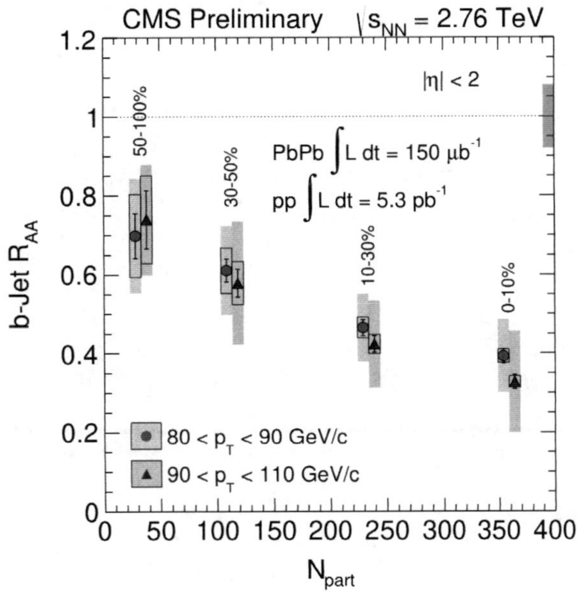

Figure 2. Nuclear modification factor (R_{AA}) as a function of centrality for b jets in two p_T bins (80-90 GeV/c, and 90-110 GeV/c) in PbPb collisions at $\sqrt{s_{NN}} = 2.76$ TeV. The centrality is quantified using N_{part}, the number of interacting nucleons in the collision.

The Strangeness in Quark Matter conference series has specifically focused on the physics of the strange (and more recently heavier) quarks. Thus far, CMS has not extracted any jet-related physics using strange or charm quarks. However, it is possible to use the high precision vertexing capability of the CMS tracker to tag jets from b quarks, using their displaced secondary vertex. Contrary to some expectations, the suppression of b jets was found to be comparable to that

Journal of Physics: Conference Series **509** (2014) 012023 doi:10.1088/1742-6596/509/1/012023

for inclusive jets [13]. Figure 2 shows the nuclear modification factor (R_{AA}) for b jets in PbPb collisions as a function of centrality. Results are shown for two different ranges in b-jet p_{T}. It may also be possible to use more differential analysis of fragmentation functions as well as comparison of both dijets with γ-jet and 2-jet with 3-jet events to determine the relative energy loss of quark and gluon jets.

In the spring of 2013, pPb collisions were performed for the first time at the LHC. Because of the different rigidities of the proton and Pb ion, these collisions were at a nucleon-nucleon center of mass energy of $\sqrt{s_{NN}} = 5.02$ TeV. The data have revealed several unexpected finding, including possible evidence for hydrodynamical collective effects in events with very high multiplicity [14, 15]. The ratio of transverse momenta for dijets emitted in pPb collisions has been analyzed for a range of centralities [16]. Since there is no pp data currently available at this same $\sqrt{s_{NN}}$, these ratios were compared to PYTHIA predictions. No modification of the dijet p_{T} ratios were found over the entire range of p_{T} studied, even for the most central 2.5% of the events.

The future prospects for using a variety of jet data to probe the hot and dense system formed in ultra-relativistic heavy ion collisions is bright indeed. Observables of interest include individual jets, both inclusive and with tagged flavor, as well as correlations of jets with other jets, photons, and charged particles. The next PbPb run, expected in late 2015 following the current LHC shut-down, will provide significantly higher integrated luminosity. The CMS L1 trigger is being upgraded to provide better selection capability for triggering on jets. Currently, that selectivity is limited because the CMS front-end electronics are configured in such a way as to make efficient jet background subtraction impossible. As a result, a large fraction of the events that fire the various jet triggers are actually due to energy from the underlying event. The upgraded trigger will allow new combinations of information from different regions of the detector, making the fast jet trigger much more efficient. This significantly reduced data stream can then be processed by the HLT for the final decision on what events to keep. With larger data sets, more selective as well as more differential analyses will be possible. Precision tomography of the quark gluon plasma using jets and other high-p_{T} probes is just beginning.

Acknowledgments

Acknowledgments We congratulate our colleagues in the CERN accelerator departments for the excellent performance of the LHC. We thank the technical and administrative staff at CERN and all of the other CMS institutes. This work was partially supported by U.S. DOE grant DE-FG02-94ER40818

References

[1] CMS Collaboration 2008 *JINST* **03** S08004
[2] https://twiki.cern.ch/twiki/bin/view/CMSPublic/PhysicsResultsHIN
[3] Chatrchyan, S., et al. (CMS Collaboration) 2011 *Phys. Rev. C* **84** 024906
[4] Chatrchyan, S., et al. (CMS Collaboration) 2012 *Phys. Lett. B* **712** 176
[5] CMS Preliminary 2013 CMS PAS HIN-12-004
[6] Chatrchyan, S., et al. (CMS Collaboration) 2012 *EPJC* **72** 1945
[7] Chatrchyan, S., et al. (CMS Collaboration) 2012 *Phys. Lett. B* **710** 256
[8] Chatrchyan, S., et al. (CMS Collaboration) 2013 *Phys. Lett. B* **718** 773
[9] CMS Preliminary 2013 CMS PAS HIN-13-006
[10] Chatrchyan, S., et al. (CMS Collaboration) 2012 *JHEP* **10** 087
[11] CMS Preliminary 2013 CMS PAS HIN-12-013
[12] Chatrchyan, S., et al. (CMS Collaboration) 2013 arXiv:1310.8651 Submitted to Phys. Rev. C
[13] Chatrchyan, S., et al. (CMS Collaboration) 2013 arXiv:1312.4198 Submitted to Phys. Rev. Lett.
[14] Chatrchyan, S., et al. (CMS Collaboration) 2013 *Phys. Lett. B* **718** 795
[15] Chatrchyan, S., et al. (CMS Collaboration) 2013 *Phys. Lett. B* **724** 213
[16] CMS Preliminary 2013 CMS PAS HIN-13-001

Results on angular correlations with ALICE

Panos Christakoglou (for the ALICE Collaboration)
Nikhef, Science Park 105, 1098 XG Amsterdam, The Netherlands
E-mail: Panos.Christakoglou@nikhef.nl

Abstract. Angular correlations are a sensitive probe of the transport properties of the system produced in nucleus–nucleus collisions. Similar studies performed in p–Pb collisions have recently revealed intriguing features as well. In this article, we review the latest results on charged and identified particle correlations obtained with the ALICE detector at the LHC in both Pb–Pb and p–Pb events.

1. Introduction

Angular correlations have emerged as a powerful tool to study the properties of the strongly interacting medium created in high energy nucleus–nucleus collisions [1]. Measurements of azimuthal correlations established the picture of a system that behaves as an almost perfect fluid with low value of dissipation i.e. low value of shear viscosity. The flow harmonics spectrum revealed that elliptic flow (v_2) is the dominant component, reflecting the elliptic shape of the nuclear overlap region and the hydrodynamic response of the medium [2, 3]. Recent experimental results [4] revealed that initial state fluctuations can generate higher order flow components, the study of which can further constrain the values of viscosity [5]. In parallel, the study of high transverse momentum (p_{T}) probes has also established the picture of a system that is dense and opaque [2]. In this article, I review the latest experimental results on angular correlations from the analysis of Pb–Pb and p–Pb collisions at $\sqrt{s_{NN}} = 2.76$ and $\sqrt{s_{NN}} = 5.02$ TeV, respectively, recorded with the ALICE detector at the LHC [6].

2. Anisotropic flow results in Pb–Pb and p–Pb

Anisotropic flow studies are based on the description of the azimuthal distribution of particles by a Fourier series $dN/d\varphi \approx 1 + \sum_{n=1}^{\infty} 2v_n \cos[n(\varphi - \Psi_n)]$ [7], where φ is the azimuthal angle of particles, and v_n are the different flow coefficients developing relative to the symmetry planes Ψ_n for each harmonic n. The first harmonic v_1, the directed flow, affected by the amount of nuclear stopping and the initial position of nucleons in the transverse plane, is a sensitive probe of the initial conditions of the system. It has been studied in ALICE relative to the deflection of the spectator nucleons (Ψ^t_{SP} and Ψ^t_{SP} for the target and projectile spectators, respectively), reconstructed with the Zero Degree Calorimeters (ZDC) [6] in Pb–Pb collisions at $\sqrt{s_{NN}} = 2.76$ TeV. Figure 1 presents the pseudo–rapidity dependence of the odd and even components of v_1 for various centrality classes [8], defined as $v_1^{odd}(\Psi_{SP}) = 0.5[v_1(\Psi^p_{SP}) + v_1(\Psi^t_{SP})]$ and $v_1^{even}(\Psi_{SP}) = 0.5[v_1(\Psi^p_{SP}) - v_1(\Psi^t_{SP})]$, respectively. The former, that originates from flow developing relative to the symmetry plane, exhibits a negative slope with little centrality dependence and a magnitude that is smaller by a factor of 3 compared to the highest (i.e. $\sqrt{s_{NN}} = 200$ GeV) RHIC energy [9]. On the other hand, v_1^{even} has a negative sign, and shows little or no dependence on η and on centrality within the current uncertainties, consistent with the expectations for a component originating from fluctuations of the initial energy density in the collision.

Journal of Physics: Conference Series **509** (2014) 012024 doi:10.1088/1742-6596/509/1/012024

It is now established, that these initial fluctuations lead to non–zero higher order harmonics that develop relative to their own symmetry plane. A step forward in understanding the nature of these fluctuations is to investigate whether they are imprinted in the correlation between the different Ψ_n. Figure 2 presents the p_{T} differential mixed harmonic correlator $\langle\cos(\varphi_a - 3\varphi_b + 2\Psi_2)\rangle$, where $\varphi_{a,b}$ are the azimuthal angles of particles a and b, that probes the correlations between the first (Ψ_1), second (Ψ_2), and third (Ψ_3) order symmetry planes via the relation $\langle\cos(\varphi_a - 3\varphi_b + 2\Psi_2)\rangle \approx v_1 v_2 v_3 \langle\cos(\Psi_1 - 3\Psi_3 + 2\Psi_2)\rangle$ [10]. The data points for all centralities, represented by different colours and symbols, exhibit non–zero correlations, indicating the existence of a 3–plane correlation. It is interesting to note that these results are in qualitative agreement with ideal hydrodynamical calculations [11] at low p_{T}, while the same calculations fail to describe the evolution at high p_{T}.

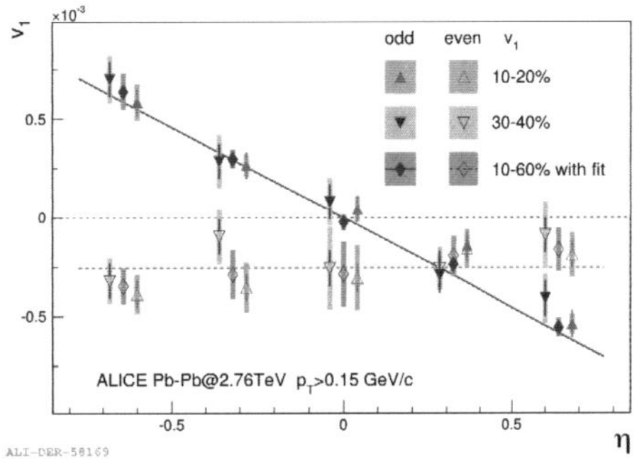

Figure 1. The η dependence of v_1^{odd} and v_1^{even} for different centrality class of Pb–Pb collisions [8].

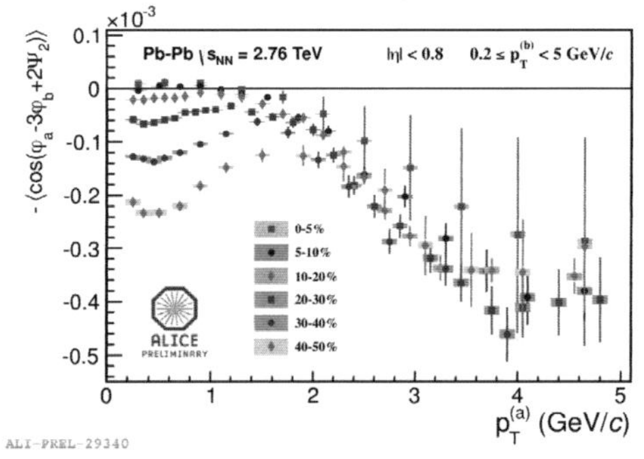

Figure 2. The p_{T} dependence of $\langle\cos(\varphi_a - 3\varphi_b + 2\Psi_2)\rangle$, probing the 3–plane correlations, for different centrality class of Pb–Pb collisions.

Studying the different order flow harmonics that develop relative to their own symmetry plane can constrain the value of shear viscosity. However, an additional constraint can be given by studying the second Fourier coefficient, the elliptic flow v_2, as a function of centrality and p_{T} for different particle species. Hydrodynamic model calculations predict a characteristic dependence of v_2 on the particle mass at low values of transverse momentum ($p_{\mathrm{T}} < 2$ GeV/c). This dependence is induced by the system's collective radial expansion, which being cumulative throughout the whole collision, has a significant contribution from both the partonic and the hadronic phases. Figure 3 presents the p_{T} differential v_2 for central Pb–Pb collisions (i.e. 10–20% centrality) for charged π, K, and $\bar{\mathrm{p}}$, $\mathrm{K^0_S}$ and Λ. The measurement was performed with the scalar product method applying a pseudo–rapidity gap of $\Delta\eta > 1$ to suppress correlations not related with the symmetry plane (i.e. non–flow effects). It is seen that the data points exhibit the characteristic mass ordering in the low p_{T} region (i.e. $p_{\mathrm{T}} \leq 2$ GeV/c), attributed to collective radial flow that shifts heavy particles to higher values of p_{T}. The experimental points are compared to hydrodynamical calculations [12] using a value of $\eta/s = 0.2$ and the Color Glass Condensate (CGC) initial conditions, represented by the curves. The model reproduces the p_{T} dependence of v_2 for pions and kaons up to $p_{\mathrm{T}} \approx 2$ GeV/c but overestimates v_2 for heavier particles.

A similar analysis has recently been performed in p–Pb collisions at $\sqrt{s_{NN}} = 5.02$ TeV and reported by ALICE [13]. It was based on studying two–particle correlations measuring particle pairs from same events separated in the pair angles $\Delta\eta$ and $\Delta\varphi$ corrected for acceptance effects and efficiency losses using mixed events. The results revealed a double–ridge structure for π–hadron, K–hadron and p–hadron correlations after subtracting the associated yield per trigger

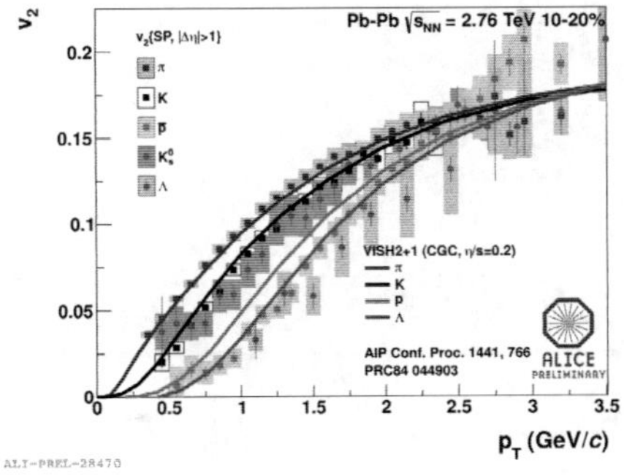

Figure 3. The p_T–differential v_2 measurement for central Pb–Pb collisions and for different particle species.

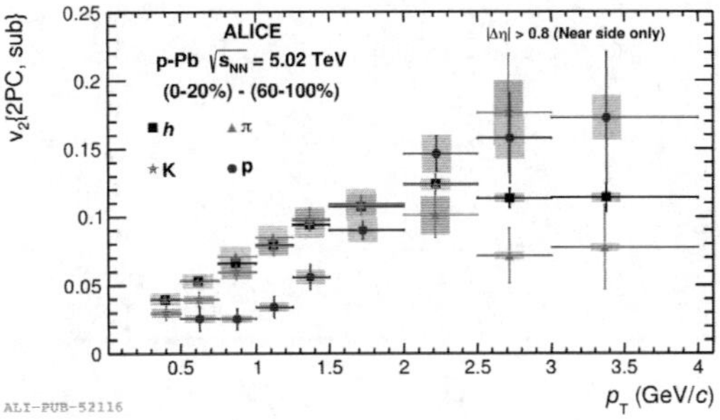

Figure 4. The p_T–differential v_2 measurement, after applying the subtraction procedure (see text for details), for p–Pb collisions and for different particle species [13].

particle in $\Delta\eta$ and $\Delta\varphi$ for the 60–100% multiplicity class from the corresponding yield from events with 20% highest multiplicity p–Pb collisions. This subtraction procedure was performed to reduce the contribution from the dominant near side jet structure. A selection of $\Delta\eta > 0.8$ was applied to reduce any residual contribution from the near side jet peak. The resulting distribution was projected on the $\Delta\varphi$ axis and was fitted with a Fourier series. Figure 4 presents the p_T differential v_2, which is the dominant harmonic, for π, K, p and charged particles. A mass ordering between π and p is observed with a hint that this ordering is also followed by K at low p_T values. The resemblance with the corresponding Pb–Pb picture (see fig. 3), where this ordering is attributed to collective effects, is striking.

3. Two-particle correlations and their charge dependence in Pb–Pb

The study of two–particle correlations described in the previous paragraph has been established as an important tool for the study of different momentum ranges e.g. the bulk, the jet dominated region, with different properties of the system being imprinted in the final observed structures. It was proposed in [14] that using this type of analysis one has the possibility to probe the modification of the jet structure and shape, originating from the interplay with the longitudinally flowing medium. This was studied in Pb–Pb collisions, estimating the $\Delta\eta$ independent effects by quantifying the long range correlations on the near side (at $\Delta\eta > 1$), and subtracting them from the short range region (at $\Delta\eta < 1$). The resulting near side jet peak is then fitted with a double Gauss function. The resulting values of $\sigma_{\Delta\eta}$ (width in $\Delta\eta$, represented by the solid curves) and of $\sigma_{\Delta\varphi}$ (width in $\Delta\varphi$, represented by the symbols) and are plotted as a function of centrality in fig. 5 for different values of p_T of the trigger and the associated particles. They show a significant centrality dependence for $\sigma_{\Delta\eta}$ but a rather mild effect for $\sigma_{\Delta\varphi}$. The data are described by AMPT calculations (not shown in the plot) and can be accommodated by models that consider the interplay of low p_T jets with the flowing medium.

The charge dependent part of these correlations has recently emerged as a key feature in many studies since they can probe the effect of local charge conservation coupled to a system that exhibits collective behaviour. This is addressed by the so called balance functions [15], defined as the difference between the unlike– and like–sign pair densities divided by the number of trigger particles. Each pair density term is corrected for detector acceptance and efficiency effects using mixed events. The balance function measures the correlation strength of balancing partners, imprinted in the width of the distribution, and it is sensitive to the time of hadronization. Figure 6 presents the centrality dependence in Pb–Pb collisions of the width of the balance function

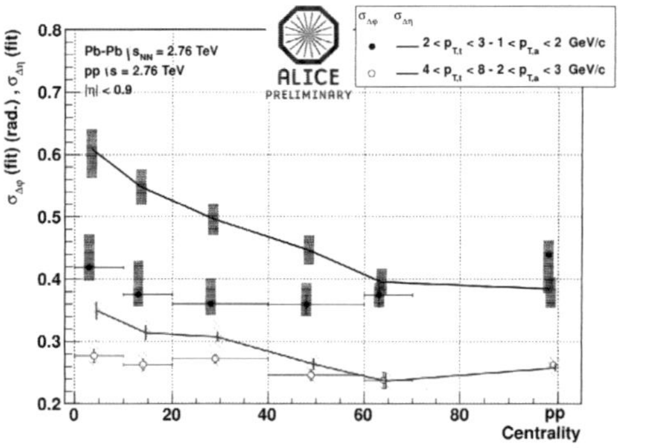

Figure 5. The centrality dependence of the width of the near–side jet–peak in $\Delta\eta$ and $\Delta\varphi$ for different ranges of p_T for the trigger and the associated particles, in Pb–Pb collisions.

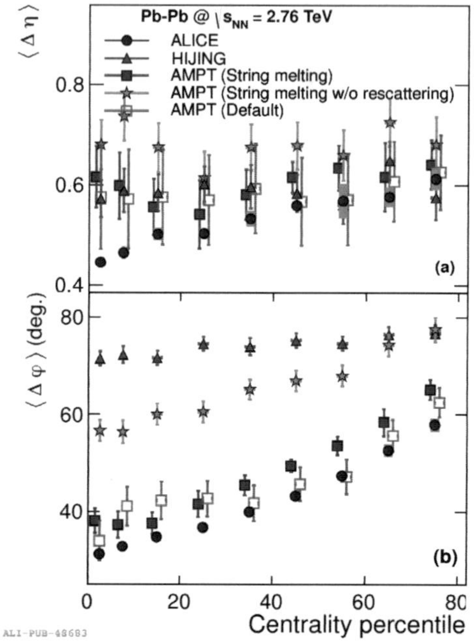

Figure 6. The centrality dependence of the width of the balance functions, $\langle\Delta\eta\rangle$ and $\langle\Delta\varphi\rangle$, in Pb–Pb collisions [16].

in $\Delta\eta$ ($\langle\Delta\eta\rangle$) and $\Delta\varphi$ ($\langle\Delta\varphi\rangle$), calculated as the weighted average from each distribution. It is seen that the width for data (red full circles) decreases for more central collisions in both representations, consistent with the picture of late stage creation of balancing charges focused by the strong radial flow. It is also interesting to note that only models that incorporate collective effects (e.g. AMPT) describe the data points in $\Delta\varphi$ but seem to fail to do so in $\Delta\eta$ [16].

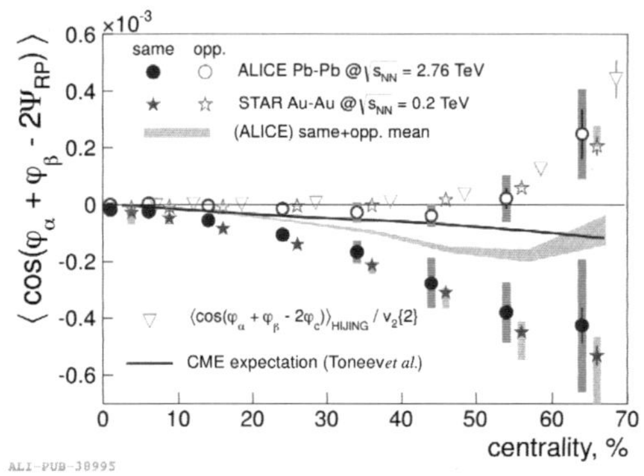

Figure 7. The centrality dependence of the charge dependent correlations of $\langle\cos(\varphi_\alpha + \varphi_\beta - 2\Psi_{RP})\rangle$ in Pb–Pb collisions [18].

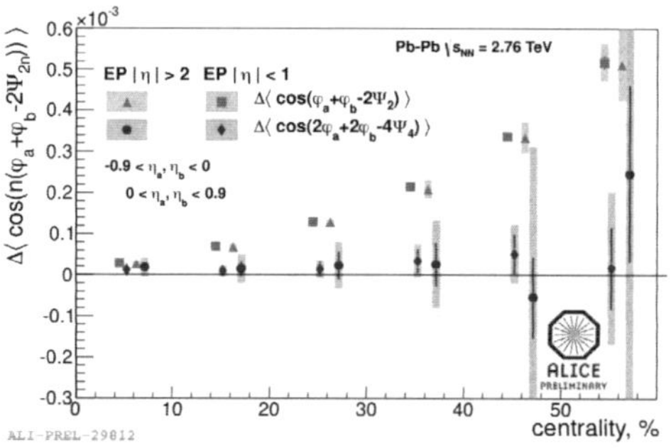

Figure 8. The centrality dependence of the differences of the charge dependent correlations of $\langle\cos(\varphi_\alpha + \varphi_\beta - 2\Psi_2)\rangle$ and $\langle\cos(2\varphi_\alpha + 2\varphi_\beta - 4\Psi_4)\rangle$ in Pb–Pb collisions.

These charge–dependent correlations relative to the collision symmetry plane, were recently

Journal of Physics: Conference Series **509** (2014) 012024 doi:10.1088/1742-6596/509/1/012024

used to investigate the possibility of observing P– and CP–odd effects in heavy–ion collisions. These effects originate from the interaction of local domains (i.e. QCD excited vacuum states) that can violate the CP invariance, with the quarks of the deconfined phase and with the strong, of the order of 10^{19} G, magnetic field that develops in peripheral collisions at the LHC energies. This is better known as the Chiral Magnetic Effect (CME), the experimental consequence of which is a charge asymmetry relative to the reaction plane. The first results from both STAR [17] and ALICE [18] are based on the study of correlations of two particle relative to the reaction plane (i.e. $\langle\cos(\varphi_\alpha + \varphi_\beta - 2\Psi_{RP})\rangle$) and show a strong charge dependent signal that develops for more peripheral events, qualitatively consistent with the CME expectations. This is illustrated in fig. 7 where the ALICE results for same and opposite charge combinations are represented by the red filled and open circles, respectively. It is seen that while the unlike–sign correlations are compatible with zero, with the exception of the most peripheral points, the like–sign ones exhibit an increasing magnitude for more peripheral events, coupled to a negative sign. It is also interesting to note that the LHC results are in quantitative agreement with the RHIC values (i.e. blue filled and open stars for the like and unlike–sign pairs, respectively) inpite the almost 14 times higher centre–of–mass energy at the LHC. The focus now turns to analyses whose aim is to disentangle the potential signal originating from the CME, from the background. For the latter, it was recently realised that the biggest contributor is the effect of local charge conservation coupled to elliptic flow modulations [19]. One way to probe the background contribution is to study the charge dependent correlations relative to the fourth harmonic symmetry plane i.e. the so–called double harmonic correlator $\langle\cos(2\varphi_\alpha + 2\varphi_\beta - 4\Psi_4)\rangle$. The contribution from the CME is argued to be little, if any at all (i.e. correlations relative to Ψ_4), while the background should create a charge dependent signal albeit scaled compared to the previous case (i.e. $\langle\cos(\varphi_\alpha + \varphi_\beta - 2\Psi_2)\rangle$, the so–called single harmonic correlator) by the ratio of v_4/v_2. Figure 8 presents the centrality dependence of the differences between the correlations of unlike and like–sign pairs for both the single (i.e. red points) and the double harmonic (i.e. blue points) correlators. While the single harmonic correlator exhibits a rather strong charge dependent signal (see also fig. 7), the double harmonic one is consistent with zero within the current large uncertainties. This could be interpreted as an indication that the effect of local charge conservation and its interplay with azimuthal anisotropy might not contribute significantly in these measurements. However, before reaching a final conclusion, a measurement of better precision needs to be performed from the experimental side, while there is an increasing need for these effects to be studied thoroughly from the theory side e.g. using models that can describe both the hydrodynamical evolution of the system but also its final state, hadronic interactions.

4. Summary

The study of angular correlations in Pb–Pb provided valuable information on the expansion of the system, driven by initial pressure gradients. During this phase it behaves as a liquid with low dissipation values. This opens up the possibility for the initial energy density fluctuations to be imprinted in the final state observables. These studies in p–Pb collisions have recently yielded surprisingly similar observations compared to the Pb–Pb results, the interpretation of which leads to rather intriguing paths.

References

[1] Heinz U and Snellings R 2013 *Annu. Rev. Nucl. Part. Sci.* **63** 123.
[2] Arsene I *et al.* [BRAHMS Collaboration] 2005 *Nucl. Phys.* A **757** 1.
Adcox K *et al.* [PHENIX Collaboration] 2005 *Nucl. Phys.* A **757** 184.
Back B B *et al.* [PHOBOS Collaboration] 2005 *Nucl. Phys.* A **757** 28.
Adams J *et al.* [STAR Collaboration] 2005 *Nucl. Phys.* A **757** 102.
[3] Aamodt K *et al.* [ALICE Collaboration] 2010 *Phys. Rev. Lett.* **105** 252302.
[4] Aamodt K *et al.* [ALICE Collaboration] 2011 *Phys. Rev. Lett.* **107** 032301.
[5] Alver B and Roland G 2010 *Phys. Rev.* C **81** 054905.

[6] Aamodt K *et al.* [ALICE Collaboration] 2008 *JINST* **3** S08002.
[7] Voloshin S and Zhang Y 1996 *Z. Phys.* C **70** 665.
[8] Abelev B I *et al.* [ALICE Collaboration] arXiv:1306.4145 [nucl-ex].
[9] Abelev B I *et al.* [STAR Collaboration] 2008 *Phys. Rev. Lett.* **101** 252301.
[10] Bhalerao R S, Luzum M, Ollitrault J Y 2011 *Phys. Rev.* C **84** 034910.
[11] Teaney D and Yan L 2011 *Phys. Rev.* C **83** 064904.
[12] Shen C *et al.* 2011 *Phys.Rev.* C **84** 044903.
[13] Abelev B I *et al.* [ALICE Collaboration], arXiv:1307.3237 [nucl-ex].
[14] Armesto N *et al.* 2005 *Phys. Rev.* C **72** 064910.
[15] Bass S A, Danielewicz P and Pratt S 2000 *Phys. Rev. Lett.* **85** 2689.
[16] Abelev B I *et al.* [ALICE Collaboration] 2013 *Phys. Lett.* B **723** 267.
[17] Abelev B I *et al.* [STAR Collaboration] 2009 *Phys. Rev. Lett.* **103** 251601.
[18] Abelev B I *et al.* [ALICE Collaboration] 2013 *Phys. Rev. Lett.* **110** 012301.
[19] Schlichting S and Pratt S 2011 *Phys. Rev.* C **83** 014913.

Journal of Physics: Conference Series **509** (2014) 012025 doi:10.1088/1742-6596/509/1/012025

Measurement of azimuthal anisotropy of hadrons in Au+Au collisions from the beam energy scan program by the PHENIX experiment at RHIC.

Yoshimasa Ikeda

RIKEN, Hirosawa 2-1, Wako, Saitama 351-0198, Japan

E-mail: y_ikeda@ribf.riken.jp

Abstract.
The elliptic flow as given by the second term of the Fourier series for the azimuthal distribution of particles with respect to the event plane is believed to carry information on the initial geometrical anisotropy. The large azimuthal anisotropy of the particle emission was observed in heavy ion collisions at RHIC in 2001. In Au+Au $\sqrt{s_{NN}} = 200$ GeV collisions, the elliptic flow of identified charged hadrons was found to scale with the number of constituent quarks, which may be an indication of the flow of constituent quarks in the QGP phase. An event plane detector, RxPN, was installed in the PHENIX experiment in 2007, in order to improve the resolution of the event plane determination. The RxPN detector was also used to make accurate measurements of the elliptic flow at lower collision energies, where the transition of the QGP might be explored. The latest elliptic flow measurement for identified hadrons in Au+Au collisions from the beam energy scan program at the PHENIX will be presented and discussed.

1. Introduction

Quarks and gluons are expected to be deconfined in a high temperature or high density. This is called Quark Gluon Plasma (QGP)[1, 2]. The QGP may have existed in the early universe according to the big bang theory or in the core of the neutron star[3, 4]. Experimentally, it is formed with a relativistic heavy ion collider. The system geometry is elliptical at the first stage of a non-central heavy ion collision which generates the asymmetry in the yield of particles as a function of azimuthal angle with respect to the event plane of an event. The magnitude of the azimuthal anisotropy of particle emission is measured as the second term of Fourier series (v_2),

$$\frac{dN}{d(\phi - \Psi)} \propto 1 + 2v_2 \cos 2(\phi - \Psi), \quad (1)$$

where ϕ is the particle emission angle and Ψ is the event plane angle of the heavy ion collision which is the direction of the short axis of the oval region of the participants. The v_2 gives information about the initial state and its expansion, possibly through the QGP phase. The measured v_2 increases with p_T at the low momentum range ($p_T < 3$ GeV/c). The rise of v_2 shifts towards higher p_T for heavier particles. A hydrodynamic model with a low viscosity reproduces the collective behavior for the particles [5]. The v_2 reaches a constant value at $p_T = 2-3$ GeV/c where the value scales with the constituent quark number and independent of the particle mass [6]. It indicates that the flow of hadrons is built up by the flow of quarks in QGP according to the quark coalescence model [7, 8]. The study of the v_2 of rare particles, with high resolution event plane measurement provides further information about the quark number dependence and quark flow. ϕ-*meson* is important because it is not only a heavy mass meson (the mass is similar to p even through it consists of two quarks), but also the hadronic re-scattering cross-section is different from that of baryons [9]. Λ v_2 should be similar to p v_2 because they are baryons and have similar mass. *Deuteron* v_2 should be larger than baryon v_2 if the quark coalescence model holds at the high p_T range. The v_2 may show a threshold behavior as a function of colliding energies if the quark number scaling of hadron v_2 is an indicator of a QGP phase. The resolution of v_2 measurement for the low energy collisions is reduced because the multiplicity is smaller than that of $\sqrt{s_{NN}} = 200$ GeV collision.

2. v_2 measurement and reaction plane resolution

When measuring the azimuthal anisotropy, particles must not be shared between reaction plane and v_2 measurement. Auto correlation effects will manifest themselves in this case, which create a large bias on v_2 measurement. Therefore, detectors which measure the event plane and v_2 should be separated in acceptance (the identified hadron v_2 is measured at a central rapidity range and the reaction plane is measured at forward rapidity range in PHENIX). Measurements of the v_2 of rare particles are limited by statistics of the identified particles and the reaction plane resolution. The reaction plane resolution is estimated with a distribution of $\Delta\Phi$ which is the difference between the measured event plane Φ and the reaction plane Ψ of each event ($\Delta\Phi = \Phi - \Psi$). The observed v_2 is reduced by the intrinsic event plane resolution of a detector. The relation between the observed $v_2^{observe}$ and the corrected v_2^{real} is as follows,

$$v_2^{observe} = v_2^{real} \times \langle\cos(2\Delta\Phi)\rangle, \tag{2}$$

The observed v_2 is corrected for the correction factor $\langle\cos 2\Delta\Phi\rangle$. This correction factor is called the reaction plane resolution. The value will be 1 if the detector has perfect event plane resolution. The statistical value of the measurement is reduced by a factor of $(\frac{1}{\cos 2\Delta\Phi})^2$ due to the reaction plane resolution.

In 2007, a new reaction plane detector RxPN was installed and v_2 was measured with two times better resolution compared to the ones measured before in $\sqrt{s_{NN}} = 200$ GeV Au+Au collisions at RHIC-PHENIX [10, 11]. It allows us to study v_2 of the rare particles or at low collision energy. In 2010, the identified hadron v_2 was measured in $\sqrt{s_{NN}} = 39$ and 62 GeV Au+Au collisions at RHIC-PHENIX. The higher event plane resolution obtained from RxNP detector compensates for the lower statistics in lower energy collisions.

3. Results

The v_2 of π, K, *(anti-)p*, *(anti-)d*, *(anti-)*Λ and ϕ are measured in $\sqrt{s_{NN}} = 200$ GeV Au+Au collisions and shown in figure 1. The v_2 increases with p_T in the low momentum range ($p_T < 2$ GeV/c). Heavier particles have smaller values of v_2 than lighter particles in this p_T range. K-meson has smaller v_2 than π-*meson* and ϕ-*meson* has smaller v_2 than K-*meson* at $p_T < 2$ GeV/c. Alternately, one can say that the v_2 is shifted towards higher p_T for heavier particles. It is an indication of collective flow of the particles. v_2 of π, K and ϕ (mesons) or p and Λ (baryons) are consistent separately with $KE_T = m_T - m_0$ (see panel (b) of Fig.1). The v_2 of the particles depends on the number of constituent quarks. It was found that the v_2 of mesons and baryons saturate at $p_T > 2$ GeV/c. The saturated values for baryons are higher than those of mesons. The d v_2 is higher than that of baryons at $p_T > 3$ GeV/c. The v_2 of the particles divided by the number of constituent quarks were plotted as a function of KE_T/n_q (see panel (c) of Fig.1). These particles are almost consistent with each other at $KE_T/n_q < 0.7$ GeV. This

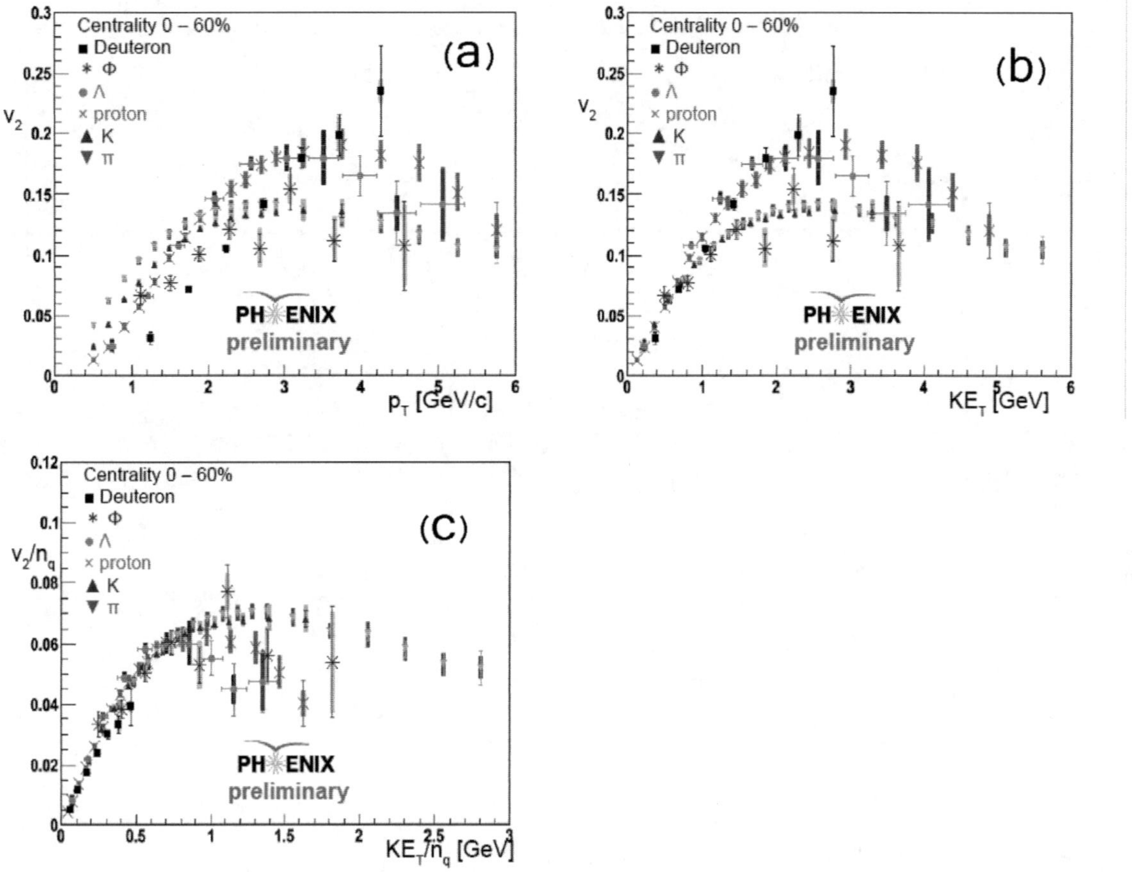

Figure 1. v_2 of identified particles (π, K, *(anti-)p*, *(anti-)*Λ, ϕ and *(anti-)d)* in Au+Au $\sqrt{s_{NN}}$ = 200 GeV. (a) shows v_2 as a function of transverse momentum p_T. (b) shows v_2 as a function of KE_T. (c) shows the v_2 with the number of constituent quarks and KE_T scaling.

v_2 dependence on the number of constituent quarks of the particle suggests the collective flow of quarks and the quark coalescence of u, d and s quarks. The v_2 scaling of the quark number and KE_T is broken at $KE_T/n_q > 0.7$ GeV. v_2 of π and p are approaching each other at the high p_T range (6 GeV/c). There could be another process generates v_2, such as jet production from hard processes and jet quenching from energy loss. In this case, the v_2 of the high p_T particle is expected to be given by the path length dependence of the jet quenching coming from the partonic energy loss.

Figure 2 shows the v_2 with the KE_T and the quark number scaling for π, K and *(anti-)p* in Au+Au lower energy collisions. The p includes not only direct p but also p from p-π decay of Λ. It is expected that v_2 of p from Λ decay is similar to v_2 of Λ because p is considerably heavier than π. Λ v_2 as a function of p_T is consistent with that of p in Au+Au $\sqrt{s_{NN}} = 200$ GeV (see panel (a) of Fig.1). The v_2 of p from Λ decay may be slightly smaller than the v_2 of the direct p. Therefore, the direct p v_2 may be slightly larger than the measured p v_2. It is expected that the number of constituent quark dependence of hadron v_2 disappears and the scaled v_2 with the quark number of baryon will be smaller than that of meson without the QGP phase. The scaled v_2 of p is not found to be smaller than that of π and K in Au+Au $\sqrt{s_{NN}} = 39$ GeV which

Journal of Physics: Conference Series **509** (2014) 012025 doi:10.1088/1742-6596/509/1/012025

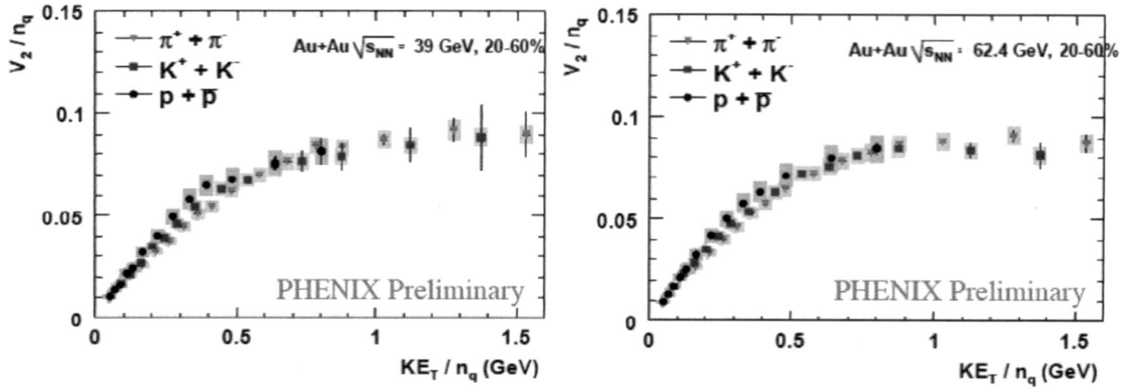

Figure 2. v_2 with the number of constituent quarks scaling of π, K and *(anti-)p* in low energy Au+Au collisions. Left picture shows those for Au+Au $\sqrt{s_{NN}} = 39$ GeV. Right picture shows those for Au+Au $\sqrt{s_{NN}} = 62$ GeV.[12]

indicates that QGP is generated already at this energy.

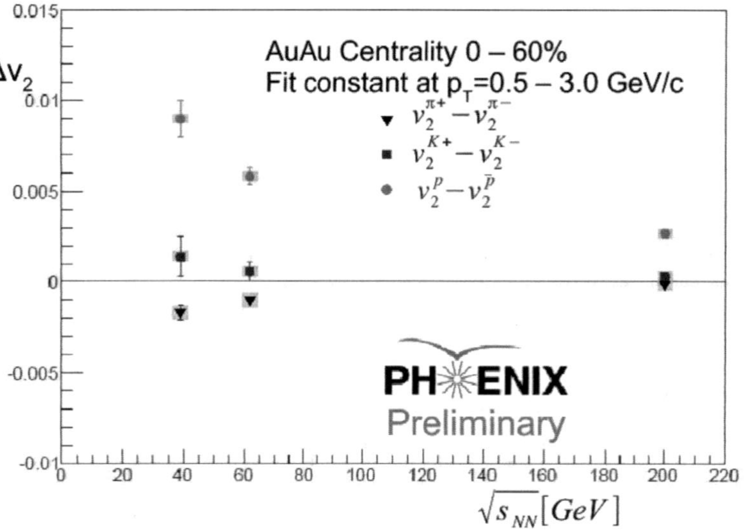

Figure 3. The difference v_2 of π, K and *(anti-)p* between positive charged hadron and negative charged hadron as a function of the collision energy at $p_T = 0 - 3.0$ GeV/c.

Figure 3 shows the difference of v_2 of π, K and *(anti-)p* for positive charge and negative charge as a function of the collision energy at $p_T = 0 - 3.0$ GeV/c. The particle (especially p) v_2 differs from the anti-particle v_2 which could be given by annihilation within the high baryon density caused by the baryon stopping of the low beam energy.

4. Summary

The elliptic flow v_2 was measured for the identified hadrons, π, K, *(anti-)p*, *(anti-)d*, *(anti-)*Λ and ϕ in Au+Au $\sqrt{s_{NN}} = 200$ GeV and π, K, *(anti-)p* in Au+Au $\sqrt{s_{NN}} = 39$ and 62 GeV with the help of the enhanced resolution obtained from the new reaction plane detector at RHIC-PHENIX. The measured v_2 of the particles are consistent with quark number scaling at $KE_T/n_q < 0.7$ GeV in Au+Au $\sqrt{s_{NN}} = 200$ GeV. This behavior is explained well by the quark coalescence mechanism for hadron production. The v_2 of π, K, *(anti-)p* are almost consistent with the quark number scaling at Au+Au $\sqrt{s_{NN}} = 39$ and 62 GeV. The measured v_2 indicates the QGP is already created at $\sqrt{s_{NN}} = 39$ GeV. Charged separated v_2 of π, K and *(anti-)p* were measured at the three collision energies. The difference v_2 between positive charge and negative charge for the charged separated v_2 of π, K and *(anti-)p* were shown as a function of the collision energy. The particle (especially p) v_2 deviates from the anti-particle v_2 in the lower energy collisions.

References

[1] Karsch F 2002 *Lecture Notes in Physics* **583** 209
[2] Karsch F, Laermann E and Peikert A 2000 *Phys. Lett. B* **478** 447
[3] Shuryak E V 1980 *Phys. Rep.* **61** 71
[4] Collins J C and Perry M J 1975 *Phys. Rev. Lett.* **34** 1353
[5] Houvinen P, Kolb P E, Heinz U W, Ruuskanen P V and Voloshin S A 2001 *Phys. Lett. B* **503** 58
[6] Adler S S *et al.* 2003 *Phys. Rev. Lett.* **91** 182301
[7] Fries R J, Muller B, Nonaka C and Bass S A 2003 *Phys. Rev. Lett.* **90** 202303
[8] Fries R J, Muller B, Nonaka C and Bass S A 2003 *Phys. Rev. C* **68** 044902
[9] Shor A 1985 *Phys. Rev. Lett.* **54** 1122
[10] Richardson E *et al.* 2010 *Nucl. Intsr. And Mass. A* **636** 99
[11] Adare A *et al.* 2012 *Phys. Rev. C* **85** 064914
[12] Huang S 2011 *27th Winter Workshop on Nuclear Dynamics* (Winter Park)

Journal of Physics: Conference Series **509** (2014) 012026 doi:10.1088/1742-6596/509/1/012026

Status and Physics Opportunities of the STAR Heavy Flavor Tracker and the Muon Telescope Detector Upgrades

Qiu Hao (for the STAR Collaboration)

Lawrence Berkeley National Lab, 1 Cyclotron Road, Berkeley, CA 94720, U.S.A.

E-mail: hqiu@lbl.gov

Abstract. The STAR Collaboration will complete the Heavy Flavor Tracker (HFT) and the Muon Telescope Detector (MTD) upgrades by 2014. HFT utilizes the state-of-art active pixel detector technology, which will greatly enhance the STAR physics capabilities by measuring heavy quark yield, collectivity and correlations via the topological reconstruction of charmed hadrons over a wide momentum range. The MTD is based on the long Multi-Gap Resistive Plate Chamber detector technology designed to measure muons penetrating the bulk of other detectors and the magnet yoke. It will enable STAR to study di-muon and electron-muon correlations and enhance heavy quarkonium studies. With the addition of these upgrades, STAR is well suited to perform precise measurements of production as well as correlations of rare probes (heavy flavors, dileptons) to systematically investigate the quark-gluon plasma properties at RHIC. For Run 13 63% of the MTD has been installed and data have been taken. Prototype PXL sectors (30% coverage) have also been installed and commissioned. Anticipated physics results and current status of these upgrades is reported.

1. Introduction

With full azimuthal particle identification ability at middle rapidity, the STAR [1] experiment made many important measurements in order to discover and study the Quark Gluon Plasma (QGP) produced at the Relativisitic Heavy Ion Collider (RHIC) [2]. In the future, heavy flavor and di-leptons will be two new physics focuses for the STAR heavy ion program.

The mass of heavy flavor quarks is significantly higher than the critical temperature at RHIC energies, Λ_{QCD} or the mass of u, d, and s quarks. As a consequence, heavy quarks are produced by hard processes early in the collision. Because heavy quarks decay weakly to light quarks within a much longer time scale than the collision, they can be treated as conserved in total number. These features make heavy quarks an ideal probe for studying the QGP medium properties at RHIC.

Leptons are produced during the whole evolution of the created matter and can traverse the medium with minimal interactions. Different kinematics of di-lepton pairs (mass and transverse momentum ranges) can selectively probe the properties of the formed matter throughout its entire evolution. The di-lepton spectra in the intermediate mass range (1.1 < M < 3.0 GeV/c^2) are expected to be directly related to the thermal radiation of the Quark-Gluon Plasma (QGP) [3, 4].

In order to better study heavy flavor and di-leptons, the STAR Collaboration will complete the Heavy Flavor Tracker (HFT) [5] and the Muon Telescope Detector (MTD) [6] upgrades by 2014. HFT will greatly enhance the measurements of open heavy flavor particles, while MTD will improve heavy quarkonium as well as di-lepton studies. In the following, the physics motivation, design, status and performance of HFT and MTD will be discussed.

2. Heavy Flavor Tracker (HFT)

2.1. *Physics motivation*

Most heavy quarks produced in heavy ion collisions end up as open heavy flavor particles. However, the rare production of heavy quarks and large combinatorial background in heavy ion collisions makes the measurements of open heavy flavor particles very difficult with current STAR detectors. Using precise silicon vertex detectors, decay vertices of open heavy flavor particles displaced from primary vertices can be measured, which will greatly reduce combinatorial background from primary tracks.

By systematically measuring the yields of ground state charmed hadrons, the total charm yield can be deduced, which provides a very important baseline for charmonium suppression and coalescence. At high p_T, the R_{CP} and R_{AA} of charm and bottom quarks can be used to study their energy loss mechanism in the QGP matter. Figure 1 shows the projected error of D^0 R_{CP} measurement using HFT with 1 billion minimum bias events. Very good precision can be achieved over a wide p_T range, a large improvement over the current STAR measurements [7]. The B meson and bottom quark yields, R_{CP}, and R_{AA} can also be measured with the HFT, for example, through the B to J/ψ decay. Unlike prompt J/ψ, J/ψ from B meson decay have a displaced vertex. This can be used to measure the contribution of J/ψ from B meson decay.

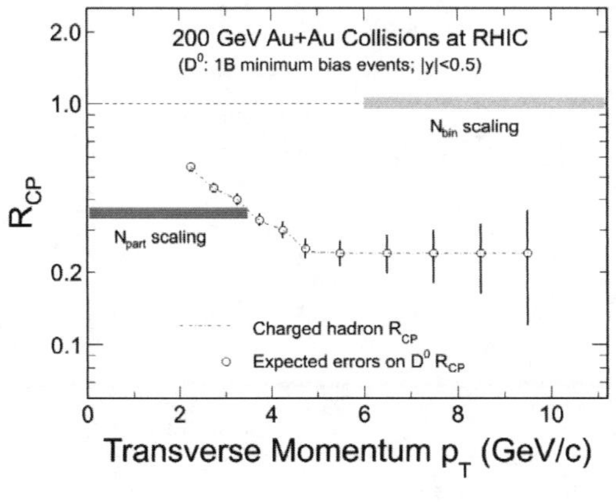

Figure 1. Expected errors on D^0 R_{CP} with HFT .

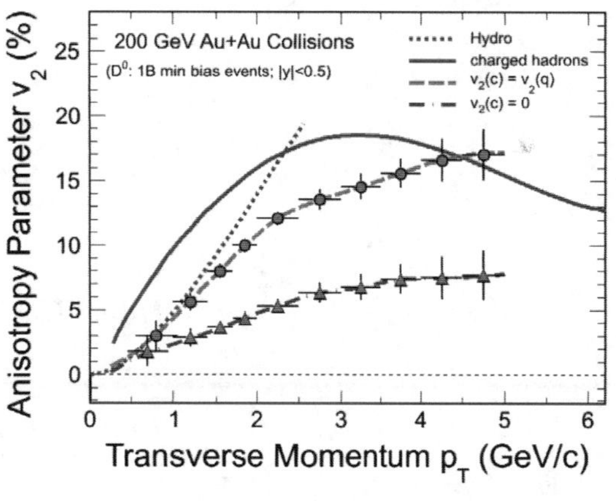

Figure 2. Expected errors on D^0 v_2 with HFT.

With much larger mass than that of light quarks, heavy quarks are more resistant to having their velocity changed, and are thus expected to thermalize slower than light partons. If charm quarks are observed to have sizable collective motion, then light partons, which dominate the medium, should be fully thermalized. The charm quark flow can be measured through D^0 v_2 and non-photonic electron v_2 [8]. The former is a more direct measurement as the latter is influenced by decay kinematics. As shown in Figure 2, with the HFT STAR will be able to measure the D^0 v_2 precise enough to distinguish between the case of full charm collectivity and no charm collective motion in the low p_T region.

With the HFT we can also study charm interactions with the medium by measuring $c\bar{c}$ ($D^0\bar{D}^0$) angular correlations. By measuring the Λ_c^+/D^0 ratio we can test the coalescence model,

Journal of Physics: Conference Series **509** (2014) 012026 doi:10.1088/1742-6596/509/1/012026

which explains the enhancement of baryon over meson ratio at intermediate p_T in the light quark sector. There are many other interesting physics topics that will be enabled by the HFT.

2.2. Design

The decay length of open heavy flavor particles is very small. For example, the $c\tau$ of D^0 mesons is about 120 microns. Thus position resolution is the primary concern for the design of the HFT. The HFT consists of 3 sub-detectors: Silicon Strip Detector (SSD), Intermediate Silicon Tracker (IST) and the PIXEL (PXL) detector, as shown in Figure 3. Table 1 shows technical details of the 3 sub-detectors. The innermost sub-detector PXL uses state-of-art active pixel detector technology to achieve a position resolution of 12 microns, which will be the most precise position detector used in collider experiments so far. The low radiation length of the HFT is crucial for the reconstruction of D^0 at low p_T, enabling charm total cross section and flow measurements.

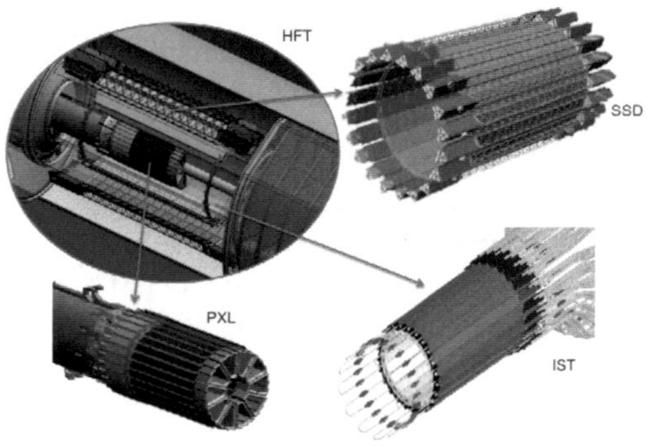

Figure 3. HFT overview.

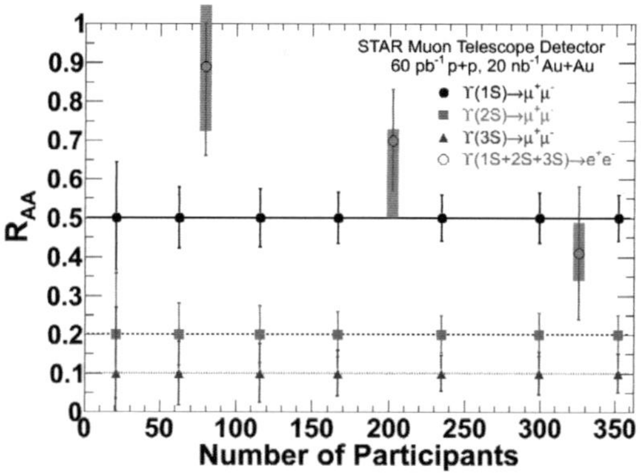

Figure 4. Expected errors for R_{AA} of different Υ states with the MTD.

Table 1. HFT technical details

Sub detector	r (cm)	Sensitive units	$\sigma_{R\phi}(\mu m)$	$\sigma_z(\mu m)$	$X/X_0(\%)$
Silicon Strip Detector	22	2 side strips with 95 μm pitch	20	740	1
Intermediate Silicon Tracker	14	600 μm $\times$ 0.6 cm strips	170	1800	< 1.5
PIXEL	2.5/8	20 μm pixel pitch	12	12	0.4/layer

2.3. Status

The PXL detector employs new technology. It is also the key detector to achieve the HFT target pointing resolution (< 50 μm for 750 MeV/c Kaons). An engineering run for a PXL prototype with 3 out of 10 sectors was carried out in 2013. The engineering run was very successful. The prototype was installed in STAR within 14 hours, and first data were obtained after only 2 days. From the data a clear TPC track and PXL hit position correlation is observed. The construction of IST and SSD is also on schedule.

3. Muon Telescope Detector (MTD)

3.1. Physics motivation

The ability to identify and trigger muons enables physics studies of both di-muon pairs and single muons. Different kinematics of di-lepton pairs (mass and transverse momentum ranges) can selectively probe the properties of the formed matter throughout its entire evolution. Single muons can be used to study the parent heavy flavor hadrons. With the TPC, TOF and BEMC detectors, STAR already has a good ability to detect electrons at mid-rapidity. However, muons have several advantages over electrons. There is no γ conversion and much less Dalitz decay contribution for muons. Muons are also less affected by radiative losses in the detector materials. Furthermore, muons can be triggered in Au+Au collisions, enabling STAR to sample the full luminosity for studies with muons.

Sequential suppression of different Υ states can be used as a QGP thermometer [9]. The MTD will enable STAR to measure the R_{AA} of different Υ states. As shown in Figure 4, the projected error for R_{AA} of different Υ states is much smaller than for current measurements of the R_{AA} for all Υ states measured through the di-electron channel.

Similarly, the J/ψ R_{AA} and v_2 measurements will have much smaller errors with MTD, especially in the low p_T region. Systematical measurements of J/ψ R_{AA} and v_2 can probe its production mechanism, especially the magnitude of $c\bar{c}$ recombination contribution, thus deduce the magnitude of the suppression due to the color screening effect [10].

The di-lepton spectra in the intermediate mass range ($1.1 < \mathrm{M} < 3.0\ GeV/c^2$) are expected to be directly related to the thermal radiation of the Quark-Gluon Plasma (QGP) [3, 4]. However, di-leptons from correlated open heavy-flavor decays also have significant contribution in the same invariant mass range. Although this two sources are difficult to separate in both di-electron and di-muon correlations, the $e - \mu$ correlation only has contribution from the correlated $c\bar{c}$ quarks. By subtracting $e - \mu$ correlation from di-electron correlation we can obtain the di-electron correlation from thermal production. Additionally, the $c\bar{c}$ correlation is important in itself to study the charm interaction with the medium. Figure 5 shows the simulated $e - \mu$ invariant mass distribution from $c\bar{c}$ quarks with a amount of data taken with MTD. The precision is good enough to distinguish two scenarios: charm correlation fully preserved as in p+p collisions, and charm de-correlated and softened due to interaction with the medium.

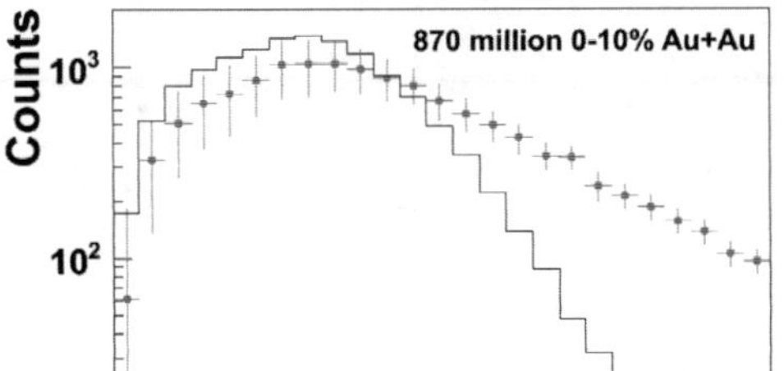

Figure 5. Simulated $e - \mu$ invariant mass from $c\bar{c}$ quarks with MTD. The red points assume correlated $c\bar{c}$ directly from PYTHIA, while the black line assumes de-correlation and soften of charm due to interaction with the medium.

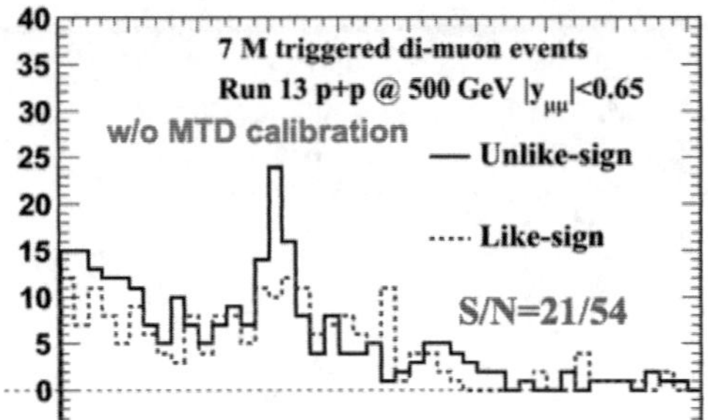

Figure 6. J/ψ signal from di-muon trigger events

3.2. Design

The MTD uses the magnet iron yoke as absorber, so that most other charged particles, except muons, will be absorbed and not reach the MTD modules placed outside the magnet yoke. The MTD employs the long Multi-gap Resistive Plate Chamber (MRPC) technique. It has 122 modules, 1464 readout strips, and 2928 readout channels.

3.3. Status

For the 2013 RHIC run, 63 % of the MTD was completed. The whole project will be finished for Run 2014. A clear correlation between TPC track projection and fired MTD channels is observed. The time resolution is measured with cosmic data to be 96 ps. The intrinsic space resolution along the beam line direction is 2.6 cm. The efficiency for $p_T > 4GeV/c$ is about 90 %. In Run 2013, a di-muon trigger was set up. Even with 7 M di-muon events out of a total of 120 M and without time calibration, a clear J/ψ signal is observed, as shown in Figure 6.

4. Summary and Outlook

In summary, STAR is conducting two major upgrades for the heavy ion program: the HFT for open heavy flavor measurements and the MTD for muon detection. These upgrades will enable or enhance a rich set of physics programs, including open and closed heavy flavor measurements, which is a clear probe for the QGP phase, and thermal radiation, which is considered as a QGP thermometer. The combination of HFT and MTD, together with the existing mid-rapidity subsystems, will make STAR the best suited detector to carry out the mission of studying the hot QCD matter properties. Construction of both detectors is going well. The technical run for the PXL prototype just finished successfully, and 63 % of MTD took data in Run 2013 with excellent performance. Both detectors will be completed for Run 2014.

In the coming 2 years, RHIC will focus on 200 GeV Au+Au, p+p and p+Au collisions for heavy ion programs. We can foresee that new physics results with HFT and MTD will greatly enhance our understanding of the QGP created at RHIC.

References

[1] Ackermann K H et al. 2003 *Nucl. Instrum. Meth.* A **499** 624
[2] Hahn H et al. 2003 *Nucl. Instrum. Meth.* A **499** 245
[3] Rapp R and Wambach J 2000 *Adv. Nucl. Phys.* **25** 1
[4] David G, Rapp R and Xu Z 2008 *Phys. Rept.* **462** 176
[5] Chasman C et al. LBNL #5509-2008.
[6] Ruan L et al 2009 *J. Phys. G* **36** 095001
[7] Tlusty D for STAR Collaboration 2013 *Nucl. Phys.* A **904** 639c
[8] Mustafa M for STAR Collaboration 2013 *Nucl. Phys.* A **904** 665c
[9] Digal S, Petreczky P and Satz H 2001 *Phys. Rev.* D **64** 094015
[10] Matsui T and Satz H 1986 *Phys. Lett.* B **178** 416

Quarks, Flow and Temperature in Spectra

T.S. Biró, Z. Szendi
MTA Wigner Research Centre for Physics, Budapest, Hungary
E-mail: biro.tamas@wigner.mta.hu, insane.zsu@gmail.com

Z. Schram
Department of Theoretical Physics, University of Debrecen, Debrecen, Hungary
MTA-DE Particle Physics Research Group, Debrecen, Hungary
E-mail: schram@phys.unideb.hu

Abstract. Strangeness in Quark Matter 2013 Theory Review

1. Introduction

A good theory summary should not consist of a direct enumeration of talks, rather a critical reflection, possibly a synthesis of different viewpoints is expected. Here we follow this concept: after identifying the key fundamental questions in the research field on quark matter five competing views will be presented as duelling alternatives. The physics behind experimental findings may not only differ from, but sometimes even contradict one another, as well as mathematical model approaches can be based on diametrally opposite philosophies. We shall exhibit a few simple theoretical exercises in order to illuminate these alternatives. By its very nature this summary is subjective.

In our view there are three fundamental questions about high energy matter: i) the degree of stohasticity in the phenomena we experience, ii) the particle – wave (particle gas – strong field) duality in the description of the acting objects, and iii) the reliability of the conjectured initial state without any direct experimental evidence. "The Lord does not play dice" - surmised Albert Einstein, but quite often it looks like that: we lack information even then, when physically it would not be prohibited to measure it. We are blind to part of the phase space by the design of our detectors (we hope only to an unimportant part). We are trying to model and understand *average* outcomes from hundreds of millions of phyiscally independent events by statistical and continuum physics calculations and we do not have much exclusive information event by event. Furthermore if we had, we possibly could not digest it. We are talking about (point) particles, but calculate with (plane wave) fields. We model initial and intermediate state fluctuations by guessing and expect to see an emerging pattern on the statistics of outcomes. But what are our choices when doing theoretical calculations? What are viable alternatives? What has been our progress lately in answering these fundamental questions? In this review we try to collect some partial answers.

Journal of Physics: Conference Series **509** (2014) 012027 doi:10.1088/1742-6596/509/1/012027

The main "idea-wise" duels occur as follows:

(i) thermalization: exponential vs power-law p_T spectra
(ii) hydrodynamics: initial state vs eos-driven flow features
(iii) hadronization: phase space vs QCD matrix element dominance
(iv) Quark Matter eos: lattice vs dual gravity methods
(v) nuclear medium: cross sections vs mean fields
(vi) simulation: particle vs field based models

In the this review we plan to discuss many of these dichotomies.

Last but not least, before starting to review theoretical contributions to this conference, let us recall Helen Caines' introductory talk about experiments: we have gathered more information from experiments, seeing more size, more time and more gradients in heavy ion collisions. From the overview of several different beam energies we learn approximate relations between yield and rapidity range,

$$\frac{dN_{\rm ch}}{d\eta} \propto \ln \sqrt{s} \sim y_{\rm beam}, \tag{1}$$

between volume and charged particle rapidity yields

$$V \approx 730{\rm fm}^3 + 2.545\frac{dN_{\rm ch}}{d\eta}, \tag{2}$$

and between time duration and charged particles per rapidity

$$\mathcal{T} \approx 0.875 \left(\frac{dN_{\rm ch}}{d\eta}\right)^{1/3}. \tag{3}$$

These formulas are approximate fits to the figure she showed[1].

2. Thermalization

The competition between thermal and non-thermal views of the intermediate "fireball" state of high-energy heavy ion collisions is reflected in the hunt for exponential spectra. Besides that any continuous spectrum can have a straight line fit on the logarithmic plot in a short enough interval, claiming thermal phenomena must be based on a thermal state, on having (at least local) equilibrium in the energy distribution of the parts of the surmised physical system. Therefore the theoretical study of possible ways of thermalization, of reaching such a state, has key importance.

Either studying the emergence or assuming the existence of a thermal state, thermal and hydrodynamical models are numerous. Just citing some of the titles from the program: "Thermalization of massive partons in anisotropic medium (Florkowski), First 3 seconds (Rafelski), Thermal model (Stachel), Strangeness balance in HIC (Kolomeitsev), Systematic Properties of the Tsallis Distribution (Cleymans), Thermalization through Hagedorn States (Beitel), Relativistic distribution function... (Grossi)," already shows the wealth and divergence among theoretical approaches. While some say that "the fireball never thermalized, but was born in a thermal state", such a statement also requires explicit, quantitative calculations[2, 3, 4, 5].

Since the hydrodynamical approach assumes local thermal equilibrium, these two theoretical tools should be treated together. We summarize the very essence of the relativistic hydro- and thermodynamics as follows.

The formulas are based on the Local Conservation of Noether Currents (LCNC),

$$\partial_\mu J^{\mu\, a} \; = \; 0. \tag{4}$$

Here $J^{\mu a}$ is a Lorentz four-vector indexed by μ ranging over all spacetime directions, while a stands generally for any further collection of indices related to the independent generators of infinitesimal symmetry transformations. More specifically the essential LCNC-s used in our theories are related to conserved charges, like baryon (B), strangeness (S) or electric charge (Q); to spacetime shifts (ν) and rotations and Lorentz-boosts (hyperbolic rotations with $\nu\rho$ antisymmetric index pairs):

$$\begin{aligned} \partial_\mu J^{\mu\,(B,S,Q)} &= 0, \\ \partial_\mu T^{\mu\,\nu} &= 0, \\ \partial_\mu M^{\mu\,\nu\rho} &= 0. \end{aligned} \tag{5}$$

In building up hydrodynamics it is essential to explore the connections between these currents. Chemistry deals with several types of charges (q_i^a):

$$J^{\mu\,a} = \sum_{i=+,-} q_i^a\, u_{(i)}^\mu \tag{6}$$

denote conductive currents, carried by the common stream, u^μ, or by several streams, $u_{(i)}^\mu$ in multicomponent fluids. There can be further directions in transport, too. Conduction or radiation[1] can be directed also transverse to the matter charge flow: This is most prominent in the energy-momentum. Its symmetric descriptive tensor can be splitted to several terms aligned and orthogonal to a fiducial flow velocity vector, u^μ:

$$T^{\mu\nu} = P^\mu u^\nu + P^\nu u^\mu + \mathfrak{T}^{\mu\nu} \tag{7}$$

with $P^\mu = eu^\mu + q^\mu$. Finally when polarisation related phenomena are considered then the angular momentum and spin density tensor has to be taken into account:

$$M^{\mu\nu\rho} = x^\mu T^{\nu\rho} - x^\nu T^{\mu\rho} + S^{\mu\nu\rho} \tag{8}$$

is anti-symmetric in all its index pairs.

Expectations and hopes by using hydordynamics in the description of quark (and sometimes hadronic) matter include

- Determination of Properties of Matter (equation of state and transport coefficients),
- Comparison with Lattice QCD,
- Determination of Initial State just after the Collision
- Key for the early dynamics...

By its nature the majority of our efforts in relativistic hydrodynamics and in the underlying kinetic theory approaches are numerical. A characteristic example of how far such efforts can go and how strong the visualising power of the numerical results can be was presented by Joannis Bouras showing the evolution of Mach cones in BAMP simulations[6, 7].

On the other hand theoretical efforts to improve the classical hydrodynamical calculations do not weaken[8, 9, 10, 11, 12, 13, 14, 15, 16, 17]. In his review talk at this conference Takeshi Kodama presented a picture of a magnifying glass revealing that in reality the flowing matter is not smooth, but very much fluctuating. This coarse graining problem opens the hydrodynamics

[1] Conduction may be viewed as phonon radiation, its speed is also limited by that of the light.

at its ultraviolet end, at small sizes towards possible improvements. At the infrared end, because of missing infinitely large reservoirs even in the heaviest ion collision, the thermodynamical treatment has to be modified by finite size effects[18, 19].

By this point we have to clarify the connection between relativistic hydrodynamics and thermodynamics. In fact, after considering the entropy four-current, S^μ, thermodynamics follows from the above described general structure of the basic system of equations. During a generic process dissipation occurs, closed systems show a tendency of approaching an equilibrium state. The net and local production of entropy is non-negative:

$$\partial_\mu S^\mu + \lambda_a \partial_\mu J^{\mu,a} \geq 0. \tag{9}$$

Here all LCNC-s (locally conserved Noether currents) have to be taken into account, each with its own Lagrange multiplier, λ_a. Classically the energy-momentum tensor plays a prominent role, its Lagrange-multiplier, $\lambda_\nu = \beta_\nu$, describes a local, but moving thermometer. Its invariant length represents a Doppler-shifted local temperature[2] [20].

The thermodynamic Gibbs potential analogue is a four-vector, an integral of the above (in a standing body it becomes p/T the pressure over the temperature):

$$S^\mu + \lambda_a J^{\mu\, a} = \Phi^\mu \tag{10}$$

is the first theorem of thermodynamics. From eqs.(9) and (10) follows the Gibbs-Duhem relation (valid in and out of local equilibrium):

$$\partial_\mu \Phi^\mu \geq J^{\mu\, a} \partial_\mu \lambda_a \tag{11}$$

Finally linear transport coefficients, η_{ab}, are defined as positive semi-definite in order to ensure the inequality (9):

$$\partial^\mu \lambda_a = -\eta_{ab} J^{\mu\, b}. \tag{12}$$

In higher order hydrodynamics the evolution (relaxation) of such coefficients is also calculated.

However, is this the truth? Do we really produce smooth, flowing and relativistic continua in high-energy experiments? Is there also a definite, sharp-valued thermodynamic temperature? Or we experience pseudo-thermalization and pseudo-flow? Some hints towards that this might be the case dates back to 1975, when it was shown that simple acceleration may produce a thermal feeling, it may lead to radiation spectra appearing as the black-body radiation [2, 3, 4, 5].

The physical reason is simple: a moving monochromatic source radiating with a single frequency, ω, is seen shifted according to the relativistic Doppler-effect:

$$\omega' = \omega \sqrt{\frac{1 - v(\tau)}{1 + v(\tau)}}. \tag{13}$$

If the source accelerates, its momentaneous velocity, $v(\tau)$ depends on the proper time along its trajectory. With a constant deceleration, $v(\tau) = v(0) - g\tau$. The spectral analysis Fourier transforms the ever changing phase by the changing Doppler factor obtaining an intensity distribution according to

$$I(\Omega) \propto \left| \int e^{i \int \omega'(\tau')d\tau'} e^{-i\Omega\tau} \, d\tau \right|^2. \tag{14}$$

This intensity divided by $\hbar\Omega$ delivers a photon number yield per invariant phase space element; observing only the high energy part of spectra also massive (but light) particles might show

[2] With the relative velocity between the local flow and the thermometer.

similar effects. Finally it is just a mathematical fact that the integral in eq.(14) is proportional to[3]

$$I(\Omega) \propto \frac{1}{e^{2\pi\Omega/g} - 1}. \tag{15}$$

By this $T = g/2\pi$ is the Unruh temperature in Planck units. For a velocity change in the order of 1 (lightspeed) in a time traversing about 1 fm, it is really in the order of 150 MeV. By this we are talking about extreme accelerations.

In our experiments, however, we cannot suppose monochromatic radiators or movements with forever constant acceleration. Towards obtaining semiclassical photon spectra from more realistic scenarios work is in progress[21]. As a preliminary a few characteristic photon rapidity distributions are shown in Figure 1 at different transverse momenta, k_T, of the photon. While long deceleration times lead to a flat rapidity distribution, like the Bjorken-flow scenario, pictures produced using short times rather resemble those obtained by the Landau-flow scenario. Not only a pseudo-temperature, but also a pseudo-flow occurs in these patterns.

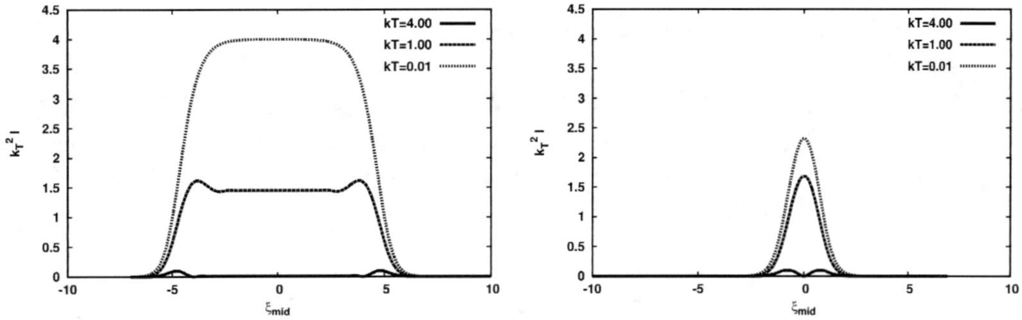

Figure 1. Numerically calculated differential photon rapidity distribution from a classical point charge moving straight with constant deceleration for a finite time. Left for long, right for short deceleration window.

Hydrodynamics is well and alive among the theories in our field embracing a list of intriguing questions to be clarified. A list of talks, like Effects of Jets in the Flow Observables (Takahashi), Thermal Equilibrium and ... Initial Condition (Kodama), Landau hydrodynamics... (Tamosiunas), Shear viscosity of hadrons... (Wiranata), Turbulence, Vorticity and Lambda Polarization (Csernai), Hydrodynamic models of particle production (Bozek), QGP viscosity and the flow ... (Song), Event-by-event correlation... (Molnar), Elliptic Flow from CGC (Scradina), Dynamical freeze-out in event-by-event hydro... (Huovinen), more than proves it.

3. Hadronization

The main question of the hadronization process is since years whether its result is phase space dominated (and in this case statistical descriptions are relevant) or elmentary QCD matrix elements play the main role (and in this case only a microdynamical approach prevails). Theoretically both factors can be studied.

We had talks among others on Monte Carlo (Werner), Strongly interacting parton-hadron matter... (Bratkovskaya), Gluon radiation by heavy quarks (Gousset), Diffusion of Non-Gaussianity ... (Kitazawa), Strangeness baryon to meson ratio (Flores), on the Excluded-

[3] It is derived by using a new integration variable, z, which satisfies $\omega'(\tau) = dz/d\tau$.

Volume Model (Tiwari), on Interpretation of strange hadron production at LHC (Petran), and on Chemical freezeout via HRG (Chatterjee).

Is a "thermal model" or a "statistical model" is a good approximation to the complicated quantum (chromo-) dynamics? By seeking an answer it can be helpful to factorize our search into several steps. Such a factorizing of the "thermo" concept should include i) finding of a scaling variable which unifies different mass hadron $p_\perp$-spectra, ii) testing the quark coalescence hypothesis, in case baryon and meson branches differ after the above step, iii) finding out the unique functional form on the quark (parton) level describing all hadron spectra, iv) testing trends with binary scaling, participant number, rapidity window, etc. v) and only in a final step interpreting the parameters in physical terms (as on/off-equilibrium, finite/infinite size, quantum/classical, phase space/matrix element dominated). This defactorization, shown on a simple example what we did on RHIC data[22], involves step by step the following questions:

(i) Is it $f(p_T, m) = f(E_v(p_t, m) - \mu(m)) = f(X)$?
(ii) Is there a quark scaling? Is it $f_h(X) = f_q^n(X/n)$ for $n = 2, 3$?
(iii) Do we see an (X-)energy distribution? $f(X) \sim (1 + a\beta X)^{-1/a} \longrightarrow e^{-\beta X}$
(iv) Are there trends with $(N_{part}, N_{bin}, P(N), \sqrt{s}, \Delta\eta\ ...)$?
(v) How to interpret the parameters $\beta, v, a = (q - 1), \ldots$?

Also thermal cosmology was introduced by Johann Rafelski. We cite his remark: "When old people make a new theory, they know all about assumptions, approximations, implied or explicit. But young people, who learn it from a textbook, believe that this were the TRUTH". We agree, scientists always should be careful.

4. Quarks off and on Lattices

We heard a number of talks about Quarkonia and Energy Loss, like Heavy Quark Energy Loss (Horowitz), ... Boltzmann vs Langevin (Das), Towards... QGP (Berrehrah), Heavy vs light flavor energy loss... (Uphoff), Jet quenching and Heavy Quarks (Renk), and about thermal behavior of quark matter, such as Free energy vs internal energy potential... (Lee), Quantum and semiclassical... (Katz), ... finite magnetic mass... (Djordjevic). One of the most interesting questions was whether AdS/CFT does well? From Thorsten Trenk's review the answer is "not quite", more precisely "clear no for both light and heavy quarks! AdS techniques predict too much suppression at LHC when tuned to RHIC and extrapolated."

So it remains to study quark matter with the more traditional lattice regularization of QCD. Comparisons between Lattice QCD results with other (non-AdS) models were presented in talks about high-density quark matter (Torrieri), Sphalerons (Chao), flavor hierarchy (Bluhm), the PNJL model (Yamazaki) or the original Nambu-Jona-Lasinio model for SU(3)f (Marty). Full blood lattice results were reported in talks about First Principles Calculation (Allton), Role of fluctuations in detecting the QCD phase transition (Redlich), and "ab initio Lattice QCD calculations" (Schmidt), while alternative approaches as a talk on Holographic descriptions of dense quark matter (Kumar) or on chiral fluid dynamics (Herold) tried to keep balance.

Most of these approaches were presented in a high-temperature context. Therefore it is the right place to demonstrate that contrary to a wide-spread false belief, high-T *does not mean* the unrestricted applicability of perturbative QCD (pQCD). We may consider the following simple example. Since pQCD relies on high Q^2 physics, we consider the thermal distribution of Q^2 in an ideal Boltzmann gas of massless partons:

$$P(Q^2) = \frac{\iiint dE_1\, dE_2\, d\cos\theta\ E_1^2 E_2^2\, e^{-\beta(E_1+E_2)}\, \delta\left(Q^2 - 2E_1E_2(1-cos\theta)\right)}{\iiint dE_1\, dE_2\, d\cos\theta\ E_1^2 E_2^2\, e^{-\beta(E_1+E_2)}}. \tag{16}$$

Journal of Physics: Conference Series **509** (2014) 012027 doi:10.1088/1742-6596/509/1/012027

This integrals can be analytically solved, the result contains some Bessel K-functions:

$$P(Q^2) = \frac{1}{64T^2}\left(\frac{Q^3}{T^3}K_1\left(\frac{Q}{T}\right) + 2\left(\frac{Q^2}{T^2}\right)K_2\left(\frac{Q}{T}\right)\right) \tag{17}$$

This Boltzmann-Gibbs Q^2 distribution is shown in Figure 2 (left).

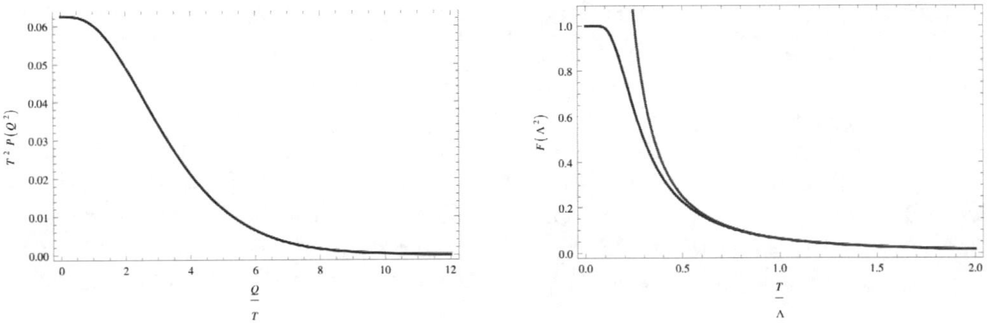

Figure 2. Thermal probability to have Q^2 four-momentum squared (left) and the integrated probability of being soft as a function of temperature (right) for a pair of massless partons.

While the width of the distribution does scale with T, the maximum probability is at $Q = 0$ at any temperature. At arbitrary high temperature non-perturbative effects prevail, only their effect may diminish in some selected (infrared safe) observables.

Regarding an order parameter of Non-Perturbativity, the thermal expectation value of "being soft" may be defined as follows:

$$F(\Lambda^2) := \langle \theta(\Lambda^2 - Q^2) \rangle = \int_0^{\Lambda^2} P(Q^2)\, dQ^2. \tag{18}$$

This definition selects out the integrated distribution function, pictured in Figure 2 (right). The upper line belongs to the perturbative approximation yielding $\Lambda^2/16T^2$. While the full result is saturated until about $T \approx \Lambda/6$, which can be considered as $T_c \approx 167$ MeV, this approximation starts to be really good only at $T \approx \Lambda$. Hadronic effects may therefore well prevail until $T \approx 1$ GeV. Between $T_1 \approx 1/6$ GeV and $T_2 \approx 1$ GeV both pQCD and non-pQCD worlds reign.

5. Highlights

Finally let us summarize theory highlights of this meeting. In our view either new theories or new aspects, or new proposals for measurable effects, perhaps new calculations on hard old problems are worth considering as highlights, those which give or promise new insights into (collective or individual, strange or less-strange) hadronic physics.

Towards a possibly new theory align efforts to derive the EoS, the transport coefficients, heat capacities (acting on the Tsallis parameter via the formula q=1+1/C), the distribution of fluctuations. Expectations and hopes are related to the determination of properties of elementary matter, to comparison of phenomenology based findings with Lattice QCD calculational results, to the very determination of an initial state just after the collision, and to a deepened theoretical understanding of the birth of hadrons.

A new aspect connected with a new proposal was peresented by Becattini and Csernai[23, 24] about the Λ-polarisation, enriching the hydrodynamical approach with vorticity and spin effects.

New calculations seem to bring new conclusions: different confinement temperature for the strange than for the non-strange sector – as it has been cited in Markus Bleicher's opening talk on theory news. Certainly a new insight was expressed in the analysis of Klaus Werner about the EPOS simulation, namely that Flow is Everywhere... Another, relatively new proposal was mediated by Krzysztof Redlich: Fluctuations carry the most important message about quark matter and about the phase transition. He found the so far most beautiful order parameter, jumping from one value to another really sharply at T_c, in form of a sophisticated ratio of susceptibilities. The proposal by Giorgio Torrieri to seek for quarkyonic signals in the v_2 coefficient against p_T plot while keeping the fluctuations was also a novelty.

Finally other topics, like the astronuclear approach in Alford's review taught us that some theories do not survive observations. This clears ground for optimism.

In summary, quoting a commercial text discovered on a bottle of cider, just replacing the word "apple" by theory, and the word "ice" by (experimental) "data" and "Bulmers original cider" by "Our original quark matter", comes as follows: *We use big theories, small theories, juicy theories and bittersweet theories to make the well-balanced and medium-sweet flavor of our original quark matter that you know and love. Enjoy poured over data for ultimate refreshment.*

Acknowledgement This work has been supported by the Hungarian National Rsearch Fund (OTKA K104260) and by the Helmholtz International Center for FAIR within the framework of the LOEWE program launched by the State of Hesse.

References

[1] ALICE Collaboration, *Centrality dependence of π, K, p production in Pb-Pb collisions at sqrt(sNN)=2.76 TeV*, arxiv 1303.0737[hep-ex].
[2] Unruh WG, *Phys. Rev. D* **14**, 870, 1976
[3] Kharzeev D, Satz H, *Phys. Lett. B* **334**, 155, 1994
[4] Casterina P, Kharzeev D, Satz H, *EPJ C* **52**, 187, 2007
[5] Biro TS, Gyulassy M, Schram Z, *Phys. Lett. B* **708**, 276, 2012
[6] Bouras I, Molnar E, Niemi H, Xu Z, El A, Fochler O, Greiner C, Rischke DH, *Phys. Rev. Lett.* **103**, 032301, 2009
[7] Cardoso N, Bicudo P, Eilhauer U, Bouras I, arxiv: 1208.0961 [hep-ph]
[8] Aguiar CE, Kodama T, Osada T, Hama Y, *J. Phys. G* **27**, 75, 2001
[9] Socolowski O, Grassi F, Hama Y, Kodama T, *Phys. Rev. Lett.* **93**, 182301, 2004
[10] Takahashi J, Tavares BM, Qian WL, Andrade R, Grassi F, Hama Y, Xu N, *Phys. Rev. Lett.* **103**, 242301, 2009
[11] Osada T, Wilk G, *Phys. Rev. C* **77**, 044903, 2007
[12] Osada T, Wilk G, *CEJP* **7**, 432, 2009
[13] Osada T, Wilk G, *Indian J. Phys* **85**, 941, 2011
[14] Biro TS, Molnar E, *EPJ A* **48**, 172, 2012
[15] Denicol GS, Koide T, Rischke DH, *Phys. Rev. Lett.* **105**, 162501, 2010
[16] Niemi H, Denicol GS, Huovinen P, Molnar E, Rischke DH, *Phys. Rev. Lett.* **106**, 212302, 2011
[17] Denicol GS, Niemi H, Molnar E, Rischke DH, *Phys. Rev. D* **85**, 114047, 2012
[18] Biro TS, *Physica A* **392**, 3132, 2013
[19] Biro TS, Barnafoldi GG, Van P, *EPJ A* **49**, 110, 2013
[20] Biro TS, Van P, *EPL* **89**, 30001, 2010
[21] Biro TS, Szendi Z, Schram Z: *Illusory Flow in Radiation from Accelerating Charge, work in progress*, 2013
[22] Biro TS, Urmossy K, Van P, Barnafoldi GG, Schram Z, *Acta Phys. Polon. B* **43**, 811, 2012
[23] Becattini F, Csernai LP, Wang DJ, arxiv: 1304.4427
[24] Csernai LP, Magas VK, Wang DJ, *Phys. Rev. C* **87**, 034906, 2013

SQM2013: Experimental Summary

Andrea Dainese

INFN – Sezione di Padova, via Marzolo 8, I-35131 Padova, Italy

E-mail: andrea.dainese@pd.infn.it

Abstract. We present an overview of the new experimental results from heavy-ion collisions discussed at the Strangeness in Quark Matter 2013 conference.

1. Introduction

The SQM2013 conference was held in Birmingham on July 22–27, 2013. This was the first major conference in the field of heavy-ion physics after the run with proton–lead collisions at a centre-of-mass energy $\sqrt{s_{\rm NN}} = 5.02$ TeV per nucleon–nucleon pair, which took place at the LHC in January–February 2013. Therefore, the new results from this run were among the most awaited for. On one hand, the measurement of heavy flavour, quarkonia and jet production in this collision system provides a control experiment to assess the role of initial-state effects on the probes of the hot and dense medium formed in nucleus–nucleus collisions. On the other hand, the unexpected observations in the sector of bulk particle production in high-multiplicity pp and p–Pb (from the first pilot run) events at the LHC raised high expectations for the outcome of more detailed studies. New results from d–Au collisions at RHIC were presented as well, providing an essential complement to the observations at the LHC.

For nucleus–nucleus collisions, the LHC and RHIC Collaborations presented new results, in particular on strangeness and heavy-flavour production, from the Pb–Pb run of 2011 at $\sqrt{s_{\rm NN}} = 2.76$ TeV, from the Au–Au runs of 2010–2011 at $\sqrt{s_{\rm NN}} = 200$ GeV and from the beam energy scan with Au–Au at $\sqrt{s_{\rm NN}}$ in the range 7.7–39 GeV. A summary of the status and main open questions before SQM2013 can be found in Refs. [1–3].

The paper is organized as follows: Section 2 presents the results from nucleus–nucleus collisions on strangeness, open heavy flavour and quarkonia; Section 3 presents the results from the measurement of hard probes production in proton–nucleus collisions; Section 4 presents the new observations in high-multiplicity p–Pb collisions at the LHC and at RHIC.

2. Nucleus–Nucleus Collisions: New Results on Strangeness and Heavy Flavour

Strangeness

The ALICE Collaboration presented the final results on the enhancement of hyperons (Λ, Ξ, and Ω) in Pb–Pb collisions at $\sqrt{s_{\rm NN}} = 2.76$ TeV with respect to pp collisions at the same energy [4, 5]. The enhancement of the per-participant yield in Pb–Pb relative to pp increases with the strangeness content and towards more central collisions, up to a factor of about six for the triply-strange Ω. In comparison with lower energy results from NA57 at the SPS and STAR at RHIC, the enhancement is found to be decreasing for increasing $\sqrt{s_{\rm NN}}$. For example, the Ω enhancement at SPS energy is of a factor of about twenty in central collisions. This energy dependence is due to the pp reference: the hyperon/π ratios show no evident energy dependence in AA collisions and a strong dependence in pp(p–Be) collisions, which is usually attributed to increasing strangeness canonical suppression towards smaller values of $\sqrt{s}$. Overall, the experimental data are consistent with a scenario in which the bulk of strange hadrons are formed from a partonic system with a large abundance of strange quarks.

Final results from the ALICE experiment were also presented for the Λ/K^0_S ratio in Pb–Pb collisions, as a function $p_{\rm T}$ and centrality [4, 6]. The baryon/meson enhancement in central collisions with respect to peripheral collisions is of a factor of about two and it is consistent with the value measured by STAR at top RHIC energy. The STAR Collaboration presented the results of this measurement from the beam energy scan: the enhancement is visible starting from $\sqrt{s_{\rm NN}} = 19.6$ GeV (consistently with the observation by NA57 at the top SPS energy of about 17 GeV) and not evident at 11.5 and 7.7 GeV. The peak of the Λ/K^0_S ratio in central collisions is at about 2.5 GeV/c at energies 19.6–200 GeV and 3.5 GeV/c at 2.76 TeV. Models including parton recombination, which was the first explanation put forward when the effect was observed at RHIC, describe the LHC data only qualitatively. Models including only radial flow, where the effect is related to the particle mass and not to its quark composition, also provide a good description. Therefore, the role of recombination in this effect is still not clarified. The ϕ meson could provide important additional information, because it is a meson and has mass similar to that of the p and Λ baryons. However, preliminary results from ALICE and final result from PHENIX at RHIC show that the ϕ nuclear modification factor $R_{\rm AA}$ has intermediate value between those of light mesons (π, K) and those of baryons (p, Λ) [4, 7, 8]. This suggests that both effects, radial flow and recombination, may play a role.

Open Heavy Flavour

The long-standing question about the quark mass dependence of parton energy loss is on the way of finding a clear answer in the LHC data. The ALICE Collaboration presented a new comparison of D meson $R_{\rm AA}$ as a function of centrality with the measurement of non-prompt J/ψ from B decay by the CMS Collaboration [9–11]. The new D meson data in the interval $8 < p_{\rm T} <$ 16 GeV/c (corresponding to average $p_{\rm T}$ of about 10–11 GeV/c) show a suppression in central collisions larger by a factor of two with respect to J/ψ with $p_{\rm T} > 6.5$ GeV/c (corresponding to approximately the same average $p_{\rm T}$ for B mesons). Considering the uncertainties of the two measurements, in central collisions, the difference between charm and beauty is observed with a significance of more than $3\,\sigma$. This is the first observation of a hadron species dependence of energy-loss induced suppression at such large $p_{\rm T}$. Comparison with models of radiative energy loss indicates that the measurements are consistent with the predicted quark mass dependence.

The comparison of D mesons and pions, presented by ALICE as a function of $p_{\rm T}$ and centrality, shows comparable suppression. A hint is seen that the suppression may be larger for pions at low $p_{\rm T}$ in central collisions, although with the present uncertainties it is not possible to draw a firm conclusion [9, 10]. These observations can be described by calculations in which radiative energy loss depends on the parton colour charge and the larger energy loss of gluons with respect to quarks is to some extent counterbalanced by the harder fragmentation of $c \to D$ with respect to $g \to \pi$ [12].

The STAR experiment presented preliminary spectra of heavy-flavour decay electrons in Au–Au collisions at $\sqrt{s_{\rm NN}} = 62.4$ GeV [13]. Lacking pp reference data at this energy from RHIC, the spectra were compared with binary scaled spectra from p$\overline{\rm p}$ at the ISR and with binary scaled perturbative QCD calculations: surprisingly, no suppression is observed in Au–Au with respect to this references. It would be important to have reference pp runs at RHIC at lower energies, in order to study the onset of the jet quenching effects, not only for heavy flavour.

New results on heavy-flavour azimuthal anisotropy v_2 at the LHC were presented by the ALICE Collaboration [9, 10, 14]. The final results for D mesons show average v_2 values of about

0.21 ± 0.04 in the interval $2 < p_{\mathrm{T}} < 6$ GeV/c. The new preliminary results on heavy-flavour decay muon v_2 at forward rapidity indicate $v_2 > 0$ in the interval $3 < p_{\mathrm{T}} < 6$ GeV/c, consistent with the measurements with electrons at central rapidity. These results suggest that charm quarks participate in the collective expansion of the medium. A similar conclusion was obtained from the PHENIX and STAR measurements at top RHIC energy.

The silicon tracker upgrades of the PHENIX and STAR detectors (already partially installed) and of the ALICE detector (planned for 2018) will give a strong boost to the heavy flavour measurements [15–17]. Precise studies of R_{AA} and v_2 separately for charm and beauty hadrons and investigation of charm hadronization mechanisms in heavy-ion collusions via baryon/meson and D_s measurements will become accessible.

Quarkonia

A new piece was added to the puzzle of J/ψ production and suppression in nucleus–nucleus collisions: STAR preliminary measurements of R_{AA} at high p_{T} (> 5 GeV/c) show a lower suppression than at low p_{T}, consistently in all centrality classes. Therefore, the p_{T} dependence of the J/ψ R_{AA} is opposite at RHIC and LHC: at RHIC the suppression decreases for increasing p_{T}, while at the LHC it increases for increasing p_{T} [13, 18, 19]. This pattern is consistent with the predictions of kinetic transport models including melting and regeneration.

The comparison of J/ψ and Υ suppression provides additional insight in the relative contributions of the melting and regeneration effects. ALICE presented new preliminary results on $\Upsilon(1S)$ R_{AA} at $p_{\mathrm{T}} > 0$ and $2.5 < y < 4$, which show a suppression of a factor of about two in minimum-bias Pb–Pb collisions [18, 20]. The measurement complements at forward rapidity the results from the CMS Collaboration (presented in [11]): taken together, these data show no evident dependence of the $\Upsilon(1S)$ R_{AA} on rapidity. On the other hand, ALICE observed a rapidity dependence for the J/ψ R_{AA}, with increasing suppression towards large y. This difference between $\Upsilon(1S)$ and J/ψ is consistent with a significant role of regeneration for the latter. Indeed, this mechanism is expected to contribute to the J/ψ yield proportionally to $(\mathrm{d}N_{c\bar{c}}/\mathrm{d}y)^2$, thus more at central than at forward rapidity.

3. Proton–Nucleus Collisions: Control Experiment for Hard Probes

The main goal of proton–nucleus collisions in the heavy-ion physics programmes at RHIC and LHC is the study of initial state and cold nuclear matter effects. On one hand, it is necessary to quantify the modification in the production of hard partons determined by the fact of having nucleons (within nuclei) instead of protons in the initial state of the hard scattering processes. This is a crucial control experiment to disentangle, in the case of nucleus–nucleus collisions, the modifications that are induced by the presence of a hot and dense QCD medium in the final state. On the other hand, the study of initial state effects is highly interesting *per se*, in the context of high-density gluon dynamics and saturation at small parton fractional momentum (Bjorken x).

For the R_{pA} measurements at the LHC, an interpolated pp reference was used, since there are no reference pp data at $\sqrt{s} = 5.02$ TeV.

High p_{T} and Jets

The nuclear modification factor R_{pA} of charged particles in minimum-bias p–Pb collisions at the LHC, measured by ALICE up to $p_{\mathrm{T}} = 20$ GeV/c, is consistent with unity in the range $2 < p_{\mathrm{T}} < 20$ GeV/c, with a central value at about 1.1 and uncertainties of about $\pm 10\%$ [21]. This suggests that particle production at medium-high p_{T} is dominated by hard scattering processes and approximately scales with the number of binary collisions in this p_{T} range.

Therefore, the large suppression observed in central Pb–Pb collisions (up to a factor of seven at 6–10 GeV/c) is an effect of the hot medium in the final state. Another method to verify binary scaling of hard scattering processes in nucleus–nucleus collisions is the measurement of high-p_T photons, $W \to \ell\nu_\ell$ and $Z \to \ell^+\ell^-$, that are produced in hard scattering processes and do not interact with the QCD medium. The CMS Collaboration presented new preliminary results on $Z \to ee$ and $Z \to \mu\mu$ R_{AA} in Pb–Pb collisions, using new reference data from the high luminosity pp run at 2.76 TeV from February this year. The results show that R_{AA} is consistent with unity, in line with the previous measurements with γ, W and Z (see [22] and references therein).

The outcome of the control experiment with p–Pb at the LHC is the same also for jet production. In minimum-bias collisions ALICE measures a charged-jet R_{pA} consistent with unity in the range $20 < p_T < 100$ GeV/c [23], with uncertainty of about $\pm 20\%$. CMS presented measurements of the di-jet momentum imbalance $p_{T,2}/p_{T,1}$ for $p_{T,1} > 120$ GeV/c and $p_{T,2} > 30$ GeV/c [24]: the distributions of this variable for p–Pb collisions in various multiplicity classes are fully consistent with the pp case. These results indicate that the large R_{AA} suppression of up to a factor of four and the large di-jet imbalance observed in central Pb–Pb collisions are produced by jet quenching in the final state.

Open Heavy Flavour

The ALICE Collaboration reported on the measurement of D mesons (D^0, D^+, D^{*+}, D_s^+) and heavy-flavour decay electrons in minimum-bias p–Pb collisions at central rapidity [9,25,26]. The D meson R_{pA}, measured in the interval $1 < p_T < 24$ GeV/c, and the electron R_{pA}, 0.5–12 GeV/c, show consistency with unity within the experimental uncertainties. Theoretical predictions based on next-to-leading order perturbative QCD supplemented with nuclear modified parton distribution function and based on the Colour Glass Condensate framework are both consistent with the data. At very low p_T, both calculations predict a suppression, by 20% and 40% for the central value, respectively, due to initial-state saturation effects that reduce the effective partonic flux. While the data are not precise enough to be sensitive to effects of this level (few 10%), they clearly indicate that, like for light-flavour particles, the strong suppression of D mesons and heavy-flavour decay electrons in central Pb–Pb collisions is due to heavy-quark energy loss.

It is interesting to note that the electron R_{pA} at the LHC has a central value of 1.4 at low p_T (with large uncertainties of 30–40%), which is similar to that of the same measurement in d–Au collisions at RHIC, where the PHENIX experiment measured a larger R_{dA} for heavy-flavour decay electrons than for pions [15]. This observation at RHIC is intriguing, especially when taken together with the enhancement measured for heavy-flavour decay muons at backward rapidity (corresponding to large x values in the Au nucleus): both R_{dA} values seem to be larger than the expectations based on nuclear modification of the parton distribution functions (anti-shadowing). It will be, therefore, very interesting to see a measurement of electrons with reduced systematic uncertainties from ALICE and the corresponding measurement for backward rapidity muons.

Quarkonia

The production of J/ψ and $\Upsilon(1S)$ in minimum-bias p–Pb collisions at $\sqrt{s_{NN}} = 5.02$ TeV was reported by the ALICE (both particles) and LHCb (J/ψ) experiments using the di-muon decay channel [18, 27, 28]. Both experiments measure di-muons at forward rapidity and for $p_T > 0$. Using p–Pb and Pb–p collisions they could cover both the backward and forward rapidity regions, probing large and small Bjorken x values in the Pb, respectively. The J/ψ R_{pA} from both experiments is consistent with unity in the backward region (large x) and shows a suppression of about 30% in the forward region (small x). The Υ R_{pA} is consistent with the J/ψ R_{pA}. Both measurements are described reasonably well by perturbative calculations with nuclear shadowing. The prediction of the Colour Glass Condensate, available only for J/ψ and only for the forward region (small x) underestimates the value of R_{pA}. For J/ψ, ALICE presented also a measurement of yield in the forward region divided by yield in the backward region, as a function of p_{T}, which has better accuracy than R_{pA}, because no pp reference is required. The pattern of the data, with a pronounced suppression at low p_{T} and a consistency with unity at high p_{T}, is not well reproduced by the calculations.

The PHENIX Collaboration presented final results on R_{dA} of J/ψ and ψ' as a function of centrality in d–Au collisions at 200 GeV [15]. While the J/ψ shows a roughly constant and moderate suppression of about 20%, the ψ' suppression increases towards more central collisions reaching down to $R_{\mathrm{pA}} \approx 0.2$–$0.4$. That is: it is 2–3 times more suppressed than the J/ψ. Since the initial-state effects related to shadowing should be the same for the two states, this observation indicates strong final-state effects for the more-loosely-bound ψ'. The nature of these final-state effects is not clear: cold nuclear matter energy loss, hadronic rescattering, ... It will be interesting to see the result of the ψ' measurement in p–Pb collisions at the LHC, which is essential for the interpretation of the J/ψ and ψ' comparison in Pb–Pb collisions.

Centrality Determination and Binary Scaling in p–Pb at the LHC

The measurements of the p–Pb nuclear modification factors presented at the conference were obtained using minimum-bias collisions. In this case, the general conclusion is that hard process scale according to the number of binary collisions for p_{T} larger than a few GeV/c. The next step is to extend the comparison with binary-scaled pp reference to p–Pb collisions in various multiplicity classes. This is especially interesting in view of the effects observed in high-multiplicity events, which will be discussed in the next Section.

The intrinsic challenge of this measurement was discussed by the ALICE Collaboration [21]. In the minimum-bias case the binary scaling variable, e.g. the average number of binary collisions, is well-defined as $\langle N_{\mathrm{coll}}\rangle = \mathrm{A}\,\sigma^{\mathrm{pp}}_{\mathrm{inel}}/\sigma^{\mathrm{pA}}_{\mathrm{inel}}$. At variance, when considering centrality classes, defined in terms of the particle multiplicity or energy measured in a given pseudo-rapidity (η) range, a model of particle production and the collision geometry is needed to define the average number of binary collisions. In the case of pA collisions the correlation between the measured multiplicity and the geometrical collision variables is very loose. This is due to the small dynamical range of the geometrical variables (e.g. the number of participants reaches barely 30 in p–Pb, to be compared with more than 400 in Pb–Pb) and to the local multiplicity bias associated with jets and long-range correlations, Consequently, there is a large overlap in terms of, e.g., distribution of binary collisions for centrality classes defined in terms of multiplicity. In addition, it was shown that the hard probes bias the determination of the binary scaling variable, if the centrality classes are defined using multiplicity measured at less than five η units from the region where the hard probes themselves are measured [21]. It is now under detailed scrutiny whether the usage of centrality estimators with a larger η gap offers a solution to reduce the bias at an acceptable level, for example below the intrinsic systematic uncertainties of the measurement.

4. Proton–Nucleus Collisions: Hints of Collectivity in High-Multiplicity Events?

The measurements of particle production and correlations in high-multiplicity p–Pb collisions at $\sqrt{s_{\mathrm{NN}}} = 5.02$ TeV have produced the most unexpected and intriguing results not only of the pA programme, but probably of the entire heavy-ion programme at the LHC. The news in this sector, including for the first time identified particles, were among the highlights of the conference.

Correlations

Two-particle correlations in azimuthal angle φ and pseudo-rapidity are measured by constructing the two-dimensional distribution of $\Delta\varphi$ vs. $\Delta\eta$ for all charged particles (called 'associated' particles) in a given $p_{\mathrm{T}}^{\mathrm{assoc}}$ range with respect to a 'trigger' particle with given $p_{\mathrm{T}}^{\mathrm{trigger}}$, where generally $p_{\mathrm{T}}^{\mathrm{trigger}} > p_{\mathrm{T}}^{\mathrm{assoc}}$. In nucleus–nucleus collisions the two-particle correlation distribution for typical p_{T} ranges of 2–4 GeV/c for trigger particles and 1–2 GeV/c for associated particles exhibits a double-ridge structure, with the two ridges extended over several units in $\Delta\eta$ and described by a dominant second-order harmonic in $\Delta\phi$, quantified by the v_2 coefficient. This structure is well consistent with hydro-dynamical collective expansion of a hot and dense system.

Shortly after the pilot p–Pb run in September 2012, two-particle correlation measurements in high-multiplicity events have revealed, in a rapid sequence: first, an extended $\Delta\eta$ ridge in the (near-side) region of $\Delta\phi \sim 0$, observed by the CMS Collaboration [31]; then, a corresponding away-side ($\Delta\phi \sim \pi$) ridge of similar size, observed by the ALICE Collaboration by subtracting the distribution of low-multiplicity events from that of high-multiplicity events [21, 29, 34, 35]. The latter observation was shortly after confirmed by the ATLAS Collaboration [30, 32]. This subtraction technique is an empirical way of eliminating the correlations with the trigger particle originating from hard scattering processes (i.e. jets), in order to emphasize the underlying long-distance correlations. The resulting double-ridge resembles the structure that, in nucleus–nucleus collisions, is interpreted in terms of collective effects induced by an expanding medium. A Fourier analysis showed values of v_2 comparable to those measured in Pb–Pb collisions with similar multiplicity [29–31]. Results from a similar analysis by the PHENIX Collaboration showed that the effect is present also in high-multiplicity d–Au collisions at $\sqrt{s_{\mathrm{NN}}} = 200$ GeV, with corresponding v_2 values even larger than observed at the LHC [33], which could result from the larger initial eccentricity of deuteron-induced with respect to proton-induced interactions. These first observations were described by models with an hydro-dynamical treatment of the final state of p–Pb collisions [36, 37], as well as by calculations based on the Colour Glass Condensate framework, where multi-gluon process in the saturated initial state can give rise to correlations in the final state [38]. It was pointed out that also colour reconnection among strings in the parton shower as implemented in the latest version of the PYTHIA generator can induce multi-particle correlations in the final state [39, 40].

An important additional piece of information was provided by the ALICE Collaboration with the measurement of the two-particle correlation using identified associated particles [34, 35]. In particular, the double-ridge was observed for pions, kaons and protons and the corresponding v_2 coefficients were extracted, as a function of p_{T} up to 4 GeV/c. The data show a clear indication of mass ordering, in particular in the comparison for pions and protons, with $v_2^p < v_2^\pi$ for $p_{\mathrm{T}} < 1.5$ GeV/c and $v_2^p > v_2^\pi$ for $2.5 < p_{\mathrm{T}} < 4$ GeV/c. The same mass ordering is observed in Pb–Pb collisions, and it is interpreted in terms for collective radial and elliptic flow. Models with collective effects in pA collisions have predicted this feature [37].

Light-Flavour Particle p_{T} Spectra

As discussed in Section 2, the baryon-over-meson ratios as a function of p_{T} exhibit an enhancement in central with respect to peripheral nucleus–nucleus collisions, which can be described in terms of radial flow or in terms of parton recombination.

The ALICE Collaboration presented new measurements of π, K, p and Λ in p–Pb collisions in various multiplicity classes [4,35,41]. The $(p+\overline{p})/(\pi^+ + \pi^-)$ and the $(\Lambda+\overline{\Lambda})/K_S^0$ ratios in the 20% highest multiplicity events in comparison with the 60–80% multiplicity class show a pattern that is reminiscent of the corresponding Pb–Pb data. In particular, both baryon-over-meson ratios appear to be higher in high multiplicity than in low multiplicity events for $p_{\mathrm{T}} > 2$ GeV/c. The p/π measurement extends only to 3 GeV/c, while the Λ/K extends to 8 GeV/c and shows that at this p_{T} the ratio is independent of event multiplicity, as also observed in Pb–Pb collisions. The Λ/K ratio was also measured in pp collisions at $\sqrt{s} = 7$ TeV as a function of p_{T} and event multiplicity. It is interesting to note that, in a given p_{T} and multiplicity interval, the ratio has similar values for different colliding systems and energies. This might indicate that the same physical mechanisms are at play in different systems. Detailed comparison with models should allow for drawing a more definite conclusion.

The PHENIX Collaboration presented the p/π^+ and $\overline{p}/\pi^-$ ratios in various centrality classes in d–Au collisions at $\sqrt{s_{\rm NN}} = 200$ GeV [8]. These data show a very similar pattern as seen at LHC energy: in central collisions both ratios are clearly higher than for peripheral collisions, in the interval $1.5 < p_{\rm T} < 4$ GeV/c. Like for the case of two-particle correlations, the data look qualitatively similar at RHIC and LHC energies.

A complementary approach for investigating the change of $p_{\rm T}$ spectra with the event multiplicity is the measurement of average $p_{\rm T}$ ($\langle p_{\rm T}\rangle$) vs. multiplicity. The CMS Collaboration presented new results on this analysis for identified pions, kaons and protons in pp collisions at and p–Pb collisions [42]. In pp collisions, $\langle p_{\rm T}\rangle$ increases roughly linearly with multiplicity. The increase is steeper for protons than for pions and kaons. In p–Pb low-multiplicity collisions, $\langle p_{\rm T}\rangle$ is the same as in pp at the same multiplicity. In high-multiplicity p–Pb, the growth of $\langle p_{\rm T}\rangle$ is less steep than in pp at the same multiplicity. The multiplicity dependence of average $p_{\rm T}$ was measured for charged particles by the ALICE Collaboration for pp, p–Pb and Pb–Pb collisions and compared with several models [21, 41]. In pp and p–Pb collisions, the increase of $\langle p_{\rm T}\rangle$ with multiplicity is much stronger than that measured in Pb–Pb collisions. For pp collisions, this behaviour is described by the PYTHIA model of hadronizing strings when multiple-parton interactions and final-state colour reconnection are included [39]. Using the pp measurement, the expectation for p–Pb and Pb–Pb under the assumption that they are an incoherent superposition of nucleon–nucleon collisions was derived. Both the Pb–Pb and the p–Pb data show a $\langle p_{\rm T}\rangle$ significantly larger than this expectation. In Pb–Pb, this is generally attributed mostly to radial flow. It is still an open question whether a similar effect is at play also in p–Pb, adding to initial-state effects, like intrinsic partonic transverse momentum ($k_{\rm T}$) broadening.

5. Conclusions and Outlook

A wealth of new and exciting experimental results were presented at the conference. At the high energy frontier, the first run period of the LHC indicates that some of the long-standing questions are on the way to find a clear answer (for example the fate of quarkonia in the medium and the mass dependence of energy loss), while new and unexpected perspectives were opened by the possible hints of collectivity in high-multiplicity proton–nucleus collisions. The RHIC experiments continue to produce new results at impressive pace, and the complementarity of the two colliders is a unique richness for the field. Just to quote one example, the parallel study of high multiplicity proton(deuteron)–nucleus collisions at such different energies is providing key information and strong constraints for theoretical interpretations.

A selection of the highlights of the conference is summarized in the following. The production of D mesons ($p_{\rm T} > 1$ GeV/c) and jets ($p_{\rm T} > 20$ GeV/c) in minimum-bias p–Pb collisions at the LHC does not exhibit significant deviations from binary scaling and also the di-jet momentum imbalance is similar to the pp case. This implies that the strong modifications observed in nucleus–nucleus collisions are mainly induced by in-medium parton energy loss. A quark mass dependence of the latter is suggested by the comparison, in Pb–Pb at the LHC, of hadrons originating from c and b quarks, which manifest the only observed hadron-species dependence of $R_{\rm AA}$ for $p_{\rm T} > 8$ GeV/c. The production of quarkonia in p–Pb collisions at the LHC shows only modest deviations from binary scaling, roughly consistent with the expected initial-state effects. Remarkably, a clear suppression is seen for ψ' particles in d–Au at RHIC. These data represent crucial constraints for the interpretation of the nucleus–nucleus measurements in terms of quarkonium melting and regeneration. In high-multiplicity proton–nucleus collisions, the main highlights were the measurements with identified particles at the LHC. These exhibit multiple qualitative similarities with the nucleus–nucleus case: namely, the hadron mass ordering of the azimuthal anisotropy modulation (in particular for v_2^{π} and v_2^{p}), the baryon-over-meson (p/π and Λ/K) enhancement (also observed at RHIC), and the dependence on hadron mass and on event multiplicity of mean p_{T}. These similarities are suggestive of a similarity in the underlying physical mechanisms in nucleus–nucleus and high-multiplicity proton–nucleus (and maybe even pp) collisions. Further studies —including the challenging centrality-dependent pA nuclear modification factor and systematic multiplicity-dependent measurements in pp— will allow us to address the question whether these mechanisms are related to multi-parton correlations in the initial or in the final state of the collision, or to the formation of a high-density final state.

This and other questions promise to keep the field very lively for more than a decade ahead of us. The RHIC experiments are being upgraded and new projects are proposed for the future. The extension of the LHC heavy programme till the mid 2020s was recently endorsed and very substantial detector upgrades are planned. The low energy and high baryon density frontier will be explored as well in the next decade, with the continuation of the SPS programme and the new facilities FAIR and NICA [43–46].

Acknowledgements

The SQM2013 conference was perfectly organized, in the fascinating atmosphere of the Birmingham University campus. I would like to thank the Organizers for proposing me the interesting (and challenging!) task of preparing the experimental summary and Federico Antinori for his advises on the talk.

References

[1] Caines H, these proceedings
[2] Hippolyte B, these proceedings
[3] Granier de Cassagnac R, these proceedings
[4] Barnby L (ALICE Collaboration), these proceedings
[5] Colella D (ALICE Collaboration), these proceedings
[6] Hanratty L (ALICE Collaboration), these proceedings
[7] Knospe A G (ALICE Collaboration), these proceedings
[8] Belmont R (PHENIX Collaboration), these proceedings
[9] Stocco D (ALICE Collaboration), these proceedings
[10] Bruna E (ALICE Collaboration), these proceedings
[11] Jo M (CMS Collaboration), these proceedings
[12] Djordjevic M , these proceedings
[13] Kikola D (STAR Collaboration), these proceedings
[14] Zhang X (ALICE Collaboration), these proceedings
[15] Jouan D (PHENIX Collaboration), these proceedings
[16] Qiu H (STAR Collaboration), these proceedings
[17] Terrevoli C (ALICE Collaboration), these proceedings
[18] Bruno G E (ALICE Collaboration), these proceedings
[19] Valencia Palomo L (ALICE Collaboration), these proceedings
[20] Khan P (ALICE Collaboration), these proceedings
[21] Morsch A (ALICE Collaboration), these proceedings
[22] Gardner M (CMS Collaboration), these proceedings
[23] Verweij M (ALICE Collaboration), these proceedings
[24] Stephans G (CMS Collaboration), these proceedings
[25] Luparello G (ALICE Collaboration), these proceedings
[26] Heide M (ALICE Collaboration), these proceedings
[27] Lakomov I (ALICE Collaboration), these proceedings
[28] Zhong L (LHCb Collaboration), these proceedings
[29] ALICE Collaboration 2013 *Phys. Lett.* B **719** 29
[30] ATLAS Collaboration 2013 *Phys. Rev. Lett.* **110** 182302
[31] CMS Collaboration 2013 *Phys. Lett.* B **724** 213
[32] Trzupek A (ATLAS Collaboration), these proceedings
[33] PHENIX Collaboration, arXiv:1303.1794
[34] Christakoglou P (ALICE Collaboration), these proceedings
[35] Milano L (ALICE Collaboration), these proceedings

Journal of Physics: Conference Series **509** (2014) 012028 doi:10.1088/1742-6596/509/1/012028

[36] Bozek P, these proceedings
[37] Werner K, these proceedings
[38] Dusling K and Venugopalan R 2013 *Phys. Rev.* D **87** 094034
[39] Sjostrand T, Mrenna S and Skands P Z 2008 *Comput. Phys. Commun.* **178** 852
[40] Ortiz A *et al* 2013 *Phys. Rev. Lett.* **111** 042001
[41] Anielski J (ALICE Collaboration), these proceedings
[42] Appelt E A (CMS Collaboration), these proceedings
[43] Fodor Z (NA61/SHINE Collaboration), these proceedings
[44] Ali Y (NA61/SHINE Collaboration), these proceedings
[45] Schmidt H R (CBM Collaboration), these proceedings
[46] Sorin A (NICA/MPD Collaboration), these proceedings

Journal of Physics: Conference Series **509** (2014) 012029 doi:10.1088/1742-6596/509/1/012029

Anisotropic flow of identified particles in Pb–Pb collisions at $\sqrt{s_{\rm NN}}$ = 2.76 TeV with the ALICE detector

You Zhou (for the ALICE Collaboration)

Nikhef, Science Park 105, 1098 XG Amsterdam, The Netherlands
Utrecht University, P.O.Box 80000, 3508 TA Utrecht, The Netherlands

E-mail: you.zhou@cern.ch

Abstract. The anisotropic flow of identified particles, including charged pion, kaon, (anti-)proton as well as strange and multi-strange particles (K^0_S, Λ, Ξ and Ω) are measured in Pb–Pb collisions at $\sqrt{s_{\rm NN}}$ = 2.76 TeV with the ALICE detector. The results are compared to measurements at RHIC and hydrodynamic model calculations. The mass splitting and the scaling with the number of constituent quarks of the anisotropic flow are also discussed.

1. Introduction

Anisotropic flow [1], described by the coefficients in the Fourier expansion of the azimuthal particle distributions with respect to the symmetry planes, is an important observable to study the properties of the hot and dense matter, the Quark Gluon Plasma (QGP), created in heavy-ion collisions [2]. The anisotropic flow of identified particles allows us to probe the freeze-out conditions of the system, *e.g.* the freeze-out temperature, and the radial flow. It was measured at SPS [3], RHIC [4,5] and could be successfully described by hydrodynamic calculations [6]. Moreover, it is very sensitive to the partonic degrees of freedom at the early time of the heavy-ion collisions. Therefore, the identified particle elliptic flow, v_2, and triangular flow, v_3, measured at the LHC is expected to play a pivotal role in constraining the initial conditions as well as the shear viscosity over entropy density (η/s) in the hydrodynamic calculations.

We report on v_2 and v_3 of identified particles measured in $\sqrt{s_{\rm NN}}$ = 2.76 TeV Pb–Pb collisions by the ALICE Collaboration. The measurements are compared to those at top RHIC energy and to ideal hydrodynamic calculations [6]. The scaling with number of constituent quarks (2 for mesons and 3 for baryons) is also tested for both v_2 and v_3.

2. Data Analysis

The data sample collected by ALICE in the first Pb–Pb run at the Large Hadron Collider was used in this analysis. About 16 million Pb–Pb events were recorded with a minimum-bias trigger, based on signals from two VZERO ($-3.7<\eta<-1.7$ and $2.8<\eta<5.1$) and on the Silicon Pixel Detector. The VZERO was also used for the determination of the collision centrality and of the symmetry planes. Charged particles are reconstructed using the Inner Tracking System and the Time Projection Chamber (TPC) with full azimuthal coverage for pseudo-rapidity range $|\eta|<0.8$. The particle identification was performed using the specific ionisation energy loss in the TPC and time-of-flight measurements in the Time-Of-Flight detector. The strange and multi-strange particles were reconstructed via their weak decay channels, which are: $K^0_S \to \pi^+\pi^-$, $\Lambda \to p\pi^-$ ($\overline{\Lambda} \to \overline{p}\pi^+$), $\Xi^- \to \Lambda\pi^-$ ($\overline{\Xi}^+ \to \overline{\Lambda}\pi^+$) and $\Omega^- \to \Lambda K^-$ ($\overline{\Omega}^+ \to \overline{\Lambda}K^+$). In order to suppress combinatorial background, additional topological cuts have been applied. Anisotropic flow was measured with the scalar product [7] and event plane [8] methods with large separation in pseudo-rapidity between the reference particles and the particle of interest ($|\Delta\eta|>1$ for SP method and $|\Delta\eta|>2$ for EP method), to reduce the correlations not associated with the symmetry plane, the so called non-flow.

3. Results

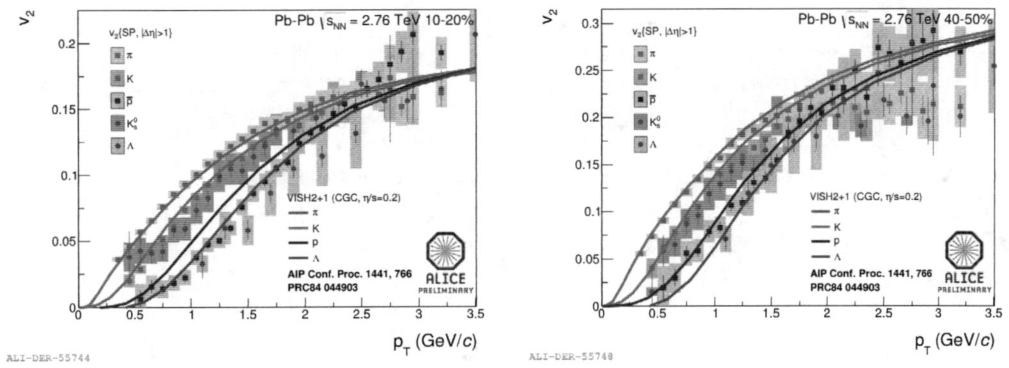

Figure 1. Identified particle $v_2(p_T)$ measured by ALICE for 10–20% (left) and 40–50% (right) centrality classes, together with viscous hydrodynamic model calculations [6]. The bars and shadows present the statistical and systematical uncertainties, respectively.

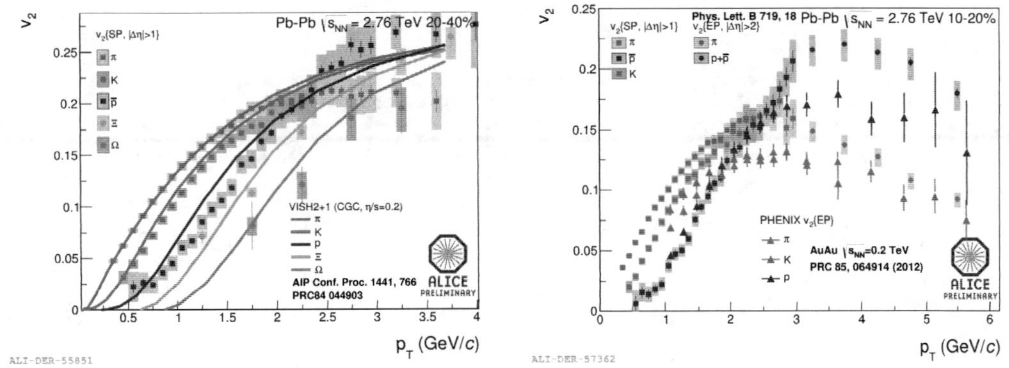

Figure 2. Ξ and Ω $v_2(p_T)$ measured by ALICE in 20–40 % centrality class (left) compared to hydrodynamic calculations [6]; pion, kaon, proton $v_2(p_T)$ measured by ALICE compared to PHENIX [9] in 10–20% centrality (right).

Figure 1 shows the elliptic flow for π, K, p($\overline{p}$), K^0_S, $\Lambda(\overline{\Lambda})$ as a function of p_T in different centrality classes along with hydrodynamical calculations [6]. A clear mass splitting is seen in the presented p_T region, more pronounced in central collisions, indicating a stronger radial flow built there. Viscous hydrodynamic model calculations (VISH2+1), using Color Glass Condensate initial condition with $\eta/s = 0.20$, quantitatively reproduce the mass splitting of v_2 for most of the particles, while a better agreement is observed in peripheral collisions. In central collisions, the hydrodynamic calculations overestimate the proton v_2, indicating that proton might freeze-out later (with larger radial flow) compared to pion and kaon. It also suggests an important role of hadronic interactions in reproducing proton v_2. A similar mass splitting can be observed for the v_2 of Ξ and Ω shown in Fig. 2 (left).

Figure 2 (right) presents a comparison of the p_{T}-differential v_2 measured in Au–Au collisions at $\sqrt{s_{\mathrm{NN}}} = 200$ GeV from PHENIX [9] and ALICE for different particles species. It is seen that the v_2 of π and K at the LHC is slightly higher compared to RHIC. However, the v_2 of proton is lower at low p_{T} but higher at high p_{T} compared to the one at RHIC, reflecting the effect of larger radial flow produced at LHC, as expected from hydrodynamical model predictions [6].

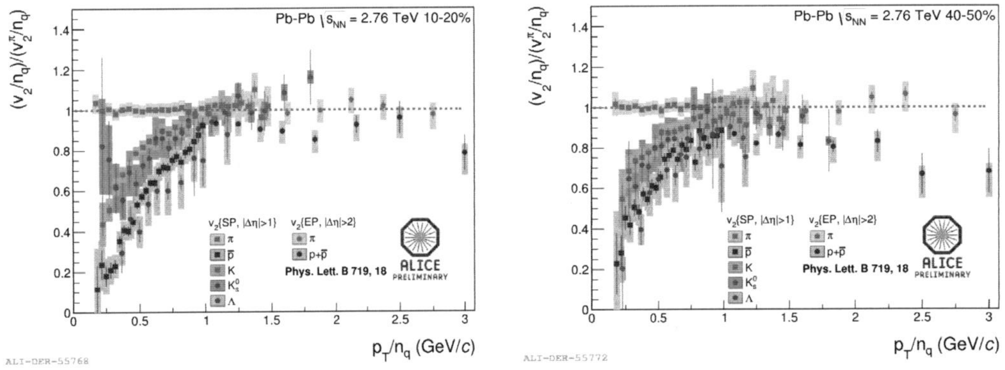

Figure 3. Ratio of identified particle v_2/n_q and v_2^{π}/n_q as a function of p_{T}/n_q in 10–20% (left) and 40–50% (right) centrality classes.

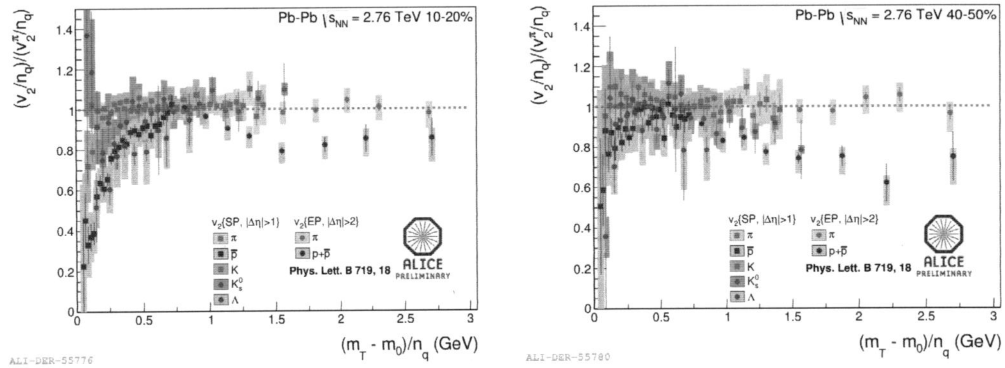

Figure 4. Ratio of identified particle v_2/n_q and v_2^{π}/n_q as a function of KE_{T}/n_q in 10–20% (left) and 40–50% (right) centrality classes.

In addition, the scaling with the number of constituent quarks observed at RHIC was used to support the picture that the collectivity already develops at the partonic level in heavy-ion collisions [4,5]. It can also serve as a test of hadron production via quark coalescence mechanism. Figures 3 and 4 show the ratio of identified particles v_2/n_q and π v_2^{π}/n_q as a function of the transverse momentum and transverse kinetic energy ($KE_{\mathrm{T}} = \sqrt{m_0^2 + p_{\mathrm{T}}^2} - m_0$)

per n_q, respectively. It appears that the scaling with n_q works only approximately at the LHC energy, within 20% at $p_T/n_q \sim 1.2$ GeV/c. For the KE_T scaling, the v_2/n_q of proton is clearly lower than that of pion for KE_T/n_q above 1 GeV/c. The scaling is violated at LHC, in contrast to the observation at top RHIC energy.

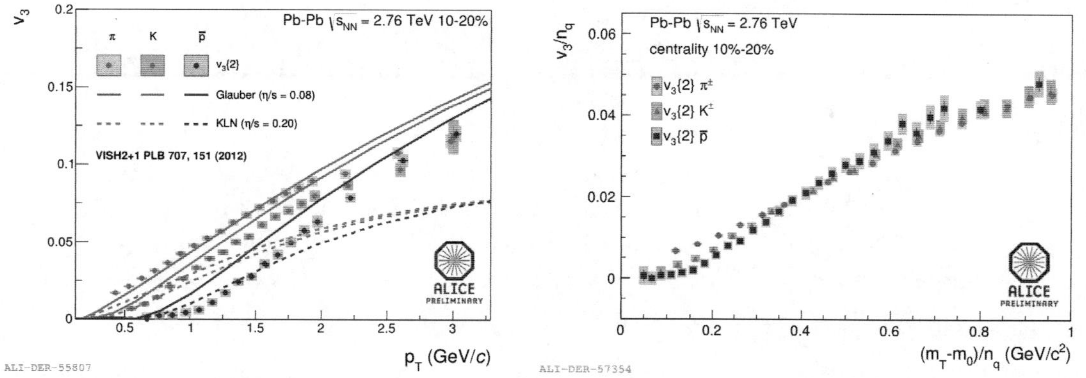

Figure 5. Identified particle $v_3(p_T)$ (left) and the KE_T/n_q scaling in 10–20% centrality.

Figure 5 (left) presents the triangular flow v_3 of identified particle as a function of transverse momentum in 10–20% centrality. The v_3 exhibits similar features as the v_2: a clear mass splitting is observed at low p_T as predicted from hydrodynamic calculations while the protons cross pions at intermediate p_T as expected from the quark coalescence mechanism. Hydrodynamic calculations can not quantitatively reproduce the v_3 results, neither using CGC initial condition with $\eta/s = 0.20$ nor with Glauber initial condition with $\eta/s = 0.08$. The KE_T scaling for v_3 works better than for v_2, but it is still only approximate as can be seen in Fig. 5 (right).

4. Summary

We review the anisotropic flow measurements of identified particles in Pb–Pb collisions at $\sqrt{s_{NN}} = 2.76$ TeV with the ALICE detector. A good agreement with hydrodynamic model calculations is found for most of the particles. Compared to the measurements at top RHIC energy, the results suggest a larger radial flow at LHC, as predicted by hydrodynamic calculations. The scaling with the number of constituent quarks in only approximate within 20% at $p_T/n_q \sim 1.2$ GeV/c. The KE_T/n_q scaling is broken over the all p_T range at the LHC energy. The v_3 shows similar features to v_2: a mass splitting at low p_T, while protons cross pions at intermediate p_T. The KE_T/n_q scaling works better for v_3 but it is still only approximate.

5. References

[1] Ollitrault J 1992 *Phys. Rev.* C **46** 229
[2] Voloshin S, Poskanzer A and Snellings R *Preprint* arXiv:0809.2949
[3] Alt C *et al.* [NA49 Collaboration] 2003 *Phys. Rev.* C **68** 034903
[4] Adams J *et al.* [STAR Collaboration] 2004 *Phys. Rev. Lett.* **92** 052302
[5] Adare A *et al.* [PHENIX Collaboration] 2007 *Phys. Rev. Lett.* **98** 162301
[6] Heinz H, Shen C and Song H 2012 *AIP Conf. Proc.* **1441** 766
[7] Adler C *et al.* [STAR Collaboration] 2002 *Phys. Rev.* C **66** 034904
[8] Poskanzer A and Voloshin S 1998 *Phys. Rev.* C **58** 1671
[9] Adare A *et al.* [PHENIX Collaboration] 2012 *Phys. Rev.* C **85** 064914

14th International Conference on Strangeness in Quark Matter (SQM2013) IOP Publishing
Journal of Physics: Conference Series **509** (2014) 012030 doi:10.1088/1742-6596/509/1/012030

Recent ATLAS results on flow measurements in lead-lead and proton-lead collisions

Adam Trzupek on behalf of the ATLAS Collaboration
The H. Niewodniczański Institute of Nuclear Physics Polish Academy of Sciences, ul. Radzikowskiego 152, 31-342 Kraków, Poland

E-mail: Adam.Trzupek@ifj.edu.pl

Abstract. Recent measurements from the ATLAS experiment of the azimuthal anisotropy of charged hadron production in the relativistic p+Pb and Pb+Pb collisions at the Large Hadron Collider (LHC) are presented. We report on distributions of event-by-event flow harmonics v_n, $n = 2-4$, for Pb+Pb collisions at energy of $\sqrt{s_{NN}}$=2.76 TeV which provide a direct measure of flow harmonic fluctuations which may be related to fluctuations in the initial geometry of the interaction region. The relative event-by-event elliptic flow fluctuations are compared to the measurement based on the cumulant approach as well as to the model predictions. We also report on measurements of the two-particle correlations in $\Delta\phi$ and $\Delta\eta$ as a function of p_T and the transverse energy (ΣE_T^{Pb}) summed over $3.1 < \eta < 4.9$ in the direction of the Pb beam in $\sqrt{s_{NN}}$ = 5.02 TeV p+Pb collisions. The recoil-corrected $\Delta\phi$-correlation exhibits flow-like modulations for all ΣE_T^{Pb} ranges and particle p_T. To study further the long-range correlations in p+Pb collisions, the elliptic flow has been measured with the cumulant approach and compared to the results from two-particle correlations. The presented p+Pb results exhibit features characteristic for collective anisotropic flow, similar to that observed in Pb+Pb collisions.

1. Introduction

The azimuthal anisotropy of hadron production is a key observable for understanding the properties of the hot and dense medium created in Pb+Pb collisions at the LHC. It is expected that this anisotropy is sensitive to conditions at the very early stage of evolution of the strongly coupled Quark-Gluon Plasma (sQGP) and is related to the spatial configuration of colliding nucleons as well as the energy density fluctuations in the initial overlap region of the two colliding nuclei [1]. The initial spatial asymmetry leads to asymmetric pressure gradients in the QGP, generating a significant azimuthal anisotropy in particle $dN/d\phi$ distributions which is usually described by means of a Fourier series with n^{th} order flow harmonic, $v_n = \langle \cos n(\phi - \Phi_n) \rangle$ where ϕ is the particle azimuthal angle and Φ_n represents the symmetry plane angle. The second order harmonic, called elliptic flow (v_2) characterizes the "elliptical" shape of the initial interaction region, while higher-order flow harmonics (v_3, v_4,...) characterize more complicated configurations.

In ATLAS [2], flow phenomena are explored with charged particles reconstructed in each event within the inner detector consisting of a silicon pixel detector and a semiconductor microstrip tracker (SCT), immersed in a 2 T axial magnetic field and covering a wide pseudorapidity range ($|\eta| <2.5$). The transverse momenta of reconstructed particles are limited by a minimum p_T of 0.5 GeV and 0.3 GeV in case of Pb+Pb and p+Pb collisions, respectively. For measurements,

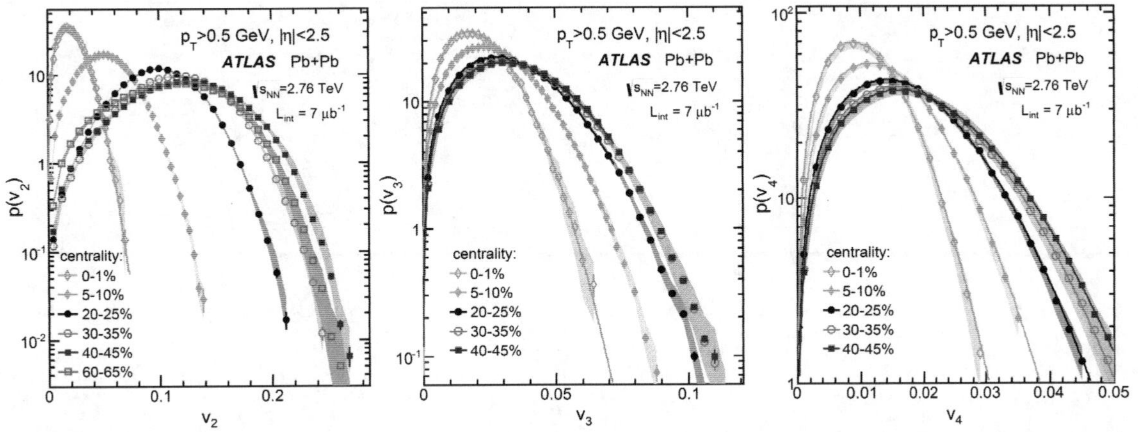

Figure 1. The event-by-event v_n distributions in selected centrality bins for $n = 2$ (left panel), $n = 3$ (middle panel) and $n = 4$ (right panel) [5]. The solid curves represent radial projections of 2D Gaussian functions with the mean rescaled to the measured $\langle v_n \rangle$.

presented in this report, a minimum bias sample of ~50M Pb+Pb collisions at an energy of $\sqrt{s_{NN}}$=2.76 TeV as well as ~2M p+Pb collisions at an energy of $\sqrt{s_{NN}}$=5.02 TeV are used.

2. Event-by-event flow harmonic distributions

A detailed analysis of flow harmonics averaged over large Pb+Pb event samples has been recently performed in the ATLAS experiment [3, 4]. Significant values of the higher order harmonics, $v_3 - v_6$, imply presence of large event-by-event flow vector fluctuations. Benefiting from the large acceptance of the ATLAS detector for measurement of charged particles, the distribution of $v_2 - v_4$ measured event-by-event were also recently obtained [5]. In this analysis, the azimuthal distribution of charged particles in each event is expanded into a Fourier series with coefficients corresponding to the single event flow vector components. However, due to finite event multiplicity, indicated by M, the absolute value of the flow vector (corresponding to the true v_n only in the limit $M \to \infty$) is smeared randomly around the true v_n. To correct for this smearing, the Bayesian unfolding procedure was applied [5]. The unfolded distributions represent the distributions of single event true flow harmonics and provide a direct measure of flow harmonic fluctuations. The unfolded distributions of v_2, v_3 and v_4, normalized to unity, are shown in Fig. 1 together with solid lines representing the radial projections of two-dimensional (2D) Gaussian functions with the mean adjusted to $\langle v_n \rangle$ from the data. Figure 1 shows that the distributions of higher order harmonics, v_3 and v_4, are consistent with the 2D Gaussian limit within the full measured centrality range. For the elliptic flow (n=2) the relative fluctuations, $\sigma_{v_2}/\langle v_2 \rangle$, where σ_{v_2} is the standard deviation, are presented in Fig. 2 as a function of centrality for three p_T ranges: $p_T > 0.5$ GeV, 0.5< p_T <1 GeV and p_T >1 GeV. It is observed that elliptic flow fluctuations strongly depend on centrality with the smallest fluctuations found in mid-central collisions ($N_{part} \approx 200$). The largest v_2 fluctuations are observed in the 2% most central collisions which are also consistent with the purely 2D Gaussian fluctuations. The relative fluctuations are nearly the same for $p_T > 1$ GeV and $0.5 < p_T < 1$ GeV.

The magnitude of event-by-event fluctuations of flow harmonics can also be estimated using the two- and four-particle cumulant method [8]. Assuming that non-flow effects and σ_{v_n} are small as compared to $\langle v_n \rangle$, then $\sigma_{v_n}/\langle v_n \rangle \approx \sqrt{(v_n\{2\}^2 - v_n\{4\}^2)/(v_n\{2\}^2 + v_n\{4\}^2)}$. The relative fluctuations from the cumulant method of the elliptic flow are shown in Fig. 3 as a function of p_T in different centrality intervals [9]. For the most central collisions (5-10%), $\sigma_{v_n}/\langle v_n \rangle$ is independent of p_T and for less central collisions increases with p_T. The relative v_2 fluctuations,

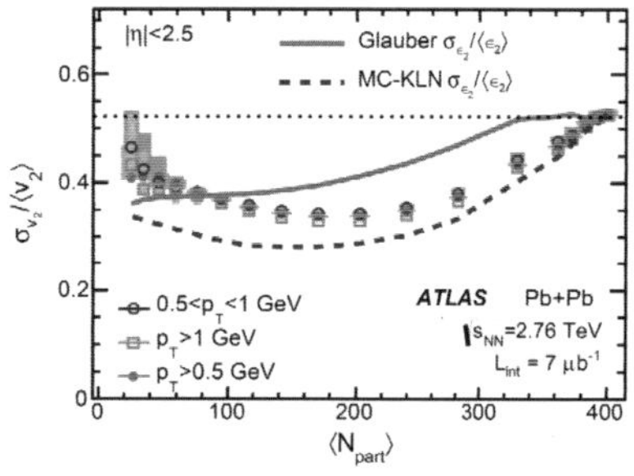

Figure 2. Centrality dependence of σ_{v2}/v_2 compared to model predictions [6, 7] and 2D Gaussian fluctuations.

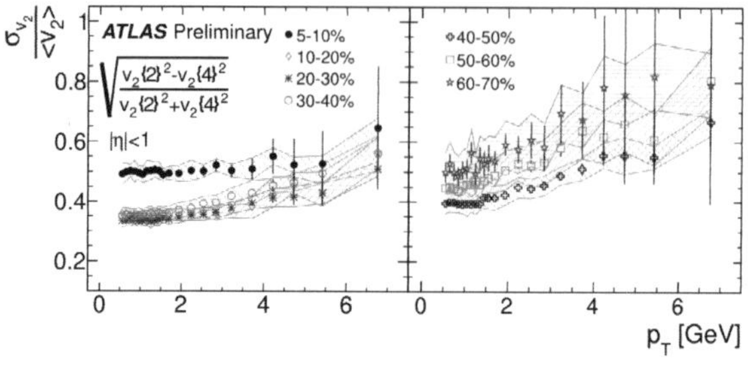

Figure 3. The p_T dependence of the relative elliptic flow fluctuations in seven centrality bins of Pb+Pb collisions from the cumulant method [9].

for $0.5 < p_T \lesssim 2$ GeV, extracted for mid-central collisions (20-30%) are at ~0.35, which is similar to that measured with event-by-event method for a corresponding centrality of N_{part} ≈200, shown in Fig. 2.

3. Azimuthal anisotropy in p+Pb collisions

An important tool to probe the collective phenomena in heavy ion collisions is the two-particle correlation function measured as a function of relative pseudorapidity ($\Delta\eta$) and azimuthal angle ($\Delta\phi$) of particle pairs. Recently, a two-particle correlation (2PC) function was obtained in ATLAS [10] for p+Pb collisions in different centrality intervals measured by the transverse energy ΣE_T^{Pb}. The 2D correlation functions for charged particles in peripheral and central collisions are shown in Fig. 4. For peripheral collisions the correlation function shows a sharp peak centered at ($\Delta\phi$,$\Delta\eta$) = (0, 0) and a broad (in $\Delta\eta$) structure at $\Delta\phi \approx \pi$ (called recoil) both predominantly originating from non-flow effects. In central collisions, in addition to the components observed in peripheral collisions, the correlation function reveals a broad (in $\Delta\eta$) structure at $\Delta\phi \approx 0$ (the "near-side ridge"). The distribution at $\Delta\phi \approx \pi$ is also broadened relative to peripheral collisions, consistent with the presence of a long-range component (the "away-side ridge").

The strength of the long-range component is commonly quantified by the "per-trigger yield", Y($\Delta\phi$), which measures the yield of particle pairs per the yield of trigger particles ($1/N_{trig}dN^{pair}/d\Delta\phi$) [11]. Figure 5 shows the Y($\Delta\phi$) distributions for $2 < |\Delta\eta| < 5$ in peripheral and central collisions as well as their difference (ΔY, solid points) which is symmetric around $\Delta\phi = \pi/2$ and consistent with flow-like modulations. The second order amplitude of these modulations, $v_2\{2PC\}$, depicted in Fig. 6 as a function of p_T is reminiscent of what is

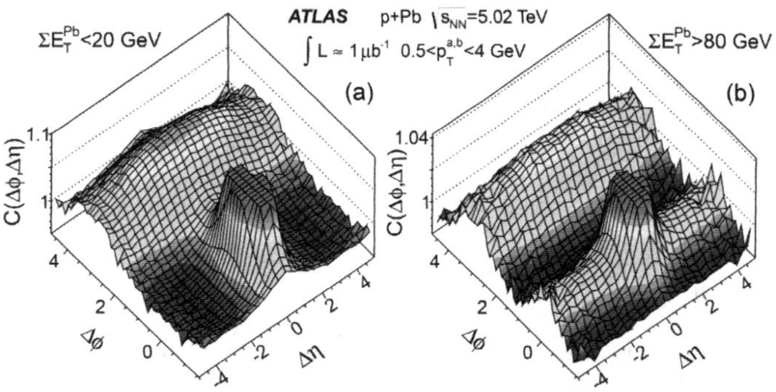

Figure 4. The two-particle correlation function in peripheral (a) and central (b) p+Pb collisions at energy of $\sqrt{s_{NN}} = 5.02$ TeV [10].

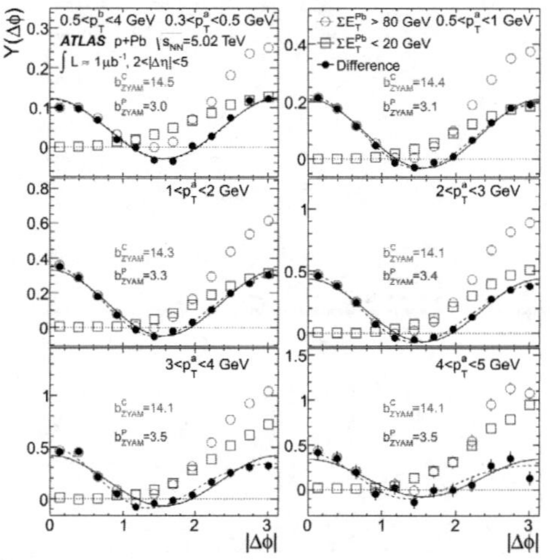

Figure 5. Per-trigger yield (PTY) for particle pairs with $2 < |\Delta\eta| < 5$ in peripheral (squares) and central (open circles) p+Pb collisions in six p_T intervals. Solid points represent the difference [10].

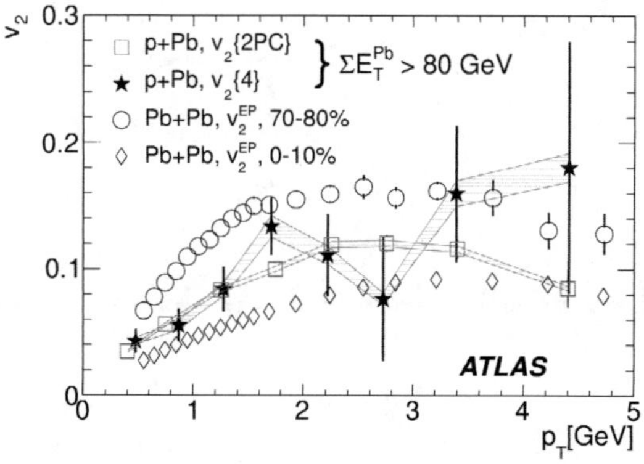

Figure 6. Comparison of the p_T dependence of the v_2 coefficient measured in p+Pb collisions with the four-particle cumulants $v_2\{4\}$ [12], and with the two-particle correlation method $v_2\{2PC\}$ [10] to v_2 obtained with the event-plane method for central and peripheral Pb+Pb collisions [3].

understood to be hydrodynamic flow in Pb+Pb collisions. To further study the collective flow in p+Pb collisions, the elliptic flow harmonics were obtained from cumulant method, $v_2\{4\}$ [12]. Figure 6 shows p_T dependence of $v_2\{4\}$ which is consistent with $v_2\{2PC\}$ and, interestingly, with a magnitude between the values of v_2 obtained with the event-plane method [3] in the most central and most peripheral centrality intervals measured for Pb+Pb collisions.

In summary, ATLAS has presented event-by-event v_2, v_3 and v_4 distributions in a wide centrality range of Pb+Pb collisions at the LHC energy of $\sqrt{s_{NN}} = 2.76$ TeV, which provide a direct insight into fluctuations in the initial geometry of the interaction region. The relative fluctuations of v_2 in the most central Pb+Pb collisions, and v_3 and v_4 within the full, measured centrality range are consistent with radially-projected 2D Gaussian distributions. In mid-central collisions, the relative fluctuations of v_2 are significantly smaller than the Gaussian limit. In $\sqrt{s_{NN}} = 5.02$ TeV p+Pb collisions, the two-particle correlation function clearly shows ridge structures resembling those observed in Pb+Pb collisions and suggesting that collective flow may also be present in p+Pb collisions. The flow interpretation of the p+Pb data is also supported by results from multi-particle azimuthal correlation measurements.

This work was supported by National Science Centre grant number 2011/01/N/ST2/ 04042.

References

[1] Ollitrault J-Y 1992 *Phys. Rev. D* **46** 229.
[2] ATLAS Collaboration 2008 *JINST* **3** S08003.
[3] ATLAS Collaboration 2012 *Phys. Lett.* **B707** 330-348.
[4] ATLAS Collaboration 2012 *Phys. Rev.* **C86** 014907.
[5] ATLAS Collaboration 2013 *JHEP* **11** 183.
[6] Miller M L, Reygers K, Sanders S J and Steinberg P 2007 *Ann. Rev. Nucl. Part. Sci.* **57** 205.
[7] Drescher H J and Nara Y 2007 *Phys. Rev.* **C75** 034905.
[8] Borghini N, Dinh P M, Ollitrault J-Y 2001 *Phys. Rev.* **C63** 054906;*Phys. Rev. C64* 054901.
[9] ATLAS Collaboration 2012 *ATLAS-CONF-2012-117* https://cdsweb.cern.ch/record/1472939.
[10] ATLAS Collaboration 2013 *Phys. Rev. Lett.* **110** 182302.
[11] PHOBOS Collaboration 2010 *Phys. Rev. Lett.* **104** 062301.
[12] ATLAS Collaboration 2013 *Phys. Lett.* **B725** 60.

Effects of Jets in the Flow Observables

Jun Takahashi, Rafael Derradi de Souza and David D. Chinellato

Instituto de Física Gleb Wataghin, Universidade Estadual de Campinas, Rua Sérgio Buarque de Holanda 777, 13083-859 São Paulo, Brazil

E-mail: rderradi@ifi.unicamp.br

Abstract. The transverse momentum anisotropy of the particles produced in heavy ion collisions is one of the most important experimental observable to investigate the collective behavior of the systems created in such collisions. Recent studies show that the complex nature of the system evolution, such as initial condition fluctuations and jets, may lead to important effects in the flow coefficients and, therefore, to misinterpretation of the results obtained. In this study, we used simulated events produced with a hydrodynamic model which allows inhomogeneous initial condition combined with proton-proton collisions produced with the Pythia event generator to create a final set of particles to be analyzed with the usual experimental flow calculation techniques. Although this simplified approach is somehow unrealistic, since it does not include the interaction of the jet with the medium, our results have shown a good agreement of the behavior of the elliptic flow coefficient as a function of the transverse momentum up to 6 GeV/c for Au+Au collisions at 200 GeV. Although each model alone is not able to describe the full range, the combination of both sets of particles as seen by the flow calculation techniques may be the key to explain the behavior observed in experimental data.

1. Introduction

One of the main observables related to the collective behavior of the system created in relativistic heavy ion collisions is the transverse momentum anisotropy of the produced particles. Ideally, one would measure the azimuthal distribution of particles with respect to the reaction plane directly, but this is not experimentally feasible and there is a need for techniques to compensate for this missing quantity based on azimuthal angle correlations of the particle momenta. In this context, recent studies have demonstrated that the system evolution may contain interesting aspects such as initial condition inhomogeneities and jets of particles originating from hard parton-parton interactions. These aspects could have important adverse effects on the flow coefficients and thus merit further investigation.

In this study, we combine two decorrelated components, one hydrodynamic-like and another jet-like part, such that the resulting event may be analyzed with common experimental calculation techniques, allowing us to explore any effects that may arise due to the inclusion of jets.

2. NeXSPheRIO + Pythia

In order to construct a simplified two component model, we combine simulated events produced with NeXSPheRIO and Pythia event generators. The NeXSPheRIO code [1, 2] is based on ideal hydrodynamics and allows for the simulation of nucleus-nucleus collisions, while Pythia is an event generator for proton-proton (pp) collisions that takes into account the leading order processes in perturbative QCD [3]. We generated events of Au+Au collisions at $\sqrt{s_{NN}} = 200$ GeV using the NeXSPheRIO code and embedded into each event an integer number α of p+p collisions generated using the Pythia code. The number of Pythia events to be included was chosen so that the resulting sum of the Pythia and NeXSPheRIO charged pion spectra according to p_t matched the experimentally measured $\pi^{\pm}$ spectra from STAR [4]. Also, we tested different Pythia configurations, namely NSD (non-single diffractive events) and QCD Jets (which prioritizes 2→2 processes), as well as different cuts on the minimum transverse momentum of the main scattering process, referred here by CKIN(3) (see reference [3] for details). Figure 1 shows the results for the 10-20% centrality class for the various event generator configurations used. Experimental points from STAR [4] are shown for comparison. The Monte Carlo configuration that best describes the spectra of charged pions was given by Pythia QCDJets with CKIN(3) = 3.5 GeV/c, for which α ranged from ∼1 to ∼9, being higher for more central collisions. It is important to note that this procedure does not take into account any type of interaction of the jet with the medium. The model still overestimates the intermediate p_t region (∼ 5 GeV/c)

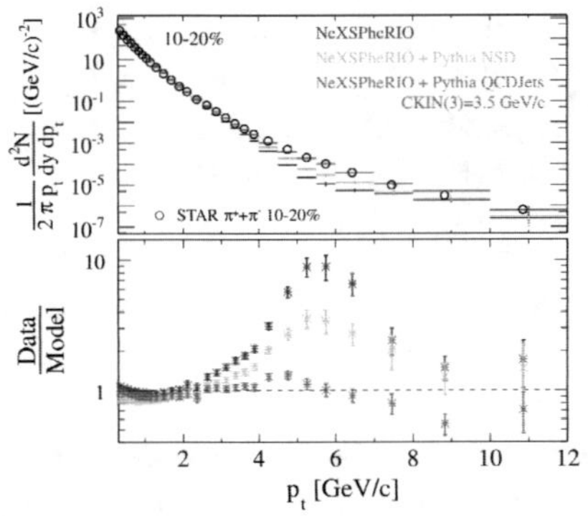

Figure 1. Invariant p_t spectra of charged pions for NeXSPheRIO+Pythia model compared to experimental results from STAR [4].

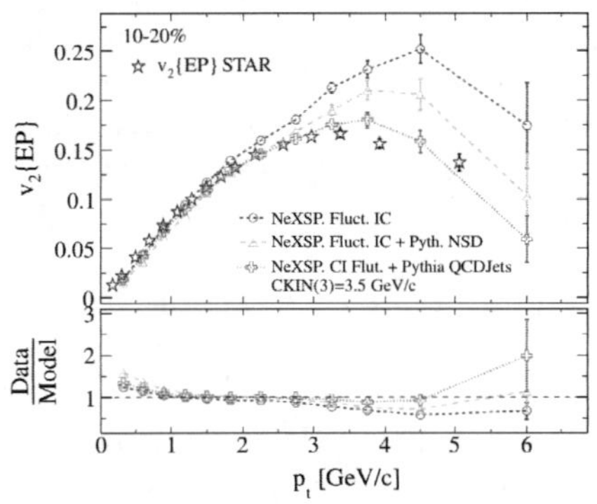

Figure 2. Differential v_2 from event plane method as a function of p_t obtained with NeXSPheRIO+Pythia model compared to experimental results from STAR [5].

and underestimates the very high p_t region. This behavior is consistent with the fact that we do not include energy loss of the jet particles to the medium, which would transfer a fraction of the momentum of these very energetic particles to lower momentum particles, thus populating the intermediate region and depleting the high p_t spectrum.

3. Effects on Flow Observables

Using a sample of simulated events generated with the NeXSPheRIO+Pythia combination, we calculated the elliptic flow of the final particles produced. The results are shown in figure 2 as a function of p_t for the centrality class 10-20%. Experimental points from STAR [5] are also shown for comparison. As can be seen, pure NeXSPheRIO particles are not able to describe the experimental results for p_t above 2.0 GeV/c. The inclusion of Pythia events with large transferred momentum decreases the values of v_2 for high p_t.

In order to explore the effects produced by the superposition of NeXSPheRIO and Pythia events, we also calculated the difference between the squares of v_2 estimates of second and fourth orders obtained with the cumulant method, namely $v_2\{2\}^2 - v_2\{4\}^2$. This quantity has been shown to be proportional to the flow fluctuation (σ_{v_2}) and the non-flow contribution

(δ) [6], the latter being mainly dominated by jets. The results obtained as a function of average number of charged particles produced at $\eta = 0$ are shown in figure 3. To illustrate

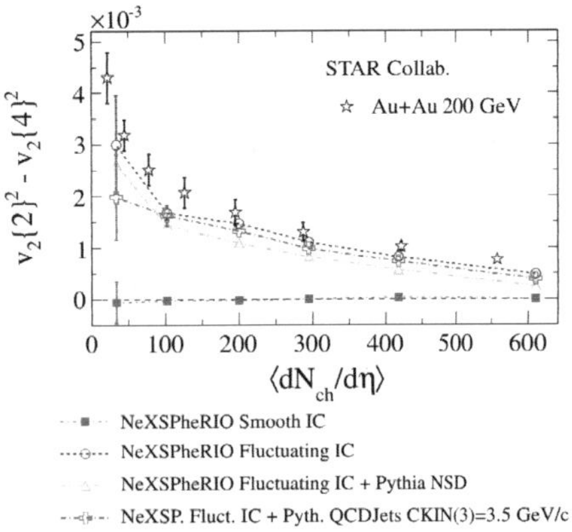

Figure 3. Difference between $v_2\{2\}^2$ and $v_2\{4\}^2$ as a function of centrality for different configurations of NeXSPheRIO+Pythia model. Experimental results from STAR [7] are shown for comparison.

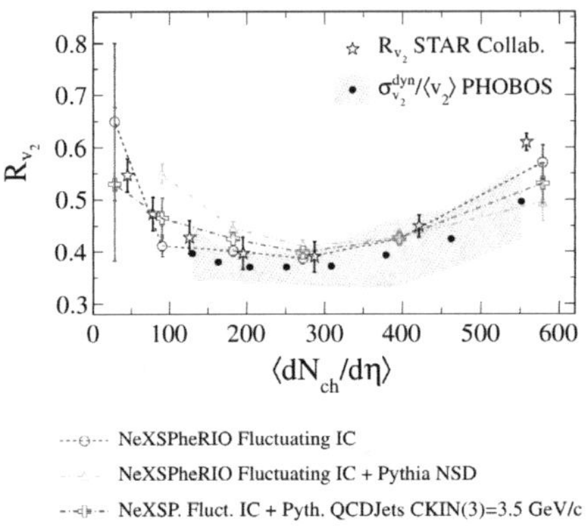

Figure 4. R_v parameter as a function of centrality for different configurations of NeXSPheRIO+Pythia model. Experimental results from STAR [7] and PHOBOS [8] are shown for comparisons.

the sensitivity of $v_2\{2\}^2 - v_2\{4\}^2$ to the fluctuations and non-flow contribution we also included the results obtained for NeXSPheRIO events generated with smoothed initial conditions, which produce values consistent with zero. However, contrary to what we expected, the addition of Pythia events did not increase the values of these parameters, suggesting that the non-flow contribution may be negligible for the integrated flow estimates. In figure 4 we show the results obtained for the R_{v_2} parameter, defined as the square-root of $(v_2\{2\}^2 - v_2\{4\}^2)/(v_2\{2\}^2 + v_2\{4\}^2)$. Again, the inclusion of Pythia particles in the NeXSPheRIO events does not seem to produce a relevant change in the behavior of R_{v_2}. For comparison, experimental results from STAR [7] and PHOBOS [8] are also shown.

4. 2-Particle Correlation

Another interesting result from particle correlations in heavy ion collision experiments is the observation of the ridge [9, 10]. Initially thought to be an effect of jets traversing the medium created in the collision, it was shown later that it could also be described within a hydrodynamic treatment considering fluctuating initial conditions [11]. However, both mechanisms may be present, and thus it is of interest to look into how the inclusion of Pythia affects the shape of the $\Delta\eta\Delta\phi$ distribuition. We define a two-particle correlation function in the same way as presented in reference [12], $C(\Delta\eta, \Delta\phi) = S(\Delta\eta, \Delta\phi)/B(\Delta\eta, \Delta\phi)$, where S refers to the same-event pair distribution and B is the combinatorial pair distribution obtained by mixing particles a and b from different events. The results obtained using pure NeXSPheRIO events (left plot) and NeXSPheRIO+Pythia events (right plot) are shown in figure 5. The correlation was calculated using charged particles with $0.5 < p_t^a < 1.5$ GeV/c (associated) and $p_t^b > 2.0$ GeV/c (trigger). In order to quantify the effects of including decorrelated Pythia events we performed a fit over slices of $\Delta\eta$ using the sum of the first five harmonics in $dN_{pairs}/d\Delta\phi \propto 1 + 2\sum V_{n,n}\cos(n\Delta\phi)$. The results are shown in figure 6 for pure NeXSPheRIO and for NeXSPheRIO+Pythia. Although Pythia particles are highly correlated within each Pythia event, the incoherent addition of a number of Pythia events to a NeXSPheRIO event drives the coefficients of the harmonic decomposition down.

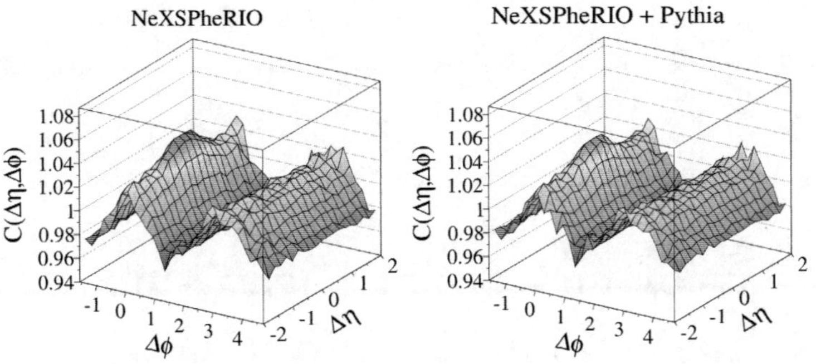

Figure 5. Two-particle correlation distribuition obtained from NeXSPheRIO with fluctuating initial condition (left plot) and from NeXSPheRIO+Pythia for QCDJets with CKIN(3)=3.5 GeV/c configuration (right plot).

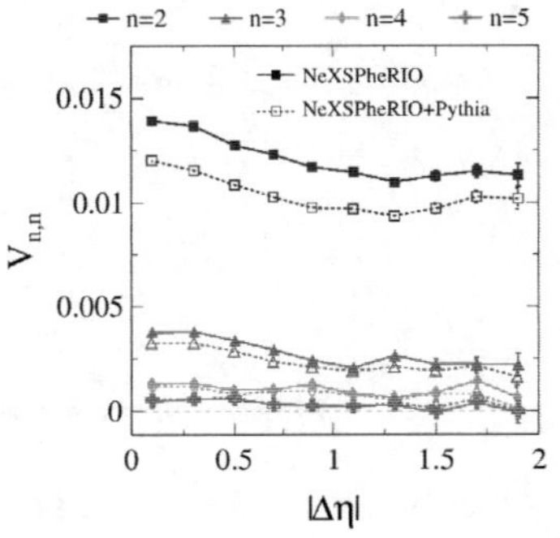

Figure 6. Results of fits over slices of the two-particle distributions for NeXSPheRIO and NeXSPheRIO+Pythia.

5. Summary and Conclusions

In this work, we have investigated a model in which a number of Pythia events is added to a single NeXSPheRIO Au+Au event and have shown that, by using a number of Pythia events so that the resulting spectra match experimental data, a better description of flow variables is also obtained. In addition, to quantify the non-flow effects over particle correlation analysis, we evaluated the difference $v_2\{2\}^2 - v_2\{4\}^2$, for integrated flow as a function of centrality as well as the harmonic decomposition of two-particle correlation distribution for a specific choice of p_t range for associated and trigger particles. The results obtained suggest that the effects over integrated quantities are very small and the observed behavior is mainly due to fluctuations in the initial condition. On the other hand, when we focus on the intermediate to high p_t region, the effects start to appear. However, it actually lowers the correlation observed, indicating that the fact that the extra added particles are added in an uncorrelated way has a stronger effect on the observables than the correlation among the few added particles. In conclusion, it is interesting to note that experimental results of v_2 are better described for higher values of p_t by adding independent jet-like proton-proton collisions. It is clear that a more detailed investigation using a similar approach may be important to understand experimental observations.

Acknowledgments

This work was supported in part by FAPESP under Project No. 2012/02895-2 and by CNPq of Brazil. The authors wish to thank Dr. Takeshi Kodama for fruitful discussions.

References

[1] Hama Y, Kodama T and Socolowski O J 2005 *Braz.J.Phys.* **35** 24–51 (*Preprint* hep-ph/0407264)
[2] Drescher H J, Liu F M, Ostapchenko S, Pierog T and Werner K 2002 *Phys. Rev. C* **65**(5) 054902
[3] Sjostrand T, Mrenna S and Skands P 2006 *JHEP* **0605** 026
[4] Abelev B *et al.* (STAR Collaboration) 2006 *Phys.Rev.Lett.* **97** 152301 (*Preprint* nucl-ex/0606003)
[5] Adams J *et al.* (STAR Collaboration) 2005 *Phys.Rev.* **C72** 014904 (*Preprint* nucl-ex/0409033)
[6] Ollitrault J Y, Poskanzer A M and Voloshin S A 2009 *Phys.Rev.* **C80** 014904 (*Preprint* 0904.2315)
[7] Agakishiev G *et al.* (STAR Collaboration) 2012 *Phys.Rev.* **C86** 014904 (*Preprint* 1111.5637)
[8] Alver B *et al.* (PHOBOS Collaboration) 2010 *Phys.Rev.Lett.* **104** 142301 (*Preprint* nucl-ex/0702036)
[9] Putschke J 2007 *Nuclear Physics A* **783** 507 – 510 ISSN 0375-9474
[10] McCumber M P (PHENIX Collaboration) 2008 *J.Phys.* **G35** 104081 (*Preprint* 0804.4319)
[11] Takahashi J *et al.* 2009 *Phys.Rev.Lett.* **103** 242301 (*Preprint* 0902.4870)
[12] Aad G *et al.* (ATLAS Collaboration) 2012 *Phys.Rev.* **C86** 014907 (*Preprint* 1203.3087)

Landau hydrodynamics for non-central heavy-ion collisions and longitudinal scaling of elliptic flow

Karolis Tamosiunas

Vilnius University, Institute of Theoretical Physics and Astronomy, Goštauto g. 12, Vilnius, 01108

E-mail: karolis.tamosiunas@tfai.vu.lt

Abstract. Landau solution of hydrodynamics is generalized for the non-central high-energy nuclear collisions. The multiparticle production after hydrodynamic expansion from the transversely asymmetric initial state shows elliptic flow formation. Moreover, obtained solution does reproduce the observed longitudinal scaling of the elliptic flow for the different collision energies at LHC and RHIC. It will be argued, that the analyticity of the solution allows us to to fit experimental data in order to obtain initial conditions.

1. Introduction

The question of particle multiplicity distribution in High-Energy Nuclear Collisions has a long-lasting history. In 1950 Fermi suggests to apply thermodynamics to particle production in high-energy collisions. In 1953 Landau improves the idea with the assumption, that system strongly interacts and particles are produced after the ideal hydrodynamic expansion. And now, this approximate and analytic solution of relativistic hydrodynamics works well not only for multiplicity distribution at LHC and RHIC energies, but with some modifications, it reproduces the experimentally observed longitudinal scaling of elliptic flow.

2. Solution of Ideal Hydrodynamics for Particle Production

The solution is inspired by the Landau hydrodynamic model of multi-particle production [1, 2] and [3], where the hydrodynamic equations for the energy-momentum conservation,

$$\partial_\mu T^{\mu\nu} = 0, \tag{1}$$

are solved by separating longitudinal and transverse expansions. It is realistic assumption, as the longitudinal expansion is much faster than transverse and starts with much thinner longitudinal profile. The equation of state of ideal relativistic gas, $P = e/3$, and the energy-momentum tensor in usual form,

$$T^{\mu\nu} = (e + P)u^\mu u^\nu - Pg^{\mu\nu}, \tag{2}$$

closes the solvable system of equations.

2.1. Longitudinal expansion

The equations of the hydrodynamic longitudinal expansion in 1+1 dimension, along z axis reads as:

$$\frac{\partial T^{00}}{\partial t} + \frac{\partial T^{0z}}{\partial z} = 0, \quad \frac{\partial T^{0z}}{\partial t} + \frac{\partial T^{zz}}{\partial z} = 0. \tag{3}$$

Solution of the equations of hydrodynamics starts by transforming relativistic velocity field components to rapidity terms, as: $u^0 = \cosh y$, $u^z = \sinh y$, and the final solution for energy density, $e(y_+, y_-)$, and rapidity, $y(y_+, y_-)$, reads as [4]:

$$e(y_+, y_-) = e_0 \exp[-4/3(y_+ + y_- - \sqrt{y_+ y_-})], \tag{4}$$

$$y(y_+, y_-) = (y_+ - y_-)/2, \tag{5}$$

while $z = t \tanh y$. The above solution of 1+1-dimensional relativistic hydrodynamics equation is connected to the solution of transverse expansion, in order to obtain multiplicities of produced particles for different rapidities.

2.2. Transverse expansion

The transverse expansion of hydrodynamic eq. 1 in polar coordinates reads as:

$$\frac{\partial T^{0r}}{\partial t} + \frac{\partial T^{rr}}{\partial r} = 0\ . \tag{6}$$

Inserting energy-momentum tensor expressions to the above equation and using ideal gas equation of state, $P = e/3$, one gets:

$$4e(u^0)^2 \frac{\partial v_r}{\partial t} + 4e(u^0)^2 \frac{\partial v_r^2}{\partial r} + \frac{\partial e}{\partial r} = 0\ . \tag{7}$$

$$t_{FO} = 2 \cosh y \sqrt{\frac{2aR_\phi}{(1 - f(R_\phi))}}, \tag{8}$$

where new function, $f(R_\phi) = e(r = R_\phi)/e(r = 0)$, is introduced as a fraction of energy density at the edge of the system with respect to the energy density at the center. The transverse and longitudinal solutions are matched at the time $t = t_{FO}$, starting the freeze-out stage, where particles stream freely to the detectors. Knowing, that $dS = su^0 dz$ at a given time within element dz and entropy density, $s = ce^{3/4}$, we express entropy change over rapidity from the energy density formula (4), as:

$$\frac{dN}{dy} \propto \frac{dS}{dy} = ce_0^{3/4} \exp[-(y_+ + y_- - \sqrt{y_+ y_-})] \frac{t_{FO}}{\cosh y}. \tag{9}$$

The transformation back to the (t, z) coordinates is: $y_+ = \ln((t + z)/\Delta)$, $y_- = \ln((t - z)/\Delta)$, while Δ is the initial thickness of the system in the beam direction, z. Also, Δ is the initial condition after which equation of state assumed to be valid and evolution equations (3) are applied. Inserting the solution for the FO time equation (8) into the entropy equation above and assuming that the number of produced particles is directly proportional to the entropy, $dN \propto dS$, one can obtain the number of particles for different rapidities at a fixed angle ϕ. In order to find particle distribution $d^2N/dyd\phi$ one needs to find $f(R_\phi)$ and Δ, what corresponds to the initial conditions.

Journal of Physics: Conference Series **509** (2014) 012032 doi:10.1088/1742-6596/509/1/012032

2.3. Initial conditions

In order to show the validity of the obtained solution, simple and transparent initial conditions will be used. In this case the widely accepted and analytically simple Wounded Nucleon (WN) model [5] will be used to parametrize initial conditions. It is based on the Woods-Saxon nuclear density parametrization [6], as follows: $\rho_A(\mathbf{r}) = \frac{\rho_0}{1+\exp(\frac{r-R_A}{d})}$, which is continuous and can be connected to the Landau equations straightforwardly. The main requirement for the initial conditions and the new function $f(R_\phi)$ is that for the central collision case, $b = 0$, the result must be equal to the original Landau one. The density of wounded nucleons in the transverse

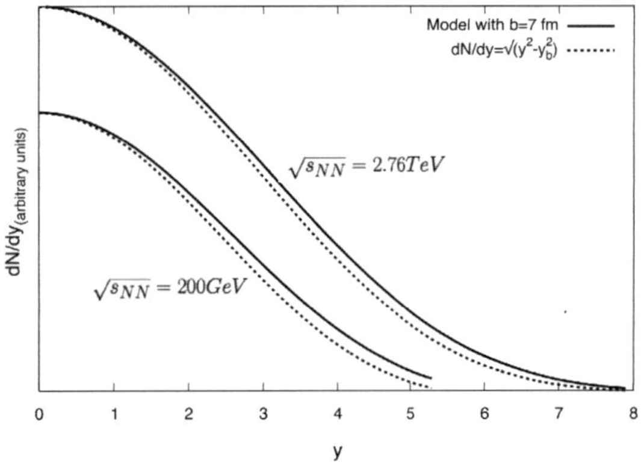

Figure 1. Rapidity distribution of produced particles in central collisions (full lines) and in peripheral collisions with impact parameter $b = 7fm$ (dashed line) for LHC and RHIC energies

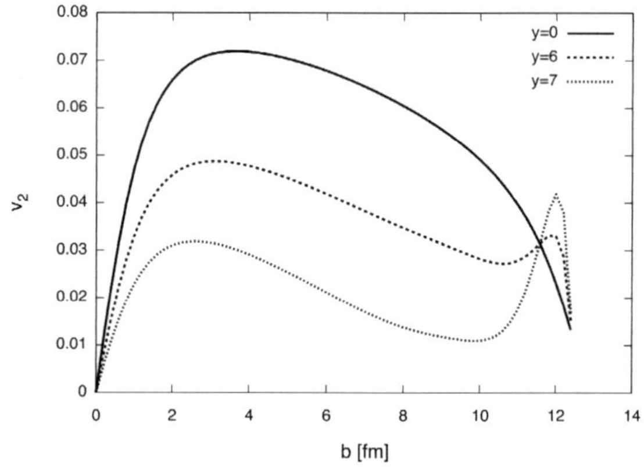

Figure 2. Elliptic flow, v_2, dependence on the impact parameter, b, for different rapidities.

plane and in polar coordinates, (r, ϕ), can be obtained by:

$$n_{WN}(r,\phi) = T_A(r,\phi)\left[1-\left(1-\frac{\sigma T_B(r,\phi)}{B}\right)^B\right] + T_B(r,\phi)\left[1-\left(1-\frac{\sigma T_A(r,\phi)}{A}\right)^A\right].$$

The thickness functions are expressed, in usual way: $T_A(r,\phi) = T_A(x - b/2, y) = \int dz\ \rho_A(\mathbf{r})$, using Woods-Saxon parametrization with $R_A = 1.12A^{1/3} - 0.86A^{-1/3}$ [fm], $d = 0.54$ [fm] and $n_0 = 0.17fm^{-3}$. Now assuming, that energy density is proportional to the Wounded Nucleon density: $e(r,\phi;b) \propto n_{WN}(r,\phi;b)$, the function $f(R_\phi)$ can be obtained. It is by definition, the ratio of energy density at the edge of the system with energy density at the center, for the fixed impact parameter b and reads as:

$$f(R_\phi) = \frac{n_{WN}(R_\phi,\phi;b) - \min(n_{WN}(R_\phi,\phi;b))}{n_{WN}(0,0;b)}. \tag{10}$$

The radius of the system, R_ϕ, is dependent on the angle ϕ and is obtained from the geometry on how two circles overlap, as: $R_\phi^2 + R_\phi b\cos\phi + \frac{b^2}{4} - R_A^2 = 0$. The term $\min(n_{WN}(R_\phi,\phi;b))$ is a minimal density at the edge of the system and is used in order to have the original Landau solution for $R_\phi = R_A$, so that $f(R_A) = 0$. In the case of Woods-Saxon, the density at the edge of the nuclei at $r = R_A$ is not zero, so the minimal value is subtracted. Now the acceleration term in (7) is the same in central collision and in Landau, but for peripheral collisions acceleration does depend on the angle ϕ. Finally, to calculate the elliptic flow one should merge equations (8), (9)

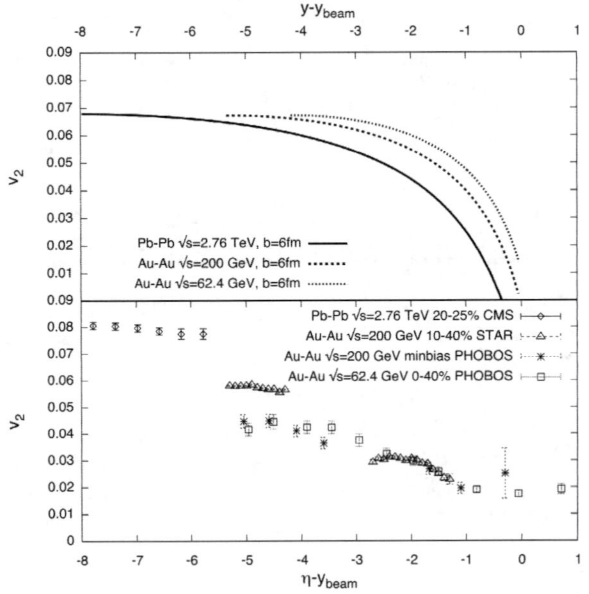

Figure 3. Elliptic flow, v_2, dependence on the rapidity, shifted by the beam rapidity, $y_{beam} = \ln(\sqrt{s_{NN}}/m_N)$, for: Pb-Pb at impact parameter $b = 6fm$ collisions at LHC energy $\sqrt{s_{NN}} = 2.76TeV$ (full-line), Au-Au at $b =$ 6fm at $\sqrt{s_{NN}} = 200GeV$ (dashed-line) and at $\sqrt{s_{NN}} = 62.4GeV$ (dotted-line) reactions used at RHIC. Data are from [7], [8] and [9]

and the definition of the elliptic flow: $v_2(y) = \frac{\int d\phi(dN/d\phi dy)\cos(2\phi)}{\int d\phi(dN/d\phi dy)}$, with the initial energy density gradient $f(R_\phi)$ and initial longitudinal thickness $\Delta(\phi)$. For the peripheral collisions initial thickness is expressed, as $\Delta(\phi) = \kappa_\phi R_A/\gamma$, where $\kappa_\phi = \sqrt{n_{WN}(R_\phi, \phi; b)/\max(n_{WN}(R_\phi, \phi; b))}$. The following means, that longitudinal expansion (3) starts with azimuthally asymmetric initial thickness, which is wider, where initial nuclear density is higher. The term $\max(n_{WN}(R_\phi, \phi; b))$ is used to have no effect of the modification for central collision case, $\kappa_\phi(b = 0) = 1$. Obtained results are shown in figure (3) in comparison with LHC and RHIC experiments.

3. Concussions

Presented analytical solution for the non-central heavy ion collisions reproduces experimentally observed elliptic flow formation and it's longitudinal scaling. Moreover, using experimental data of produced charged particles for different rapidities in Event-on-Event basis, $d^2N/d\eta d\phi$, it will be possible to find the initial energy density gradient function in transverse plane, $f(R_\phi)$, and the initial thickness, $\Delta(\phi)$.

Acknowledgments

Author acknowledges support from the Lithuanian Research Council, Ref. No: PRO-06/2012

References

[1] Landau L D 1953 *Izv. Akad. Nauk SSSR* **17** 51
[2] Belenkij S Z and Landau L D 1956 *Nuovo Cim. Suppl.* **3S10** 15
[3] Tamosiunas K 2011 *Eur. Phys. J.* A **47** 121 (*Preprint* `DOI 10.1140/epja/i2011-11121-5`)
[4] Wong C Y 2008 *Phys. Rev. C* **78** 054902 (*Preprint* `arXiv:0808.1294`)
[5] A Bialas M B and Czyz W 1976 *Nucl. Phys. B* **111** 461
[6] Woods R and Saxon D 1954 *Phys. Rev.* **95** 577
[7] CMS collaboration 2013 *Phys. Rev. C* **87** 014902 (*Preprint* `arXiv:1204.1409`)
[8] Abelev B I *et al.* 2008 *Phys. Rev. C* **77** 54901
[9] Alver B *et al.* (PHOBOS) 2007 *Phys. Rev. Lett.* **98** 242302

Journal of Physics: Conference Series **509** (2014) 012033 doi:10.1088/1742-6596/509/1/012033

Charge asymmetry dependency of $\pi/$K anisotropic flow in U+U $\sqrt{s_{NN}}$ = 193 GeV and Au+Au $\sqrt{s_{NN}}$ = 200 GeV collisions at STAR

Qi-Ye Shou (for the STAR Collaboration)

Shanghai Institute of Applied Physics, Chinese Academy of Sciences, Shanghai, China
Brookhaven National Laboratory, Upton, NY, USA

E-mail: qyshou@rcf.rhic.bnl.gov

Abstract. In this paper we present STAR's measurements of v_2 and v_3 for charged kaons and pions at low transverse momentum range ($0.15 < p_T < 0.5$ GeV/c), as a function of event charge asymmetry (A_{ch}) in both U+U collisions at $\sqrt{s_{NN}}$ = 193 GeV and Au+Au collisions at $\sqrt{s_{NN}}$ = 200 GeV. We found that in 20% − 60% centrality, the ratio of slope parameters of $\Delta v_3(A_{ch})$ to that of $\Delta v_2(A_{ch})$, is $+3\sigma$ below the predicted value (1/3) from local charge conservation at freeze-out, indicating that it is unlikely that such effect can have a significant contribution to the splitting of elliptic flow of charged pions as a function of A_{ch}. We also observed a smaller $\Delta v_2(A_{ch})$ slope for kaons relative to pions. Our measurements serve as important consistency checks for the phenomena suggested as the consequence of the chiral magnetic wave.

1. Introduction

Theoretical study [1] indicates that, the electric quadrupole deformation, induced by chiral magnetic wave (CMW) at finite baryon density in Quark Gluon Plasma, will lead to a difference in elliptic flow of low momentum hadrons, by increasing v_2 of negatively charged hadrons and decreasing v_2 of positively charged ones. The magnitude of this difference is predicted to be proportional to the charge asymmetry parameter A_{ch}, defined as

$$A_{ch} \equiv \langle \frac{N_+ - N_-}{N_+ + N_-} \rangle,$$

here N_+ and N_- represent number of positive particles and number of negative particles in one event respectively. Such charge asymmetry dependency of π elliptic flow has been observed in Au+Au collisions at $\sqrt{s_{NN}}$ = 200 GeV by the STAR experiment at the Relativistic Heavy Ion Collider [2] [3]. The trend and the magnitude of the slope of $\Delta v_2(A_{ch})$ for $\pi^\pm$, to the first order, are consistent with the theoretical calculation. This observation has been used as an evidence for CMW [4] [5], which stems from the interplay of Chiral Magnetic Effects (CME) [6] [7] [8] and Chiral Separation Effects (CSE) [9] [10]. Nevertheless, recently it is argued that the local charge conservation at freeze-out, together with the characteristic shape of $v_2(\eta)$ and $v_2(p_t)$ at STAR, may also contribute to the elliptic flow splitting as a function of A_{ch} [11]. The effect of which has been qualitatively studied by hydrodynamic model. In particular, it is predicted [11] that with this alternative mechanism, a similar effect for higher flow harmonics v_3 may also exist, and the $\Delta v_3(A_{ch})$ slope for $\pi^\pm$ should be $\sim 1/3$ of the $\Delta v_2(A_{ch})$ slope. Here the v_3, known as triangular flow, is the third harmonic coefficient of the Fourier expansion of the final momentum-space azimuthal anisotropy, defined in a way similar to that of elliptic flow v_2 [12].

In this paper, we present 1) experimental measurement of v_2 and v_3 difference between negatively charged and positively charged pions, at low p_T as a function of A_{ch} in Au+Au $\sqrt{s_{NN}} = 200$ GeV collisions at STAR, in order to test the proposed mechanism due to local charge conservation at freeze-out. 2) the same measurement in U+U $\sqrt{s_{NN}} = 193$ GeV as the CMW effect is also expected to exist. 3) A_{ch} dependency of Δv_2 for kaons, as they are suggested [1] to have a weaker effect than that of pions due to hadronic effects. The latter two measurements also serve as important consistency checks for the phenomena due to CMW.

2. Data sets and analysis details

2.1. Data sets and cuts

In this analysis, ~328M and ~555M mininum-bias triggered events, from year 2010 and year 2011 respectively, in Au+Au $\sqrt{s_{NN}} = 200$ GeV, as well as ~280M mininum-bias triggered events in U+U $\sqrt{s_{NN}} = 193$ GeV at STAR experiment [13] have been used. To reject background events, all events are required to have primary vertex along beam direction (V_z) within 30 cm from the detector center, and primary vertex radius (V_r) within 2 cm from the beam center.

All charged particles are firstly required to have pseudorapidity within unity ($|\eta| < 1$). For the calculation of charged asymmetry, charged particles within transverse momentum window, $0.15 < p_T < 12$ GeV/c, are used, with the exception of low p_T protons($p_T < 0.4$ GeV/c) which are mostly due to beam-pipe knockout interactions. The same p_T cut was also applied on antiprotons to match that used for protons. The A_{ch} is calculated event by event and according to which the whole sample is divided into five sub-groups, in which each has almost the same number of events, for each centrality bin. Due to finite detecting efficiency, the observed A_{ch} is usually larger than real A_{ch}. To take this into account, HIJING events [14] have been used to convolute with detector acceptance and efficiency to derive the real A_{ch} from the observed A_{ch}.

For the flow analysis, the cut on Distance of the Closest Approach to primary vertex of π/K, dca < 1 cm, is applied to reject secondary particles, and cut on $0.15 < p_T < 0.5$ GeV/c is used to guarantee that tracks have good particle identification and almost the same mean p_T. Time projection chamber (TPC) [15] and time-of-flight (TOF) detector[16] are used to identify particles. In this analysis, only tracks satisfy both criteria, 1) within 2σ below/above theoretical dE/dx (tracks' average energy loss per unit length) curves, 2) within m^2 windows, $0 < m_\pi^2 < 0.1$ ($0.15 < m_K^2 < 0.35$) for π(K), are selected as particles of interest.

2.2. Q-cumulants method

The Q-cumulants method [17] is used to calculate v_2 and v_3. In this method, all multi-particle cumulants are expressed in terms of various expressions with respect to flow vectors (Q-vectors). For example, two-particle correlations for single event and all events can be obtained by

$$\langle 2' \rangle = \frac{p_n Q_n^* - m_q}{m_p M - m_q},\ \langle\langle 2' \rangle\rangle = \frac{\sum_{i=1}^{N} (w_{\langle 2' \rangle})_i \langle 2' \rangle_i}{\sum_{i=1}^{N} (w_{\langle 2' \rangle})_i},$$

here p_n, q_n, Q_n^* are different flow vectors, while m_p and m_q are number of particles in particles of interest group and reference particle group, respectively. Using differential second order cumulant $d_n\{2\} = \langle\langle 2' \rangle\rangle$, one can estimate differential flow by

$$v_n'\{2\} = \frac{d_n\{2\}}{\sqrt{c_n\{2\}}},$$

here c_n denotes reference flow. More calculation details are presented in [17]. In our analysis, a η-gap of 0.3 is applied to reduce non-flow correlation [18] when calculating reference flow. One significant advantage of the Q-cumulants method is that, using flow vectors, a fast and accurate calculation without exhaustly looping over all particle combinations can be provided.

Journal of Physics: Conference Series **509** (2014) 012033 doi:10.1088/1742-6596/509/1/012033

3. Results

3.1. The slope of $\Delta v_3(A_{ch})$ for $\pi^{\pm}$ versus centrality

In Figure 1, The slope parameter of $\Delta v_3(A_{ch})$ is presented as a function of centrality for $\pi^{\pm}$. The $\Delta v_3(A_{ch})$ slope parameter is extracted from the linear relationship between the flow difference $v_3(\pi^-)$ - $v_3(\pi^+)$ and the charge asymmetry A_{ch}. The solid blue circles, which represent $\Delta v_2(A_{ch})$ slope, are taken from previous measurement [2] at STAR as reference. It is observed that the trend of $\Delta v_3(A_{ch})$ slope in Au+Au collisions (red triangles) is very similar to that of $\Delta v_2(A_{ch})$, however the magnitude of the former is smaller than the latter and are even negative in peripheral and central bins. Combining data points from 20% – 60% centrality, as well as data from both Au + Au and U + U collisions, the ratio of slope parameters of $\Delta v_3(A_{ch})$ to that of $\Delta v_2(A_{ch})$, is 0.026 ± 0.098, which is consistent with zero and is $+3\sigma$ below the predicted value (1/3) from local charge conservation at freeze-out.

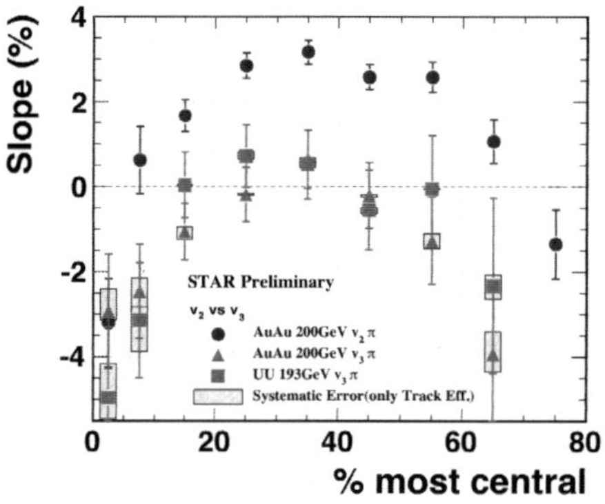

Figure 1. $\Delta v_3(A_{ch})$ slope parameters for $\pi^{\pm}$ in both Au+Au collisions (red triangles) and U+U collisions (green squares) are shown in all centrality bins, compared with $\Delta v_2(A_{ch})$ slope (blue dots) parameters in Au+Au collisions

3.2. The slope of $\Delta v_2(A_{ch})$ for $\pi^{\pm}$ versus centrality in U+U collisions

In figure 2 (Left) the slope parameters of $\Delta v_2(A_{ch})$ for $\pi^{\pm}$ is presented as a function of centrality for U+U collisions. They match well with the result in Au+Au collisions. This measurement is the first step of checking such effect in U + U collisions. However, whether this effect is in any way influenced by the special geometrical set up of U+U collisions has to be answered by detailed, future studies.

3.3. The slope of $\Delta v_2(A_{ch})$ for $K^{\pm}$ versus centrality

As mentioned above, the flow difference for charged kaons is supposed to have a weaker A_{ch} dependency than that of pions. In Figure 2 (Right), the same measurement for kaons is presented. Combining data points from 20% – 60% centrality, as well as data from both Au + Au and U + U collisions, the ratio of slope parameters of $\Delta v_2(A_{ch})$ for kaons and pions, is 0.57 ± 0.15. The observation of a smaller slope for kaons than that for pions is also consistent with the expectation in Ref. [1].

4. Summary

We have presented v_2 and v_3 of low momentum charged π, as well as v_2 of low momentum charged K, as a function of event charge asymmetry (A_{ch}) in both U+U collisions at $\sqrt{s_{NN}}$ =

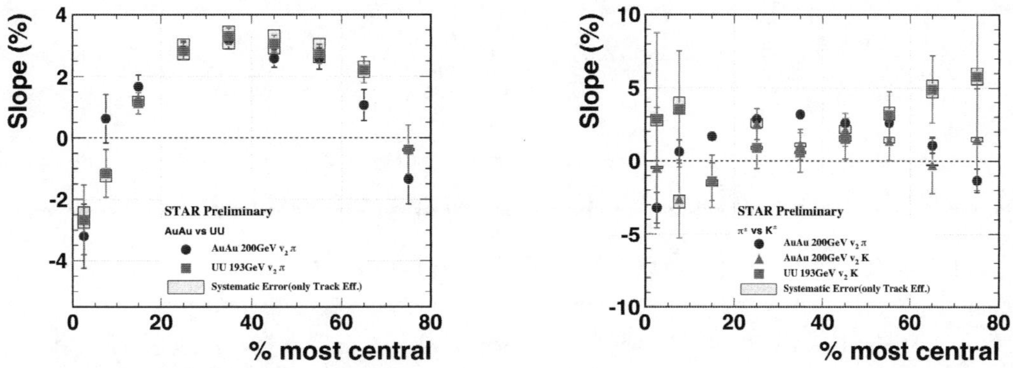

Figure 2. (Left) $\Delta v_2(A_{ch})$ slope parameters for $\pi^{\pm}$ in U+U MinBias collisions, compared with which in Au+Au collisions. (Right) $\Delta v_2(A_{ch})$ slope parameters for $K^{\pm}$, compared with that of $\pi^{\pm}$

193 GeV and Au+Au collisions at $\sqrt{s_{NN}} = 200$ GeV. We found that in 20% − 60% centrality, for pions, the ratio of slope parameters of $\Delta v_3(A_{ch})$ to that of $\Delta v_2(A_{ch})$, is consistent with zero and is $+3\sigma$ below the predicted value (1/3) from local charge conservation at freeze-out. It is unlikely that the local charge conservation at freeze-out, when convoluted with the characteristic shape of $v_2(\eta)$ and $v_2(p_t)$ at STAR, has a significant contribution to the splitting of elliptic flow of charged pions as a function of charge asymmetry. For the same centrality range, the $\Delta v_2(A_{ch})$ slope for kaons is smaller than that for pions, with a ratio of former to latter of 0.57 ± 0.15. This observation is also consistent with theoretical expectation. Our measurements serve as important consistency checks for the phenomena suggested as the consequence of the chiral magnetic wave.

5. Acknowledgement

Qi-Ye Shou was partially supported by the US Department of Energy under Grants No. DE-AC02-98CH10886 and No. DE-FG02-89ER40531.

References

[1] Burnier Y, Kharzeev D E, Liao J and Yee H U 2011 *Phys. Rev. Lett* **107** 052303
[2] Ke H W (for the STAR Collaboration) 2012 *J. Phys.: Conf. Ser.* **389** 012035
[3] Wang G (for the STAR Collaboration) 2013 *Nuclear Physics A* **904-905** 248c-255c
[4] Yee H U 2013 *Nuclear Physics A* **904-905** 310c-317c
[5] Stephanov M and Yee H Y, arXiv:1304.6410
[6] Kharzeev D E, McLerran L D, Warringa H J 2008 *Nuclear Physics A* **803** 227-253
[7] Kharzeev D E 2006 *Phys. Lett. B* **633** 260-264
[8] Kharzeev D E and Zhitnitsky A 2007 *Nucl. Phys. A* **797** 67-79
[9] Son D T and Zhitnitsky A 2004 *Phys. Rev. D* **70** 074018
[10] Metlitski M A and Zhitnitsky A 2005 *Phys. Rev. D* **72** 045011
[11] Bzdak A and Bozek P 2013 *Physics Letters B* **726** 239-243
[12] Poskanzer A M and Voloshin S A 1998 *Phys. Rev. C* **58** 1671-1678
[13] STAR Collaboration 2003 *Nucl. Instrum. Meth. A* **499** 624-632
[14] Wang X N and Gyulassy M 1994 *Comput. Phys. Commun.* **83** 307
[15] STAR Collaboration 2003 *Nucl. Instrum. Meth. A* **499** 659
[16] STAR Collaboration 2003 *Nucl. Instrum. Meth. A* **558** 419
[17] Bilandzic A, Snellings R and Voloshin S 2011 *Phys. Rev. C* **83** 044913
[18] Pandit Y (for the STAR Collaboration) 2013 *J. Phys.: Conf. Ser.* **420** 012038

14th International Conference on Strangeness in Quark Matter (SQM2013) IOP Publishing
Journal of Physics: Conference Series **509** (2014) 012034 doi:10.1088/1742-6596/509/1/012034

Influence of the in-medium kaon potential on kaon production in heavy ion collisions

P Srisawad[1,4], A Harfield[1], S Sombun[1], T Katukum[1], O Ketsungnoen[1,4], Y M Zheng[2], A Limphirat[3,4] and Y Yan[3,4]

[1] Faculty of Science, Naresuan University, Phitsanulok 65000, Thailand
[2] China Institute of Atomic Energy, P.O. Box 275(18), 102413 Beijing, China
[3] School of Physics, Suranaree University of Technology, Nakhon Ratchasima 30000, Thailand
[4] Thailand Center of Excellence in Physics (ThEP), CHE, Bangkok 10400, Thailand

E-mail: pornrads@gmail.com

Abstract.
The inclusive cross section and polar angle distribution of K^+ mesons in Ni + Ni collisions at incident energy 1.93 AGeV are studied using the Quantum Molecular Dynamics model based on the covariant kaon dynamics, and compared with experimental KaoS data from GSI . The calculated K^+ yields better fit to experiment data for a soft nuclear equation of state (compression modulus K = 200 MeV) when including the in-medium K^+ potential (its value at saturation density is $U_k(\rho_0) \approx 30MeV$). The calculated results for the polar angle distribution of K^+ in Ni + Ni collisions at 1.93 AGeV are slightly more forward and back direction than KaoS data.

1. Introduction

Kaons production in heavy ion collisions is a relevant topic for theoretical and experimental studies. Properties of kaons in dense hadronic matter are important for a better understanding of both the possible restoration of chiral symmetry in dense hadronic matter and the properties of nuclear matter at high density. Kaons are created in binary collisions of baryons and mesons (E_{beam} = 1.58 GeV for $NN \rightarrow K\Lambda N$ and E_{beam} = 2.5 GeV for $NN \rightarrow NNK\bar{K}$, the secondary processes $\Delta N \rightarrow KYN$, $\pi N \rightarrow KY$, and $\pi Y \rightarrow KN$). In the lowest order approximation to the chiral Lagrangian, the kaon (antikaon) potential has an attractive scalar and a repulsive (attractive) vector part. The calculations by Zheng et al.[1] demonstrated that the new FOPI data on the kaon in-plane flow [2] are best described by using the in-medium kaon potential given by the Brown and Rho parametrization [3] ($U_k(\rho_0) \approx$ 30MeV, where $\rho_0 = 0.16 fm^{-3}$).

In this work we calculated the inclusive cross section and polar angle distribution of K^+ mesons in Ni + Ni collisions at incident energies of 1.93 AGeV. Nucleons are described by the Quantum Molecular Dynamics (QMD) model. Kaons are treated within the covariant kaon dynamics [1]. For the nuclear force we use the standard momentum dependent Skyrme interactions corresponding to a soft and a hard nuclear equation of state (EOS)(compression modulus K = 200 MeV and K = 380 MeV respectively).

2. Kaons in Dense Matter

The natural framework to study the interaction between pseudoscalar mesons and baryons at low energies is chiral perturbation theory (ChPT). From the chiral Lagrangian the field equations for the $K^{\pm}$ mesons are derived from the Euler-Lagrange equations. [4]

$$\left[\partial_\mu \partial^\mu \pm \frac{3i}{4f_\pi^*} j_\mu \partial^\mu + \left(m_K^2 - \frac{\Sigma_{KN}}{f_\pi^{*2}} \rho_s\right)\right] \phi_{K^\pm}(x) = 0 \quad (1)$$

In Eq. (1) j_μ is the baryon four-vector current, ρ_s is the baryon scalar density,f_π^* is the in-medium pion decay constant. Introducing the kaonic vector potential

$$V_\mu = \frac{3}{8f_\pi^{*2}} j_\mu \quad (2)$$

Eq. (1) can be rewritten in the form [5]

$$\left[(\partial_\mu \pm iV_\mu)^2 + m_K^{*2}\right] \phi_{K\pm}(x) = 0 \quad (3)$$

Thus, the vector field is introduced by minimal coupling into the Klein-Gordon equation. The effective mass m_K^* of the kaon is then given by

$$m_K^* = \sqrt{m_K^2 - \frac{\Sigma_{kN}}{f_\pi^{*2}} \rho_s + V_\mu V^\mu} \quad (4)$$

where $m_K = 0.496\ GeV$ is the bare kaon mass.
The $K^{\pm}$ single-particle energies are expressed as

$$\omega_{K^\pm}(\mathbf{k}) = \pm V^0 + \sqrt{\mathbf{k}^{*2} + m_K^{*2}} \quad (5)$$

where $\mathbf{k}^* = \mathbf{k} \mp V$ is the kaon effective momentum,$V^\mu = (V^0, \mathbf{V})$, the kaon vector field is introduced by minimal coupling into the Klein-Gordon with opposite signs for K^+ and K^-, and m_K^* is the kaon effective (Dirac) mass. The kaon (antikaon) potential $U_{K^\pm}$ is defined as

$$U_{K^\pm}(\mathbf{k}) = \omega_{K^\pm}(\mathbf{k}) - \sqrt{\mathbf{k}^2 + m_{\mathbf{K}}^2} \quad (6)$$

Following Ref.[1], we use the Brown and Rho parameterization $\Sigma_{KN} = 450\ MeV$, $f_\pi^{*2} = 0.6f_\pi^2$ for the vector field and $f_\pi^{*2} = f_\pi^2$ for the scalar part given by $-\Sigma_{KN}/f_\pi^{*2}\rho_s$.This accounts for the fact that the enhancement of the scalar part using f_π^{*2} is compensated by higher-order corrections in the chiral expansion. For the nuclear forces we use the standard momentum dependent Skyrme interactions corresponding to a soft (hard) equation of state (EOS) (the compression modulus K = 200 MeV for a soft and K = 380 MeV for hard EOS). For the determination of the kaon mean field we adopt the corresponding covariant scalar-vector description of the nonlinear $\sigma\omega$ model. Up to saturation density the BR potential is $(U_k(\rho_0) \approx 30MeV)$.

3. Results and discussions

The inclusive invariant cross sections of K^+ mesons as a function of the kaon energy for ${}^{58}_{28}Ni +^{58}_{28} Ni$ at incident energy of 1.93 AGeV are given in Fig.1. In this figure full circles are KaoS data [6], empty and half empty circles are calculated results with and without the in-medium kaon potential, the left (right) panel shows the results calculated by using the soft (hard) EOS. We can see clearly from this figure that calculated results with the in-medium kaon potential and by using the soft EOS better fit to experimental data.

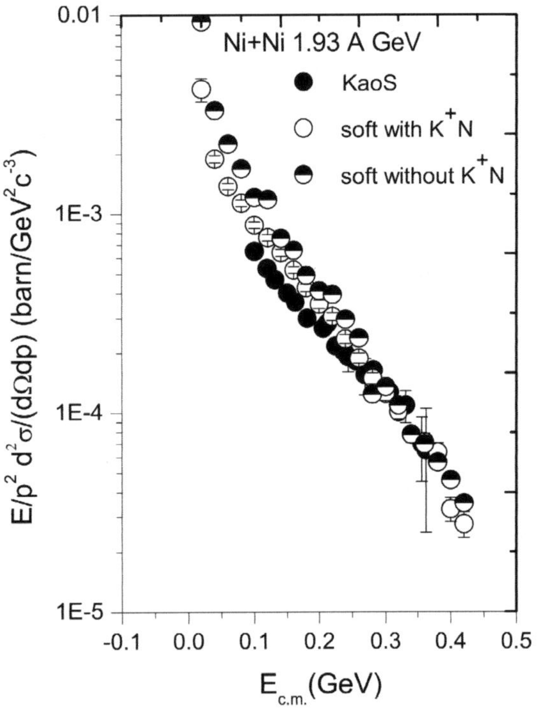

Figure 1. The inclusive invariant cross sections of K^+ mesons as a function of the kaon energy for $^{58}_{28}Ni + ^{58}_{28}Ni$ at 1.93 AGeV, in which full circles are the KaoS data [6], empty and half empty circles are calculated results with and without the in-medium kaon potential. The left (right) panel shows the results calculated by using the soft (hard) EOS.

Figure 2 shows the polar angle distribution of K^+ for $^{58}_{28}Ni +^{58}_{28} Ni$ collisions at 1.93 AGeV. The distributions are normalized to 1 for $\cos\theta_{c.m.} = 0$. The data are taken from Ref. [6]. It is seen from this figure that the calculated distributions are rather flat. The polar angle distribution is not affected by the in-medium kaon potential. Because the effects of rescattering and K^+ nucleus potential do not add linearly for the polar angle distribution of K^+. It is also shown that the polar angle distribution is not affected by the nuclear EOS.

4. Conclusions

The inclusive invariant cross sections and the polar angle distribution of K^+ in reactions $^{58}_{28}Ni +^{58}_{28} Ni$ at 1.93 AGeV are analyzed within the QMD model based on the covariant kaon dynamics, and compared to the KaoS data. Our calculated results with a repulsive in-medium K^+N potential (Its value at saturation density ρ_0 is $U_k(\rho_0) \approx 30MeV$) better fit to experimental data. This means that in order to describe reasonably data the medium effect of K^+ mesons should be included in the kaon dynamics. On the other hand, The inclusive invariant cross sections of K^+ meson turn out to be sensitive to the nuclear equation of state and to be preferred to the soft equation of state. However, the polar angle distribution of K^+ mesons is not sensitive to the equation of state.

We can conclude that the physics of kaon production close to threshold in A + A collisions is quantitatively understood for most of the observables of kaon production. To make progress in

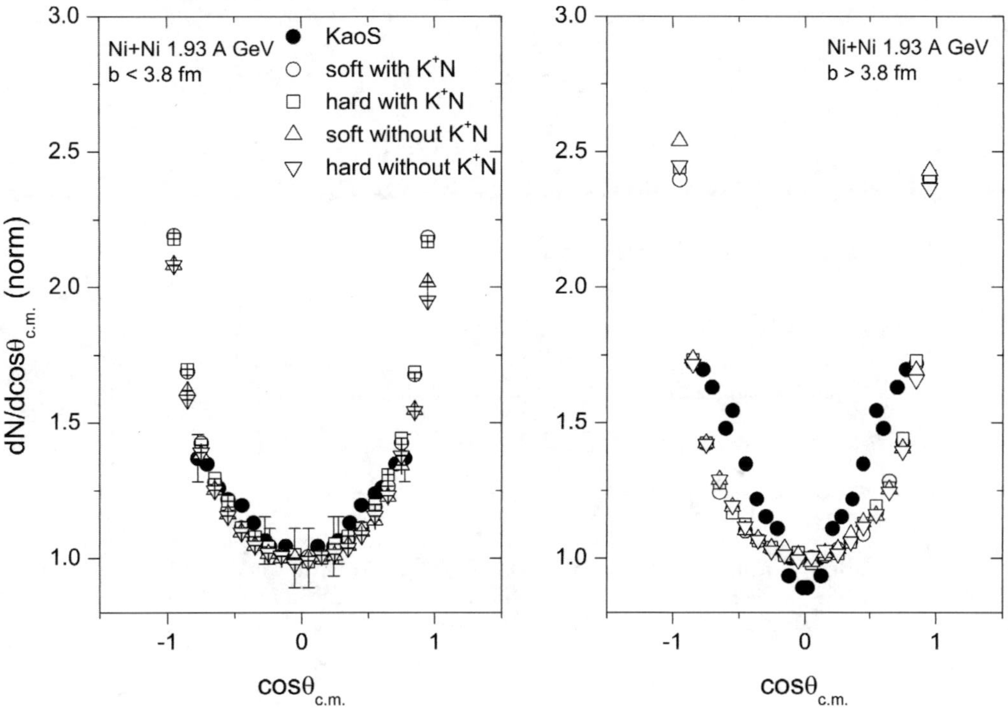

Figure 2. The polar angle distribution of K^+'s for ${}^{58}_{28}Ni + {}^{58}_{28}Ni$ collisions at 1.93 AGeV. Full circles are the experimental data. Empty symbols are calculated results. Left and right part show the calculated results for impact parameter $b < 3.8fm$ and $b > 3.8fm$ respectively. Data are taken from KaoS Collaborations [6].

understanding the strength of the K^+ nucleon interaction in the nuclear medium it is necessary to reach a consistent description of all sensitive observables in a heavy ion reactions.

Acknowledgments

This work was supported in part by Faculty of Science, Naresuan University. The computing resources have been provided by the National e-Science Infrastructure Consortium of Thailand.

5. References

[1] Zheng Y M, Fuchs C, Faessler A, Shekhter K, Yan Y and Kobdaj C 2004 *Phys. Rev. C* **69** 034907
[2] Herrmann H 1999 *Prog. Part. Nucl. Phys.* **42** 187
[3] Brown G E and Rho M 1996 *Nucl. Phys. A* **596** 503
[4] Li G Q and Ko C M 1995 *Nucl. Phys. A* **594** 460
[5] Fuchs C 2006 *Prog. Part. Nucl. Phys.* **56** 1
[6] Forster A et al., 2007 *Phys. Rev. C* **75** 024906

14th International Conference on Strangeness in Quark Matter (SQM2013) IOP Publishing
Journal of Physics: Conference Series **509** (2014) 012035 doi:10.1088/1742-6596/509/1/012035

Phenomenology of quarkyonic matter in heavy-ion collisions

Giorgio Torrieri[a,b], Sascha Vogel[a], Stefano Lottini[c]

[a]FIAS, Goethe Universität, Ruth-Moufang-Str. 1, 60438 Frankfurt am Main, Germany
[b] Pupin Physics Laboratory, Columbia University, 538 West 120th Street, New York, NY 10027, USA
[c]NIC, DESY, Platanenallee 6, 15738 Zeuthen, Germany.

Abstract. We give an overview of the possible phase structure of dense hadronic matter with $T \leq T_c$ and $\mu_Q \geq \Lambda_{\mathrm{QCD}}$, arguing for the possibility of a percolating transition in the T, μ_Q, N_c space. We show that matter after this phase transition has many properties associated with "quarkyonic" matter. Finally, we show how the existence of this phase can be experimentally investigated using electromagnetic signatures. The details of these calculations are in [1, 2, 3, 4].

1. Quarkyonic matter

The study of nuclear matter at moderate ($T \sim 0 - 180$ MeV) temperature and large baryochemical potential ($\mu_Q = \mu_B/3 \sim \Lambda_{\mathrm{QCD}} = 250$ MeV) has recently enjoyed new vigorous theoretical and experimental interest.

From the experimental side, this is due to the start of programs specifically aimed at exploring low-energy collisions with the latest detector technology [5, 6, 7].

From the theoretical side, it was realized that this regime presents both potential for very interesting physics and unique challenges. The latter come from the breakdown of most of the techniques used to study QCD: lattice gauge theory presents the well-known sign problem at finite chemical potential. Several approaches have been invented to deal with this, but the results are neither conclusive nor precise enough to draw any conclusions about the property of matter in the experimentally interesting region. Effective field theory is also problematic, since the typical momentum exchange is $\sim \mu_Q$, while the "fundamental scale of the theory" is $\Lambda_{\mathrm{QCD}} \sim \mu_Q$. Hence, we expect effective field theories to be unreliable. The effect on deconfinement, a non-perturbative phenomenon, on chiral symmetry in the critical region adds an extra complication which is little understood [8].

This ambiguity leaves room for qualitatively new phenomena, and even new phases of matter, to arise. A recent proposal of this kind is quarkyonic matter [9, 10]. The reason for conjecturing the existence of a new phase boils down to comparing the quark-hole screening with the gluon-gluon antiscreening at large chemical potential.

The "bag model phase diagram", positing that deconfinement occurs when there is one hadron per hadronic size ($T \sim \Lambda_{\mathrm{QCD}}$ or $\mu_Q \sim \Lambda_{\mathrm{QCD}}$, small box in Fig. 1) is incompatible with the expected running in momentum space of the gluon self-energy (right side of Fig. 1). If confinement is broken when the screening by quark-hole pairs, proportional to $\mu_Q^2 N_c N_f$ at the Fermi surface (which decreases the effective coupling), overpowers anti-screening by gluon loops

($\sim N_c^2$), then the low-temperature deconfinement point scales *at least* as $\sim \Lambda_{\rm QCD}\sqrt{N_c/N_f}$. (The bottom-right diagram of Fig. 1 shows higher loops yielding a $(N_c/N_f)^{z>1/2}$ scaling [2].)

Thus, in contrast to bag model intuition, the phase diagram at $N_c \to \infty$ looks like the one in Fig. 1, where as $N_c \to \infty$ the deconfinement line becomes flat. At the same time, the transition to "nuclear matter", with the baryonic density as order parameter, becomes infinitely sharp since the baryon mass is proportional to $N_c\Lambda_{\rm QCD}$. Therefore baryons drop out of the confined vacuum partition function entirely, but continue to be present at $\mu_Q \geq \Lambda_{\rm QCD}$.

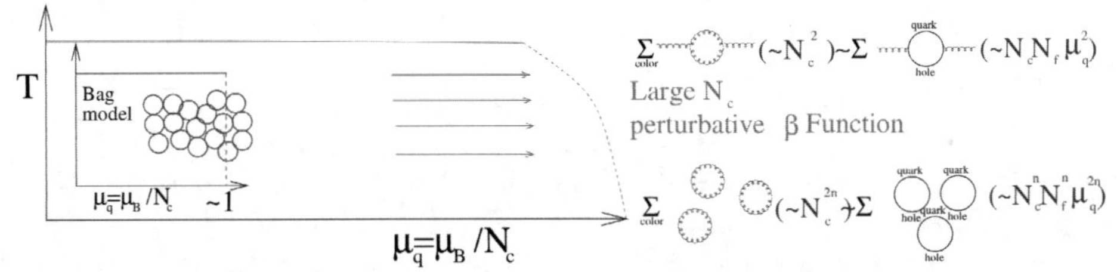

Figure 1. *Left:* Expected QCD phase diagram in the large-N_c limit, with the dashed deconfinement line parametrically located at $\mu_Q \propto \sqrt{N_c/N_f}$ (see text). In the insert, the phase diagram resulting from the bag model. *Right:* N_c- and N_f-dependences in the gluon self-energy (top diagrams), and some higher-order contributions (bottom).

Hence unless there are *non-perturbative* contributions to the running of quark-quark interactions, which could in principle bring the critical μ_Q for deconfinement down to $N_c^0\Lambda_{\rm QCD}$, "nuclear matter" at $\mu_Q \sim \Lambda_{\rm QCD}$ should, at large N_c, be in the confined phase. In configuration space, however, inter-quark distance is proportional to $N_c^{-1/3}$: for large enough N_c, then, one should be in the confinement regime yet somehow neighboring quarks should be so close that asymptotic freedom applies. The authors of [10] proposed a solution to this seeming contradiction by postulating matter in this regime exhibits quark-like degrees of freedom deep inside the Fermi surface (and hence a scaling $\sim N_c^1$ for the pressure) but baryonic excitations on the surface. Dense matter at $\Lambda_{\rm QCD} < \mu_Q < \sqrt{N_c/N_f}\Lambda_{\rm QCD}$, with features of asymptotic freedom in configuration space but features of confinement in momentum space is known as "quarkyonic".

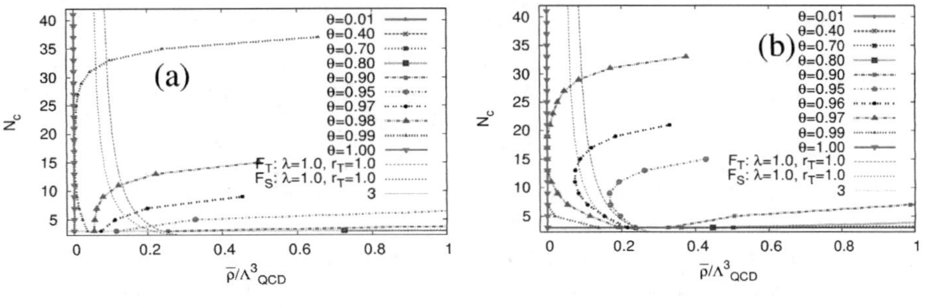

Figure 2. The percolation (lines only) and deconfinement (lines and points) transition lines as a function of the baryon density ρ_B and the number of colors N_c. $F_{T,S}$ are two quark propagators, and $\theta = T/T_c(\mu_Q = 0)$.

This is an interesting idea, but how much of quarkyonic dynamics survives at $N_c = 3$ and $N_f = 2, 3$ is an open question. It has long been known [4] that there are significant *qualitative* differences between the $N_c \to \infty$ limit and $N_c = 3$. Baryons in the $N_c \to \infty$ regime have an excitation energy $\ll \Lambda_{\mathrm{QCD}}$ and strong binding (binding energy scales as the baryon mass). The critical point for the nuclear liquid-gas phase transition is at $T, \mu_Q \sim \Lambda_{\mathrm{QCD}}$. Neither of these holds true in the real world, by at least an order of magnitude.

As argued in [1, 2], this indicates that the large-N_c limit is separated from the real world by a percolation-type phase transition. The quarkyonic matter transition line is therefore bound to be curved in N_c as well as T, μ_B space, the former being accessible only on the lattice. Indeed, some aspects of quarkyonic matter (the "Skyrme crystal" phase) are likely to be captured by the liquid-gas phase transition in the physical world. Other aspects, most importantly the appearance of quark degrees of freedom, could manifest themselves in our world [1] provided the deconfinement phase transition is far enough on the baryochemical potential axis.

In [1] it has been shown that, for a wide variety of reasonable propagators at a fixed baryonic number density of $\rho_B = \Lambda_{\mathrm{QCD}}^3/8$, a percolation transition is found as N_c is varied. If one identifies the percolation transition with the quarkyonic phase, deconfinement and percolation are indeed separate, and they cover different regions not only in T and μ_B, but also along N_c. Reference [2] further demonstrated that, as shown in Fig. 2, the critical density of percolation generally drops as N_c is raised, while that of deconfinement rises. Hence, in the "low-N_c" limit deconfinement occurs before percolation, and hence the percolating phase, which requires baryonic wave functions, does not occur. In the "large-N_c" limit, on the other hand, percolation is physically realized at densities of $\mathcal{O}(1)$ baryons per baryonic size, below deconfinement. The critical regime separating these two is at $N_c \simeq \mathcal{O}(2-8)$, and hence it is far from clear whether the quarkyonic phase occurs in our world. This makes the development of a quarkyonic matter *phenomenology* of considerable interest.

2. Phenomenology of quarkyonic matter

Looking for quarkyonic matter in experiment presents some of the same difficulties inherent in looking for Quark-Gluon plasma. The system is dynamically evolving, and includes both a high-density phase (which could be quarkyonic) and a low-density phase which should be identical to a weakly interacting hadron gas.

In addition, the quarkyonic phase should include both "hadronic" characteristics (since hadrons continue to exist, and dynamics at "slow" scales should be hadronic), and "quark characteristics", emerging via pQCD-like quark-hole interactions.

Perhaps the most specific signature of quarkyonic matter is given by electromagnetic interactions: Quark-hole scattering diagrams should, by the quarkyonic hypothesis, be essentially the same as quark-antiquark scatterings in asymptotically free pQCD. However, as shown in Fig. 3, quark wavefunctions should not be the same as free particle wavefunctions, but should contain a form factor which reflects the fact that quarks are still localized in baryons (there is no Fermi surface for antiquarks and gluons, and hence they should not exist as asymptotically free states in the confined phase).

Photon [3] and dilepton [2] spectra are both sensitive the the earliest, densest phase of the evolution of the system. They are also amenable to a rigorous calculation in the quarkyonic Ansatz (defined by the above assumptions), and details of the spectra can be used to distinguish between the "trivial" form factor of asymptotically free QGP and the quarkyonic form factor.

In Fig. 4 [3] we show the radial distribution, and the angular distribution (parametrized in azimuthal harmonic coefficients v_2) of Bremsstrahlung photons (the reaction hole-$q \to \gamma q$ hole). As can be seen, the photons exhibit a steeper p_T spectrum than QGP, due to the localization of quarks in "slow" baryons. In addition, v_2 oscillates chaotically event-by-event around zero, since it is generated not by collective flow but by the anisotropies of the baryonic wavefunction.

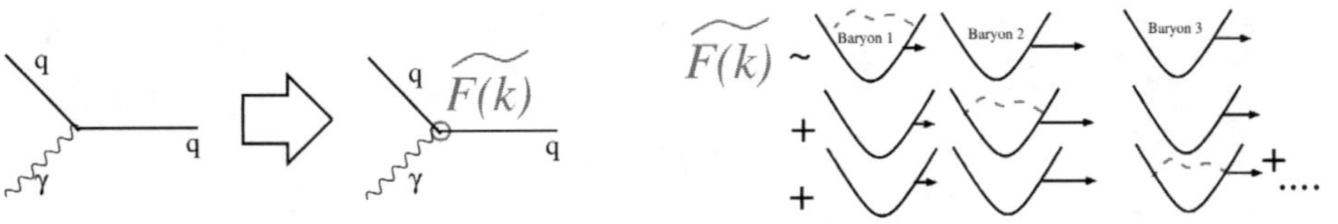

Figure 3. The form factor for γq terms in scattering diagrams, encoding the baryon distribution, originating from the presence of a quark Fermi sphere.

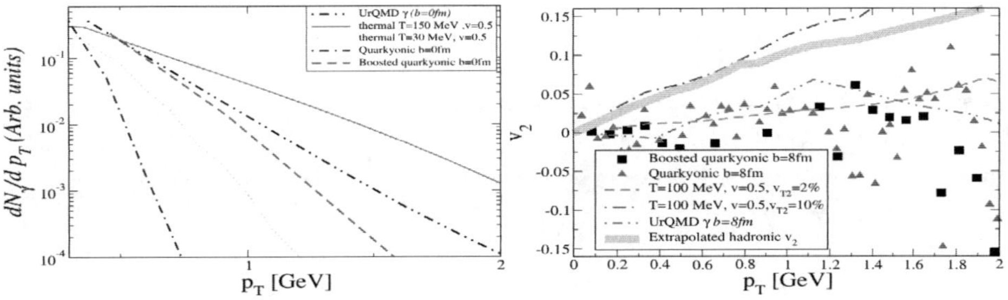

Figure 4. The p_T distribution and v_2 of Bremsstrahlung photons calculated with pQCD diagrams and the form factor in Fig. 3. The left panel shows the transverse momentum spectrum while the right panel shows the elliptic flow. While the effect of the hadronic phase and resonance decays have to be considered (see [11] for the background), these plots show the way to a possible experimental distinction between a quarkyonic-dominated system and other types of matter

In conclusion, we have argued that, in addition to the deconfinement and chiral phase transition, the formation of a Fermi sphere of quarks could be associated with a percolation phase transition which curves on the number-of-colors axis in addition to energy- and number-density, effectively separating a large-N_c from a small-N_c limit at finite chemical potential. We argued that it is difficult to say conclusively where the physical $N_c = 3$ is located w.r.t. the percolation transition, but that experimentally electromagnetic signatures are our best bet in distinguishing quarkyonic matter from the more conventional states of QCD matter.

We acknowledge the financial support received from the Helmholtz International Centre for FAIR within the framework of the LOEWE program (Landesoffensive zur Entwicklung Wissenschaftlich-Ökonomischer Exzellenz) launched by the State of Hesse. GT also acknowledges support from DOE under Grant No. DE-FG02-93ER40764.

References

[1] Torrieri G and Vogel S, 2013 *Phys. Rev. Lett.* **111**, 012301
[2] Lottini S and Torrieri G 2011 *Phys. Rev. Lett.* **107**, 152301
[3] Lottini S and Torrieri G 2013 *Phys. Rev.* C **88**, 024912
[4] Torrieri G, Lottini S, Mishustin I and Nicolini P 2012 *Acta Phys. Polon. Supp.* **5**, 897
[5] Staszel P [CBM Collaboration] 2010 *Acta Phys. Polon. B* **41**, 341
[6] Sissakian A and Sorin A [NICA Collaboration], 2009 *J. Phys.* G **36**, 064069
[7] Odyniec G, 2009 *Acta Phys. Polon.* B **40**, 1237
[8] Casher A 1979 *Phys. Lett.* B **83**, 395
[9] Fukushima K and Sasaki C 2013 *Prog. Part. Nucl. Phys.* **72** 99
[10] McLerran L and Pisarski R 2007 *Nucl. Phys.* A **796**, 83
[11] Dumitru A, Bleicher M, Bass S, Spieles C, Neise L, Stoecker H and Greiner W 1998 *Phys. Rev.* C **57**, 3271

Strongly interacting parton-hadron matter in- and out-off equilibrium

E.L. Bratkovskaya[1,2], O. Linnyk[3], W. Cassing[3], V.P. Konchakovski[3] and V. Ozvenchuk[2,4]

[1] Institute for Theoretical Physics, Goethe University, Frankfurt am Main, Germany
[2] Frankfurt Institut for Advanced Studies and Institute for Theoretical Physics, Goethe Universität, Frankfurt am Main, Germany
[3] Institut für Theoretische Physik, Universität Gießen, Gießen, Germany
[4] Subatech, UMR 6457, IN2P3/CNRS, Université de Nantes, École des Mines de Nantes, Nantes, France

E-mail: Elena.Bratkovskaya@th.physik.uni-frankfurt.de

Abstract. We study the equilibrium properties of strongly-interacting infinite parton-hadron matter, characterized by the transport coefficients such as shear and bulk viscosity and electric conductivity, and the non-equilibrium dynamics of heavy-ion collisions within the Parton-Hadron-String Dynamics (PHSD) transport approach, which incorporates explicit partonic degrees of freedom in terms of strongly interacting quasiparticles (quarks and gluons) in line with an equation of state from lattice QCD as well as the dynamical hadronization and hadronic collision dynamics in the final reaction phase. We discuss in particular the possible origin for the strong elliptic flow v_2 of direct photons observed at RHIC energies.

1. Introduction

One of the important findings from the heavy-ion experiments at ultra-relativistic energies was that the produced quark gluon plasma behaves as a strongly-interacting almost perfect fluid unlike a weakly-interacting gas [1, 2]. Recent relativistic viscous hydrodynamic calculations - using the Israel-Stewart framework - require a very small shear viscosity to entropy density ratio η/s of $0.08 - 0.24$ in order to reproduce the elliptic flow v_2 data at RHIC (cf. [3]). There is strong evidence from atomic and molecular systems that η/s should have a minimum in the vicinity of the phase transition (or rapid crossover) between the hadronic matter and the quark-gluon plasma [4], and that the ratio of bulk viscosity to entropy density ζ/s should be maximum or even diverge at a second-order phase transition [5]. It is also important to know the electromagnetic properties of the QGP since they influence the electromagnetic radiation in terms of photons and dileptons. The recent observation by the PHENIX Collaboration [6] that the elliptic flow $v_2(p_T)$ of 'direct photons' produced in minimal bias Au+Au collisions at $\sqrt{s_{NN}} = 200$ GeV is comparable to that of the produced pions was a surprise and in contrast to the theoretical expectations of small v_2 [7, 8, 9, 10].

In this contribution we present the highlights of the PHSD results [11, 12] on transport coefficients – the shear and bulk viscosities and electric conductivity – for 'infinite' parton-hadron matter in equilibrium as well as an example for non-equilibrium dynamics in heavy-ion collisions, in particular the direct photon spectra and collective flow v_2 (cf. Ref. [13]).

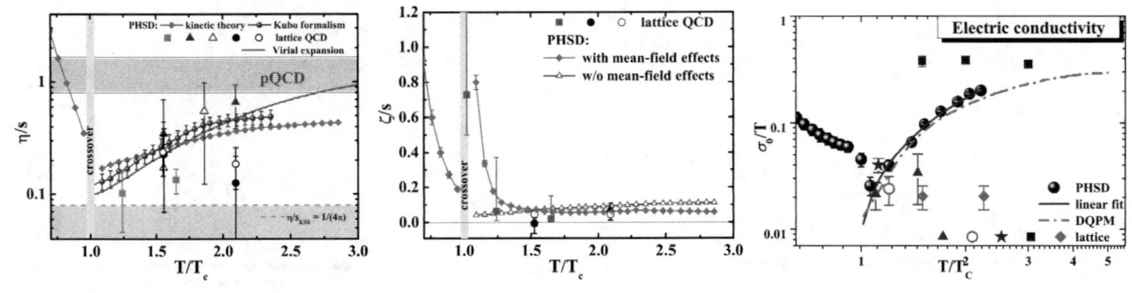

Figure 1. The PHSD results for the shear (left) and bulk (middle) viscosities of partonic and hadronic matter - as well as the electric conductivity (right) - as functions of temperature T/T_c.

2. PHSD model

The PHSD transport approach[14, 15]is based on generalized transport equations on the basis of the off-shell Kadanoff-Baym equations [16, 17] for Green's functions in phase-space representation (in first order gradient expansion beyond the quasiparticle approximation). The approach consistently describes the full evolution of a relativistic heavy-ion collision from the initial hard scatterings and string formation through the dynamical deconfinement phase transition to the strongly-interacting quark-gluon plasma (sQGP) as well as hadronization and the subsequent interactions in the expanding hadronic phase as in the Hadron-String-Dynamics (HSD) transport approach [18]. The partonic dynamics is based on the Dynamical QuasiParticle Model (DQPM) [19], which describes QCD properties in terms of single-particle Green's functions (in the sense of a two-particle irreducible (2 PI) approach) and reproduces lattice QCD results – including the partonic equation of state – in thermodynamic equilibrium (cf. review Ref. [20]).

3. Transport coefficients

In Fig. 1we present the PHSD results for the shear and bulk viscosities of partonic and hadronic matter - as well as the electric conductivity - as functions of temperature T/T_c ($T_c = 158$ MeV). In order to study the dynamical hadronic and partonic systems in equilibrium we have performed the PHSD simulations in a finite box with periodic boundary conditions. The ratio of the shear viscosity to entropy density $\eta(T)/s(T)$ from PHSD shows a minimum (with a value of about 0.1) close to the critical temperature T_c, while it approaches the perturbative QCD (pQCD) limit at higher temperatures in line with lattice QCD results. For $T < T_c$, i.e. in the hadronic phase, the ratio η/s rises fast with decreasing temperature due to a lower interaction rate of the hadronic system and a significantly smaller number of degrees-of-freedom.

The bulk viscosity $\zeta(T)$ – evaluated in the relaxation time approach – is found to strongly depend on the effects of mean fields (or potentials) in the partonic phase. We find a significant rise of the ratio $\zeta(T)/s(T)$ in the vicinity of the critical temperature T_c, which is also in agreement with that from lQCD calculations. This rise has to be attributed to mean-fields (or potential) effects that in PHSD are encoded in the temperature dependence of the quasiparticle masses, which is related to the infrared enhancement of the resummed (effective) coupling $g(T)$.

We also find that the dimensionless ratio of the electric conductivity over temperature σ_0/T rises above T_c approximately linearly with T up to $T = 2.5T_c$, but approaches a constant above $5T_c$, as expected qualitatively from perturbative QCD (pQCD) (cf. Ref. [12] for details). Our findings imply that the QCD matter even at $T \approx T_c$ is a much better electric conductor than Cu or Ag (at room temperature). We note that our result for σ_0/T close to T_c is in agreement with the most recent lQCD calculations which is important for our further discussion since the photon emission rate from the QGP or the hadronic system is controlled by the electric conductivity.

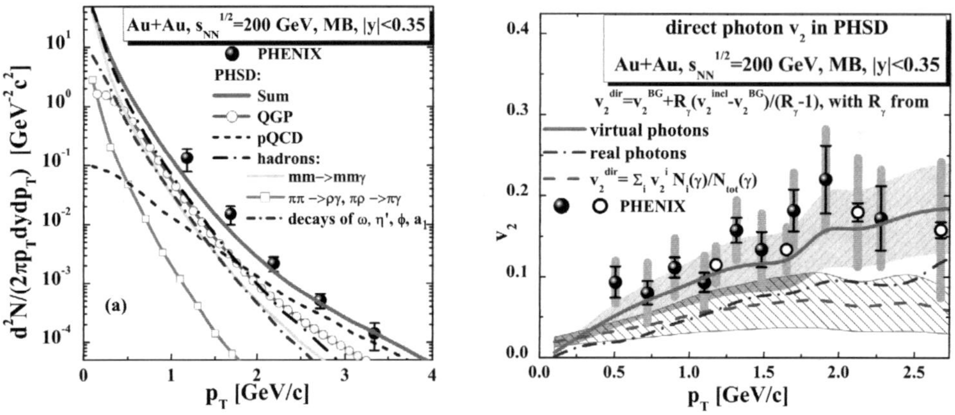

Figure 2. Left hand side: Direct photons (sum of all photon production channels except the π- and η-meson decays) from the PHSD approach (red solid line) in comparison to the data of the PHENIX Collaboration [21, 22] for minimal bias collisions of Au+Au at $\sqrt{s_{NN}} = 200$ GeV (black symbols). **Right hand side:** Elliptic flow of direct photons (hadron decays excluded) in the PHSD approach for minimal bias Au+Au collisions at $\sqrt{s_{NN}} = 200$ GeV in comparison to the data from Refs. [6, 23]. The results from the PHSD are displayed by the solid red line, equation (2), and by the dashed green line, by applying Eq. (1).

4. Photon production in heavy-ion collisions

In Ref [13] we have applied the PHSD approach to photon production in Au+Au collisions at $\sqrt{s_{NN}} = 200$ GeV and studied the transverse momentum spectrum and the elliptic flow v_2 of photons from hadronic and partonic production channels. We found that the PHSD calculations reproduce the transverse momentum spectrum of direct photons - measured by the PHENIX Collaboration in Refs. [21, 22] - as shown in the l.h.s. of Fig. 4. Furthermore, the PHSD also describes the data on the elliptic flow of inclusive photons [13]. In order to extract the flow of 'direct photons' from the inclusive one, the hadron decay background has to be subtracted. This can be done by two procedures:

Procedure 1. We calculate the direct photon v_2 (in PHSD) by summing up the elliptic flow of the individual channels contributing to the direct photons, using their contributions to the spectrum as the relative p_T-dependent weights, $w_i(p_T)$, i.e.

$$v_2(\gamma^{dir}) = \sum_i v_2(\gamma^i) w_i(p_T) = \sum_i v_2(\gamma^i) N_i(p_T) / \sum_i N_i(p_T), \quad (1)$$
$$\text{where } i = (q\bar{q} \to g\gamma, qg \to q\gamma, \pi\pi/\rho \to \rho/\pi\gamma, mm \to mm\gamma, \text{pQCD}).$$

The index i denotes the binary production channels, both the partonic quark-gluon interaction channels and the meson reactions which cannot be separated presently experimentally by model-independent methods. The direct photon elliptic flow calculated in this way is presented in Fig. 2 (r.h.s.) by the dashed green line and clearly underestimates the PHENIX data from Ref. [6] in line with Refs. [7, 8, 9, 10].

Procedure 2. The experimental collaboration has extracted the elliptic flow of direct photons $v_2(\gamma^{dir})$ from the measured inclusive photon $v_2(\gamma^{incl})$ by subtracting the hadron decay sources (π_0, η, ω, η', ϕ, a_1) as follows [6]:

$$v_2(\gamma^{dir}) = \left(R_\gamma v_2(\gamma^{incl}) - v_2(\gamma^{BG})\right) / (R_\gamma - 1) = v_2(\gamma^{BG}) + \frac{R_\gamma}{R_\gamma - 1}(v_2(\gamma^{incl}) - v_2(\gamma^{BG})) \quad (2)$$

where $R_\gamma = N^{incl}/N^{BG}$ denotes the ratio of the inclusive photon yield to that of the "background" (i.e. the photons stemming from the decays of π_0, η, ω, η', ϕ and a_1 mesons),

$v_2(\gamma^{BG})$ is the elliptic flow of the background photons; one can assume $v_2(\gamma^{BG}) \approx v_2(\pi_0)$ [13]. The ratio R_γ was obtained experimentally in Ref. [22] by analyzing the yield of dileptons with high transverse momentum $p_T > 1$ GeV and low invariant mass M. We recall here that we have studied the dilepton production at the top RHIC energy within the PHSD approach in Ref. [24]. The PHSD results reproduce well the PHENIX and STAR dilepton data differentially in the invariant mass M and transverse momentum p_T, only underestimating the excess observed by PHENIX at low M and low p_T. Note, however, that for the relatively high transverse momenta of dileptons ($p_T > 1$ GeV) the agreement of the PHSD calculations with the PHENIX data is quite good. We obtain $R_\gamma \approx 1.2$ by analyzing the yield of dileptons in the invariant mass window $M = 0.15 - 0.3$ GeV. Alternatively, we can use the calculated inclusive *real* photon spectrum to find the ratio $R_\gamma = N(\gamma)^{incl}/N(\gamma)^{BG}$. In this case we obtain $R_\gamma \approx 1.08$ from the real photons in PHSD. The difference between the values of R_γ – extracted from the dilepton spectra and the real photon spectra – is caused by the fact that in the dilepton mass window $M = 0.15 - 0.3$ GeV the background from the pion decays effectively "dies out", while the pion decay contribution is prominent for $M \to 0$.

Following the procedure of equation (2) in the PHSD, we obtain the red solid line in Fig. 2 (r.h.s.), if we use the ratio R_γ from the virtual photons in the invariant mass window $M = 0.15 - 0.3$ GeV, and the blue dash-dotted line, if we use the ratio R_γ from the calculated real photon spectrum. The two lines differ by about a factor of two. The difference between the two extraction procedures for the direct photon flow $v_2(p_T)$ can be attributed to different definitions for the ratio of the inclusive and background photons (R_γ).

In conclusion, our findings imply that there is presently no clear signal for 'unconventional physics' (beyond the strong interaction on the partonic and hadronic level) in the photon data from the PHENIX Collaboration within error bars. The strong v_2 of direct photons - which is comparable to the hadronic v_2 - in PHSD is attributed to hadronic channels, i.e. to meson binary reactions which are not subtracted in the data. On the other hand, the strong v_2 of the 'parent' hadrons, in turn, stems from the interactions in the QGP. Accordingly, the presence of the QGP shows up 'indirectly' in the direct photon elliptic flow.

References

[1] Gyulassy M and McLerran L D 2005 *Nucl. Phys.* A **750** 30
[2] Shuryak E V 2005 *Nucl. Phys.* A **750** 64
[3] Romatschke P and Romatschke U 2007 *Phys. Rev. Lett.* **99** 172301
[4] Csernai L P Kapusta J I and McLerran L D 2006 *Phys. Rev. Lett.* **97** 152303
[5] Kharzeev D and Tuchin K 2008 *JHEP* **09** 093
[6] Adare A *et al.* 2012 *Phys. Rev. Lett.* **109** 122302
[7] Chatterjee R, Frodermann E S, Heinz U W, Srivastava D K 2006 *Phys. Rev. Lett.* **96** 202302
[8] Liu F M, Hirano T, Werner K, Zhu Y 2009 *Nucl. Phys.* **A830** 587C
[9] Dion M, Gale C, Jeon S, Paquet J F, Schenke B *et al.* 2011 *J. Phys.* **G38** 124138
[10] Chatterjee R, Holopainen H, Helenius I, Renk T, Eskola K J 2013 *Phys. Rev.* **88** 034901
[11] Ozvenchuk V, Linnyk O, Gorenstein M I, Bratkovskaya E L and Cassing W 2013 *Phys. Rev.* **C87** 064903
[12] Cassing W, Linnyk O, Steinert T, Ozvenchuk V 2013 *Phys. Rev. Lett.* **110** 182301
[13] Linnyk O, Konchakovski V, Cassing W, Bratkovskaya E L 2013 *Phys. Rev.* **C88** 034904
[14] Cassing W and Bratkovskaya E L 2009 *Nucl. Phys.* A **831** 215
[15] Cassing W and Bratkovskaya E L 2008 *Phys. Rev.* **C78** 034919
[16] Kadanoff L P and Baym G 1962 *Quantum Statistical Mechanics*, Benjamin, New York
[17] Juchem S Cassing W and Greiner C 2004 *Phys. Rev.* **D69** 025006; 2004 *Nucl. Phys.* **A743** 92
[18] Cassing W and Bratkovskaya E L 1999 *Phys. Rept.* **308** 65
[19] Cassing W 2007 *Nucl. Phys.* **A795** 70; *Nucl. Phys.* **A791** 365
[20] Cassing W 2009 *Eur. Phys. J. ST* **168** 3
[21] Adare A *et al.* 2010 *Phys. Rev.* **C81** 034911
[22] Adare A *et al.* 2010 *Phys. Rev. Lett.* **104** 132301
[23] Tserruya I 2013 *Nucl. Phys.* **A904-905** 225c
[24] Linnyk O, Cassing W, Manninen J, Bratkovskaya E, Ko C M 2012 *Phys. Rev.* **C85** 024910

The Production of K^0's in p+p Reactions

Jia-Chii Berger-Chen, Laura Fabbietti for the HADES collaboration
Excellence Cluster Universe, TU München, Boltzmannstr. 2, 85748 Garching b. München
E-mail: jia-chii.chen@tum.de, laura.fabbietti@ph.tum.de

Abstract. We present the exclusive analysis of the reactions $p+p \to Y + \Delta^{++}/(p+\pi^+) + K^0$, where Y stands for the Λ or Σ^0 hyperon. The proton-proton measurement was performed with the HADES setup at GSI, Darmstadt, at a kinetic beam energy of 3.5 GeV. A dedicated sideband technique allows to reproduce the non-strange background and its kinematics in the selected data sample. Therefore, it is possible to obtain the background subtracted missing mass distribution $\mathrm{MM}(p\pi^+\pi^+\pi^-)$, where we can separate the Λ from the Σ^0 channel. From a Monte Carlo simulation of possible reactions we see contributions also by the reactions $p+p \to \Sigma^+/\Sigma(1385)^+ + p + K^0$. The ongoing analysis should provide exclusive cross sections of the mentioned reactions and their angular distributions.

1. Introduction

The study of kaons in heavy ion collisions is of particular interest, as they are not absorbed by the surrounding medium and thus carry information about the most interesting high-density stage of the collision. Such reactions are commonly interpreted in view of transport models. In this context, cross-sections of kaon production in nucleon-nucleon collisions are needed as an input for these models. A review on this topic can be found in [1]. While the 3-body strangeness production (e.g. $p+p \to Y+p+K$ with $Y = \Lambda, \Sigma$) is rather well studied in a wide energy range, 4-body reactions (e.g. $p+p \to Y+p+\pi+K$) are not that well known experimentally as well as theoretically. Especially for lower energies ($\sqrt{s} < 3.35$ GeV) only one measurement exists for the channel $p+p \to \Lambda+p+\pi^+ + K^0$ at $\sqrt{s} = 2.97$ GeV [2], but none for the analogous Σ^0 channel. At higher energies several bubble chamber experiments have been performed [3, 4, 5, 6]. All these experiments suggest considerable contribution from an intermediate Δ^{++} resonance, which decays to the measured p and π^+. Since the appearance of an intermediate resonance would modify the kinematics of its daughter particles, it is necessary to quantify its contribution and extract informations about the dynamics of such a reaction. These considerations motivate our study of the exclusive channels $p+p \to Y + \Delta^{++}/(p+\pi^+) + K^0$ at $\sqrt{s} = 3.18$ GeV. Besides, our data provides informations on the reactions $p+p \to \Sigma^+ + p + K^0$ and $p+p \to \Sigma(1385)^+ + p + K^0$.

2. The Experiment

The **H**igh-**A**cceptance **D**i-**E**lectron **S**pectrometer (HADES) is a fixed target experiment located at GSI Hemholtzzentrum in Darmstadt (Germany). The beam is delivered by the SIS18 synchrotron, which can be heavy ions of 1-2 AGeV or protons up to 3.5 GeV kinetic beam energy. The detector setup covers almost the full azimuthal range, while the acceptance in the polar angles reaches from 18° to 85° degrees. The momentum resolution is Δ p/p $\approx$ 3%.

The most important components for the current analysis are following: Four layers of Multiwire-Drift-Chambers (MDCs) deliver together with the superconducting magnet (toroidal field) informations on the particle momentum. Furthermore, particle identification is possible knowing additionally the energy loss in the MDCs. A Time-Of-Flight wall at the end of the setup allows for online triggering. Here, the LVL1 trigger required at least three hits (M3) in this Time-Of-Flight wall to reduce contributions from elastic reactions.

In this report, we discuss the measurements performed in collisions of a proton beam with 3.5 GeV kinetic energy on a liquid hydrogen target. In total, 1.14×10^9 events have been collected.

3. Exclusive Analysis

As already mentioned, it is important to understand the production of strangeness in elementary nucleon-nucleon processes. One of the uncertainties here are the contributions of resonances, which can change the dynamics of a reaction. Therefore, this exclusive analysis focuses on reactions containing a K^0 in the final state associated with an intermediate $\Delta(1232)^{++}$ resonance. In the end, the goal is to extract the cross sections of the exclusive channels and to determine angular distributions, which can serve as an input or as a check for transport models.

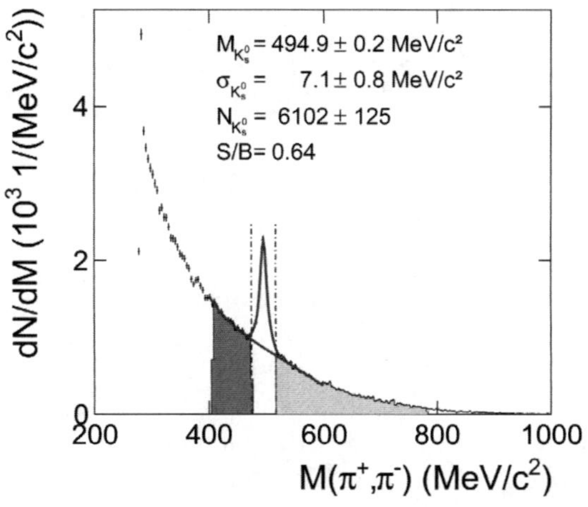

Figure 1. Invariant mass of a (π^+,π^-) pair after secondary vertex cuts for the selected event sample. The blue dashed-dotted lines show the three σ region around the K^0_S signal. The red and green area indicate the chosen low and high mass sideband sample respectively.

3.1. Event selection

The main channels that are investigated are $p+p \rightarrow \Lambda/\Sigma^0 + p + \pi^+ + K^0$, where the short-lived component K^0_S is reconstructed via its decay into π^+ and π^-. The event selection has been chosen such as to enhance events with these reactions in the analysed data set, namely by selecting events with exactly the four charged particles proton, π^+, π^+ and π^-, which are identified with the energy loss information from the MDCs. This way the data sample will also include contributions from further reactions like $p+p \rightarrow \Sigma^+ + p + K^0$ and $p+p \rightarrow \Sigma(1385)^+ + p + K^0$, which need to be considered. As we are only interested in reactions with a K^0_S in the final state, we introduce a cut on the $\pi^+\pi^-$-invariant mass. The resulting distribution is shown in figure 1 after applying the following off-vertex cuts to decrease combinatorial background: (1) distance between the two pion tracks ($d_{\pi^+-\pi^-} < 7$ mm), (2) distance between the primary reaction vertex and the secondary decay vertex ($d(K^0_S - V) > 25$ mm), (3) distance of closest approach to the primary vertex for the two pion tracks ($\mathrm{DCA}_{\pi^+} > 7$ mm, $\mathrm{DCA}_{\pi^-} > 7$ mm). This invariant mass spectrum is fitted with the sum of two Gaussians for the signal and a polynomial and a Landau function for the description of the background. A three σ cut around the K^0_S signal is then used for further analysis, where σ is the averaged standard deviation of the two Gaussians.

3.2. Background Determination

As can be seen in figure 1 considerable background contribution is left under the K^0_S signal. This background mainly comes from the remaining contribution of non-strange reactions in our data sample, for example $p+p \to p+\pi^+ + \pi^+ + \pi^- + n$ or $p+p \to p+p+\pi^+ + \pi^+ + \pi^- + \pi^-$ and many others. A good description of this background is necessary. Therefore, a dedicated sideband analysis has been developed. For that we explicitly define a sample of background events by selecting two-track combinations having a $(\pi^+\pi^-)$-invariant mass in the mass region adjacent to the K^0_S signal. This sample is marked as red and green areas in figure 1 as low and high mass sample, respectively. To obtain a correct description of the background remaining after the K^0_S mass cut, the kinematics of this background has to be reproduced by the sideband sample. That means the momentum distribution of the K^0_S background needs to be modeled. To find this distribution, a cut on the missing mass $\mathrm{MM}(p\pi^+\pi^-) < 1100$ MeV/c^2 has been applied to reject all channels with a K^0_S in the final state. By simultaneously fitting the momentum distributions of the low and high mass sample to the one of the K^0_S background, we can determine the relative contribution of both sideband samples. The result of this fit is depicted on figure 2, where the magenta histogram corresponding to the sum of low and high mass sample is compared to the K^0_S background in black. One can see a very good description of the background kinematics.

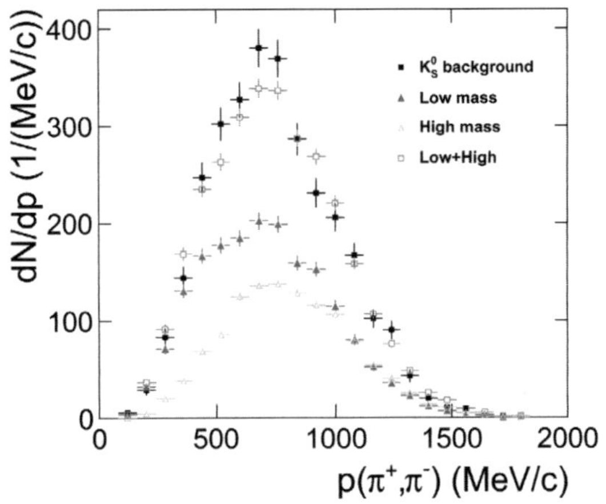

Figure 2. Momentum distribution of the K^0_S background. Applied cuts are a three σ cut around the K^0_S peak in the $\mathrm{IM}(\pi^+\pi^-)$, and a cut on $\mathrm{MM}(p\pi^+\pi^-) < 1100$ MeV/c^2. Relative weights of low mass and high mass sideband samples were extracted from a fit. The sum of both is shown in magenta.

3.3. Reaction Separation

As mentioned above this analysis focuses on the reactions $p+p \to \Lambda/\Sigma^0 + p + \pi^+ + K^0$. The only difference is the associated hyperon. Thus, one can plot the missing mass $\mathrm{MM}(p\pi^+\pi^+\pi^-)$, which should show two peaks at the Λ and the Σ^0 mass. Figure 3 illustrates this missing mass distribution with the following conditions: The cut on the K^0_S has been applied; additionally, a cut was used on the missing mass $\mathrm{MM}(p\pi^+\pi^-) > 1270$ MeV/c^2 to reject the channel $p+p \to \Sigma^+ + p + K^0$; the background as defined in the last section was subtracted. From the remaining missing mass distribution we see a clear Λ and Σ^0 peak, which allows us to distinguish between both contributions. The plotted coloured curves in the same figure are Monte Carlo simulations of the mentioned K^0 production channels. Some weakly contributing K^0 production channels like $p+p \to p+n+K^+ + \bar{K}^0$ are summed up and referred to as "other K^0_S channels". To this end, all channels have been simulated with isotropic angular distributions except for the $p+p \to \Sigma(1385)^+ + p + K^0$ and the $p+p \to \Sigma^+ + p + K^0$ reactions. For the first reaction we assume the same anisotropy as extracted for the reaction $p+p \to \Sigma(1385)^+ + n + K^+$ in the same data set [7]. For the Σ^+ reaction the anisotropy was adopted from [8] as found at $p_{beam} = 3059$ MeV/c.

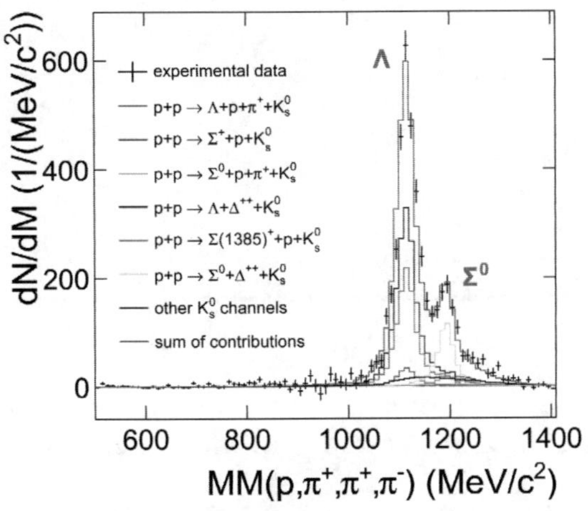

Figure 3. Missing mass of a (p,π^+,π^+,π^-) track combination with a cut on MM($p\pi^+\pi^-$) > 1270 MeV/c^2 after K^0_S preselection. The background is subtracted using the sideband sample. The grey histogram corresponds to the sum of simulated contributions.

4. Summary and Outlook

This report has shown that from the selected data sample of 4-body reactions we are able to model the non-strange background via a sideband analysis. This allows us to study the remaining K^0 production channels, mainly the reactions $p+p \to \Lambda+p+\pi^++K^0$ and $p+p \to \Sigma^0+p+\pi^++K^0$ to extract their contribution. Here we want to determine the contribution of an intermediate Δ^{++} by studying the invariant mass of p and π^+. Different kinematics are expected for these channels. Additionally, we can study the channels $p + p \to \Sigma(1385)^+ + p + K^0$ and $p + p \to \Sigma^+ + p + K^0$, which are also present in the selection. These channels are expected to exhibit anisotropic prodution. While our simulations take this into account, a quantitative study of the angular distributions of the Δ^{++} reactions remains an open issue, which are expected to show anisotropic behaviour as well. We have also shown that a separation of these two reactions is possible by cutting on the Λ or the Σ^0 mass in the missing mass distribution MM($p\pi^+\pi^+\pi^-$). Finally, one can extract the cross sections of the exclusive reactions by fitting all the simulated K^0 channels to relevant experimental observables. These information can then serve as an input for transport models or used to cross-check the available calculations.

Acknowledgment

The authors gratefully acknowledge the support of the TUM Graduate School's Faculty Graduate Center Physics at Technische Universität München, Germany, LIP Coimbra, Coimbra (Portugal) PTDC/FIS/113339/2009 SIP JUC Cracow, Cracow (Poland) N N202 286038 28-JAN-2010 NN202198639 01-OCT-2010 FZ Dresden-Rossendorf (FZD), Dresden (Germany) BMBF 06DR9059D TU Mnchen, Garching (Germany) MLL Mnchen DFG EClust 153 VH-NG-330 BMBF 06MT9156 TP5 GSI TMKrue 1012 NPI AS CR, Rez, Rez (Czech Republic) MSMT LC07050 GAASCR IAA100480803 USC - S. de Compostela, Santiago de Compostela (Spain) CPAN:CSD2007-00042 Goethe-University, Frankfurt (Germany) HA216/EMMI HIC for FAIR (LOEWE) BMBF:06FY9100I GSI F&E.

References

[1] Hartnack, Oeschler, Leifels, Bratkovskaya, Aichelin 2012 *Phys. Rep.* **510** 119-200
[2] Nekipelov et al. 2007 *J. Phys.* G **34** 627
[3] Biermann et al. 1966 *Phys. Rev.* **147** 922-931
[4] Alexander et al. 1967 *Phys. Rev.* **154** 1284-1304
[5] Klein et al. 1970 *Phys. Rev.* D **1** 3019-3030
[6] Firebaugh et al. 1968 *Phys. Rev.* **172** 1354-1369
[7] Agakishiev et al. 2012 *Phys. Rev.* C **85** 035203
[8] M. Abdel-Bary et al. 2012 *Eur. Phys. J.* A **48** 23

Effect of quark gluon plasma on charm quark produced in relativistic heavy ion collision

Mohammed Younus[a,1], **Dinesh K. Srivastava**[a,2] **and Steffen A. Bass**[b]
[a]Variable Energy Cyclotron Centre, 1/AF Bidhan Nagar, Kolkata 700064, India
[b]Duke University, Department of Physics, 139 Science Drive, Durham 27708, NC, U.S.A.
E-mail: [1]younus.presi@gmail.com, [2]dinesh@vecc.gov.in
E-mail: [b]bass@phy.duke.edu

Abstract. Charm quarks are produced mainly in the pre-equilibrium stage of heavy ion collision and serve as excellent probes entering the thermalized medium. They come out with altogether different momenta and energies and fragments into D-mesons and decay into non-photonic electrons which are observed experimentally. Here we present the effect of QGP on charm quark production using two different models: first one based on Wang-Huang-Sarcevic model of multiple scattering of partons and the second one is based on Parton Cascade Model with Boltzmann transport equation used for charm quark evolution in QGP.

1. Introduction

Charm quarks are produced mostly in the early period after heavy ion collision when quarks and gluons with high momenta interact with momentum transfer $Q^2 > 2m_c$ (mass of charm). Thus charm production rate may be controlled using perturbative QCD techniques. Also, being produced small in number as compared to massless gluons and light quarks, they remain separated from the bulk properties of the deconfined system and may serve as the probes to quark gluon plasma.

2. Production of charm and its energy loss in quark gluon plasma

At first let us discuss briefly production of charm quarks:

2.1. charm quarks production

The differential cross-section for $c\bar{c}$ production in proton-proton collision can be shown to be

$$\frac{d\sigma}{dy_1 dy_2 d^2p_T} = 2x_1x_2\sum_{ij}\left[f_i^{(1)}(x_1,Q^2)f_j^{(2)}(x_2,Q^2)\frac{d\hat{\sigma}_{ij}(\hat{s},\hat{t},\hat{u})}{d\hat{t}} + f_j^{(1)}(x_1,Q^2)f_i^{(2)}(x_2,Q^2)\frac{d\hat{\sigma}_{ji}(\hat{s},\hat{u},\hat{t})}{d\hat{t}}\right]/(1+\delta_{ij})\,, \quad (1)$$

where i and j are the interacting partons and f_i and f_j are the partonic structure functions, and x_1 and x_2 are the momentum fractions of the parent nucleons carried by the interacting

Journal of Physics: Conference Series **509** (2014) 012038 doi:10.1088/1742-6596/509/1/012038

partons [1].

$$\frac{d\sigma}{d\hat{t}} = \frac{|M|^2}{16\pi\hat{s}^2} \tag{2}$$

where the invariant amplitude $|M|^2$ is given by [2].

For heavy ion collisions the p_T spectrum for heavy quark production is given by

$$\frac{dN}{d^2p_T dy} = T_{AA}(b)\frac{d\sigma}{d^2p_T dy} \tag{3}$$

where for collisions at different centrality, $T_{AA}(b)$ can be calculated using Glauber formalism. We account for higher order corrections by taking a constant K-factor ≈ 2.5

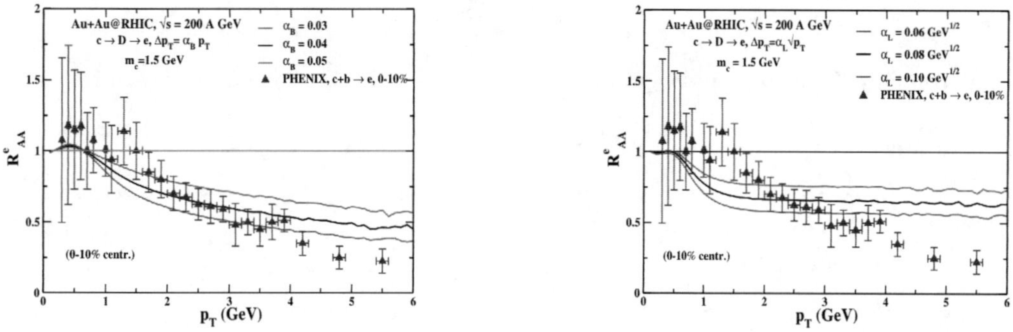

Figure 1. $R_{AA}(p_T)$ of single non-photonic electron at RHIC energy of $\sqrt{s} = 200$ A GeV.

2.2. Energy Loss of charm

The medium effect on charm p_T spectrum has been obtained using two different models. The first one deals with energy loss of charm via momentum loss per collision through multiple collisions of charm with medium partons. The calculations are partly inspired by Wang-Huang-Sarcevic model of multiple scattering of jet partons with medium partons [3]. The momentum loss per collision for $'i^{th}'$ charm is defined by

$$\Delta p_i = \alpha(p_i)^{\beta}\,, \alpha = \alpha_L(\beta = 0.5)\,, \text{and}\, \alpha_B(\beta = 1.0) \tag{4}$$

The no. of collisions for given charm momentum is calculated using Poisson distribution. Finally nuclear modification factor, $'R_{AA}(p_T)'$ and azimuthal anisotropy, $'v_2(p_T)'$ have been calculated for single non-photonic electrons and D mesons and results compared with recent data from RHIC and LHC experiments.

The second model used for charm evolution is Parton Cascade Model [4]. One of its implementation commonly called VNI/BMS-box mode has been utilized. The calculations are based on Microscopic Boltzman transport equation given by

$$p^{\mu}\frac{\partial F_k(x,\vec{p})}{\partial x^{\mu}} = \sum_{processes:i} C_i[F], \tag{5}$$

where $F_k(x,\vec{p})$ is the single particle phase space distribution and the collision term on r.h.s. is a non-linear functional of phase space distribution terms inside an integral.

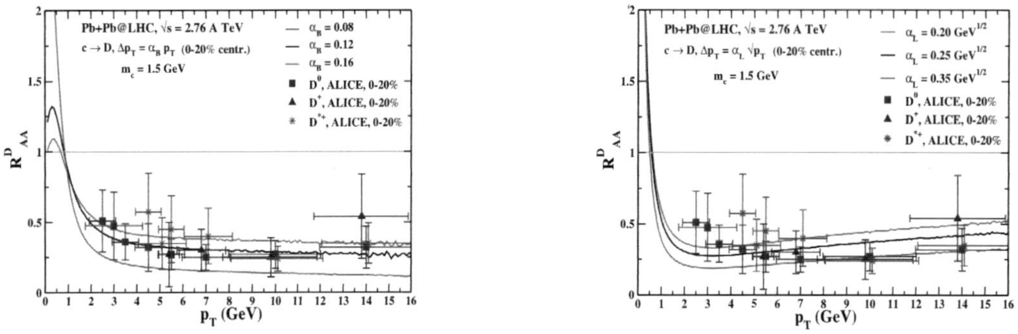

Figure 2. $R_{AA}(p_T)$ of D mesons at LHC energy of $\sqrt{s} = 2.76$ A TeV

We have included the matrix elements for all $2 \rightarrow 2$ binary elastic scattering processes for charm interaction with gluons or light quarks(u, d, s) and $2 \rightarrow n$ process for radiative (brehmsstrahlung) corrections after each scattering. The elastic processes included are

$$\begin{aligned} cg &\rightarrow cg, \\ cq(\bar{q}) &\rightarrow cq(\bar{q}). \end{aligned} \tag{6}$$

The corresponding differential scattering cross section is defined to be,

$$\frac{d\hat{\sigma}}{dQ^2} = \frac{1}{16\pi(\hat{s} - M_c^2)^2} \sum |\mathcal{M}|^2. \tag{7}$$

The invariant transition amplitude, $|M|^2$ for elastic scattering which can be calculated or obtained from [5].

In our calculations, radiative corrections are included in form of time-like branching of the probe charm into a final charm and a shower of radiated partons. In time-like branching the probe charm after each scattering may split into a daughter charm with different momentum and a gluon. Together with this, we have also included coherent emission of gluons commonly called Landau-Pomeranchuk-Migdal(LPM) effect in our radiative energy loss formalism.

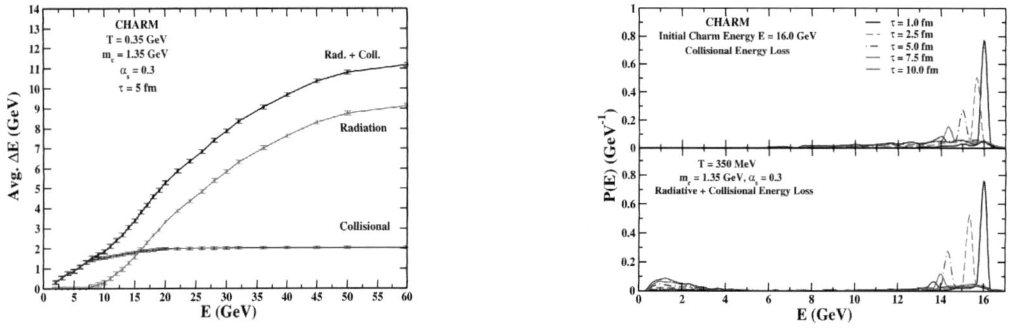

Figure 3. (left)Average energy loss of different energy charms, (right) energy profile of a 16.0 GeV charm at different time.

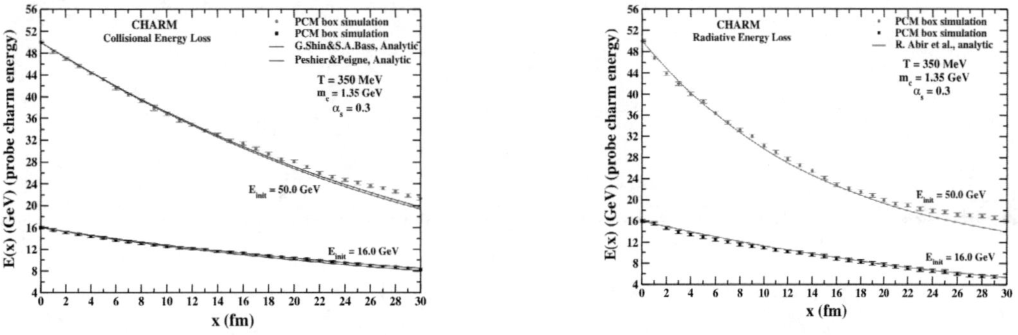

Figure 4. Energy of probe charm with distance traveled for collisional loss(left) only and radiative energy loss(right) only

3. Results

From the model calculation of multiple scattering of charm in QGP, we have obtained R_{AA} and v_2 of charm at both LHC and RHIC collision energies, for two different mechanisms for momentum loss per collision. Our calculations for $\alpha = \alpha_B(\beta = 1.0)$ explains the experimental data mostly in lower and mid-p_T region for both RHIC and LHC temperatures. However, the other mechanism of $\alpha = \alpha_L(\beta = 0.5)$ seems to explain the data for higher charm momentum, p_T region where the R_AA data for D-mesons and non-photonic electrons seem to rise just like lighter mesons, see Fig. 1 and, Fig. 2. v_2 results from our calculations however do not match experimental data well enough at RHIC but it clearly shows different trend between two mechanisms we have included in our calculations.

From the second model of PCM-VNI/BMS, we have calculated average energy loss of charms of different energies(E_c). The collisional loss seems to dominate below E_c <16.0 GeV while radiative loss takes over and dominates beyond E_c ≥16.0 GeV. We also showed energy profile of a 16.0 GeV charm tracked through different evolution time. The collisional loss shows a shift in the peak with a long tail-like structure unlike results for charm quarks,from recent calculations with Langevin equation(+ hydrodynamical background) under similar conditions, Fig. 3. In last two plots, Fig. 4 we have shown probe charm energy evolution for different distances it travels in QGP at a fixed temperature of 350 MeV. We used, two different initial charm energy of 16.0 and 50.0 GeVs respectively and studied the evolution of energy for collisional loss radiative loss, separately. The PCM simulation results are compared with some of the recent analytical calculation of charm quark energy loss per unit length it travels in QGP [6].

References

[1] Younus M and Srivastava D K 2010 *J. Phys.* **G***: Nucl. Part. Phys.* **37**, 115006 and references therein; Younus M and Srivastava D K 2012 *J. Phys.* **G***: Nucl. Part. Phys.* **39**, 025001; Jamil U and Srivastava D K 2012 *J. Phys.* **G***: Nucl. Part. Phys.* **37**, 085106 and references therein.

[2] Combridge B L 1979 *Nucl. Phys.* **B 151**, 429.

[3] Wang X-N, Huang Z and Sarcevic I 1996 **Phys. Rev. Lett. 77**, 231; Younus M and Srivastava D K 2012 *J. Phys.* **G***: Nucl. Part. Phys.* **39**, 095003; Müller B 2003 *Phys. Rev.* **C 67**, 061901.

[4] Younus M, Coleman-Smith C E, Bass S A and Srivastava D K 2013 Systematic Study of Charm Quark Energy Loss using Parton Cascade Model *Preprint- arXiv: 1309.1276v2/nucl-th* and references therein; Geiger K and Müller B 1992 *Nucl. Phys.* **B 369**, 600; Srivastava D K and Geiger K 1998 *Phys. Lett.* **B 422**, 39; Shin G R, Bass S A and Müller 2010 *J. Phys.* **G***: Nucl. Part. Phys.* **37**, 105112.

[5] Svetitsky B 1988 *Phys. Rev.* **D 37**, 2484.

[6] Peigné S and Peshier A 2008 *arXiv:0802.4362v1[hep-ph]*, Abir R, Jamil U, Mustafa M G and Srivastava D K 2012 *Phys. Lett.* **B 715**, 183.

Gluon radiation by heavy quarks

Joerg Aichelin, Pol Gossiaux, and Thierry Gousset

Subatech, CNRS/IN2P3, Université de Nantes, Ecole des Mines de Nantes
4 rue Alfred Kastler, 44307 Nantes cedex 3, France

E-mail: gousset@subatech.in2p3.fr

Abstract. The gluon emission cross section of a heavy quark colliding a light parton from the plasma is computed in pQCD at leading order. It is possible to compute the complete energy dependence of the result and therefore to assess the range of applicability in energy of standard high-energy approximation.

1. Introduction

It is generally assumed that in ultrarelativistic heavy ion collisions a plasma of quarks and gluons is produced, and then quickly expands and hadronizes. The main objective of present experiments at ultrarelativistic heavy ion colliders is to study this plasma. The study of the plasma properties is difficult because the initial momentum distribution of the light quarks is not known and that before being detected the system passes the chiral/confinement phase transition in which hadrons are formed. Hadrons interact on the way to the detectors what modifies their spectrum. Therefore soft light hadrons do not carry information on the early stage of the plasma.

The situation is much better for heavy quarks (c and b) which are created in hard processes. Here perturbative QCD allows for the calculation of the production cross sections (in contradistinction to light quarks) and these cross sections have also been measured. Also the details of the chiral/confinement phase transition are less important than for light quarks because, due to its mass, the momentum of the heavy quark determines the momentum of the open charm hadrons. In addition, the momentum distribution at production and at the transition is very different from that expected if the heavy quarks are in thermal equilibrium with the plasma of light quarks and gluons. Therefore the modification of the initial momentum distribution by the interaction of the heavy quarks with the plasma carries information on the plasma properties.

The interaction of the heavy quark with the plasma has two parts, elastic collisions and radiative collisions. For the first a model was developped [1] in which the cross section of the elementary interactions are calculated by perturbative QCD with a running coupling constant and an infrared behavior adjusted so as to match hard thermal loop calculations. Embedding these cross sections in the hydrodynamical description of the expanding plasma of Heinz and Kolb it was shown that the collisional energy loss underpredicts the measured energy loss of heavy mesons at large momenta as well as their elliptic flow by roughly a factor of two.

The present work extends our pQCD calculation toward the calculation of the radiative energy loss, following the Gunion-Bertsch approach to light quark radiation phenomenology in high-energy collisions [2]. We compute the gluon emission cross section of a heavy quark (instead of a light quark in [2]) colliding a light parton from the plasma. The heavy quark is supposed to be relativistic but with an intermediate energy so that coherence effects are not dominant. We keep track of the full energy dependence of the result in order to assess the range of applicability in energy of the standard high-energy approximation. The latter has the advantage of leading to very compact expression for the various integrated cross sections and also of making more transparent the discussion of physical phenomenons. We discuss in particular the occurrence of the dead cone effect.

2. Energy dependence

The quantity $\int x\,(d\sigma/dx)\,dx$, which is related to the radiated energy loss $dE_{\rm rad}/dz$ normalized to the density of scatterers ρ and the quark energy E_Q, was recently computed for a heavy quark traversing a hot QCD medium [3]. The result is shown in Fig. 1 for a charm quark $m_c = 1.3$ GeV and a bottom quark $m_b = 4.5$ GeV.

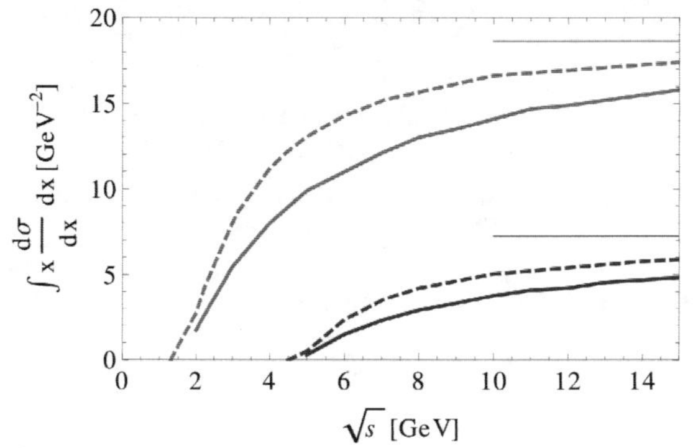

Figure 1. Energy dependence of $\int x\,(d\sigma/dx)\,dx$ for charm (upper curves) and bottom (lower curves). Solid lines show the full energy dependence and dashed lines correspond to an approximation where the influence of finite energy is taken into account only in phase space boundaries. Thin straight lines show the asymptotic results. The screening scale is taken as $\mu = 0.4$ GeV.

It can be seen that at respectively $E_c \sim 10$ GeV and $E_b \sim 40$ GeV, 50% of the respective asymptotic behaviors are reached. This indicates first that the finite energy constraint is effective in the high-energy range of relevance nowadays, say $E_Q \approx p_T \sim 10 - 100$ GeV, especially for the b sector. A simplified model in which the fully differential cross section is approximated by its high-energy limit ($s \to \infty$) but phase space limitation is taken into account upon integration is capable of explaining typically half of the deviation between the full result and the asymptotic limit.

A second consequence of the above observation is that high-energy phenomenons are at least indicative of the physics at work in the range $10 - 100$ GeV. The high-energy limit expressions can thus serve as basis for the discussion of the full result.

The asymptotic behavior roughly scales as

$$\int x \frac{d\sigma}{dx} dx \propto \frac{1}{\mu\, m_Q}$$

giving a suppression of heavy quark radiation $\propto 1/m_Q$ but also showing a sensitivity to the details of screening. This behavior results from contributions from both below $x_M = \mu/m_Q$ ('hard' collisions) and above ('soft' collisions) to the x-integral.

The fractional energy loss spectrum $x\,d\sigma/dx$ is displayed in Fig. 2 for charm (left) and bottom (right). The top curve on each plot is the asymptotic result. It approaches the finite energy results best in the intermediate x-range, slightly below $x_M = \mu/m_Q$, which is also the range of relevance for quenching phenomenology [4].

Below we give some details on the origin of radiation in the high-energy limit.

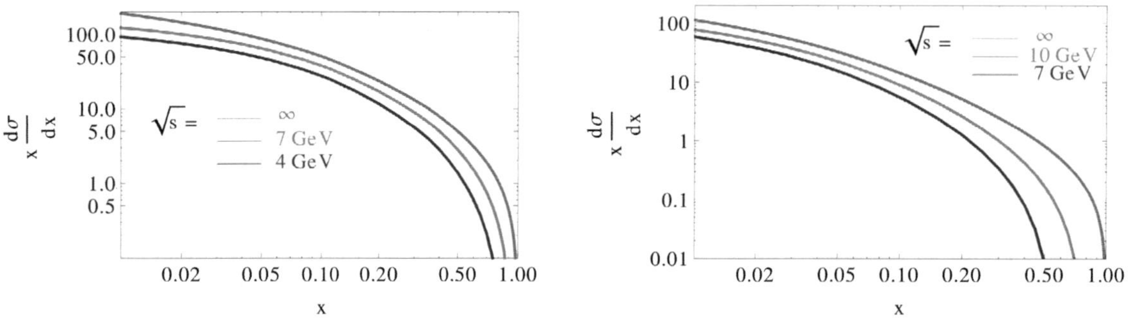

Figure 2. Charm (left) and bottom (right) fractional energy loss spectrum, $x\,d\sigma/dx$, for various $\sqrt{s}$, together with the high-energy approximation $\sqrt{s} \to \infty$. $\mu = 0.4$ GeV.

3. High-energy approximation

Let us first explain how the high-energy approximation is realized. We consider the collision of a heavy quark Q onto a massless parton q with the production of a gluon g:

$$Q(P) + q(q) \to Q(P') + q(q') + g(k)$$

where 4-momenta are indicated between brackets. The heavy quark momentum may be parameterized as

$$P = p + \frac{m_Q^2}{s - m_Q^2} q$$

where p is a second light-like vector, in addition to q the light-like momentum of the initial massless quark, and chosen such that $2\,p \cdot q = s - m_Q^2$. The gluon momentum may be written as

$$k = x\,p + y\,q + \vec{k}_\perp,$$

where x is the light-cone momentum fraction already encountered and $\vec{k}_\perp$ the transverse momentum of the gluon which is a second quantity of interest for radiation. $y \equiv y(x, \vec{k}_\perp)$ is fixed by imposing $k^2 = 0$. The last relevant momentum is the momentum transfer to the light quark $\ell = q - q'$ and we will also write $\ell^2 = t$.

At large s the fully differential cross section factorizes

$$\frac{d\sigma}{dx d^2k_\perp d^2\ell_\perp} \approx \frac{d\sigma_{\text{el}}}{d^2\ell_\perp} \times P_g(x, \vec{k}_\perp, \vec{\ell}_\perp),$$

where high-energy means that $s - m_Q^2 \gg |t|, \vec{k}_\perp^{\,2}$ [3]. Then the s-dependence disappears from the cross section, be it either differential or integrated. The reason for this simplification comes from $d\sigma_{\text{el}}/d^2\ell_\perp \propto 1/t^2$ at large $|t|$ and $P_g \propto 1/k_\perp^4$ at large $k_\perp$ making the cross section insensitive to s at large s. In return the cross section is sensitive to low momentum transfer and in particular to the Debye screening effect in the plasma. At small t we use the prescription

$$\frac{d\sigma_{\text{el}}}{d^2\ell_\perp} \propto \frac{1}{t^2} \to \frac{1}{(t - \mu^2)^2}$$

with μ related to the Debye mass m_D. This sets a natural scale for $\ell_\perp \sim \mu$ in the cross sections.

Now

$$P_g \propto \left(\frac{\vec{k}_\perp}{\vec{k}_\perp^{\,2} + x^2 m_Q^2} - \frac{\vec{k}_\perp - \vec{\ell}_\perp}{(\vec{k}_\perp - \vec{\ell}_\perp)^2 + x^2 m_Q^2} \right)^2 \tag{1}$$

and there are different regimes depending on the comparison between $\ell_\perp$ and $x\, m_Q$. If $\ell_\perp \gg x\, m_Q$ we have a hard scattering regime, with a behavior indicated in Fig. 3 (left).

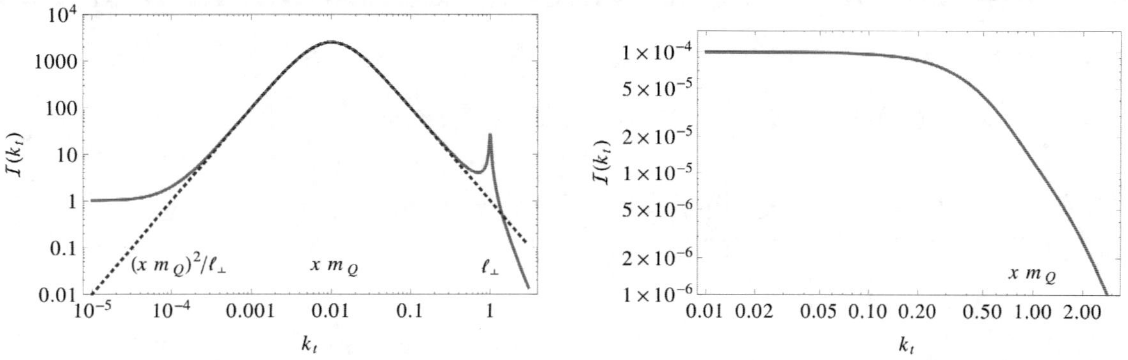

Figure 3. Gluon distribution P_g (averaged over the angle between $\vec{k}_\perp$ and $\vec{\ell}_\perp$) for $\ell_\perp = 1$ and $x\, m_Q = 0.01$ (hard scattering, left) and $\ell_\perp = 0.01$ and $x\, m_Q = 1$ (soft scattering, right).

In this regime, there is no interference between the two terms in the bracket of eq. (1) at intermediate $\vec{k}_\perp$ (where the second term is negligible), as can be seen in the plot where the dotted curve is simply the resulting P_g, eq. (1), when dropping the second term. As a consequence, $P_g \propto 1/\vec{k}_\perp^{\,2}$ in the range $x\, m_Q < k_\perp < \ell_\perp$, and

$$\int P_g\, d^2k_\perp \sim \ln \frac{\ell_\perp}{x\, m_Q}.$$

The suppression of radiation for $k_\perp < x\, m_Q$ generates a dead cone. Replacing $\ell_\perp$ by μ the range $x < x_M \equiv \mu/m_Q$ in $d\sigma/dx$ is identified.

If $\ell_\perp \ll x\, m_Q$ (soft scattering regime) there is a strong interference at all $\vec{k}_\perp$ as shown in Fig. 3 (right), leading to

$$\int P_g\, d^2k_\perp \sim \frac{\ell_\perp^2}{x^2\, m_Q^2}.$$

Replacing $\ell_\perp$ by μ leads to the identification of the $x > x_M$ range in $d\sigma/dx$. In this regime, the mass dependence is not reducible to the dead cone phenomenon.

References

[1] Gossiaux P B and Aichelin J 2008 *Phys. Rev.* C **78** 014904; 2009 *J. Phys.* G **36** 064028
[2] Gunion J F and Bertsch G 1982 *Phys. Rev.* D **25** 746
[3] Aichelin J, Gossiaux P B and Gousset T 2013 *Preprint* arXiv:1307.5270 [hep-ph]
[4] Baier R, Dokshitzer Y L, Mueller A H and Schiff D 2001 *JHEP* **0109** 033

Journal of Physics: Conference Series **509** (2014) 012040 doi:10.1088/1742-6596/509/1/012040

System-size dependence of charged kaon and pion production in nucleus-nucleus collisions at beam energies of 40*A* GeV and 158*A* GeV

H. Ströbele for the NA49 collaboration

Goethe-Universität Frankfurt

E-mail: `stroebel@ikf.uni-frankfurt.de`

Abstract. We present the system size (centrality) dependence of the mean transverse mass, of the K/π ratio, and the width of the rapidity distributions. Except for the latter a steep increase with centrality is observed for small systems followed by a weak rise or even saturation for higher centralities at both energies. This behavior is compared to calculations using transport models (UrQMD and HSD), a percolation model and the core-corona approach.

The NA49 collaboration has published data on the system-size dependence of pion and kaon [1], hyperon (Λ and Ξ) [2], and (anti-)proton [3] yields and distributions in Pb+Pb collisions at 40*A* GeV and 158*A* GeV beam energy. Results on pions, kaons, ϕ mesons and Λ hyperons in C+C and Si+Si at 158*A* GeV were reported in Ref. [4]. Pion and kaon production was also studied in p+p interactions at 158 GeV/c [5, 6].

In this contribution we summarize the findings for the system size dependence of the kaon and pion observables $\langle m_T \rangle - m_0$, of the K/$\pi$ ratio, and the width of the rapidity distributions RMS_y. In addition to the centrality selected Pb+Pb collisions semi-central C+C and Si+Si collisions are considered. These new data are published in [1].

The recorded minimum bias Pb+Pb collisions were divided into five consecutive centrality bins C0 - C4 with C0 being the most central one. The centrality selection is based on the forward going energy of projectile spectators measured in a downstream (or zero degree) calorimeter VCAL positioned behind a suitably adjusted collimator. We quantify centrality by the fraction of cross-section according to intervals of forward going spectator energy. For each centrality interval a characteristic quantity, the mean value of "wounded nucleons" $\langle N_w \rangle$ is calculated. A nucleon is considered wounded if its interaction occurs in the nuclear overlap volume. After applying quality cuts to the sample of all reconstructed tracks, pion spectra were obtained by subtracting from the yield of all negatively charged particles the yield of K^-, $\bar{p}$, e^-, and of particles from the decay of strange particles and from secondary interactions. π^+ were identified in a combined dE/dx and TOF analysis only near midrapidity Their yields in full phase space were calculated from the π^+/π^- ratios near midrapidity assuming that their spectra have the same shapes. Kaons were identified using the energy loss (dE/dx) information from the TPCs and time-of-flight measurements in the phase space covered by the TOF walls. For the analysis based on dE/dx information alone, raw yields of K^+ and K^- were extracted from fits to the dE/dx distributions in bins of laboratory and transverse momenta. The resulting distributions were corrected for geometrical acceptance, in-flight decay of kaons, and reconstruction efficiency.

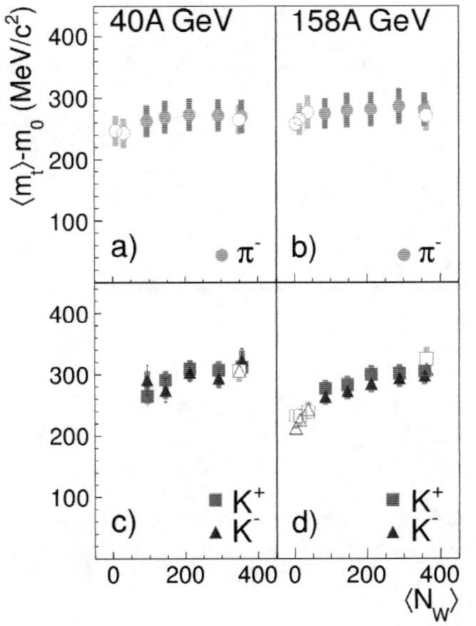

Figure 1. (Color online) Mean transverse mass as extracted from the transverse mass spectra at mid-rapidity for π^-, K$^+$ and K$^-$ versus number of participating nucleons. Open symbols show results from p+p, C+C, Si+Si and central Pb+Pb collisions. Except for the new results for C+C and Si+Si collisions at 40*A* GeV beam energy the data are taken from [5, 4, 7, 8, 6]. Filled symbols represent the new centrality selected Pb+Pb data. The thin vertical bars indicate the statistical errors, which are mostly not visible. The thick shaded bars indicate the systematic errors.

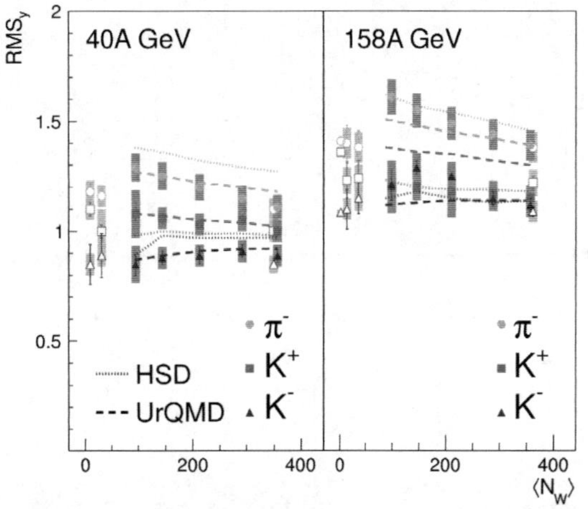

Figure 2. (Color online) Width of the rapidity distributions described by the RMS values for π^-, K$^+$ and K$^-$ versus mean number of participating nucleons. Open symbols show results from p+p, C+C, Si+Si and central Pb+Pb collisions. Except for the new results for C+C and Si+Si collisions at 40*A* GeV beam energy the data are taken from [5, 4, 7, 8]. Filled symbols represent the new centrality selected Pb+Pb data. Curves show results from HSD [9, 10] and UrQMD2.3 [11] simulations as indicated in the figure. The thin vertical bars indicate the statistical errors, which are mostly not visible. The thick shaded bars indicate the systematic errors.

Transverse mass spectra at mid-rapidity were determined for kaons (combined dE/dx and TOF analysis) and pions for different centrality classes. For a model independent study of the transverse mass spectra the average transverse mass $\langle m_{\mathrm{T}} \rangle - m_0$ was calculated. To account for the small unmeasured high m_{T} part of the kaon spectra, which contributes to m_{T} only at the percent level, the spectra were extrapolated by suitable exponential functions. Figure 1 shows the dependence of the resulting $\langle m_{\mathrm{T}} \rangle - m_0$ on the mean number of wounded nucleons $\langle N_{\mathrm{w}} \rangle$ for the investigated collision systems at 40*A* GeV and 158*A* GeV beam energy. Results from central Pb+Pb [7, 8], p+p at 158 GeV/*c* [6], C+C and Si+Si collisions [4] at 158*A* GeV are also shown (by open symbols).

Owing to the mass difference and well known radial flow effect, $\langle m_{\mathrm{T}} \rangle - m_0$ values are larger for kaons than for pions. For central collisions [7, 8] mean transverse masses do not change significantly from 40*A* GeV to 158*A* GeV. For pions, if any, only a weak centrality dependence is observed at 40*A* GeV and none at 158*A* GeV beam energy, while kaons show an increase of mean transverse masses towards central Pb+Pb collisions. This increase is particularly pronounced when comparing to the much smaller C+C and Si+Si collision systems with $\langle N_{\mathrm{w}} \rangle < 60$.

Transverse momentum spectra were measured in bins of rapidity in order to extract the rapidity dependence of p_T-integrated yields. The measured p_T-spectra were extrapolated into

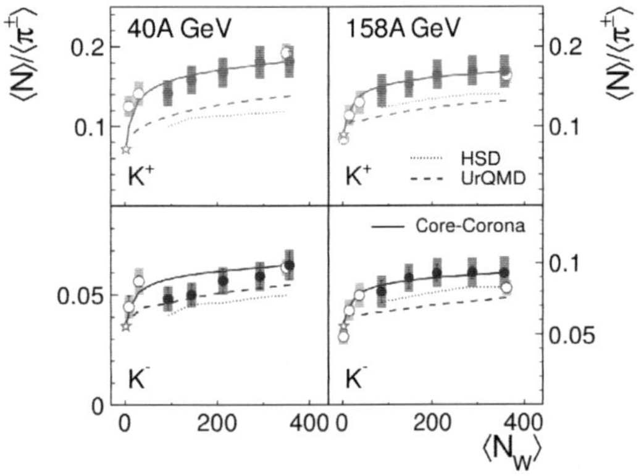

Figure 3. (Color online) Kaon yields divided by mean number of pions as a function of centrality. Closed symbols show the new results from centrality selected Pb+Pb collisions, open symbols p+p data from NA49 [5, 6] as well as C+C, Si+Si (at 158A GeV [4]) and central Pb+Pb collisions from [7, 8]). The stars denote the values obtained from the parametrization "B" in reference [12]. Curves show results from HSD [9, 10] and UrQMD2.3 [11] simulations and the core-corona model (the latter for the Pb+Pb collision system only,[13]). The thin vertical bars indicate the statistical errors, which are mostly not visible. The thick shaded bars indicate the systematic errors.

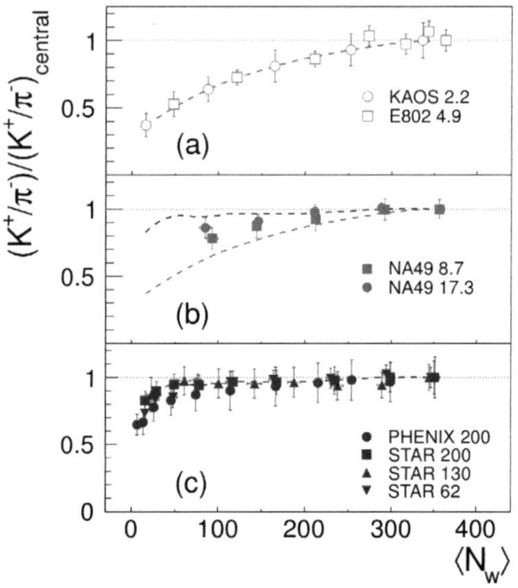

Figure 4. (Color online) Centrality dependence of the double ratio of the K^+ to π^- multiplicities scaled to the ratios from the corresponding most central measurements for minimum bias Pb+Pb (Au+Au) collisions at SIS [14] and AGS [15, 16] energies (a), at SPS energies (b) and at RHIC energies [17, 18] (c). The dashed curves in panels (a) and (c) are meant to guide the eye and are also drawn as a reference in panel (b). Only statistical errors are shown.

unmeasured regions by a single exponential. A double exponential had to be used for pions in Pb+Pb collisions because of the concave shape of their p_T-spectra. The rapidity distributions of p_T-integrated yields for the particles, collision systems and energies under study are well described by two Gaussians of equal widths σ which are displaced symmetrically around mid-rapidity by a constant y_0:

$$\frac{\mathrm{d}N}{\mathrm{d}y} = C\left[\exp\left(-\frac{(y-y_0)^2}{2\sigma_y^2}\right) + \exp\left(-\frac{(y+y_0)^2}{2\sigma_y^2}\right)\right]. \tag{1}$$

Using this functional form the measured rapidity spectra can be extrapolated into the unmeasured regions, and particle yields in full phase space can be extracted.

The widths of the rapidity distributions are quantified by $RMS_y = \sqrt{\sigma_y^2 + y_0^2}$. Figure 2 presents the system-size dependence of the widths of the rapidity distributions. While kaons show no significant change of the width as a function of the number of wounded nucleons, rapidity distributions of pions are significantly wider in peripheral than in central Pb+Pb collisions. In fact, it turns out that the width of the pion rapidity distribution in central Pb+Pb collisions is the same as in C+C and Si+Si reactions as well as in p+p interactions at 158 GeV/c. Only non-central Pb+Pb collisions deviate from this common behavior, exhibiting a wider distribution.

The widening of the pion rapidity distribution for peripheral collisions is probably due to pion production in interactions of participants with spectator matter, an idea which had been already invoked to explain the proton rapidity spectra in the same data sets [3].

One observable used to quantify the strangeness enhancement in A+A collisions with respect to p+p collisions is the K/π ratio. We present the $K^+/\pi^\pm$ and $K^-/\pi^\pm$ ratios at both energies in Fig. 3. ($\langle\pi^\pm\rangle = 0.5\cdot(\langle\pi^-\rangle + \langle\pi^+\rangle)$.) The normalized yields of all kaons show a steep increase for small system sizes followed by a slow rise or even saturation for higher centralities. A closer look reveals that at 158*A* GeV the K/π ratios in the small collision systems C+C and Si+Si seem to follow the trend set by the centrality dependence of the Pb+Pb data which is well described by the core-corona approach, whereas at 40*A* GeV the normalized kaon abundance in peripheral Pb+Pb collisions is lower than expected from the extrapolation of the K/π ratio in C+C and Si+Si collisions (most pronounced for the negatively charged kaons). The Pb+Pb data in Fig. 3 are compared to transport models (HSD and UrQMD) as well as the core-corona approach.

In Fig. 4 we summarize the K^+/π^- ratios for centrality selected Au+Au (Pb+Pb) collisions at different beam energies. The ratios at the different centralities were scaled to the ratios measured in the most central collisions. The observed saturation of strangeness production at SPS energy and above can be understood in the context of statistical models by the approach to a grand-canonical description. Increasing the energy available for particle production no longer changes the multiplicities of strangeness carrying particles relative to pions [7, 8], which means that strangeness becomes fully saturated. The situation is different at low (SIS) energies. Here the K^+/π^- ratio does not saturate and its dependence on the number of wounded nucleons is rather linear. The results obtained at 40*A* GeV are intermediate between those at top SPS and AGS energies. We thus observe a smooth transition between both scenarios which occurs in the SPS energy region.

Acknowledgments
I thank the organizers of SQM2013 and the NA49 collaboration for the opportunity to give this talk.

References

[1] Anticic T *et al* 2012 (NA49 Collaboration) *Phys. Rev.* C **86** 054903
[2] Anticic T *et al* 2009 (NA49 Collaboration) *Phys. Rev.* C **80** 034906
[3] Anticic T *et al* 2011 (NA49 Collaboration) *Phys. Rev.* C **83** 014901
[4] Alt C *et al* 2005 (NA49 Colaboration) *Phys. Rev. Lett.* **94**, 052301
[5] Alt C *et al* 2006 (NA49 Collaboration) *Eur. Phys. J.* C **45** 343
[6] Anticic T *et al* 2010 (NA49 Collaboration) *Eur. Phys. J.* C **68** 1
[7] Afanasiev S V *et al* 2002 (NA49 Collaboration) *Phys. Rev.* C **66** 054902
[8] Alt C *et al* 2008 (NA49 Collaboration) *Phys. Rev.* C **77** 024903
[9] Ehehalt W and Cassing W 1996 *Nucl. Phys.* A **602** 449
[10] Cassing W and Bratkovskaya E L 1999 *Phys. Rep.* **308**, 65
[11] Petersen H, Bleicher M, Bass S A and Stöcker H arXiv:0805.0567 [hep-ph]
[12] Rossi A M *et al* 1975 *Nucl. Phys.* B **84** 269
[13] Blume C 2010 *J. Phys.: Conf. Ser.* **230** 012003
[14] Förster A *et al* 2005 *J. Phys. G* **31** S693
[15] Ahle L *et al* 1999 *Phys. Rev.* C **60** 044904
[16] Wang F, Liu L, Sorge H, Xu N and Yang J 2000 *Phys. Rev.* C **61** 064904
[17] Adler S S *et al* 2004 (Phenix Collaboration) *Phys. Rev.* C **69** 034909
[18] Abelev B I *et al* 2009 (STAR Collaboration) *Phys. Rev.* C **79** 034909
[19] Cleymans J, Oeschler H and Redlich K 1999 *Phys. Rev.* C**59** 1663

Journal of Physics: Conference Series **509** (2014) 012041 doi:10.1088/1742-6596/509/1/012041

Production of π/K/p in pp and Pb-Pb collisions measured with ALICE

Marek Chojnacki for the ALICE Collaboration
Niels Bohr Institute, Blegdamsvej 17, 2100 Copenhagen Denmark
E-mail: `Marek.Chojnacki@cern.ch`

Abstract. The ALICE experiment is a dedicated heavy-ion physics detector at the LHC, with unique capabilities for the study of identified particle production. In this paper, an overview of ALICE results on the measurement of pion, kaon and proton transverse momentum (p_T) spectra at mid-rapidity in proton-proton and Pb-Pb collisions at different energies is presented. The particle identification is performed using the following detectors: the Time Projection Chamber (TPC), the Inner Tracking System (ITS), the Time of Flight system (TOF) and a Ring-imaging Cherenkov system (HMPID).

1. Introduction

The ALICE detector [1, 2, 3] features multiple particle identification systems including: the Time Projection Chamber (TPC), the Inner Tracking System (ITS), the Time of Flight system (TOF) and a Ring-imaging Cherenkov detector (HMPID). This combination of detectors, along with the excellent tracking capabilities of ALICE, provides the opportunity to measure the production of pions, kaons and protons over a broad transverse momentum (p_T) range, from 100 MeV/c up to 20 GeV/c. Particle identification at low p_T (below $\sim$ 1 GeV/c) is performed using the energy loss of particles in the ITS. The TOF velocity measurement contributes to the identification for the p_T range between 0.5 GeV/c and 3-5 GeV/c (depending on the particle type and the colliding system). At high p_T (up to 20 GeV/c), particles are identified using the HMPID or the relativistic rise of the energy loss distribution in the TPC.

2. Results

Since November 2009, the LHC has produced collisions of different systems (pp, p-Pb and Pb-Pb) at several energies. For all of those systems, the ALICE collaboration has measured pion, kaon and proton production as a function of p_T at mid-rapidity. A detailed description of the analyses can be found in [4, 5, 6].

Figure 1 shows the p_T dependent $(K^+ + K^-)/(\pi^+ + \pi^-)$ and $(p + \bar{p})/(\pi^+ + \pi^-)$ ratios in pp collisions at $\sqrt{s} = 2.76$ and 7 TeV with comparisons to some theoretical predictions. Within the systematic uncertainties, both ratios are consistent for two energies. The NLO based calculations of [7] do not reproduce the measured ratios. The PHYTIA Monte Carlo generator (tune Perugia 2011) does not fully reproduce the ratios, underestimating both ratios at intermediate p_T.

The results for Pb-Pb are summarized in [6]. Figure 2 shows the p_T spectra measured by ALICE for the 0-5% most central Pb-Pb collisions. The spectra are compared to RHIC results from Au-Au collisions at $\sqrt{s_{NN}}$ =200 GeV/c [8, 9] and theoretical models. Since the

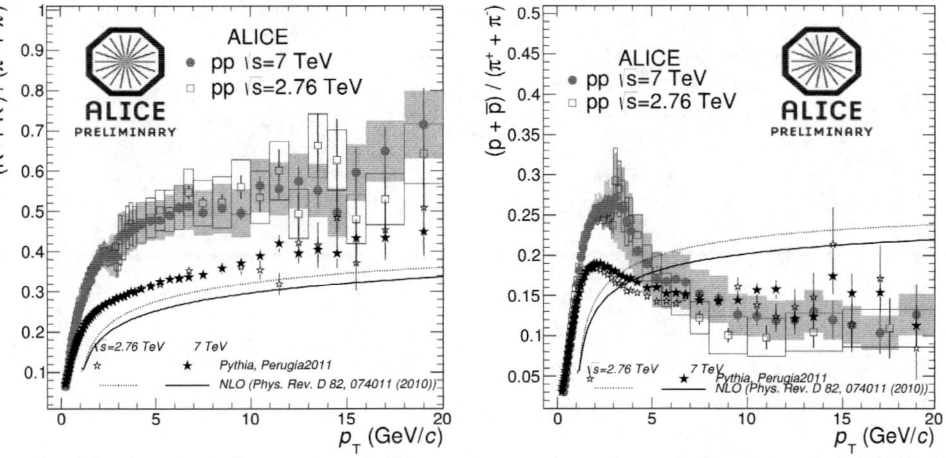

Figure 1. Ratios of kaon, proton and antiproton $p_{\rm T}$ spectra to the pion $p_{\rm T}$ spectra in pp collisions at $\sqrt{s}$ = 2.76 and 7 TeV compared to PYTHIA Monte Carlo predictions and NLO calculations.

ALICE spectra for positive and negative particles are compatible within systematic uncertainties, the combined spectra (summed charge states) are used. The ALICE results show harder distributions compared to the RHIC results which indicates a significantly stronger radial flow at LHC energies. The comparison to hydrodynamical models shows that VISH2+1 [10] describes the pion and kaon spectra below $p_{\rm T}$ 1.5 GeV/c well, but it does not describe the proton spectra. The lack of the hadronic phase in this model could explain the discrepancy with the shape of the proton spectra. This hypothesis is supported by comparison to the HKM Model [11], which better agrees with the data. This model uses a hadronic cascade model to describe the system evolution after the hydrodynamic phase until the final decoupling. The Kraków model [12], which also shows a good agreement with data, uses non-equilibrium corrections due to the bulk viscosity which changes the effective T_{ch}. The EPOS model, which uses breakup of flux tubes created during initial hard scatterings, describes the spectra shapes in a broad $p_{\rm T}$ range.
The ratios of the production yields at mid-rapidity were compared (see Figure 3) to the RHIC results and thermal model predictions [14, 15], (for μ_B= 1 MeV and T_{ch} = 164 MeV or 170 MeV). The baryochemical potential μ_B vanishes because the yields of antiparticles and particles are equal within uncertainties. The thermal models do not reproduce the proton to pion ratio. To better understand this discrepancy, the production yields of different particles were also compared with the thermal model predictions (see Figure 4). It seems that a simple equilibrium thermal model cannot describe the data. Possible solutions for this unexpected observation include: baryon annihilation after chemical freeze-out [16], non-equilibrium statistical hadronization [17], flavour hierarchy in the QCD phase transition [18], missing higher mass resonance states in the equilibrium thermal model[19].

Figure 5 shows the pion, kaon and proton nuclear modification factors (R_{AA}) for the 0-5% and 60-80% centrality bins. At lower $p_{\rm T}$ a difference in the suppression pattern of different particle species can be seen, with the heavier particles being less suppressed. This reflects the "baryon anomaly" also observed e.g. in the Λ/K^0_S ratio [20] and could be a consequence of radial flow or quark coalescence. At higher $p_{\rm T}$, above 8 GeV/c, the R_{AA} for all particle species are consistent within uncertainties. Theoretical calculations based on the separate treatment of radiative and collisional energy losses [21] can reproduce the measured R_{AA} in this region. Any difference between R_{AA} for different particle species at high $p_{\rm T}$ could be interpreted as a

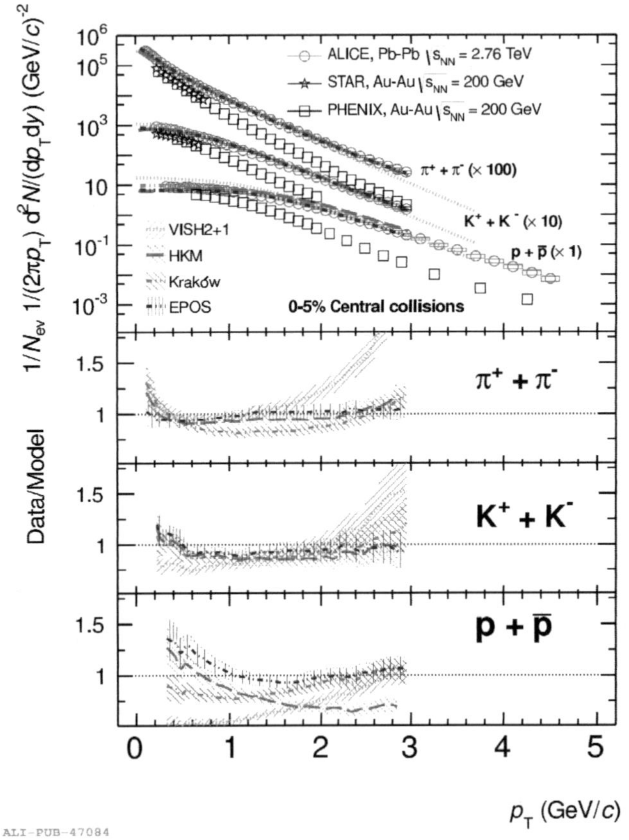

Figure 2. Pion, kaon and proton spectra in the most central Pb-Pb collisions at $\sqrt{s_{NN}}$ =2.76 TeV compared to the results from lower energies and model predictions.

difference in the fragmentation functions within the medium. However, no strong constraints on the fragmentation functions can be extracted from the present data.

3. Conclusion

The ALICE collaboration has successfully measured the p_{T} dependence of pion, kaon and proton production at several energies for different colliding systems. The ALICE results for pp collisions show no energy dependence in hadronic ratios as function of p_{T}. In Pb-Pb collisions, p_{T} spectra indicate a stronger radial flow than at lower (RHIC) energies. The proton production is significantly lower than thermal model predictions with a chemical freeze-out temperature T_{ch} = 160 - 170 MeV. At higher p_{T}, current measurements indicate that the medium does not significantly affect the parton fragmentation.

References

[1] ALICE Collaboration 2004 *J. Phys.* G **30** 1517-1763
[2] ALICE Collaboration 2006 *J. Phys.* G **32** 1295-2040
[3] ALICE Collaboration 2008 *J. Instrum.* **3** SO8002
[4] ALICE Collaboration 2011 *Eur. Phys. J.* C **71**(6) 1655
[5] ALICE Collaboration 2012 *Phys. Rev. Lett.* **109** 252301
[6] ALICE Collaboration 2013 *Phys. Rev.* C **88** 044910

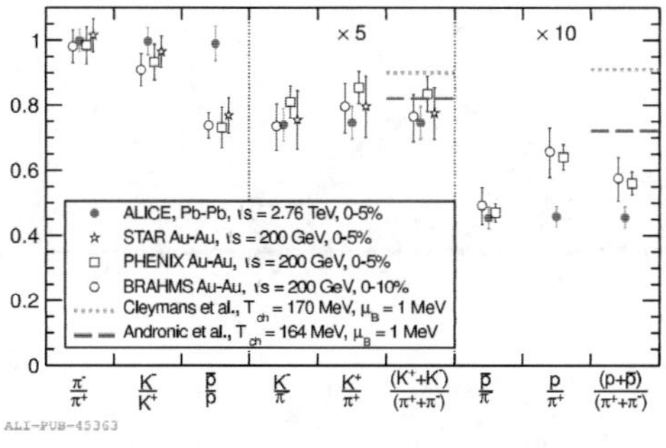

Figure 3. Ratios of pion, kaon and proton yields in the most central (0-5%) Pb-Pb collisions at $\sqrt{s_{NN}}$ =2.76 TeV compared to predictions from thermal models.

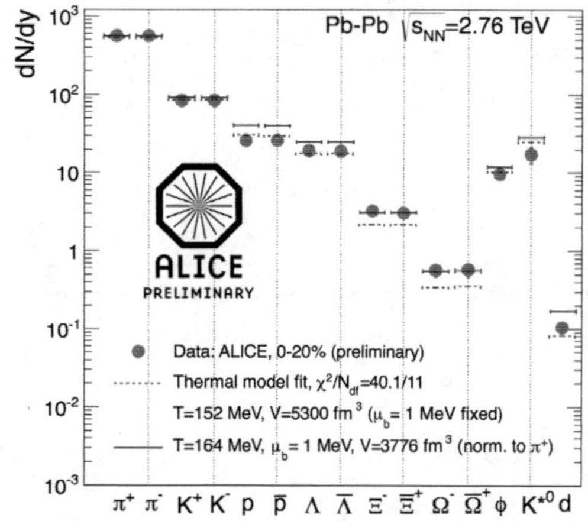

Figure 4. The production yields of different particles in the most central (0-20%) Pb-Pb collisions at $\sqrt{s_{NN}}$ =2.76 TeV compared to predictions from thermal models.

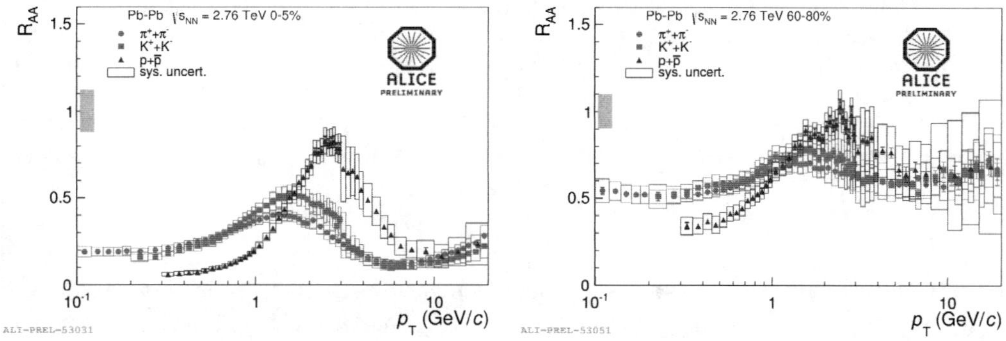

Figure 5. Pion, kaon and proton nuclear modification factors in Pb-Pb collisions at $\sqrt{s_{NN}}$ = 2.76 TeV in two centrality bins.

[7] Sassot R, Zurita P and Stratmann M 2010 *Phys. Rev.* D **82** 074011
[8] STAR Collaboration 2009 *Phys. Rev.* C **79** 034909
[9] PHENIX Collaboration 2004 *Phys. Rev.* C **69** 034909
[10] Shen C, Heinz U, Huovinen P and Song H 2011 *Phys. Rev.* C **84** 044903
[11] Karpenko I, Sinyukov Y and Werner K 2013 *Phys. Rev.* C **87** 024914
[12] Bożek P 2012 *Phys. Rev.* C **85** 034901
[13] Werner K, Karpenko I, Bleicher M, Pierog T and Porteboeuf-Houssais S 2012 *Phys. Rev.* C **85** 064907
[14] Andronic A, Braun-Munzinger P and Stachel J 2009 Phys. Lett. B **673** 142
[15] Cleymans J, Kraus I, Oeschler H, Redlich K and Wheaton S 2006 *Phys. Rev.* C **74** 034903
[16] Becattini F, Bleicher M, Kollegger T, Schuster T, Steinheimer J and Stock R 2013 *Phys. Rev. Lett.* **111** 082302
[17] Petran M, Letessier J, Petracek V and Rafelski J 2013 *Phys. Rev.* C **88** 034907
[18] Ratti C, Bellwied R, Cristoforetti M and Barbaro M 2012 *Phys. Rev.* D **85** 014004
[19] Stachel J *these proceedings*
[20] ALICE Collaboration 2013 *Phys. Rev. Lett.* **111** 222301
[21] Djordjevic M and Djordjevic M *Preprint* arXiv:1307.4714

Journal of Physics: Conference Series **509** (2014) 012042 doi:10.1088/1742-6596/509/1/012042

Kaon femtoscopy in $\sqrt{s_{\rm NN}}$=200 GeV central Au+Au collsions at STAR

Róbert Vértesi (for the STAR Collaboration)
Nuclear Physics Institute ASCR, 25068 Řež, Czech Republic
E-mail: robert.vertesi@ujf.cas.cz

Abstract. Three-dimensional analyses of the pion source revealed a heavy, non-Gaussian tail in the direction of the pair transverse momentum. The interpretation of these pion sources in terms of pure hydrodynamical evolution is, however, complicated by the strong contribution of feed-down from long-lived resonances to the source. On the other hand, kaons provide a much cleaner probe of the expanding fireball. Here we present a recent three-dimensional kaon correlation analysis using Cartesian harmonics decomposition technique. In contrary to pions, the three-dimensional source function of kaons is largely Gaussian. Comparison with thermal simulations and a hydrokinetic model show that resonance decays, as well as non-zero emission duration and/or rescattering in the hadronic phase play an important role. The analysis of the three dimensional extent of the kaon source w.r.t. the pair transverse momentum favors the hydrokinetic model over the exact $m_{\rm T}$-scaling featured by perfect hydrodynamical models.

1. Introduction

Bose-Einstein correlations (BEC) between bosons emitted incoherently carry information about the space-time extent of the emitting source. Femtoscopy studies the size and evolution of the hot, strongly interacting medium created in high energy heavy ion collisions using BEC measurements. Analysis of two-pion correlation functions with source imaging [1] revealed a long-range, non-Gaussian component in the pion source function [2]. The extraction of the three-dimensional (3D) pion source function using the Cartesian surface-spherical harmonic decomposition technique [3, 4] in conjunction with model comparisons has permitted the decoupling of the spatio-temporal observable into its spatial and temporal aspects, and the latter into source lifetime and emission duration [5, 6]. However, an interpretation of pion correlations in terms of pure hydrodynamic evolution is complicated by significant contributions from later stages of the reaction, such as decays of long-lived resonances and anomalous diffusion from rescattering [7, 8]. A purer probe of the fireball decay can be obtained with kaons, which suffer less contribution from long-lived resonances. The lower yields, however, make kaon measurements more difficult. A one-dimensional kaon source image measurement by PHENIX recently reported an even heavier tail than in pions [9], but this measurement corresponds to a fairly broad range of the pair transverse momentum $2k_{\rm T}$, which makes the interpretation complicated. A different aspect of the fireball expansion can be addressed by studying the $k_{\rm T}$-dependence of the source size. NA49 data at SPS energies, as well as PHENIX data at relatively higher $k_{\rm T}$ values, showed a scaling behavior between pions and kaons, as expected from perfect hydrodynamics models [10, 9].

2. Data analysis

We present recent STAR 3D analyses [11] of the shape and k_T-dependent size of the kaon source at mid-rapidity, extracted from BEC using low transverse momentum like–sign kaon pairs in $\sqrt{s_{NN}}$=200 GeV central Au+Au collisions. The kaon source shape was analyzed using 4.6 million 0–20% central events from 2004, and 16 million 0–20% central events from 2007. The k_T–dependent analysis was carried out using 6.6 million 0–30% central events from 2004. The position of the primary vertex along the beam line was constrained to be $|z|<30$ cm. Kaon tracks with a low transverse momentum ($0.1<p_T<1.0$ GeV/c) were selected in the STAR Time Projection Chamber (TPC) [12], within a 0.5 T quasi-uniform magnetic field provided by the solenoid surrounding it. The identification of kaons is based on fractional energy loss dE/dx. In the source shape analysis, candidates were required to fulfill the kaon hypothesis to the extent of $|n\sigma_K| < 2$ and to be rejected as pions or electrons as $|n\sigma_\pi| > 3$ or $|n\sigma_e| > 2$, where $n\sigma_X$ is the deviation from the normalized distribution of particle type X at a given momentum. Only pairs with $0.2<k_T<0.36$ GeV/c were accepted. In the k_T–dependent analysis, kaon pairs were collected in two bins: $0.2<k_T<0.36$ GeV/c and $0.36<k_T<0.48$ GeV/c, with a slightly different kaon selection of $-1.5<n\sigma_K<2.0$ and $-0.5<n\sigma_K<2.0$, respectively. Since proton and kaon bands are well separated in both k_T bins, protons are also effectively removed by these cuts.

From an experimental point of view, the 3D correlation function is defined as the ratio of the 3D relative momentum distribution of charged kaon pairs to pairs from mixed events,

$$C(\mathbf{q}) = \frac{N_{\text{same}}(\mathbf{q})}{N_{\text{mixed}}(\mathbf{q})}, \tag{1}$$

where $2\mathbf{q}$ is the difference between the momenta of the two particles in the pair center-of-mass system (PCMS). The source function $S(\mathbf{r})$ is the probability of emitting a pair with a spatial separation $\mathbf{r}$. It is linked to the correlation function via the Koonin-Pratt equation [13, 14, 15],

$$C(\mathbf{q}) - 1 \equiv R(\mathbf{q}) = \int K(\mathbf{q}, \mathbf{r}) S(\mathbf{r}) d\mathbf{r}, \tag{2}$$

where the kernel can be expressed in terms of the relative wave function, $K(\mathbf{q}, \mathbf{r}) = |\phi(\mathbf{q}, \mathbf{r})|^2 - 1$, and encodes the quantum correlation as well as final state interactions. Source imaging methods numerically invert the above equation without any assumption for the source shape [1, 3]. In the 3D case when the source and correlation functions are handled direction-dependently, the longitudinally co-moving system (LCMS) is often used: The coordinate axes x-y-z form a right-handed *out-side-long* Cartesian coordinate system oriented so that the z-axis is parallel to the beam direction and x points in the direction of the pair total transverse momentum. The correlation and source functions can be expanded in Cartesian harmonics basis elements $A^l_{\alpha_1 \ldots \alpha_l}(\Omega_\mathbf{q})$ [3, 4], where $l = 0, 1, \ldots$, $\alpha_i = x, y$ or z, $\Omega_\mathbf{q}$ ($\Omega_\mathbf{r}$) is the solid angle in $\mathbf{q}$ ($\mathbf{r}$) space, as

$$R(\mathbf{q}) = \sum_{l, \alpha_1 \ldots \alpha_l} R^l_{\alpha_1 \ldots \alpha_l}(q) \, A^l_{\alpha_1 \ldots \alpha_l}(\Omega_\mathbf{q}) \quad \text{and} \quad S(\mathbf{r}) = \sum_{l, \alpha_1 \ldots \alpha_l} S^l_{\alpha_1 \ldots \alpha_l}(r) A^l_{\alpha_1 \ldots \alpha_l}(\Omega_\mathbf{r}) \,. \tag{3}$$

The limited statistics of charged kaons prevented us from carrying out an imaging study. We found, however, that all the moments except for l=0 and l=2 are consistent with zero, and that the direction independent correlation function R(q) agrees with R^0 within statistical errors. Therefore a 3D Gaussian function,

$$S^G(r_x, r_y, r_z) = \frac{\lambda}{(2\sqrt{\pi})^3 R_x R_y R_z} \exp[-(\frac{r_x^2}{4R_x^2} + \frac{r_y^2}{4R_y^2} + \frac{r_z^2}{4R_z^2})], \tag{4}$$

was applied as a trial function for the source, where R_x, R_y and R_z are the characteristic radii of the source in the *out*, *side* and *long* directions, and λ represents the overall correlation strength.

While λ may be sensitive to feed-down from long-lived resonance decays, remainder sample contamination or track splitting/merging not removed by purity and track quality cuts, the radii are virtually independent of these effects. Technically, the fit is carried out as a simultaneous fit on the even independent moments up to $l=4$, yielding $\chi^2/ndf=1.7$ in the source shape analysis, $\chi^2/ndf=1.1$ and $\chi^2/ndf=1.3$ in the $0.2<k_T<0.36$ GeV/c and $0.36<k_T<0.48$ GeV/c bins in the k_T-dependent analysis, respectively. The shape assumption was tested using a double Gaussian trial function and by pushing the fit parameters to the edges of their errors. Other systematic errors were obtained under varying conditions including magnetic field, data collection periods, charge and various sample purity selections. The systematic errors are largely governed by the limited statistics available.

3. Results

The source function profiles in the x, y and z directions are shown on Figure 1 (circles). The two solid curves around the Gaussian source function profiles represent the error band arising from the statistical and systematic errors on the 3D Gaussian fit. Note that the latter becomes important for large r values only. The 3D pion source functions from PHENIX [5] are shown for comparison purposes (squares). While the Gaussian radii are similar, there is a strinking difference between the source shapes of the two particle species, especially in the *out* direction. Note that the PHENIX and STAR pion measurements are fully consistent [6]. We have used the STAR tune of the Therminator Blast-Wave model [16, 17] to gain a better understanding of this difference. The simulation reproduces the source function profiles with emission duration $\Delta\tau=0$ (solid upward-pointing triangles). However, with resonance contribution switched off, Therminator gives a distribution that is narrower than the measurement (empty triangles). Also note that the pion source function is reproduced with Therminator only when non-zero emission duration is assumed [5]. Recent simulations with the hydrokinetic model (HKM) [18] show good agreement in the *side* direction, although it is slightly over the measurements at larger radii in *out* and *long* (downward-pointing triangles).

Figure 2 shows the dependence of the Gaussian radii in LCMS ($R_{out}=R_x/\gamma$, $R_{side}=R_y$ and $R_{long}=R_z$; γ is the Lorentz boost in the outward direction from the LCMS to the PCMS frame) as a function of the transverse mass $m_T = (m^2 + k_T^2)^{1/2}$. PHENIX kaon data [9] are also shown. The perfect fluid hydrodynamics calculations from the Buda-Lund model [19] and the HKM with Glauber initial conditions [20] are plotted for comparison purposes. While pions are well described by the Buda Lund model in the whole interval shown [19], low-k_T kaons in the *long* direction seem to favor HKM over the Buda-Lund model, suggesting that the m_T-scaling is broken in the *long* direction.

4. Summary

We have accounted for the first model-independent extraction of 3D kaon source by the STAR Collaboration [11], at mid-rapidity in $\sqrt{s_{NN}}=200$ GeV central Au+Au collisions, using low-k_T kaon pair correlations and the Cartesian surface-spherical harmonic decomposition technique. No significant non-Gaussian tail has been observed. Comparison with the Therminator model calculations indicates that, although the transverse extent of the source is similar to pions, the shape and the size in the longitudinal direction is very different. This can be attributed to resonance decays, but also indicates that kaons and pions may be subject to different freeze-out dynamics. Although the Gaussian radii follow m_T-scaling in the outward and sideward directions, the scaling appears to be broken in the longitudinal direction. Thus the hydro-kinetic predictions [20] are favored over pure hydrodynamical model calculations. The agreement of HKM simulations with both the source shape and the radii suggests that rescattering in the hadronic phase may also give a non-negligible contribution to the observed source characteristics.

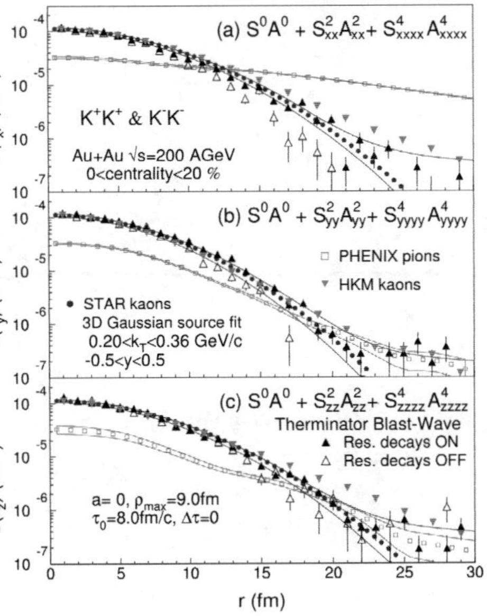

Figure 1. Kaon source function profiles extracted from the data (solid circles) compared to 3D pion source function (squares) from PHENIX [5], and to the Therminator (triangles pointing upwards) and HKM (triangles pointing downwards) models.

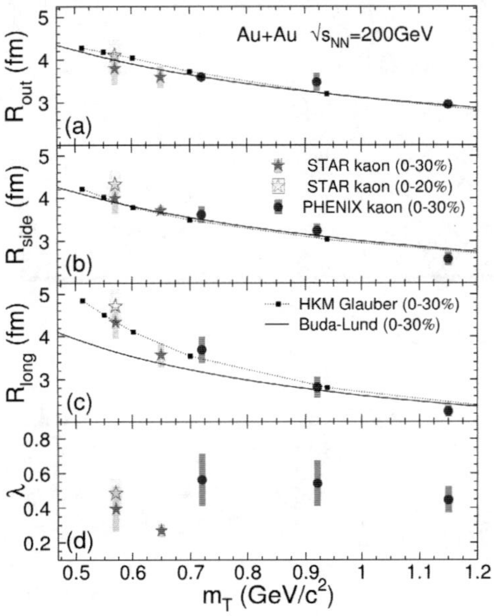

Figure 2. Transverse mass dependence of Gaussian radii and the λ for the 30% most central Au+Au collisions (solid stars). PHENIX data are also plotted (dots). Squares are HKM, solid curves are Buda-Lund model calculations. The 20% most central data are also shown for comparison (open stars).

This research was supported in part by the grant CZ.1.07/2.3.00/20.0207 of the ESF programme "Education for Competitiveness Operational Programme", the grant LA09013 of the Ministry of Education of the Czech Republic and by the Hungarian OTKA grant NK 101438.

References

[1] Brown D A and Danielewicz P 1997 *Phys. Lett. B* **398** 252
[2] Adler S S *et al.* (PHENIX Collaboration) 2007 *Phys. Rev. Lett.* **98** 132301
[3] Danielewicz P and Pratt S, 2005 *Phys. Lett. B* **618** 60
[4] Danielewicz P and Pratt S, 2007 *Phys. Rev. C* **75** 034907
[5] Afanasiev S *et al.* (PHENIX Collaboration) 2008 *Phys. Rev. Lett.* **100** 232301
[6] Chung P (STAR Collaboration) 2011 *Phys. Part. Nucl. Lett.* **8** 1019
[7] Vértesi R (PHENIX Collaboration) 2007 *Proc. of 23rd Winter Workshop on Nuclear Dynamics, Big Sky, Montana, USA* (Budapest: EP Systema)
[8] Csanád M, Csörgő T and Nagy M 2007 *Braz. J. Phys.* **37** 1002
[9] Afanasiev S *et al.* (PHENIX Collaboration) 2009 *Phys. Rev. Lett.* **103** 142301
[10] Afanasiev S V *et al.* (NA49 Collaboration) 2003 *Phys. Lett. B* **557** 157
[11] Adamczyk L *et al.* (STAR Collaboration) 2013 *Phys. Rev. C* **88** 034906
[12] Ackermann K H *et al.* (STAR Collaboration) 2003 *Nucl. Instrum. Meth. A* **499** *624*
[13] Koonin S E 1977 *Phys. Lett. B* **70** 43
[14] Lisa M A, Pratt S, Soltz R and Wiedemann U, 2005 *Ann. Rev. Nucl. Part. Sci.* **55** 357
[15] Lednicky R 2009 *Phys. Part. Nucl.* **40** 307
[16] Kisiel A, Taluc T, Broniowski W and Florkowski W, 2006 *Comput. Phys. Commun.* **174** 669
[17] Kisiel A *et al.* 2006 *Phys. Rev. C* **73** 064902
[18] Shapoval V M, Sinyukov Y M and Karpenko I A 2013 *Phys. Rev. C* **88** 064904
[19] Csanád M and Csörgő T, 2008 *Acta Phys. Polon. Supp.* **1** 521
[20] Karpenko I A and Sinyukov Y M 2010 *Phys. Rev. C* **81** 054903

Journal of Physics: Conference Series **509** (2014) 012043 doi:10.1088/1742-6596/509/1/012043

Study of h-h and V^0-h angular correlations in Pb–Pb collisions at $\sqrt{s_{NN}} = 2.76$ TeV

Marek Bombara on behalf of the ALICE Collaboration
Pavol Jozef Šafárik University, Šrobárova 2, Košice 041 80, Slovakia
E-mail: Marek.Bombara@cern.ch

Abstract. Two-particle angular correlations provide a powerful tool to study jets and their modification in ultra-relativistic heavy-ion collisions. The study of the particle species dependence of correlation structures as a function of transverse momentum provides additional information on particle production mechanisms. In this contribution we present studies of a near-side jet-like correlation peak in h-h correlations in Pb–Pb collisions at $\sqrt{s_{NN}} = 2.76$ TeV. We observe significant broadening of the jet-like peak that increases with decreasing p_{T}^{trig} and p_{T}^{assoc} and with increasing centrality and additional broadening of the jet-like peak in $\Delta\eta$ with respect to broadening in $\Delta\phi$ in most central collisions.

The prospects of expanding these studies to include strange particles (Λ, $\overline{\Lambda}$ and K^0_S) is explored.

1. Introduction

In Pb–Pb collisions at LHC the jet shapes and fragmentation functions show modifications at low p_{T} (large jet cone radii) in comparison to pp collisions but the high p_{T} core is unchanged [1]. It was also predicted for heavy-ion collisions that the jet shape can be deformed by a longitudinally flowing medium [2] and therefore we might observe jets with asymmetrical cones in central heavy-ion collisions. Jets highly modified by such mechanisms might be difficult to reconstruct by jet-finding algorithms. In addition, full jet reconstruction in central heavy-ion collisions is restricted to the jet high p_{T} core due to fluctuations in the bulk. Since two-particle correlations are less sensitive to these effects, they provide a complementary approach to full jet reconstruction.
Quark and gluon jets can be distinguished (on a statistical basis) due to differences in their multiplicity and width [3] and also baryon and meson content [4]. Baryon production in gluon jets is enhanced with respect to quark jets. Studying K^0_S-h and Λ-h ($\overline{\Lambda}$-h), denoted V^0−h (i.e. V^0−hadron), correlations could provide samples enhanced in gluon or quark jets. An advantage of V^0−h correlations is that it is possible to get a clean sample even at high p_{T} ($\sim$20 GeV/c) via reconstruction of the daughter tracks ($K^0_S \rightarrow \pi^+\pi^-$, $\Lambda \rightarrow p\pi^-$ and $\overline{\Lambda} \rightarrow \overline{p}\pi^+$).

2. Analysis

The ALICE experiment is designed to study hot and dense nuclear matter created in heavy-ion collisions [5]. In the h-h analysis, 15×10^6 Pb–Pb collisions taken in 2010, at $\sqrt{s_{NN}} =$ 2.76 TeV and 55×10^6 pp collisions at $\sqrt{s} = 2.76$ TeV taken in 2011 were investigated. The centrality of the Pb–Pb collision was estimated using the VZERO detector and tracking was performed using information from the TPC (Time Projection Chamber). In the correlation

Journal of Physics: Conference Series **509** (2014) 012043 doi:10.1088/1742-6596/509/1/012043

technique we select a high p_{T} (trigger) particle in a specific p_{T}^{trig} interval and associated particles with $p_{\mathrm{T}}^{assoc} < p_{\mathrm{T}}^{trig}$. Calculating angular differences $\Delta\phi = \phi_{assoc} - \phi_{trig}$ and $\Delta\eta = \eta_{assoc} - \eta_{trig}$ and plotting these two variables as 2D histogram (same-event pairs), we obtain a jet-like peak sitting on a triangular shaped background. The triangular shape can be reproduced by mixed pairs, where the trigger and the associated particles come from different events. The acceptance corrected distribution can be estimated as:

$$\frac{\mathrm{d}^2 N^{raw}}{\mathrm{d}\Delta\phi \mathrm{d}\Delta\eta}(\Delta\phi, \Delta\eta) = \frac{N_{pair}^{same}(\Delta\phi, \Delta\eta)}{N_{pair}^{mixed}(\Delta\phi, \Delta\eta)}\beta \tag{1}$$

The normalization factor β was chosen in such a way that the $N_{pair}^{mixed}(\Delta\phi, \Delta\eta)$ is 1 at $\Delta\phi = \Delta\eta = 0$. There were also additional corrections applied for the single track efficiency and contamination correction (from secondary particles) and also for track merging/splitting.

The corrected angular correlations can be quantified by the per-trigger associated particle yield:

$$\frac{\mathrm{d}^2 N}{\mathrm{d}\Delta\phi \mathrm{d}\Delta\eta}(\Delta\phi, \Delta\eta) = \frac{1}{N_{\mathrm{trig}}}\frac{\mathrm{d}^2 N_{\mathrm{assoc}}}{\mathrm{d}\Delta\phi \mathrm{d}\Delta\eta} \tag{2}$$

for p_{T}^{trig} and p_{T}^{assoc} intervals. We can usually see the near-side peak located at (0,0) and the away-side peak (recoiling jet) located at $\Delta\phi \approx \pi$ and spread out in $\Delta\eta$. Both peaks sit (in Pb–Pb) on a flow-modulated background (more pronounced for low p_{T}^{trig} and p_{T}^{assoc} intervals). To estimate this background we assume that the flow is independent of $\Delta\eta$ within (-1.8,1.8) and the jet does not contribute to $|\Delta\eta| > 1.0$. The signal is then extracted by subtracting the scaled side bands $1 < |\Delta\eta| < 1.6$ (η-gap method).

3. Results for jet shape analysis in h-h correlations

The near-side peak in h-h correlations was fitted with a sum of two 2D Gaussians centred at $\Delta\phi = \Delta\eta = 0$. The fit parameters were used to calculate σ in $\Delta\phi$ and in $\Delta\eta$. The results for five p_{T}^{trig} and p_{T}^{assoc} intervals in pp collisions and in Pb–Pb collisions with five centrality bins are shown in fig. 1 and 2. As can be seen in fig. 1, the $\sigma_{\Delta\phi}$ is, within our uncertainties, independent of centrality, and decreases with increasing p_{T}^{assoc} and p_{T}^{trig}. The $\sigma_{\Delta\eta}$ (fig. 2) increases as collisions become more central, but decreases as p_{T}^{assoc} and p_{T}^{trig} increase.

Another way of studying the jet-like peak is the kurtosis ($\mathrm{K}_{\Delta\phi}$, $\mathrm{K}_{\Delta\eta}$) = μ_4/μ_2^2 - 3 (note: μ_n - n^{th} moment about the mean): which is a measure of the peakedness of the distribution (Laplace: K=3, Gaussian: 0, semicircle: -1, uniform: -1.2). Both kurtoses (fig. 3 and 4) decrease going to lower p_{T} (the peaks are less sharp) and decrease in more central events. Interestingly, for the lowest p_{T} interval in $\Delta\eta$ the peak is almost flat at the top.

4. Outlook for studying V^0-h correlations

In V^0-h analysis we have used 30×10^6 Pb–Pb collisons at $\sqrt{s_{NN}} = 2.76$ TeV, triggered on centrality, taken in 2011. V^0 candidates were selected using track topological cuts and $|y| < 0.75$. Invariant mass distributions for most central events (0-10%) are shown in the fig. 5 and 6. V^0 candidates in the shaded area were used for the correlation analysis. For the p_{T}^{trig} interval we selected V^0 candidates with $6 < p_{\mathrm{T}}^{trig} < 15$ GeV/c and for associated charged unidentified particles we chose 3 GeV/c $< p_{\mathrm{T}}^{assoc} < p_{\mathrm{T}}^{trig}$. Fig. 7 and 8 present the acceptance corrected correlations for $\mathrm{K}_{\mathrm{S}}^0$-h and Λ-h, respectively. The plots are not efficiency corrected. Very clear jet-like peaks can be seen well above background, indicating that the analysis is feasible.

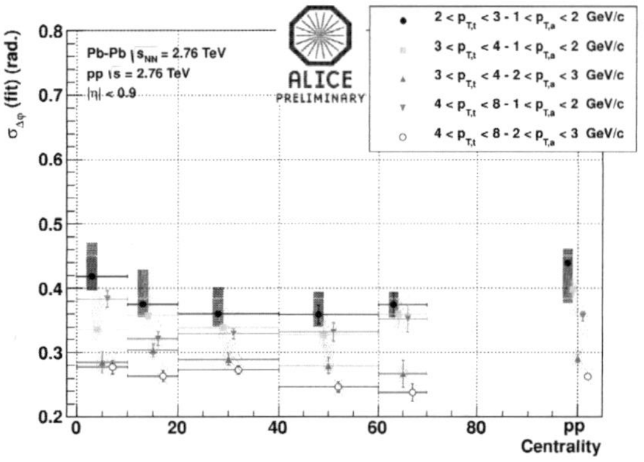

Figure 1. RMS in $\Delta\phi$ in h-h correlations for five different p_{T}^{trig} and p_{T}^{assoc} intervals as a function of centrality in Pb–Pb collisions. Results for pp collisions are on the right.

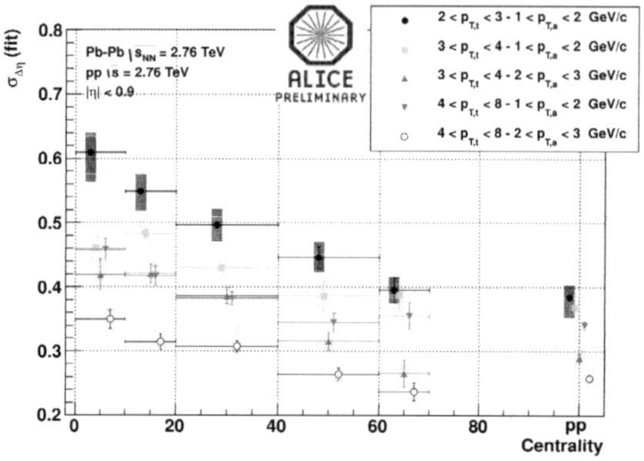

Figure 2. RMS in $\Delta\eta$ in h-h correlations for five different p_{T}^{trig} and p_{T}^{assoc} intervals as a function of centrality in Pb–Pb collisions. Results for pp collisions are on the right.

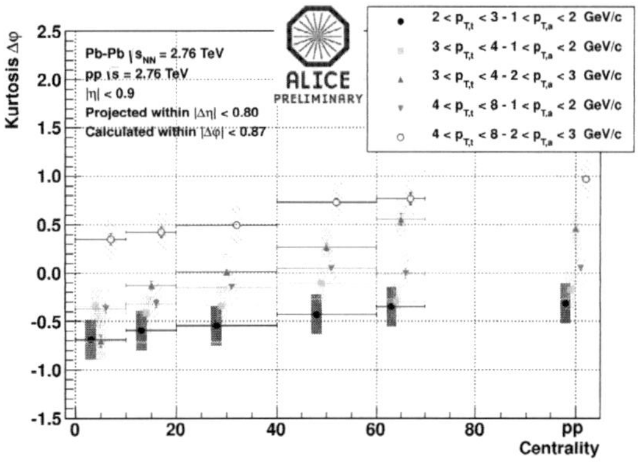

Figure 3. Excess kurtosis in $\Delta\phi$ in h-h correlations for five different p_{T}^{trig} and p_{T}^{assoc} intervals as a function of centrality in Pb–Pb collisions. Results for pp collisions are on the right.

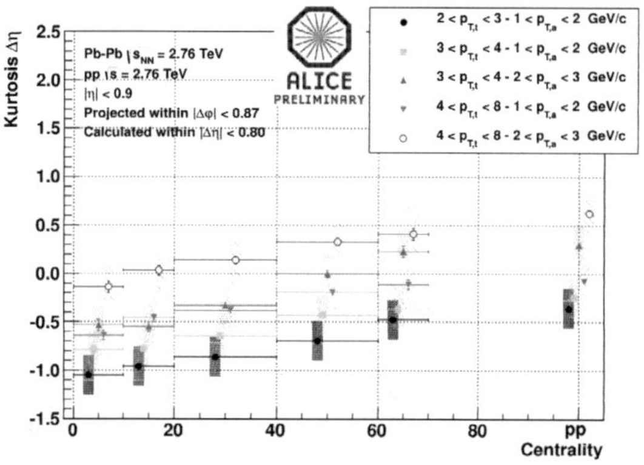

Figure 4. Excess kurtosis in $\Delta\eta$ in h-h correlations for five different p_{T}^{trig} and p_{T}^{assoc} intervals as a function of centrality in Pb–Pb collisions. Results for pp collisions are on the right.

5. Summary

Properties of the jet-like peaks in di-hadron correlations were investigated. We observe a significant broadening at lower p_{T} intervals of trigger and associated particles and in more central Pb–Pb collisions. Jet-like peak widths in $\Delta\phi$ show no centrality dependence while the widths in $\Delta\eta$ increase in central Pb-Pb collisions. This might be an indication of interaction of jets with longitudinal flow. In the kurtosis study we see the peak narrowing at high trigger and associated p_{T} and widening in central Pb-Pb collisions. V^0-h studies can be used for tagging quark and gluon jets on a statistical basis and therefore could be an interesting alternative for measuring colour charge dependent energy loss. Clear jet-like peaks in V^0-h correlations show the feasibility of these measurements in central Pb-Pb collisions.

Journal of Physics: Conference Series **509** (2014) 012043 doi:10.1088/1742-6596/509/1/012043

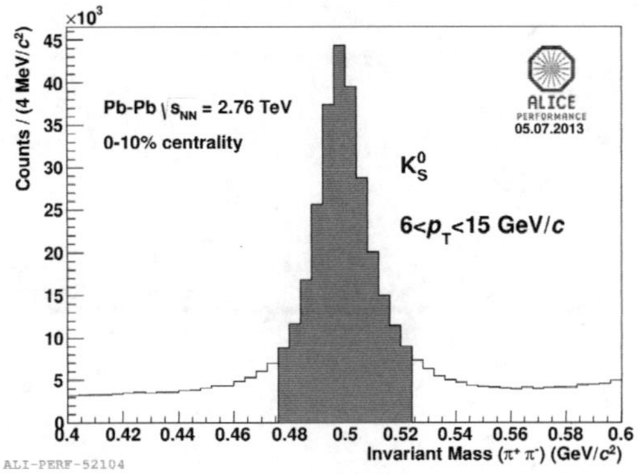

Figure 5. Invariant mass distribution for $\pi^+\pi^-$ pairs

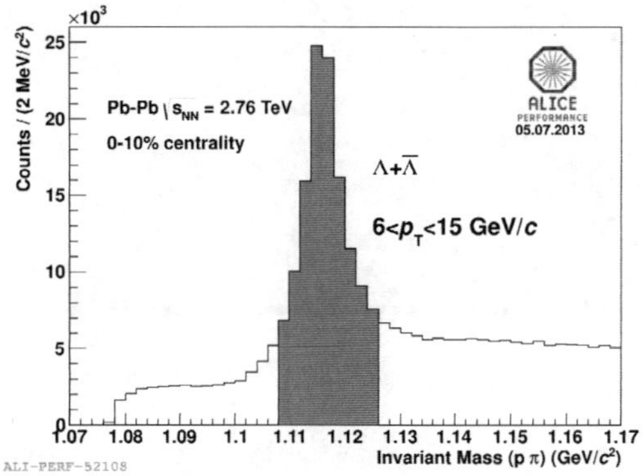

Figure 6. Invariant mass distribution for p π pairs

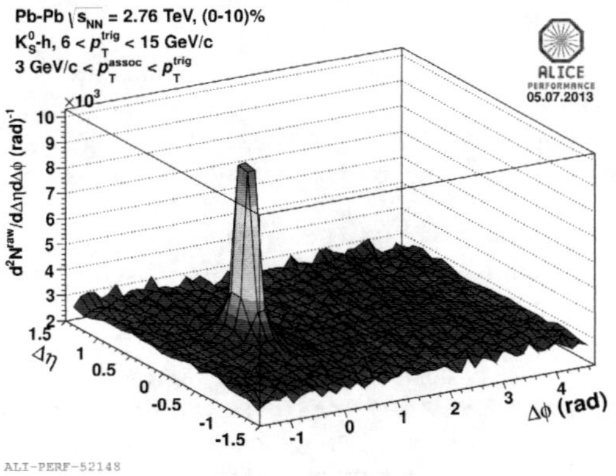

Figure 7. Acceptance-corrected K^0_S-h

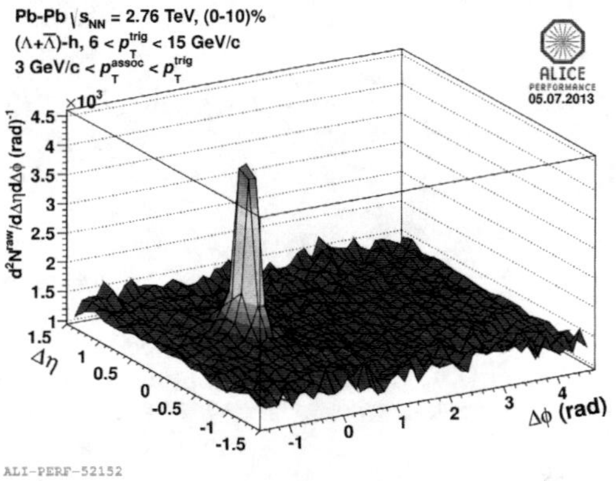

Figure 8. Acceptance-corrected $(\Lambda+\overline{\Lambda})$-h

References

[1] Veres G *et al.* (CMS Collaboration) 2013 *Nuclear Physics* A **904** 146
[2] Armesto N, Salgado C A and Wiedemann U A 2004 *Phys. Rev. Lett.* **93** 242301
[3] Gallicchio J and Schwartz M D 2011 *Phys. Rev. Lett.* **107** 172001
[4] Ackerstaff K *et al.* (OPAL Collaboration) 1999 *Eur. Phys. J. C* **8** 241
[5] Aamodt K *et al.* (ALICE Collaboration) 2008 *JINST* **3** 08002

Journal of Physics: Conference Series **509** (2014) 012044 doi:10.1088/1742-6596/509/1/012044

Multiplicity fluctuations of identified hadrons in p+p interactions at SPS energies.

Maja Maćkowiak-Pawłowska[1,2] and Andrzej Wilczek[3]
for the NA61 Collaboration

[1]Goethe University, Max-von-Lause-Str. 1, 60438, Frankfurt am Main, Germany
[2]Warsaw University of Technology, ul. Koszykowa 75, 00662 Warsaw, Poland
[3]University of Silesia, ul. Bankowa 12, 40007 Katowice, Poland

E-mail: majam@if.pw.edu.pl, awilczek@us.edu.pl

Abstract. Study of energy and system size fluctuations of identified hadrons is one of the key goals of NA61/SHINE at the CERN SPS. Results may allow to discover the critical point (CP) of strongly interacting matter as well as to uncover properties of the onset of deconfinement (OD). But fluctuations exhibit numerous other sources starting from most basic ones like volume effects and conservation laws.

NA49 seems to observe fluctuations related to CP in collisions of medium size nuclei at top SPS energy. However, this result will remain inconclusive until systematic data on energy and system size dependence will be available. Moreover, fluctuations in p+p as well as in Pb+Pb interactions should be better understood.

In this contribution results on multiplicity fluctuations of identified hadrons in p+p interactions at the CERN SPS energies will be presented. The NA61 data will be compared with the corresponding results from central Pb+Pb collisions of NA49 in the common acceptance region of both experiments. Moreover, predictions of models (EPOS, UrQMD and HSD) for both reactions will be tested.

1. Introduction

It is a well established fact that matter exists in different states. One of the most important goals of heavy-ion collision (HIC) experiments is to study the phase diagram of strongly interacting matter by finding possible phase boundaries and establishing their properties. In particular, high energy HIC experiments want to study the deconfined matter (QGP) and the transition between QGP and confined hadrons. The NA61/SHINE experiment [1] is a fixed target experiment in the European Organization for Nuclear Research (CERN) at the Super Proton Synchrotron (SPS). The study of energy and system size dependence of fluctuations of identified hadrons is one of the goals of the NA61/SHINE ion program. The aim of this program is to search for the critical point (CP) [2] of strongly interacting matter as well as to uncover properties of the onset of deconfinement (OD) [3, 4]. The predecessor of NA61/SHINE, the NA49 experiment [5] seems to observe fluctuations possibly related to the CP in collisions of medium size nuclei at the top energy of SPS [6]. However, measured fluctuations are affected by numerous effects not related to CP or OD like volume fluctuations and conservation laws. Thus, this result will remain inconclusive until systematic data on energy and system size dependence will be available. These data are being recorded and analysed by NA61/SHINE.

2. NA61/SHINE identified hadron fluctuation studies

Addressed, multiplicity and chemical fluctuations of identified hadrons produced in strong and electromagnetic processes were measured in inelastic p+p interactions at $\sqrt{s_{NN}} =$ 7.6, 8.7, 12.3, and 17.3 GeV. Particle identification is based on energy loss measurements in the relativistic rise region. Fluctuations of charged pions ($\pi = \pi^+ + \pi^-$), kaons ($K = K^+ + K^-$) and protons + antiprotons as well as of positively charged hadrons were studied via first and second (pure and mixed) moments of identified particle multiplicity distributions. For the completeness, figure 1 shows energy dependence of mean multiplicity of identified all charged and positively charged hadrons in the analysis acceptance. Note that the mean multiplicities are not corrected for possible biases and should not be used for direct comparison with models.

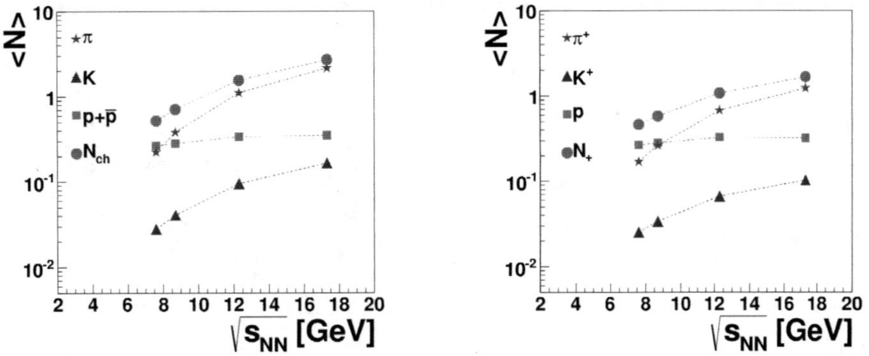

Figure 1. The energy dependence of mean multiplicity of identified charged (*left*) and positively charged (*right*) hadrons in p+p interactions in the acceptance used for the fluctuation analysis.

Second moments of identified hadron multiplicity distributions were obtained using the identity method [7, 8, 9]. Presented results include the statistical uncertainty and a first estimate of systematic uncertainty (still under studies are detector effects and influence of feed-down).

The multiplicity fluctuations were addressed using the scaled variance measure, ω_i of the multiplicity distribution. It is defined as $\omega = \frac{\langle N_i^2\rangle - \langle N_i\rangle^2}{\langle N_i\rangle}$, where $\langle N_i\rangle$ and $\langle N_i^2\rangle$ are the mean multiplicity and the second moment of the multiplicity distribution of particles of type i, respectively. The scaled variance is an intensive measure [10], i.e. it is independent of the number of wounded nucleons in the Wounded Nucleon Model [11] or volume in the Grand Canonical Ensemble, but it depends on their fluctuations. The latter feature makes it difficult to compare results from p+p interactions with those from nucleus+nucleus collisions. For the Poisson distribution $\omega = 1$.

Figure 2 (*left*) shows the scaled variance of π, K and $p+\bar{p}$. The scaled variance of all considered particle types increases with increasing energy. Rich data on the charged particle multiplicity distribution in the full phase space obey the KNO scaling [12, 13]. From this scaling follows a linear increase of the scaled variance with the mean multiplicity of charged particles visible as an increase of ω_π with collision energy (see the bottom left panel of Fig. 2). The scaled variance of unidentified charged hadrons measured within the acceptance chosen for this analysis is shown in the same panel by purple diamonds. The increase with collision energy is weaker than that measured in the full phase space, but agreement between $\omega_{N_{ch}}$ and ω_π as well as between N and N_π indicates that KNO scaling may also apply to pion production.

Figure 2 (*right*) shows the energy dependence of ω_p, ω_{K^+}, and ω_{π^+}. The scaled variance of identified positively charged hadrons is smaller than the one measured for the sum of charges. Moreover, in the case of positively charged hadrons, instead of an increase with the collision energy the scaled variance is approximately independent of energy.

The suppressed (<1) values of $\omega_{p+\bar{p}}$ and ω_p are probably due to the baryon number conservation [14]. In p+p interactions, the net baryon number is two, thus, fluctuations of the sum of charges is dominated by protons [14]. Values of ω_K equal to or above 1 are probably connected with the strangeness conservation, which leads to a correlation between

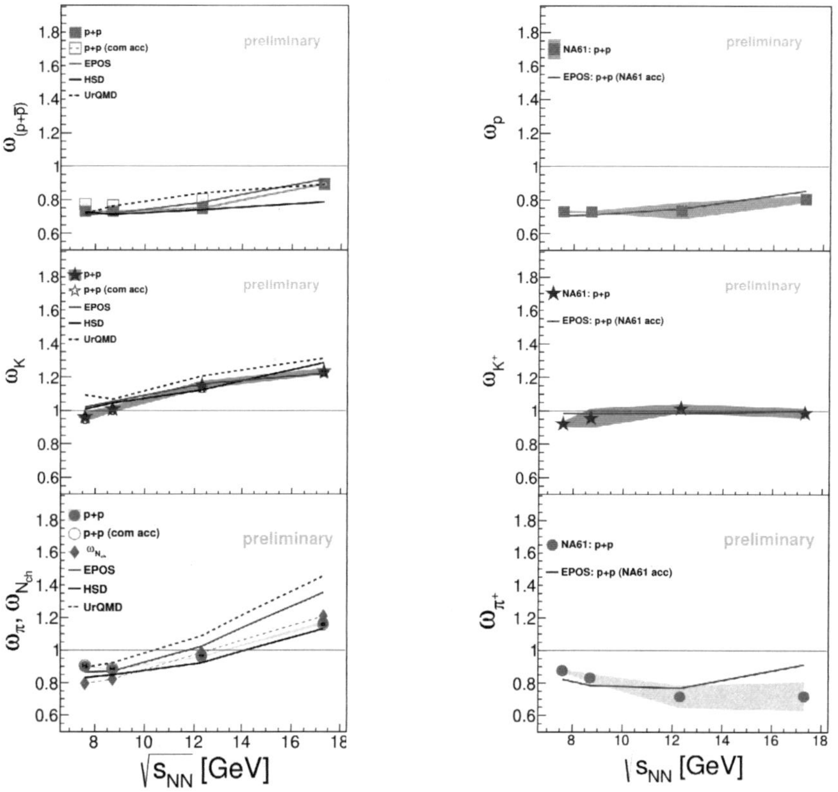

Figure 2. The energy dependence of the scaled variance in p+p interactions: $\omega_{p+\bar{p}}$, ω_K, ω_π (*left*), ω_p, ω_{K^+}, and ω_{π^+} (*right*). Full symbols denote the fluctuation results within the NA61 acceptance. Open symbols denote the results within the common phase space region of NA61 and NA49. Model predictions are presented for the NA61/SHINE acceptance.

the K^+ and K^- productions. It is supported by the fact that $\omega_{K^+} < \omega_K$ [15]. It should be underlined that the influence of a specific charge conservation on the scaled variance depends on the volume of the system, mean number and mass of charge carriers, and the fraction of measured particles and anti-particles. Measured quantities are approximately reproduced by the EPOS [16, 17], UrQMD [18, 19] and HSD [20, 21, 22] models. In order to compare results for p+p and 3.5% of most central Pb+Pb collisions , the strongly intensive measure, $\Phi_{ij} = \frac{\sqrt{<N_i><N_j>}}{<N_i+N_j>} \times \left(\sqrt{\Sigma^{ij}} - 1\right)$ [23, 10], defined for two hadron types, i and j, was chosen, where $\Sigma^{ij} = [< N_i > \omega_j + < N_j > \omega_i - 2(< N_i N_j > - < N_i >< N_j >)]/ < N_i + N_j >$.
As a strongly intensive measure, Φ_{ij} is not only independent of number of wounded nucleons or volume but also of their fluctuations. For independent particle emission, $\Phi_{ij} = 0$. Figure 3 shows the energy dependence of Φ_{ij} of two hadron types: $\pi(p+\bar{p})$, πK, and $(p+\bar{p})K$ as well as only for positively charged particles. Full symbols refer to the NA61 respectively NA49 acceptance. Open symbols display results for the common phase-space region of NA49 and NA61. Differences between the individual and common acceptance are small. For $\pi(p+\bar{p})$ there is a dip at intermediate energies. The energy dependence is similar in the case of positively charged hadrons. A similar but weaker effect is visible in Pb+Pb interactions. In the case of πK and $(p+\bar{p})K$ there is an increase of Φ_{ij} with collision energy, which is not visible in positively charged hadrons. The increase of $\Phi_{\pi K}$ is not visible in Pb+Pb collision but $\Phi_{\pi K} \geq 0$ at all considered energies in both reactions. The opposite energy dependencies between p+p and Pb+Pb interactions are visible in $\Phi(p+\bar{p})K$ (see bottom panel of Fig. 3 (*left*)).

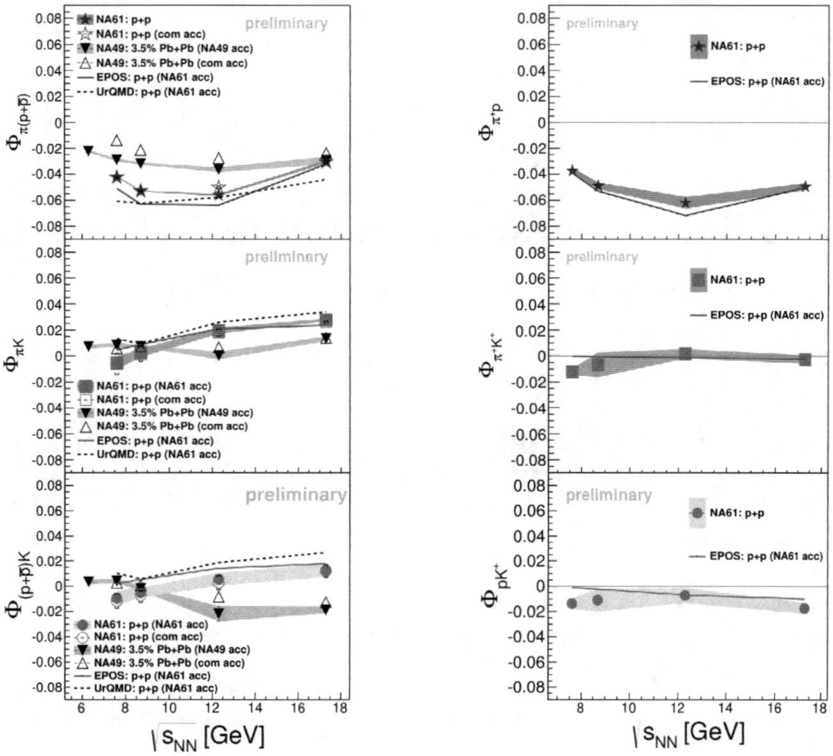

Figure 3. The energy dependence of $\Phi_{\pi(p+\bar{p})}$, $\Phi_{\pi K}$, $\Phi_{(p+\bar{p})K}$ (*left*), Φ_{π^+p}, $\Phi_{\pi^+K^+}$, and Φ_{pK^+} (*right*) in p+p interactions. Full symbols denote fluctuation results within the NA61 or NA49 acceptance. Open symbols denote results within the common phase space region of NA61 and NA49. Model predictions are presented for the NA61/SHINE acceptance.

References

[1] Alekseev G A *et al.* 2001 *Nucl. Instr. and Meth.* A **461** 381
[2] Fodor Z and Katz S D 2004 *JHEP* **0404** 050
[3] Gaździcki M and Gorenstein M I 1999 *Acta Phys. Polon.* B **30** 2705
[4] Alt C *et al.* (NA49 Coll.) 2008 *Phys. Rev.* C **77** 024903
[5] Afanasiev S *et al.* (NA49 Coll.) 1999 *Nucl. Instrum. Meth.* A **430** 210
[6] Grebieszkow K for the NA49 Coll. 2009 *Nucl. Phys.* A **830** 546c–50c
[7] Gaździcki M, Grebieszkow K, Maćkowiak M and Mrówczyński S 2011 *Phys. Rev.* C **83** 054907
[8] Gorenstein M I 2011 *Phys. Rev.* C **84** 024902
[9] Rustamov A and Gorenstein M I 2012 *Phys. Rev.* C **86** 044906
[10] Gorenstein M I and Gazdzicki M 2011 *Phys. Rev.* C **84** 014904
[11] Bialas A, Bleszynski M and Czyz W 1976 *Nucl. Phys.* B **111** 461
[12] Koba Z, Nielsen H B and Olesen P 1972 *Nucl. Phys.* B **40** 317
[13] Heiselberg H 2001 *Phys. Rept.* **351** 161–94
[14] Becattini F, Keranen A, Ferroni L and Gabbriellini T 2005 *J. Phys. Conf. Ser.* **27** 164.
[15] M. Gazdzicki M 1999 *Eur. Phys. J.* C **8** 131
[16] Werner K 2007 *Phys. Rev. Lett.* **98** 152301
[17] Drescher H J, Hladik M, Ostapchenko S, Pierog T and Werner K 2001 *Phys. Rept.* **350** 93–289
[18] Bass S A, *et al.* 1998 *Prog. Part. Nucl. Phys.* **41** 225-370
[19] Bleicher M *et al.* 1999 *J. Phys. G Nucl. Part. Phys.* **25** 1859–96
[20] Ehehalt W and Cassing W 1996 *Nucl. Phys.* A **602** 449–86
[21] Geiss J, Cassing W and Greiner C 1998 *Nucl. Phys.* A **644** 107–38
[22] Cassing W and Bratkovskaya E L 1999 *Phys. Reports* **308** 65–233
[23] Gazdzicki M and Mrowczynski S 1992 *Z. Phys.* C **54** 127

Journal of Physics: Conference Series **509** (2014) 012045 doi:10.1088/1742-6596/509/1/012045

Nuclear modification factor and elliptic flow of muons from heavy-flavour decays in Pb–Pb collisions at $\sqrt{s_{\mathrm{NN}}} = 2.76$ TeV

Xiaoming Zhang for the ALICE Collaboration

Lawrence Berkeley National Laboratory, 1 Cyclotron Road, Berkeley, CA 94720, U.S.A.
Institute of Particle Physics, Central China Normal University, Wuhan, China

E-mail: `xmzhang@lbl.gov`

Abstract. We report on the latest results from ALICE on the measurement of the nuclear modification factor and elliptic flow of muons from heavy-flavour decays at forward rapidity in Pb–Pb collisions at $\sqrt{s_{\mathrm{NN}}} = 2.76$ TeV. A strong suppression of muons from heavy-flavour decays is observed at high transverse momentum, p_{T}, in the most central collisions compared to the binary-scaled expectation from pp events at the same energy, while a positive elliptic flow is observed at intermediate p_{T} in semi-central collisions. Comparisons with model predictions are presented.

1. Introduction

ALICE [1] is the dedicated heavy-ion experiment at the LHC, optimized to investigate the properties of strongly-interacting matter in extreme conditions of temperature and energy density where the formation of Quark Gluon Plasma (QGP) is expected. Heavy quarks (charm and beauty) are regarded as efficient probes of the properties of the QGP as they are created on a very short time scale in initial hard parton scattering processes and subsequently interact with the medium. Open heavy-flavour hadrons are expected to be sensitive to the energy density of the system through the in-medium energy loss of heavy quarks. In the high p_{T} region, the suppression of the heavy-flavour yields quantified by means of the nuclear modification factor, R_{AA}, defined as the ratio of the yield measured in Pb–Pb collisions to that observed in pp collisions scaled by the number of binary nucleon-nucleon collisions, is used to study the heavy quark in-medium energy loss mechanisms. Due to the color charge effect, the radiative energy loss of gluons should be larger than that of quarks, and due to the dead-cone effect [2] heavy quark energy loss may be reduced with respect to that of light quarks. The heavy-flavour elliptic flow, the second order coefficient of the Fourier expansion of particle azimuthal distributions (v_2), should carry complementary information on the medium transport properties. It is expected to provide insights on the degree of thermalization of heavy quarks in the deconfined medium and on the path-length dependence of parton energy loss in the low and high p_{T} regions, respectively.

In the following, we focus on open heavy-flavour measurements at forward rapidity via single muons detected in the ALICE muon spectrometer (pseudo-rapidity coverage: $-4 < \eta < -2.5$) which consists of a thick front absorber, a beam shield, a dipole magnet, five tracking stations and two trigger stations behind an iron wall. Muons are identified by requiring that the track

Journal of Physics: Conference Series **509** (2014) 012045 doi:10.1088/1742-6596/509/1/012045

candidate in the tracking system is matched with the track reconstructed in the muon trigger system. This condition allows to reject most of the punch-through hadrons that are stopped in the iron wall. Then, acceptance cuts are applied. Furthermore, the correlation between the track momentum and the geometrical distance of closest approach to the primary vertex is used in order to remove fake tracks and tracks from beam-gas interactions. The remaining background after these selection cuts consists of muons from decays of primary light hadrons (π and K, mainly), called decay muons in the following. In addition to the muon spectrometer, for the analyses presented here, the VZERO detector comprising two of scintillator arrays, the two Zero Degree Calorimeters (ZDC), the Silicon Pixel Detector (SPD) and the Time Projection Chamber (TPC) are used. Collisions were classified according to their centrality, determined from the sum of the amplitudes of the signals in the VZERO detector and defined in terms of percentiles of the total hadronic Pb–Pb cross section.

2. Nuclear modification factor of muons from heavy-flavour decays

The study of in-medium effects with the R_{AA} observable requires the measurement of the pp reference. The latter was obtained from the analysis of the pp data sample at $\sqrt{s} = 2.76$ TeV collected in 2011 [3, 4]. The Pb–Pb data sample collected in 2010 with minimum bias trigger events is used to obtain the p_{T} distributions of muons from heavy-flavour decays for different centrality selections. The background contribution that needs to be subtracted is evaluated by

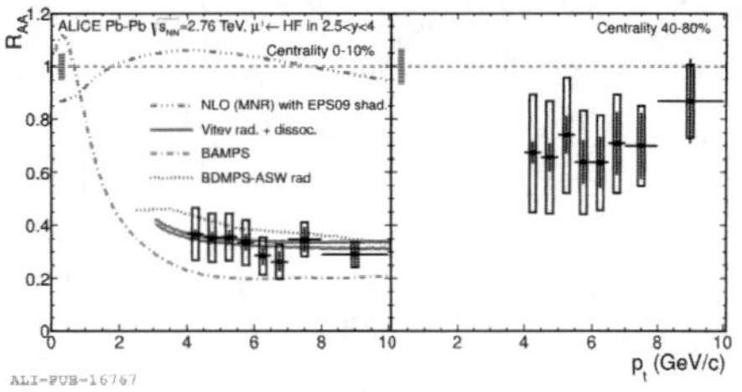

Figure 1. p_{T}-differential R_{AA} of muons from heavy-flavour decays in 0–10% (left) and 40%–80% (right) centrality classes; statistical (bars), systematic (empty boxes) and normalization (filled boxes) uncertainties are shown. Data from the most central collisions are compared to models including shadowing alone and in-medium energy loss.

extrapolating to forward rapidity the π and K distributions measured in pp and Pb–Pb collisions in the ALICE central barrel [1], and generating the decay muons through simulations of the decay kinematics and interaction with the front absorber [4]. Figure 1 shows the p_{T}-differential R_{AA}, in $4 < p_{\mathrm{T}} < 10$ GeV/c, in the 10% most central collisions (left) and in the 40%–80% centrality class (right). A stronger suppression is observed in central collisions compared to peripheral collisions, with no significant p_{T} dependence within uncertainties. The suppression reaches a factor 3–4 in the 10% most central collisions. The R_{AA} measured in central collisions is compared to model predictions (Fig. 1, left). In addition to final state effects where in-medium energy loss would be dominant, initial state effects could influence the R_{AA} measurement. In the kinematic range relevant for heavy-flavour production, the main expected effect is nuclear shadowing. This effect was estimated by means of perturbative QCD calculations [5] and the EPS09 NLO parameterization of nuclear PDFs [6]. These calculations, shown as grey-dotted-dashed curve in the left panel of Fig. 1, indicate that, in the range $4 < p_{\mathrm{T}} < 10$ GeV/c, nuclear shadowing alone cannot explain the observed suppression at forward rapidity. The other predictions shown in Fig. 1 refer to the R_{AA} calculations with models implementing radiative energy loss (BDMPS-APW [8]), radiative and dissociation energy loss with in-medium hadronization [9], and a partonic transport model (BAMPS), which includes collisional energy loss and accounts effectively for radiative processes by scaling the binary cross section with a correction factor tuned at RHIC energies [7]. They describe the data reasonably well within uncertainties. The measured suppression at forward rapidity is similar to that reported for D mesons [10] and electrons from heavy flavour decays [11] measured at mid-rapidity with ALICE.

3. Elliptic flow of muons from heavy-flavour decays

We report on the first measurement of the elliptic flow of muons from heavy-flavour decays at forward rapidity ($2.5 < y < 4$) and high p_{T} ($p_{\mathrm{T}} > 3$ GeV/c) in Pb–Pb collisions at $\sqrt{s_{\mathrm{NN}}} = 2.76$ TeV. The analysis uses the 2011 Pb–Pb data sample and is based on central and semi-central trigger events. Various analysis techniques which exhibit different sensitivity to non-flow effects and flow fluctuations have been exploited. In the following, we discuss results obtained with the scalar product [12], 2- and 4-particle Q-cumulants [13], and Lee-Yang zeros with sum and product generating functions [14]. The reference particle flow is determined from TPC standalone tracks. In order to determine the elliptic flow of muons from heavy-flavour decays ($v_2^{\mu\leftarrow\mathrm{HF}}$), one needs to subtract background contribution from that of the inclusive muons. This is done according to:

$$v_2^{\mu\leftarrow\mathrm{HF}} = \frac{v_2^{\mu} - f^{\mu\leftarrow\mathrm{K}\pi} \cdot v_2^{\mu\leftarrow\mathrm{K}\pi}}{1 - f^{\mu\leftarrow\mathrm{K}\pi}}, \quad (1)$$

where v_2^{μ} is the measured inclusive muon elliptic flow and $v_2^{\mu\leftarrow\mathrm{K}\pi}$ is the elliptic flow of decay muons. $f^{\mu\leftarrow\mathrm{K}\pi}$ is decay muon fraction that contributes to the inclusive muon yield. The v_2 of decay muons is estimated by extrapolating to forward rapidity the elliptic flow of charged hadrons measured in $|\eta| < 2.5$ by the ATLAS collaboration [15] and performing fast detector simulations in order to obtain the decay muon v_2. The extrapolation is done in various p_{T} intervals with a 2nd oder polynomial. Several other functions (polynomials up to 3rd order, gaussian) are used to evaluate the systematic uncertainty, and they give the maximum $\sim 4\%$ uncertainty in the final results. The fraction of decay muons varies between 5–15%, depending

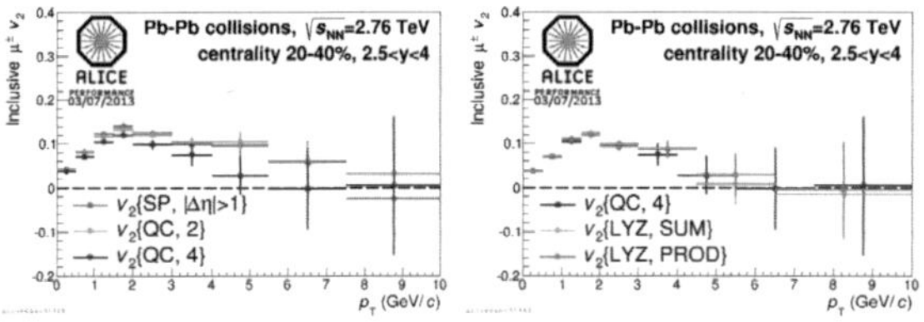

Figure 2. p_{T}-differential inclusive muon v_2 from various flow methods. Only statistical uncertainties are displayed.

on p_{T} and centrality [4]. Figure 2 presents the p_{T}-differential inclusive muon v_2 from 2- and multi-particle correlation flow methods in the centrality class 20–40%. Scalar product and 2-particle Q-cumulants give compatible results within uncertainties (Fig. 2, left). A similar behaviour is also observed when comparing results with 4-particle Q-cumulants and Lee-Yang zeros, indicating that non-flow contributions are negligible at 4th order (Fig. 2, right). There is a systematic decrease of the v_2 magnitude with 4-particle Q-cumulants as compared to 2-particle Q-cumulants and scalar product, which might be attributed to initial flow fluctuations as already reported for charged hadrons measured with ALICE at mid-rapidity [16]. Figure 3 (left) displays the v_2 of muons from heavy-flavour decays in $3 < p_{\mathrm{T}} < 10$ GeV/c as a function of the collision centrality, obtained via 2-particle Q-cumulants. A positive v_2 is measured in semi-central collisions (20–40% centrality class) with a significance larger than 3σ, and the

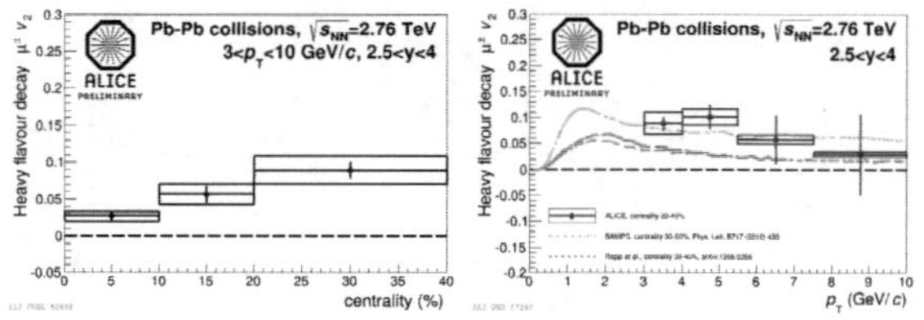

Figure 3. Left: v_2 of muons from heavy-flavour decays in $3 < p_{\mathrm{T}} < 10$ GeV/c as a function of the collision centrality. Right: p_{T}-differential v_2 of muons from heavy-flavour decays in the 20–40% centrality class compared to model predictions (see text). Statistical (bars) and systematic (empty boxes) uncertainties are shown. The results are obtained via 2-particle Q-cumulants.

magnitude of v_2 decreases as the collision centrality increases. The p_{T}-differential v_2 of muons from heavy-flavour decays in the 20–40% centrality is compared to transport model predictions, BAMPS [7] and Rapp *et al* [17] (two orange lines show the band of uncertainty), in Fig. 3 (right). The data are well reproduced by the BAMPS calculations within uncertainties, while they are underestimated by the Rapp *et al* predictions obtained with a heavy-flavour transport model based on collisional processes where the interactions of heavy quarks with the medium proceed via resonance formation. Finally, it is worth noticing that the measured v_2 is comparable in magnitude with that measured at mid-rapidity for electrons from heavy-flavour decays [18].

4. Conclusion

The production and propagation of muons from heavy-flavour decays has been measured at forward rapidity in Pb–Pb collisions with ALICE. The R_{AA} results provide evidence for strong in-medium energy loss of heavy quarks in central collisions. We have presented the first measurement of the v_2 of muons from heavy-flavour decays at forward rapidity, which is found positive in the region $3 < p_{\mathrm{T}} < 10$ GeV/c for semi-central collisions.

Acknowledgments

This work is supported partly by the NSFC Key Grant 11020101060, IRG11221504, 11375071 and 1005044, the 973 Grant of MOST of China 2013CB837803 and CCNU Key grant CCNU13F026.

References

[1] Aamodt K *et al* [ALICE Collaboration] 2008 *JINST* **3** S08002
[2] Dokshitzer Y L and Kharzeev D E 2001 *Phys. Lett.* B **519** 199 (*Preprint* hep-ph/0106202)
[3] Abelev B *et al* [ALICE Collaboration] 2012 *Phys. Lett.* B **708** 265 (*Preprint* arXiv:1201.3791)
[4] Abelev B *et al* [ALICE Collaboration] 2012 *Phys. Rev. Lett.* **109** 112301 (*Preprint* arXiv:1205.6443)
[5] Mangano M L *et al* 1992 *Nucl. Phys.* B **373** 295
[6] Eskola K *et al* 2012 *JHEP* **0904** 065 (*Preprint* arXiv:0902.4154)
[7] Uphoff J, Fochler O, Xu Z, and Greiner C 2012 *Phys. Lett.* B **717** 430 (*Preprint* arXiv:1205.4945)
[8] Armesto N *et al* 2005 *Phys. Rev.* D **71** 054027 (*Preprint* hep-ph/0501225)
[9] Sharma R *et al* 2009 *Phys. Rev.* C **80** 054902 (*Preprint* arXiv:0904.0032)
[10] Abelev B *et al* [ALICE Collaboration] 2012 *JHEP* **1209** 112 (*Preprint* arXiv:1203.2160)
[11] Thomas D for the ALICE Collaboration 2014 these proceedings.
[12] Adler C *et al* [STAR Collaboration] 2002 *Phys. Rev.* C **66** 034904 (*Preprint* nucl-ex/0206001)
[13] Bilandzic A, Snellings R, and Voloshin S 2011 *Phys. Rev.* C **83** 044913 (*Preprint* arXiv:1010.0233)
[14] Bhalerao R S *et al* 2003 *Nucl. Phys.* A **727** 373 (*Preprint* nucl-th/0310016)
[15] Aad G *et al* [ATLAS Collaboration] 2012 *Phys. Lett.* B **707** (2012) 330 (*Preprint* arXiv:1108.6018)
[16] Aamodt K *et al* [ALICE Collaboration] 2010 *Phys. Rev. Lett.* **105** 252302 (*Preprint* arXiv:1011.3914)
[17] He M, Fries R J, and Rapp R 2012 *Phys. Rev.* C **86** 014903 (*Preprint* arXiv:1106.6006)
[18] Dubla A for the ALICE Collaboration 2014 these proceedings

Journal of Physics: Conference Series **509** (2014) 012046 doi:10.1088/1742-6596/509/1/012046

Heavy Quark Energy Loss and Bulk Physics

W. A. Horowitz

Department of Physics, University of Cape Town, Private Bag X3, Rondebosch 7701, South Africa

E-mail: wa.horowitz@uct.ac.za

Abstract. We show that the low-p_T data and high-p_T data from RHIC and LHC give a contradictory picture of the relevant dynamics of a quark-gluon plasma. While the bulk observables imply a strongly coupled fluid best described by AdS/CFT the rare probes imply a weakly coupled gas best described using thermal pQCD. These conclusions are based on leading order theoretical calculations, and we speculate that more sophisticated calculations that include higher order effects might provide a resolution to the current puzzle.

1. Introduction

The goal of high-energy nuclear physics is the detailed understanding both theoretically and experimentally of the many-body dynamics of QCD. This goal is an immodest one: there are few examples of a detailed theoretical understanding of even many-body QED. However, there are many reasons to be interested in the emergent properties of QCD: the nuclear force is fundamental, the universe was in a state of quark-gluon plasma (QGP) a microsecond after the Big Bang, QCD is the only non-Abelian theory for which we have both experimental and theoretical access to many-particle physics, and it turns out that there appear to be surprising connections between QGP physics and disparate other fields of physics such as cold atoms and high temperature superconductivity.

Experimentalists can measure an enormous number of observables at the Relativistic Heavy Ion Collider (RHIC) and the Large Hadron Collider (LHC) as we attempt to tease out Nature's nuclear secrets. Broadly speaking, two of the most important classes of observables are those that are related to particles that have "low" transverse momentum, $p_T \lesssim 2$ GeV/c, and those that are related to particles with a large scale, such as heavy quarks and/or those with "high" transverse momentum, $p_T \gtrsim 5$ GeV/c.

The consensus view of the community is that the low-p_T observables are well understood within a nearly perfect fluid paradigm [1]. Both naïve [2] and sophisticated, next-to-leading order [3] applications of pQCD predict a viscosity much larger than that suggested by the hydrodynamics models. On the other hand, leading order strong coupling calculations utilizing the AdS/CFT correspondence yield predictions within a factor of 2 of the measured QGP η/s [4]. The low-p_T observables thus exclusively suggest that the many-body dynamics of QGP at temperatures accessible at colliders is best understood as a strongly-coupled fluid.

It is natural then to ask: how do predictions of observables related to hard probes based on this strong-coupling paradigm compare to data?

2. Application of AdS/CFT to High-p_T Observables

In answering this question, we will assume that the high momentum particles are strongly coupled to a strongly coupled medium. Assuming all couplings are large corresponds to the not unreasonable assumption that any influence from weak coupling dynamics—from, say, running coupling in QCD—is small compared to those from the strong coupling dynamics.

Within this strong coupling paradigm the leading order calculation of the energy loss for a heavy quark is relatively straightforward [5, 6] and corresponds to a drag, $dp_T/dt = -\mu\, p_T$, $\mu = \pi T^2\sqrt{\lambda}/2M_Q$. We show in Fig. 1 that one may well describe the suppression of non-photonic electrons at RHIC using this energy loss model [7], but not simultaneously the suppression of D mesons at LHC. However this model only takes into account the mean heavy quark energy loss. As is shown in [8], fluctuations in the mean energy loss are large and grow quickly, $\propto \gamma^{5/2}$; we are currently working on quantifying the importance of these fluctuations in energy loss observables.

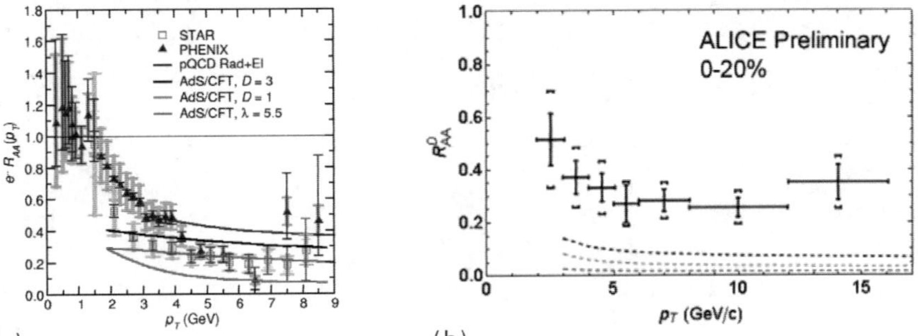

(a) (b)

Figure 1: (a) AdS/CFT predictions [7] for non-photonic electron decay products of heavy c and b quarks at RHIC [9, 10] and (b) D meson suppression predictions from AdS/CFT [11] at LHC [12].

It is less clear how to apply the AdS/CFT correspondence to the energy loss of light quarks and gluons. Nevertheless, although there are large uncertainties associated with the predictions, it appears that models based on naïve leading order calculations also overpredict the suppression; see Fig. 3 (a) and [13].

3. Appliction of pQCD to High-p_T Observables

On the other hand we may choose to assume that the high-p_T probe is weakly coupled to a weakly coupled QGP. In this scenario, the weak coupling to the plasma might be justified by the presence of a large scale in the problem, such as the quark mass or the transverse momentum of the probe. Assuming the medium is weakly coupled to itself is less justified. If one computes the running coupling at the first Matsubara frequency, then $g(2\pi T) \approx 2$, which is not much less than 1. Nevertheless we find that a comparison of predictions from an energy loss model [14, 15] based on these assumptions does an excellent job of qualitatively describing all single particle observables at both RHIC and LHC to within approximately a factor of 2, Fig. 2. We include here a new, very favorable comparison[1] between the WHDG energy loss model [14, 15] and the recent $R_{AA}^{B\to J/\psi}(N_{part})$ results from CMS [16].

It is worth emphasizing that the WHDG energy loss model includes both radiative and elastic energy loss, but is based on only the leading order in α_s contributions to these channels. The uncertainty estimates shown in Fig. 2 are due only to the rigorous extraction of the proportionality constant that we assume connects the thermalized QGP medium density and

[1] We wish to thank Zaida Conesa Del Valle and Andrea Dainese for fragmenting our B meson predictions to J/ψ mesons for the comparison with the CMS results.

the observed multiplicity [17]; we do not include theoretical uncertainty estimates for higher order effects. In particular, running coupling effects are very likely important, possibly leading to a more quantitative agreement with data [18]. The inclusion of running coupling and higher order corrections in collinearity, softness, and eikonality, also clearly very important [19, 20], are current works in progress.

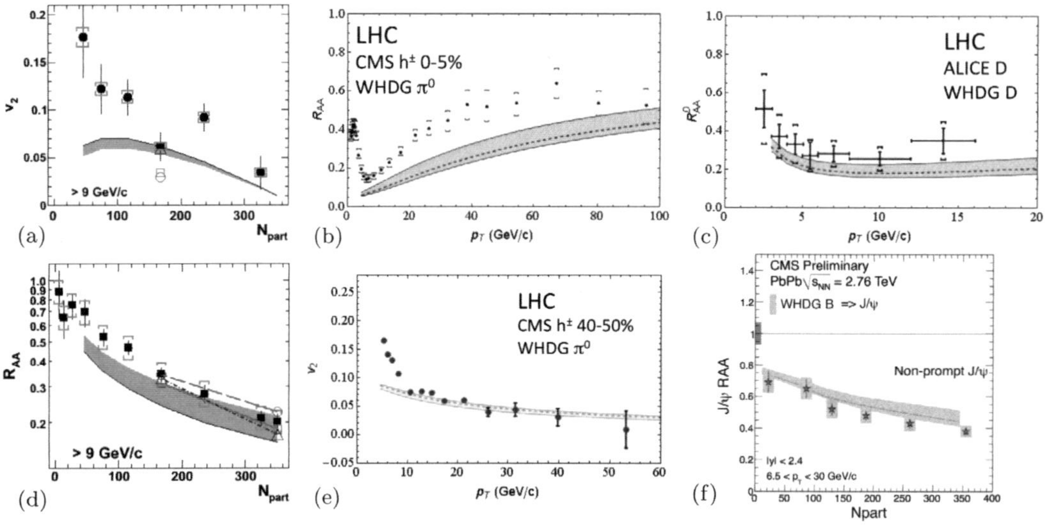

Figure 2: Constrained zero parameter WHDG predictions compared to data for (a) $v_2^{\pi^0}(N_{part})$ at RHIC [14, 21], (b) 0-5% centrality $R_{AA}(p_T)$ for light flavors at LHC [15, 22], (c) $R_{AA}^{D}(p_T)$ at 0-20% centrality at LHC [12, 13], (d) $R_{AA}^{\pi^0}(N_{part})$ at RHIC [14, 21], (e) $v_2(p_T)$ at LHC for light flavors at 40-50% centrality [13, 23], and (f) $R_{AA}^{B\to J/\psi}(N_{part})$ at LHC [11, 16].

4. Towards a Hybrid AdS/CFT and pQCD Model

One would very much like to reduce the very large theoretical uncertainties associated with light flavor strong coupling energy loss, as shown, for example, in Fig. 3 (a). A better comparison between the light flavor energy loss and data might come from a calculation of the energy-momentum tensor associated with the high-p_T probe. There is an additional reason for investigating the energy-momentum tensor of probes in AdS/CFT: we show in Fig. 3 (b) the surprising sensitivity to the stopping distance on the off-shellness of the initial conditions of the high-p_T probe. One might hope to create a hybrid energy loss model that better approximates the relevant physics in a heavy ion collision by matching the results from pQCD from time $t = 0$ up to some finite time such as t_{therm} to a subsequent evolution using AdS/CFT methods. Matching the calculations in AdS/CFT and pQCD requires computing a gauge invariant observable in both regimes; the energy-momentum tensor seems a natural choice.

5. Conclusions

The inclusion of higher order corrections are necessary for a quantitative extraction of QGP properties within the paradigm of pQCD energy loss. However, it is important to recognize that this pQCD paradigm, which is shown here to be very successful in qualitatively describing the features of energy loss observables at RHIC and LHC, is in contradiction to the strong coupling paradigm implied by data associated with low-p_T observables. Naïve leading order applications of AdS/CFT techniques to energy loss observables leads to an oversuppression compared to

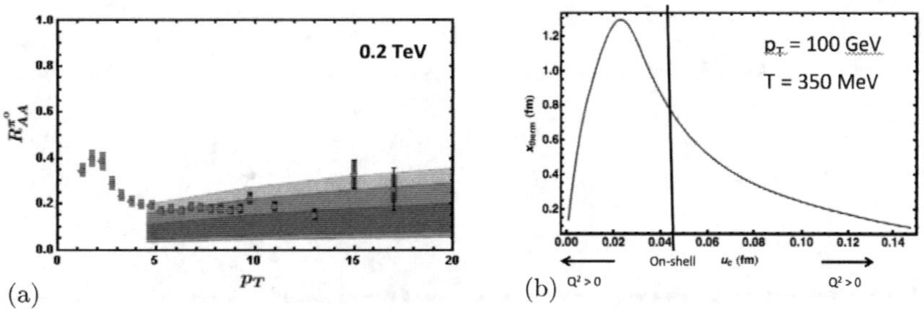

Figure 3: (a) Predictions from a simple Bragg peak model for AdS/CFT light flavor energy loss [13] compared to PHENIX data at RHIC [21]. (b) The stopping distance in AdS/CFT for light flavors depends strongly on the initial conditions.

data. Perhaps the success of the perturbative calculations is mainly due to a relatively correct description of the very early time physics. Then possibly a hybrid model in which 1) fluctuations in the AdS/CFT energy loss are included and 2) there is a better understanding of AdS/CFT initial conditions and light flavor energy loss will lead to a consistent description of high-p_T data within the AdS/CFT paradigm, and, therefore, we will have a consistent picture of the dynamics of QGP in heavy ion collisions.

References

[1] Heinz U, Shen C and Song H 2012 *AIP Conf.Proc.* **1441** 766–770 (*Preprint* 1108.5323)
[2] Danielewicz P and Gyulassy M 1985 *Phys.Rev.* **D31** 53–62
[3] Chen J W, Deng J, Dong H and Wang Q 2013 *Phys.Rev.* **C87** 024910 (*Preprint* 1107.0522)
[4] Kovtun P, Son D and Starinets A 2005 *Phys.Rev.Lett.* **94** 111601
[5] Gubser S S 2006 *Phys.Rev.* **D74** 126005 (*Preprint* hep-th/0605182)
[6] Herzog C, Karch A, Kovtun P, Kozcaz C and Yaffe L 2006 *JHEP* **0607** 013
[7] Horowitz W and Gyulassy M 2008 *Phys.Lett.* **B666** 320–323 (*Preprint* 0706.2336)
[8] Gubser S S 2008 *Nucl.Phys.* **B790** 175–199 (*Preprint* hep-th/0612143)
[9] Dion A (PHENIX collaboration) 2009 *Nucl.Phys.* **A830** 765C–768C (*Preprint* 0907.4749)
[10] Bielcik J (STAR Collaboration) 2006 *Nucl.Phys.* **A774** 697–700
[11] Horowitz W 2012 *AIP Conf.Proc.* **1441** 889–891 (*Preprint* 1108.5876)
[12] ALICE Collaboration 2012 *JHEP* **09** 112 (*Preprint* 1203.2160)
[13] Horowitz W and Gyulassy M 2011 *J.Phys.* **G38** 124114 (*Preprint* 1107.2136)
[14] Wicks S, Horowitz W, Djordjevic M and Gyulassy M 2007 *Nucl.Phys.* **A784** 426–442
[15] Horowitz W and Gyulassy M 2011 *Nucl.Phys.* **A872** 265–285 (*Preprint* 1104.4958)
[16] Mironov C (CMS Collaboration) 2013 *Nucl.Phys.* **A904-905** 194c–201c
[17] Adare A *et al.* (PHENIX Collaboration) 2008 *Phys.Rev.* **C77** 064907 (*Preprint* 0801.1665)
[18] Buzzatti A and Gyulassy M 2013 *Nucl.Phys.A904-905* **2013** 779c–782c
[19] Horowitz W 2008 (*Preprint* 0806.3092)
[20] Horowitz W and Cole B 2010 *Phys.Rev.* **C81** 024909 (*Preprint* 0910.1823)
[21] Adare A *et al.* (PHENIX) 2010 *Phys. Rev. Lett.* **105** 142301 (*Preprint* 1006.3740)
[22] Chatrchyan S *et al.* (CMS) 2012 *Eur. Phys. J.* **C72** 1945 (*Preprint* 1202.2554)
[23] Chatrchyan S *et al.* (CMS Collaboration) 2012 *Phys.Rev.Lett.* **109** 022301

Journal of Physics: Conference Series **509** (2014) 012047 doi:10.1088/1742-6596/509/1/012047

Heavy-flavor azimuthal correlations of D mesons

Marlene Nahrgang[1,2,3], **Jörg Aichelin**[3], **Pol Bernard Gossiaux**[3], **Klaus Werner**[3]

[1] Department of Physics, Duke University, Durham, North Carolina 27708-0305, USA

[2] Frankfurt Institute for Advanced Studies (FIAS), Ruth-Moufang-Str. 1, 60438 Frankfurt am Main, Germany

[3] SUBATECH, UMR 6457, Université de Nantes, Ecole des Mines de Nantes, IN2P3/CNRS. 4 rue Alfred Kastler, 44307 Nantes cedex 3, France

E-mail: marlene.nahrgang@phy.duke.edu

Abstract. Observables of heavy-quark azimuthal correlations in heavy-ion collisions are new and promising probes for the investigation of the in-medium energy loss. We explore the potential of these observables to discriminate the collisional and radiative contributions within a hybrid EPOS+MC@sHQ transport approach.

1. Introduction

Traditional heavy-quark observables like the nuclear modification factor $R_{\rm AA}$ and the elliptic flow v_2 are thoroughly measured at RHIC [1] and the LHC [2, 3, 4]. The results suggest a significant energy loss at high transverse momentum p_T and partial thermalization with the QGP medium at low p_T. Many theoretical approaches including energy loss via gluon bremsstrahlung [5], elastic scatterings [6] or a mixture of both processes are able to reproduce the $R_{\rm AA}$ and v_2 data within numerical simulations, for which a rescaling of the interaction cross sections or of the diffusion coefficient and a coupling to a background fluid dynamical medium of light partons is necessary [7, 8, 9, 10]. It remains, however, a challenge to reproduce both of these observables within a single setup. Besides the energy loss mechanism (elastic, radiative or a mixture of both) also the modeling of the fluid dynamical medium and the details of the coupling between the medium and the heavy quarks via an interpretation of the equation of state affect the numerical results substantially [11, 12]. It is therefore of crucial importance to use fluid dynamical models, which are well tested in the light sector, and to properly couple the both sectors.

We show that investigating heavy-flavor azimuthal correlations in addition to the $R_{\rm AA}$ and the v_2 offers the possibility to discriminate the different contributions to the energy loss, radiative or collisional. We use our approach of heavy-quark propagation, MC@sHQ coupled to a $3+1$ d ideal fluid dynamical evolution, which is subsequent to EPOS initial conditions. EPOS in its integral version, i. e. including a hadronic afterburner, successfully reproduces light-hadron observables [13]. The heavy quarks are initialized via the p_T-distributions from FONLL [14], then propagated via the Boltzmann equation taking the local temperature and fluid velocities from the background medium and finally hadronized via coalescence at low-p_T and fragmentation, which dominates at high-p_T [8, 9, 10, 15, 12].

Our procedure is the following: We will fix a global and temperature-independent K-factor by reproducing the LHC data of the high-p_T D-meson $R_{\rm AA}$ within the experimental errors for

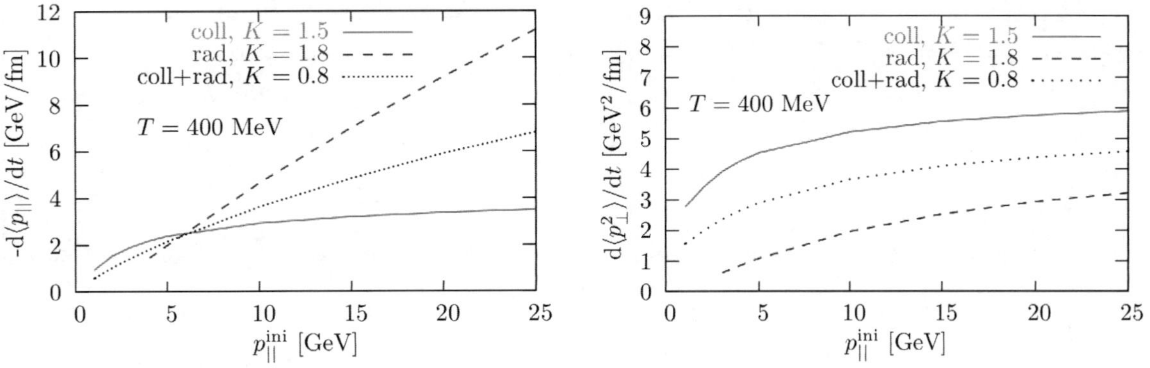

Figure 1. The drag coefficient and the average perpendicular deflection for charm quarks.

each interaction mechanism. We will see that the collisional and radiative energy loss contribute about the same to the overall energy loss. With these respective K-factors we will then evaluate the D-meson v_2 and the azimuthal correlations of $D\bar{D}$-pairs.

2. Properties of the interaction

In a first step we analyze the drag coefficient and the average deflection of the charm quark with respect to its initial direction in an infinite, static medium at the given temperature $T = 400$ MeV. For this purpose we initialize the charm quarks with $\vec{p}^{\text{ini}} = (0, 0, p_{||}^{\text{ini}})$. The drag coefficient, which describes the average parallel momentum loss, is then given by $A = -\mathrm{d}\langle p_{||}\rangle/dt$. We observe in figure 1(a) that for the purely radiative energy loss the increase of A with the charm quark momentum is much faster than for the purely collisional energy loss. The combination of collisional and radiative corrections lies between the two pure mechanisms. There is, however, a region for small and intermediate momenta, where the drag coefficient A for the purely collisional energy loss mechanism is larger than for the purely radiative one.

For the study of azimuthal correlations it is important to understand how much momentum perpendicular to its initial direction $p_\perp^2 = p_x^2 + p_y^2$ a charm quark acquires on average during the evolution. This quantity $\mathrm{d}\langle p_\perp^2\rangle/dt$ is shown in figure 1(b). Over the whole momentum range it is distinctively different for the three interaction mechanisms. It is largest for the purely collisional energy loss and smallest for the radiative energy loss. From this we expect that the initial azimuthal correlations will be broadened more efficiently by the purely collisional than by the radiative scatterings.

Due to approximations in the implementation the purely radiative scenario is not applicable for small transverse momenta $p_T < 3$ GeV.

3. Traditional observables

For a decoupling temperature $T_c = 155$ MeV we evaluate the D-meson R_{AA} in central $0 - 7.5$ % and v_2 in mid-peripheral $30 - 50$ % Pb+Pb collisions at $\sqrt{s} = 2.76$ TeV, see figure 2. The K-factors for each interaction mechanism are tuned such that the experimental R_{AA} is reproduced within the errors above $p_T \simeq 10$ GeV. This is possible for all three interaction mechanisms, although in figure 2(a) one is inclined to find a slightly better agreement for the purely collisional energy loss scenario. The v_2, figure 2(b), is also reasonably well reproduced by all three scenarios including their respective K-factors. Again, the purely collisional interaction is slightly closer to the central data points. One expects, however, a significant contribution to the D-meson v_2 from the final hadronic phase, which is not included in the current study.

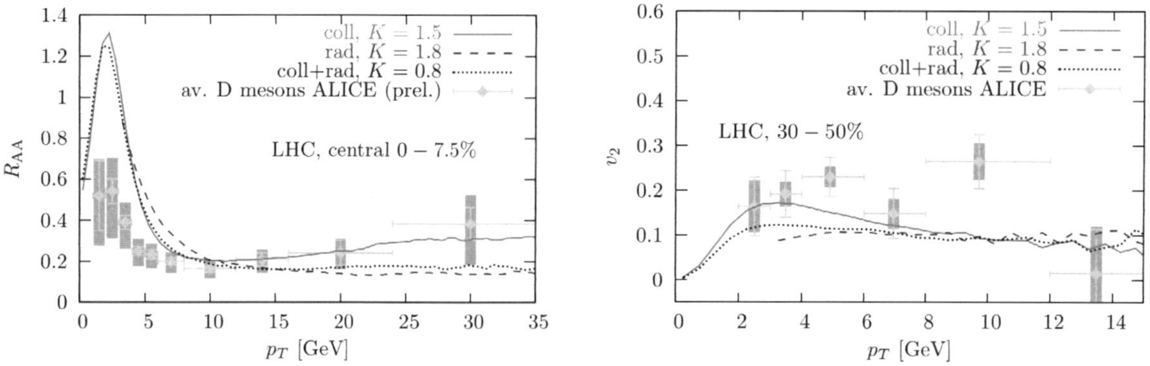

Figure 2. The nuclear modification factor R_{AA} and the elliptic flow v_2 of average D-mesons. ALICE data of averaged D mesons, which includes D^0, D^+ and D^{*+}, is from [3] and [4].

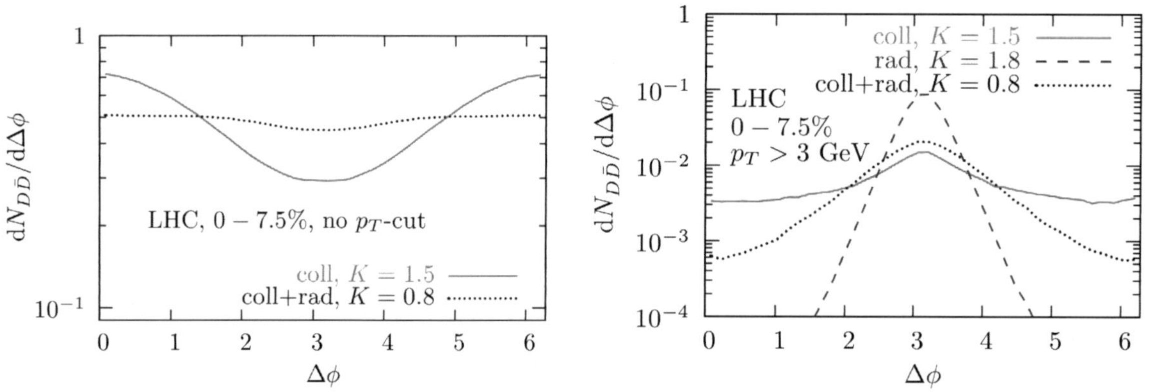

Figure 3. Azimuthal correlations of $D\bar{D}$-pairs with and without a p_T-cut.

4. Azimuthal correlations

Due to the next-to-leading order (NLO) production processes already in proton-proton collisions the azimuthal distributions will show a broadening of the back-to-back peak at $\Delta\phi = \pi$ and in particular the gluon splitting process leads to an additional enhancement at $\Delta\phi = 0$. The theoretical and experimental investigations for $D\bar{D}$ azimuthal correlations even in elementary reactions do not constitute conclusive evidence on the precise shape for initial $c\bar{c}$ azimuthal distributions. The higher mass of the bottom quarks makes similar calculations more easy. In a recent publication [15] we showed that by using $b\bar{b}$ initial distributions from the NLO+parton shower event generator MC@NLO [16] we obtained the same qualitative results as in a simple leading-order back-to-back initialization. In the present study we still initialize the $c\bar{c}$-pairs according to the p_T-distributions from FONLL [14] and strictly back-to-back.

In figure 3 the azimuthal correlations of $D\bar{D}$-pairs, which were initially produced together, are shown for all pairs in the left plot 3(a) and for pairs, where both transverse momenta are above 3 GeV in the right plot 3(b). One clearly sees that the overall yield is dominated by the low-p_T pairs, which show signs of thermalization with the flowing background medium – depending on the interaction mechanism. While the distribution is flat for the collisional+radiative scenario, one even observes a final correlation around $\Delta\phi = 0$ for the purely collisional scenario, which stems from the radial flow of the medium, pushing a $c\bar{c}$-pair towards smaller opening angles.

This is the so-called "partonic wind" effect [17]. When performing a cut in p_T to concentrate on intermediate and higher p_T this picture is confirmed: the purely collisional interaction is most efficient in washing out the initial correlations during the evolution in the medium. The purely radiative scenario is least efficient, the residual correlation at $\Delta\phi = \pi$ are most pronounced over a small background of isotropized pairs. This is a natural consequence of the larger average $p_\perp^2$ for the purely collisional energy loss mechanism than for any other of the energy loss mechanisms as discussed above.

We thus conclude that a precise measurement of heavy-flavor azimuthal correlations will give insight into the nature of the interaction between heavy-quarks and the QGP beyond what we know from the traditional observables. Implementing more realistic exclusive initial distributions and studying quantitative correlation observables, which are closer to the experimentally accessible ones, is work in progress.

Acknowledgments

We are grateful for support from the Hessian LOEWE initiative Helmholtz International Center for FAIR, ANR research program "hadrons @ LHC" (grant ANR-08-BLAN-0093-02), TOGETHER project Région Pays de la Loire and I3-HP.

References

[1] STAR Collab., *Phys. Rev. Lett.* **98** (2007) 192301; Erratum-ibid. **106** (2011) 159902; PHENIX Collab., *Phys. Rev. C* **84** (2011) 044905; Dong X [STAR Collaboration], *Nucl. Phys. A*904-905 **2013** (2013) 19c.
[2] ALICE Collab., *JHEP* **09** (2012) 112; Dainese A [ALICE Collaboration], arXiv:1212.0995 [nucl-ex].
[3] Conesa del Valle Z [ALICE Collaboration], *Nucl. Phys. A*904-905 **2013** (2013) 178c.
[4] Abelev B *et al.* [ALICE Collaboration], arXiv:1305.2707 [nucl-ex].
[5] Gyulassy M and Wang X N, Nucl. Phys. B **420** (1994) 583; Wang X N, Gyulassy M, and Plümer M, *Phys. Rev. D* **51** (1995) 3436; Baier R, Dokshitzer Y L, Peigné S, and Schiff D, *Phys. Lett. B* **345** (1995) 277; Baier R, Dokshitzer Y L, Müller A H, Peigné S, and Schiff D, *Nucl. Phys. B* **483** (1997) 291; *Nucl. Phys. B* **484** (1997) 265; Zakharov B G, *JETP Lett.* **63** (1996) 952; *JETP Lett.* **64** (1996) 781; *JETP Lett.* **65** (1997) 615; *JETP Lett.* **73** (2001) 49; *JETP Lett.* **78** (2003) 759; *JETP Lett.* **80** (2004) 617; Gyulassy M, Levai P, and Vitev I, *Phys. Rev. Lett.* **85** (2000) 5535; *Nucl. Phys. B* **571** (2000) 197; *Nucl. Phys. B* **594** (2001) 371; Dokshitzer Y L and Kharzeev D E, *Phys. Lett. B* **519** (2001) 199; Arnold P B, Moore G D, and Yaffe L G, *JHEP* 0011 (2000) 001; *JHEP* 0305 (2003) 051; Armesto N, Salgado C A, and Wiedemann U A, *Phys. Rev. D* **69** (2004) 114003; *Phys. Rev. C* **72** (2005) 064910; Zhang B W, Wang E and Wang X N, *Phys. Rev. Lett.* **93** (2004) 072301.
[6] Bjorken J D, Fermilab preprint Pub-82/59-THY (1982); Peshier A, *Phys. Rev. Lett.* **97** (2006) 212301; Peigne S and Peshier A, *Phys. Rev. D* **77** (2008) 114017.
[7] Moore G D and Teaney D, *Phys. Rev. C* **71** (2005) 064904; Uphoff J, Fochler O, Xu Z and Greiner C, *Phys. Rev. C* **82** (2010) 044906; Alberico W M, Beraudo A, De Pace A, Molinari A, Monteno M, Nardi M and Prino F, *Eur. Phys. J. C* **71** (2011) 1666; Uphoff J, Fochler O, Xu Z and Greiner C, *Phys. Rev. C* **84** (2011) 024908; Lang T, van Hees H, Steinheimer J and Bleicher M, arXiv:1211.6912 [hep-ph]; Cao S, Qin G Y and Bass S A, arXiv:1308.0617 [nucl-th].
[8] Gossiaux P B and Aichelin J, *Phys. Rev. C* **78** (2008) 014904.
[9] Gossiaux P B, Aichelin J, Gousset T and Guiho V, *J. Phys. G* **37** (2010) 094019.
[10] Gossiaux P B, Nahrgang M, Bluhm M, Gousset T and Aichelin J, *Nucl. Phys. A* **904-905** (2013) 992.
[11] Gossiaux P B, Vogel S, van Hees H, Aichelin J, Rapp R, He M and Bluhm M, [arXiv:1102.1114 [hep-ph]].
[12] Nahrgang M, Aichelin J, Gossiaux P B and Werner K, arXiv:1305.6544 [hep-ph].
[13] Werner K, Karpenko I, Pierog T, Bleicher M and Mikhailov K, *Phys. Rev. C* **82** (2010) 044904; Werner K, Karpenko I, Bleicher M, Pierog T and Porteboeuf-Houssais S, *Phys. Rev. C* **85** (2012) 064907; Werner K, Bleicher M, Guiot B, Karpenko I and Pierog T, arXiv:1307.4379 [nucl-th].
[14] Cacciari M, Greco M and Nason P, *JHEP* **9805** (1998) 007; Cacciari M, Frixione S and Nason P, *JHEP* **0103** (2001) 006; Cacciari M, Frixione S, Houdeau N, Mangano M L, Nason P and Ridolfi G, *JHEP* **1210** (2012) 137; www.lpthe.jussieu.fr/ cacciari/fonll/fonllform.html.
[15] Nahrgang M, Aichelin J, Gossiaux P B and Werner K, arXiv:1305.3823 [hep-ph].
[16] www.hep.phy.cam.ac.uk/theory/webber/MCatNLO/; Frixione S and Webber B R, *JHEP* **0206** (2002) 029; Frixione S, Nason P and Webber B R, *JHEP* **0308** (2003) 007.
[17] Zhu X, Xu N and Zhuang P, *Phys. Rev. Lett.* **100** (2008) 152301.

Journal of Physics: Conference Series **509** (2014) 012048 doi:10.1088/1742-6596/509/1/012048

Heavy Flavor Suppression: Boltzmann vs Langevin

S. K. Das[1,2], F. Scardina [1,2], S. Plumari[1,2] and V. Greco[1,2]

[1] Department of Physics and Astronomy, University of Catania, Via S. Sofia 64, I- 95125 Catania, Italy

[2] Laboratori Nazionali del Sud, INFN-LNS, Via S. Sofia 62, I-95123 Catania, Italy

E-mail: santosh@lns.infn.it

Abstract. The propagation of heavy flavor through the quark gluon plasma has been treated commonly within the framework of Langevin dynamics, i.e. assuming the heavy flavor momentum transfer is much smaller than the light one. On the other hand a similar suppression factor R_{AA} has been observed experimentally for light and heavy flavors. We present a thorough study of the approximations involved by Langevin equation by mean of a direct comparison with the full collisional integral within the framework of Boltzmann transport equation. We have compared the results obtained in both approaches which can differ substantially for charm quark leading to quite different values extracted for the heavy quark diffusion coefficient. In the case of bottom quark the approximation appears to be quite reasonable.

1. Introduction

The experimental efforts at Relativistic Heavy Ion Collider (RHIC) and Large Hadron Collider (LHC) energies is to create and study the properties of matter called quark gluon plasma (QGP). The heavy flavors, namely, charm and bottom quarks are particularly playing a vital role to serve this purpose. The most common approach to study heavy flavors propagation in QGP is the Fokker-Planck [1, 2, 3, 4, 5, 6, 7, 8, 9, 10, 11, 12, 13] equation that can be realized from the Boltzmann equation which constitutes a significant simplification *i.e.* the heavy flavor momentum transfer is small or the scattering are sufficiently forward peaked. Thus it is a approximation of the full collision term of the Boltzmann equation. Such an approximation is expected to be asymptotically valid for $m/T \rightarrow \infty$, here we present a first study of the validity of such an approximation for b and c quark at temperature typically reached in ultrarelativistic heavy ion collisions at LHC.

2. Relativistic Langevin Equation

The relativistic Langevin equations of motion corresponds to evolve the particles in the phase space according to the following equations [1],

$$dx_i = \frac{p_i}{E} dt,$$
$$dp_i = -\Gamma p_i dt + C_{ij} \rho_j \sqrt{dt} \quad (1)$$

where Γ and C_{ij} describe the drag force and the stochastic force in terms of independent Gaussian-normal distributed random variables ρ, which obey $< \rho_i \rho_j >= \delta_{ij}$ and $< \rho_i >= 0$, respectively. To study the momentum evolution of charm and bottom quarks, we use the Ito discretization.

3. Relativistic Boltzmann Transport Equation

The relativistic Boltzmann equation can be written as follows [14, 15]:

$$p^\mu \partial_\mu f(x,p) = \mathcal{C}(x,p) \tag{2}$$

where $\mathcal{C}(x,p)$ is the Boltzmann collision integral, the main ingredient of the cascade codes.

If we define $\omega(p,k)$, the rate of collisions which change the heavy quark momentum from p to $p-k$, then we have [2]

$$\mathcal{C}(x,p) = \int d^3k \left[\omega(p+k,k)f(p+k) - \omega(p,k)f(p)\right] \tag{3}$$

The first term in the integrand represents the gain of probability through collisions which knock the heavy quark into the volume element, and the second term represents the loss out of that volume element. If we expand $\omega(p+k,k)f(p+k)$ around k,

$$\omega(p+k,k)f(p+k) \approx \omega(p,k)f(p) + k\frac{\partial}{\partial p}(\omega p) + \frac{1}{2}k_i k_j \frac{\partial^2}{\partial p_i \partial_j}(\omega p) \tag{4}$$

We obtain the Fokker Planck/Langevin Equation. In the following we will discuss a first comparison between the full Boltzmann equation and the Langevin equation coming from Eq. 4.

4. Numerical results and discussion

The basic ingredients required for solving the Langevin equation are the drag and diffusion coefficients and the initial momentum distributions of heavy quarks. We are using $f^c_{t=0} = 4.1 * 10^2/(0.7 + 0.09p)^{15.44}$ and $f^b_{t=0} = 1/(57.74 + p^2)^{5.04}$ as the initial momentum distribution of the charm and bottom quarks respectively. For the sake of comparison, we are solving the Langevin equation in a box where the bulk consists of only gluon at T=0.4 GeV. The elastic collisions of heavy quarks with gluon has been considered within the pQCD framework to calculate the drag and diffusion coefficients. In the present calculation we used $\alpha_s = 0.35$ and the Debye screening mass as $m_D = gT$. It can be mentioned that the experimental data on the nuclear suppression factor and the elliptic flow of nonphotonic electrons provide the best agreement to the experimental data for the spatial diffusion coefficient $D_x = 6/2\pi T$ [12]. Keeping this in mind we enhanced the pQCD cross sections by a factor 2 to match our diffusion coefficients as $D_x = 6/2\pi T$, so that we could extract conclusion of phenomenological interest from our calculation. On the other hand the relativistic Boltzmann equation also solved in a identical environment as the Langevin equation with the enhanced pQCD cross sections by a factor 2, in a Montecarlo cascade [14] based on the stochastic interpretation of the transition rate discussed in Ref [15].

In Fig 1 the ratio of Langevin to Boltzmann spectra for the charm quark has been displayed as a function of momentum at different times to quantify how much the ratio deviates from 1. We started the simulation at $t = 0$ with the same initial distribution for both Lanvevin and Boltzmann equations which leads to a ratio 1. So any deviation from 1 would reflect how much the Langevin deviations differ from the Boltzmann as a approximation. From Fig 1 it is observed that for $t = 4$ fm the deviation of Langevin from Boltzmann is around 25% and for $t = 6$ fm the deviation is around $35\% - 40\%$ at $p = 7$ GeV, which suggests Langevin approach overestimates the interaction considerably due to approximation it involved. we have seen that the effect can be significantly larger or smaller depending on the value of the drag and diffusion coefficients and/or on the angular depndence of the collisions. The same quantity is depicted in Fig 2 as a function momentum at different times for bottom quark. From Fig 2 it is shown that the ratio is almost unity at all the times considered in the manuscript. Therefore, for bottom quark the Langevin approach is really a good approximation of Boltzmann equation.

Journal of Physics: Conference Series **509** (2014) 012048 doi:10.1088/1742-6596/509/1/012048

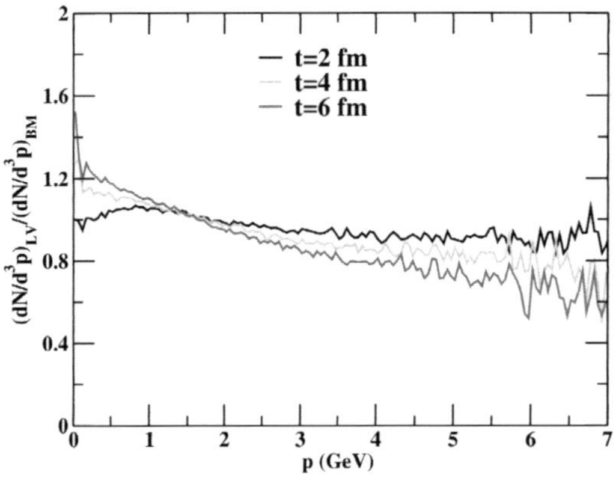

Figure 1. Ratio between the Langevin (LV) and Boltzmann (BM) spectra for charm quark as a function of momentum at different time

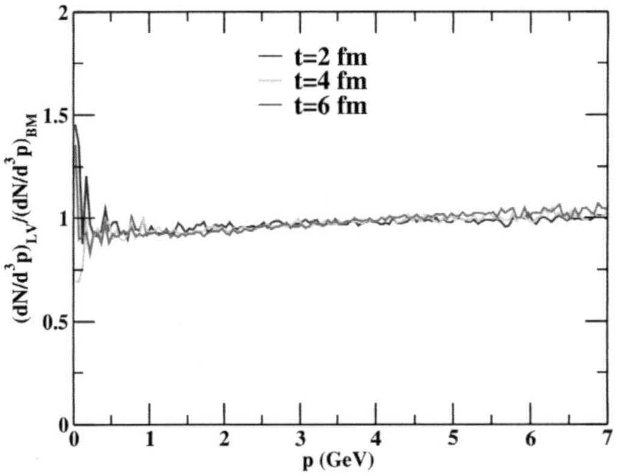

Figure 2. Ratio between the Langevin (LV) and Boltzmann (BM) spectra for bottom quark as a function of momentum at different time

We calculate the nuclear suppression factor, R_{AA}, using our initial $t = 0$ and final $t = t_f$ charm quark distribution as $R_{AA} = \frac{f_{t=f}}{f_{t=0}}$. The nuclear suppression factor, R_{AA}, has been displayed in Fig 3 as a function of momentum from both Langevin and Boltzmann side at $t = 6$ fm. From Fig 3 it is observed that the nuclear suppression factor differ substantially from Langevin to Boltzmann. Since the diffusion coefficient is very important for the phenomenological study, it will be very useful from phenomenological point of view to study how much the diffusion coefficient change from Langevin side to reproduce the same nuclear suppression factor of Boltzmann equation. It is observed that (Fig 3) we need to change (reduced) the diffusion coefficient of Langevin equation by 30% to get similar nuclear suppression factor as of Boltzmann equation, although it is not anyway possible to reproduce the same momentum dependence. Heavy flavor supression within Boltzmann tranport approach can be found in Ref. [16](see also Ref. [17]) .

5. SUMMARY AND CONCLUSIONS

We present a thorough study of the approximations involved by Langevin equation by mean of a direct comparison with the full collisional integral within the framework of Boltzmann

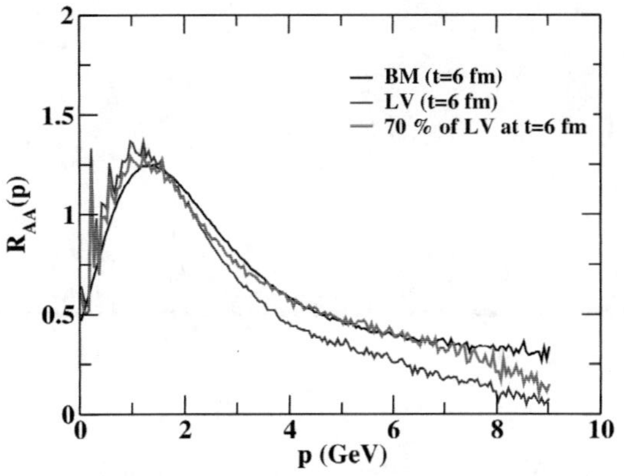

Figure 3. The nuclear suppression factor, R_{AA} as a function of momentum from the Langevin (LV) equation and Boltzmann (BM) equation for charm quark in a box at T=0.4 GeV

transport equation in a box where the bulk consists of only gluon at T=0.4 GeV. We found that the Langevin approach is a good approximation for bottom quark where as for charm quark Langevin approach deviates of about a 30% at intermediate momentum for a time evolution of 5-6 fm/c typical of the QGP created in heavy-ion collisions respect to the spectra calculated from the solution of the full Boltzmann transport equation. It has also been found that to get a similar suppression factor form both the approach we need to reduce the diffusion coefficient of the Langevin approach by around 30%. These results can have significant effects on the heavy ion phenomenology at RHIC and LHC energies. Calculation for a realistic fireball evolution as created in ultra-relativistic heavy-ion collisions is undergoing.

Acknowledgements

We acknowledge the support by the ERC StG under the QGPDyn Grant n. 259684

References

[1] Rapp R and Hees H van 2010 Hwa R C, Wang X N (Ed.) Quark Gluon Plasma 4, *World Scientific*, **111** [arXiv:0903.1096 [hep-ph]]
[2] Svetitsky B 1988 *Phys. Rev.* D **37**, 2484
[3] Mustafa M G, Pal D and Srivastava D K 1998 *Phys. Rev.* C **57**, 889
[4] Moore G D, Teaney D 2005 *Phys. Rev.* C **71** 064904
[5] Hees H van, Greco V and Rapp R 2006 *Phys. Rev.* C **73**, 034913
[6] Hees H van, Mannearelli M, Greco V and Rapp R 2008 *Phys. Rev. Lett.* **100** 192301
[7] Gossiaux P B, Aichelin J 2008 *Phys. Rev.* C **78** 014904 ; Gossiaux P B, Aichelin J, Gousset T and Guiho V 2010 *J. Phys.* G **37** 094019
[8] Akamatsu Y, Hatsuda T, Hirano T 2009 *Phys. Rev.* C **79** 054907
[9] Das S K, Alam J, Mohanty 2009 *Phys. Rev.* C **80** 054916 ; Majumdar S, Bhattacharyya T, Alam J, Das S K 2012 *Phys. Rev.* C **84** 044901
[10] Alberico W M *et al.* 2011 *Eur. Phys. J.* C **71** 1666; Alberico W M *et al.* 2013 *Eur. Phys. J.* C **73** 2481
[11] Young C, Schenke B, Jeon S and Gale C 2012 *Phys. Rev.* C **86**, 034905
[12] Cao S, Bass S A Bass 2011 *Phys. Rev.* C **84** 064902
[13] Lang T, Hees H, Steinheimer J and Bleicher M, arXiv:1208.1643 [hep-ph]
[14] Ferini G, Colonna M, Toro M Di and Greco V, 2009 *Phys. Lett.* **B670**,325; Greco V, Colonna M, Toro M Di and Ferini G, 2009 *Progr. Part. Nucl. Phys.* **62**, 562
[15] Xu Z, Greiner C 2005 *Phys. Rev.* C **71**, 064901
[16] Uphoff J, Fochler O, Xu Z and Greiner C 2011 *Phys. Rev.* C **84** 024908
[17] Younus M, Coleman-Smith C E,Bass S A and Srivastava D K, arXiv:1309.1276 [nucl-th]

Journal of Physics: Conference Series **509** (2014) 012049 doi:10.1088/1742-6596/509/1/012049

Shear viscosity of a multi-component hadronic system

A. Wiranata[1,2], V. Koch[2], M. Prakash[3] and X.N. Wang[1,2]
[1]Institute of Particle Physics and Key Laboratory of Quark and Lepton Physics (MOE), Central China Normal University, Wuhan 430079, China
[2]Lawrence Berkeley National Laboratory, Nuclear Science Division, MS 70R0319, Berkeley, CA 94720, USA
[3]Department of Physics and Astronomy, Ohio University, Athens, OH, 45701

Abstract. The shear viscosity η and entropy density s of a hadron gas with zero baryon number density are calculated using the Chapman-Enskog and virial expansion approaches, respectively. Interactions are included via the K-matrix parametrization of cross sections preserving the unitarity of the S-matrix. In the four component mixture ($\pi - K - N - \eta$), a total of 57 resonances up to 2 GeV mass are included. Interactions forming resonances reduce the magnitude of η and increase s, both effects serving to progressively reduce η/s as the temperature nears the QCD phase transition temperature.

1. Introduction

Studies of strongly interacting matter at high temperature T have revealed the qualitative behavior of the temperature dependence of the shear viscosity η and its ratio with entropy density s [1, 2]. The ratio η/s decreases with increasing T in the hadronic sector. Perturbative QCD calculations [3] predict an η/s that increases logarithmically with T in the quark-gluon plasma (QGP). This ratio, conjectured to be bound by a limit of $\simeq 1/(4\pi)$ (in units of $\hbar/k_B$) from AdS/CFT calculations [4], likely reaches a minimum around the QCD phase transition temperature T_c [5]. To date, most hydrodynamical model studies of heavy-ion collisions at RHIC/LHC have assumed a constant η/s, whereas some studies have incorporated hadronic cascade models [6] following hydrodynamic evolution to explore the effect of a large η in the hadronic phase. Here, we present results of η/s in a multi-component mixture of hadrons so that its effect on observed collective flow in heavy-ion collisions may be assessed.

2. Formalism

We employ the Chapman-Enskog approach, generalized to include relativistic kinematics, for the calculation of the shear viscosity. In this approach, the off-equilibrium distribution function can be written as $f(x,p) = f^0(x,p)\,[1 + \phi(x,p)]$, where $f^0(x,p)$ is the equilibrium distribution function and $\phi(x,p)$ is the deviation function which contains shear and bulk viscosities as well as heat conductivity [7]. We follow closely the formalism developed in [8] considering only number conserving elastic processes. In the context of hadronic interactions, our recent calculations have been detailed in [9, 10]. The magnitude of η is inversely proportional to the momentum transport cross sections of the various constituent particles in the system. Large cross sections, characteristic of a strongly interacting system, naturally lead to small viscosities. For all but the lightest particles, first principle calculations (or data) of hadronic interactions, particularly those involving massive resonances do not exist. One therefore often uses empirical parameterizations of these hadronic cross sections. To assess the impact of different parameterizations of cross sections on the shear viscosity, we have examined three forms obtained from (i) measured phase shifts (where available) [2], (ii) the Breit-Wigner parametrization [11], and (iii) the $K-$matrix [12] parametrization. Here, we show results ensuing from the $K-$matrix parametrization. Leading contributions to the thermal properties from interactions between hadrons are calculated using the second virial coefficient deduced from two-body phase shifts according to the procedures described in [13, 14, 15].

3. Results

In a pion gas, 8 resonant channels lead to $\pi\pi$ final states [16]. With increasing T, all of these resonances influence how η and s vary with T as exemplified by the two cases: (i) $\pi\pi \to \rho \to \pi\pi$ and (ii) $\pi\pi \to$ All Channels $\to \pi\pi$ (Fig. 1). Relative to the case when only the $\rho-$ resonance is considered (dashed curves), η decreases (due to increased transport cross sections) and s increases (due to the increased number of (spin and isospin) degrees of freedom). The corresponding ratios of η/s are shown in Fig. 2. As $T \to T_c = 155\pm5$ MeV from lattice QCD simulations [17], η/s decreases more rapidly when all channels are included (solid curves) than when only the $\rho-$ channel is considered (dashed curves). The inclusion of several resonances not only causes a reduction in η, it also results in a significant increase in s. Both of these effects render the ratio η/s to become progressively small as $T \to T_c$.

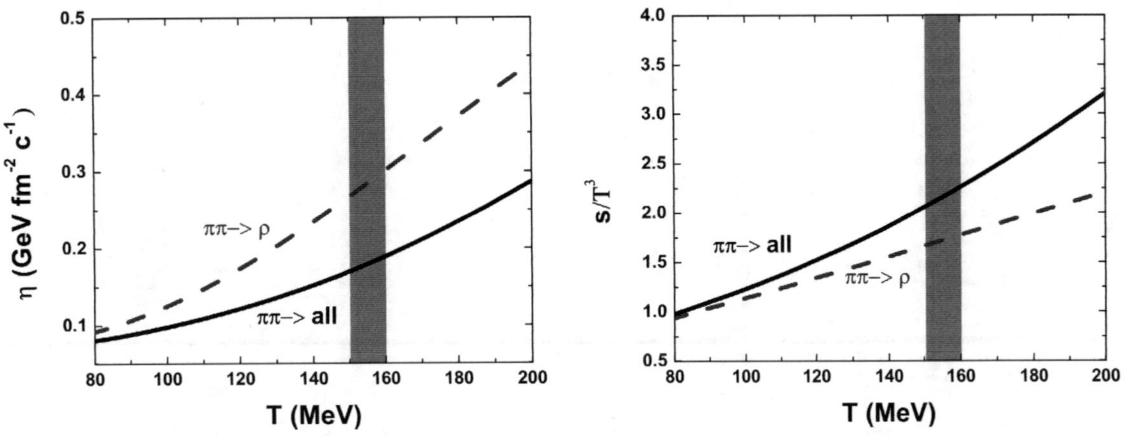

Figure 1. Results of η and s/T^3 for a single component pions gas. Dashed curves correspond to the case when only the $\rho-$resonance is considered. Solid curves show results when all possible resonances are included. The transition temperature, T_c, is indicated by the rectangular box [17].

In the four-component mixture comprising of $\pi - K - N - \eta(548)$, we include the dominant resonances produced in binary elastic interactions among the various constituents. To understand how the inclusion of increasing number of particles (and their associated resonances) determines the magnitudes of η and η/s, we also show results for the two-component mixtures of $\pi - K$ and $\pi - N$, as well as for the three-component mixture of $\pi - K - N$ together with results for the $\pi - K - N - \eta(548)$ mixture. The interaction measure $I = (\epsilon - 3P)/T^4$ calculated using the virial expansion method is compared with the lattice result of the Budapest-Wuppertal (BW) collaboration [17] (left panel of Fig. 3). To be consistent with the ingredients of transport

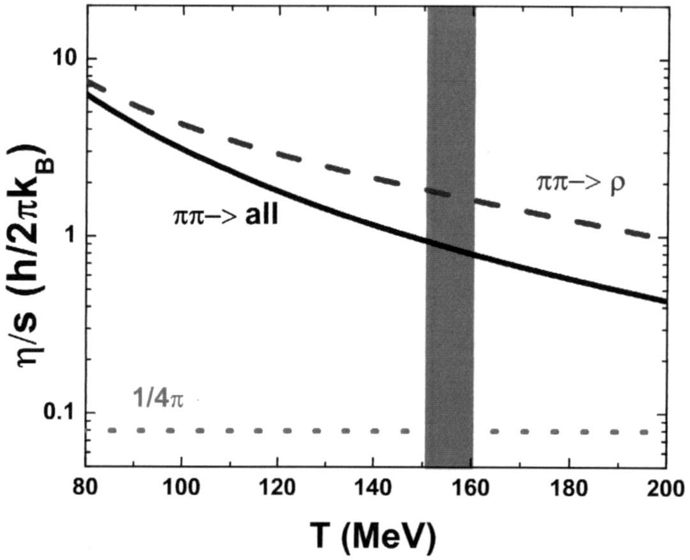

Figure 2. Same as Fig. 1, but for the ratio η/s.

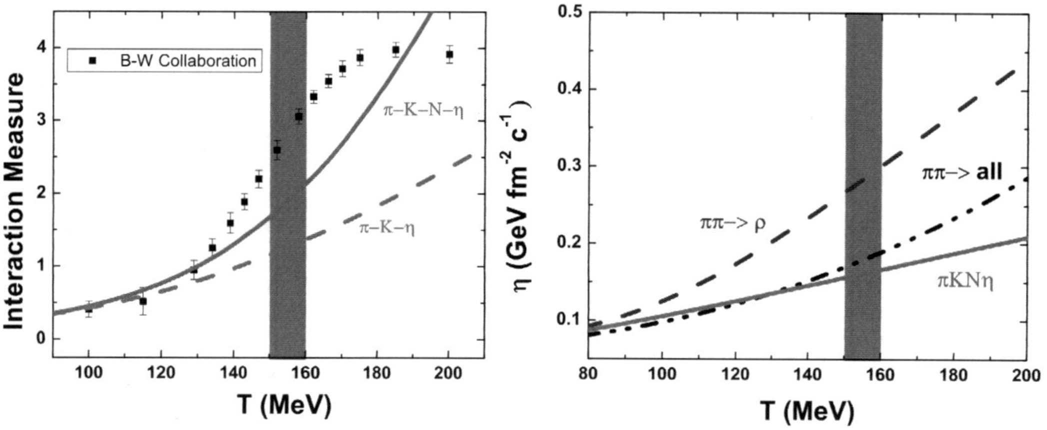

Figure 3. Left panel: Interaction measure I versus temperature T from lattice calculations (data points) with 3-quarks, $m_\pi = 135$ MeV and lattice spacing, $N_t = 8$ (BMW Collaboration). Right panel: Shear viscosity versus temperature. The transition temperature, T_c, is shown by the rectangular box [17].

calculations, only those resonances that are formed in the $\pi - K - \eta - N$ mixture are included in the virial expansion approach. The number of resonances included in the case of the $\pi - K - \eta$ mixture is 21, whereas 57 resonances are included in the $\pi - K - \eta - N$ mixture. The additional resonances present in the four component mixture improve the agreement with the lattice result up to 140 MeV. The inclusion of additional mesons and baryons more massive than realized in the $\pi - K - \eta - N$ mixture is expected to improve agreement with the current lattice results (as supported by results of calculations with all resonances in the PDG book) up to the phase transition temperature T_c [17]. The progressive decreasing in the magnitude of the shear viscosity with increasing temperature as more and more resonances are included is readily apparent from

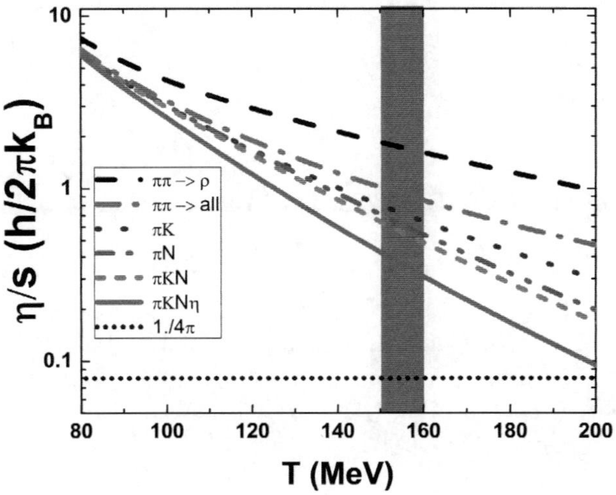

Figure 4. Shear viscosity to entropy density ratio of interacting hadrons. Results are for a single component system (π gas), two component mixtures ($\pi - K$ and $\pi - N$), a three component mixture ($\pi - K - N$) and a four component mixture ($\pi - K - \eta(548) - N$). The horizontal curve at $1/(4\pi)$ is the AdS/CFT result.

Fig. 3. The inclusion of even more resonances than considered in this work is, therefore, likely to reduce the magnitude of η even further as T_c is approached (see also [18]).

The results of η/s are presented in Fig. 4. The role of increasing number of resonances is evident even in the case of a single component pion gas. In addition to decreasing η, resonances increase s. Both these effects serve to decrease η/s with increasing temperature. A similar trend is also observed in the case of binary mixtures as seen from the results for the $\pi - K$ and $\pi - N$ mixtures. Results for the three ($\pi - K - N$) and four ($\pi - K - \eta(548) - N$) component systems, highlight the increasing role of the enhanced entropy density in these systems as the heaviest resonances are not very effective in transferring momentum in a direction perpendicular to that of fluid flow. For reference, the AdS/CFT result of $1/(4\pi)$ is also shown in this figure. From the trends seen in these results, we infer that the inclusion of additional mesons and baryons will further decrease η/s as T_c is approached though not to the level of the AdS/CFT result.

This work is supported by the NSFC under grant No. 11221504 & No. 1350110493, and by the US DOE under contracts #DE-AC02-05CH11231 and #DE-FG02-93ER-40756.

References

[1] Kapusta J I 2010 *Viscous Properties of Strongly Interacting Matter at High Temperature* (Relativistic Nuclear Collisions Vol I) ed R Stock (Landolt-Bornstein New Series) pp 23
[2] Prakash M, Prakash M, Venugopalan R, and Welke R 1993 *Phys. Rep.* **227** 331
[3] Arnold P , Moore G, and Yaffe L G 2000 *JHEP* **0011** 001
[4] Kovtun P K, Son D, and Starinets A O 2005 *Phys. Rev. Lett.* **94** 111601
[5] Csernai L P, Kapusta J I and McLerran L D 2006 *Phys. Rev. Lett.* **97** 152303
[6] Gale C, Jeon S, Schenke B, Tribedy P, and Venugopalan R 2013 *Phys. Rev. Lett.* **110** 012302
[7] Leeuwen W van, Kox A J, and Groot S de 1975 *Physica* **79A** 233
[8] Leeuwen W van 1975 *Physica* **81A** 249
[9] Wiranata A and Prakash M 2012 *Phys. Rev. C* **85** 054908
[10] Wiranata A, Koch V, Prakash M, and Wang X N 2013 *Phys. Rev. C* **88** 044917
[11] Bass S A, Gyulassy M, Stöcker H, and Greiner W 1999 *J Phys. G* **25** R1
[12] Chung S U*et al* 1995 *Ann. Physik* **4** 404
[13] Dashen R, Ma S, and Bernstein H J 1969 *Phys. Rev.* **187** 349
[14] Welke G, Venugopalan R, and Prakash M 1990 *Phys. Lett. B* **245** 137
[15] Venugopalan R and Prakash M 1992 *Nuclear Physics A* **A546** 718
[16] Beringer J *et al* 2012 *Phys. Rev. D* **86** 010001
[17] Borsanyi S *et al* 2010 *JHEP* **1011** 077
[18] Hostler J N, Noronha J, and Greiner C 2009 *Phys.Rev.Lett.* **103** 172302

Journal of Physics: Conference Series **509** (2014) 012050 doi:10.1088/1742-6596/509/1/012050

Flavor-specific behavior of conserved charge fluctuations at the QCD confinement transition

M Bluhm[1], P Alba[1], W M Alberico[1], R Bellwied[2] and C Ratti[1]

[1] Dipartimento di Fisica Teorica, Università degli Studi di Torino & INFN, Sezione di Torino, via Pietro Giuria 1, I-10125 Torino, Italy

[2] Physics Department, University of Houston, Houston, TX 77204-5005, USA

E-mail: mbluhm@to.infn.it

Abstract. We propose measurements of fluctuation observables, which are sensitive to the chemical freeze-out of the matter produced in heavy-ion collisions, in order to discriminate a potential flavor-specific behavior of light and strange quarks at the QCD confinement transition. This complements our former considerations which were restricted to the strange quark sector only.

1. Introduction

The QCD confinement transition from a state with deconfined, colored and (almost) massless partonic degrees of freedom to a state containing color-neutral and massive hadrons and resonances, in which the partons are confined, is known from first-principles to be an analytic crossover at vanishing net baryon-number density [1]. Accordingly, this phase transformation of QCD matter takes place over a broad temperature T-interval, where an exact value for the pseudo-critical temperature T_c cannot be pinned down unambiguously. In fact, depending on the considered order-parameter-like observable, values of $T_c = (154 \pm 9)$ MeV [2], or $T_c = (147 \pm 5)$ MeV and $T_c = (155 \pm 6)$ MeV [3] are found.

Within large-scale world-wide efforts, the phase diagram of QCD matter is sought to be explored experimentally. Envisaged goals are the investigation of the confinement/deconfinement transition in the laboratory and a determination of the properties of deconfined matter and of the nature of its degrees of freedom. In the relativistic heavy-ion collisions (HIC) being performed at the LHC (Large Hadron Collider) and at RHIC (Relativistic Heavy Ion Collider), indeed, the deconfined phase of QCD matter is transiently produced. Observables suitable for indicating this are, for example, fluctuations in conserved charges, which are sensitive to the microscopic structure of the matter [4]. The conserved charges relevant in HIC are the net baryon-number, net electric-charge and net strangeness. A rapidly changing, non-monotonous behavior in the fluctuations of these charges can provide clear signatures for the confinement/deconfinement transition. For example, in [5] the ratio of fourth- to second-order cumulants of net electric-charge fluctuations was proposed as a possible observable.

2. Indications for a flavor-hierarchy at confinement

In the gauge field theory of the strong interaction, QCD, the conserved charges are instead the net quark-numbers of individual quark flavors. Fluctuations in these charges can be quantified in the

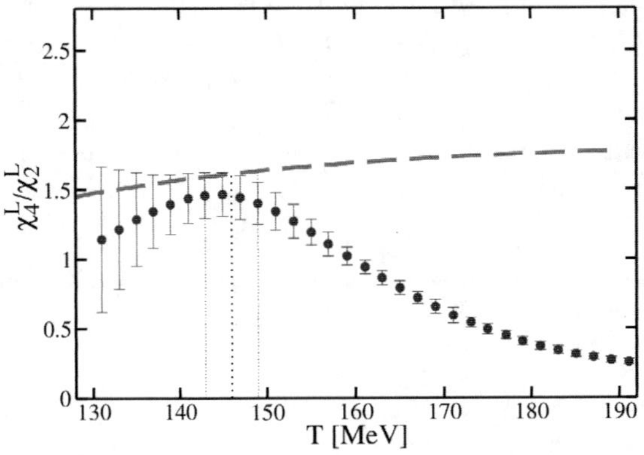

Figure 1. Comparison of the HRG model result (long-dashed curve) for χ_4^L/χ_2^L with the continuum-extrapolated lattice QCD data at physical quark masses (circles) from [8]. The change in the behavior of the χ_4^L/χ_2^L-data with T as indicator for the onset of the deconfinement of light quarks happens at about $T \simeq (146 \pm 3)$ MeV (dotted band). Given the large errors in the data from [8] within the transition region, this is also where the HRG model starts to deviate significantly from lattice QCD.

grand canonical ensemble description of the system by susceptibilities and are therefore accessible from first-principles by means of lattice QCD calculations. Quark-number susceptibilities are related to the equilibrium pressure P by appropriate derivatives reading

$$\chi_{lmn}^{uds} = \left. \frac{\partial^{l+m+n}(P/T^4)}{\partial(\mu_u/T)^l \partial(\mu_d/T)^m \partial(\mu_s/T)^n} \right|_{\mu_u=\mu_d=\mu_s=0}, \tag{1}$$

where μ_u, μ_d and μ_s represent the up, down and strange quark chemical potentials associated with the net quark-numbers of up, down and strange quarks, respectively.

Recent continuum-extrapolated lattice QCD results for χ_2^u and χ_2^s at physical quark masses showed a clear difference in the behavior of the net quark-number fluctuations with T between up and strange quarks [6]. This observation was interpreted in [7] as signal for a flavor-hierarchy at the QCD confinement transition. In order to shed more light onto this possibility, higher-order susceptibilities were studied in lattice QCD simulations in [8]. In particular, the volume-independent ratio of fourth- to second-order quark-number susceptibilities χ_4^f/χ_2^f of flavor f was found to depict a non-monotonous behavior with T, which occurs for light (up and down) quarks at a smaller temperature than for strange quarks, cf. [8]. Relating μ_u and μ_d via $\mu_u = \mu_d = \mu_L/2$ to the light quark chemical potential μ_L, the light quark-number susceptibilities χ_2^L and χ_4^L can be obtained from up and down quark-number susceptibilities as $\chi_2^L = (\chi_2^u + 2\chi_{11}^{ud} + \chi_2^d)/4$ and $\chi_4^L = (\chi_4^u + 4\chi_{31}^{ud} + 6\chi_{22}^{ud} + 4\chi_{13}^{ud} + \chi_4^d)/16$. The corresponding lattice QCD results [8] for χ_4^L/χ_2^L as a function of T are shown in Fig. 1.

Moreover, particular combinations of higher-order cumulants of fluctuations in the conserved charges and of correlations among them were considered in [8] which probe directly the degrees of freedom carrying the quantum numbers of light or of strange quarks. For a system composed of uncorrelated hadrons and resonances as described by the Hadron Resonance Gas (HRG) model, which properly describes QCD thermodynamics in the hadronic phase, some of these combinations exactly vanish. Thus, any deviation of the corresponding lattice QCD data from this HRG model prediction may be interpreted as the onset of a liberation of quarks. Again, a difference in the behavior between light and strange quarks was observed [8], where the model-dependent result obtained for the liberation temperature of strange quarks coincided with the observations formerly made in [9]. Thus, the first-principle results reported in [8] provided further indications for a flavor-specific behavior in the confinement transition of QCD.

In the HRG model, the equilibrium pressure reads as

$$P(T,\mu_B,\mu_Q,\mu_S) = \sum_j (-1)^{B_j+1} \frac{d_j T}{(2\pi)^3} \int d^3\vec{k} \, \ln\left[1 + (-1)^{B_j+1} e^{-\left(\sqrt{\vec{k}^2+m_j^2}-\mu_j\right)/T}\right], \tag{2}$$

Journal of Physics: Conference Series **509** (2014) 012050 doi:10.1088/1742-6596/509/1/012050

where the sum runs over all the hadronic and resonance states included in the model. In Eq. (2), d_j, m_j and μ_j denote the degeneracy factor, the mass and the chemical potential of the excitation j, where the latter is given by $\mu_j = B_j\mu_B + Q_j\mu_Q + S_j\mu_S$ with B_j, Q_j and S_j as quantum numbers of j in baryonic charge, electric charge and strangeness, respectively, while μ_B, μ_Q and μ_S denote the chemical potentials associated with the corresponding charge-numbers being conserved in HIC. Fluctuations in these charges can be quantified by the susceptibilities χ_{lmn}^{BQS} readily obtained from Eq. (1) by replacing (u, d, s) by (B, Q, S). Relating the chemical potentials via $\mu_B = \mu_u + 2\mu_d$, $\mu_Q = \mu_u - \mu_d$ and $\mu_S = \mu_d - \mu_s$, cf. [10], the ratio χ_4^L/χ_2^L can be determined from the susceptibilities calculable in the HRG model as

$$\frac{\chi_4^L}{\chi_2^L} = \frac{\left(81\chi_4^B + 108\chi_{31}^{BS} + 54\chi_{22}^{BS} + 12\chi_{13}^{BS} + \chi_4^S\right)}{4\left(9\chi_2^B + 6\chi_{11}^{BS} + \chi_2^S\right)} \,. \tag{3}$$

The corresponding result for a HRG model containing excitations up to a mass of 2 GeV is shown in Fig. 1 to agree fairly well with the lattice QCD data from [8] for $T \lesssim 146$ MeV.

3. Verifying a possible flavor-hierarchy experimentally

The ratios of those susceptibilities, which can be associated with fluctuations in the conserved charges, form in principle measurable, volume-independent observables. They can be related to higher-order moments of multiplicity distributions [11], which are fixed at the chemical freeze-out of the matter produced in HIC. Assuming that the chemical freeze-out takes place subsequent to the confinement transition, a flavor-specific behavior should leave a fingerprint in the higher-order moments of flavor-sensitive particle-identified multiplicity distributions. Hence, a flavor-hierarchy at confinement should be testable experimentally.

In [12], we proposed a measurement suitable for determining the freeze-out characteristics of strange quarks. Here, we want to extend this idea to the light quark sector. Given the spectrum of hadrons identifiable with the ALICE detector at the LHC, we propose to determine for an ensemble composed of K^+, K^-, p and $\overline{p}$ the statistical moments of the combined multiplicity distribution $N_L = \frac{3}{2}N_p + \frac{1}{2}N_{K^+} - \frac{3}{2}N_{\overline{p}} - \frac{1}{2}N_{K^-}$. In this ensemble, pions are disregarded because they do neither contribute to the net light quark-number nor to its fluctuations. By comparison of the product between kurtosis κ and variance σ^2 of this distribution with the susceptibility ratio χ_4^L/χ_2^L determined via Eq. (3) from the pressure in Eq. (2), where the sum is restricted to the above proposed ensemble, a freeze-out temperature specific to light quarks might be deducible from the fluctuation observables κ and σ^2 in a model-dependent way, see Fig. 2 (left).

For the strange quark sector, instead, we suggested in [12] to consider an ensemble composed of K^+, Λ^0, Ξ^- and Ω^- as well as their anti-particles. By determining $\kappa\sigma^2$ for the combined multiplicity distribution $N_s = N_{K^-} + N_{\Lambda^0} + 2N_{\Xi^-} + 3N_{\Omega^-} - N_{K^+} - N_{\overline{\Lambda}^0} - 2N_{\overline{\Xi}^+} - 3N_{\overline{\Omega}^+}$ and comparing with $\chi_4^s/\chi_2^s \equiv \chi_4^S/\chi_2^S$ obtained from Eq. (2) restricted to this ensemble, the corresponding freeze-out temperature for strange quarks may be deduced, see Fig. 2 (right). A possible difference in the freeze-out behavior between light and strange quarks found from the fluctuation observables would provide an experimental verification for a flavor-hierarchy in the QCD confinement transition.

For a meaningful comparison of experimental results with the HRG model, however, strong resonance decays have to be included into the considerations. This will be discussed elsewhere. Moreover, detector acceptance cuts have to be taken into account. They can be included into the HRG model approach by replacing in Eq. (2) the measure $d^3\vec{k} \to k_T\sqrt{k_T^2 + m_j^2}\cosh(y)\, dk_T\, dy\, d\phi$ and $\sqrt{\vec{k}^2 + m_j^2} \to \sqrt{\vec{k}_T^2 + m_j^2}\cosh(y)$, where k_T is the transverse momentum and y the rapidity of the identified particles. The influence of acceptance cuts on the HRG model results is shown in Fig. 2 (solid curves). The applied cuts read $|y| < 0.5$ as well as $0.2 < k_T(\mathrm{GeV/c}) < 2$ and $0.5 < k_T(\mathrm{GeV/c}) < 3$ for the light and strange quark specific ensembles, respectively.

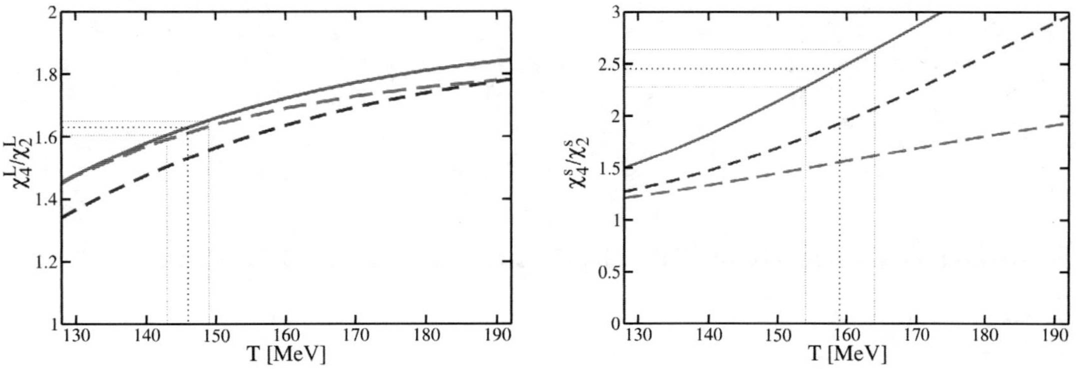

Figure 2. Determination of the freeze-out temperature for light (left panel) and strange quarks (right panel) from comparing fluctuation observables. Possible results depend on the potentially measured $\kappa\sigma^2$-values for the specific multiplicity distributions N_L and N_s of primary particles (dotted bands). Long-dashed and short-dashed curves show the HRG model results for χ_4^f/χ_2^f ($f = L, s$), where excitations of mass up to 2 GeV or only the particles contained in the specific ensembles are considered in Eq. (2), respectively, while the solid curves depict the impact of detector acceptance cuts on the specific ensemble results.

Acknowledgments
The work of C. Ratti and M. Bluhm is supported by funds provided by the Italian Ministery of Education, Universities and Research under the FIRB Research Grant RBFR0814TT.

References

[1] Aoki Y, Endrodi G, Fodor Z, Katz S D and Szabo K K 2006 The order of the quantum chromodynamics transition predicted by the standard model of particle physics *Nature* **443** 675-8
[2] Bazavov A *et al* 2012 The chiral and deconfinement aspects of the QCD transition *Phys. Rev.* D **85** 054503
[3] Borsanyi S, Fodor Z, Hoelbling C, Katz S D, Krieg S, Ratti C and Szabo K K 2010 Is there still any T_c mystery in lattice QCD? Results with physical masses in the continuum limit III *J. High Energy Phys.* JHEP09(2010)073
[4] Asakawa M, Heinz U W and Müller B 2000 Fluctuation probes of quark confinement *Phys. Rev. Lett.* **85** 2072-5
[5] Ejiri S, Karsch F and Redlich K 2006 Hadronic fluctuations at the QCD phase transition *Phys. Lett.* B **633** 275-82
[6] Borsanyi S, Fodor Z, Katz S D, Krieg S, Ratti C and Szabo K K 2012 Fluctuations of conserved charges at finite temperature from lattice QCD *J. High Energy Phys.* JHEP01(2012)138
[7] Ratti C, Bellwied R, Christoforetti M and Barbaro M 2012 Are there hadronic bound states above the QCD transition temperature? *Phys. Rev.* D **85** 014004
[8] Bellwied R, Borsanyi S, Fodor Z, Katz S D and Ratti C 2013 Is there a flavor hierarchy in the deconfinement transition of QCD? *Preprint* arXiv:1305.6297 [hep-lat]
[9] Bazavov A *et al* 2013 Strangeness at high temperatures: from hadrons to quarks *Phys. Rev. Lett.* **111** 082301
[10] Cheng M *et al* 2009 Baryon number, strangeness and electric charge fluctuations in QCD at high temperature *Phys. Rev.* D **79** 074505
[11] Karsch F and Redlich K 2011 Probing freeze-out conditions in heavy ion collisions with moments of charge fluctuations *Phys. Lett.* B **695** 136-42
[12] Alba P, Alberico W M, Bluhm M, Ratti C and Bellwied R 2013 Flavor hierarchy in the confinement transition of QCD *Proc. of Science* PoS(CPOD2013)060

Journal of Physics: Conference Series **509** (2014) 012051 doi:10.1088/1742-6596/509/1/012051

Quark-hadron phase transition in a three flavor PNJL model

Kanako Yamazaki[1], **T. Matsui**
Institute of Physics, University of Tokyo, Komaba, Tokyo, 153-8902, Japan
E-mail: [1] kyamazaki@nt1.c.u-tokyo.ac.jp

Abstract.
We study the quark-hadron phase transition by using a three flavor Nambu-Jona-Lasinio model with the Polyakov loop at zero chemical potential, extending our previous work with two flavor model. We show that the equation of state at low temperatures is dominated by pions and kaons as collective modes of quarks and anti-quarks. As temperature increases, mesonic collective modes melt into the continuum of quark and anti-quark so that hadronic phase changes continuously to the quark phase where quark excitations dominate pressure.

1. Introduction

Studying the quark-hadron phase transition is one of the most fundamental problem in modern nuclear physics. Although many works have been done on the QCD phase transitions by effective models and lattice calculations, there are still many uncertainties especially in the intermediate region between hadronic phase and quark phase. The goal of this work is to study the behavior of the equation of state in order to investigate how the degrees of freedom change from those of hadrons to quarks and gluons with increasing temperature. We work with an effective model written in quark and anti-quark fields and take into account correlations between quarks and anti-quarks to describe mesons. Since we wish to describe the chiral and de-confining transitions simultaneously, we use the Nambu-Jona-Lasinio model with the Polyakov loop (PNJL model)[1]. In this report, we present a brief account of the results of calculations, extending our previous work with the two flavor model [2], with strangeness degree of freedom. It is shown that the equation of state is transformed continuously from that of a gas of light hadrons (pions and kaons) at low temperatures, to a gas of light quarks (u-, d- and s-quarks) and their anti-particles with gluon excitation at high enough temperatures. Full account of this work will be reported elsewhere [3].

2. Model setting

We start with the Lagrangian of a three flavor NJL model in external temporal color gauge fields:

$$\mathcal{L} = \sum_{i,j=1}^{3} \bar{q}_i (i\not{D} - \hat{m})_{ij} q_j + \mathcal{L}_4 + \mathcal{L}_6 \tag{1}$$

where

$$\mathcal{L}_4 = G\sum_{a=0}^{8}\left[(\bar{q}\lambda^a q)^2 + (\bar{q}i\gamma_5\lambda^a q)^2\right] \tag{2}$$

and

$$\mathcal{L}_6 = -K\left[\det \bar{q}(1+\gamma_5)q + \det \bar{q}(1-\gamma_5)q\right] \tag{3}$$

for three flavor light quarks, $\bar{q} = (\bar{q}_1, \bar{q}_2, \bar{q}_3) = (\bar{u}, \bar{d}, \bar{s})$. $D_\mu = \partial_\mu + gA_0\delta_{\mu,0}$ where A_0 is the temporal component of gauge fields, $A_0 = -iA_4$. The gauge field is not treated as a dynamical variable; it plays a role of an external parameter like imaginary color-dependent chemical potential for quarks, and we will take a statistical average over A_4 later to define the PNJL model. $\hat{m}$ is a 3 by 3 mass matrix, giving bare masses of u, d and s quarks. The partition function is given by

$$Z(T, A_4) = \int [dq][d\bar{q}]\exp\left[\int_0^\beta d\tau \int d^3x\mathcal{L}(q, \bar{q}, A_4)\right]. \tag{4}$$

With the interaction terms, we build pseudo scalar mesons and scalar mesons from this model. For pseudo scalar mesons, there are nine mesons, three kinds of π, four kinds of K, η and η'. They make a nonet in SU(3) flavor classification. In the chiral limit, all mesons becomes the massless Nambu-Goldstone modes. The 6-point interaction, Eq.(3), breaks the axial U(1) symmetry and makes η^0 massive η' meson. In addition, the SU(3) flavor symmetry breaking by non-vanishing bare quark masses reproduces the physical mass splitting of strange and non-strange mesons. There appear also nine scalar mesons, σ, four kinds of κ, three kinds of a_0, and f_0 [4, 5]. They also form a nonet, although not all of them have been confirmed by experiments.

The three flavor NJL type models have not only 4-point interaction of fermion fields but also 6-point interaction. We replace the 4 and 6 point interactions by Yukawa couplings to bosonic auxiliary fields, extending the method of Hubbard and Stratonovich [6, 7] to 6-point interaction, and perform integral over the quarks field, converting the original Lagrangian of quark fields to that of scalar mesons ϕ_a and pseudo scalar mesons π_a.

The simplest way to calculate the equation of state by the PNJL model is to perform the mean field approximation. However in this approximation, meson fields are treated as uniform back ground fields, in other words, thermal fluctuations of mesons are not included under the mean field approximation even at low temperatures. We therefore must go beyond the mean field approximation, taking mesonic correlation into account to obtain the equation of state of a meson gas.

In order to calculate a pressure of mesonic correlation, we expand an effective action up to the second order of fluctuations around a stationary point given by the non-vanishing uniform ϕ_0 field. All other components of the scalar fields and pseudo scalar fields are set to zero at the stationary point.

$$I(\phi, \pi, A_4) = I_0 + \frac{1}{2}\left.\frac{\delta^2 I}{\delta\phi_a\delta\phi_b}\right|_{\phi=\phi_0}\phi_a\phi_b + \frac{1}{2}\left.\frac{\delta^2 I}{\delta\pi_a\delta\pi_b}\right|_{\phi=\phi_0}\pi_a\pi_b \cdots \tag{5}$$

and the thermodynamic potential is written by

$$\Omega(T, A_4) = T\ \ I_0 + \frac{1}{2}\mathrm{Tr}_M\ln\frac{\delta^2 I}{\delta\phi_a\delta\phi_b} + \frac{1}{2}\mathrm{Tr}_M\ln\frac{\delta^2 I}{\delta\pi_a\delta\pi_b}\ \ . \tag{6}$$

3. Equation of state

3.1. Mean field approximation

Under the mean field approximation, pressure is given by

$$p_{MF}(T) = p_{MF}^0 + 2 \times 3 \sum_f \frac{d^3p}{(2\pi)^3} \frac{\mathbf{p}^2}{3E_f} f_\Phi(E_f) - \mathcal{U}(T, \Phi) \tag{7}$$

where the first term is a vacuum pressure that doesn't depend on temperature explicitly, the second term depends on the temperature through the quark distribution function $f_\Phi(E_f)$. The third term is an effective potential of gluon, given by hand in order to give the pressure of gluon at high temperatures. The effect of the Polyakov loop appears in the second and third terms. In the quark distribution function, the Polyakov loop works to suppress quark excitations at low temperatures.

We show the pressure scaled by T^4 as a function of temperature in Fig.(1). The red and the blue lines are calculated by the PNJL model: the red line contains the contribution from an effective potential $\mathcal{U}(T, \Phi)$ while the blue one doesn't so that all contributions of the blue line at high temperatures come only from quarks. The pink line is calculated by the NJL model. Since the NJL model has no mechanism for the quark confinement, quark excitations are not forbidden even in the hadronic phase at low temperatures. In the PNJL model, there is a phase cancellation of the distribution functions of three different color quark states at low temperature. This cancellation is however not complete and the color neutral excitations with three colored quarks are still allowed. It looks like a baryonic excitation but the number of degrees of freedom is not what is expected for baryons [2].

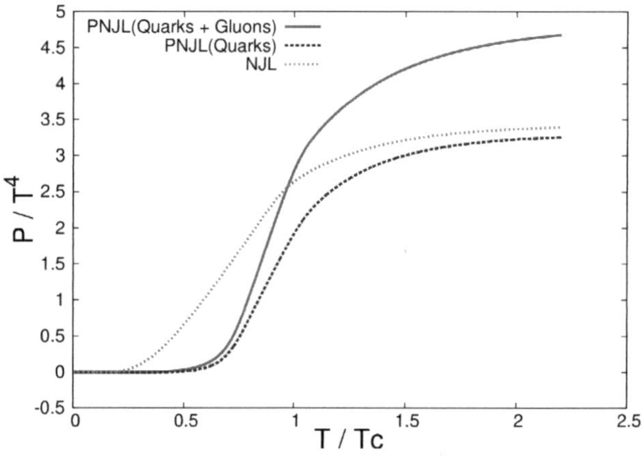

Figure 1. Pressures under the mean field approximation.

3.2. Mesonic correlation

In order to obtain the pressure from mesonic correlations, we calculate the second and the third terms of Eq.(6). From the relation between the thermodynamic potential and the pressure, $p = -\Omega/V$, the pressure of mesonic correlations is given by

$$\begin{aligned} p_M = -\sum_n \int \frac{d^3q}{(2\pi)^3} \Big\{ & 3\ln\mathcal{M}_\pi(\omega_n, q) + 4\ln\mathcal{M}_K(\omega_n, q) + \ln\mathcal{M}_\eta(\omega_n, q) + \ln\mathcal{M}_{\eta'}(\omega_n, q) \\ & + \ln\mathcal{M}_\sigma(\omega_n, q) + 4\ln\mathcal{M}_\kappa(\omega_n, q) + 3\ln\mathcal{M}_{a_0}(\omega_n, q) + \ln\mathcal{M}_{f_0}(\omega_n, q) \Big\} \end{aligned} \tag{8}$$

where

$$\mathcal{M}_M(\omega_n, q) = \frac{1}{2K'_M} - \Pi_M(\omega_n, q). \tag{9}$$

with K'_M the effective coupling consisting of the four point coupling G and the six point coupling K with appropriate weights for each meson. Π_M may be decomposed into two terms as indicated in Fig.2. In Eq.(8), the first four terms are contributions of pseudo scalar mesons and the last four terms of scalar mesons. The pre-factors of π, K, κ and a_0 are degeneracies of these mesons.

Figure 2. Meson self energy, Π_M.

The behavior of the pressure is shown in Fig.(3). At low temperatures, the pressure is dominated mesonic correlations especially pions and kaons. As temperature increases, it approaches quark mean field pressure and corresponds to it at high temperatures. It means that mesons as collective modes at low temperatures melt as temperature increases and resolve to quarks finally.

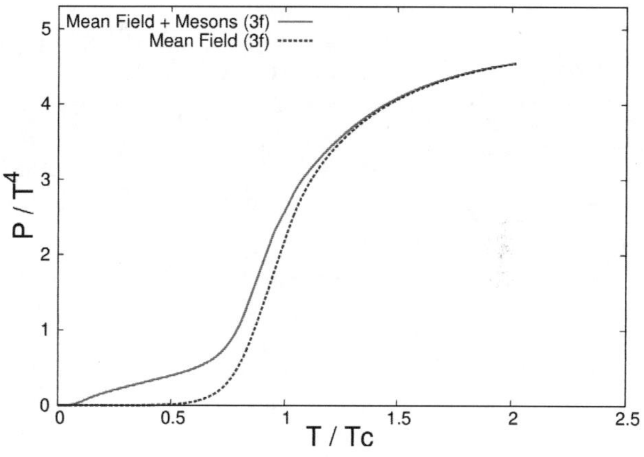

Figure 3. Pressure scaled by T^4 as a function of temperature. Red line: pressure from both contributions from the mean field and mesonic correlations. Blue line: pressure without mesonic correlations. The pseudo critical temperature T_c is decided in the mean field level.

4. Conclusion

We study the quark-hadron phase transition by using a three flavor PNJL model which contains the order parameters of both the chiral phase transition and the de-confining phase transition. Under the flavor SU(3) symmetry breaking, four kinds of pseudo scalar mesons(π, K, η and η') and scalar mesons(σ, κ, a_0, and f_0) appear with different masses and the equations of state are dominated by mesonic correlations, especially pions and kaons, at low temperatures. As temperature increases, the contribution from mesonic correlations decreases and the equation of state is dominated by quarks and antiquark excitations, with gluon excitations added by hand. Mesons melt into the continuum of quark-antiquark excitations and the degrees of freedom change from those of hadrons to quarks and gluons.

Acknowledgements

We thank members of Komaba Nuclear Theory Group for their interests in this work. KY's work has been supported by the University of Tokyo Grants for Ph.D Research, Research Assistant for Creation of the Research Core in Physics, and School of Science Grants for PhD Students. TM's work has been supported by the Grant-in-Aid # 25400247 of MEXT, Japan.

References

[1] Fukushima K 2004 *Phys. Lett.* B **591** 227
[2] Yamazaki K and Matsui T 2013 *Nucl. Phys.* A **913** 19
[3] Yamazaki K and Matsui T 2013 Quark-hadron phase transition in a three flavor PNJL model for interacting quarks *Preprint* hep-ph/13104960
[4] Ishida S, Ishida M, Ishida T, Takamatsu K and Tsuru T 1997 *Prog. Theor. Phys.* **98** 621
[5] Fariborz A H, Jora R and Schechter J 2009 *Phys. Rev.* D **79** 074014
[6] Hubbard J 1956 *Phys. Rev. Lett.* **3** 77
[7] Stratonovich R L 1957 *Doklady Akad. Nauk SSSR.* **115** 1097

Journal of Physics: Conference Series **509** (2014) 012052 doi:10.1088/1742-6596/509/1/012052

Transport coefficients from the Nambu-Jona-Lasinio model for $SU(3)_f$

R. Marty[1], **E. Bratkovskaya**[1], **W. Cassing**[2], **J. Aichelin**[3], **H. Berrehrah**[1]

[1] Frankfurt Institut for Advanced Studies and Institute for Theoretical Physics, Johann Wolfgang Goethe Universität, Frankfurt am Main, Germany
[2] Institut für Theoretische Physik, Universität Gießen, Gießen, Germany
[3] Subatech, UMR 6457, IN2P3/CNRS, Université de Nantes, École des Mines de Nantes, Nantes, France

E-mail: marty@fias.uni-frankfurt.de

Abstract. We calculate the shear $\eta(T)$ and bulk viscosities $\zeta(T)$ as well as the electric conductivity $\sigma_e(T)$ and heat conductivity $\kappa(T)$ within the Nambu-Jona-Lasinio (NJL) model for 3 flavors as a function of temperature as well as the entropy density $s(T)$, pressure $P(T)$ and speed of sound $c_s^2(T)$. We compare the results with other models such as the Polyakov-Nambu-Jona-Lasinio (PNJL) model and the dynamical quasiparticle model (DQPM) and confront these results with lattice QCD data whenever available. This work is based on Ref. [1].

1. Introduction

In order to study the expansion of the partonic plasma created in ultra-relativistic heavy-ion collisions [2], it is very helpful to calculate and compare thermodynamic properties as well as transport coefficients in equilibrium as a function of the temperature T. These results are not easily obtained from lattice QCD calculations and one has to consider suitable effective models in addition to achieve a more transparent picture.

It is the purpose of this contribution to evaluate these quantities in the Nambu-Jona-Lasinio (NJL) model [3, 4, 5] and to compare with similar approaches such as the Polyakov-Nambu-Jona-Lasinio (PNJL) [6, 7] and the dynamical-quasiparticle model (DQPM) [8, 9] as well as with the available lattice QCD data.

2. Equation of state

In order to compute the equation of state of strongly interacting matter, we use thermal distribution functions for partons (i.e. Fermi-Dirac and Bose-Einstein distributions). In the NJL model we compute the pressure P and energy density ϵ (cf. Fig. 1) as a function of temperature T using the relations based on the definition of the stress-energy tensor $T^{\mu\nu}$ for non-interacting particles at quark chemical potential μ:

$$T^{\mu\nu}(T,\mu) = g \int_0^\infty \frac{d^3p}{(2\pi)^3} f(E) \frac{p^\mu p^\nu}{E}. \tag{1}$$

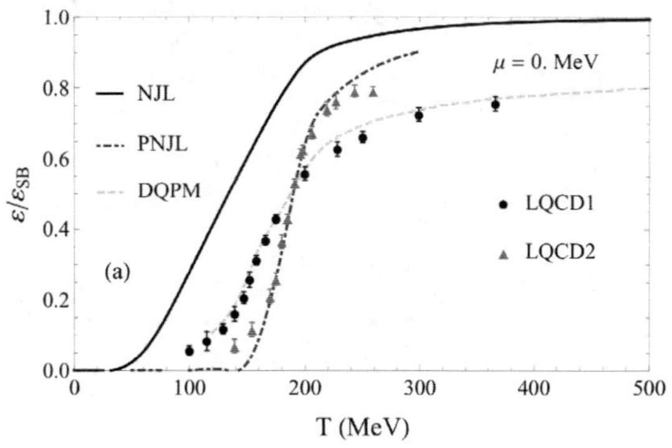

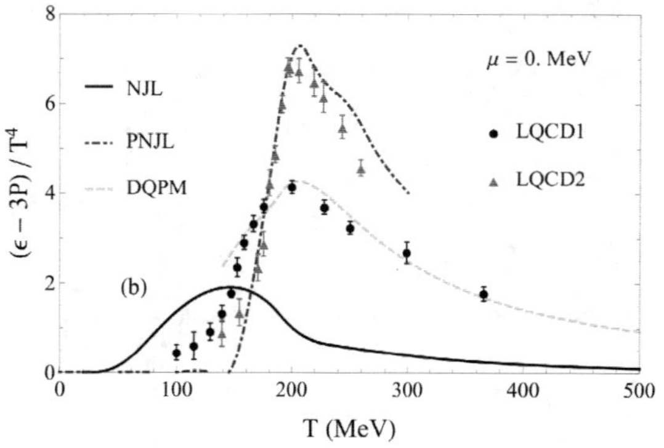

Figure 1. Energy density ε normalized to the energy density in the SB limit (a) and the trace anomaly $(\varepsilon - 3P)/T^4$ (b) as a function of temperature T from different models compared to the LQCD1 data from Ref. [10] and LQCD2 data from [11]. The PNJL results have been taken from Ref. [12].

Note that in Fig. 1, which displays the resulting energy density ε relative to the energy density in the Stefan Boltzmann (SB) limit, the SB limit is different for the NJL model and for PNJL/DQPM/LQCD due to the lack of the gluon degrees of freedom. We recall that the results of the Polyakov-NJL model differ from the NJL results due to the explicit gluon potential $\mathcal{U}$ [12] which reflects the pressure from the gluons. The critical temperature T_c is also different between the NJL, PNJL and LQCD since the parameters of the NJL are fixed at T=0.

In Fig. 1(b) we display the interaction measure – known in LQCD as the trace anomaly – in comparison with LQCD and find again the NJL model not to be in good agreement with the lattice data from Ref. [10, 11]. Note that the PNJL [12] and DQPM calculations agree well with the LQCD data, however, both have been adjusted explicitly to different LQCD results, i.e. either to those from Ref. [10] (DQPM) or from Ref. [11] (PNJL).

3. Shear viscosity

The shear viscosity η is defined using the relaxation time τ in the dilute gas approximation for interacting particles as [13]:

$$\eta(T,\mu) = \frac{1}{15T} g_g \int_0^\infty \frac{d^3p}{(2\pi)^3} \tau_g f_g \frac{\mathbf{p}^4}{E_g^2} + \frac{1}{15T}\frac{g_q}{6} \int_0^\infty \frac{d^3p}{(2\pi)^3} \left[\sum_q^{u,d,s} \tau_q f_q + \sum_{\bar{q}}^{\bar{u},\bar{d},\bar{s}} \tau_q f_{\bar{q}} \right] \frac{\mathbf{p}^4}{E_q^2}.$$

We find that the shear viscosity over entropy density ratio $\eta/s(T)$ in the NJL model shows a temperature dependence $\propto T^{-1}$ for high temperatures. Nevertheless, the order of magnitude is in agreement with the lattice QCD data from 1.2 T_c up to 1.5 T_c. The T^{-1} behavior of the viscosity in the NJL implies to go beyond the Kovtun-Son-Starinets (KSS) bound [14]: $(\eta/s)_{\rm KSS} = 1/4\pi$ above $T \sim 1.7T_c$ which limits the applicability of the NJL model.

4. Bulk viscosity

The bulk viscosity defined in Ref. [13] reads in the relaxation time approximation (RTA)

$$\begin{aligned} \zeta(T,\mu) =& \frac{1}{9T} g_g \int_0^\infty \frac{d^3p}{(2\pi)^3} \tau_g f_g \frac{1}{E_g^2} \left[\mathbf{p}^2 - 3c_s^2 \left(E_g^2 - T^2 \frac{dm_g^2}{dT^2} \right) \right]^2 \\ &+ \frac{1}{9T}\frac{g_q}{6} \int_0^\infty \frac{d^3p}{(2\pi)^3} \left[\sum_q^{u,d,s} \tau_q f_q + \sum_{\bar{q}}^{\bar{u},\bar{d},\bar{s}} \tau_q f_{\bar{q}} \right] \frac{1}{E_q^2} \left[\mathbf{p}^2 - 3c_s^2 \left(E_q^2 - T^2 \frac{dm_q^2}{dT^2} \right) \right]^2. \end{aligned} \quad (2)$$

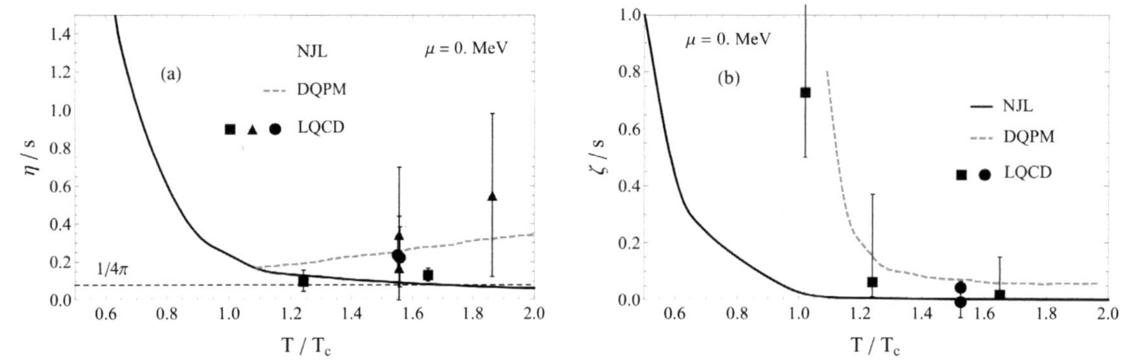

Figure 2. Shear viscosity η/s (a) as a function of T/T_c compared to the LQCD data points from Ref. [15] (square), [16] (triangle) and [17] (circle) and the result from the DQPM (dashed line); the bulk viscosity to entropy density ratio ζ/s (b) as a function of T/T_c compared to the LQCD data points from Ref. [18] (square) and [17] (circle) and the DQPM (dashed line).

The bulk viscosity over entropy density ζ/s from the NJL model is displayed in Fig. 2(b) and shows a very different temperature dependence than η/s. Indeed, for high temperatures we find the limit $\zeta/s \to 0$. Moreover, the behavior around T_c shows a peak in LQCD as well as in the DQPM. This peak is not seen in the NJL model (or shifted to much lower temperatures). This can be easily explained by the fact that the T-dependence of the masses of the degrees of freedom plays an important role for the bulk viscosity (2).

5. Electric conductivity

The electric conductivity for charged particles – known as the Drude-Lorentz conductivity for a classical gas – is defined as [19, 20]:

$$\sigma_e(T,\mu) = \sum_q \frac{e_q^2\, n_q(T,\mu)\, \tau_q(T,\mu)}{m_q(T,\mu)}, \qquad \text{with} \qquad e_q^2 = \frac{4\pi}{137} e^2, \tag{3}$$

with $q = u, d, s, \bar{u}, \bar{d}, \bar{s}$, and $e = +2/3$ or $-1/3$ denoting the quark electric charge fractions.

Fig. 3(a) shows that for the DQPM as well as the NJL model the dimensionless ratio of the electric conductivity over T is approximately linearly in T for $T \geq T_c$ up to about 2 T_c [20]. Both results are in a reasonable agreement with the present lattice QCD results although there is quite some uncertainty in the LQCD extrapolations.

6. Heat conductivity

The heat conductivity κ is another quantity of interest that describes the heat flow in interacting systems [21, 22] and only recently has regained interest in the context of relativistic heavy-ion collisions [23, 24].

The heat conductivity for charged particles is defined using the specific heat c_V and the relaxation time [25]:

$$\kappa(T,\mu) = \frac{1}{3} v_{\rm rel}\; c_V(T,\mu)\; \sum_q \tau_q(T,\mu), \tag{4}$$

with $q = u, d, s, \bar{u}, \bar{d}, \bar{s}$. For our purpose we assume $v_{\rm rel} \simeq 1$ in the NJL model because the masses of quarks decrease with temperature T whereas the mean momentum increases.

Fig. 3(b) displays the dimensionless quantity κ/T^2 for both models. Whereas the DQPM shows a slightly rising ratio for $T_c < T < 2T_c$ the NJL model predicts a rapid decrease with T for $T > T_c$. In this case no LQCD results are presently available.

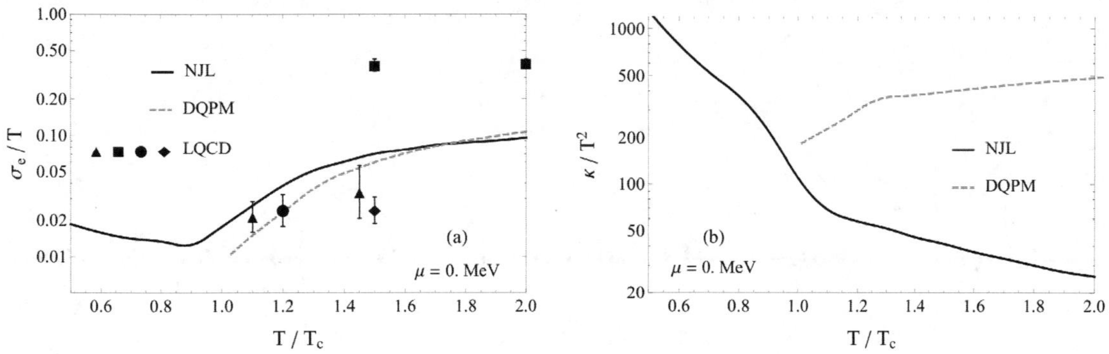

Figure 3. The electric conductivity σ_e/T (a) as a function of T/T_c compared to the LQCD data points from Ref. [26] (triangle), [27] (diamond), [28] (square) and [29] (circle) and the results from the DQPM (dashed lines); heat conductivity κ/T^2 (b) as a function of T/T_c in the NJL model and the DQPM.

7. Conclusion

In this study we have calculated thermodynamic properties of the NJL model for three flavors such as the energy density and pressure as a function of temperature T up to a few times the critical temperature T_c. Furthermore, we have calculated the shear η and bulk ζ viscosity as well as the electric conductivity σ_e and heat conductivity κ as a function of T and compared to corresponding results from the DQPM, from the PNJL model and lattice QCD results when available.

References

[1] Marty R, Bratkovskaya E, Cassing W, Aichelin J and Berrehrah H 2013 (*Preprint* 1305.7180)
[2] Marty R and Aichelin J 2013 *Phys. Rev. C* **87**(3) 034912
[3] Klevansky S 1992 *Rev.Mod.Phys.* **64** 649–708
[4] Vogl U and Weise W 1991 *Progress in Particle and Nuclear Physics* **27** 195 – 272 ISSN 0146-6410
[5] Rehberg P, Klevansky S and Hüfner J 1996 *Nucl.Phys.* **A608** 356–388
[6] Fukushima K 2004 *Phys.Lett.* **B591** 277–284
[7] Ratti C, Rößner S, Thaler M and Weise W 2007 *Eur.Phys.J.* **C49** 213–217
[8] Peshier A 2004 *Phys. Rev. D* **70**(3) 034016
[9] Cassing W 2009 *The European Physical Journal Special Topics* **168** 3–87 ISSN 1951-6355
[10] Borsanyi S, Endrodi G, Fodor Z, Jakovac A, Katz S D *et al.* 2010 *JHEP* **1011** 077
[11] Cheng M, Ejiri S, Hegde P, Karsch F, Kaczmarek O *et al.* 2010 *Phys. Rev. D* **81**(5) 054504
[12] Costa P, Ruivo M, de Sousa C and Hansen H 2010 *Symmetry* **2** 1338–1374
[13] Chakraborty P and Kapusta J 2011 *Phys.Rev.* **C83** 014906
[14] Kovtun P K, Son D T and Starinets A O 2005 *Phys. Rev. Lett.* **94**(11) 111601
[15] Meyer H B 2007 *Phys.Rev.* **D76** 101701
[16] Nakamura A and Sakai S 2005 *Phys. Rev. Lett.* **94**(7) 072305
[17] Sakai S and Nakamura A 2007 *PoS* **LAT2007** 221
[18] Meyer H B 2008 *Phys.Rev.Lett.* **100** 162001
[19] Reif F 1965 *Fundamentals of Statistical and Thermal Physics* (McGraw-Hill)
[20] Cassing W, Linnyk O, Steinert T and Ozvenchuk V 2013 *Phys. Rev. Lett.* **110**(18) 182301
[21] Israel W and Stewart J 1979 *Annals of Physics* **118** 341 – 372 ISSN 0003-4916
[22] de Groot S, van Leeuwen W and van Weert C 1980 *Relativistic Kinetic Theory: Principles and Applications*
[23] Denicol G S, Niemi H, Molnár E and Rischke D H 2012 *Phys. Rev. D* **85**(11) 114047
[24] Greif M, Reining F, Bouras I, Denicol G S, Xu Z and Greiner C 2013 *Phys. Rev. E* **87**(3) 033019
[25] Heiselberg H and Pethick C 1993 *Phys.Rev.* **D48** 2916–2928
[26] Ding H T, Francis A, Kaczmarek O, Karsch F, Laermann E *et al.* 2013 (*Preprint* 1301.7436)
[27] Aarts G, Allton C, Foley J, Hands S and Kim S 2007 *Phys. Rev. Lett.* **99**(2) 022002
[28] Gupta S 2004 *Phys.Lett.* **B597** 57–62
[29] Brandt B B, Francis A, Meyer H B and Wittig H 2013 (*Preprint* 1302.0675)

Journal of Physics: Conference Series **509** (2014) 012053 doi:10.1088/1742-6596/509/1/012053

Diffusion of non-Gaussianity in heavy ion collisions

Masakiyo Kitazawa, Masayuki Asakawa, Hirosato Ono
Department of Physics, Osaka University, Toyonaka, Osaka 560-0043, Japan
E-mail: kitazawa@phys.sci.osaka-u.ac.jp

Abstract. We investigate the time evolution of higher order cumulants of bulk fluctuations of conserved charges in the hadronic stage in relativistic heavy ion collisions. The dynamical evolution of non-Gaussian fluctuations is modeled by the diffusion master equation. Using this model we predict that the fourth-order cumulant of net-electric charge is suppressed compared with the recently observed second-order one at ALICE for a reasonable parameter range. Significance of the measurements of various cumulants as functions of rapidity window to probe dynamical history of the hot medium created by heavy ion collisions is emphasized.

1. Introduction

Bulk fluctuations of conserved charges, especially higher order cumulants characterizing non-Gaussianity, are believed to be promising experimental observables to reveal thermodynamic properties of the hot medium created by relativistic heavy ion collisions [1, 2]. These observables are experimentally measured by event-by-event analysis, and active analyses of cumulants have been performed at RHIC [3] and LHC [4]; In particular, fourth-order cumulants are measured with a good statistics at RHIC. Numerical analyses of higher order cumulants of conserved charges in equilibrium have been also carried out in lattice QCD Monte Carlo simulations.

An important property of the conserved-charge fluctuations in heavy ion collisions which has been clarified by the recent experimental results is that the experimentally-observed fluctuations are not those of the equilibrated medium at some stage in the hadronic medium. In fact, suppression of the net-electric charge fluctuation observed at ALICE and especially its strong rapidity window dependence [4] clearly show that the fluctuation is not equilibrated at LHC energy. Moreover, a comparison of the fluctuations observed at STAR with the cumulants measured on the lattice [5] also indicates that the fluctuations observed at RHIC are not consistent with the equilibrated values at chemical freezeout. These results show that an appropriate description of the non-equilibrium nature of fluctuations owing to dynamical evolution of the hot medium is inevitably needed to investigate their thermodynamic properties using the fluctuation observables, especially those of conserved charges.

In the present study, we investigate the time evolution of higher order cumulants of conserved charges in the hadronic stage using the diffusion master equation [6]. After hadronization, the fluctuations approach the equilibrated distribution in the hadronic medium, which is approximately given by the Skellam distribution [7]. Our approach can describe this feature of the non-Gaussian fluctuations, while the time evolution of the average and Gaussian fluctuations are consistent with the ones in the stochastic diffusion equation [8]. By analyzing the time evolution of higher order cumulants in this model, we show that the dependence of the cumulants of conserved charges on the size of rapidity window to count the particle number in experiments show characteristic behaviors reflecting the dynamical history of the time evolution of the hot medium. In particular, we predict that the fourth-order cumulants of conserved charges will be suppressed compared with the second ones at LHC energy for a wide range of parameters.

2. Stochastic formalism to describe non-Gaussianity in diffusive systems

In relativistic heavy ion collisions with sufficiently large $\sqrt{s_{\rm NN}}$, the hot medium created at mid-rapidity has an approximate boost invariance. In a sufficiently large space-time scale where hydrodynamic equations at first order are applicable, the average of the net number of a conserved charge per unit coordinate-space rapidity, $n(\eta,\tau)$, follows the diffusion equation

$$\partial_\tau n(\eta,\tau) = D\partial_\eta^2 n(\eta,\tau), \qquad (1)$$

with the coordinate-space rapidity η, proper time τ, and the diffusion constant D. Assuming that the kinetic freezeout takes place at a certain proper time $\tau_{\rm fo}$, the experimentally-observed conserved-charge number at mid-rapidity at RHIC and LHC is given by $Q(\tau) = \int_{-\Delta\eta/2}^{\Delta\eta/2} d\eta n(\eta,\tau)$ at $\tau = \tau_{\rm fo}$ with the size of the rapidity window to count the particle number $\Delta\eta$.

In order to describe fluctuations around the solution of Eq. (1), one may employ a stochastic model, in which the time evolution of the deterministic part satisfies Eq. (1). A choice of such stochastic models is the theory of hydrodynamic fluctuations [9], in which the hydrodynamic equations are promoted to Langevin equations with stochastic terms representing fast random forces. It is known that these stochastic equations well describe Gaussian fluctuations in fluids. However, extension of this formalism to treat non-Gaussian fluctuations is nontrivial. In fact, one can show that the stochastic force in the theory of hydrodynamic fluctuations for Markov process is of Gaussian [6]. Using this property it is shown that all higher order cumulants of $Q(\tau)$ vanish in equilibrium unless $D(\tau)$ is explicitly dependent on n. This property is not welcome to describe non-Gaussianity in heavy ion collisions, because higher order cumulants are expected to increase toward nonzero equilibrated values in the hadronic medium [1, 3, 4].

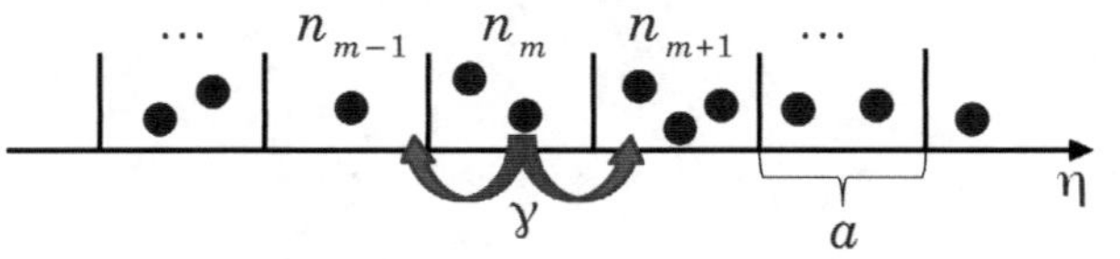

Figure 1. System described by the diffusion master equation Eq. (2).

In the present study, instead of directly extending the theory of hydrodynamic fluctuations, we investigate the time evolution of higher order cumulants starting from a microscopic model. In this study, as such a model we consider a simple one-dimensional system composed of Brownian particles. Instead of tracking the motion of each Brownian particle separately, however, we represent the system as follows (See, Fig. 1) [6]. First, the coordinate η is divided into discrete cells with an equal length a. Second, we consider a single species of particle for the moment, and denote the number of particles in each cell, labeled by an integer m, as n_m, and the probability that each cell contains n_m particles as $P(\mathbf{n},\tau)$ with $\mathbf{n} = (\cdots, n_{m-1}, n_m, n_{m+1}, \cdots)$. Finally, we assume that each particle moves to adjacent cells with a probability γ per unit proper time. The probability $P(\mathbf{n},\tau)$ then follows the differential equation

$$\partial_\tau P(\mathbf{n},\tau) = \gamma \sum_m [(n_m+1)\{P(\mathbf{n}+\mathbf{e}_m-\mathbf{e}_{m+1},\tau) + P(\mathbf{n}+\mathbf{e}_m-\mathbf{e}_{m-1},\tau)\} - 2n_m P(\mathbf{n},\tau)], \quad (2)$$

which is referred to as *diffusion master equation*, where $\mathbf{e}_m$ is the vector that all components are zero except for the mth one, which takes unity. One can show that the average density and Gaussian fluctuation of $n(\eta,\tau)$ in Eq. (2) in the continuum limit, $a \to 0$, agree with those in the stochastic diffusion equation [8] with $D = \gamma a^2$ [6].

3. Solution of diffusion master equation

Now, we solve the time evolution of cumulants for the stochastic process Eq. (2). In order to simplify the problem, in the following we limit our attention to the time evolution in an infinitely long system without boundaries. Since we are interested in the solution in the continuum limit, $a \to 0$, we represent the particle numbers n_m by a function $n(\eta, \tau)$. After some algebra [6], one finds that the cumulants of $Q(\tau)$ with the fixed initial condition $n(\eta, 0) = M(\eta)$ are given by

$$\langle (Q(\tau))^n \rangle_c = \int_{-\infty}^{\infty} d\eta M(\eta) H_X^{(n)}(\eta), \tag{3}$$

with

$$H_X^{(1)}(z) = I_X(z/\Delta\eta), \quad H_X^{(2)}(z) = I_X(z/\Delta\eta) - I_X(z/\Delta\eta)^2, \tag{4}$$

$$H_X^{(3)}(z) = I_X(z/\Delta\eta) - 3I_X(z/\Delta\eta)^2 + 2I_X(z/\Delta\eta)^3, \tag{5}$$

$$H_X^{(4)}(z) = I_X(z/\Delta\eta) - 7I_X(z/\Delta\eta)^2 + 12I_X(z/\Delta\eta)^3 - 6I_X(z/\Delta\eta)^4, \tag{6}$$

and $I_X(z) = \int_{-1/2}^{1/2} dx \int dq/(2\pi) e^{-X^2q^2} e^{iq(x+z)}$, where $\Delta\eta$ and τ dependences are encoded in the dimensionless parameter $X = \sqrt{D\tau}/\Delta\eta$. We note that $\sqrt{2D\tau}$ is the mean diffusion length of the Brownian particles at τ.

In order to examine the time evolution of conserved charges in heavy ion collisions, one must extend the above result to general initial conditions containing fluctuations. We also extend the result to the system with two particle species with densities $n_1(\eta, \tau)$ and $n_2(\eta, \tau)$, and consider cumulants of the difference, $Q_{(\mathrm{net})}(\tau) = \int_{-\Delta\eta/2}^{\Delta\eta/2} d\eta (n_1(\eta, \tau) - n_2(\eta, \tau))$, in order to compare the results with the cumulants of net charge numbers. In the following, we limit our attention to the solution for the initial conditions which satisfies spatial uniformity and locality with vanishing net-charge number. In this case, second- and fourth-order cumulants of $Q_{(\mathrm{net})}(\tau)$ are given by

$$\langle Q_{(\mathrm{net})}^2 \rangle_c = \Delta\eta [M_{(\mathrm{tot})}]_c (1 - F_X^{(2)}), \tag{7}$$

$$\langle Q_{(\mathrm{net})}^4 \rangle_c = \Delta\eta \{3[M_{(\mathrm{tot})}^2]_c (F_X^{(2)} - 2F_X^{(3)} + F_X^{(4)}) + [M_{(\mathrm{tot})}]_c (1 - 7F_X^{(2)} + 12F_X^{(3)} - 6F_X^{(4)})\}, \tag{8}$$

with $F_X^{(n)} = \int_{-\infty}^{\infty} dz [I_X(z)]^n$, and $M_{(\mathrm{net}),(\mathrm{tot})}(\eta) = M_1(\eta) \mp M_2(\eta)$, respectively. In Eq. (8), $[M_{(\mathrm{tot})}^2]_c$ is the fluctuation of the total number of the particles per unit rapidity at the initial condition. This quantity is not constrained by the conservation laws and strongly depends on the hadronization mechanism [6]. We thus treat this quantity as a parameter that characterizes the hadronization mechanism. From Eqs. (7) and (8), one can check that the distribution of $Q_{(\mathrm{net})}$ approaches a Skellam one with $\lim_{\tau\to\infty} \langle Q_{(\mathrm{net})}^{2n} \rangle_c = \Delta\eta [M_{(\mathrm{tot})}]_c$. The time evolution of the Gaussian fluctuation, Eq. (7), is equivalent with the one in the stochastic diffusion equation [8].

4. Time evolution of cumulants and $\Delta\eta$ dependence

Now, let us consider the cumulants of conserved charges in relativistic heavy ion collisions. To make the argument simple, we assume that a boost invariant system with local equilibration is realized just above the critical temperature of the deconfinement transition. We further assume that the fluctuations of conserved charges vanish at this time, reflecting the small fluctuations in the deconfined phase [1] and the local charge conservations. Due to the diffusion in the hadronic phase, the fluctuations keep on approaching the equilibrated values in the hadronic medium until kinetic freezeout at $\tau = \tau_{\mathrm{fo}}$. Provided that this diffusion process is well described by the diffusion master equation, $\Delta\eta$ dependence of the cumulants of conserved charges at kinetic freezeout are given by Eqs. (7) and (8) with $\tau = \tau_{\mathrm{fo}}$.

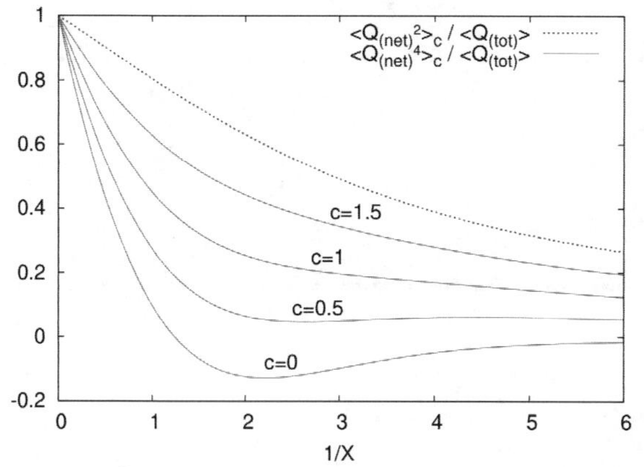

Figure 2. Second and fourth order cumulant of $Q_{(\mathrm{net})}$ as a function of $1/X$ with the fixed initial condition.

In Fig. 2, we show the $1/X$ dependences of Eqs. (7) and (8). Since $1/X$ is proportional to $\Delta\eta$ with fixed τ, this result can directly be compared with the $\Delta\eta$ dependence of the cumulants in experiments. The result for the fourth-order is shown with several values of the parameter $c = [M^2_{(\mathrm{tot})}]_\mathrm{c}/[M_{(\mathrm{tot})}]_\mathrm{c}$, which is the quantity sensitive to hadronization mechanism [6].

In the figure, one finds that $\langle Q^4_{(\mathrm{net})}\rangle_\mathrm{c}$ is suppressed compared with $\langle Q^2_{(\mathrm{net})}\rangle_\mathrm{c}$ in the parameter range $c < 1.5$, while the behavior of $\langle Q^4_{(\mathrm{net})}\rangle_\mathrm{c}$ depends sensitively on the value of c. This result indicates that $\langle (N_\mathrm{Q}^{(\mathrm{net})})^4\rangle_\mathrm{c}$ at ALICE, which has not been measured yet, will be suppressed compared with the $\langle (N_\mathrm{Q}^{(\mathrm{net})})^2\rangle_\mathrm{c}$ which has been already measured [4], while the statement is altered for large c. The same conclusion is also anticipated for the relation between the baryon number cumulants [10], $\langle (N_\mathrm{B}^{(\mathrm{net})})^2\rangle_\mathrm{c}$ and $\langle (N_\mathrm{B}^{(\mathrm{net})})^4\rangle_\mathrm{c}$. Our results also indicate that experimental measurements of not only the magnitudes of various cumulants at a fixed $\Delta\eta$ but also their $\Delta\eta$ dependence enable us to explore various aspects of the time evolution of the hot medium and the hadronization mechanism in the experiments. In particular, these analyses would enable us to estimate the magnitude of the parameter c, which is sensitive to the hadronization mechanism.

References

[1] Koch V 2008 Hadronic Fluctuations and Correlations *Preprint* arXiv:0810.2520 [nucl-th]

[2] Asakawa M, Ejiri S, and Kitazawa M 2009 Third moments of conserved charges as probes of QCD phase structure *Phys. Rev. Lett.* **103** 262301

[3] Adamczyk L, *et al.* [STAR Collaboration] 2013 Energy Dependence of Moments of Net-proton Multiplicity Distributions at RHIC *Preprint* arXiv:1309.5681 [nucl-ex]

[4] Abelev B, *et al.* [ALICE Collaboration] 2013 Net-Charge Fluctuations in Pb-Pb collisions at $\sqrt{s_{NN}} = 2.76$ TeV *Phys. Rev. Lett.* **110** 152301

[5] Mukherjee S and Wagner W 2013 Deconnement of strangeness and freeze-out from charge uctuations *Preprint* arXiv:1307.6255 [nucl-th]

[6] Kitazawa M, Asakawa M, Ono H 2013 Nonequilibrium time evolution of higher order cumulants of conserved charges and event-by-event analysis *Preprint* arXiv:1307.2978 [nucl-th]

[7] F. Karsch and K. Redlich 2011 Probing freeze-out conditions in heavy ion collisions with moments of charge fluctuations *Phys. Lett.* **B695** 136

[8] Shuryak E V and Stephanov M A 2001 When can long range charge fluctuations serve as a QGP signal? *Phys. Rev. C* **63** 064903

[9] Kapusta J I, Müller B and Stephanov M 2012 Relativistic Theory of Hydrodynamic Fluctuations with Applications to Heavy Ion Collisions *Phys. Rev. C* **85** 054906

[10] Kitazawa M and Asakawa M 2012 Revealing baryon number fluctuations from proton number fluctuations in relativistic heavy ion collisions *Phys. Rev. C* **85** 021901R; *ibid.* 2012 Relation between baryon number fluctuations and experimentally observed proton number fluctuations in relativistic heavy ion collisions *Phys. Rev. C* **86** 024904

Turbulence, Vorticity and Lambda Polarization

L.P. Csernai[1], **F. Becattini**[2,3], **and D.J. Wang**[1,3]

[1] Dept. of Physics and Technology, Univ. of Bergen, Allegaten 55, 5007 Bergen, Norway
[2] University of Florence and INFN Florence, Italy
[3] Frankfurt Institute for Advanced Studies (FIAS), Johann Wolfgang Goethe University, Germany

E-mail: csernai@ift.uib.no

Abstract. Present highest energy heavy ion experiments show the dominance of the fluid dynamical processes. The possibility of new collective phenomena, rotation and turbulence occurs, which were known only in macroscopic systems up to now. These cand be detected with a new type of polarization measurements.

In recent years studying heavy ion collisions at LHC energies in a high resolution 3+1D fluid dynamical model we observed that the flow leads to collective rotation [1] in peripheral events arising from the initial angular momentum [2].

This rotation leads to a change of the flow pattern, and the antiflow, as characterized by the v_1 parameter, weakens and peaks at angles, which point more forward rapidities.

The v_1 becomes very small at high, LHC energies, less than one per cent, thus its measurement is not easy. Still the trend that the antiflow peak moves to more forward rapidities is apparent in experiments.

At high energies the role of random fluctuations is significant, and these random fluctuations are becoming strong, exceeding the magnitude of the collective v_1 flow component. Flow harmonics were identified up to high multipolarities, having a maximum amplitude at v_3, in central heavy ion collisions. In these collisions due to symmetry reasons all collective flow harmonics should have vanished. This in itself indicates the low viscosity of the underlying matter, which enables large random fluctuations.

The stability of the flow implies a given amount of viscosity. This stability against turbulence is characterized by the dimensionless, Reynolds number: $Re = l_1 u_1/\nu$, where $\nu = \eta/\rho$ is the kinetic viscosity, η is the shear viscosity, and ρ is the mass (or energy) density, u_1 is the characteristic flow velocity, and l_1 is the characteristic length scale of the flow pattern.

Kovtun, Son and Starinets have shown in a recent work that in certain field theories, that are dual to black branes in higher space-time dimensions, the ratio $\eta/s = 1/4\pi$ [3]. Interestingly, this bound is obeyed by $\mathcal{N} = 4$ supersymmetric $SU(N_c)$ Yang-Mills theory in the large N_c limit. Subsequently in another recent work with [4] it was pointed out that η/s has a minimum at the critical point of the QGP - Hadronic Matter phase transition. At pressures below the critical pressure there is a discontinuity, and η/s increases, and at pressures above it there is a broad smooth increase again from the minimum.

Apart of the fluctuatios from low viscosity, the phase transition leads to critical fluctuations also. These fluctuations, and their multipolarity spectrum can identify and characterize the phase transition. This is discussed in large number of publications, including [5, 6]. These fluctuations can shed light on the critical properties of the phase transition.

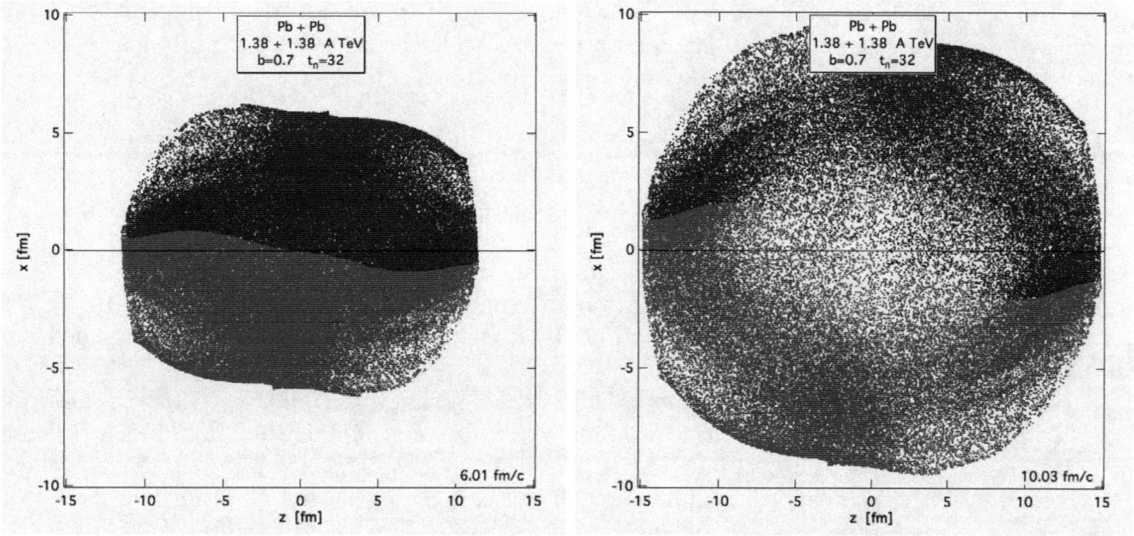

Figure 1. (color online) The position of computational "marker particles" which represents the moton of the net baryon charge in the raction plane for a periheral Pb+Pb reaction. At the initial configuration of the fluid dynamical development the top (projectile side) half of the marker particles were painted red, while the lower (target side) half was painted blue. We can see that with time a non-linear wave develops with increasing amplitude. This is the Kelvin Helmholtz Instability.

A consequence of the low viscosity is turbulence. The most typical type of turbulence is the Kelvin Helmholtz Instability (KHI) which is well known in macroscopic systems, as breaking waves in strong wind or in the air with strong wind shear in different vertical layers. It was observed in low viscosity numerical fluid dynamical model calculations that the KHI may occur in heavy ion collisions under favorable conditions [7]. See Fig. 1. The development of KHI is dependent on several conditions, discussed in ref. [7], but most importantly low viscosity is required. Thus, this phenomenon is a direct measure of the low viscosity.

Turbulent phenomena may occur also from random fluctuations. We have seen that these fluctuations are significant. In ref. [8] vorticity arising from random fluctuations in the transverse plane were studied, and a maximum vorticity amplitude of 0.2 c/fm was found. Due to the fact that this vorticity is a product of random fluctuation in a sample of a large number of collisions the average integrated vorticity should vanish.

On the other hand the vorticity in the reaction plane is due to the initial angular momentum, thus the plane of the rotation is the reaction plane, $[x, z]$ plane, and the rotation axis is the y axis. Due to the high angular momentum, $\approx 10^6 \hbar c$, at LHC energies, the vorticity vector is large and pointing into one direction, to the y direction. The amplitude of the vorticity reaches a maximum of 3 c/fm [9], which is more than an order of magnitude larger than the vorticity arising from random fluctuations.

The observation of the collective phenomena is not a trivial problem, although it is practiced for more than three decades. The present emphasis of random fluctuations has extended the evaluation methods to those classes of symmetries, which would not occur in case of collisions without fluctuations.

Without fluctuations, the heavy ion collision is mirror symmetric across the reaction plane, as well as reflection symmetric with respect to the center of mass (c.m.) point. Thus in the earlier, low energy experiments the longitudinal distribution along the z axis was antisymmetric across the c.m. point or with respect to the c.m. point of the rapidity coordinate. From the mirror symmetry across the reaction plane it follows that the distribution as function of the azimuth angle should be symmetric, so only cosine functions can be present in the in azimuthal angle Fourier series.

For random fluctuations these restrictions do not apply. The azimuthal multipole and longitudinal analysis should contain even and odd functions. This does not lead to any problem in exactly central collisions, where global flow patterns do not lead to specific angles.

The precise identification of the c.m. of the observed participant system would still be necessary for "exactly" central collisions, because random azimuthal fluctuations may be correlated with the longitudinal momentum in the c.m. system. Unfortunately this is not done yet in recent experiments, although longitudinal fluctuations of the participant c.m. momentum are not negligible at high beam energies.

Several collective processes in peripheral collision are strongly correlated with the symmetry axes of the collision, such as the reaction plane and the c.m. momentum or rapidity. These include the directed flow, all odd flow harmonics, rotation, and all polarization effects, which are correlated with the reaction symmetries. If these symmetry axes and the c.m. are not identified, then these processes are not measurable or are not measured accurately.

For example if we intend to measure polarization of particles which are arising from the collective rotation in the reaction plane then it is necessary to identify the plane, the direction of the rotation, i.e. which is the projectile and target side, and the position of the rotation axis, i.e. the c.m. momentum of the participants. This is possible to do, as described e.g. in ref. [10]. If this is not done with a given precision than the measurement of these collective phenomena is not possible.

In view of these general comments we discussed the effect of rotation on the directed flow, which turned out very sensitive and subject to significant perturbations from random fluctuations [1]. As the v_1 becomes small ($\sim 1\%$ or less) at RHIC and LHC energies, other measurements were studied which are more sensitive to rotation and turbulence.

The polarization is one of the consequences of rotation and local vorticity, which arises in a stratified shear flow. This can happen due to equipartition of orbital rotation and spin due to local thermalization. The Λ polarization in the participant centre-of-mass frame, as a function of its momentum, depends on the vorticity of the inverse temperature four-vector, $\beta^\mu(x) = (1/T(x))u^\mu(x)$ as shown in ref. [11]. The fact that the peripheral areas of the flow have lower temperatures, leads to an increased thermal vorticity in these areas, thus the thermal vorticity governing polarization is more uniform than the simple vorticity generated just from the velocity field. This also led to the interesting phenomenon that the polarization not necessarily increasing with increasing anglar momentum at high energy collisions, because at the same time the temperature is also strongly increasing.

It is shown that about 8-10% Λ or $\bar{\Lambda}$ polarization can be reached in peripheral heavy ion collisions [12]. This is observable for Λ or $\bar{\Lambda}$ emitted into the $\pm X$ direction in the reaction plane and the both the Λ or $\bar{\Lambda}$ will be polarized into the $-y$ direction (i.e. transverse polarization). This is in contrast to electromagnetic effects where the Λ and $\bar{\Lambda}$ polarizations would be opposite. Of course to perform this measurement the Event by Event identification of the reaction plane, its direction, and the c.m. momentum are necessary. Otherwise the overall integration of all directions, eliminates the effect.

Although early measurements at RHIC were negative, thses were averaged over all azimuth! We propose a selective measurement in the reaction plane (in the $\pm x$ direction) in the event by event (EbE) c.m. frame. Statistical error is much reduced now, so a significant effect is expected

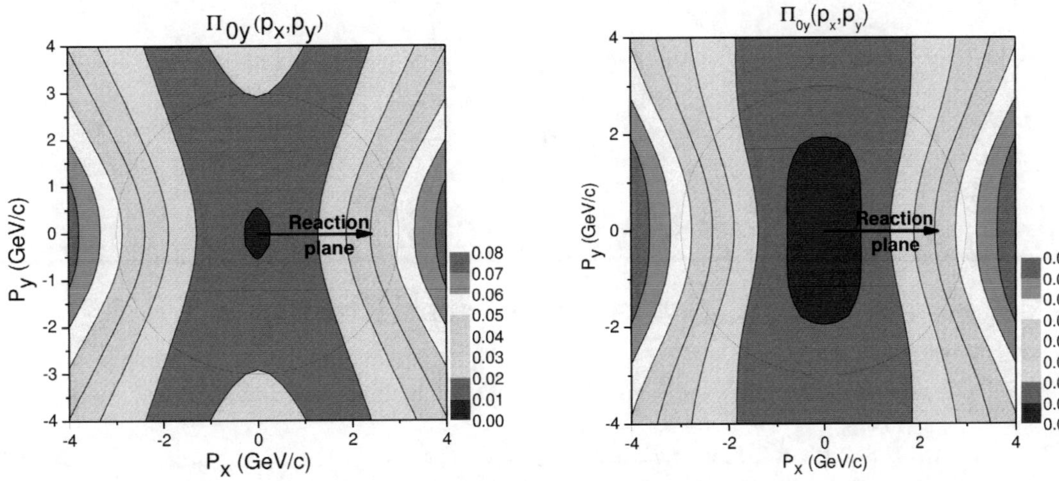

Figure 2. (color online) The polarization of Λ and $\bar{\Lambda}$ due to thermal equipartition with with local vorticity is slightly stronger at RHIC (right) than at LHC (left), due to the much higher temperatures at LHC.

around $p_x \approx 3$ GeV/c. See Fig. 2.

Another sensitive method of studying heavy ion reactions is the Hanbury Brown, and Twiss (HBT) effect. This is widely used for identifying the size of the emitting source, and also the radial expansion speed, which also influences the observed effective radii. Furthermore, with the azimuthal HBT method the shape of the emission ellipsoid can also be determined.

Recently, a Differential HBT (DHBT) method was worked out, which is similarly sensitive to rotation, and has advantages compared to the flow harmonics type of analysis. The estimated signal is of the order of 10% also [13]. As in the previous case this also requires the identification of reaction geometry.

This is not a fundamental problem and at lower energies this was done earlier. Recently the shift of focus towards analysis of fluctuations, diverted attention from this task, but we hope that the experimental teams will pay attention to this task.

In conclusion, we have a good perspective to continue the research along the lines and successes of the earlier research in this field.

References

[1] Csernai LP Magas VK Stöcker H and Strottman DD 2011 *Phys. Rev.* C **84** 024914
[2] Vovchenko V Anchishkin D and Csernai LP 2013 *Phys. Rev.* C **88** 014901
[3] Kovtun PK Son DT and Starinets AO 2005 *Phys. Rev. Lett.* **94** 111601
[4] Csernai LP Kapusta JI and McLerran LD 2006 *Phys. Rev. Lett.* **97** 152303
[5] Csernai LP Mocanu G and Neda Z 2012 *Phys. Rev.* C **85** 068201
[6] Wang DJ Csernai LP Strottman D Anderlik C Cheng Y Zhou DM Yan YL Cai X and Sa BH 2012 *Eur. Phys. J.* A **48** 168
[7] Csernai LP Strottman DD and Anderlik C 2012 *Phys. Rev.* C **85** 054901
[8] Floerchinger S and Wiedemann UA 2011 *JHEP* **11** 100; and *J. Phys.* G **38** 124171
[9] Csernai LP Magas VK and Wang DJ 2013 *Phys. Rev.* C **87** 034906
[10] Csernai LP Eyyubova G Magas VK 2012 *Phys. Rev.* C **86** 024912
[11] Becattini F Chandra Del Zanna L and Grossi E 2013 *Ann. Phys.* **338** 32
[12] Becattini F Csernai LP and Wang DJ 2013 *Phys. Rev.* C **88** 034905
[13] Csernai LP Velle S Wang DJ 2013 arXiv:1305.0396v2 [nucl-th]

14th International Conference on Strangeness in Quark Matter (SQM2013) IOP Publishing
Journal of Physics: Conference Series **509** (2014) 012055 doi:10.1088/1742-6596/509/1/012055

Relativistic distribution function for particles with spin at local thermodynamical equilibrium

F.Becattini [1,2,3], **V. Chandra**[2], **L. Del Zanna**[1,2] **and E. Grossi**[1,2]
[1] Universitá di Firenze, Florence Italy.
[2] INFN Sezione di Firenze, Florence, Italy.
[3] Universität Frankfurt and FIAS, Frankfurt am Main, Germany.

E-mail: becattini@fi.infn.it, vinod.chandra@fi.infn.it, ldz@arcerti.astro.it and grossi@fi.infn.it

Abstract. We present an extension of relativistic single-particle distribution function for weakly interacting particles at local thermodynamical equilibrium including spin degrees of freedom, for massive spin 1/2 particles. We infer, on the basis of the global equilibrium case, that at local thermodynamical equilibrium particles acquire a net polarization proportional to the vorticity of the inverse temperature four-vector field. The obtained formula for polarization also implies that a steady gradient of temperature entails a polarization orthogonal to particle momentum. The single-particle distribution function in momentum space extends the so-called Cooper-Frye formula to particles with spin 1/2 and allows to predict their polarization, and particularly of Lambda hyperons, in relativistic heavy ion collisions at the freeze-out.

1. Introduction

The single-particle distribution function is the main quantity in kinetic theory and its form at local thermodynamical equilibrium for relativistic, weakly interacting, gases is well known. For spinless particles, it is simply the Bose-Einstein distribution with x-dependent values of temperature and chemical potential (which can be defined as Bose-Jüttner distribution):

$$f(x,p) = \frac{1}{\mathrm{e}^{\beta(x)\cdot p - \xi(x)} - 1} \tag{1}$$

where $\beta = \frac{1}{T_0}u$ is the inverse temperature four-vector,T_0 being the proper temperature measured by a comoving thermometer with the four-velocity u and $\xi = \mu_0/T_0$ is the ratio between the proper chemical potential μ_0 and T_0. The above formula has, as straightforward consequence, the invariant momentum spectrum at local thermodynamical equilibrium, the so-called Cooper-Frye formula [1]:

$$\varepsilon\frac{\mathrm{d}N}{\mathrm{d}^3\mathrm{p}} = \int_\Sigma \mathrm{d}\Sigma_\mu p^\mu f(x,p) \tag{2}$$

where Σ is a space-like 3-dimensional hypersurface. This formula is widespreadly used in e.g. relativistic heavy ion collisions to calculate hadronic spectra at the end of the hydrodynamical stage.

The distribution (1), multiplied by a degeneracy factor $(2S+1)$, is also used for particles with spin (for fermions replacing the -1 with $+1$ in the denominator) being understood that f

means the *total* particle density in phase space, i.e. summed over polarization states. However, in general, particles may not evenly populate the various polarization states and one may then wonder what is the appropriate extension of (1) in this case. Indeed, in this work, we will answer this question and provide a generalization of (1) and (2) including the spin degrees of freedom. We will argue that a non-even population of the polarization states arises when the inverse temperature four-vector field has a non-vanishing antisymmetric part of its gradient and calculate the polarization vector for massive spin 1/2 particles. Phenomenologically, this extension may have several interesting applications. For instance, it would make possible to predict the value of particle polarization in relativistic heavy ion collisions [2, 3, 4, 5, 6, 7] at the hydrodynamical decoupling, provided that local thermodynamical equilibrium applies to spin degrees of freedom as well. The detailed analysis and calculation can be found in [8].

2. Single particle distribution function with spin

At global thermodynamical equilibrium with finite angular momentum density the density operator is well known [9, 10] and reads:

$$\widehat{\rho} = \frac{1}{Z} \exp[-\widehat{H}/T + \mu\widehat{Q}/T + \boldsymbol{\omega}\cdot\widehat{\mathbf{J}}/T]\mathsf{P}_V \tag{3}$$

where $\boldsymbol{\omega}$ is a constant fixed vector whose physical meaning is that of an angular velocity and T is the global temperature, that is the temperature of a thermostat in contact with the system or that measured by a thermometer at rest with respect to the external inertial observer. The P_V operator is the projector operator onto localized states [11] which is needed in order to avoid the relativistic singularity at $r = c/\omega$. For this distribution, it has been shown the single-particle distribution function of an ideal relativistic Boltzmann gas of particles with spin S can be obtained from just statistical mechanics arguments [11]:

$$f(x,p)_{rs} = \mathrm{e}^{\xi}\,\mathrm{e}^{-\beta\cdot p}\,\frac{1}{2}\left(D^S([p]^{-1}\mathsf{R}_{\hat\omega}(i\omega/T)[p]) + D^S([p]^{\dagger}\mathsf{R}_{\hat\omega}(i\omega/T)[p]^{\dagger-1})\right)_{rs} \tag{4}$$

where $\beta = \frac{1}{T}(1, \boldsymbol{\omega}\times\mathbf{x})$ is the inverse temperature four-vector; λ is the fugacity; $[p]$ is the SL(2,C) matrix corresponding to the Lorentz transformation taking the time unit vector $\hat{t}$ into $\hat{p}$ (so-called standard transformation); D^S stands for the $(S,0)$ irreducible representation of SL(2,C); R is the SL(2,C) corresponding of a rotation, which is calculated for an imaginary angle $i\omega/T$.

The extension to Fermi-Dirac statistics of this formula is not a straightforward calculation like in the non-rotating case, hence we make an *ansatz* about this extension which reproduces the usual Fermi-Dirac distribution in the non-rotating case and, at the same time, has the (4) for $S = 1/2$ case as its Boltzmann limit.

$$\begin{aligned} f(x,p)_{rs} &= \frac{1}{2m}\bar{u}_r(p)\left(\exp[\beta\cdot p - \xi]\exp\left[-\frac{1}{2}\varpi_{\mu\nu}\Sigma^{\mu\nu}\right] + I\right)^{-1} u_s(p) \\ \bar{f}(x,p)_{rs} &= -\frac{1}{2m}\bar{v}_s(p)\left(\exp[\beta\cdot p + \xi]\exp\left[\frac{1}{2}\varpi_{\mu\nu}\Sigma^{\mu\nu}\right] + I\right)^{-1} v_r(p) \end{aligned} \tag{5}$$

whereas ϖ is the covariant form for the angular velocity:

$$\varpi_{\mu\nu} = (\omega/T)(\delta^1_\mu\delta^2_\nu - \delta^1_\nu\delta^2_\mu) = \sqrt{\beta^2}\Omega_{\mu\nu} \tag{6}$$

and $\Omega_{\mu\nu}$ turns out to be the acceleration tensor of the Frenet-Serret tetrad of the β field lines [11]. The above second equality holds for a rigid velocity field only [11, 12].

The expression of the single particle distribution function for a fermion field, at local thermodynamical equilibrium, in principle, should be calculated using the density operator for local themodynamical equilibrium [13]. At the lowest order of approximation, however, we know that the single particle distribution function must yield the same formal expression at global thermodynamical equilibrium with space-time dependent intensive thermodynamics functions, that is space-time dependent β, ϖ and ξ. Hence, the single-particle distribution functions (4) for $S = 1/2$ case become:

$$f(x,p)_{rs} = \frac{1}{2m}\bar{u}_r(p)\left(\exp[\beta(x)\cdot p - \xi(x)]\exp\left[-\frac{1}{2}\varpi(x)_{\mu\nu}\Sigma^{\mu\nu}\right] + I\right)^{-1} u_s(p)$$
$$\bar{f}(x,p)_{rs} = -\frac{1}{2m}\bar{v}_s(p)\left(\exp[\beta(x)\cdot p + \xi(x)]\exp\left[\frac{1}{2}\varpi(x)_{\mu\nu}\Sigma^{\mu\nu}\right] + I\right)^{-1} v_r(p) \quad (7)$$

where β is the inverse temperature four-vector, ξ is the ratio between comoving chemical potential and temperature, $\Sigma^{\mu\nu} = (i/4)[\gamma^\mu, \gamma^\nu]$ are the generators of Lorentz transformations of 4-components spinors, $u(p)$ and $v(p)$ are the spinors solutions of the free Dirac equation. While the general physical meaning of the fields β and ξ can be easily inferred from the equilibrium limit (β is the local inverse temperature four-vector field and ξ the ratio between the comoving chemical potential $\mu_0(x)$ and the local comoving temperature $T_0(x) = 1/\sqrt{\beta^2}$), $\varpi_{\mu\nu}(x)$'s expression cannot be uniquely obtained from the equilibrium distribution. The reason of this ambiguity is that at rotational equilibrium, the tensor ϖ is:

$$\varpi_{\mu\nu} = \sqrt{\beta^2}\Omega_{\mu\nu} \quad (8)$$

where $\Omega_{\mu\nu}$ is the acceleration tensor of the Frenet-Serret tetrad of the β field lines [11] and [12]:

$$\varpi_{\mu\nu} = -\frac{1}{2}(\partial_\mu\beta_\nu - \partial_\nu\beta_\mu) \quad (9)$$

that is the relativistic generalization of angular velocity of fluid. In a nonequilibrium situation, the right hand sides of eqs. (9) and (8) differ and it is not obvious which one applies, perhaps neither. However, if the system is not too far from equilibrium, ϖ cannot be too distant from the right hand side of (9). Particularly, the difference must be of the 2nd order in the gradients of the β field (for instance: $(\partial\cdot\beta)(\partial_\mu\beta_\nu - \partial_\nu\beta_\mu)$ or second order gradients like $\partial_\mu\partial_\nu\beta^\lambda$) which vanish at equilibrium [12]. For the lowest-order formulation of relativistic hydrodynamics, the expression (9) is sufficient to determine the expression of spin-related quantities.

3. Polarization

The main consequence of the distribution functions (7) is that spin 1/2 particles get polarized at local thermodynamical equilibrium. The polarization four-vector, for a particle with mass m and four-momentum p is defined as:

$$\Pi_\mu = -\frac{1}{2}\epsilon_{\mu\rho\sigma\tau}S^{\rho\sigma}\frac{p^\tau}{m} \quad (10)$$

where $S^{\rho\sigma}$ is the mean value of the *total* angular momentum operator of the single particle. However the Levi-Civita tensor makes the orbital part of the total angular momentum irrelevant, so we are left with the spin-tensor density in the phase space $\mathcal{S}^{0,\rho\tau}(x,p)$ contribution only:

$$\langle\Pi_\mu(x,p)\rangle = -\frac{1}{2}\frac{1}{\sum_r f_{rr}}\epsilon_{\mu\rho\sigma\tau}\frac{\mathrm{d}\mathcal{S}^{0,\rho\sigma}(x,p)}{\mathrm{d}^3p}\frac{p^\tau}{m} \quad (11)$$

The *canonical* spin tensor can be computed using (7):

$$\mathcal{S}^{\lambda,\mu\nu}(x) = \frac{1}{2}\int\frac{\mathrm{d}^3\mathrm{p}}{2\varepsilon}\sum_{rs}\left(f_{rs}(x,p)\bar{u}_s(p)\{\gamma^\lambda,\Sigma^{\mu\nu}\}u_r(p)\right) - \left(\bar{f}_{rs}(x,p)\bar{v}_r(p)\{\gamma^\lambda,\Sigma^{\mu\nu}\}v_s(p)\right) \quad (12)$$

Not far from equilibrium, using the (9), expanding for small value of $\varpi^{\mu\nu}$, the polarization vector becomes:

$$\langle \Pi_\mu(x,p) \rangle \simeq \frac{1}{16} \epsilon_{\mu\rho\sigma\tau} \left(1 - n_F\right) \left(\partial^\rho \beta^\sigma - \partial^\sigma \beta^\rho\right) \frac{p^\tau}{m} = \frac{1}{8} \epsilon_{\mu\rho\sigma\tau} \left(1 - n_F\right) \partial^\rho \beta^\sigma \frac{p^\tau}{m} \tag{13}$$

where $n_F = 1/(\exp(\beta(x) \cdot p - \xi(x)) + 1)$ is the Fermi-Dirac distribution function.

It may be of interest, e.g. for relativistic heavy ion collisions, to calculate the space-integrated mean polarization vector. For a three-dimensional spacelike hypersurface Σ, one has:

$$\langle \Pi_\mu(p) \rangle \equiv \frac{\displaystyle\int \mathrm{d}\Sigma_\lambda \frac{p^\lambda}{\varepsilon} (-1/2) \epsilon_{\mu\rho\sigma\tau} \frac{\mathrm{d}\mathcal{S}^{0,\rho\sigma}}{\mathrm{d}^3\mathrm{p}} \frac{p^\tau}{m}}{\displaystyle\int \mathrm{d}\Sigma_\lambda \frac{p^\lambda}{\varepsilon} \sum_r f_{rr}(x,p)} \simeq \frac{1}{8} \epsilon_{\mu\rho\sigma\tau} \frac{p^\tau}{m} \frac{\displaystyle\int \mathrm{d}\Sigma_\lambda \, p^\lambda \, n_F (1 - n_F) \partial^\rho \beta^\sigma}{\displaystyle\int \mathrm{d}\Sigma_\lambda \, p^\lambda n_F} \tag{14}$$

For antiparticles, one gets the same formula, with $\bar{n}_F = 1/(\exp(\beta(x) \cdot p + \xi(x)) + 1)$ replacing n_F. Here an important comment is in order: the fact that local thermodynamical equilibrium implies the same orientation for the polarization vector of particles and antiparticles (unlike e.g. in the electromagnetic field) is a general outcome and does not depend on the introduced approximations. It stems from the fact that the spin tensor, as well as the angular momentum, is a charge-conjugation even operator, or, more simply stated, that thermal and mechanical effects do not "see" the internal charge of the particles.

This formula (14) may be used to predict the polarization of particles, in particular Λ hyperons, produced at the freeze-out (primary particles) in a relativistic heavy-ion collision, after the hydrodynamical evolution. This polarization is proportional to the thermal vorticity (9) at freeze-out hypersurface and could give an observable contribution [15]. The observation of this effect would be a confirmation of the achievement the local thermodynamical equilibrium also to spin degrees of freedom and would also indicate that vorticity and circulation may persist up to freeze-out.

References

[1] Cooper F and Frye G 1974 *Phys. Rev.* D **10** 186
[2] Baznat M, Gudima K, Sorin A and Teryaev O 2013 *Phys. Rev.* C **88** 061901
[3] Huang X G, Huovinen P and Wang X N 2011 *Phys. Rev* . C **84** 054910
[4] Gao J H, Chen S W, Deng W T, Liang Z T, Wang Q and Wang X N 2008 *Phys. Rev.* C **77** 044902
[5] Gao J H, Liang Z T, Pu S, Wang Q and Wang X N 2012 *Phys. Rev. Lett.* **109** 232301
[6] Betz B, Gyulassy M and Torrieri G 2007 *Phys. Rev.* C **76** 044901
[7] Csernai L, Magas V K and Wang D J 2013 *Phys. Rev.* C **87** 034906
[8] Becattini F, Chandra V, Del Zanna L and Grossi E 2013 *Annals Phys.* **338** 32
[9] Landau L and Lifshitz L 1980 *Statistical Physics, Part 1* (Oxford, England: Pergamon Press)
[10] Vilenkin A 1980 *Phys. Rev.* D **21** 2260
[11] Becattini F and Tinti L 2010 *Annals Phys.* **325** 1566
[12] Becattini F 2012 *Phys. Rev. Lett.* **108** 244502
[13] Becattini F and Tinti L 2011 *Phys. Rev.* D **84** 025013
[14] Becattini F and Piccinini F 2008 *Annals Phys.* **323** 2452
[15] Becattini F, Csernai L and Wang D J 2013 *Phys. Rev.* C **88** 034905

Hyperon Polarization in Heavy ion Collisions

C. C. Barros Jr.

Depto de Física - CFM - Universidade Federal de Santa Catarina - Florianópolis - SC - CP. 476 - CEP 88.040 - 900 - Brazil

E-mail: barros@fsc.ufsc.br

Abstract. The STAR collaboration has measured the Λ and anti-Λ polarizations in 200 GeV Au-Au collisions [1]. These results can be understood in terms of a model [2], [3], that we proposed recently, based on the hydrodynamical model, and taking into account the effect of the final-state interactions (that occur in the hadronic phase) between the hyperons and other produced particles. These final interactions are described in terms of chiral effective lagrangians, that consider many hadronic processes. This model describes quite well the antihyperon polarization data obtained in proton-nucleus collisions, and now we extended it to study nucleus-nucleus collisions, with a very good accord. Theoretical results obtained with other models will also be discussed. The perspectives of hyperon and antihyperon polarization at LHC is another subject of interest.

Since the discovery of significant polarization of the Λ particles produced in 100 GeV p-Be collisions by Bunce [4], hyperon polarization has been found to be a very challenging subject, as, at the time it was a totally surprising result. This fact, unexpected both experimentally and theoretically has been confirmed by further experiments, and this puzzle has been complicated when the polarizations of the other hyperons and antihyperons have been measured [5]-[12].

Hyperon polarization may be quite well described by parton-based models [13]-[15], but antihyperon polarization not. In [2], we proposed a model [2] that was able to describe successfully the antihyperon polarization in terms of final-state interactions that occur in the hadronic phase of such collisions, in a mechanism based in the hydrodynamics.

Recently, at RHIC, in 200 GeV Au-Au collisions, the Λ and $\overline{\Lambda}$ polarizations have been measured [1], as functions of the transverse momentum, in the range $0 < p_t < 5$ GeV, and as functions of the pseudorapidity, in the range $-1.5 < \eta < 1.5$. In this region, the final polarization for both particles may be considered consistent with zero. Some models deal with this problem [16]-[18]. Other models [19], [20] propose that some polarization should be produced due to the mechanical processes.

In this work, we show the results obtained in [3], with a model that we used to calculate antihyperon polarization in p-A collisions, now in the study of the Au-Au collisions performed at RHIC. This model is based in the hydrodynamical aspects of such collisions, and depends on the velocity distribution of the fluid formed during the collision. Then, we will use it in order to obtain the average polarization, taking into account the $\pi\Lambda$ and $\pi\overline{\Lambda}$ final interactions.

We will consider that the velocity distribution of such a fluid may be given by the expression

$$u^0\frac{d\rho}{d^3u} = A\left[e^{-\beta(\alpha-\alpha_0)^2} + e^{-\beta(\alpha+\alpha_0)^2}\right]e^{-\beta_t\xi^2} \quad , \tag{1}$$

that is written in terms of its longitudinal (α) and transversal (ξ) rapidities. That means that the formed fluid expands in the incident nuclei direction (α), and also in the transverse direction (ξ). We may visualize this fluid geometrically as a hot expanding cylinder. The constants β, β_t and α_0 are parameters that describe the shape of this distribution, and are determined by calculating the distributions of the produced particles, and, comparing them with the RHIC experimental data for the transverse momentum p_t [23] and pseudorapidity (η) distributions [24].

This objective may be achieved, making a convolution of the fluid velocity distribution (1) with the particle distribution inside these fluid elements, that may be considered a Bose distribution as most of the produced particles are pions. We will consider

$$\frac{dN}{d\vec{p}_0} = \frac{N_0}{\exp\left(E_0/T\right) - 1} \tag{2}$$

with the temperature $T \sim m_\pi$, and $\vec{p}_0$ and E_0 are the momentum and energy of the pions inside one fluid element. The observed distributions of particles are given by [21], [22]

$$E\frac{dN}{d\vec{p}} = C \int \left[e^{-\beta(\alpha-\alpha_0)^2} + e^{-\beta(\alpha+\alpha_0)^2}\right] \frac{e^{-\beta_t\xi^2} E_0(\alpha,\xi,\phi)}{\exp\left(E_0(\alpha,\xi,\phi)/T\right) - 1} \times \sinh\xi \, \cosh\xi \; d\alpha \; d\xi \; d\phi \;, \tag{3}$$

where ϕ is the azimuthal angle. Fig. 1 and Fig. 2 show the resulting particle distributions compared with the experimental data.

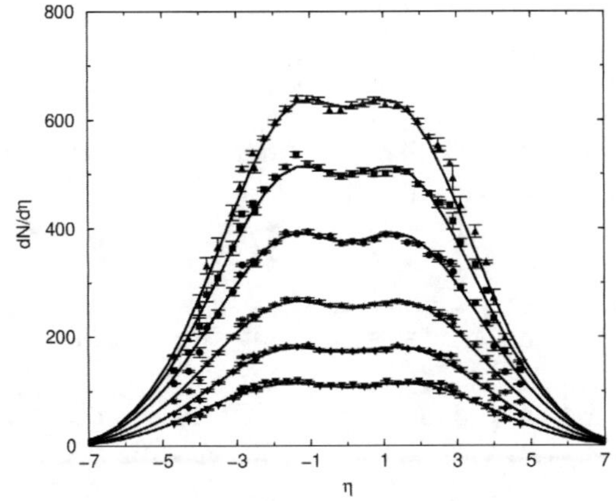

Figure 1. Distributions $dN/d\eta$, for many centralities. From the top, 0-5%, 5-10%, 10-20%, 20-30%, 30-40%, 40-50%. We compare our results (solid lines), with the experimental data from [24] (points).

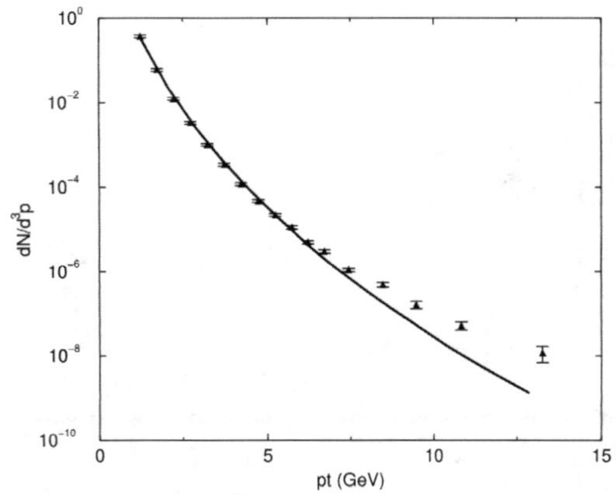

Figure 2. Comparison of the calculated distribution dN/d^3p as function of p_t with the experimental data from [23].

The final-state interactions may be described by effective chiral lagrangians, as was done in [25]-[27], where the resonance $\Sigma^*(1385)$ in the intermediate state is a key element.

With the knowledge of the velocities distribution and of the final interactions, we are able to calculate the average polarization of the produced particles in the same way as in [2]. The average polarization may be calculated by the expression [3], [2]

$$\langle\vec{P}\rangle = \frac{\int \left(\vec{P}' \; d\sigma/dt\right) \mathcal{G} \; d\alpha \; d\xi d\phi \; d\vec{\Lambda}_0' d\vec{\pi}_0'}{\int \left(d\sigma/dt\right) \mathcal{G} \; d\alpha \; d\xi d\phi \; d\vec{\Lambda}_0' d\vec{\pi}_0'} \;. \tag{4}$$

The factor $\mathcal{G}$ that appears in eq. (4) contains the statistical weighs of the production of the particles and the ones relative to the expansion of the fluid, and can be written as

$$\mathcal{G} = \frac{(d\rho/d^3u)\Lambda_0'^2\pi_0'^2}{(\exp(E'_{\pi_0}/T)-1)(\exp(E'_0/T)+1)} \times \delta\left(E'_0 + E'_{\pi_0} - E' - \sqrt{m_\pi^2 + (\vec{\pi}'_0 + \vec{\Lambda}'_0 - \vec{\Lambda}')^2}\right) \quad , \tag{5}$$

where $d\rho/d^3u$, is given by (1).

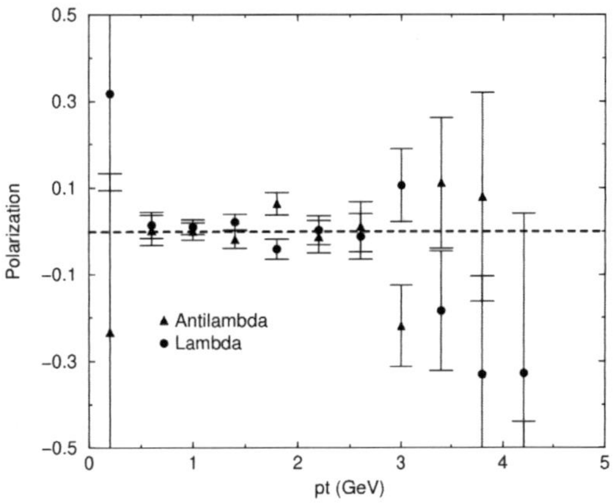

Figure 3. Calculated Λ and $\overline{\Lambda}$ polarizations as functions of the transverse momentum (dashed line) compared with the RHIC data [1].

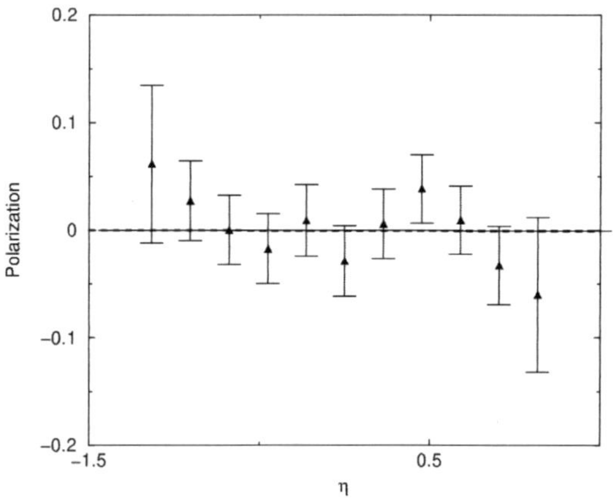

Figure 4. Calculated Λ and $\overline{\Lambda}$ polarizations as functions of η (solid line) compared with the RHIC data [1].

With this procedure we obtained the results shown in Fig. 3 and Fig. 4. As we can see, the resulting polarization is very small (smaller than 1%) for all values of the centrality, and are in good accord with the experimental data for both Λ and $\overline{\Lambda}$.

In view of these results, if the hyperon polarization were to be measured at LHC in heavy ion collisions, we expect that the effect due to the final state interactions will be totally washed out in average, for Λ, $\overline{\Lambda}$ and for the other hyperons, as the energy of the system is much higher. If some polarization effect could be found (averaging the data in a different way, for example) it would be due to global aspects of the system, in processes similar to the ones proposed in [19], [20] and [28].

Another intriguing possibility is the hyperon polarization in pA collisions at LHC. In this case we imagine that it is possible that some polarization could survive considering the final state interactions, but the exact results are not available yet.

Acknowledgments
I would like to acknowledge FAPESC for the financial support.

References

[1] Abelev B I *et al* 2007 *Phys. Rev. C* **76** 024915
[2] Barros Jr C C and Hama Y 2008 *Int. J. Mod. Phys. E* **17** 371
[3] Barros Jr C C and Hama Y 2011 *Phys. Lett. B* **699** 74
[4] Bunce G *et al* 1976 *Phys. Rev. Lett.* **36** 1113

[5] Heller K *et al* 1977 *Phys. Lett.* **68B** 480, Heller K *et al* 1978 *Phys. Rev. Lett.* **41** 607, Erhan S *et al* 1979 *Phys. Lett.* **82B** 301
[6] Lundberg B *et al* 1989 *Phys. Rev. D* **40** 3557
[7] Wilkinson C *et al* 1981 *Phys. Rev. Lett.* **46** 803
[8] Deck L *et al* 1983 *Phys. Rev. D* **28** 1
[9] Ramerika R *et al* 1986 *Phys. Rev. D* **33** 3172
[10] Adamovich M I *et al* 1995 *Z. Phys. A* **350** 379
[11] Ho P M *et al* 1990 *Phys. Rev. Lett.* **65** 1713, Ho P M *et al* 1991 *Phys. Rev. D* **44** 3402
[12] Morelos A *et al* 1993 *Phys. Rev. Lett.* **71** 2172, Morelos A *et al* 1995 *Phys. Rev. D* **52** 3777
[13] Andersson B Gustafson G and Ingelman G 1979 *Phys. Lett.* **85B** 417
[14] DeGrand T A and Miettinen H I 1981 *Phys. Rev. D* **24** 2419
[15] Troshin S M and Tyurin N E 1997 *Phys. Rev. D* **55** 1265
[16] Panagiotou A D 1986 *Phys. Rev. C* **33** 1999
[17] Betz B Gyulassy M and Torrieri G 2007 *Phys. Rev. C* **76** 044901
[18] Liang Z and Wang X 2005 *Phys. Rev. Lett.* **94** 102301
[19] Becattini F Piccinini F and Rizzo J 2008 *Phys. Rev. C* **77** 024906
[20] Becattini F Piccinini F 2008 *Ann. Phys.* **323** 2452
[21] Hama Y 1978 *Nuovo Cim.* **46A** 569
[22] Hama Y and Navarra F S 1992 *Z. Phys. C* **53** 501
[23] Adler S S *et al* 2003 *Phys. Rev. Lett.* **91** 241803
[24] Bearden I G *et al* 2002 *Phys. Rev. Lett.* **88** 202301
[25] Barros Jr C C and Hama Y 2001 *Phys. Rev. C* **63** 065203
[26] Barros Jr C C and Robilotta M R 2006 *Eur. Phys. J. C* **45** 445
[27] Barros Jr C C 2003 *Phys. Rev. C* **68** 034006
[28] Csernai L P, Magas V K and Wang D J 2013 *Phys. Rev. C* **87** 034906

eRHIC: A Precision Tool for Studying Nuclear Structure

Thomas Burton

Brookhaven National Lab, Upton, NY 11973, USA

E-mail: tpb@bnl.gov

Abstract. eRHIC is a proposed electron-ion collider to be located at Brookhaven National Lab. This high-luminosity electron-hadron machine will provide unprecedented precision and versatility in studying features of QCD and the structure of nucleons and nuclei. We describe some key features of the eRHIC proposal, focusing on its potential for understanding the gluon structure of nuclei.

1. Introduction

eRHIC is proposed as a future electron-ion collider (EIC) to be located at Brookhaven National Lab. It would expand the existing RHIC collider complex with a polarised, high-intensity and variable-energy electron beam. This upgrade, providing capability for electron-proton and electron-nucleus collisions, would open up new frontiers in the precision study of QCD, nuclear structure and spin physics. eRHIC will utilise much of the existing RHIC infrastructure and investment, while incorporating novel new technologies, such as a pair of energy-recovery linacs to accelerate the electron beam. The upgraded facility will build on the successes of both the RHIC programme, and of the only previous electron-proton collider, HERA. The significantly higher luminosity of eRHIC (up to $\sim$ 500 times that of HERA), coupled with its ability to run with a wide variety of ion species at variable beam energy will allow unparalleled precision and flexibility in studying the structure of nuclear matter.

2. Electron-hadron collisions

RHIC has operated as a highly successful hadron-hadron collider for over a decade now. One may therefore ask the question: why perform *electron*-hadron collisions? There are three key features of electron-hadron collisions that distinguish them from hadron-hadron collisions. Firstly, they eliminate "spectator" backgrounds that may interfere with signal. Secondly, they make it simple to distinguish initial- and final-state effects, as interactions between "beam" and "target" do not occur before the hard interaction as they do in hadron-hadron collisions. Thirdly, and perhaps most compellingly, they provide direct access to parton-level kinematics via deeply inelastic scattering (DIS).

DIS (figure 1(a)) involves the interaction of an electron with a quark inside a hadron via the exchange of a virtual photon. It is such a powerful process for understanding nuclear structure because of its direct, but experimentally simple, access to the photon-quark kinematics. Consider a DIS interaction, where an electron, of four-momentum k, interacts with a quark inside a

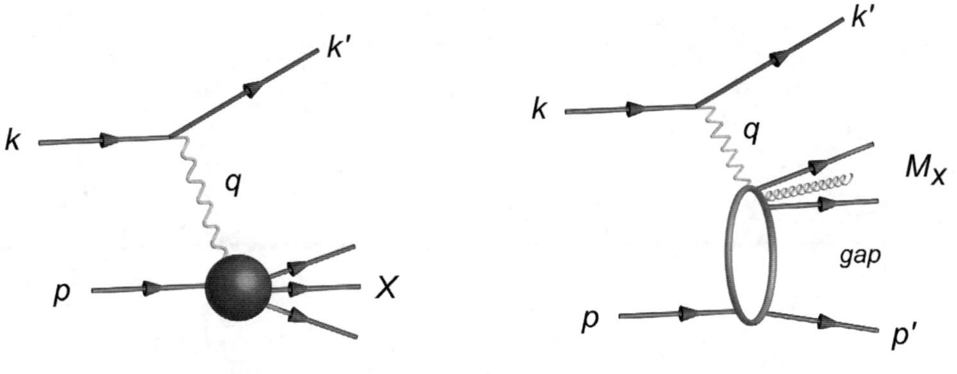

(a) DIS (b) Diffractive DIS

Figure 1. Schematic view of a DIS interaction between an electron (blue) and a nucleon or nucleus (red) via exchange of a virtual photon (yellow). In diffractive DIS (b) there is also an exchange of a colour-neutral state, such as a pair of gluons (green).

hadron, p, via the exchange of a virtual photon, q. All the information about the photon-quark interaction is encapsulated in a few invariant quantities:

$$Q^2 = -q^2 = -(k - k')^2 \qquad x = \frac{Q^2}{2p \cdot q} \qquad y = \frac{q \cdot p}{k \cdot p}. \tag{1}$$

Q^2 is the squared momentum transfer in the interaction. The Bjorken variable, x, can be interpreted as the fraction of the proton's momentum carried by the struck parton. The variable y is a measure of the inelasiticity of the interaction. Significantly, all these quantities may be obtained simply by measuring the angle and energy of the scattered electron. They are also related to one-another and the squared centre-of-mass energy, s, of the collision via

$$Q^2 = sxy. \tag{2}$$

The eRHIC physics programme is very diverse, spanning a wide variety of measurement using both electron-proton and electron-nucleus collisions. The goals and potential of the programme, and the details of the accelerator itself, are summarised in the EIC whitepaper document [1]. Here, we shall select just two topics of particular interest, in the electon-nucleus programme, to highlight the capabilities of eRHIC: gluon saturation and nuclear imaging.

3. Saturation

Quarks exist in nature only as tightly bound states. However, the overwhelming majority of the mass of these states, such a protons and neutrons, is provided not by the mass of the quarks themselves, but by the QCD interaction binding them. The partonic composition of the proton shows explosive growth in the gluon density as small values of the Bjorken parameter x. This growth cannot continue indefinitely, and so must be tamed at some sufficiently small x. Such taming can be accounted for by non-linear evolution equations. In addition to the splitting of gluons included in linear evolution schemes (such as DGLAP), non-linear evolution incorporate the re-merging of small-x gluons into larger-x gluons. As gluon density increases, recombination competes with splitting until the growth in gluon density is restrained. This is referred to as gluon *saturation.*

Results from HERA show no clear signature of saturation in e-p collisions for x down to $\sim 10^{-6}$. As can be seen from equation 2, access to low x (at a given Q^2 and y) requires high

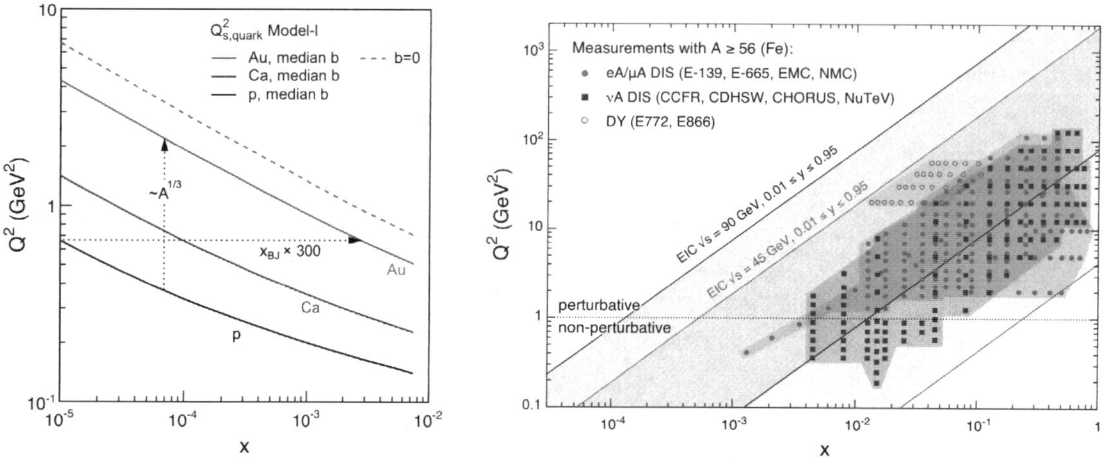

(a) (b)

Figure 2. (a) Nuclear amplification of the saturation scale makes it accessible at larger x, and therefore lower energy. (b) eRHIC kinematics for e-A collisions compared to existing world data

centre-of-mass energy. While HERA reached $s \sim 300$ GeV, eRHIC will be a lower energy-machine. Theoretical predictions for the onset of saturation suggest an e-p machine with $s \sim$ 1-2 TeV would be required. How, then, can eRHIC investigate the onset of saturation at low x? The answer lies in using *nuclear* beams. In a high-energy collision, the electron probe sees a nucleus Lorentz-contracted into a disk. It therefore "sees" a higher gluon density. This results in a nuclear amplification of the saturation scale, Q_s^2, below which saturation manifests, approximately as follows:

$$Q_s^2(x) \sim \left(\frac{A}{x}\right)^{1/3}, \qquad (3)$$

for a nucleus of mass A. This increases the x below which saturation appears by a factor of several hundred (see figure 2(a)). Thus, saturation is accessible at an e-A machine with $s \sim 100$ GeV, as opposed to a TeV-scale e-p machine.

In addition, eRHIC will possess the ability to run at different energies, allowing a very wide span of x and Q^2 to be mapped out. This will allow the saturated and usaturated regimes, and the onset of saturation, to be studied in great detail (figure 2(b)).

3.1. Diffractive collisions

Of particular interest in saturation studies are *diffractive* interactions, shown in figure 1(b). These interactions differ from a simple DIS interaction by the exchange of a colour-neutral object, such as a pair of gluons, with the scattered nucleus either remaining intact, or subsequently breaking up due to excitation. Diffractive collisions are ideal for studying gluons because their cross section increases approximately as the *square* of the gluon density. The collisions have a clear experimental signature, whereby a hadronic final state is produced with a large angular separation, often referred to as a *rapidity gap*, from the scattered nucleus. A particularly clean channel for study is *exclusive vector meson production*, where the hadronic final state is a single vector meson like a ϕ or J/ψ. The cross section has two components; coherent, in which the scattered nucleus remains intact; and incoherent, in which the nucleus breaks apart. Simulated differential cross sections for exclusive diffractive production of ϕ mesons at eRHIC are shown

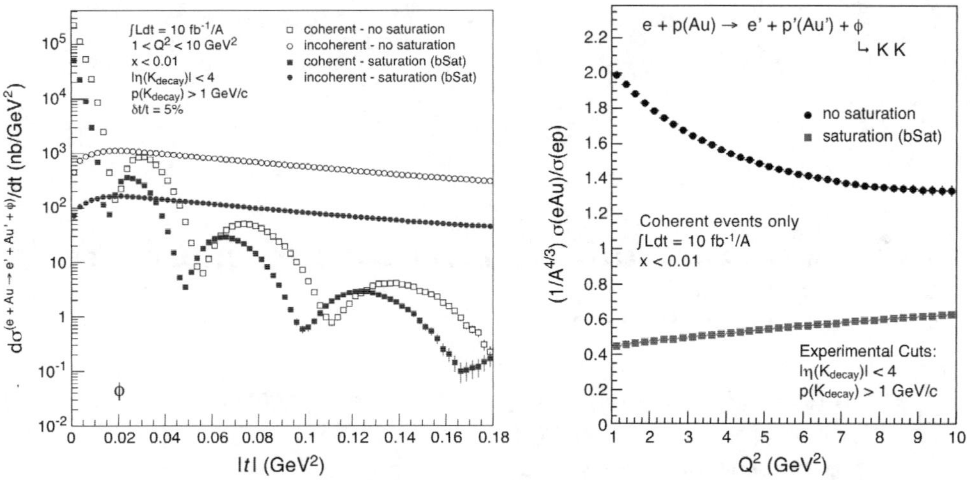

(a) (b)

Figure 3. (a) Simulated differential cross sections with respect to t for $e + Au \rightarrow e' + Au' + \phi$ [2]. Solid points show results with a saturation model, while hollow points show results with a non-saturation model. (b) The coherent cross section ratio between e-A and e-p collisions is highly sensitive to the presence of saturation.

in 3(a).

Diffractive collisions are characterised by an additional kinematic variable, t, the four-momentum transferred to the nucleon or nucleus in the scattering. The shape of the coherent differential cross section with respect to t displays a shape similar to a classical diffraction pattern. By detecting neutrons emitted close to the beam as a signature of nuclear breakup, it will be possible to disentangle the coherent and incoherent contributions.

Figure 3(b) shows the utility of diffractive collisions in studying saturation, as the cross section is strongly sensitive to the presence of saturation. The predictions when saturation effects are included differ significantly from the non-saturation prediction, and it is well within the eRHIC precision to separate them.

4. Nuclear Imaging

In addition to their use in understanding saturation, diffractive interactions have great potential for imaging nuclei. Parton distribution functions (PDFs) yield a 1-dimensional view of the nucleus: they give a picture only in terms of x. The momentum transfer t involved in a diffractive collisions is the conjugate variable of the impact parameter, b, of the interaction. Consequently, a Fourier transform of the coherent differential cross-section, $d\sigma/dt$, yields the distribution of partons in impact parameter space. Therefore, one is able to form a "3-dimensional" view of the nucleus: simultaneously in x and in transverse spatial profile, b. Given the sensitivity of diffractive interactions to gluons, this would allow the gluon density of the nucleus to be imaged in exquisite detail. Furthermore, the incoherent cross section is related to the "lumpiness" of the initial gluon density, providing information about fluctuations and "hot-spots" in the nuclear gluon distribution.

References

[1] Accardi A, Albacete J, Anselmino M, Armesto N, Aschenauer E *et al.* 2012 (*Preprint* 1212.1701)
[2] Toll T and Ullrich T 2013 *Phys. Rev. C* **87** 024913

Azimuthal correlation of charm quark pair produced in heavy ion collision

Mohammed Younus[1] **and Dinesh K. Srivastava**[2]
1/AF, Bidhan Nagar, Kolkata 700064, India
E-mail: [1]younus.presi@gmail.com
E-mail: [2]dinesh@vecc.gov.in

Abstract. It is known that heavy quarks are produced in pairs. There always exists a correlation in azimuthal angle in transverse momentum plane between the members in a pair. The quark and anti-quark in a pair may lose different amount of energy while traveling through QGP and it is expected that the correlation of $Q\bar{Q}$ will alter considerably. Also in a complete different picture, collective flow of the medium can change the correlation of heavy quark pair. Here we present the effect of charm quark energy loss based on Wang-Huang-Sarcevic model on its correlation. The flow feature is shown using a naive model based on Cautle-Páic model.

1. Introduction

Charm quarks are always produced in pairs in agreement with the conservation of heavy flavour quantum no. in relativistic heavy ion collision. The produced charm pairs($c\bar{c}$), show correlation in azimuthal angle($C(\Delta\phi)$), in transverse momentum plane. It is known that at leading order. c and $\bar{c}$ will be exactly back-to-back($\Delta\phi$=π), with $\mathbf{p_{Tc}}$=-$\mathbf{p_{T\bar{c}}}$. While at next-to-leading order, as one of the produced final state charm may emit a gluon or one of the final state gluons from hard scattering process may split into $c\bar{c}$ pair, this back-to-back correlation is now modified with distribution now extended throughout, $\Delta\phi$=0–π. Again if charm pair undergoes medium modification due to energy loss or due to the collective flow of the medium, then the azimuthal correlation will alter considerably. Thus azimuthal correlation of heavy quark pair is fast emerging as one of the possible signature for the QGP properties.

2. Charm pair production

The correlation of heavy quarks produced in pp collisions is defined as

$$E_1\,E_2\,\frac{d\sigma}{d^3p_1\,d^3p_2}=\frac{d\sigma}{dy_1\,dy_2\,d^2p_{T1}\,d^2p_{T2}}=C\,, \tag{1}$$

where y_1 and y_2 are the rapidities of heavy quark and anti-quark and $\mathbf{p_{Ti}}$ are the respective momenta. One can now calculate [1]

$$\frac{d\sigma_{pp}}{dy_1dy_2d^2p_{T1}d^2p_{T2}} = 2x_ax_b\sum_{ij}\left[f_i^{(a)}(x_a,Q^2)f_j^{(b)}(x_b,Q^2)\frac{d\hat{\sigma}_{ij}(\hat{s},\hat{t},\hat{u})}{d\hat{t}} + f_j^{(a)}(x_a,Q^2)f_i^{(b)}(x_b,Q^2)\frac{d\hat{\sigma}_{ij}(\hat{s},\hat{u},\hat{t})}{d\hat{t}}\right]/(1+\delta_{ij})\;, \tag{2}$$

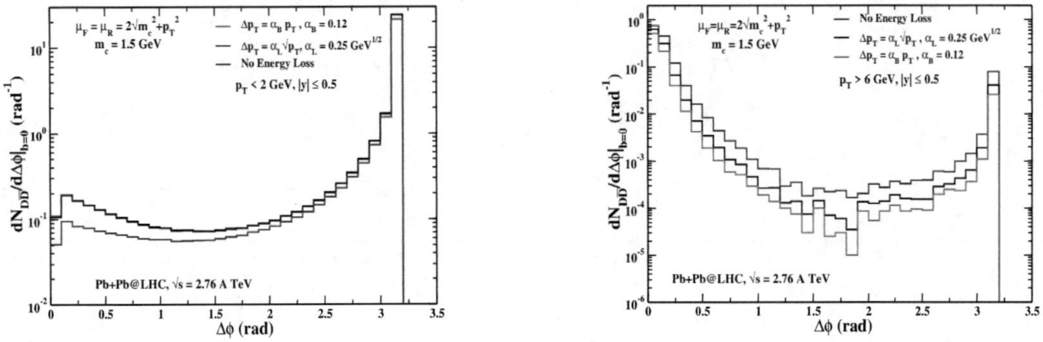

Figure 1. Effect of energy loss on azimuthal correlation of $D\overline{D}$ pair for two different charm momentum region

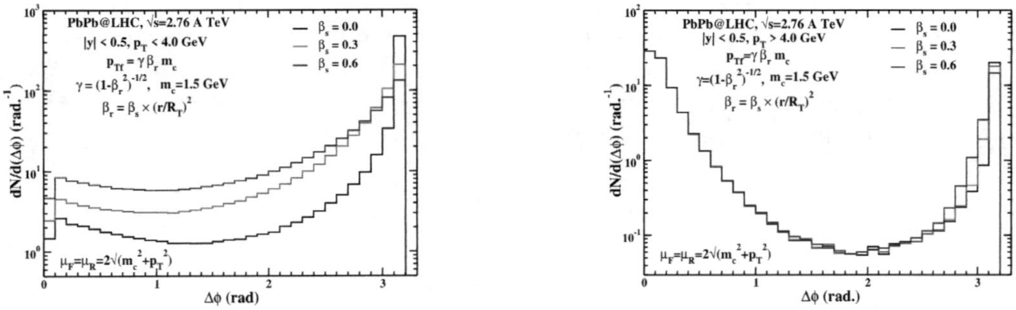

Figure 2. Effect of flow on azimuthal correlation of $c\bar{c}$ for $p_T < 4.0$ GeV(left) and $p_T > 4.0$ GeV

where $p_{T1,2}$ and $y_{1,2}$ are the momenta and rapidities of produced charm and anti-charm and x_a and x_b are the fractions of the momenta carried by the partons from their interacting parent hadrons. We have included all leading and next-to-leading order diagrams for charm quark pair production and obtained the cross-section using NLO-pQCD calculations [2].

The Eq. 1 is used to calculate the correlation of heavy quarks from initial fusion in proton-proton collision. The azimuthal correlation of heavy quark for Pb+Pb collision at given impact parameter is given by

$$E_c\, E_{\bar{c}}\, \frac{dN_{AA}}{d^3p_c\, d^3p_{\bar{c}}} = T_{AA} E_c\, E_{\bar{c}}\, \frac{d\sigma_{pp}}{d^3p_c\, d^3p_{\bar{c}}} \tag{3}$$

For lead on lead collisions at LHC, we have used $T_{AA} = 292$ fm^{-2} for b = 0 fm We have used CTEQ5M structure function. The factorization, renormalization, and fragmentation scales are chosen as $2\sqrt{m_c^2 + p_T^2}$ and the charm quark mass, m_c has been taken as 1.5 GeV.

3. Effect of energy loss and medium flow on charm quark azimuthal correlation and discussions of results:

The medium effect on charm p_T spectrum has been obtained using two different models. The first model shows the effect of momentum loss of charm quark per collision through multiple scattering with medium partons. Partly inspired by Wang-Huang-Sarcevic model, [3], the momentum loss per collision for 'i^{th}' charm with momentum, $|\mathbf{p_T}|=p_i$, is calculated by

$$\Delta p_i = \alpha(p_i)^\beta\,, \alpha = \alpha_L(\beta = 0.5)\,, \text{and}\, \alpha_B(\beta = 1.0) \tag{4}$$

The no. of collisions for given charm momentum is calculated using Poisson distribution(with mean free path,λ $\sim$1 fm) and the net momentum loss is then calculated and included in the Eq. 3 to obtain the correlation for D mesons, Fig. 1, after adding fragmentation calculations to the final charm spectra.

The second model shows the effect of collective flow of the medium on the angular correlation, if the heavy quarks are assumed to be thermalized. In order to estimate the effect of the flow on the correlation of the heavy quarks, we use a toy model used earlier by [4, 5] and proceed as follows. We first give a random orientation and position to the charm quark-pairs at co-ordinates, (x, y), randomly chosen according to the probability:

$$P = \frac{\int\int\, dx\, dy\, T_A(x, y, b = 0) T_B(x, y, b = 0)}{T_{\mathrm{AB}}(b = 0)}\,, \tag{5}$$

where T_i is the transverse density profile of the nucleus with uniform density, and $T_{\mathrm{AB}}(b = 0)$ is the nuclear thickness for impact parameter, $b = 0$. Then we add vectorially, the flow momentum $\mathbf{p_{Tf}}= p_{Tf}\ (\mathbf{r}/r)$ to the momentum of the heavy quark $\mathbf{p_T}$.

We use the blast-model [6] to write p_{Tf} as

$$p_{Tf} = \gamma \beta_r m_Q\,, \tag{6}$$

where

$$\beta_r = \beta_s \times \left(\frac{r}{R_T}\right)^2\,. \tag{7}$$

and $r = \sqrt{x^2 + y^2}$. We show our results in Fig. 2 for two ranges of p_T of the charm quarks, $p_T < 4$ GeV and $p_T > 4$ GeV.

4. Conclusions

We studied heavy quark azimuthal correlation and showed that it affected by both energy loss and flow of the medium and thus it may serve as one of the observables of quark gluon plasma properties. However much improvement in terms of calculations and statistics must be done and will be addressed in future.

References

[1] Younus M, and Srivastava D K 2012 *J. Phys.* **G**: *Nucl. Part. Phys.* **39**, 025001 and references therein; Nahrgang M, Aichelin J, Gossiaux P B, and Werner K 2013 Azimuthal correlations of heavy quarks in Pb+Pb collisions at LHC ($\sqrt{s} = 2.76$ TeV) Preprint: *arXiv:1305.3823 [hep-ph]*.

[2] Mangano M L, Nason P, and Ridolfi G 1992 *Nucl. Phys.* **B 373**, 295.

[3] Wang X-N, Huang Z, and Sarcevic I 1996 **Phys. Rev. Lett. 77**, 231; Muller B 2003 *Phys. Rev.* **C 67**, 061901(R). Younus M, and Srivastava D K 2013 *J. Phys.* **G**: *Nucl. Part. Phys.* **40**, 065004 and references therein.

[4] Cuautle E, and Paić G 2006 Proton/pion ratios and radial flow in pp and peripheral heavy ion collisions Preprint: *arXiv:hep-ph/0604246v2*.

[5] Tsiledakis G, Appelshäuser H, Schweda K, and Stachel J 2011 *Nucl. Phys.* **A 858**, 86.

[6] Schnedermann E, Sollfrank J, and Heinz U W 1993 *Phys. Rev.* **C 48**, 2462-75.

Relativistic hydrodynamics on graphics processing units

Jan Sikorski[1], Sebastian Cygert[2], Joanna Porter-Sobieraj[2], Marcin Słodkowski[1], Piotr Krzyżanowski[1], Natalia Książek[1] and Przemysław Duda[1]

[1] Faculty of Physics, Warsaw University of Technology, Koszykowa 75, 00-662 Warsaw, PL
[2] Faculty of Mathematics and Information Science, Warsaw University of Technology, Koszykowa 75, 00-662 Warsaw, PL

E-mail: `sikorski.jan@gmail.com`

Abstract. Hydrodynamics calculations have been successfully used in studies of the bulk properties of the Quark-Gluon Plasma, particularly of elliptic flow and shear viscosity. However, there are areas (for instance event-by-event simulations for flow fluctuations and higher-order flow harmonics studies) where further advancement is hampered by lack of efficient and precise 3+1D program. This problem can be solved by using Graphics Processing Unit (GPU) computing, which offers unprecedented increase of the computing power compared to standard CPU simulations. In this work, we present an implementation of 3+1D ideal hydrodynamics simulations on the Graphics Processing Unit using Nvidia CUDA framework. MUSTA-FORCE (MUlti STAge, First ORder CEntral, with a slope limiter and MUSCL reconstruction) and WENO (Weighted Essentially Non-Oscillating) schemes are employed in the simulations, delivering second (MUSTA-FORCE), fifth and seventh (WENO) order of accuracy. Third order Runge-Kutta scheme was used for integration in the time domain. Our implementation improves the performance by about 2 orders of magnitude compared to a single threaded program. The algorithm tests of 1+1D shock tube and 3+1D simulations with ellipsoidal and Hubble-like expansion are presented.

1. Numerical algorithms

The equations of hydrodynamics can be written as a system of conservative equations

$$\frac{\partial u}{\partial t} + \sum_i \frac{\partial f_i(u)}{\partial x_i} = 0 \tag{1}$$

where the sum runs over all spacial dimensions, and $u = u(x)$ represents a conserved variable, which include the energy density, momenta densities and conserved charge densities. The equation is discretized on a cartesian grid into an ordinary differential equation

$$\frac{\mathrm{d}u}{\mathrm{d}t} = \sum_i \frac{1}{|e_i|}\left(f_i\left(u\left(x - \frac{e_i}{2}\right)\right) - f_i\left(u\left(x + \frac{e_i}{2}\right)\right)\right) \equiv L(u) \tag{2}$$

where e_i are the primitive vectors, i.e. vectors between two adjacent cells in the positive direction of dimension i. It is then integrated using a standard, third order Runge-Kutta algorithm. The final discretized operator L depends on a number of neighbouring cells on the lattice in all spacial directions.

1.1. *MUSTA-FORCE scheme*

In order to obtain a general and accurate algorithm for estimating L we use a hybrid MUSTA (MUlti-STAge) approach [1, 2]. It utilizes a centered flux in a predictor-corrector loop, solving the intercell Riemann problem numerically, without using a priori information about waves that propagate in this system.

The algorithm is extended with the MUSCL scheme, which uses linear interpolation inside the cells for second order accuracy. To reduce oscillations that arise in the vicinity of strong gradients, we used a variety of slope limiting methods.

1.2. *WENO scheme*

Due to high numerical cost and complexity of the MUSTA algorithm the finite difference WENO scheme was also implemented [3, 4, 5, 6, 7]. It uses polynomial interpolations from multiple stencils, a convex combination of which, with suitably non-linear coefficients, lead to a scheme that stays high order in smooth regions of the solution and discards stencils containing shocks, thus avoiding spurious oscillations.

In our program the fifth (WENO5) and seventh (WENO7) order accurate schemes were implemented.

2. Technical aspects

The code is written using the NVIDIA CUDA programming framework, which produces binaries for NVIDIA graphics processing units. We compared our implementation of the MUSTA-FORCE algorithm with an equivalent implementation for the central processing unit. Using contemporary hardware the GPU code executed approximately 200 times faster.

3. Results

We found that MUSTA-FORCE algorithm is more computationally expensive and gives less accurate description of shock waves compared to WENO algorithms. Thus only results obtained with WENO schemes are presented.

Figures 1 and 2 show a comparison of the WENO5 scheme simulation with an analytical solution to the shock tube problem. The results from WENO7 reproduce shock front better at the cost of slight oscillations in the flat regions.

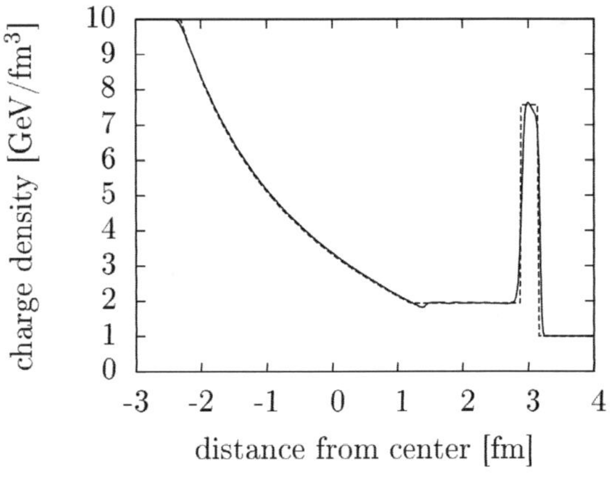

Figure 1. Sod shock tube, charge density.

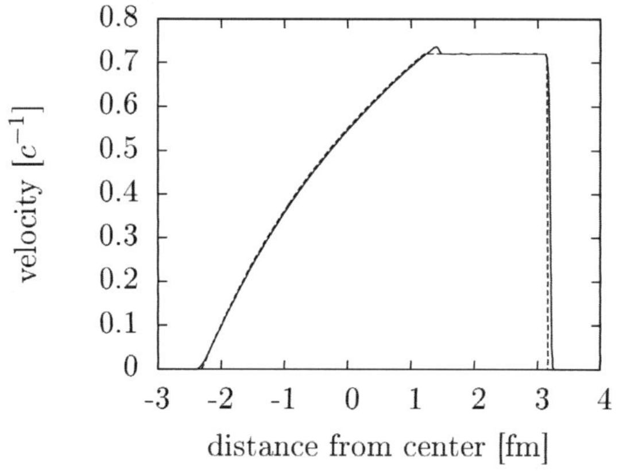

Figure 2. Sod shock tube, velocity.

The following two simulations of ellipsoidal flow solution [8] and Hubble-like solution were done in full 3+1D. On figures 3–6 one dimensional sections are shown. Both WENO5 and WENO7

Journal of Physics: Conference Series **509** (2014) 012059 doi:10.1088/1742-6596/509/1/012059

shown similiar results in these tests—higher order simulations were slightly more accurate in both cases.

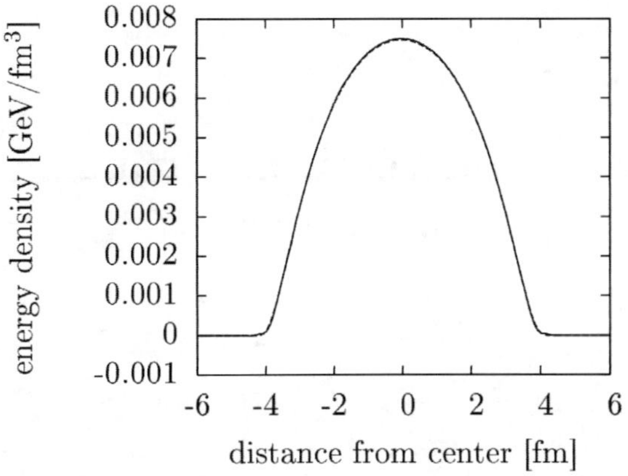

Figure 3. Ellipsoidal flow, energy density (WENO7).

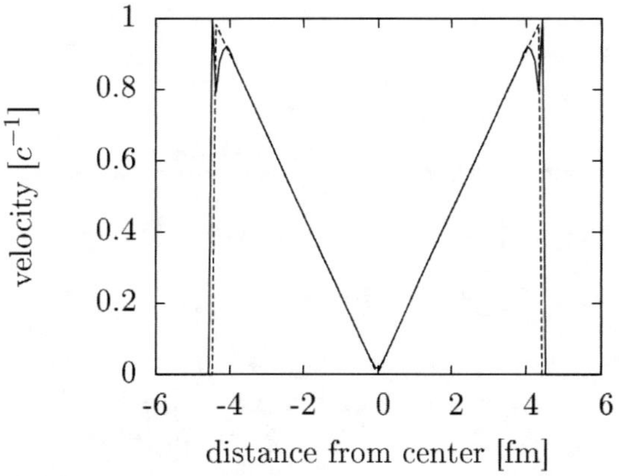

Figure 4. Ellipsoidal flow, velocity (WENO7).

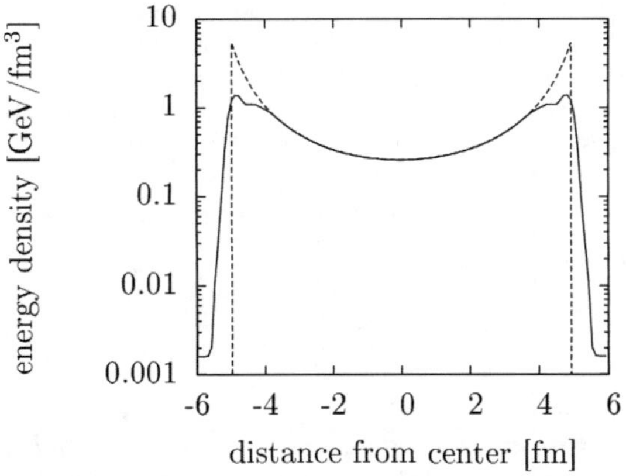

Figure 5. Hubble-like flow, energy density (WENO5).

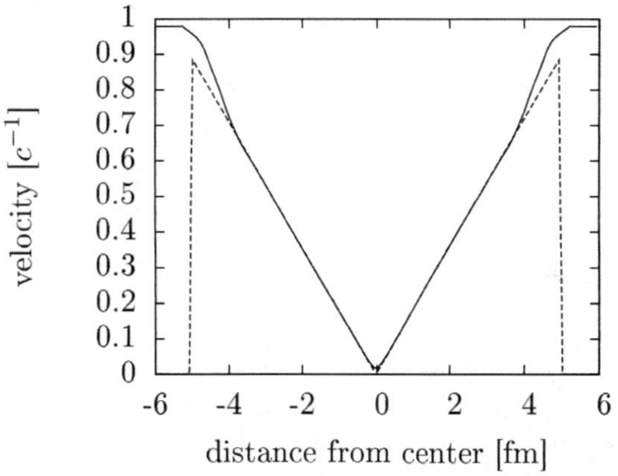

Figure 6. Hubble-like flow, velocity (WENO5).

References

[1] Toro E F 2006 *Appl. Numer. Math.* **56** 1464–1479 ISSN 0168-9274 URL `http://dx.doi.org/10.1016/j.apnum.2006.03.022`

[2] Titarev V A and Toro E F 2005 *International Journal for Numerical Methods in Fluids* **49** 117–147 ISSN 1097-0363 URL `http://dx.doi.org/10.1002/fld.980`

[3] Shen Y and Zha G 2008 (American Institute of Aeronautics and Astronautics) chap A Robust Seventh-Order WENO Scheme and ITS Applications Aerospace Sciences Meetings 0 URL `http://dx.doi.org/10.2514/6.2008-757`

[4] Balsara D and Shu C W 2000 *Journal of Computational Physics* **160** 405–452 ISSN 00219991 URL `http://dx.doi.org/10.1006/jcph.2000.6443`

[5] Shu C W 2001 *International Journal of Computational Fluid Dynamics* **17** 107–118

[6] Jiang G S, Shu C W and L I 1995 *J. Comput. Phys* **126** 202–228

[7] Qiu J and Shu C W 2002 *SIAM J. Sci. Comput.* **24** 2185–2198 ISSN 1064-8275 URL `http://dx.doi.org/10.1137/S1064827502412504`

[8] Sinyukov Y M and Karpenko I A 2005 (*Preprint* `nucl-th/0505041`)

Journal of Physics: Conference Series **509** (2014) 012060 doi:10.1088/1742-6596/509/1/012060

Measurements of the correlation between electrons from heavy-flavour hadron decays and light hadrons with ALICE at the LHC

E. Pereira De Oliveira Filho on behalf of the ALICE collaboration

Universidade de São Paulo, Departamento de Física Nuclear
Travessa R da rua do Matão 187
05508900, São Paulo, Brasil

E-mail: epereira@cern.ch

Abstract. In relativistic heavy-ion physics two-particle correlations provide a very useful tool to investigate the Quark-Gluon Plasma (QGP). This observable is sensitive to several of the properties of the QGP such as resonances, interaction of partons with the medium and collective effects (e. g. elliptic flow). In the present work, the correlation function between electrons from heavy-flavour hadron decays and light hadrons was measured in pp and Pb-Pb collisions (central and semi-central). Furthermore, in pp collisions the relative beauty contribution to the total cross section of electrons from heavy-flavour decays was estimated by comparing the measured correlation with Monte-Carlo templates.

1. Introduction

In relativistic heavy-ion collisions, two-particle correlation is a powerful tool to investigate the properties of the Quark-Gluon Plasma (QGP). This observable is sensitive to the near and away side jet structure, resonances and collective effects [1, 2, 3]. Which effect dominates the correlation is strongly determined by the kinematic region of the investigated particles, as well as the event class as discussed in [3].

Heavy quarks (charm and beauty) are of particular importance in this context. They are well-suited probes of the QGP, since they are produced during the initial stages of the collisions through hard parton-parton scattering processes. Subsequently, they experience the entire evolution of the system and they might participate in the collective expansion of the medium. One of the common approaches in the study of heavy quarks is the measurement of electrons from the decay of hadrons containing such particles. This is the technique exploited in the present work.

In order to address the QGP properties, a given observable is often studied as a function of the event class (centrality) of the collision and in various collision systems. The correlation function of electrons from heavy-flavour decays and light hadrons was measured with ALICE in pp and central and semi-central Pb-Pb collisions. Furthermore, for the case of pp collisions, the relative beauty contribution to the total cross section of electrons from heavy-flavour decays was estimated by comparing the measured correlation function with Monte-Carlo templates.

2. Electron Identification and Background Reconstruction

At the LHC, ALICE is dedicated to the study of heavy-ion collisions [4]. ALICE has excellent capabilities for heavy-flavour measurements thanks to several subsystems which allow vertexing and track reconstruction with high resolution and particle identification (PID) in a wide momentum range [4]. The Time Projection Chamber (TPC) and the ElectroMagnetic Calorimeter (EMCal) are the main PID detectors used in the present analysis. Fig. 1 (left) illustrates how electron identification can be performed using TPC and EMCAL (combined) information. The electron band can be seen around $E/p = 1$ and TPC N$\sigma = 0$, where E is the energy measured in the EMCal and p the track momentum. The TPC Nσ is defined as the measured specific energy loss (dE/dx) relative to the expected value, for a given mass assumption (the electron mass in this case).

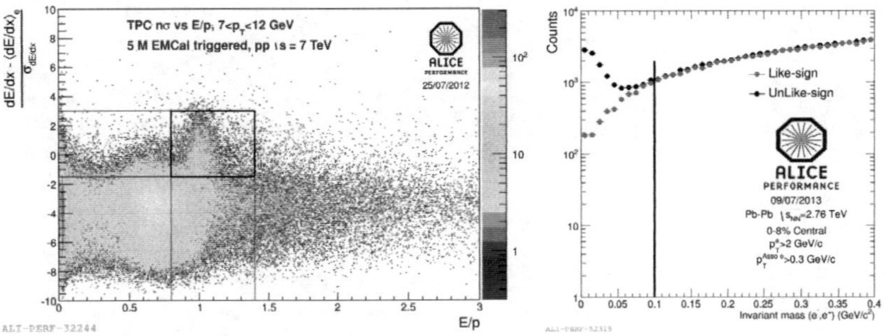

Figure 1: Electron identification using TPC and EMCal information (left panel). Invariant Mass of $e^{\pm}e^{\mp}$ (black points) and $e^{\pm}e^{\pm}$ (red points) pairs, used to reconstruct electrons from gamma conversions in the detector material and from light meson Dalitz decays (right panel).

After the identification, electrons not coming from heavy-flavour hadron decays should be removed from the sample. This background is mainly composed by electrons from photon conversions in the detector material and light meson (e. g. π^0 and η mesons) Dalitz decays. This contribution can be statistically estimated through the invariant mass of $e^{\pm}e^{\mp}$ pairs where the combinatorial background is determined using like-sign pairs ($e^{\pm}e^{\pm}$). Fig. 1 (right) shows the background reconstruction in Pb-Pb collisions.

3. Azimuthal correlation analysis and results

The azimuthal ($\Delta\phi$) correlation function between a trigger and an associated particle is defined in Eq. 1 [1]. For the case of electrons from heavy-flavour decays, the background is removed as discussed in Section 2. The remaining hadron contamination in the electron sample is removed by scaling down the di-hadron correlation distribution and subtracting it from the one of inclusive electrons. The final correlation function is obtained after correcting for the tracking efficiency and for the Mixed Event Correlation, which accounts for acceptance effects [1].

$$C_{t-a}(\Delta\phi) = \left(\frac{dN_{t-a}}{N_{t-a}d(\Delta\phi)}\right)^{Same-Event} \Bigg/ \left(\frac{dN_{t-a}}{N_{t-a}d(\Delta\phi)}\right)^{Mixed-Event} \tag{1}$$

Fig. 2 shows the correlation function of electrons from heavy-flavour hadrons decay and light hadrons for the kinematic range defined by $6 < p_{\mathrm{T}}^{e} < 8$ GeV/c and $8 < p_{\mathrm{T}}^{e} < 10$ GeV/c with $4 < p_{\mathrm{T}}^{h} < 6$ GeV/c, for central and semi-central Pb-Pb collisions at $\sqrt{s_{\mathrm{NN}}} = 2.76$ TeV, as well as for pp collisions at $\sqrt{s} = 7$ TeV. Fig. 3 (left) shows the modification I_{AA} of the near-side yield in heavy-ion collisions relative to pp collisions defined as:

$$I_{\mathrm{AA}}(p_{\mathrm{T}}^{e}, p_{\mathrm{T}}^{h}) = \left(\int_{-\frac{\pi}{2}}^{\frac{\pi}{2}} C(\Delta\phi; p_{\mathrm{T}}^{e}, p_{\mathrm{T}}^{h}) d(\Delta\phi)\right)^{Pb-Pb} \Bigg/ \left(\int_{-\frac{\pi}{2}}^{\frac{\pi}{2}} C(\Delta\phi; p_{\mathrm{T}}^{e}, p_{\mathrm{T}}^{h}) d(\Delta\phi)\right)^{pp} \tag{2}$$

The relative beauty contribution to the total cross section of electrons from heavy-flavour decays is shown in Fig. 3 (right), for pp collisions at $\sqrt{s} = 2.76$ TeV and 7 TeV, obtained by fitting the near-side correlation to Monte-Carlo templates obtained with PYTHIA [5].

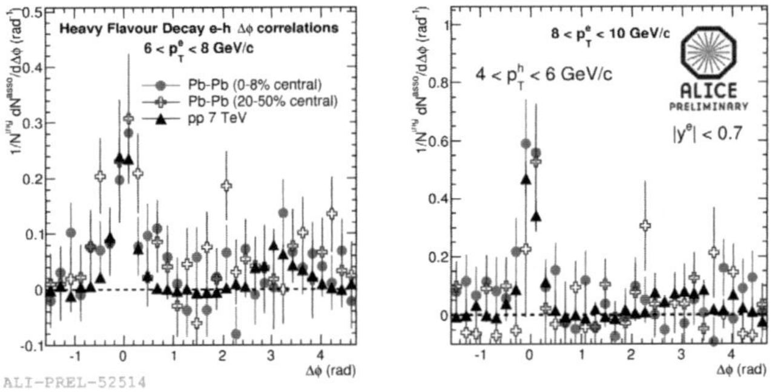

Figure 2: Correlation function between electrons from heavy-flavour decays and light hadrons in pp ($\sqrt{s} = 7$ TeV) and Pb-Pb ($\sqrt{s_{\mathrm{NN}}} = 2.76$ TeV) collisions.

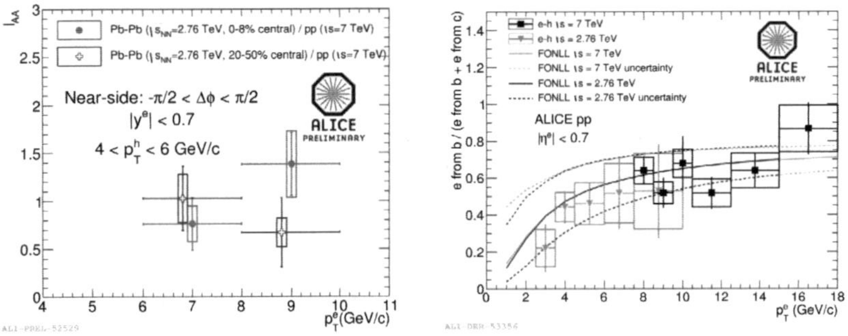

Figure 3: Yield modification in Pb-Pb relative to pp collisions (left panel). Relative beauty contribution to the total cross section of electrons from heavy-flavour decays, as a function of p_{T}, in pp collisions (right panel).

4. Conclusions

The measured I_{AA} is consistent with unity within statistical and systematic uncertainties. Indeed, the small statistics available does not allow stronger conclusions, as well as it does not exclude a difference in the correlation between heavy-ion and pp collisions.
The relative beauty contribution to the cross section of electrons from heavy-flavour decays is consistent with FONLL pQCD calculations [6] within the uncertainties for both collision energies of the present analysis.

Acknowledgments

The author thanks all his colleagues from the ALICE Collaboration, University of São Paulo and Utrecht University, for every help, advice and very useful discussions. He also thanks the "Conselho Nacional de Desenvolvimento Científico e Tecnológico - CNPq", for the financial support.

References

[1] B. Abelev, et al. (ALICE Collaboration), Phys. Lett. B 719 (2013) 29-41.
[2] B. Abelev, et al. (ALICE Collaboration), Phys. Rev. Lett. 108, 092301 (2012).
[3] B. Abelev, et al. (ALICE Collaboration), Phys. Lett. B 708 (2012) 249-264.
[4] B. Abelev, et al. (ALICE Collaboration), J. Instrum. 3, S08002 (2008).
[5] T. Sjostrand, S. Mrenna, Comput. Phys. Commun. 178 (2008) 852-867.
[6] M. Cacciari, S. Frixione, N. Houdeau, M. L. Mangano, P. Nason, et al., JHEP 1210 (2012) 137.

Journal of Physics: Conference Series **509** (2014) 012061 doi:10.1088/1742-6596/509/1/012061

Study of b-jet tagging performance in ALICE

Linus Feldkamp, for the ALICE Collaboration
Westfälische Wilhelms-Universität Münster, Institut für Kernphysik, Wilhelm-Klemm Straße 9, 48149 Münster, Germany
E-mail: l.feldkamp@cern.ch

Abstract. We present the current status and Monte Carlo study based performance estimates of b-jet tagging using ALICE, as obtained using both impact parameter as well as secondary vertex methods. We also address the prospects of the identification of electrons from heavy-flavour hadron decays to obtain b-jet enhanced jet samples.

1. Introduction

Charm and beauty quarks, produced in the early stage of heavy-ion collisions, are ideal probes to study the characteristics of the hot and dense deconfined medium (Quark-Gluon Plasma) formed in these collisions. The radiative energy loss of high energy partons interacting with the medium is expected [1][2] to be larger for gluons than for quarks, and to depend on the quark mass, with beauty quarks losing less energy than charm quarks, light quarks and gluons. Therefore a comparison of the modification in the momentum distribution or possibly in the jet shape of b-jets with that of light flavour or c-jets in Pb-Pb collisions relative to pp collisions allows one to investigate the mass dependence of the energy loss. It also allows one to study the redistribution of the lost energy and possible modifications to b-quark fragmentation in the medium. We present a MC study of the b-jet tagging performance in ALICE for pp collisions at $\sqrt{s} = 7$ TeV using two different algorithms, both exploiting the long lifetime and high mass of B mesons to discriminate b-jets from jets from lighter partons.

1.1. Monte Carlo sample and jet reconstruction

In this analysis two sets of Pythia 6 [3] (+GEANT 3 [4]) simulations of proton-proton collisions at $\sqrt{s} = 7$ TeV were used: 1.5×10^7 events with a c-$\bar{\text{c}}$ or b-$\bar{\text{b}}$ pair and 8.0×10^7 minimum bias events. Jets were reconstructed using FastJet [5] v2.4.2 with the anti-k_{T} algorithm [6], a anti-k_{T} cone radius of $R = 0.4$ and a minimum track p_T of 150 MeV/c. The jet reconstruction was done both on reconstructed tracks and at generator level, matched by the minimal $\Delta R = \sqrt{(\eta_{Jet,Gen.} - \eta_{Jet,Rec.})^2 + (\varphi_{Jet,Gen.} - \varphi_{Jet,Rec.})^2}$. At generator level neutral particles were considered to correct to the full jet energy. Jets with $p_{\mathrm{T}} > 10$ GeV/c and a charged area, given by FastJet, larger than $0.6\pi R^2$ were taken into account for further analysis steps. The jet flavour is determined by the flavour of the leading p_{T} parton in the cone of the matched generator level jet. In case it is a gluon, the daughters are also considered to take gluon splitting into account.

1.2. Track counting algorithm

The *track counting algorithm* [7] exploits the large $r\varphi$-impact parameters, $d_0 = |\vec{d_0}|$, of B-meson decay products to identify b-jets. The signed $r\varphi$-impact parameter, $d_0 = sign(\vec{d_0} \cdot \vec{p}^{Jet})d_0$, is calculated for each track in the jet cone, where $\vec{d_0}$ is pointing away from the primary vertex. The tracks are required to have a $p_T > 1$ GeV/c, a hit in the innermost layer of the ITS and a Distance of Closest Approach between track and jet smaller than 0.07 cm. To tag a jet as a b-jet, the d_0 parameter of each track in the jet is ordered by size, and the N-th largest d_0 is compared to check if it is larger than a threshold value a. In the analysis presented here N=3 and the threshold value a was varied in such was to keep the same b-tagging efficiency for different jet p_T bins. The tagging efficiency for a jet assigned to a given parton, c, b, LFg=(u,d,s,g), can be defined as:

$$\epsilon_i(\mathrm{p}_\mathrm{T}^{Jet}) = \frac{N_i^{tagged}(\mathrm{p}_\mathrm{T}^{\mathrm{Jet}})}{N_i(\mathrm{p}_\mathrm{T}^{\mathrm{Jet}})} \quad \text{with } i = \text{c,b, LFg(=u,d,s,g)} \tag{1}$$

Tagging efficiencies for c- and b-jets and the ratio ϵ_c/ϵ_b (at constant ϵ_b) for the N=3 discriminator were obtained as a function of the transverse momentum of the jet at kinematic level, including both the p_T contributions of charged and neutral final state particles. To allow a comparison of the efficiencies of charm-, beauty- and LFg-Jets the ratio $\frac{\epsilon_i}{\epsilon_b}|_{\epsilon_b(\mathrm{p_T})\text{fixed}}$ is used. In Figure 1 we show the results for the case $\epsilon_b(\mathrm{p_T}) = 0.1$, which corresponds to threshold a on the 3rd most displaced track of about 0.01 cm. It may be noted that a is varied with p_T in order to achieve a constant ϵ_b.

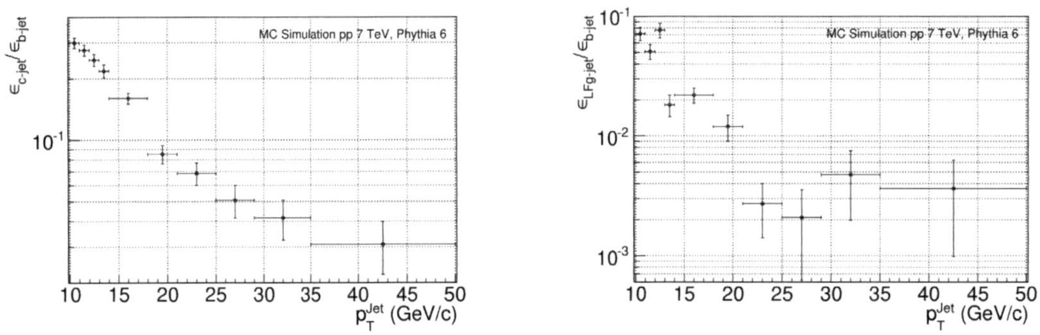

(a) (b)

Figure 1: c-jet vs. b-jet performance for N=3, **(a)**: $\epsilon_c(\mathrm{p_T})/\epsilon_b(\mathrm{p_T})$, **(b)**: $\epsilon_{LFg}(\mathrm{p_T})/\epsilon_b(\mathrm{p_T})$ at $\epsilon_b(\mathrm{p_T}) = 0.1$.

At $\epsilon_b = 0.1$ the efficiency ratio ϵ_c/ϵ_b, obtained from the b- and c-quark enhanced Monte Carlo sample, is decreasing between 20 GeV/c and 50 GeV/c from $\sim$ 0.08 to $\sim$ 0.03 for the N=3 discriminator.

The light-flavour and gluon jets have typically tracks with smaller impact parameters with respect to charm jets. Thus the $\epsilon_{LFg}/\epsilon_b$ efficiency ratio, obtained from the minimum bias sample, is approximately one order of magnitude smaller, below 0.005 above 20 GeV/c.

1.3. Secondary vertex reconstruction

In the *secondary vertex algorithm* [8] secondary vertices are reconstructed combining 2 or 3 jet tracks, approximating the tracks as straight lines in the vicinity of the primary vertex. The presence of vertices with a displaced topology and with a large invariant mass is used as a criterion to tag jets from b. In particular, the discriminator variables used are the vertex invariant mass and the decay length in the transverse plane, $L_{xy} = L_{xy} \cdot sign(\vec{L}_{xy} \cdot \vec{\mathrm{p}}_{\mathrm{T,Jet}})$.

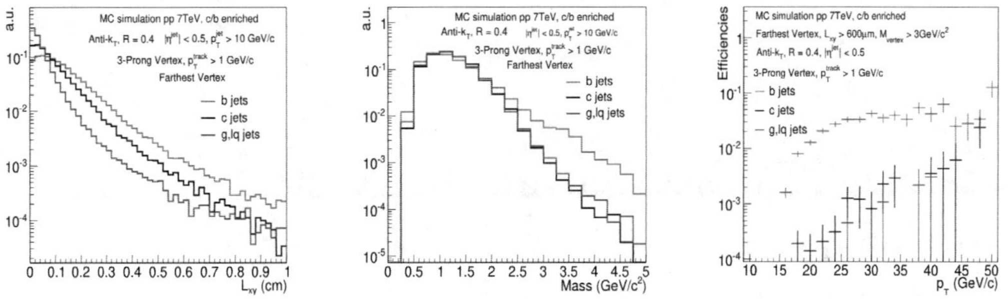

(a) (b) (c)

Figure 2: (**a**): $L_{xy} = L_{xy} \cdot sign(\vec{L}_{xy} \cdot \vec{p}_{\mathrm{T,Jet}})$ using the most displaced vertex in the jet, (**b**): Vertex invariant mass for the most displaced vertex in the jet, (**c**): Tagging efficiencies: $L_{xy} > 600$ μm, $M_{\mathrm{Vertex}} > 3$ GeV/c^2

L_{xy} and the vertex invariant mass distributions are shown in Figure 2a and 2b. In Figure 2c the tagging efficiency for a given combination of cut values is shown exemplarily.

1.4. Electron Identification

The performance of b-jet tagging may further improve by using electron triggered events, as semi-electronic decays of heavy-flavour hadrons have branching ratios of the order of 10% [9]. The inclusive electron spectrum contains a significant contribution from heavy-flavour hadron decays above momenta of a few GeV/c with the beauty-decay contribution becoming dominant above ~ 5 GeV/c [10]. Thus using both the TPC and the electromagnetic calorimeter (EMCal) for electron identification in the analysis and the high-p_{T} EMCal triggers a significant increase in b-jet statistics may be achieved by applying a TPC-dE/dx cut of $]-1\sigma, 3\sigma[$ around the expected electron dE/dx, with σ corresponding to the gaussian dE/dx-resolution and requiring a EMCal cluster energy E over track momentum p ratio of $0.8 < E/p < 1.2$.

2. Conclusions

Flavour-dependent tagging efficiencies for the track counting and the secondary vertex algorithm have been estimated using simulated pp events at $\sqrt{s} = 7$ TeV. Beside further optimisation these efficiencies can be used as input to estimate the contamination of the raw tagged sample of b-jets using FONLL [11] or POWHEG [12]. High-p_T electron triggers promise a significant increase in statistics. A measurement of the b-jet spectrum in pp and p-Pb collisions down to ≈ 20 GeV/c and of the b-jet spectrum in Pb-Pb collisions down to $30 - 40$ GeV/c might become feasible in the future.

References

[1] Salgado C A, Wiedemann U A (2003) (*Preprint*: hep-ph/0302184)
[2] Dokshitzer Yu L, Kharzeev D E (2001)*Phys. Lett.* **B519** 199-206
[3] Sjostrand T, Mrenna S, Skands P (2006) *JHEP* **0605** 026
[4] Brun, R. et al. (1987) CERN-DD/EE/84-1
[5] Cacciari M, Salam G P, Soyez G (2011) (*Preprint*: arXiv:1111.6097 [hep-ph])
[6] Cacciari M, Salam G P, Gregory G (2008) *JHEP* **0804** 063
[7] Rizzi A, Palla F, Segneri G (2006) CERN-CMS-NOTE-2006-019.
[8] Aad G et al. (2009) ATL-PHYS-PUB-2009-018
[9] Beringer J et al. (2012) *Phys. Rev.* **D86**, 010001
[10] Abelev B et al. (2013) *Phys. Lett.* **B721** *Issues* 1-3 *Pages* 13-23
[11] Cacciari M, Greco M, Nason P (1998) *JHEP* **9805** 007
[12] Frixione S, Nason P, Ridolfi G (2007) *J. High Energy Phys.* **0709** 126

Journal of Physics: Conference Series **509** (2014) 012062 doi:10.1088/1742-6596/509/1/012062

Elliptic flow of heavy-flavour decay electrons in Pb-Pb collision at $\sqrt{s_{\rm NN}} = 2.76$ TeV with ALICE

Andrea Dubla for the ALICE Collaboration

Department of Physics, Utrecht University, Princetonplein 5, 3584 CC Utrecht, The Netherlands

E-mail: andrea.dubla@cern.ch

Abstract. We present the elliptic flow v_2 of electrons from the semi-leptonic decays of heavy-flavour hadrons at mid-rapidity for semi-central (20-40% centrality) Pb-Pb collisions at $\sqrt{s_{\rm NN}} =$ 2.76 TeV measured with the ALICE detector. A positive v_2 is observed at low $p_{\rm T}$ in semi-central collisions. Comparisons with previous measurements obtained at lower collision energy at RHIC in Au-Au collisions at $\sqrt{s_{\rm NN}} = 0.2$ TeV and with theoretical models are also presented.

1. Introduction

Heavy-quarks, i.e. charm and beauty, are mainly produced in hard scattering processes in the early stages of high energy nucleus-nucleus collisions and they experience the full evolution of the medium formed in such collisions, the Quark-Gluon Plasma (QGP). Therefore, heavy-flavour hadrons are well suited probes to investigate the properties of the medium. Due to their large masses, charm and beauty quarks are expected to lose less energy than light quarks and gluons when traversing the medium due to the dead-cone effect [1] and also to have a longer relaxation time [2]. In this contribution we present the measurement of elliptic flow v_2 of electrons from the semi-leptonic decays of heavy-flavour hadrons in semi-central (20-40% centrality) Pb-Pb collisions at $\sqrt{s_{\rm NN}} = 2.76$ TeV with ALICE. The elliptic flow is the second Fourier coefficient of the azimuthal distribution of particle momenta in the transverse plane with respect to the reaction plane. At low $p_{\rm T}$, the v_2 of the electrons from heavy-flavour hadron decays is sensitive to the degree of thermalization of charm and beauty quarks in the deconfined medium. At higher $p_{\rm T}$, the measurement of v_2 carries information on the path length dependence of in-medium parton energy loss. The analysis was carried out with different methods that have different sensitivity to non-flow contributions, namely event plane [3], scalar product [4] and Q-Cumulants [5]. We compare the experimental result to several theoretical models and with measurements from Au-Au collisions at $\sqrt{s_{\rm NN}} = 0.2$ TeV at RHIC [7].

2. Data analysis

The data analyzed were recorded with the ALICE detector [8] during the 2010 and 2011 LHC runs with Pb-Pb collisions at $\sqrt{s_{\rm NN}} = 2.76$ TeV. The VZERO detector provides the MB trigger, centrality determination, which is used online to enhance the sample of central and semi-central events, and event plane determination. In addition to the sample from the semi-central trigger (8×10^6 events in the centrality range 20-40%), a sample of 1.3×10^6 events in 20-40% centrality triggered by the Electro-Magnetic Calorimeter (EMCal) is used. Electron candidates are selected and identified using different approaches. In the low $p_{\rm T}$ region ($p_{\rm T} < 6$ GeV/c) information from the TPC (Time Projection Chamber) and the TOF (Time Of Flight) detectors is used, while the EMCal contributes to electron identification for $1.5 < p_{\rm T} < 13$ GeV/c. Particle identification in the TPC is based on the measurement of the specific energy loss dE/dx in the detector gas, while in the TOF the charged particles are separated via the measurement of their time of flight from the interaction point to the detector. Electron identification in the EMCal is based on the measurement of the E/p ratio, where E is the energy of the EMCal cluster that was matched to the prolongation of the track with momentum p reconstructed in the TPC and ITS detectors. Electrons deposit their total energy in the EMCal and due to their small mass the E/p ratio should be equal to unity.

The sample of identified electron candidates is mainly composed of two distinct sources: electrons from the semi-leptonic decays of hadrons containing a charm or beauty quark, and photonic electrons, originating mainly from Dalitz decays of π^0 and η mesons and from photon conversion. In order to extract the heavy-flavour signal, the background contributions have to be subtracted from the inclusive electron spectra and v_2. In this analysis this task is accomplished using two different methods. In the first method a cocktail of electron spectra from background sources is calculated using a Monte Carlo event generator of hadron decays [6]. In the second method the electron background is reconstructed using the invariant mass method. The latter makes use of the fact that an electron from a photonic source is always accompanied by a positron, forming an electron-positron pair with small invariant mass. In the current analysis pairs with a mass below 70 MeV/c^2 are tagged as photonic.

The heavy-flavour electron $v_2^{\rm HFe}$ was then obtained as:

$$v_2^{\rm HFe} = \frac{(1+R_{\rm SB})v_2^e - v_2^{\gamma e}}{R_{\rm SB}} \qquad (1)$$

where v_2^e is the inclusive electron v_2, $(1 + R_{\rm SB})$ is the ratio of inclusive electrons to the estimated background electrons and $v_2^{\gamma e}$ is the v_2 coefficient for the background sources estimated from the cocktail. The $(1 + R_{\rm SB})$ was estimated with both methods, cocktail and invariant mass method as a function of $p_{\rm T}$. The ratio of the $p_{\rm T}$ spectra comparing the two different methods is shown in in Fig. 1[1] and a good agreement is observed within statistical and systematic uncertainties. The $v_2^{\gamma e}$ estimated for the background with the cocktail is reported in Fig. 2.

3. Results and conclusions

The transverse momentum dependence of v_2 of heavy-flavour decay electrons is shown in Fig. 3[1] for the different methods used in the analysis, which are found to be consistent within the uncertainties. We observe positive v_2 at low $p_{\rm T}$ in semi-central collisions. The positive v_2 has a contribution from heavy-quark collective flow at low $p_{\rm T}$ and may be due to the path-length dependence of the energy loss mechanism. In Fig. 4 the experimental results are compared with measurements by PHENIX in Au-Au collisions at RHIC at $\sqrt{s_{\rm NN}} = 0.2$ TeV [7] and with various transport model calculations. The v_2 measured at the LHC is similar in magnitude to that measured by PHENIX at lower collision energy. The model calculations shown in Fig. 4 incorporate parton transport based on the Boltzmann approach that includes the coupling of the quarks with an expanding medium via collisional and radiative processes (BAMPS) [9]; heavy-quarks transport with in-medium resonance scattering and coalescence (Rapp et. al) [10]; and heavy-quarks transport (Langevin equation) with collisional energy loss (POWLANG) [11].

[1] Fig. 1 and Fig. 3 are updated versions of the plots shown in the conference. The v_2 results in the transverse momentum intervals $8 < p_{\rm T} < 10$ GeV/c and $10 < p_{\rm T} < 13$ GeV/c were added for the second order cumulants and for the scalar product methods.

The BAMPS calculation is in good agreement with the v_2 measurement while the calculations by Rapp et al. and POWLANG underestimate the v_2.

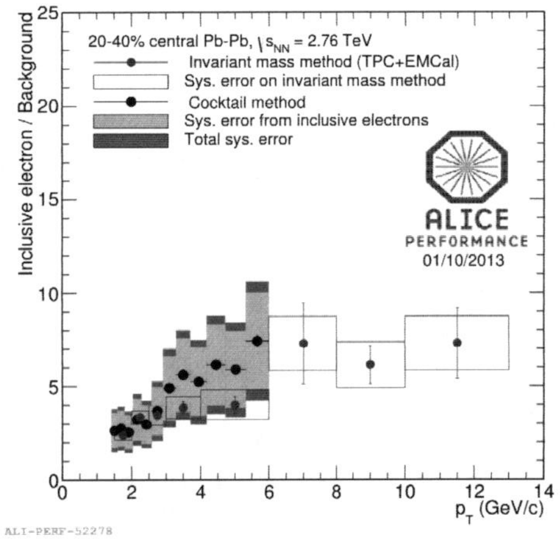

Figure 1. Inclusive over background electron ratio estimated with the cocktail method (black marker) and with the invariant mass method (blue marker).

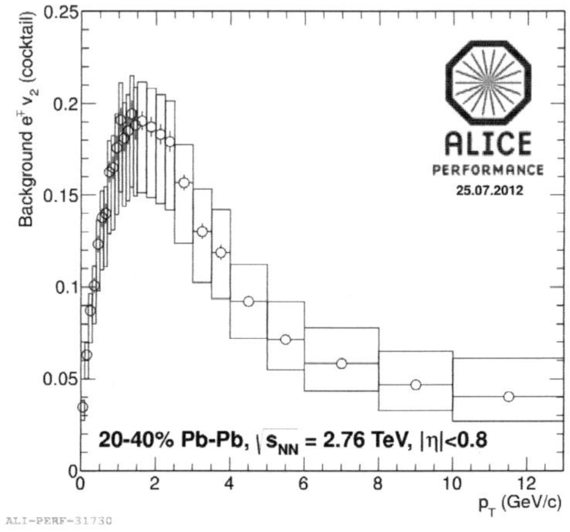

Figure 2. Background electron v_2 estimated with the cocktail method for Pb-Pb collisions in the 20-40% centrality range.

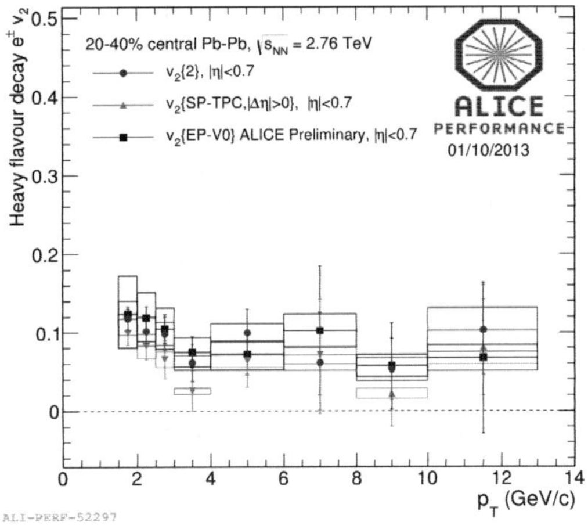

Figure 3. Comparison of heavy flavor decay electron v_2 extracted with the second order cumulants (blue marker), scalar product (red marker) and event plane (black marker) methods.

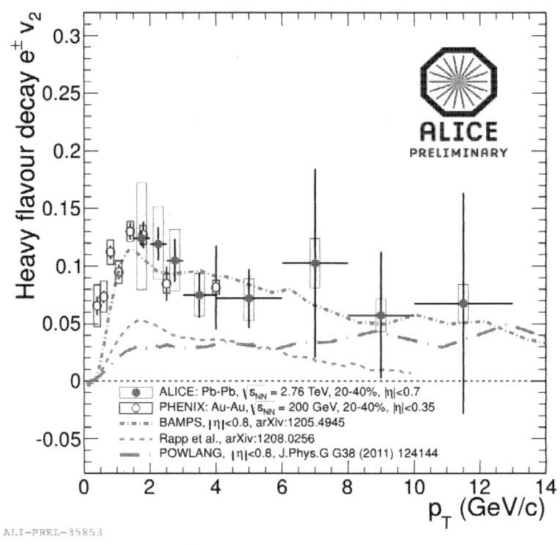

Figure 4. Heavy-flavour decay electrons v_2 measured with the event plane method in 20-40% Pb-Pb collisions at $\sqrt{s_{\rm NN}} = 2.76$ TeV (red marker) compared to different models specific for LHC and with PHENIX results in Au-Au collisions at $\sqrt{s_{\rm NN}} = 0.2$ TeV (blue marker).

4. References

[1] Dokshitzer Y L and Kharzeev D E arXiv:hep-ph/0106202

[2] Moore G D and Teaney D *Phys. Rev.* C **71** (2005) 064904
[3] Poskanzer A M and Voloshin S A *Phys. Rev.* C **58**, 1671
[4] C. Adler et al. [STAR Coll.] *Phys. Rev.* C **66**, 03 4904 (2002)
[5] Bilandzic A, Snellings R, Voloshin S arxiv.org/abs/1010.0233
[6] B. Abelev et al. [ALICE Coll.] arXiv:1205.5423
[7] A.Adare et al. [PHENIXColl.] *Phys.Rev.* C **84**, 04 4905 (2011)
[8] K. Aamodt et al. [ALICE Coll.] *JINST* 3, 2008
[9] Uphoff J, Fochler O, Xu Z and Greiner C arXiv:1205.4945
[10] He M, Fries R J and Rapp R arXiv:1106.6006v1
[11] Monteno M et al. *J. Phys.* G **38** 124144

Charm decay as a source of multistrange hadrons

M Petráň[1], J Letessier[2], V Petráček[3] and J Rafelski[1]

[1] Department of Physics, University of Arizona, Tucson, AZ 85719, USA
[2] Laboratoire de Physique Théorique et Hautes Energies, Université Paris 6, F-75005
[3] Czech Technical University in Prague, Brehova 7, 115 19 Praha 1, Czech Republic

Abstract. We describe a newly formulated approach to account for charm production and decay in Statistical Hadronization approach. Considering Pb–Pb collisions at $\sqrt{s_{NN}} = 2.76$ TeV at LHC we show that charm hadron decays can be a significant contributor to the multistrange hadron abundance. We discuss the magnitude of expected effects as a function of charm yield.

1. Charm Production and Decay in Heavy Ion Collisions

We demonstrate that in a precise analysis of hadron abundances within, e.g., the statistical hadronization model (SHM), it is necessary at LHC to allow for contribution to hadron yields from charm decays. In order to accomplish this we need to know: 1) total charm quark yield in QGP at hadronization; 2) charm hadronization pattern into hadrons; and 3) charm hadron decay distribution.

Initial state charm production can be estimated by folding the *pp* cross section with the number of initial nucleon-nucleon collisions. To be specific, we recall that in central Au–Au collisions at RHIC, the total charm cross–section has been measured to be $d\sigma_{c\bar{c}}/dy|_{y=0} = 175\pm12\pm23\mu b$, and shown to scale with the number of binary collisions [1]. Thus, in most central 0%–5% centrality bin RHIC reactions, one finds $N_{cc}^{\rm RHIC} \equiv dN_{c+\bar{c}}/dy \simeq 4.3$ for $y \in (-0.5, 0.5)$ measured in the CM frame.

In Pb–Pb collisions at $\sqrt{s_{NN}} = 2.76$ TeV at LHC, theoretical considerations lead to $N_{cc} = 246\pm154$ (123 ± 77 pairs), where the error is due to uncertainty in charm production cross section [2]. It is believed that most of these charm pairs survive the parton thermalization and the sequel QGP evolution to the hadronization stage. However, if the prediction of such high abundance is true, one must expect some significant charm reannihilation. This can be checked in the near future within our kinetic flavor evolution model [3].

At QGP hadronization, all surviving charm hadronizes, producing heavy particles with hidden, open single charm, and multiple charm content. After traveling at most a few hundred μm, all these particles have decayed. For nearly all decay channels, it is practically impossible to experimentally distinguish which final state hadrons originate in a charm decay. Considering that, the hadron yields one measures contain input from post-hadronization charm decay. If all particles were fed by charm decay with a strength that is similar to the SHM model production, the contribution of charm would simply be a shift in overall normalization. The principal objective we pursue here is to show that this is not the case. We show that charm and strangeness are strongly correlated, and thus charm replaces some of strangeness production in QGP. Given N_{cc}, we predict yields of all charmed hadrons using the statistical hadronization method.

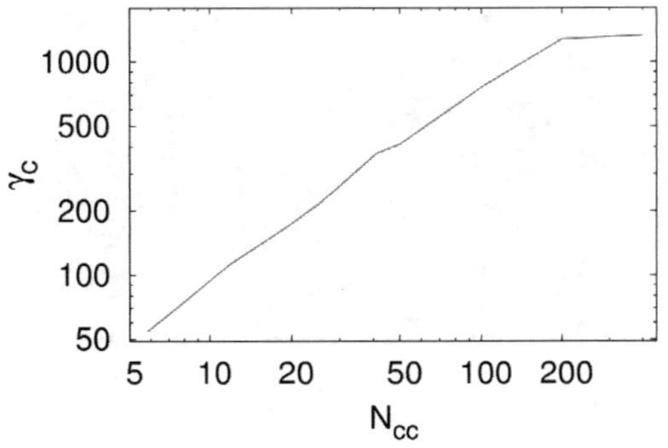

Figure 1. The charm phase space occupancy as a function of total charm present at hadronization.

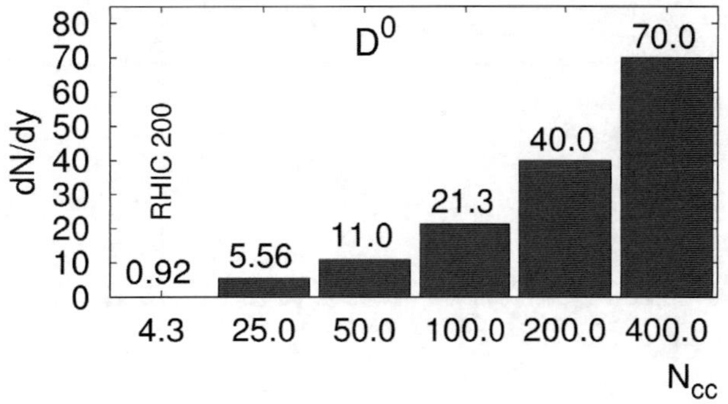

Figure 2. The D^0 meson yield as a function of total charm abundance N_{cc} at hadronization.

2. SHARE with CHARM

We have developed a new program predicting the contribution of charm hadron decays and allowing a fit with charm feed accounted into all hadron abundance. This program comprises a CHARM module which adds charm decay hadron multiplicity into SHARE, the statistical hadronization model implementation we use [4, 5]. The new SHARE with CHARM utility uses N_{cc} as an additional fit parameter when analyzing hadron production in heavy–ion collisions. SHARE with CHARM further allows the differentiation of the temperature of charm hadronization T_c from the temperature T of other hadron production. However, for reason of space we refrain here from showing results with this refinement.

SHARE with CHARM functions as follows: for a given N_{cc} in the first step we compute the distribution of charm among all charm hadrons following the principles of statistical hadronization at a prescribed temperature T_c. To describe the normalization of charm yield, the phase space occupancy γ_c has been introduced. Given N_{cc}, γ_c is solved for using methods described in [6]. Charm production is not thermal, but is due to the initial hard parton scattering. Therefore it is expected to be strongly out of chemical equilibrium. In Figure 1 we see the magnitude of γ_c as a function of N_{cc} for hadronization temperature $T_c = T$. We note that for $N_{cc} > 50$, the magnitude of $\gamma_c > 500$ implies that some of the charm produced will reannihilate. This effect may explain how the large initial predicted yield can be consistent with the more realistic yield obtained in direct experimental effort, as we now describe.

Figure 2 shows the expected dN_{D^0}/dy yield (number above each column) for a prescribed value of N_{cc}. Charmed hadrons in general show dependence on N_{cc} proportional to the number of constituent (anti-)charm quarks with small variations due to changes in statistical parameters such as T and γ_s arising when a fit is performed. The ALICE experiment at LHC has the capability of recognizing the charm decay in specific channels and thus can measure the charmed hadron yields. The D^0 meson $p_\perp$–spectrum has been presented [7] for the 0-20% centrality bin. Due to the uncertain low $p_\perp$ bin, the yield dN/dy cannot be reliably integrated. However, the other points allow us to estimate by extrapolation to $p_\perp = 0$ a yield $dN/dy|_{D^0} \in (1.3, 9.0)$.

Considering the results presented in Figure 2, this implies a 5 times larger total yield $N_{cc} \simeq 6 - 45$ charm and anti-charm quarks. Scaling from 0–20% to the 0–5% centrality by a factor 1.35 we find $N_{cc}^{\mathrm{LHC}} \in (10, 60)$ as the best direct experiment-based estimate, much less than the central point of the initial production evaluation $N_{cc} = 246 \pm 154$ [2]. Still, we choose to show expected yields for up to a value $N_{cc} = 400$, and the yield of D^0 mesons clearly is not compatible even with $N_{cc} = 100$.

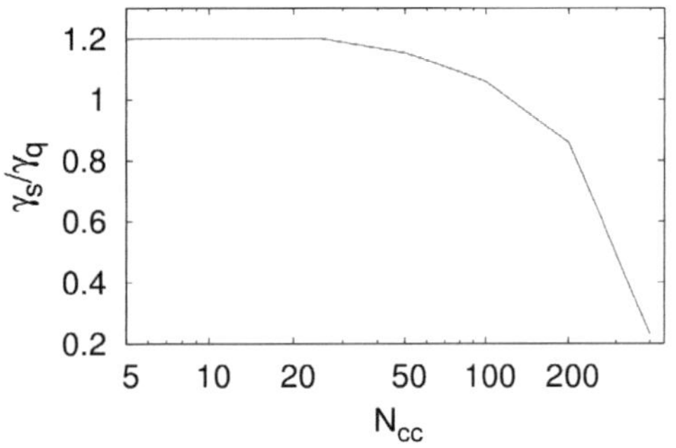

Figure 3. Strangeness over light quark phase space occupancy as a function of total charm present at hadronization.

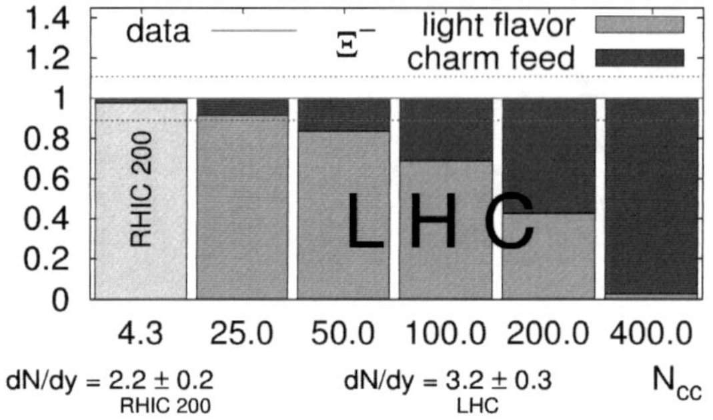

Figure 4. The relative contribution of charm decays to the yield of Ξ^- as a function of charm abundance, see text for details.

In the next SHARE with CHARM step, each of the produced charm hadrons is allowed to decay. Based on experimental decay data [8], symmetry principles, and plausibility arguments, we prepared a complete charmed hadron decay table. This task took much effort, required to study of many additional theoretical references, and further improvements will certainly be necessary. One must see our present effort as a first step to be as complete as the current understanding of charm hadron decays.

In the final SHARE with CHARM step, the produced hadrons and their resonances are added to the SHARE initial SHM yields obtained for a set of chosen SHM parameters. Hadron resonances are decayed; the total hadron yields are compared to the experimental data input field, and χ^2 computed. An improved set of statistical parameters is proposed by MINUIT routine embedded in SHARE. The cycle of computations is carried out until SHARE concludes that the best local minimum of χ^2 is achieved.

3. Charm decay feed

We quantify the charm hadron decay contributions in the final hadron yields using previous analysis of central Au–Au collisions at RHIC [9] and for LHC, we perform our study using the same input data as in our recent analysis of Pb–Pb collisions at LHC [10, 11]. In order to quantify the expected effect in the absence of a measured charm hadron yield at LHC we perform a parametric study prescribing a value of $N_{cc} \in (0, 400)$ while fitting the yields of observed hadrons. Charm feed–down of light hadrons generally has a different pattern than the light hadron production in QGP hadronization. Therefore, although the charm contribution may be hidden within the experimental errors of light hadrons, a very large abundance introduces a tension in the SHM fit of multiple hadron species with different strangeness content.

In Figure 3, we show the change of the statistical parameter γ_s/γ_q, characterizing the strangeness present at time of hadronization, resulting from the fit. For the value $N_{cc} = 0$, the results of our prior fit apply [10, 11]. There is a nominal change up to $N_{cc} < 50$. For $N_{cc} > 50$ the charm decay feed replaces in a significant way the strangeness abundance, and for $N_{cc} = 400$ strangeness in QGP at hadronization would be practically erased. Naturally, this is impossible, and it is clear that $N_{cc} > 150$ is an unpalatable result also inconsistent with the D^0 spectral results as we discussed.

During our study, we confirm a previous result [6], that charm decays contribute substantially to multistrange hadrons, ϕ, Ξ and Ω. In Figure 4, we show as an example the relative contribution to Ξ^- yield as a function of total charm abundance N_{cc} normalized to the experimental value indicated at the bottom for RHIC and LHC. The red dotted lines represent the experimental error, the blue part shows the contribution to Ξ^- yield from charm decays. One can see that at RHIC-200, the charm contribution presents an insignificant part of the total yield. However, at LHC-2760, the contribution to Ξ^- yield (and similarly other multistrange particles) exceeds the experimental error for $N_{cc} > 25$ (that is 13 or more pairs).

We further study the effect of charm on the thermal parameters of the SHM fit. Volume dV/dy, chemical freeze-out (hadronization) temperature T and light hadron phase space occupancy γ_q show a very slow, but steady decrease within their respective errors. As discussed, charm decays are a significant source of strangeness and hence the strange phase space occupancy γ_s decreases significantly upon the introduction of a relatively large charm yield. This is the cause why multistrange particles are affected the most by charm decays.

We investigated the effect of including the charm degree of freedom on the bulk properties of the particle source. We found that the physical bulk properties, such as energy density ε, entropy density σ, and pressure P remain unchanged. Therefore, the charm contribution to particle yields does not affect our previous findings about universality of hadronization condition [10].

4. Outlook

Using the new SHM program, SHARE with CHARM, we found that in RHIC Au–Au collisions at 200 GeV, the amount of charm has very little direct effect on hadron yields, whereas in LHC Pb–Pb collisions at 2.76 TeV, charm has an observable effect on hadron yields. We have shown that the present best estimate of the total $c+\bar{c}$ yield $N_{cc} \simeq 250 \pm 150$ seems too large compared to the reported D^0 spectrum. We have shown that $N_{cc} \simeq 100$ (50 pairs) will contribute nearly 3sd to the yield of Ξ and thus analysis of the hadron yields should include charm decays once such a yield is confirmed.

A precise yield of at least one of the charmed hadrons is necessary for an analysis of hadron yields with SHARE with CHARM to be possible. Without charmed hadrons we find that a fit of N_{cc} does not lead to a clear χ^2 minimum: we find a nearly totally flat χ^2 as a function of N_{cc}. This is of course good news in the sense that any acceptable in size measured charmed hadron yield will "work". If and when we have more than one charmed hadron yield measured directly, we can determine the difference in hadronization temperature $\delta T \equiv T_c - T$. With more than two charmed hadron yields available we begin to test the hypothesis that statistical hadronization applies to charmed hadron production. We noted that based on future experiment further refinements in our charm hadron decay tables are needed. To conclude, there is still much to do in preparation for the charm hadron era of relativistic heavy ion collisions.

Acknowledgments

This work has been supported by a grant from the U.S. Department of Energy, grant DE-FG02-04ER41318, Laboratoire de Physique Théorique et Hautes Energies, LPTHE, at University Paris 6 is supported by CNRS as Unité Mixte de Recherche, UMR7589.

References

[1] Tlusty D [STAR Collaboration] 2013 *Nucl. Phys.* A **904-905**, 639c
[2] Nelson R E, Vogt R and Frawley A D 2013 *Phys. Rev.* C **87**, 014908 and Vogt R, private communication
[3] Letessier J and Rafelski J 2007 *Phys. Rev.* C **75**, 014905
[4] Torrieri G *et al.* 2005 *Comput. Phys. Commun.* **167**, 229
[5] Torrieri G, Jeon S, Letessier J and Rafelski J 2006 *Comput. Phys. Commun.* **175**, 635.
[6] Kuznetsova I and Rafelski J 2007 *Eur. Phys. J.* C **51**, 113
[7] Abelev B *et al.* [ALICE Collaboration] 2012 JHEP **1209**, 112
[8] Beringer J *et al.* [Particle Data Group Collaboration] 2012 *Phys. Rev.* D **86**, 010001
[9] Rafelski J, Letessier J and Torrieri G 2005 *Phys. Rev.* C **72**, 024905
[10] Petran M and Rafelski J 2013 *Phys. Rev.* C **88**, 021901(R)
[11] Petran M, Letessier J, Petracek V and Rafelski J 2013 *Phys. Rev.* C **88**, 034907

Color reconnection and flow-like patterns in pp collisions.

I.A. Maldonado-Cervantes[1], E. Cuautle[1], Guy Paic.[1], A. Ortiz Velasquez[2] and Peter Christiansen.[2]

[1]Instituto de Ciencias Nucleares, Universidad Nacional Autónoma de México. Apartado Postal 70-543, México Distrito Federal 04510, México

[2]Lund, University, Department of Physics, Division of Particle Physics, Box 118, SE-221 00, Lund, Sweden.

E-mail: ivonne@nucleares.unam.mx

Abstract. Data collected at the LHC are confronted with the possible existence of flow in pp collisions. Here we present a study inside the framework of PYTHIA 8, showing that it contains implicit flow-like effects coming from multiple hard subcollisions and color string formation between initial and final partons from independent hard scatterings: color reconnection. We present studies with strange hadron observables in pp collisions at 7 TeV. Studies have been done both for minimum bias and multiplicity intervals in events with and without color reconnection to isolate the flow-like effect.

1. Introduction

The existence of something that is called flow has been taken as a proof of collective behavior of partons and hadrons. In central heavy ion collisions the transverse pressure gradient causes a transverse hydrodynamic expansion. This effect is transmitted to hadrons via a boost by the local velocity field. The transverse flow shifts the emitted particles to higher transverse momentum p_{T}, affecting the baryon to meson ratio as a function of p_{T} which exhibits an enhancement of the ratio and its position is pushed to higher momenta when one goes from peripheral to central Pb-Pb collisions [1]. In pp collisions at $\sqrt{7}$ TeV, the behavior of different ratios of particles could be related with flow. The ratios $(p^+ + p^-)/(\pi^+ + \pi^-)$ and $(\Lambda^0 + \bar{\Lambda}^0)/(2 \times K_s^0)$ exhibits a small enhancement, and it looks like there is no or little energy dependence [2] which is qualitatively well reproduced by PYTHIA version 8.17 [3] tune 4C [4], even though in general PYTHIA underestimates the production of strange particles [5].

The aim of this work is to show that the enhancement in the baryon to meson ratio is attributed to color reconnection, a mechanism present in PYTHIA 8.

2. Color reconnection

Color reconnection (CR) is a microscopic mechanism in which final partons are connected by color string, in such a way that the total string length becomes as short as possible [6]. One string connecting two partons follows the movement of the partonic end points. The effect of this movement is a common boost of the string fragments. With CR two partons from independent hard scattering at mid-rapidity can color reconnect and make a large transverse boost (Fig. 1).

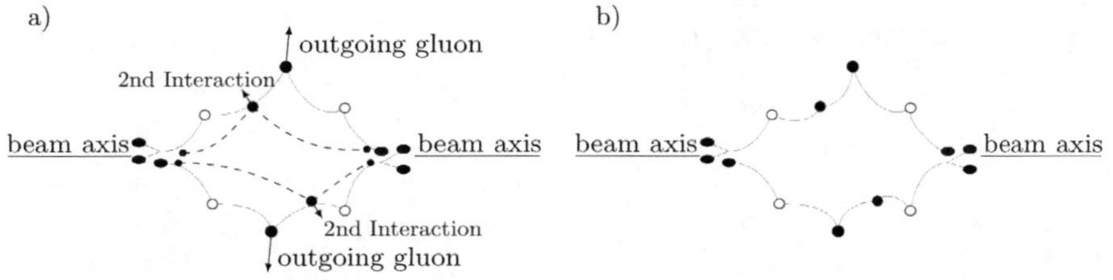

Figure 1. The figure illustrates the color reconnection mechanism in the string fragmentation model. The outgoing partons which are color connected to the projectile and target remnants (continuous lines) are reconnect with the partons in the second hard scattering (in dashed lines). (a) Color reconnected string (b).

3. Studies to isolate the flow-like effect.

The parameter which controls CR is the reconnection range (RR) which enters in the probability to merge a hard scale $\hat{p}_T$ system with one of a harder scale, tune 4C uses the value $RR = 1.5$ which gives a good description of mean transverse momentum as a function of multiplicity.

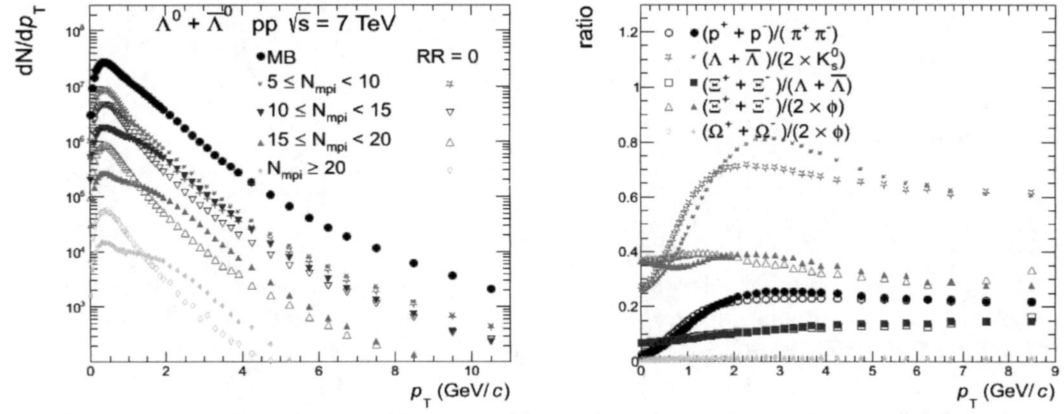

Figure 2. Spectra (left) and baryon to meson ratios (right) with RR = 1.5 (solid points) and without CR, RR = 0 (open points). We observe that CR typically produces a bump around 2.5 − 3 GeV/c.

Comparison with and without CR. To study the effect we simulate MB events with PYTHIA 8, tune 4C, with (RR = 1.5) and without (RR = 0) CR. Left side of Fig. 2 shows the evolution of the p_{T} spectra of Λ for ranges of the number of Multiple Parton Interactions (MPI), from this it is observed a harder spectra when CR are included. Also shown in the right side of Fig. 2 different baryon to meson ratios as well as heavy to light meson ratios as a function of p_{T}, all of them shown a bump around 2.5−3 GeV/c which diminishes when CR is turned off. The position of the maximum is shifted to larger p_{T} values when CR is included, however after $p_{\mathrm{T}} > 7$ Gev/c, the particles ratio are unaffected by CR. A qualitatively similar evolution of the spectral shapes and particles ratios with collision centralities has been observed in heavy ion collisions [7].
MPI and Particle ratios. In reference [8] We shown that the baryon to meson ratios are enhanced in events with increased MPI activity. In events with more than 20 MPI's the peak is also pushed

up to higher p_T, a characteristic effect of flow. On the contrary, in events with less than 5 MPI's the ratio behaves like in simulations without CR.

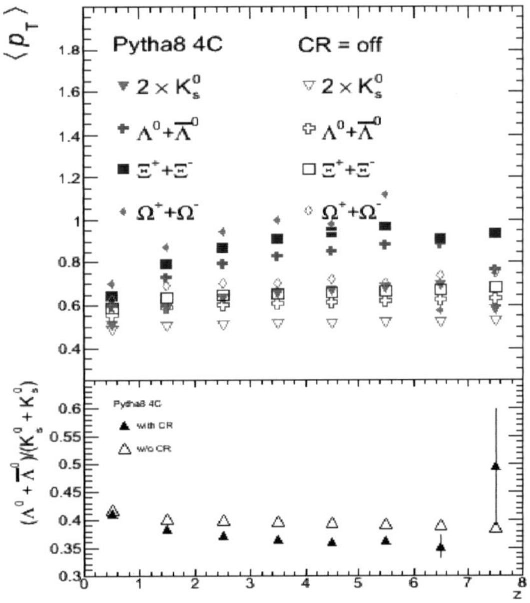

Figure 3. The behavior of the $\langle p_T \rangle$ as a function of multiplicity($z = dN/d\eta/\langle dN/d\eta \rangle$) for strange hadrons.

$\langle p_T \rangle$ as a function of multiplicity. At LHC energies the CMS collaboration has published p_T spectra for pions, kaons and protons as a function of the track multiplicity. They found that $\langle p_T \rangle$ increases with multiplicity, this can only be described by PYTHIA when CR is included.

In Fig. 3 we observe an increase of $\langle p_T \rangle$ with multiplicity when CR is turned on, while it is essentially flat when color reconnection is turned off. The stronger increase of $\langle p_T \rangle$ for heavy hadrons mimics a flow-like effect due to the string boost.

Energy dependence. If the center of mass energy is increased in the simulations the MPI activity increases and the peak is shift further out in p_T and the magnitude increases. So the effect of CR becomes more prominent at higher energies.

4. Summary

The CR scheme used in the PYTHIA 8 event generator exhibits a qualitatively new feature that has potentially important consequence on our understanding of the details of the pp collisions. The "flow" mechanism present in PYTHIA 8 does not require thermalization or a medium to be formed. This could lead us to make predictions that are different from hydrodynamics for small systems as pp and p-Pb.

Acknowledgments

Support for this work has been received by CONACyT under grant numbers 101597 and PAPIIT-UNAM under grant number IN108414.

References

[1] Ortiz Velasquez A (for the ALICE Collaboration) 2013 *Nucl. Phys.* **A904-905** 763c
[2] Aamodt K *et al* (ALICE Collaboration) 2011 *Eur. Phys. J C* **71** 1594
[3] Sjöstrand T, Mrenna S and Skands P 2008 *Computer Phys. Commun.* **178** 852
[4] Corke R and Sjöstrand T 2011 *JHEP* **1103** 32
[5] Abelev B *et al* (ALICE Collaboration) 2012 *Eur. Phys. J C***72** 2183
[6] Sjöstrand T, Mrenna S and Skands P 2006 *JHEP* **0605** 026
[7] Aamodt K *et al* (ALICE Collaboration) 2013 K^0_s and Λ production in Pb-Pb collisions at $\sqrt{S_{NN}} = 2.76$ TeV *e-print* arXiv:1307.5530v1
[8] Ortiz Velasquez A, Christiansen P, Cuautle Flores E, Maldonado Cervantes I A and Paić G (2013) *Phys. Rev. Lett.* **111** 042001

Journal of Physics: Conference Series **509** (2014) 012065 doi:10.1088/1742-6596/509/1/012065

The QCD phase transition in a fully dynamical model of heavy-ion collisions

Christoph Herold[1,2], Marlene Nahrgang[2,3], Igor Mishustin[2,4], Marcus Bleicher[1,2]

[1] Institut für Theoretische Physik, Goethe-Universität, Max-von-Laue-Str. 1, 60438 Frankfurt am Main, Germany
[2] Frankfurt Institute for Advanced Studies (FIAS), Ruth-Moufang-Str. 1, 60438 Frankfurt am Main, Germany
[3] Department of Physics & CTMS, Duke University, Durham, NC 27708, USA
[4] Kurchatov Institute, National Research Center, 123182 Moscow, Russia

E-mail: herold@fias.uni-frankfurt.de

Abstract. Experimental signals for a possible QCD critical point and first-order phase transition are strongly influenced by the rapid nonequilibrium dynamics during a heavy-ion collision. In order to estimate and understand these effects we study the cooling through the phase transition within a nonequilibrium chiral fluid dynamics model. The order parameters for the chiral and deconfinement transition are explicitly propagated, taking into account dissipation and fluctuation stemming from the interaction with a quark-antiquark fluid. In studies of single events, we demonstrate how the formation of domains in net-baryon density at the first-order phase transition leads to a clear enhancement of higher flow harmonics. For the detection of the critical point it is crucial that the relevant signal survives the rapid dynamics. We observe critical slowing down and long-wavelength fluctuations in the vicinity of the critical point.

1. Introduction

Exploring the phase structure of strongly interacting matter is one of the primary goals of heavy-ion physics. The implementation of quantum chromodynamics (QCD) on the lattice predicts a crossover transition at small baryochemical potential $\mu_{\rm B}$ [1] with a pseudo-critical temperature in the range between $150 - 170$ MeV [2]. The concept of a critical point (CP) and a first-order phase transition at large values of $\mu_{\rm B}$ is mainly supported by mean-field effective models [3] and functional methods such as Dyson-Schwinger equations [4] or renormalization group techniques [5]. Experimentally, fluctuations are of crucial importance for the detection of any of these transition scenarios. At the CP, a diverging correlation length in equilibrium would lead to diverging event-by-event fluctuations of conserved quantities [6]. On the other hand, a different type of fluctuation is expected at a dynamical nonequilibrium first-order phase transition, where spinodal instabilities could foster spatial inhomogeneities within single events [7, 8]. We address these issues within the nonequilibrium chiral fluid dynamics model trying to find hints for both critical behavior and spinodal decomposition.

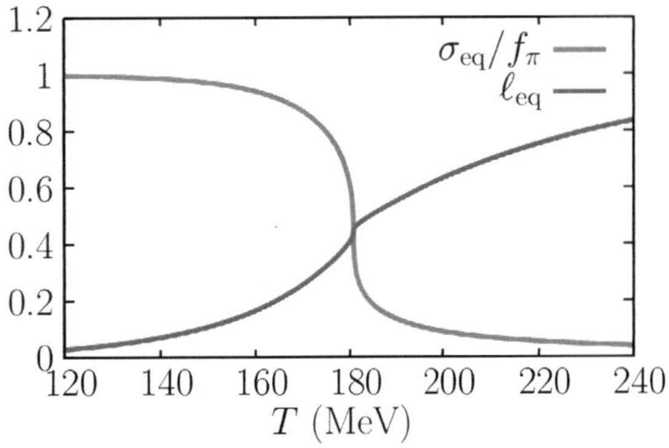

Figure 1. Equilibrium values of the sigma and Polyakov loop fields for a critical point scenario with $g = 3.52$.

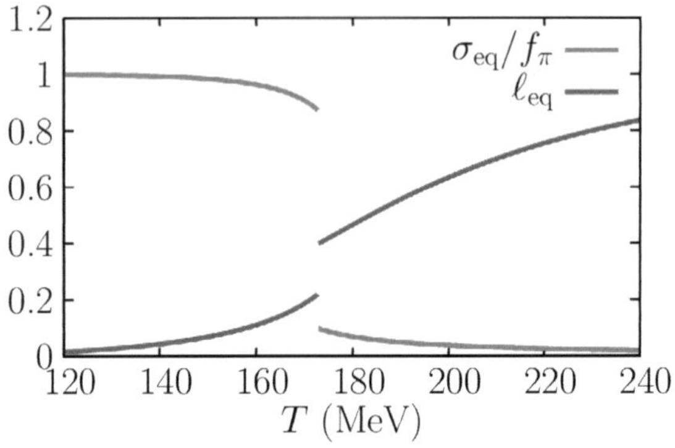

Figure 2. Equilibrium values of the sigma and Polyakov loop fields for a scenario with a first-order phase transition with $g = 4.7$.

2. Coupling fluid dynamics to chiral fields and Polyakov loop

Starting from the Polyakov-Quark-Meson model [9], we obtain the dynamics of the sigma field, the chiral order parameter, by adding damping and stochastic noise to the usual Euler-Lagrange equation

$$\partial_\mu \partial^\mu \sigma + \eta_\sigma(T)\partial_t \sigma + \frac{\partial V_{\mathrm{eff}}}{\partial \sigma} = \xi_\sigma \ . \tag{1}$$

This can be explicitly derived using the two-particle irreducible effective action approach [10], which gives a temperature-dependent damping coefficient $\eta_\sigma(T)$ vanishing only around the CP, and a dissipation-fluctuation relation. Here, V_{eff} is the grand canonical potential

$$V_{\mathrm{eff}} = U(\sigma) + \mathcal{U}(\ell) + \Omega_{\mathrm{q\bar{q}}} \ . \tag{2}$$

It is composed of the chiral potential U describing spontaneous breaking of chiral symmetry, the effective Polyakov loop potential $\mathcal{U}$ and the mean-field quark-antiquark contribution $\Omega_{q\bar{q}}$ from integrating out the quark degrees of freedom in the path integral formulation of the partition function.

A rigorous derivation of the equation of motion for the Polyakov loop is not possible. This quantity is only defined in Euclidean time and we do not understand the real-time equivalent. All we can do is make some reasonable phenomenological estimate. From the mean-field approximation we would have $\partial V_{\mathrm{eff}}/\partial \ell = 0$, which forces ℓ to be equal to its thermal expectation value at all times. To gain an independent dynamical description we augmented the equation of motion with damping and noise terms, giving [11]

$$\eta_\ell \partial_t \ell + \frac{\partial V_{\mathrm{eff}}}{\partial \ell} = \xi_\ell \ . \tag{3}$$

Thereby, we found our results to be independent of the choice of η_ℓ for a wide range of applications. We chose the value of $\eta_\ell = 5/\mathrm{fm}$. Both noise fields ξ_σ and ξ_ℓ are assumed to be Gaussian and white.

The quark fluid is described by an ideal stress-energy tensor $T_{\mathrm{q}}^{\mu\nu} = (e+p)u^\mu u^\nu - p g^{\mu\nu}$, where pressure and energy density are determined via $p = -\Omega_{\mathrm{q\bar{q}}}$ and $e = T\partial p/\partial T - p + \mu \partial p/\partial \mu$. The coupling to the order parameter fields σ and ℓ is achieved through source terms S_σ^ν and S_ℓ^ν,

$$\partial_\mu T_{\mathrm{q}}^{\mu\nu} = S_\sigma^\nu + S_\ell^\nu \ , \tag{4}$$

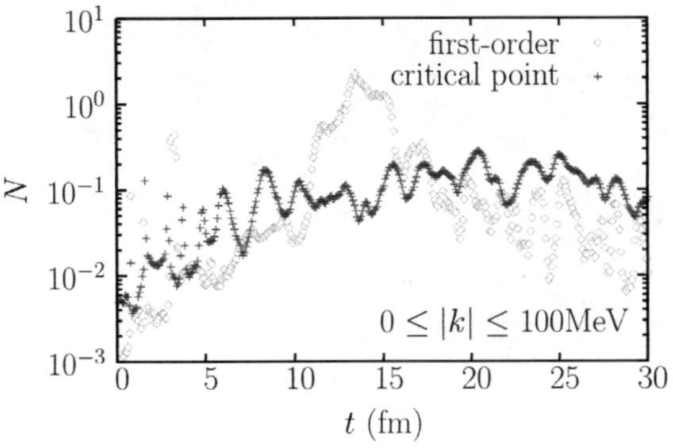

Figure 3. Fluctuation intensity of the sigma field for $0 \leq |k| \leq 100$ MeV.

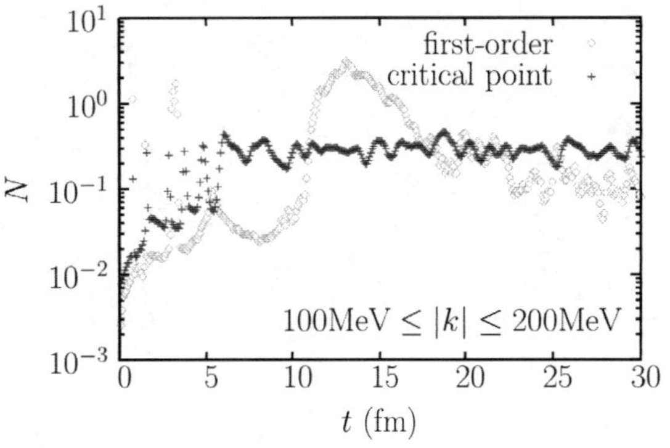

Figure 4. Fluctuation intensity of the sigma field for $100 \leq |k| \leq 200$ MeV.

which ensures conservation of the total energy of the coupled system [11, 12]. It is important to note here, that the local pressure of the fluid depends on the local values of the fields, so we do not use an equilibrium equation of state. Conservation of the total quark number is ensured by the continuity equation $\partial_\mu n^\mu = 0$.

3. Enhancement of low-momentum modes at the critical point

We investigate the dynamics of the order parameter fields after a temperature quench. For this purpose we put the system in a box of finite size with periodic boundary conditions. In this setup, pressure gradients are negligibly small and the dynamics is dominated by the Langevin dynamics of the fields. We restrict ourselves to the case of vanishing baryochemical potential and vary the strength of the transition by changing the quark-meson coupling g. In principle this quantity has to be chosen such that $g\sigma$ in vacuum reproduces the constituent quark mass, resulting in $g \sim 3.3$ and a crossover chiral and deconfinement transition. For a value of $g = 3.52$ we obtain a CP and for larger values of g the transition is of first-order type. Figs. 1 and 2 illustrate this through the equilibrium values $\sigma_{\rm eq}$ and $\ell_{\rm eq}$ as a function of temperature.

We let the system in the box relax in the vicinity of the respective transition temperature for the case of a CP and a first-order phase transition. During the evolution we focus on the behavior of sigma field fluctuations which we analyze through their intensity. This quantity is given by a combination of creation and annihilation operators which in equilibrium resembles the number of excited particles from the quantum field. It reads

$$\frac{\mathrm{d}N}{\mathrm{d}^3k} = \frac{a_k^\dagger a_k}{(2\pi)^3 2\omega_k} = \frac{\omega_k^2|\delta\sigma_k|^2 + |\partial_t\sigma_k|^2}{(2\pi)^3 2\omega_k} , \tag{5}$$

with the Fourier transformed field $\delta\sigma_k$ and the time derivative $\partial_t\sigma_k$. The corresponding energy of the k-the mode is given by $\omega_k = \sqrt{m_\sigma^2 + \vec{k}^2}$. The intensities integrated over the low-momentum ranges from 0 to 100 MeV and from 100 to 200 MeV are plotted in Figs. 3 and 4, respectively. In both figures we can make out significant differences between the two types of transition: For the CP, the field needs less time to reach the equilibrium value, nevertheless, fluctuations remain strong here with a clear enhancement compared to the first-order phase transition towards the end of the evolution. On the other hand, we find large fluctuations when the system undergoes the first-order phase transition, induced by spinodal instabilities around the potential barrier separating the two competing minima. After equilibration, fluctuations are significantly weaker than near the CP.

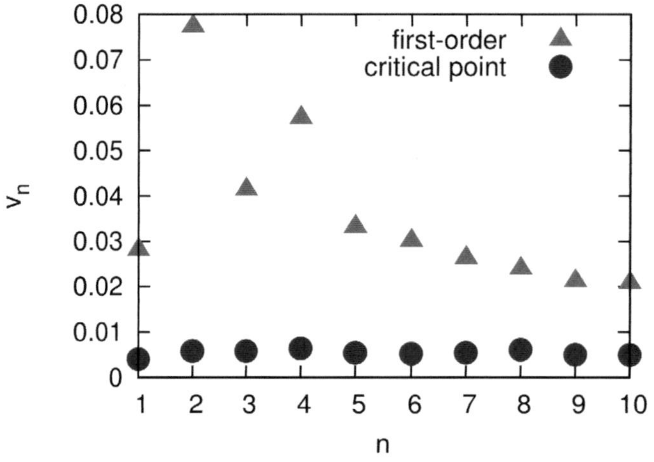

Figure 5. Event-averaged harmonic coefficients from the azimuthal net-baryon number distribution. We find a strongly anisotropic expansion for at first-order transition, while small coefficients indicate a homogeneous spherical shape at the critical point.

4. Anisotropic flow at large baryon densities

In order to study what happens during a heavy-ion collision, we consider a freely expanding blob of hot and dense quark matter cooling through the first-order phase transition at large μ_B. Here, we find that a metastable phase dynamically fragments into small droplets, leading to inhomogeneities in the azimuthal baryon number distribution within single events [13]. Averaged over many events, we find a clear enhancement of higher harmonics in comparison with an evolution through the CP, cf. Fig. 5. Note here that we used spherical initial conditions instead of a more realistic ellipsoidal shape to illustrate the mere influence of the phase transition.

In the future, we aim at investigating more realistic scenarios by introducing initial conditions from transport models and a hadronic freeze-out. We expect that the enhancement of the harmonic coefficients at the first-order phase transition leads to strongly enhanced flow.

Acknowledgements

The authors thank Stefan Leupold and Carsten Greiner for fruitful discussions and Dirk Rischke for providing the SHASTA code. This work was supported by the Hessian LOEWE initiative Helmholtz International Center for FAIR. I. M. acknowledges partial support from the grant NSM-215.2012.2 (Russia).

References

[1] Aoki Y, Endrodi G, Fodor Z, Katz S D and Szabo K K, *Nature* **443** (2006) 675
[2] Borsanyi S *et al.* [Wuppertal-Budapest Collaboration], *JHEP* **1009** (2010) 073
[3] Scavenius O, Mocsy A, Mishustin I and Rischke D H, *Phys. Rev. C* **64** (2001) 045202
[4] Fischer C S and Luecker J, *Phys. Lett. B* **718** (2013) 1036
[5] Herbst T K, Pawlowski J M and Schaefer B J, *Phys. Rev. D* **88** (2013) 014007
[6] Stephanov M A, Rajagopal K and Shuryak E V, *Phys. Rev. D* **60** (1999) 114028
[7] Scavenius O and Dumitru A, *Phys. Rev. Lett.* **83** (1999) 4697
[8] Steinheimer J and Randrup J, *Phys. Rev. Lett.* **109** (2012) 212301
[9] Schaefer B J, Pawlowski J M and Wambach J, *Phys. Rev. D* **76** (2007) 074023
[10] Nahrgang M, Leupold S, Herold C and Bleicher M, *Phys. Rev. C* **84** (2011) 024912
[11] Herold C, Nahrgang M, Mishustin I and Bleicher M, *Phys. Rev. C* **87** (2013) 014907
[12] Nahrgang M, Herold C, Leupold S, Mishustin I and Bleicher M, *J. Phys. G* **40** (2013) 055108
[13] Herold C, Nahrgang M, Mishustin I and Bleicher M, arXiv:1304.5372 [nucl-th]

Study of freeze-out dynamics in STAR at RHIC Beam Energy Scan Program

Sabita Das (for the STAR collaboration)
Institute of physics, Bhubaneswar - 751005, INDIA
Brookhaven National Laboratory, Upton, NY, 11973-5000, USA
E-mail: sabita@rcf.rhic.bnl.gov

Abstract. The STAR detector at RHIC, due to its large uniform acceptance and excellent particle identification capabilities, has measured a variety of hadron species ($\pi^{\pm}$, $K^{\pm}$, p, $\bar{p}$, K_S^0, Λ, $\bar{\Lambda}$, Ξ^-, $\bar{\Xi}^+$) produced in Au+Au collisions at $\sqrt{s_{NN}}$ = 7.7, 11.5, 27 and 39 GeV. These data are part of the Beam Energy Scan (BES) program at RHIC and provide an opportunity to measure the yields and transverse momentum spectra (p_T) of the particles produced in the collisions. The corresponding measurements allow to study the freeze-out properties and dynamics of heavy ion collisions. A statistical thermal model analysis of particle production in BES energies in both grand canonical and strangeness canonical ensembles, is used to extract the chemical freeze-out parameters. The p_T spectra, particle ratios, and the energy and centrality dependence of freeze-out parameters determined from the thermal fit of particle ratios are discussed.

1. Introduction
Relativistic heavy-ion collisions provide the opportunity to study strongly interacting nuclear matter at different thermodynamic conditions. Quantum Chromodynamics (QCD), a fundamental theory to describe the interactions of quarks and gluons, has anticipated the transition from hadronic matter to a new state of matter called Quark-Gluon Plasma (QGP) [1] phase at high temperature and high energy density (≈ 1 GeV/fm^3). The QCD phase diagram is characterized by the temperature (T) and the baryon chemical potential (μ_B) or the (net) baryon density (n_B), and it contains the information about the phase boundary that separates the QGP and hadronic phases [2,3]. Finite temperature lattice QCD calculations [4] predict a cross-over from hadronic to QGP phase at vanishing baryon chemical potential and large T while several QCD-based calculations [5] show that at lower T and μ_B a first-order phase transition may take place. The point in the QCD phase diagram, where the first order phase transition ends would be the QCD critical point [6]. The BES program at RHIC has been carried out with the specific aim to explore several features of QCD phase diagram such as to search for the phase boundaries and the location of QCD critical point by colliding nuclei at several center-of-mass energies. When heavy ions collide at ultra-relativistic energies, the state of the system at the time of the final interaction is reflected by the observed single particle spectra. The process of hadron decoupling from an interacting system is called freeze-out. Freeze-out are of two types - kinetic and chemical freeze-out. Chemical freeze-out occurs at a temperature (T_{ch}) when inelastic collisions cease and the particle yields become fixed and thermal (kinetic) freeze-out occurs at a temperature (T_{kin}) when elastic collisions cease and particle transverse

Journal of Physics: Conference Series **509** (2014) 012066 doi:10.1088/1742-6596/509/1/012066

momenta spectra are fixed.

In this paper we discuss the identified transverse momentum spectra and particle ratios produced in Au+Au collisions at $\sqrt{s_{NN}} = 27$ GeV. Similar measurements at other BES center-of-mass energies $\sqrt{s_{NN}} = 7.7$, 11.5, and 39 GeV are reported in [7]. Statistical thermal models, such as THERMUS [8] have been proven to be successful in describing the particle production in heavy-ion collisions [9, 10, 11]. Experimental particle ratios are used in a statistical thermal model in both grand canonical ensemble (GCE) and strangeness canonical ensemble (SCE) approach to extract various chemical freeze-out parameters such as chemical freeze-out temperature (T_{ch}), baryon chemical potential (μ_B), strangeness chemical potential (μ_S) and strangeness saturation factor (γ_S). In this study we have used mid-rapidity particle ratios that include measured yields for charged pions ($\pi^{\pm}$), charged kaons ($K^{\pm}$), protons (p, $\bar{p}$), K^0_S, lambdas (Λ, $\bar{\Lambda}$) and cascades (Ξ^-, $\bar{\Xi}^+$) [7, 12]. The energy and centrality dependence of extracted chemical freeze-out parameters in Au+Au collisions at above BES energies are studied.

2. Results

2.1. Identified particle spectra and ratios

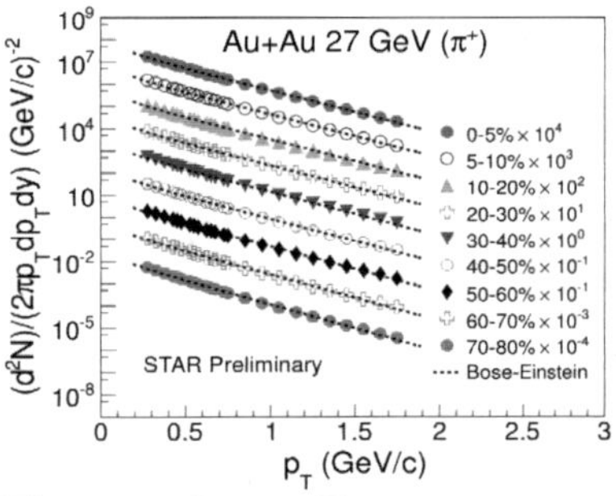

Figure 1. Transverse momentum spectra for charged pion at mid-rapidity ($|y| < 0.1$) in Au+Au collisions at $\sqrt{s_{NN}} = 27$ GeV. Errors shown here are statistical only. The lines represents the Bose-Einstein fits to the p_T-spectra.

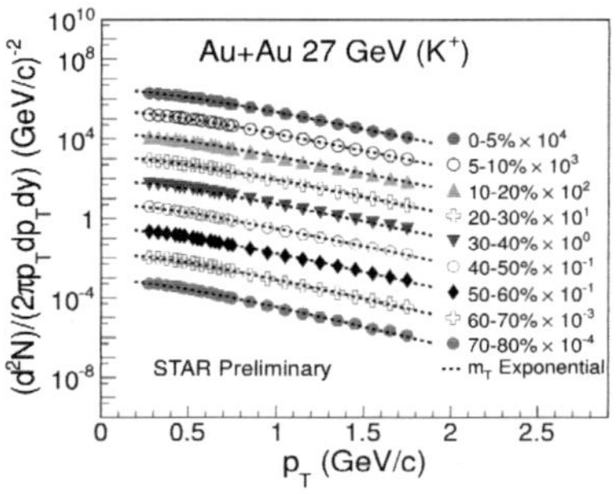

Figure 2. Transverse momentum spectra for charged kaon at mid-rapidity ($|y| < 0.1$) in Au+Au collisions at $\sqrt{s_{NN}} = 27$ GeV. Errors shown here are statistical only. The lines represents the m_T exponential fits to the p_T-spectra.

Figures 1 and 2 show the transverse momentum spectra of π^+ and K^+ respectively in Au+Au collisions at $\sqrt{s_{NN}} = 27$ GeV in different centralities at mid-rapidity $|y| < 0.1$. The p_T-integrated pion yields obtained from the Bose-Einstein fit to p_T-spectra have been corrected for feed-down from weak decays. The m_T-exponential fit is used to obtain the p_T-integrated kaon yields. The collision-energy dependence of the anti-particle to particle ratios in central heavy-ion collisions for π^-/π^+ and K^-/K^+ are shown in Fig. 3. The π^-/π^+ ratio at lower beam energies have values larger than unity, which could be due to significant contributions from resonance decays (such as from Δ baryons). The $\bar{p}/p$ ratio, not shown in this paper, increases with increasing collision energy and approaches unity for top RHIC energies [21]. This indicates that at higher beam energies, the p ($\bar{p}$) production at mid-rapidity is dominated by pair production. The K^-/K^+ ratio approaches unity as collision energy increases, indicating the dominance of kaon-pair production while at lower BES energies associated production of K^+ dominates.

2.2. Chemical freeze-out

At chemical freeze-out, inelastic collisions among the particles stop and hadron yields are fixed. In Au+Au collisions at $\sqrt{s_{NN}} = 27$ GeV, the measured mid-rapidity particle ratios including yields of π, K, p, K^0_S, Λ and Ξ are used in THERMUS. The thermal model fit of other BES

Journal of Physics: Conference Series **509** (2014) 012066 doi:10.1088/1742-6596/509/1/012066

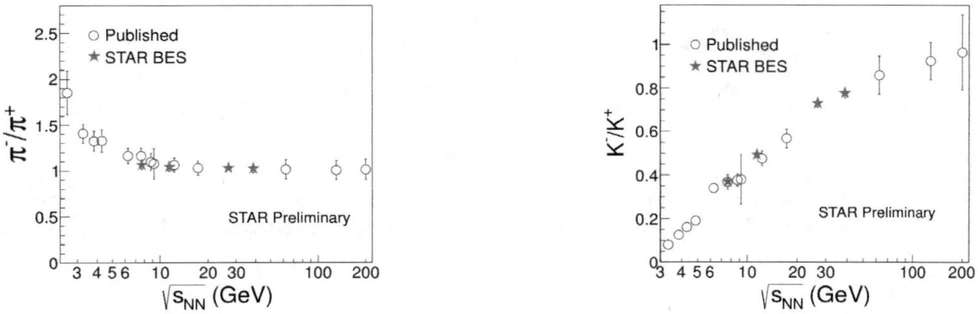

Figure 3. π^-/π^+ and K^-/K^+ ratios for 0–5% centrality in Au+Au collisions at $\sqrt{s_{NN}} = 27$ GeV compared with the other BES energies and previous published results from AGS, SPS, and RHIC [14-22] respectively. Errors are the quadratic sum of statistical and systematic uncertainties.

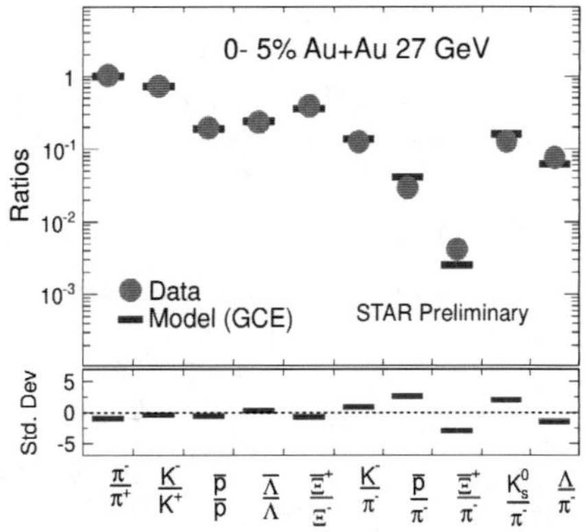

Figure 4. Statistical thermal model [8] fit to experimental mid-rapidity particle ratios using grand-canonical ensemble for 0–5% centrality in Au+Au at $\sqrt{s_{NN}} = 27$ GeV.

energies at $\sqrt{s_{NN}}$ = 7.7, 11.5, and 39 GeV is discussed in [13]. The π, K, p yields are measured at rapidity $|y| < 0.1$ and those for K^0_S, Λ , Ξ are measured for $|y| < 0.5$. The errors on particle ratios including yields of π, K, p, K^0_S, Λ, and Ξ, are the quadratic sum of statistical and systematic uncertainties. Proton and anti-proton yields have not been corrected for feed-down contributions. The Λ yields have been corrected for the feed-down contributions from Ξ and Ξ^0 weak decays [12]. The errors on freeze-out parameters are obtained from the THERMUS model and errors on particle ratios are treated as independent errors. Figure 4 shows the statistical thermal model fit to experimental particle ratios for 0–5% centrality in Au+Au collisions at $\sqrt{s_{NN}}$ = 27 GeV. The data show a good agreement with the fit to model having $\chi^2/d.o.f = 23.8/6$. Figure 5(a) shows that the T$_{ch}$ increases with increasing collision energy. Figure 5(b) shows that the μ_B decreases with increasing collision energy and it increases when going from peripheral to central collisions at lower energies. We observe a centrality dependence of chemical freeze-out curve (T$_{ch}$ vs. μ_B) at BES energies which was not observed at higher energies of Au+Au 200 GeV [10].

In contrast to GCE approach where all quantum numbers are conserved on average, the THERMUS model also allows for a strangeness canonical ensemble where only the strangeness quantum number is required to be conserved exactly where as baryon and charge quantum numbers are conserved on an average. It is observed that in peripheral collisions, T$_{ch}$ and μ_B

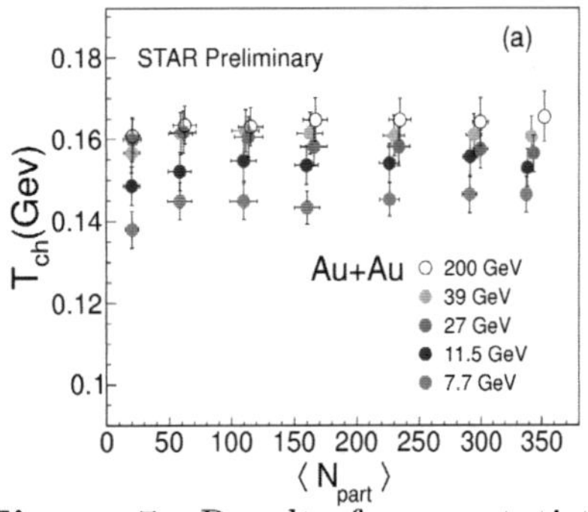

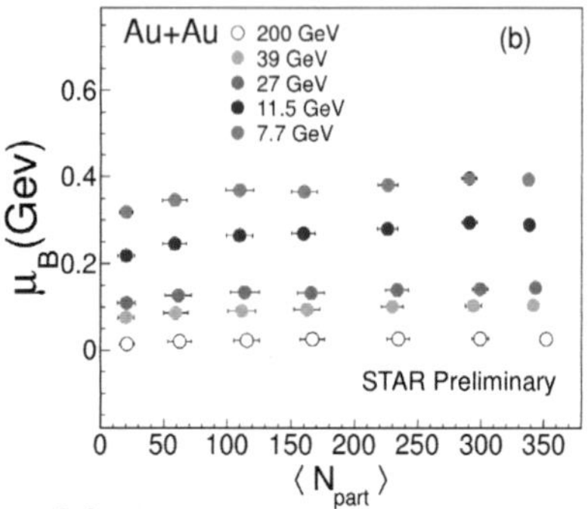

Figure 5. Results from a statistical thermal model fit [8] for Au+Au at $\sqrt{s_{NN}}$ = 7.7, 11.5, 27 and 39 GeV. Chemical freeze-out temperature, baryon chemical potential, and strangeness saturation factor are shown as a function of the average number of participating nucleons. The 200 GeV results are taken from Ref. [10].

follow a different behavior in GCE and SCE at lower BES energies [13]. Further studies by using the yields for the thermal fit towards a more quantitative analysis for all the BES energies are ongoing.

3. Conclusions

The study of the hadron particle ratios at BES energies in STAR within the framework of a statistical model yields the following conclusions. A centrality dependence of the chemical freeze-out parameters is observed at the lower energies. Baryon chemical potential (μ_B) range extends from 20 to 400 MeV in the QCD phase diagram in the new BES measurements at $\sqrt{s_{NN}} = 7.7$, 11.5, 27 and 39 GeV. A detailed study using particle yields and different choices of ensembles in the statistical model is ongoing: this is expected to provide a better understanding of the freeze-out dynamics.

References

[1] Adams J *et al* (STAR Collaboration) 2005 *Nucl.Phys.* A **757** 102
[2] Braun-Munzinger P *et al* 2011 *arXiv:1101.3167*
[3] Mohanty B 2009 *Nucl. Phys.* A **830** 899C
[4] Aoki Y *et al* 2006 Nature **443** 675
[5] Ejiri S 2008 *Phys. Rev.* D **78** 074507; Bowman E S and Kapusta J I 2009 *Phys. Rev.* C **79** 015202
[6] Gupta S *et al* 2011 *Science* **332** 1525
[7] Kumar L (STAR Collaboration) 2011 *J. Phys.* G *Nucl. Part. Phys.* **38** 124145
[8] Cleymans J *et al* 2009 *Computer Physics Communications* **180** 84
[9] Cleymans J *et al* 2005 *Phys.Rev.* C **71** 054901
[10] Aggarwal M M *et al* (STAR Collaboration) 2011 *Phys. Rev.* C **83** 024901; Kumar L (STAR Collaboration) 2013 *Nucl. Phys.* A **904-905** 256C
[11] Cleymans J *et al* 2006 *Phys. Rev.* C **73** 034905
[12] Zhu X (STAR Collaboration) 2012 *Acta Phys. Polon.* B **Proc. Supp. 5** 213-218
[13] Das S (STAR Collaboration) 2013 *Nucl. Phys.* A **904-905** 891C
[14] Ahle L*et al* (E866 Collaboration and E917 Collaboration) 2000 *Phys. Lett.* B **490** 53
[15] Klay J L *et al* (E895 Collaboration) 2002 *Phys. Rev. Lett.* **88** 102301
[16] Ahle L *et al* (E866 Collaboration and E917 Collaboration) 2000 *Phys. Lett.* B **476** 1
[17] Afanasiev S V *et al* (NA49 Collaboration) 2002 *Phys. Rev.* C **66** 054902
[18] Alt C *et al* (NA49 Collaboration) 2008 *Phys. Rev.* C **77** 024903
[19] Anticic T *et al* (NA49 Collaboration) 2004 *Phys.Rev.* C **69** 024902
[20] Abelev B I *et al* (STAR Collaboration) 2009 *Phys. Rev.* C **79** 034909
[21] Adams J *et al* (STAR Collaboration) 2004 *Phys. Rev. Lett.* **92** 112301
[22] Abelev B I *et al* (STAR Collaboration) 2010 *Phys. Rev.* C **81** 024911

Beam energy scan using a viscous hydro+cascade model

Iu.A. Karpenko[1,2], M. Bleicher[1,3], P. Huovinen[1] and H. Petersen[1]

[1] Frankfurt Institute for Advanced Studies, Ruth-Moufang-Straße 1, 60438 Frankfurt am Main, Germany

[2] Bogolyubov Institute for Theoretical Physics, 14-b, Metrolohichna str., 03680 Kiev, Ukraine

[3] Institute for Theoretical Physics, Johann Wolfgang Goethe Universität, Max-von-Laue-Str. 1, 60438 Frankfurt am Main, Germany

E-mail: `karpenko@fias.uni-frankfurt.de`

Abstract. Following the experimental program at BNL RHIC, we perform a similar "energy scan" using 3+1D viscous hydrodynamics coupled to the UrQMD hadron cascade, and study the collision energy dependence of pion and kaon rapidity distributions and m_T-spectra, as well as charged hadron elliptic flow. To this aim the equation of state for finite baryon density from a Chiral model coupled to the Polyakov loop is employed for hydrodynamic stage. 3D initial conditions from UrQMD are used to study gradual deviation from boost-invariant scaling flow. We find that the inclusion of shear viscosity in the hydrodynamic stage of evolution consistently improves the description of the data for Pb-Pb collisions at CERN SPS, as well as of the elliptic flow measurements for Au-Au collisions in the Beam Energy Scan (BES) program at BNL RHIC. The suggested value of shear viscosity is $\eta/s \geq 0.2$ for $\sqrt{s_{NN}} = 6.3 \ldots 39$ GeV.

1. Introduction

During the recent years there has been considerable progress in modelling of bulk matter dynamics in ultrarelativistic heavy ion collisions at BNL Relativistic Heavy Ion Collider (RHIC) and CERN Large Hadron Collider (LHC). The state-of-the-art theoretical descriptions are based on viscous hydrodynamics coupled to a hadron cascade. They successfully describe bulk observables such as radial flow, elliptic and higher order flow harmonics [1], femtoscopy [2] etc. The idea of the present study is to apply a hybrid (or hydro+cascade) model to heavy ion collisions at lower collision energies to study how well the model can describe the existing data from CERN Super Proton Synchrotron (SPS) and recent results from Beam Energy Scan program at RHIC.

2. Model description

The lower the collision energy, the worse the approximation of boost-invariant scaling flow is. Thus it is important to employ initial conditions with non-trivial rapidity dependence. We use the Ultrarelativistic Quantum Molecular Dynamics (UrQMD) model [3] for the description of initial stage dynamics. The two nuclei are initialized according to Wood-Saxon distributions and the binary interactions are taken into account until a hypersurface at constant $\tau = \sqrt{t^2 - z^2}$. We average over many UrQMD events to get smooth distribution of particles. The energy and momentum of particles is then converted to energy and momentum densities of the fluid.

In addition to energy/momentum densities, initial baryon and charge densities are non-zero and obtained from UrQMD, evolved in the hydro stage and accounted for in the particlization procedure. On the other hand, net strangeness density is set to zero.

The hydrodynamic equations are solved in Milne ($\tau - \eta$) coordinates. The starting time of hydrodynamic evolution is chosen to be $\tau_0 = 2R/\sqrt{(\sqrt{s}/2m_N)^2 - 1}$, where R is a radius of nucleus and m_N is a nucleon mass. This corresponds to the time when the two nuclei have passed through each other.

For beam energy scan, an equation of state (EoS) for finite baryon density must be used. We employ the Chiral model based EoS [5], which features correct asymptotic degrees of freedom, i.e. quarks and gluons at high temperature and hadrons in the low-temperature limits, crossover-type transition between confined and deconfined matter for all values of μ_B and qualitatively agrees with lattice QCD data at $\mu_B = 0$.

The hydrodynamic code used solves the equations of relativistic viscous hydrodynamics in Israel-Stewart framework [4]. In particular we solve the following equations for the shear stress tensor:

$$\langle u^\gamma \partial_{;\gamma} \pi^{\mu\nu} \rangle = -\frac{\pi^{\mu\nu} - \pi^{\mu\nu}_{\text{NS}}}{\tau_\pi} - \frac{4}{3}\pi^{\mu\nu}\partial_{;\gamma}u^\gamma \tag{1}$$

where $\partial_{;\gamma}$ denotes covariant derivative and the brackets $\langle A^{\mu\nu} \rangle$ denote the symmetric, traceless and orthogonal to u^μ part of $A^{\mu\nu}$. For the purpose of the current study we consider only the effects of shear viscosity, fixing bulk viscosity to zero, $\zeta/s = 0$. We do not include the baryon/electric charge diffusion either. For viscous hydro simulations, we initialize the shear stress tensor to zero. The relaxation time for shear, τ_π, is taken as $\tau_\pi = 3\eta/(sT)$.

The transition from fluid to particle description (so-called particlization) is made at the constant energy density $\epsilon_{\text{sw}} = 0.5$ GeV/fm^3 when the medium has already hadronized (as in previous studies at $\sqrt{s_{NN}} = 200$ A GeV RHIC and LHC energies [2]). Cornelius subroutine [6] is used to calculate the 3-volume elements $d\sigma_\mu$ of the transition hypersurface. It is important to note that while we use fixed energy density as transition criterion for all collision energies, the value of net baryon density on this surface is non-uniform and its average increases with decreasing collision energy. This corresponds to the increase of average temperature and decrease of baryon chemical potential with increasing collision energy, resembling the results for the collision energy dependence of chemical freezeout parameters from thermal model studies [7]. Also, since Chiral EoS deviates from free hadron-resonance gas (HRG) in hadronic phase, we switch to free HRG EoS when sampling the particles according to Cooper-Frye prescription. We recalculate the energy density, pressure, flow velocity and corresponding thermodynamical quantities from energy-momentum tensor using free HRG EoS, and employ them when sampling. Finally, we use the same corrections to the local equilibrium distribution functions for all hadron species:

$$f_i(x,p) = f_{i,\text{eq}}(x,p)\left[1 + (1 \mp f_{i,\text{eq}})\frac{p_\mu p_\nu \pi^{\mu\nu}}{2T^2(\epsilon + p)}\right] \tag{2}$$

The scatterings and decays happening after particlization are then treated with UrQMD code.

3. Results

First we simulate Pb-Pb collisions at energies $E_{\text{lab}} = 158, 80, 40, 30$ and 20 A GeV (i.e. $\sqrt{s_{NN}} = 17.3, \ldots, 6.3$ GeV), and compare m_T and dN/dy distributions with the data from NA49 collaboration. Note that the only variable parameter is the shear viscosity to entropy density ratio η/s in hydro stage, with the other parameters fixed as described above.

With a given criterion for fluid to particle transition, the duration of hydro phase decreases from about 6.5 fm/c (at $\sqrt{s_{NN}} = 17.3$ GeV) to just about 4.5 fm/c (at $\sqrt{s_{NN}} = 6.3$ GeV), while the duration of pre-hydrodynamic stage increases according to the Eq. 2. Consequently, at the lowest SPS energy ($\sqrt{s_{NN}} = 6.3$ GeV) the lifetimes of pre-hydro (4.1 fm/c) and hydro stages are comparable.

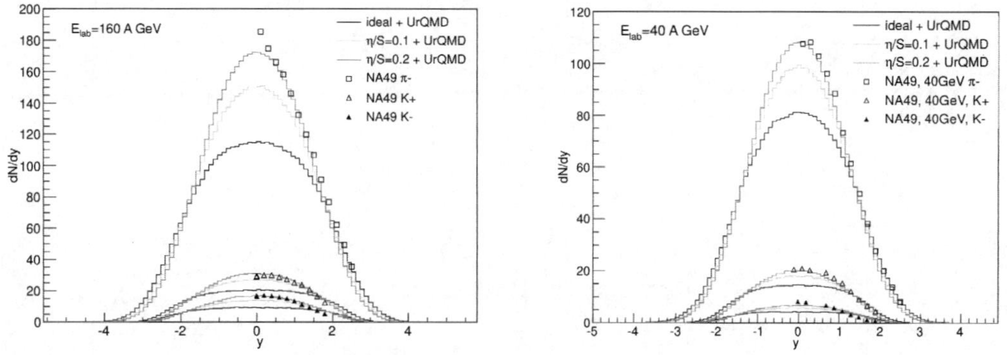

Figure 1. Rapidity distributions for π^-,K^+ and K^- for Pb-Pb collisions at $E_{\rm lab} = 158$ and 40 A GeV (corresponding to $\sqrt{s_{NN}} = 17.3$ and 6.3 GeV). Model calculations are compared to NA49 data [9]

In the case of zero shear viscosity in hydro phase we underestimate the particle yields and radial flow at midrapidity. However, the inclusion of shear viscosity in hydro phase increases the yields at midrapidity due to viscous entropy production, see Fig. 1. One can see that shear viscosity makes the dN/dy profile narrower. The longitudinal expansion is weaker, and the expansion tends to be more spherical, as seen in the comparison of the p_T-spectra, Fig. 2: p_T-spectra become flatter, which is due to stronger transverse expansion and larger radial flow.

Next we calculate charged hadron elliptic flow at collision energies $\sqrt{s_{NN}} = 7.7, 27$ and 39 A GeV to compare with the results from RHIC BES, Fig. 3. Elliptic flow for the case of $\eta/s = 0$ overestimates the data, while choosing $\eta/s = 0.2$ greatly improves agreement with it. Here we do not try to fit the data and only demonstrate the model results for $\eta/s = 0, 0.1$ and 0.2. So one can conclude that a consistent description of v_2, dN/dy, and p_T-spectra requires a value of η/s which is somewhat larger than 0.2, especially for lower energy points. The $\eta/s = 0.2$ also makes $v_2(p_T)$ almost independent of the collision energy.

We conclude that the shear viscosity in hydrodynamic phase is a key component for a hybrid model to better describe the data in the low energy region. The suggested value of the effective shear viscosity is $\eta/s \geq 0.2$ for both Pb-Pb collisions at SPS and Au-Au in BES program at RHIC, which is at least twice higher than the typical value of $\eta/s \approx 0.08$ [11] obtained for $\sqrt{s_{NN}} = 200$ GeV RHIC energy and Monte Carlo Glauber initial state.

Acknowledgements

IK and HP acknowledge the financial support by the ExtreMe Matter Institute EMMI and Hessian LOEWE initiative. HP acknowledges funding by the Helmholtz Young Investigator Group VH-NG-822. Computational resources have been provided by the Center for Scientific Computing (CSC) at the Goethe-University of Frankfurt.

References

[1] Hirano T, Huovinen P, Murase K and Nara Y 2013, *Prog.Part.Nucl.Phys.* **70** 108-158
[2] Karpenko Iu A, Sinyukov Yu M, Werner K 2013, *Phys.Rev.* C **87** 24914

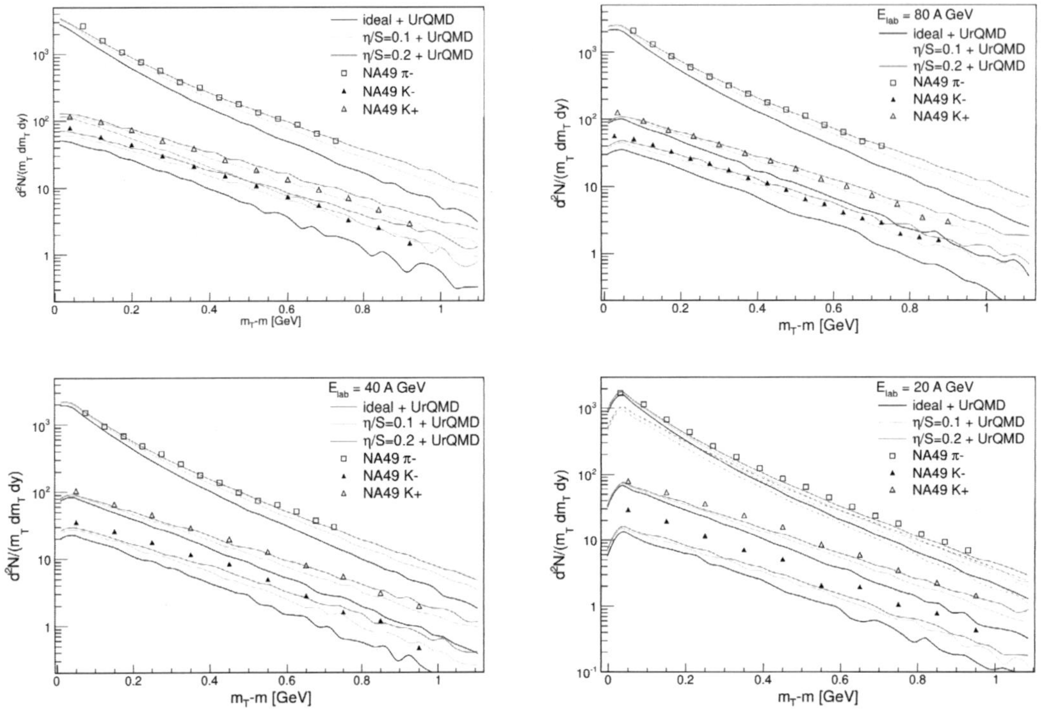

Figure 2. m_T-spectra for π^-,K^+ and K^- for Pb-Pb collisions at $E_{\rm lab} = 158, 80, 40, 20$ A GeV ($\sqrt{s_{NN}} = 17.3, 12.3, 8.8, 6.3$ GeV, respectively). Model calculations are compared to NA49 data [8, 9]. Dashed lines on bottom right plot: model calculations for π^+.

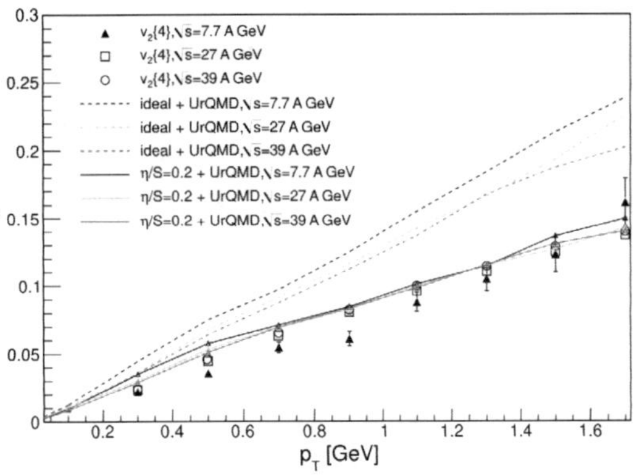

Figure 3. p_T-differential elliptic flow of all charged hadrons for Au-Au collisions at $\sqrt{s_{NN}} = 7.7, 27$ and 39 GeV. The data are from STAR collaboration [10]

[3] Bass S A et al. 1998, *Prog.Part.Nucl.Phys.* **41** 255-369
[4] Israel W and Stewart J M 1979, *Ann. Phys.* **118**, 341
[5] Steinheimer J, Schramm S and Stocker H 2011, *J.Phys.G* **38** 35001
[6] Huovinen P and Petersen H 2012, *Eur.Phys.J.* A **48** 171
[7] Andronic A, Braun-Munzinger P, Stachel J 2006, *Nucl.Phys.* A **772** 167-199
[8] C. Alt et al. (NA49 Collaboration) 2008, *Phys.Rev.* C **77** 24093
[9] S. V. Afanasiev et al. (NA49 Collaboration) 2002, *Phys.Rev.* C **66** 54902
[10] L. Adamczyk et al. (STAR Collaboration) 2012, *Phys.Rev.* C **86** 54908
[11] Song H, Bass S A, Heinz U, Hirano T, Shen C 2011 *Phys.Rev.Lett.* **106** 192301

The elliptic flow and the shear viscosity of the QGP within a kinetic approach

S. Plumari[1,2], A. Puglisi[2], F. Scardina[1,2], V. Greco[1,2] and L.P. Csernai[3]

[1] Department of Physics and Astronomy, University of Catania, Via S. Sofia 64, I-95125 Catania (Italy)
[2] Laboratorio Nazionale del Sud, INFN-LNS, Via S. Sofia 63, I-95125 Catania (Italy)
[3] Dept. of Physics and Technology, University of Bergen (Norway)

E-mail: salvatore.plumari@ct.infn.it

Abstract. We use a relativistic transport approach to study the role of a temperature dependent shear viscosity to entropy density ratio, $\eta/s(T)$, on the build-up of the elliptic flow, v_2. The recent results from $\sqrt{s_{NN}} = 62.4$GeV at RHIC up to 2.76 TeV at LHC have shown an intriguing property of the $v_2(p_T)$, which appears to be nearly invariant with energy. We show that in our transport approach this surprising behavior can be described by a particular temperature dependence of $\eta/s(T)$, typical of matter that undergoes a phase transition or a cross-over, with a rise and fall and the minimum close to critical temperature T_c.

1. Introduction

The experimental results from RHIC and recently from LHC have provided a lot of information about the physics of quark-gluon plasma (QGP). In particular in these experiments has been shown that the elliptic flow $v_2 = \langle cos(2\,\varphi_p)\rangle = \langle (p_x^2 - p_y^2)/(p_x^2 + p_y^2)\rangle$, which is a measure of the momentum anisotropy of the emitted particle, is the largest ever seen in uRHICs [1, 2]. On the other hand theoretical calculations within viscous hydrodynamics [3] or transport approach [4, 5, 6] have shown that this large value of v_2 is consistent with a very low shear viscosity to entropy density ratio η/s close to the conjectured lower bound for a strongly interacting system in the limit of infinite coupling, $\eta/s = 1/4\pi$ [7]. However recent measurements of higher harmonics $v_n = \langle cos(n\,\varphi_p)\rangle$ with $n > 2$ seems to confirm a small value of $\eta/s \sim 1 - 3$ [8, 9]. In itself a small value of $\eta/s \sim 1/4\pi$ it is not an evidence of the creation of a QGP phase but rather a phenomenological estimation of its temperature dependence which could give us more information of how the matter created in these collisions undergoes a phase transition [10, 11]. On the other hand, recent results of both the STAR Collaboration at RHIC [12] and the ALICE at LHC [2] have shown a surprising property of the $v_2(p_T)$ that it appears invariant in a very wide colliding energy range of $62.4\,\text{GeV} \leq \sqrt{s_{NN}} \leq 2.76\,\text{TeV}$. It is important to explain the invariance of $v_2(p_T)$ and how it is related to η/s. We show that the invariance of $v_2(p_T)$ in the energy range $62.4\,\text{GeV} \leq \sqrt{s_{NN}} \leq 2.76\,\text{TeV}$ is caused by a fall and rise of the $\eta/s(T)$ as one would expect if the created matter undergoes a phase transition.

2. Relativistic Boltzmann Transport approach at fixed η/s

In this section we introduce a Relativistic Boltzmann Transport (RBT) approach at fixed η/s ratio. It is inspired by the success of the hydrodynamical approach that has shown the key role played by the η/s. Within this approach it is possible to make a direct comparison to viscous hydrodynamics because instead of focusing on specific microscopic details we fix the cross section in order to have the wanted η/s and it provides a tool to directly estimate the viscosity of the plasma in a wider range of η/s and p_T.

To solve the RBT with fixed $\eta/s(T)$, we determine locally in space and time the total cross section σ_{tot} according to the Chapmann-Enskog theory. For a pQCD inspired cross section, $d\sigma/dt \sim \alpha_s^2/(t-m_D^2)^2$, typically used in parton cascade approaches[4, 13, 5], this gives:

$$\eta/s = \frac{1}{15}\langle p\rangle\, \tau_\eta = \frac{1}{15}\frac{\langle p\rangle}{g(a)\sigma_{tot}\rho}\,, \tag{1}$$

where $a = m_D/2T$, with m_D being the screening mass regulating the angular dependence of the cross section σ_{tot}. The $g(a)$ function is the proper function accounting for the pertinent relaxation time $\tau_\eta^{-1} = g(a)\sigma_{tot}\rho$ associated to the shear transport coefficient and it is given by $g(a) = 50^{-1}\int dy y^6\left[(y^2+3^{-1})K_3(2y) - yK_2(2y)\right]h\left(a^2y^{-2}\right)$ with K_n-s being the Bessel functions and the function h relating the transport cross section to the total cross section $\sigma_{tr}(s) = \sigma_{tot}\, h(m_D^2/s)$ and $h(\zeta) = 4\zeta(1+\zeta)[(2\zeta+1)ln(1+1/\zeta) - 2]$. As shown in Ref. [14] the eq.(1) correctly describes the relation between $\eta/s \leftrightarrow T, \sigma(\theta), \rho$ of the system. This approach is similar to the one [15],where it is shown to converge to viscous hydrodynamic in 1+1D.

In the calculations shown in the next section we have used longitudinal boost invariant initial conditions. The initial $dN/d\eta$ have been chosen in order to reproduce the final $dN_{ch}/d\eta(b)$ at mid rapidity as observed in the experiments at RHIC and LHC energies. In coordinate space the partons are initially distributed according to the Glauber model. In the momentum space the distribution is thermal up to $p_T = 2\,\mathrm{GeV}$ and at larger p_T we include the spectrum of non-quenched minijets according to standard NLO-pQCD calculations. We have used the relation $(\tau A_T)^{-1}dN_{ch}/d\eta \propto T^3$ to determine how the maximum temperature, T_{m0} scales with the collision energy. The initial time, τ_0, is fixed by uncertainty relation between the initial average thermal energy and the initial time $T_{m0}\tau_0 \approx 1$. As also done in hydrodynamics we fix $T_{m0} = 340MeV$ and $\tau_0 = 0.6fm/c$ for $Au+Au$ at $\sqrt{s} = 200$ GeV. which implies $T_{m0} = 290$ MeV and 560 MeV respectively for $\sqrt{s} = 62.4$ GeV and 2.76 TeV. The temperature profile scales with the energy density as $T(\vec{r}) = T_{m0}(\epsilon(\vec{r})/\epsilon(0))^{1/4}$.

3. The $v_2(p_T)$ at different beam energies and $\eta/s(T)$

In this section we discuss a connection between the different regions of $\eta/s(T)$ phase diagram in Fig.1 and the value of the $v_2(p_T)$ at different beam energies and how it could be related to the intriguing invariance of $v_2(p_T)$ with the collision energies as observed in the experiments [2, 12]. From a theoretical point of view there are several indications that η/s should have a typical behavior of phase transition with a 'U' shape. In Fig.1 it is shown the $\eta/s(T)$ used in our calculations. To have a match with the estimates of η/s in the chiral perturbation theory for a meson gas [16, 17], shown by down-triangles in Fig. 1, we have considered an η/s increasing linearly at low temperature $T < T_0 = 1.2T_c$ which also determines kinetic freeze out (f.o.). We however consider an η/s increasing up to 0.40 at $T = 0.8T_c$ a value which is also comparable to the estimate of η/s extrapolated from heavy-ion collisions at intermediate energies (HIC-IE diamonds in Fig. 1)[18, 19]. On the other hand at higher temperature lattice QCD (lQCD) calculations have shown large error bars for η/s and it is not possible to infer a clear temperature dependence in the QGP phase. Therefore for $T > 1.2T_c$ we have considered two cases: one with a linear dependence $4\pi\eta/s = T/T_0$ (red solid line in fig.1) in agreement with lQCD data of Ref.

[20] (up-triangles) and quasi-particle model prediction (orange band) where $\eta/s \sim T^{1-1.5}$ [21]. The other one with a quadratic dependence $4\pi\eta/s = 3.64(T/T_0 - 1) + (T/T_0)^2$ (blue dashed

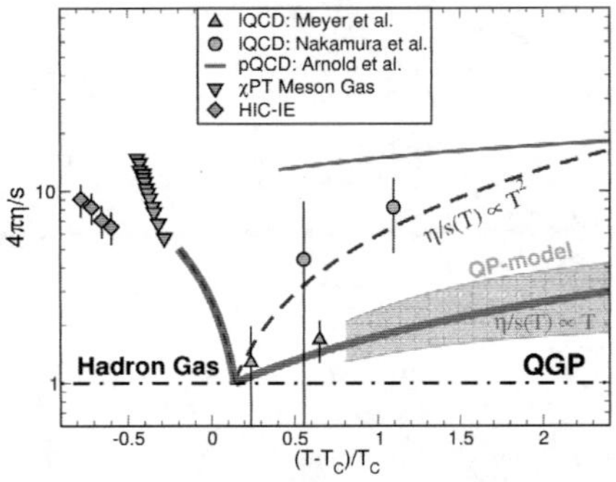

Figure 1. Different temperature dependent parametrizations for η/s. The orange area takes into account the quasi-particle model predictions for η/s [21]. The different lines indicate different T dependencies assumed in the simulation of heavy-ion collision while with the green line it is shown the result for the pQCD calculation [22]. Symbols are as in the legend. See the text for more details.

line) resembling the lQCD in quenched approximation in Ref. [23] as given also in [24]. We also consider a common case of a constant η/s at its conjectured minimum value $1/4\pi$. In Fig. 2(a-d) show the differential elliptic flow $v_2(p_T)$ for the three different beam energies at RHIC and LHC at the same centrality $10 - 20\%$. In Fig. 2a the results are shown for $\eta/s = 1/4\pi$ all over the evolution of the system. As clearly shown in such a case it is not possible to predict an invariance $v_2(p_T)$ in the range of collision energies explored, it breaks at the LHC energy (dot-dashed line) with a breaking up of about 20%. In this case where $\eta/s = const$, the $v_2(p_T)$ at LHC energies is smaller than that at lower energies and this is an implicit effect coming from the different initial p_T distributions. At LHC energies the initial spectrum is flatter than that at lower energies and for fixed η/s it produces a smaller $v_2(p_T)$,for details see [25]. In Fig. 2b we study the effect of a temperature dependent $\eta/s(T)$ with an η/s increasing in the cross over region and constant at $1/4\pi$ in the QGP phase. Comparing Fig. 2a and 2b we can see that the effect of an increasing η/s at lower temperature is to reduce the elliptic flow at lower collision energy, the $v_2(p_T)$ becomes more sensitive to the value of η/s at low temperature. This different behavior is determined by the different initial temperature and consequent lifetime of the fireball. In fact at RHIC energies such a lifetime is about $4 - 6\,\mathrm{fm/c}$ while at LHC it is about $10\,\mathrm{fm/c}$. Therefore at RHIC the elliptic flow has not enough time to fully develop in the QGP phase, while at LHC the lifetime is long enough to let the v_2 develop almost completely in the QGP phase and it becomes sensitive to the value of η/s in the QGP phase, similar result has been found in Ref. [26]. From this reasoning one has the hint that an invariant $v_2(p_T)$ can be caused by a specific T-dependence of η/s that balances the suppression due to the viscosity above and below T_c where a minimum in η/s should occur. In Fig.s 2c and 2d we have shown the effect of a temperature dependence of η/s for $T > T_C$ in the QGP phase. Comparing Fig. 2c and 2d we see that the rapidly increasing $\eta/s(T)$ affects more the system created at LHC and this generates again a larger splitting of the $v_2(p_T)$ among the different energies. This means that also a strong T-dependence in the QGP phase is in contrast with the observed $v_2(p_T)$ invariance. In Fig. 2d it is shown how it is possible to reproduce an almost perfect invariance (within a 5%) with the collision energy if we consider a linear temperature dependence $4\pi\eta/s = T/T_0$, according to the solid red line in Fig. 1. In Fig. 2d the symbols are the experimental results for the $v_2[4]$ measured at RHIC and LHC energy, data taken from [2, 12]. However, we notice that the initial conditions are such that the p_T experimental spectra are fairly well reproduced for all the cases considered except for the case of $\eta/s \sim T^2$ at LHC energy. In such a case the distribution of mini jets cannot be discarded and the final p_T spectra remain too flat if the $\eta/s(T) \sim T^2$ and one would exclude this case.

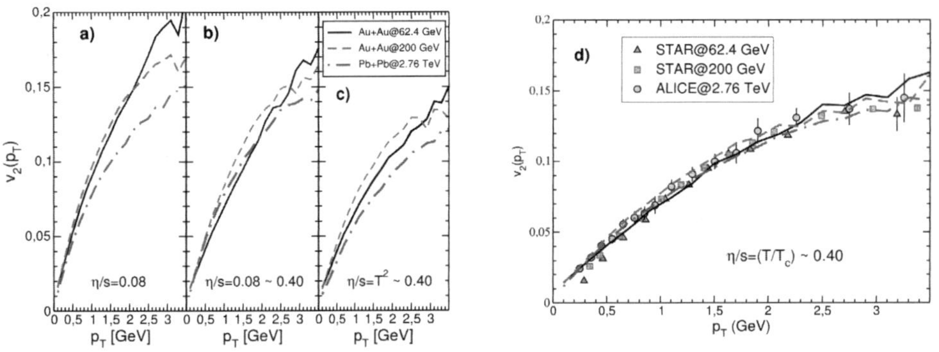

Figure 2. $v_2(p_T)$ at mid rapidity for $10\% - 20\%$ collision centrality. The solid line, the dashed line and the dot dashed line refer to: $Au+Au$ at $\sqrt{s} = 62.4$ GeV and $\sqrt{s} = 200$ GeV and $Pb+Pb$ at $\sqrt{s} = 2.76$ TeV, respectively for different behavior of $\eta/s(T)$ as indicated in the labels.

4. Conclusions

We have studied within a transport approach the effect of a temperature dependent η/s on the generation of the $v_2(p_T)$ from RHIC to LHC collision energies. We find that depending on the range of energies explored the suppression of the $v_2(p_T)$ due to the viscosity of the medium has different contamination coming from hadronic phase or QGP phase. We have shown that it is possible to have a nearly invariant $v_2(p_T)$ if η/s has a 'fall and rise' temperature behavior typical of a phase transition.

5. Acknowledgements

V.G. and F.S. acknoweledge the financial support of the ERC-StG under the QGPDyn Grant.

6. References

[1] Adams J *et al.* (STAR Collaboration) 2005 *Nucl.Phys.* **A757** 102–183 (*Preprint* nucl-ex/0501009)
[2] Aamodt K *et al.* (ALICE Collaboration) 2010 *Phys.Rev.Lett.* **105** 252302 (*Preprint* 1011.3914)
[3] Romatschke P and Romatschke U 2007 *Phys.Rev.Lett.* **99** 172301 (*Preprint* 0706.1522)
[4] Ferini G, Colonna M, Di Toro M and Greco V 2009 *Phys.Lett.* **B670** 325–329 (*Preprint* 0805.4814)
[5] Xu Z and Greiner C 2009 *Phys.Rev.* **C79** 014904 (*Preprint* 0811.2940)
[6] Plumari S and Greco V 2012 *AIP Conf.Proc.* **1422** 56–61 (*Preprint* 1110.2383)
[7] Kovtun P, Son D and Starinets A 2005 *Phys.Rev.Lett.* **94** 111601 (*Preprint* hep-th/0405231)
[8] Staig P and Shuryak E 2011 *Phys.Rev.* **C84** 044912 (*Preprint* 1105.0676)
[9] Cifarelli L, Csernai L and Stocker H 2012 *Europhys.News* **43N2** 29–31
[10] Csernai L P, Kapusta J and McLerran L D 2006 *Phys.Rev.Lett.* **97** 152303 (*Preprint* nucl-th/0604032)
[11] Lacey R A *et al.* 2007 *Phys.Rev.Lett.* **98** 092301 (*Preprint* nucl-ex/0609025)
[12] Adamczyk L *et al.* (STAR collaboration) 2012 *Phys.Rev.* **C86** 054908 (*Preprint* 1206.5528)
[13] Plumari S, Baran V, Di Toro M, Ferini G and Greco V 2010 *Phys.Lett.* **B689** 18–22 (*Preprint* 1001.2736)
[14] Plumari S, Puglisi A, Scardina F and Greco V 2012 *Phys.Rev.* **C86** 054902 (*Preprint* 1208.0481)
[15] Huovinen P and Molnar D 2009 *Phys.Rev.* **C79** 014906 (*Preprint* 0808.0953)
[16] Prakash M, Prakash M, Venugopalan R and Welke G 1993 *Phys.Rept.* **227** 321–366
[17] Chen J W, Li Y H, Liu Y F and Nakano E 2007 *Phys.Rev.* **D76** 114011 (*Preprint* hep-ph/0703230)
[18] Danielewicz P, Barker B and Shi L 2009 *AIP Conf.Proc.* **1128** 104–111 (*Preprint* 0903.2560)
[19] Schmidt W, Katscher U, Waldhauser B, Maruhn J, Stocker H *et al.* 1993 *Phys.Rev.* **C47** 2782–2793
[20] Meyer H B 2007 *Phys.Rev.* **D76** 101701 (*Preprint* 0704.1801)
[21] Plumari S, Alberico W M, Greco V and Ratti C 2011 *Phys.Rev.* **D84** 094004 (*Preprint* 1103.5611)
[22] Arnold P B, Moore G D and Yaffe L G 2003 *JHEP* **0305** 051 (*Preprint* hep-ph/0302165)
[23] Nakamura A and Sakai S 2005 *Phys.Rev.Lett.* **94** 072305 (*Preprint* hep-lat/0406009)
[24] Niemi H, Denicol G, Huovinen P, Molnar E and Rischke D 2012 *Phys.Rev.* **C86** 014909 (*Preprint* 1203.2452)
[25] Plumari S, Greco V and Csernai L 2013 (*Preprint* 1304.6566)
[26] Niemi H *et al.* 2011 *Phys.Rev.Lett.* **106** 212302 (*Preprint* 1101.2442)

Journal of Physics: Conference Series **509** (2014) 012069 doi:10.1088/1742-6596/509/1/012069

Higher moments of net kaon multiplicity distributions at RHIC energies for the search of QCD Critical Point at STAR

Amal Sarkar (for the STAR collaboration)
Indian Institute of Technology Bombay (IITB)
Mumbai, India - 400076
E-mail: amal@rcf.rhic.bnl.gov

Abstract. In this paper we report the measurements of the various moments mean (M), standard deviation (σ), skewness (S) and kurtosis (κ) of the net-Kaon multiplicity distribution at mid rapidity in Au+Au collisions from $\sqrt{s_{NN}}$ =7.7 to 200 GeV in the STAR experiment at RHIC in an effort to locate the critical point in the QCD phase diagram. These moments and their products are related to the thermodynamic susceptibilities of conserved quantities such as net baryon number, net charge, and net strangeness as also to the correlation length of the system. A non-monotonic behaviour of these variables indicate the presence of the critical point. In this work we also present the moment products Sσ and $\kappa\sigma^2$ of net-Kaon multiplicity distribution as a function of collision centrality and energy. The energy and the centrality dependence of higher moments of net-Kaons and their products have been compared with its Poisson expectation and with simulations from AMPT which does not include the critical point. From the measurement at all seven available beam energies, we find no evidence for a critical point in the QCD phase diagram.

1. Introduction

One of the fundamental goals of the heavy-ion collision experiment is to map the QCD phase diagram as a function of temperature and baryon-chemical potential (μ_B) [1]. Finite temperature lattice QCD calculations at baryon chemical potential $\mu_B = 0$ suggest a crossover from a system with hadronic degrees of freedom to a system where the relevant degrees of freedom are quarks and gluons [2]. Several QCD based calculations find the quark-hadron phase transition to be of the first order at large μ_B [3]. The point in the QCD phase plane (T vs μ_B) where the first order phase transition become continuous is called the QCD critical point (CP) [4]. This results in long range correlation and fluctuation at all length scales. Such properties of state open possibilities for distinct experimental signatures which can be used to discover the critical point. Current theoretical calculations are highly uncertain about location of the CP as Lattice calculations at finite μ_B present numerical challenges in computing. The correlation length (ξ) and the magnitude of the fluctuations of the conserved quantities diverge at the critical point [5] but because of the finite size and time effects, ξ takes values in the range of $2-3$fm [6]. Higher non Gaussian moments such as skewness, S ($\propto \xi^{4.5}$), and kurtosis, $\kappa(\propto \xi^7)$ can provide much better handle in location of CP as they are much more sensitive than variance ($\propto \xi^2$) to the correlation length [7]. To cancel out the volume dependence of the higher moments, moment products, such as, $S\sigma(\propto \xi^{2.5})$ and $\kappa\sigma^2(\propto \xi^5)$ are constructed [8]. Lattice calculations and QCD-based models show that moments of net-conserved (baryons (B), strangeness (S) and charge (Q)) distributions are related to its conserved number susceptibilities ($\chi_x = \frac{\langle(\Delta N_x)^2\rangle}{VT}$; where, x represents B, S and Q, & V is the volume) [2, 8]. Volume independent moment products $S\sigma$ and $\kappa\sigma^2$ are the ratio of third ($\chi_x^{(3)}$) to second $\chi_x^{(2)}$ order and fourth $\chi_x^{(4)}$ to second $\chi_x^{(2)}$ order susceptibilities [2, 8]. Close to the critical point, models predict the distributions of the conserved quantities to be non-Gaussian and susceptibilities to diverge causing $S\sigma$ and $\kappa\sigma^2$ to deviate from being constants and have large values [7, 8]. Experimentally measuring event-by-event all produced conserved quantities is very difficult. However, net proton, kaons and charge can be measured to serve proxy for B, S, & Q. We calculate event-by-event the net-Kaon multiplicity ($\Delta N_K = N_{K^+} - N_{K^-}$) to obtain the net-strangeness distribution. In this paper, we report the measurement of the moments of the net-Kaon multiplicity distributions as a function of baryon chemical potential which was varied from 410 to 20 MeV by changing the $\sqrt{s_{NN}}$ from 7.7, 11.5, 19.6, 27, 39, 62.4 to 200 GeV in Au+Au collisions [9] by the STAR collaboration [10]. $S\sigma$ and $\kappa\sigma^2$ have been calculated as a function of collision centrality.

2. Analysis Details

The STAR (Solenoidal Tracker At RHIC) experiment at Brookhaven National Laboratory provides excellent particle identification and large uniform acceptance at mid-rapidity for all the $\sqrt{s_{NN}}$ [9]. Ionization energy loss dE/dx of charged particles in the Time Projection Chamber (TPC) was used to identify charged kaons by comparing it to the theoretical (parameterized) expectation along with a mass square cut from the information of Time-of-Flight (TOF) detector [9]. The analysis was carried out event-by-event with identified positively charged kaons (K^+) and negatively charged kaons (K^-) in full azimuthal coverage for collisions occurring within 30 cm of the TPC center along the beam line. The ΔN_K measurements are carried out within the pseudo-rapidity range $|\eta| < 0.5$ in the range $0.2 < p_T < 1.6$ GeV/c. To reduce the contamination from secondary kaons, we required each $K^+(K^-)$ track to have a minimum p_T of 0.2 GeV/c and a distance of closest approach (DCA) to the primary vertex of less than 1 cm. To suppress the contamination from other particles we use $|n\sigma_{Kaon}| < 2.0$,

$$n\sigma_x = \frac{1}{R}\log\frac{\langle dE/dx\rangle|_{measured}}{\langle dE/dx\rangle|_{expected}} \tag{1}$$

Where, R is the $\langle dE/dx\rangle$ resolution of TPC. For centrality selection we have used uncorrected charged particle multiplicity within $0.5 < |\eta| < 1.0$, to avoid the self-correlation. For each centrality, the average numbers of participants ($\langle N_{part}\rangle$) are obtained by Glauber model calculations. The results are corrected for the finite centrality bin width effects [11].

3. Results

The raw net-Kaon ($\Delta N_K = N_{K^+} - N_{K^-}$) multiplicity distribution which has not been corrected for reconstruction efficiency for various collision centralities (70-80%, 30- 40%, and 0-5%) in Au+Au collisions at $\sqrt{s_{NN}}$ = 7.7 - 200 GeV at mid-rapidity ($|\eta| < 0.5$), are shown in the Figure 1(a). The mean of net-Kaon (ΔN_K) distribution shifts towards zero with increase in $\sqrt{s_{NN}}$. The four moments (M, σ, S, and κ) of the ΔN_K distributions at various collision energies are plotted as a function of average number of participants $\langle N_{\rm part}\rangle$ in Figure 1(b).The moments are calculated from the ΔN_K distributions after correcting for the finite centrality bin width effects [11]. M shows a linear variation with $\langle N_{\rm part}\rangle$ and increases as $\sqrt{s_{NN}}$ decreases. σ increases with $\langle N_{\rm part}\rangle$ as also with beam energies. S is positive and decreases as $\langle N_{\rm part}\rangle$ increases for a given collision energy. The values also decrease as $\sqrt{s_{NN}}$ increases. This indicates that the distributions become symmetric for more central collision and for higher beam energies. κ

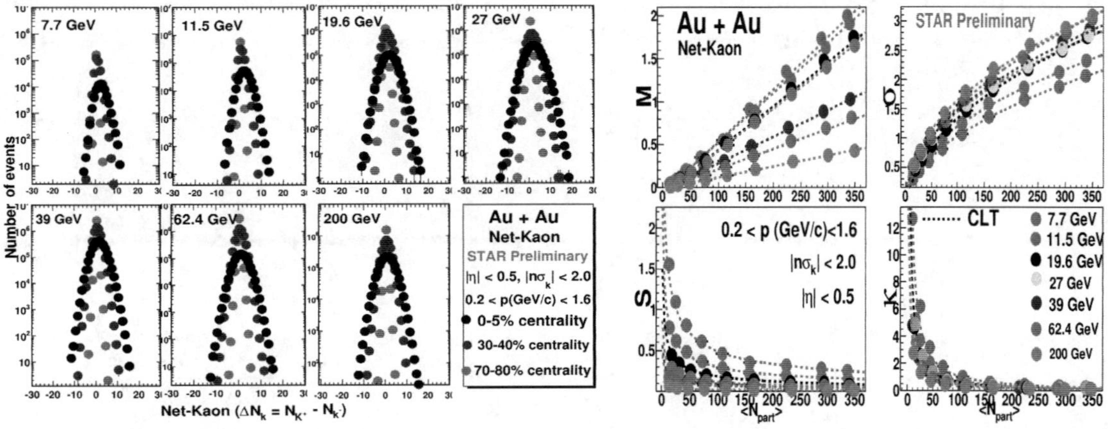

Figure 1. (left panel)ΔN_K multiplicity distribution in Au+Au collisions at $\sqrt{s_{NN}}$ = 7.7 - 200 GeV for various collision centralities at mid-rapidity ($|\eta| < 0.5$) (right panel) Centrality dependence of moments of ΔN_k distributions for Au+Au collisions at $\sqrt{s_{NN}}$ = 7.7 - 200 GeV. The doted lines are the expected values from the central limit theorem. Error bars are statistical.

decreases as $\langle N_{\text{part}} \rangle$ increases, and also decreases as $\sqrt{s_{NN}}$ increases with beam energy. The fitted dotted lines are expected values from the central limit theorem (CLT), which varies as volume's x, $\sqrt{x}$, $\frac{1}{\sqrt{x}}$ and $\frac{1}{x}$ respectively. Error bars are statistical only, calculated from Delta theorem [12]. All results shown in this paper are not corrected for tracking efficiency. The centrality dependence Sσ and $\kappa\sigma^2$ are shown in Figure 2 and Figure 3 compared with the baseline from Poisson statistics for the seven available beam energies. The Poisson baseline has been calculated from the mean value of the N_{K+} and N_{K-} distribution. There is a weak centrality dependence of Sσ value. However, the variation is within 15%. Sσ value is greater than the Poisson baseline for beam energy below 200 GeV. Sσ increases with the decrease in collision energy. Volume independent product $\kappa\sigma^2$ is independent of centrality with in 10%. In

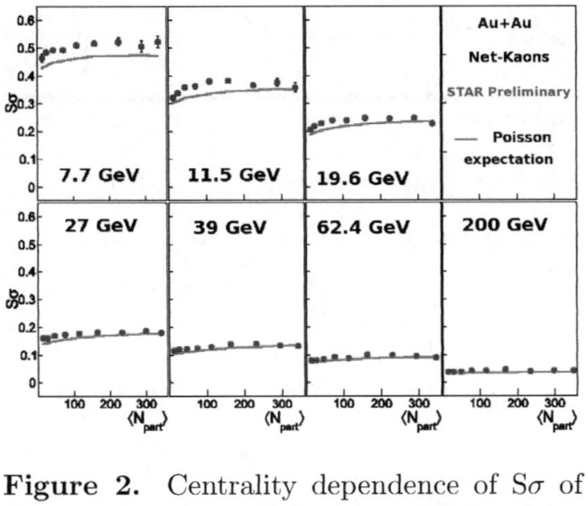

Figure 2. Centrality dependence of Sσ of ΔN_k in Au+Au collisions at $\sqrt{s_{NN}}$ = 7.7 - 200 GeV compared to its Poisson expectation.

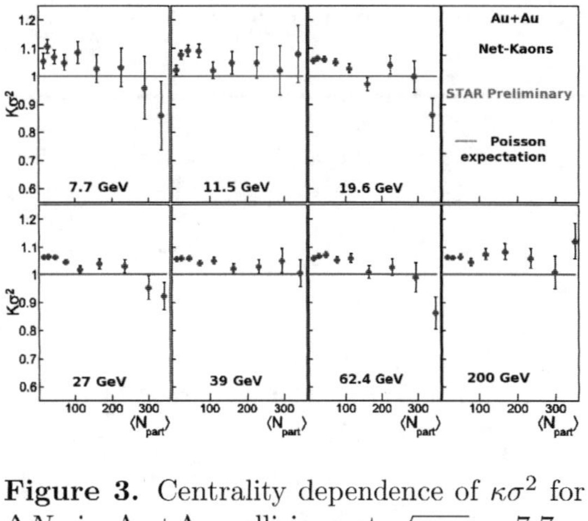

Figure 3. Centrality dependence of $\kappa\sigma^2$ for ΔN_k in Au+Au collisions at $\sqrt{s_{NN}}$ = 7.7 – 200 GeV compared to its Poisson expectation.

Figure 4 and Figure 5 Sσ and $\kappa\sigma^2$ as a function of energy for the top central $0-5\%$ Au+Au collisions are shown. The data as can be seen in the figure matches well with the Poisson baseline as well as with the AMPT model (without CP).

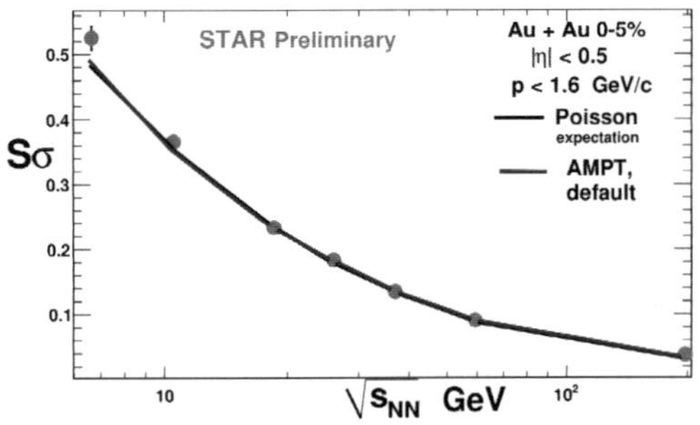

Figure 4. $\sqrt{s_{NN}}$ dependence of Sσ for net-Kaon distributions measured at RHIC for the top central $0-5\%$ Au+Au collisions for available energies. The results are compared to AMPT and with the Poisson expectation.

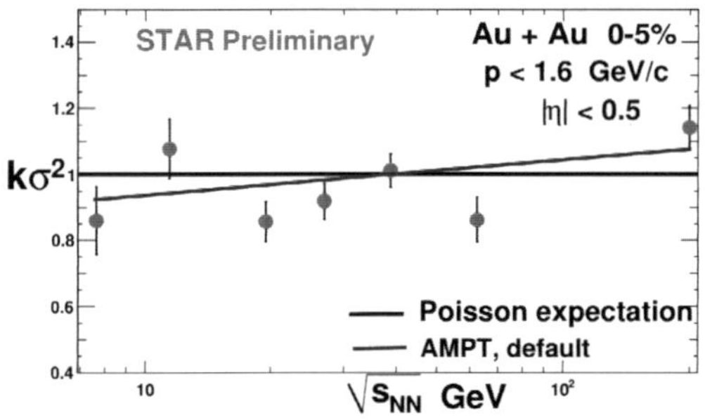

Figure 5. $\sqrt{s_{NN}}$ dependence of $\kappa\sigma^2$ for net-Kaon distributions measured at RHIC for the top central $0-5\%$ Au+Au collisions for available energies. The results are compared to AMPT and with the Poisson expectation.

4. Summary

The first measurements of kurtosis (κ), skewness (S), and variance (σ^2) of net-Kaon multiplicity distributions at mid-rapidity ($|\eta| < 0.5$) in $0.2 < p_T < 1.6$ GeV/c in Au+Au collisions for baryon chemical potentials (μ_B) between 410 and 20 MeV have been shown. The centrality dependence of moments follows the central limit theorem. Sσ and $\kappa\sigma^2$ values have been found to be independent of centrality as a function of $\langle N_{\text{part}} \rangle$ for all available collision energies within 15% and are consistent with expectations from AMPT model (which does not have the critical point). No significant enhancement of moment products compared to the Poisson has been observed at measured energies.

References

[1] STAR Collaboration, Adams J *et al* 2005 *Nucl. Phys.* A **757** 102
[2] Aoki Y *et al* 2010 *arXiv*:1007.2613; Cheng M *et al* 2009 *Phys. Rev.* D **79** 074505
[3] Ejiri S 2008 *Phys. Rev.* D **78** 074507; Bowmanand E S and Kapusta J I 2009 *Phys. Rev.* C **79** 015202
[4] Stephanov M A 2004 *Prog. Theor. Phys. Suppl.* **153** 139
[5] STAR Collaboration, Aggarwal M M *et al* 2010 *arXiv*:1007.2613; Stephanov M A 2011 *Phys. Rev. Lett.* **107** 052301
[6] Berdnikov B *et al* 2000 *Phys. Rev.* D **61** 105017
[7] STAR Collaboration, Aggarwal M M *et al* 2010 *Phys. Rev. Lett.* **105** 022302; Stephanov M A 2009 *Phys. Rev. Lett.* **102** 032301
[8] Athanasiou C *et al* 2010 *Phys. Rev. D* **82** 074008; Gupta S 2009 *arXiv:* 0909.4630; Karsch F *et al* 2011 *Phys. Lett.* B **695** pp 136-142
[9] STAR Collaboration, Abelev B I *et al* 2010 *Phys. Rev.* C **81** 024911; STAR Collaboration, Ackermann K H *et al* 2006 *Nucl. Instrum. Meth.* A **558** pp 419-429; STAR Collaboration, Abelev B I *et al* 2006 *Nuclear Physics* A vol 774 pp 956-958
[10] Hatta Y and Stephanov M A 2003 *Phys. Rev. Lett.* **91** 102003
[11] Luo X 2011 *J. Phys.: Conf. Series* **316** 445 012003
[12] Luo X 2012 *J. Phys.* G **39** 025008

Journal of Physics: Conference Series **509** (2014) 012070 doi:10.1088/1742-6596/509/1/012070

Probing the QCD phase diagram with measurements of ϕ-meson production and elliptic flow in heavy-ion collisions at STAR

Md. Nasim (for the STAR Collaboration)

National Institute of Science Education and Research (NISER), Bhubaneswar-751005, India.

E-mail: nasim@rcf.rhic.bnl.gov/mnasim2008@gmail.com

Abstract. We present measurements of ϕ-meson production and elliptic flow (v_2) at mid-rapidity in Au + Au collisions at $\sqrt{s_{NN}}$ = 7.7 - 200 GeV. Energy dependence of nuclear modification factor ($R_{\rm CP}$) of ϕ meson is presented. The ϕ-meson $R_{\rm CP}$ has a value ≥ 1.0 for $\sqrt{s_{NN}}$ $\leq$ 39 GeV. The Ω/ϕ ratios are also presented and show a different trend at the intermediate transverse momentum (p_T) for $\sqrt{s_{NN}}$ = 11.5 GeV compared to higher beam energies. The number-of-constituent quark (NCQ) scaling of v_2 has been studied at various beam energies. The NCQ scaling holds for particles and anti-particles separately including the ϕ meson for $\sqrt{s_{NN}} \geq 19.6$ GeV, which can be considered as an evidence of partonic collectivity. We observe that ϕ-meson v_2 falls off the trend from the other hadrons at highest measured p_T values by 1.8σ and 2.3σ at $\sqrt{s_{NN}}$ = 7.7 and 11.5 GeV, respectively.

1. Introduction

The ϕ vector meson is the lightest bound state of a strange (s) quark and a anti strange ($\bar{s}$) quark. It has a mass of 1.01945 $\pm$ 0.00002 GeV/c^2 which is comparable to the mass of lightest baryons p (0.938 GeV/c^2) and Λ (1.115 GeV/c^2) [1]. The life time of ϕ meson is $\sim$ 42 fm/c. ϕ meson mostly decay outside the fireball because of its longer life time, therefore its daughters do not have much time to re-scatter in the hadronic phase. The hadronic interaction cross-section of ϕ meson is expected to have a small value and hence seems to freeze out early compared to other lighter hadrons (π, K and p) [2]. Therefore its production should be less affected by the later stage hadronic interactions in the evolution of the system formed in heavy-ion collisions.

The elliptic flow parameter v_2 is a good tool for studying the system formed in the early stages of high energy collisions at RHIC [3]. It describes the momentum anisotropy of particle emission from non-central heavy-ion collisions. It is defined as the second harmonic coefficient of the Fourier decomposition of azimuthal distribution with respect to the reaction plane angle (Ψ) and can be written as

$$v_2 = \langle \cos(2(\phi - \Psi)) \rangle, \tag{1}$$

where ϕ is emission azimuthal angle [4]. Although elliptic flow is an early time phenomenon, its magnitude might still be affected by the later stage hadronic interaction. Since the hadronic interaction cross section of ϕ meson is smaller than the other hadrons, its v_2 remain almost unaffected by the later stage interaction [5, 6]. Therefore ϕ-meson v_2 reflect the collective motion of the partonic phase. This makes ϕ meson a clean probe for the study of the QCD phase diagram in the Beam Energy Scan (BES) program at the Relativistic Heavy Ion Collider (RHIC) [7].

2. Data Sets and Analysis Details

The results presented here are based on data collected from Au+Au collisions at $\sqrt{s_{NN}}$= 7.7, 11.5, 19.6, 27, 39, 62.4 and 200 GeV with the STAR detector for minimum bias trigger in 2010 and 2011. The cuts on primary vertex position along the longitudinal beam direction (V_z) is 30 cm for 200 GeV; 40 cm for 39 and 62.4 GeV; 50 cm for 11.5 GeV; 70 cm for 7.7, 19.6 and 27 GeV data set. An additional cut on vertex radius (< 2 cm) has been used to reject contamination from beam pipe interaction. The Time Projection Chamber (TPC) and Time Of Flight (TOF) detectors with full azimuthal coverage were used for particle identification in the central pseudo-rapidity (η) region ($|\eta| < 1.0$). For spectra analysis we have used only the TPC whereas for elliptic flow analysis both TPC and TOF detectors have been used. ϕ mesons are identified using the invariant mass technique from their decay to $K^+ + K^-$ (branching ratio $\sim$ 49.04 $\pm$ 0.6 %). The mixed event technique has been used for combinatorial background estimation [8]. The η-sub event plane method with a η gap of $|\Delta\eta| < 0.05$ has been used for elliptic flow measurements [4].

3. Results

3.1. Transverse momentum spectra

We present $R_{\rm CP}(0-10\%/40-60\%)$ measurement of ϕ mesons at mid-rapidity ($|y| < 0.5$) in Au+Au collisions at $\sqrt{s_{NN}}$ = 7.7 - 39 GeV. The $R_{\rm CP}$ is defined as the ratio of the particles yield in the central to peripheral collisions normalized by number of inelastic binary collisions (N_{bin}). The value of N_{bin} is calculated from the Monte Carlo Glauber simulation [9]. The $R_{\rm CP}$ will be equal to one if the nucleus-nucleus collision is simply a superposition of nucleon-nucleon collision. Deviation of $R_{\rm CP}$ from the unity would imply the contribution from the nuclear medium effects. Because of the energy loss of the partons traversing the high density QCD medium, the $R_{\rm CP}$ of ϕ meson goes below unity at 200 GeV [2]. One can see from Fig. 1, at the intermediate p_T (> 1.5 GeV/c) $R_{\rm CP} \geq 1.0$ for $\sqrt{s_{NN}} \leq 39$ GeV. This indicates that, at low energy the parton energy loss effect is less important.

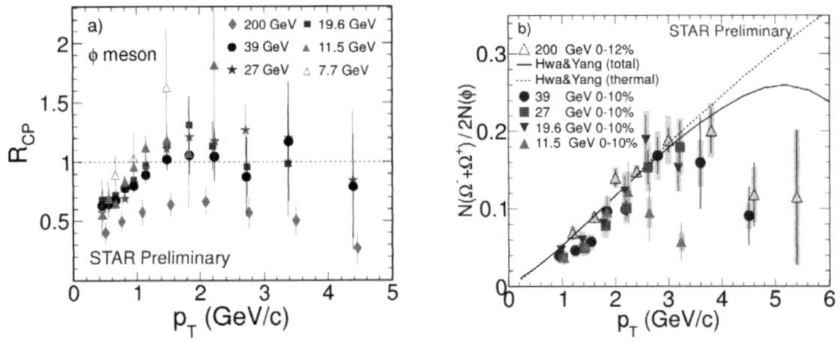

Figure 1. (Color online) Left panel : $R_{\rm CP}$ as function of p_T in Au+Au collisions at various beam energies. The $R_{\rm CP}(0-05\%/40-60\%)$ at $\sqrt{s_{NN}}$ = 200 GeV are taken from previous STAR measurement [10]. Error bars are only statistical uncertainties. Right panel : $(\Omega^- + \overline{\Omega}^+)/2\phi$ as a function of p_T. Green bands are the systematic error and vertical lines are statistical error.

The panel (b) of Fig. 1 shows the baryon-to-meson ratio, $\mathrm{N}(\Omega^- + \overline{\Omega}^+)/2\mathrm{N}(\phi)$, as a function of p_T in Au + Au collisions from $\sqrt{s_{NN}}$ = 11.5 GeV to 200 GeV. The data points for 200 GeV are taken from Ref. [10]. The dashed lines are the results from the recombination model calculations with thermal strange quarks [11]. In Au+Au central collisions at $\sqrt{s_{NN}}$ = 200 GeV, the ratios of $\mathrm{N}(\Omega^- + \overline{\Omega}^+)/2\mathrm{N}(\phi)$ in the intermediate p_T range are explained by the recombination model with thermal strange quarks. The ratios $\mathrm{N}(\Omega^- + \overline{\Omega}^+)/2\mathrm{N}(\phi)$ for $\sqrt{s_{NN}} \geq 19.6$ GeV show similar trend. But at $\sqrt{s_{NN}}$ = 11.5 GeV, the ratio at the highest measured p_T shows a deviation from the trend of other energies. The value of χ^2/ndf for deviation between 11.5 and 19.6 GeV is $\sim$ 8.3/2 for $p_T > 2.4$ GeV/c. This may suggest a change in Ω and/or ϕ production mechanism at $\sqrt{s_{NN}}$ = 11.5 GeV.

3.2. Elliptic flow

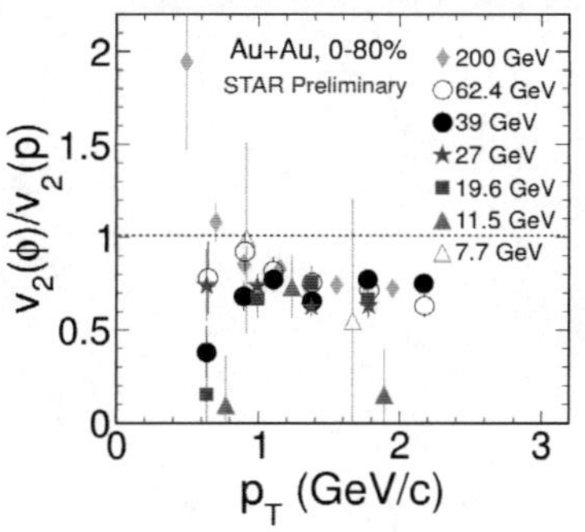

Figure 2. (Color online) $v_2(\phi)/v_2(p)$ ratio as function of p_T in the Au+Au collision at various beam energies for 0-80% centrality [13, 14]. Error bars are only statistical uncertainties.

Figure 2 shows $v_2(\phi)/v_2(p)$ ratios as function of p_T for all beam energies from $\sqrt{s_{NN}}$ = 7.7 GeV to 200 GeV. We observe that at low p_T the $v_2(\phi)/v_2(p)$ ratio decreases with decrease in beam energy. This could be due to less partonic collectivity at low energy, as the ϕ v_2 mostly reflect collectivity from the partonic phase. On the other hand, if we look at 200 GeV, the situation is different. One can see that at top RHIC energy $v_2(\phi)/v_2(p) > 1.0$ at low p_T (< 0.7 GeV/c). This can be explained due to later stage hadronic interaction effect on proton v_2 as predicted in the theoretical (hydro + hadron cascade) model [12].

Figure 3 shows v_2 divided by number-of-constituent quark (n_q) as function of $(m_T - m_0)/n_\mathrm{q}$, where $m_T(=\sqrt{p_T^2 + m_0^2})$ is the transverse mass and m_0 is the rest mass of hadron, at $\sqrt{s_{NN}}$ = 7.7, 11.5 and 19.6 GeV [13, 14]. The NCQ scaling holds fairly well for $\sqrt{s_{NN}} \geq 19.6$ GeV for particles and anti-particles separately including ϕ meson (results for $\sqrt{s_{NN}} > 19.6$ GeV are not shown here). This could be considered as a signature of partonic collectivity [15]. However, we observe that ϕ-meson v_2 falls off the trend from the other hadrons at highest measured p_T values by 1.8σ and 2.3σ at $\sqrt{s_{NN}}$ = 7.7 and 11.5 GeV, respectively. Due to the small hadronic interaction cross-section, v_2 of ϕ meson mostly reflect collectivity from the partonic phase [5, 6]. So the small magnitude of ϕ-meson v_2 at $\sqrt{s_{NN}} \leq 11.5$ GeV could reflect that hadronic interaction are more important, however we need more statistics to make a strong conclusion.

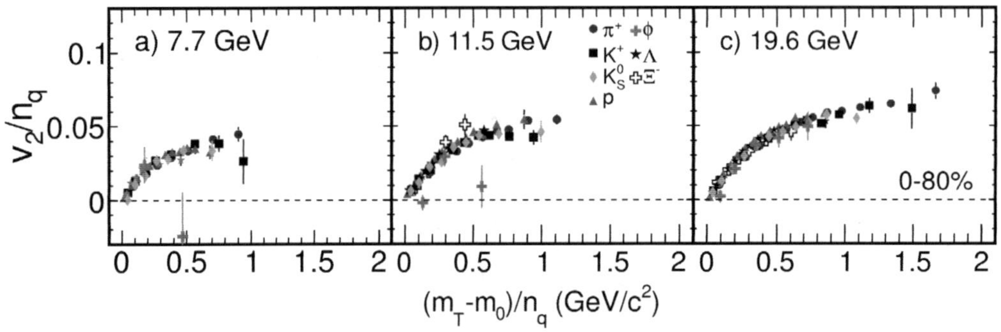

Figure 3. (Color online) The elliptic flow (v_2) scaled by number-of-constituent quark (n_q) as a function of $(m_T - m_0)/n_q$ for selected particles in the Au+Au collision at various beam energies for 0-80% centrality [13, 14]. Error bars are only statistical uncertainties.

4. Summary

We report the study of ϕ-meson production and elliptic flow at mid-rapidity in Au + Au collisions at $\sqrt{s_{NN}}$ = 7.7 - 200 GeV recorded by the STAR detector. At the intermediate p_T, the nuclear modification factor (R_{CP}) of ϕ increases with decreasing beam energies, indicating that the partonic energy loss effect becomes less important at lower beam energies. The ratios of $(\Omega^- + \overline{\Omega}^+)/2\phi$ in the intermediate p_T range show a different trend at 11.5 GeV compared to those for the higher beam energies. This may suggest a change of particle production mechanism at lower beam energy. The NCQ scaling holds for $\sqrt{s_{NN}} \geq 19.6$ GeV for particle and anti-particle separately. We observe that ϕ-meson v_2 show deviation ($\sim 2\sigma$) from the other hadrons at highest measured p_T values at $\sqrt{s_{NN}} \leq$11.5 GeV and also the $v_2(\phi)/v_2(p)$ ratios at low p_T decreases with decrease in beam energy. This may indicate the contribution to the collectivity from partonic phases decreases at lower beam energies.

5. Acknowledgements

Financial assistance from DST International Travel Support and DST SwarnaJayanti project, Government of India is gratefully acknowledged.

References

[1] Beringer J *et al* 2012 *Phys. Rev.* D **86** 010001
[2] Adams J *et al* (STAR Collaboration) 2005 *Nucl. Phys.* A **757** 102
[3] Kolb P F *et al* 2003 *Nucl. Phys.* A **715** 653c
[4] Poskanzer A M and Voloshin S A 1998 *Phys. Rev.* C **58** 1671
[5] Mohanty B and Xu N 2009 *J. Phys.* G **36** 064022
[6] Nasim M, Mohanty B, and Xu N 2013 *Phys. Rev.* C **87**, 014903
[7] Aggarwal M M *et al* (STAR Collaboration) *Preprint* nucl-ex/10072613
[8] Abelev B I *et al* (STAR Collaboration) 2009 *Phys. Lett.* B **673** 183
[9] Abelev B I *et al* (STAR Collaboration) 2009 *Phys. Rev.* C **79** 034909
[10] Abelev B I *et al* (STAR Collaboration) 2009 *Phys. Rev.* C **49** 064903
[11] Hwa R C and Yang C B 2007 *Phys. Rev.* C **75** 054904
[12] Hirano T *et al* 2008 *Phys. Rev.* C **77** 044909
[13] Adamczyk L *et al* (STAR Collaboration) 2013 *Phys. Rev.* C **88** 014902
[14] Adamczyk L *et al* (STAR Collaboration) 2013 *Phys. Rev. Lett.* **110** 142301
[15] Molnar D and Voloshin S A 2003 *Phys. Rev. Lett.* **91** 092301

Electroweak boson production in heavy-ion collisions with CMS.

Michael Gardner on behalf of the CMS Collaboration
University of California, Davis, USA
E-mail: michael.david.gardner@cern.ch

Abstract. The Compact Muon Solenoid (CMS) is fully equipped to measure leptonic decays of electroweak probes in the high multiplicity environment of nucleus-nucleus collisions. Electroweak boson production is an important benchmark process at hadron colliders. Precise measurements of W and Z production in heavy-ion collisions can help to constrain nuclear PDFs as well as serve as a standard candle of the initial state in PbPb collisions at the LHC energies. The inclusive and differential measurements of the Z boson yield in the muon and electron decay channel will be presented, establishing that no modification is observed with respect to pp collisions, scaled by the number of incoherent nucleon-nucleon collisions.

1. Introduction

W and Z bosons were first observed by the UA1 and UA2 experiments at CERN thirty years ago in proton-antiproton collisions at $\sqrt{s} = 540$ GeV [1]. Since then a succession of electron-positron, proton-proton (pp) and proton-antiproton collider experiments have been used to measure their properties, such as mass, decay width and production cross sections at a variety of center-of-mass energies. Now for the first time, due to the large center-of-mass energy and luminosities offered at the LHC, the Z and W boson production can be studied in nucleus-nucleus collisions. Based on the first lead-lead (PbPb) collisions corresponding to 7.2 μb^{-1} integrated luminosity, the CMS collaboration reported first results on the Z $\rightarrow \mu\mu$ [2] and W $\rightarrow \mu\nu$ [3] processes, showing electroweak bosons are essentially unmodified by the hot and dense medium created in heavy-ion collisions, often referred to as the quark-gluon plasma (QGP).

Z-boson production is reported in this proceedings with the second set of PbPb collisions that occurred in late 2011 and accounts for 150 μb^{-1} integrated luminosity. In this document, the second set of pp data at the same energy in the center-of-mass, recorded in 2013, is used as a reference in order to calculate the nuclear modification factor, R_{AA}. The increases in statistics of both the pp and PbPb samples allow more detailed study of the transverse momentum (p_T), rapidity (y) and collision centrality dependencies of the Z-boson yields. The dimuon channel is analyzed with an improved reconstruction algorithm with respect to Ref. [7], and electrons are used for the first time to reconstruct the Z decays in the heavy-ion environment in CMS [8].

Z bosons, once produced, decay within the medium with a typical lifetime of 0.1 fm/c. Since leptons pass freely through the strongly-interacting medium, dileptons from Z bosons can serve as a reference to the processes expected to be heavily modified in the QGP, such as quarkonium production, or the production of an opposite-side jet in Z+jet processes [4]. However, in heavy-ion collisions, Z-boson production can be slightly affected by initial-state effects. This modification is expected to be about 3% from the different mix of protons and neutrons in AA collisions (isospin effect) [5]. Energy loss and multiple scattering of the initial partons could produce modifications of the order of 3% [6]. In addition, the parton distribution functions can be depleted in nuclei, which could modify the Z-boson yield by at most 20% [5].

2. Lepton reconstruction and Signal Extraction

2.1. Muon reconstruction in CMS

With respect to the previous analysis reported in Ref. [7], a new muon offline reconstruction algorithm is used for the PbPb data, with increased efficiency due to an improvement in the track reconstruction. The single-muon reconstruction efficiency is raised from $\simeq$ 85% to $\simeq$ 98%, for muons from Z decays, reaching the level of the algorithm used for pp collisions. Background muons from cosmic rays and heavy-quark semileptonic decays are rejected by requiring a transverse (longitudinal) distance of closest approach of less than 0.2 (5.0) mm from the measured vertex. These selection criteria further reject about 2% of real muons for both pp and PbPb collisions.

2.2. Electron reconstruction in CMS

The electron reconstruction uses information from the pixel and strip tracker, and the electromagnetic calorimeter (ECAL). Electrons produced in pp or PbPb collisions traversing the silicon layers of the tracker can emit bremsstrahlung photons and deposit energy in the ECAL with a significant spread in the azimuthal direction. An ECAL clustering algorithm and in particular the building of so-called *superclusters*, was used for collecting and estimating the proper energy of photons in the heavy-ion environment as in Ref. [9]. Pixels facing these superclusters are looked for, in order to initiate the building of trajectories in the inner tracker. A dedicated algorithm which takes into account the bremsstrahlung emission is used [10].

The reconstruction algorithm results in a reconstruction efficiency of 85% for electrons from Z decays. Further selection criteria reduce this efficiency by about 10%. For the pp sample, the standard algorithms and identification criteria presented in Ref. [11] are used, resulting in a reconstruction efficiency of 98% and a further reduction of 5% due to selection criteria.

2.3. Signal Extraction

Figures 1 and 2 show the dimuon and dielectron invariant mass spectra in the 60–120 GeV/c^2 mass range after applying selection criteria in pp and PbPb collisions. Only leptons with $p_T > 20$ GeV/c and $|\eta| < 2.4$ and 1.44 for the muons and the electrons, respectively, are considered. The dimuon rapidity is limited to $|y| < 2$, while the dielectron rapidity is kept in the $|y| < 1.44$ range. Overlaid histograms are from the Monte Carlo (MC) simulation.

In the 60–120 GeV/c^2 mass range of the PbPb sample, 1022 events are found with opposite-charge (OC, black solid circles) muons and not a single same charge (SC, black open squares) muon pair, while the pp sample has 830 OC and 1 SC pair. In the restricted dielectron rapidity range, 328 (388) OC electrons are measured in the PbPb (pp) data, with 27 (17) SC pairs resulting from higher fake and charge mis-identification rates. The SC pairs provide a measure of uncorrelated combinatorial background: negligible (1/1000 level) in the muon channel, and 8% (4)% in the electron channel. The number of Z candidates is taken as the OC–SC difference. Other sources of background are found to be negligible and covered by systematic uncertainties.

3. Results

Based on pp and PbPb Z-boson yields at the same energy in the center-of-mass, we compute the nuclear modification factors R_{AA} for both electron and muon channels as a function of the Z-boson p_T and rapidity, and event centrality, as the following: $R_{\mathrm{AA}} = N_{\mathrm{AA}}/T_{\mathrm{AA}} \times \sigma_{\mathrm{pp}}$, where

Journal of Physics: Conference Series **509** (2014) 012071 doi:10.1088/1742-6596/509/1/012071

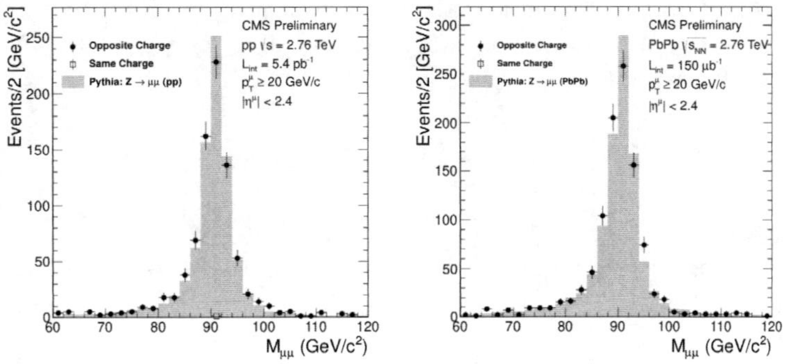

Figure 1. Dimuon invariant mass spectra for muons with $|\eta^{\mu}| < 2.4$ and $p_T^{\mu} > 20$ GeV/c, and dimuons with $|y| < 2.0$ in pp (left) and in PbPb (right) collisions. Full black circles are opposite-charge pairs, open black squares are same-charge pairs. Superimposed is the MC simulation from PYTHIA pp $\to$ Z $\to \mu^+\mu^-$ embedded in HYDJET for the PbPb case.

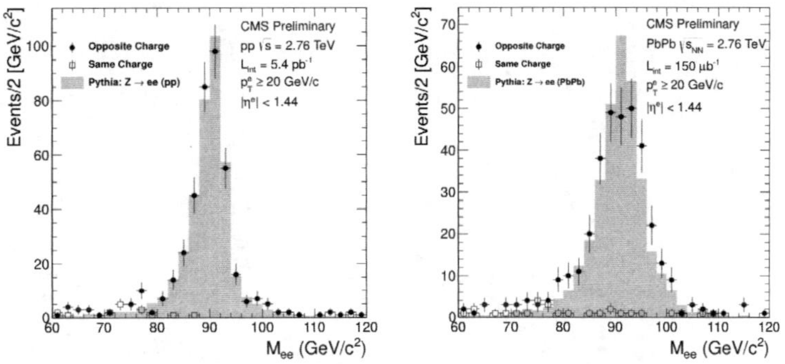

Figure 2. Dielectron invariant mass spectra for electrons with $|\eta^{e}| < 1.44$ and $p_T^{e} > 20$ GeV/c, and dielectrons with $|y| < 1.44$ in pp (left) and in PbPb (right) collisions.

$N_{\rm AA}$ are the Z boson yields per event estimated from PbPb collisions corrected for acceptance and efficiency measured in simulation, and $\sigma_{\rm pp}$ the differential cross sections estimated from pp collisions by dividing the corrected Z-boson yields by the integrated luminosity. The nuclear overlap function $T_{\rm AA}$, computed with a Glauber model [12] is proportional to the number of elementary nucleon-nucleon binary collisions $N_{\rm coll} = T_{\rm AA} \times \sigma_{\rm NN}$ where the latter is an assumed inelastic nucleon-nucleon cross section that cancels to first order in $T_{\rm AA}$.

The data are divided in different independent ranges: 6 in event centrality, 8 (5) in dimuon (dielectron) rapidity, and 7 in Z-boson p_T. Figure 3 shows the obtained AA as a function of the Z-boson p_T, rapidity and event centrality (translated to the average number of participants ($N_{\rm part}$), through the Glauber model mentioned above). The $T_{\rm AA}$ uncertainties, displayed as vertical green bands are common to all points when centrality-integrated, and to the dielectron and dimuon points when centrality-binned. The pp luminosity uncertainty, common to all points is displayed as a vertical grey band.

Within our statistical and systematic uncertainties (dominated by uncertainty in the efficiency from a data-driven Tag and Probe method) the nuclear modification factors in both the electron and the muon channels as a function of p_T, rapidity and event centrality are compatible with unity and no strong deviations are observed. The integrated R_{AA} values are $1.06 \pm 0.05(stat) \pm 0.11(syst)$ in the dimuon channel and $1.08 \pm 0.09(stat) \pm 0.14(syst)$ for dielectrons, and the differential values are shown in Figure 3.

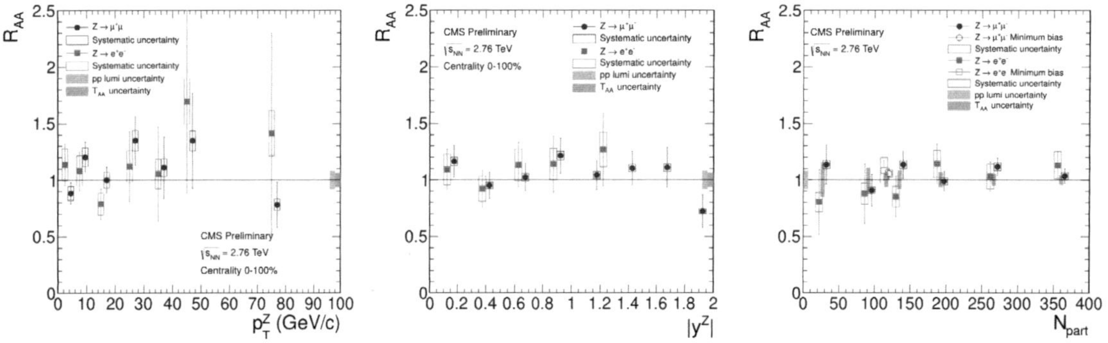

Figure 3. Nuclear modification factor (R_{AA}) extracted from PbPb and pp collisions at $\sqrt{s_{NN}} = 2.76$ TeV for Z $\rightarrow e^+e^-$ and Z $\rightarrow \mu^+\mu^-$ as a function of Z-boson p_T (left), rapidity (center) and number of participants in the collision (right).

4. Conclusions

The Z-boson R_{AA} has been measured from PbPb and pp collisions at $\sqrt{s_{NN}} = 2.76$ TeV. This was done as a function of Z-boson transverse momentum, rapidity and event centrality in both the dimuon and dielectron decay channels with 150 μb^{-1} integrated luminosity. No significant deviation is observed, underlying that within uncertainties, no nuclear effects are observed.

References

[1] UA1 collaboration *Experimental Observation of Lepton Pairs of Invariant Mass Around* $95 - GeV/c^2$ *at the CERN SPS Collider*, *Phys.Lett. B* **126** (1983) 398-410

[2] CMS collaboration *Study of Z boson production in PbPb collisions at nucleon-nucleon center of mass energy = 2.76 TeV*, *Phys.Rev.Lett.* **106** (2011) 212301, arXiv:1102.5435 [nucl-ex]

[3] CMS collaboration *Study of W boson production in PbPb and pp collisions at* $\sqrt{s_{NN}} = 2.76$ *TeV*, *Phys.Lett. B* **715** (2012) 66, arXiv:1205.6334 [nucl-ex]

[4] Kartvelishvili V, Kvatadze R and Shanidze R *On Z and Z+jet production in heavy ion collisions*, *Phys.Lett. B* **356** (1995) 589, arXiv:hep-ph/9505418

[5] Paukkunen H and Salgado C A *Constraints for the nuclear parton distributions from Z and W production at the LHC*, *JHEP* **1103** (2011) 071, arXiv:101.5392 [hep-ph]

[6] Neufeld R B, Vitev I and Zhang B W *Toward a determination of the shortest radiation length in nature*, *Phys.Lett. B* **704** (2011) 590-595. arXiv:1010.3708 [hep-ph]

[7] CMS collaboration *Study of Z boson production with 150* μb^{-1} *integrated PbPb luminosity at* $\sqrt{s_{NN}} =$ 2.76 *TeV*, *CMS PAS HIN-12-008* **HIN-2012/008** (2012)

[8] CMS collaboration *The CMS experiment at the CERN LHC*, *JINST* 0803:S08004 (2008)

[9] CMS collaboration *Measurement of isolated photon production in pp and PbPb collisions at* $\sqrt{s_{NN}} = 2.76$ *TeV*, *Phys.Lett. B* **710**, (2012) 256-277, arXiv:1201.3093 [nucl-ex]

[10] Adam W et al *Reconstruction of Electrons with the Gaussian-Sum Filter in the CMS Tracker at the LHC*, *J Phys. G: Nucl. Part. Phys.* **31** (2005) N9-N20, arXiv:physics/0306087

[11] CMS collaboration *Electron reconstruction and identification at sqrt(s) = 7 TeV*, *CMS PAS EGM-10-004* **EGM-2010/004** (2010)

[12] Miller M L et al *Glauber modeling in high energy nuclear collisions*, *Ann. Rev. Nucl. Part. Sci.* **57** (2007) 205, arXiv:nucl-ex/0701025

Investigation of Mach cones and the corresponding two-particle correlations in a microscopic transport model

I. Bouras[1], B. Betz[1], Z. Xu[2] and C. Greiner[1]

[1] Institut für Theoretische Physik, Johann Wolfgang Goethe-Universität, Max-von-Laue-Str. 1, D-60438 Frankfurt am Main, Germany

[2] Department of Physics, Tsinghua University, Beijing 100084, China

E-mail: bouras@th.physik.uni-frankfurt.de

Abstract. We investigate the evolution Mach cones in ultra-relativistic heavy-ion collisions within a microscopic transport model. Using smooth initial conditions and central collisions the jet-medium interaction is studied using highly-energetic jets and various values of the shear viscosity over entropy density ratio, η/s, of the matter. We observe the formation of Mach cones for small shear viscosity over entropy density ratio, η/s, while for larger values of η/s the characteristic structures smear out and eventually vanish. We extract the final azimuthal two-particle correlation from the final gluon distribution. A double-peak structure shows up if, in a single event, the jet propagates in the direction opposite to the radial flow. This indicates that the contribution from the head shock and from the diffusion wake is superimposed by the radial flow and the contribution of the Mach-cone wings can show up. Considering the superposition of many different jet paths in a central heavy-ion collision, a double-peak structure also appears. The double-peak structure then originates mostly from a superposition of deflected jet-induced Mach cones with the contribution originating from head shock and diffusion wake. A large value of the shear viscosity over entropy density ratio destroys the double-peak structure for any scenario.

1. Introduction

In relativistic heavy-ion collisions at the relativistic heavy-ion collider (RHIC) and the Large Hadron Collider (LHC) a new state of matter, the Quark-Gluon Plasma (QGP), is supposed to be created. In these collisions highly energetic partons propagate through the hot and dense medium and rapidly lose their energy and momentum as the energy is deposited in the medium. Measurements of two- and three-particle correlations in heavy-ion collisions show a complete suppression of the away-side jet, whereas for lower p_T a double peak structure is observed in the two-particle correlation function [1]. One possible origin of these structures were assumed to be the interaction of fast partons with the soft matter which generates collective motion of the medium in form of Mach cones [2, 3, 4, 5, 6, 7, 8, 9]. Although today the most promising explanation for this phenomenon is due to fluctuating initial conditions [10], we address here the question whether the jet-induced Mach cones can still be considered as possible candidates.

In a previous work [11] we demonstrated in BAMPS the transition of Mach cones from perfect fluid limit to the highly viscous regime by adjusting the shear viscosity over entropy density ratio η/s. For this purpose we used a simplified setup of a static box in order to investigate the pure

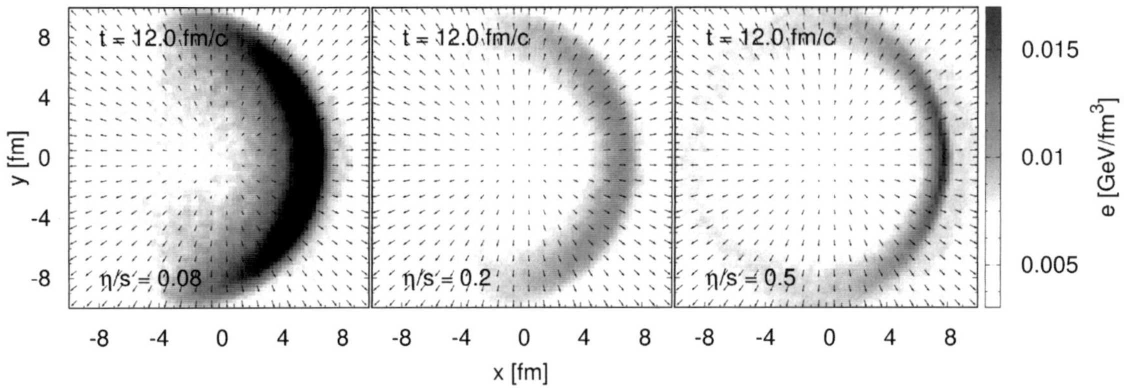

Figure 1. Preliminary results. Shape of a Mach cone in a single event in central ultrarelativistic heavy-ion collisions within BAMPS. We show a snapshot at $t = 12$ fm$/c$ for different values of the shear viscosity over entropy density ratio, η/s. The initial jet energy is $E_{\rm jet} = 20$ GeV. Depicted is the LRF energy density within a specific range. As an overlay we show the velocity profile as arrows with a scaled length arrow.

evolution of a Mach cone. Using two different source terms with an infinite energy reservoir we have shown, that the double peak structure in extracted two-particle correlations appears only, if the source deposits only energy, but no momentum. However, this only holds if in addition the energy deposition is large enough. Furthermore, a finite shear viscosity over entropy density ratio η/s can destroy this double peak structure.

Although this source term with only energy but no momentum deposition was able to explain under some circumstances the double peak structure in extracted two-particle correlations, the physical motivation for this source term is weak, since its properties does not fit in the usual picture of a jet assumed in heavy-ion collisions. Therefore, the origin of such a double peak structure observed in heavy-ion collision is most probably not the source alone, but maybe medium effect, such as the expansion of the medium. Furthermore, due to the infinite energy in our previous setups, the source never stopped, which also may have an effect of the final results.

In this proceeding we demonstrate first preliminary results using a more realistic setup in order to describe the evolution of jet-induced Mach cones in relativistic collisions of heavy-ions using the microscopic transport model BAMPS. For this purpose we use smoothed Glauber initial conditions for binary collisions only in the transverse direction [12]. For this study we want to focus on the impact of the longitudinal and radial flow on the jet only and thus we neglect additional effects like local density fluctuations. Moreover, we consider only central $Au + Au$ collisions ($b = 0$ fm) at RHIC energies, which results in neglecting effects originating from elliptic flow, v_2, or higher harmonics. We further study the influence of the shear viscosity over entropy density ratio, η/s, on the final results. We use only binary collisions with an isotropic angle distribution. The jet is initialized at midrapidity with $E_{\rm jet} = 20$ GeV and looses its energy only via binary scatterings.

Since the microscopic transport model BAMPS has no effective hadronization process implemented yet, the final particle distribution is obtained by stopping the simulation at a certain time and extracting the hydrodynamic quantities as well as the two-particle correlations from the final gluon-momentum distribution.

In Fig. 1 we discuss our first scenario of a single jet event a central heavy-ion collision. The initial position of the jet is at $x = -4$ fm and $y = 0$ fm with initially only momentum in x-direction. We perform calculations with different values of the shear viscosity over entropy

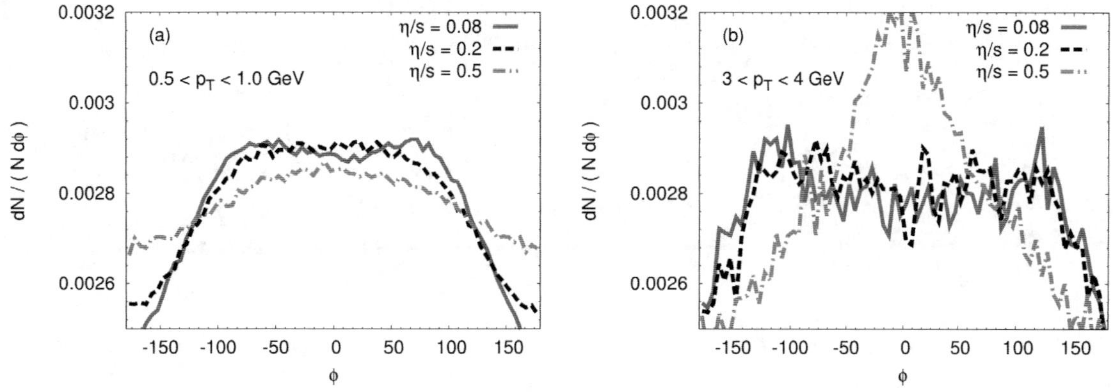

Figure 2. Preliminary results. Two-particle correlations, $\mathrm{d}N/(N\mathrm{d}\phi)$, extracted from BAMPS calculations using several values of η/s in a single event. The results are shown for several regions of p_T and at fixed time $t = 12$ fm/c.

density ratio which reflect possible strengths for the medium interactions expected in HIC. Depicted is the LRF energy density within a specific range. As an overlay we show the velocity profile as arrows with a scaled length arrow. We show a snapshot at $t = 12$ fm/c (Note that the energy density is obviously lower than the expected energy density for deconfined matter. The further evolution when the decondined matter should have already hadronized when the critical energy density is reached does not affect ponderable our final conclusions taken from these calculations). We clearly observe the development of a shock wave in form of Mach cone for a small value of the shear viscosity over entropy density ratio, $\eta/s = 0.08$. However, the shock front is strongly curved, which is due to the strong jet quenching. Increasing the value of η/s results into a smearing out of the profile, while for $\eta/s = 0.5$ the Mach cone structure is very diffuse and vanishes. The vanishing Mach cone structure is analogous to the work done in [11].

As demonstrated in [11] the strong contribution of head shock and diffusion wake in a static box scenario prevents the observation of a double-peak structure originating from the Mach cone wings. However, in this scenario shown in Fig. 1 of a single event, the initial jet propagates in opposite direction to the radial flow of the expanding medium in the heavy-ion collision. This interplay of the medium with the jet strongly affects the final distribution of particles and is the strongest for small values of the shear viscosity. Thus, the contribution of head shock and diffusion wake is reduced.

For this purpose we inspect the extracted normalized azimuthal particle distribution, $\mathrm{d}N/(N\mathrm{d}\phi)$, in Fig. 2. Using different cuts in p_T we demonstrate that a double-peak structure develops for several regions of p_T, and a rather small value for the shear viscosity over entropy density ratio, η/s. For $\eta/s = 0.08$ and $0.5 < p_T < 1$ GeV, as shown in the left panel of Fig. 2, a double-peak structure is observable. We assume that the contribution of head shock and diffusion wake is indeed compensated by the radial flow and the contribution from the Mach cone wings show up. The appearing peaks are approximately at $\phi \approx \pm 70^\circ$. However, for larger transverse momentum, i.e., $3 < p_T < 4$ GeV, and for $\eta/s = 0.08$ and 0.2 again a double-peak structure appears but the peaks move to approximately at $\phi \approx \pm 120^\circ$. In this case we suggest that this contribution comes from the region of the diffusion wake, where matter flows into the region of lower pressure and energy density. Finally, the head-shock contribution for $\eta/s = 0.5$ is very strong and not reduced by the radial flow.

In Fig. 3 we show the results, where we average over all possible jets with randomly chosen

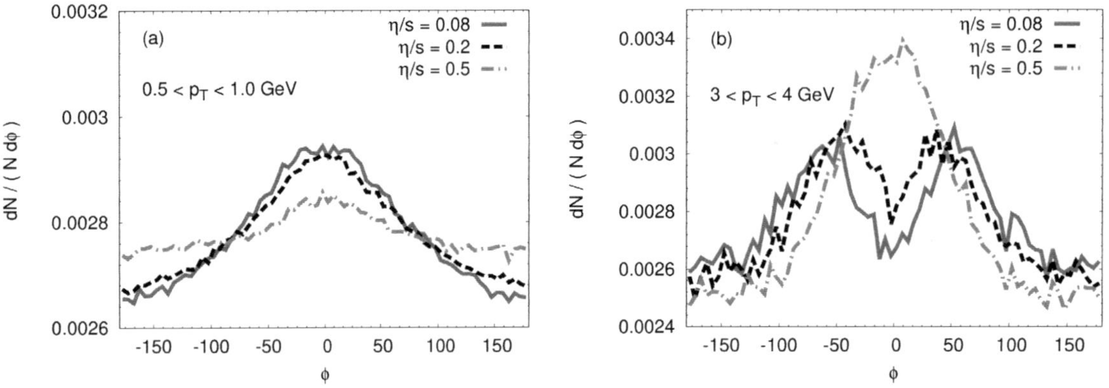

Figure 3. Preliminary results. Two-particle correlations, $dN/(Nd\phi)$, extracted from BAMPS calculations using several values of η/s. In this scenario we average over several jet paths, i.e., we consider many different events. The results are shown for several regions of p_T and at fixed time $t = 12$ fm/c.

starting positions. In contrast to the above scenario for a single event this scenario gets closest to the experimental situation as many different events are considered. In the left panel of Fig. 3 we show the normalized azimuthal particle distribution for $p_T = 0.5 - 1$ GeV. We do observe a double-peak structure for any value of the shear viscosity over entropy density ratio, η/s. In contrast, for $p_T = 3 - 4$ GeV, as demonstrated in the right panel of Fig. 3, a double-peak structure appears for sufficiently low $\eta/s = 0.08 - 0.2$. The peaks are approximately at $\phi \approx 50°$. In this scenario the double-peak structure originates mostly from a superposition of deflected jet-induced Mach cones with the contribution originating from head shock and diffusion wake [13].

Acknowledgements
The authors are grateful to the Center for the Scientific Computing (CSC) at Frankfurt for the computing resources. I. B. is grateful to HGS-Hire. This work was supported by BMBF and the Helmholtz International Center for FAIR within the framework of the LOEWE program launched by the State of Hesse.

References

[1] Wang F (STAR) 2004 *J. Phys.* **G30** S1299–S1304 (*Preprint* nucl-ex/0404010)
[2] Stoecker H 2005 *Nucl. Phys.* **A750** 121–147 (*Preprint* nucl-th/0406018)
[3] Scheid W, Mller H and Greiner W 1974 *Phys.Rev.Lett.* **32** 741–745
[4] Baumgardt H G *et al.* 1975 *Z. Phys.* **A273** 359–371
[5] Satarov L M, Stöcker H and Mishustin I N 2005 *Phys. Lett.* **B627** 64–70 (*Preprint* hep-ph/0505245)
[6] Betz B, Noronha J, Torrieri G, Gyulassy M, Mishustin I *et al.* 2009 *Phys.Rev.* **C79** 034902 (*Preprint* 0812.4401)
[7] Bouras I *et al.* 2010 *J. Phys. Conf. Ser.* **230** 012045 (*Preprint* 1004.4615)
[8] Bouras I *et al.* 2009 *Phys. Rev. Lett.* **103** 032301 (*Preprint* 0902.1927)
[9] Bouras I *et al.* 2010 *Phys. Rev.* **C82** 024910 (*Preprint* 1006.0387)
[10] Schenke B, Jeon S and Gale C 2011 *Phys.Rev.Lett.* **106** 042301 (*Preprint* 1009.3244)
[11] Bouras I, El A, Fochler O, Niemi H, Xu Z *et al.* 2012 *Phys.Lett.* **B710** 641–646 (*Preprint* 1201.5005)
[12] Kolb P, Heinz U W, Huovinen P, Eskola K and Tuominen K 2001 *Nucl.Phys.* **A696** 197–215 (*Preprint* hep-ph/0103234)
[13] Betz B, Noronha J, Torrieri G, Gyulassy M and Rischke D H 2010 *Phys.Rev.Lett.* **105** 222301 (*Preprint* 1005.5461)

ATLAS high-p_{T} measurements in lead-lead collisions

Zvi Citron on behalf of the ATLAS Collaboration
Weizmann Institute of Science, Rehovot Israel
E-mail: zvi.citron@weizmann.ac.il

Abstract. Measurements of high-p_{T} processes in lead-lead collisions can provide insight into the physics responsible for the quenching of jets in the quark gluon plasma. Results are presented from ATLAS measurements of single jet suppression, both inclusively and as a function of the azimuthal angle of the jets with respect to the elliptic flow plane, single jet fragmentation, and gamma-jet correlations. Implications of the results are discussed. Results are also presented for measurements of photon and Z boson production that provide essential tests on calculations of hard scattering rates using binary collision scaling.

Lead-lead collisions at $\sqrt{s_{NN}} = 2.76$ TeV at the Large Hadron Collider (LHC) provide the opportunity to study strongly interacting matter at the highest temperatures achieved in a laboratory setting. The ATLAS experiment has a robust heavy-ion program to take advantage of this opportunity. In the 2010 and 2011 LHC runs, yielding approximately 10 μb^{-1} and 0.15 nb^{-1}, respectively, the ATLAS experiment has made a set of measurements that form an emerging picture of the hot dense matter created in heavy ion collisions. These measurements include hard probes such as photons [1, 2], Z bosons [3, 4], and single and di-jet measurements [5, 6, 7, 8]. Some of the present discussion appears in [9].

The ATLAS detector [10] at the LHC covers nearly the entire solid angle around the collision point. It consists of an inner tracking detector surrounded by a thin superconducting solenoid, electromagnetic and hadronic calorimeters, and a muon spectrometer incorporating three superconducting toroid magnet systems. The inner-detector system (ID) is immersed in a 2 T axial magnetic field and provides charged particle tracking in the pseudorapidity range $|\eta| < 2.5$. The high-granularity silicon pixel detector covers the vertex region and is surrounded by the silicon microstrip tracker and transition radiation tracker. The calorimeter system covers the range $|\eta| < 4.9$. Within the region $|\eta| < 3.2$, electromagnetic calorimetry is provided by barrel and endcap high-granularity lead liquid-argon (LAr) calorimeters, with an additional thin LAr presampler covering $|\eta| < 1.8$. Forward calorimeters (FCal) are located in the range $3.1 < |\eta| < 4.9$. The muon spectrometer comprises separate trigger and high-precision tracking chambers that measure the deflection of muons in a magnetic field generated by superconducting air-core toroids. The precision tracking chamber system covers the region $|\eta| < 2.7$ with trigger coverage in the range $|\eta| < 2.4$.

High momentum particles allow us to probe the properties of the dense medium created in heavy ion collisions, and how color sensitive objects interact with it. Measurements of inclusive single particle spectra as well as two particle correlations have suggested that jets are being "quenched" in the medium, and much effort has been made to understand the mechanisms of this quenching. The suppression of jet production may be quantified using the nuclear modification factor, $R_{\rm CP}$:

$$R_{\rm CP} = \frac{\langle N_{\rm coll}\rangle(P)}{\langle N_{\rm coll}\rangle(C)} \frac{(1/N_{evt,C})d^2N_C/d\eta dp_{\rm T}}{(1/N_{evt,P})d^2N_P/d\eta dp_{\rm T}} \quad (1)$$

where C and P refer to central and peripheral event classes, respectively, and $\langle N_{\rm coll}\rangle$ is the mean number of binary nucleon-nucleon collisions in a given centrality class. Figure 1 shows the $R_{\rm CP}$ of jets as a function of the mean number of nucleons participating in a Pb+Pb collision, $\langle N_{\rm part}\rangle$, clearly demonstrating the suppression of the overall jet yield. The suppression observed is independent of the jet $p_{\rm T}$ within the experimental uncertainties.

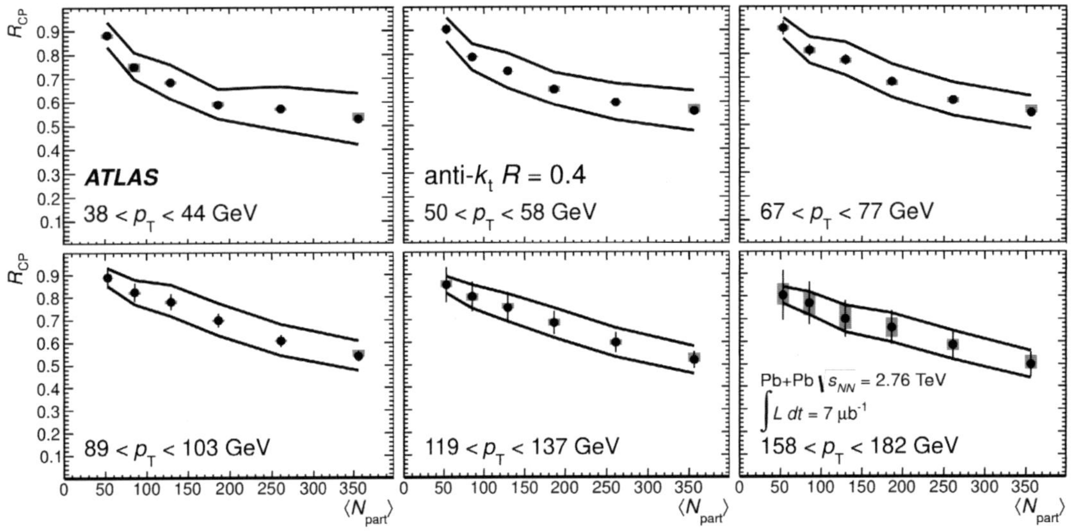

Figure 1. $R_{\rm CP}$ as a function of $\langle N_{\rm part}\rangle$ for R = 0.4 anti-k$_t$ jets in six $p_{\rm T}$ bins. The jet spectra are unfolded to correct for any distortions due to detector resolution and underlying event fluctuations. The error bars indicate statistical errors from the unfolding; the shaded boxes indicate point-to-point systematic errors that are only partially correlated. The solid lines indicate systematic errors that are fully correlated between all points. The horizontal errors indicate systematic uncertainties on $\langle N_{\rm part}\rangle$. Figure is from reference [5].

Further understanding of how jets interact with the medium is provided by considering the azimuthal distribution of jets. The di-jet energy asymmetry [8] and jet $R_{\rm CP}$ show that there is significant quenching, however this is averaged over all in-medium path lengths the jets may be probing. In order to directly examine the path length dependence of jet modification the distribution of jets relative to the reaction plane of the collision was measured, which reflects the anisotropy of the reaction region in which the medium formed. Similarly defined to the v_2 used to measure the flow, v_2^{jet} measures anisotropy of jet yields relative to the amount of medium they traverse. Figure 2 shows a significant non-zero v_2^{jet} as a function of $p_{\rm T}$.

To study the mechanisms of the quenching we may measure the structure of the reconstructed jets themselves. A possible mechanism of jet suppression involves radiated gluons from the leading parton being shunted out of the jet cone due to multiple scattering with the medium, which would lead to a dependence on the jet cone size of the jet $R_{\rm CP}$. This trend is seen in the $R_{\rm CP}$ measured for jets reconstructed with different radii [5].

To further understand the structure of the jets and possible mechanisms of quenching, we

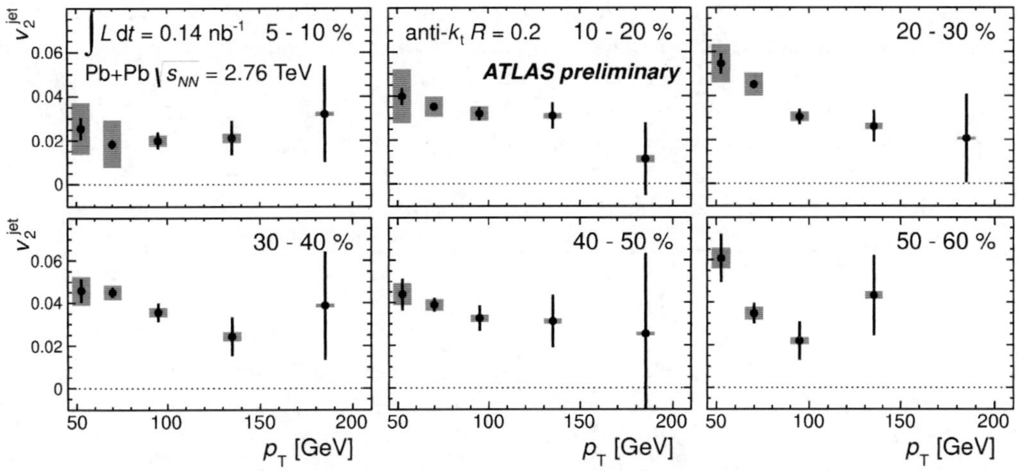

Figure 2. v_2^{jet} as a function of p_T for several centrality bins. The error bars on the points indicate statistical uncertainties while the shaded bands represent systematic uncertainties. Figure is from reference [7].

may consider the fragmentation function of the jets and its central to peripheral ratio:

$$D(z) = \frac{{p_T}^{particle}}{{p_T}^{jet}} cos(\sqrt{\Delta\eta^2 + \Delta\phi^2}), R_{D(z)} = \frac{D(z)|_{central}}{D(z)|_{peripheral}} \quad (2)$$

where particle refers to reconstructed tracks inside the reconstructed jet cone with $\Delta\eta$ and $\Delta\phi$ the distance between the particle and jet directions. The ratio $R_{D(z)}$ reflects the increasing modification of the jet substructure with centrality. Figure 3 shows $R_{D(z)}$ for three centrality intervals, and indicates that within the experimental uncertainties the leading particle at high z is unmodified but rather that only lower momentum particles at mid z are.

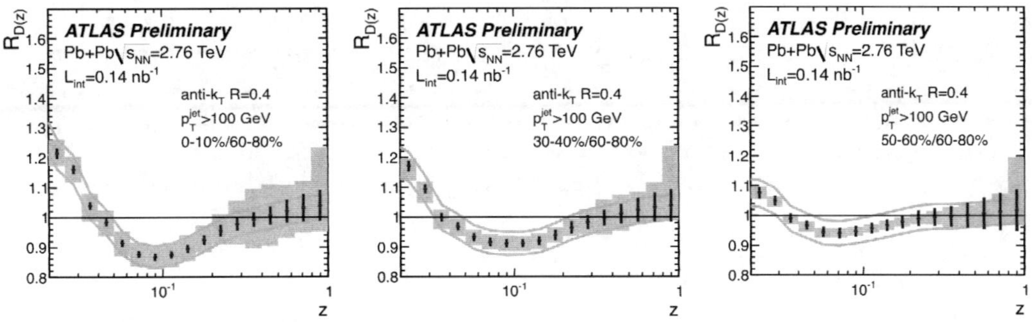

Figure 3. Ratios of $D(z)$ for three bins in collision centrality to those in peripheral (60-80%) collisions, $R_{D(z)}$, for $R = 0.4$ jets. The error bars on the data points indicate statistical uncertainties while the shaded bands indicate systematic uncertainties that are uncorrelated or partially correlated between points. The solid lines indicate systematic uncertainties that are 100% correlated between points. Figure is from reference [6].

In contrast to the modification of color sensitive objects discussed above, color neutral electroweak bosons are expected to be unmodified in the medium. To confirm this as well as the $\langle N_{\rm coll}\rangle$ scaling assumptions, we measure photons and Z bosons. The boson yields are compared to pQCD models of production that include no modification due to the presence of the medium or to the nuclear parton distribution functions, and show good agreement [2, 3].

The color neutral bosons provide an unbiased baseline from which to measure jet modification in the medium. In a boson+jet event the boson allows direct access to the initial hard scattering which produced both the boson and the jet, and is thus a useful baseline from which to evaluate the jet modification. We may consider the azimuthal distribution of boson+jet pairs and the momentum fraction $p_{\rm T}^{\rm jet}/p_{\rm T}{}^{boson}$. As in the di-jet case the azimuthal distribution is consistent between central Pb+Pb events and an unmodified pQCD model, however selecting azimuthally correlated events there is a reduction in the mean $p_{\rm T}$ of the jet compared to the boson $p_{\rm T}$. These events have been studied for the Z boson and photon+jet cases [4, 2]. The means of the $p_{\rm T}^{\rm jet}/p_{\rm T}{}^{boson}$ and the boson+jet yield normalized by the number of bosons are shown in Figure 4 for photons from both the data and simulation. Although the uncertainties are large (especially in the Z boson case) these measurements show a qualitative result consistent with jet quenching and hold promise for future precision studies.

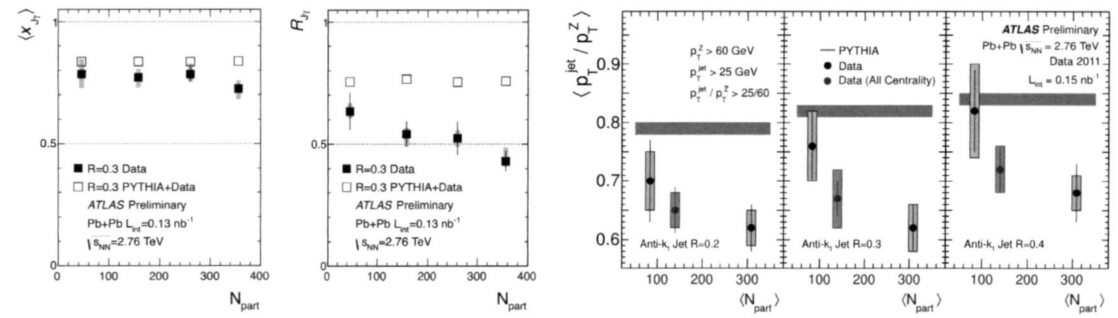

Figure 4. Far left: The mean energy fraction $\langle x_{J_\gamma}\rangle = \langle p_{\rm T}^{\rm jet}/p_{\rm T}{}^{photon}\rangle$ from fully corrected and unfolded distributions calculated as a function of jet radius ($R = 0.2$ and $R = 0.3$), and for each radius as a function of $\langle N_{\rm part}\rangle$. Center: The integrated yield of jets per photon R_{J_γ}. The figures are from reference [2]. Far right: The extracted means from the $p_{\rm T}^{\rm jet}/p_{\rm T}^{Z}$ distributions of the fully corrected data plotted as a function of $\langle N_{\rm part}\rangle$. The systematic uncertainties are largely correlated across the $\langle N_{\rm part}\rangle$ bins. The width of the PYTHIA band represents its uncertainty. The points at $\langle N_{\rm part}\rangle = 140$ refer to 0-80% centrality. The figures are from reference [4]. In all figures bars represent statistical uncertainties, and shaded boxes systematic uncertainties. The data are compared to PYTHIA events embedded into minimally biased data events.

References

[1] ATLAS Collaboration 2012 ATLAS-CONF-2012-051, https://cdsweb.cern.ch/record/1451913
[2] ATLAS Collaboration 2012 ATLAS-CONF-2012-121, https://cdsweb.cern.ch/record/1473135
[3] ATLAS Collaboration 2013 *Phys. Rev. Lett.* **110** 02231
[4] ATLAS Collaboration 2012 ATLAS-CONF-2012-119, https://cds.cern.ch/record/1472941
[5] ATLAS Collaboration 2013 *Phys.Lett.* B **719** 220-241
[6] ATLAS Collaboration 2012 ATLAS-CONF-2012-115, http://cds.cern.ch/record/1472936
[7] ATLAS Collaboration 2012 ATLAS-CONF-2012-116, http://cds.cern.ch/record/1472938
[8] ATLAS Collaboration 2010 *Phys. Rev. Lett.* **105** 252303
[9] Citron Z (on behalf of the ATLAS Collaboration) 2013 *J. Phys.: Conf. Series* **455** 012011
[10] ATLAS Collaboration 2008 JINST **3** S08003
[11] Cacciari M, Salam G P, and Soyez G 2008 JHEP **0804** 063

14th International Conference on Strangeness in Quark Matter (SQM2013) IOP Publishing
Journal of Physics: Conference Series **509** (2014) 012074 doi:10.1088/1742-6596/509/1/012074

Tomography of QGP with jet asymmetries

Guo-Liang Ma

Shanghai Institute of Applied Physics, Chinese Academy of Sciences, P.O. Box 800-204, Shanghai 201800, China

E-mail: glma@sinap.ac.cn

Abstract. Within a multi-phase transport (AMPT) model, the transverse momentum asymmetries for γ-jet and dijet are studied in Pb+Pb collisions at $\sqrt{s_{NN}}$ = 2.76 TeV. A large asymmetry is produced by strong interactions between jets and partonic matter. It is demonstrated that hadronization and hadronic rescatterings have little effects on the formed asymmetries. The asymmetry evolution function discloses that final asymmetry is driven by both initial asymmetry and partonic jet energy loss. The γ-jet imbalance ratio, $x_{j\gamma}$, is sensitive to both production position and passing direction of γ-jet, which could enable a detail tomographic study on the formed partonic matter by selecting $x_{j\gamma}$.

1. Introduction

Jets, initially produced in hard scatterings, are thought as powerful probes to investigate the properties of the deconfined quark-gluon plasma (QGP), because they interact and loss energy when they pass the medium through in high-energy heavy-ion collisions [1]. Recently, the measurements based on full jet reconstruction further provide us a comprehensive understanding in jet energy loss mechanisms [2, 3, 4]. On the other hand, more information about the QGP medium is believed to be extracted with jet tomography, such as by using dijets or γ-jets [5, 6]. Compared to dijets, γ-jets have an innate advantage which makes γ-jets probe the QGP medium more deeply than dijets owing to photon's electromagnetic interactions only with the medium. In this work, both transverse momentum (p_T) asymmetries for reconstructed dijets and γ-jets are investigated within a multi-phase transport (AMPT) model for Pb+Pb collisions at $\sqrt{s_{NN}}$ = 2.76 TeV [7, 8]. With the help of their asymmetries, a detail γ-jet tomography is proposed to study different initial configurations between jets and the partonic matter.

2. Jet reconstruction in the AMPT model

The AMPT model with string melting scenario is implemented in this work [9]. It consists of four main stages of high-energy heavy-ion collisions: the initial condition, parton cascade, hadronization, and hadronic rescatterings. In order to study the transport behaviors of jets, a dijet ($p_T \sim$ 120 GeV/c) or a γ-jet ($p_T \sim$ 60 GeV/c) is triggered in the initial condition based on HIJING model. The high-p_T primary partons pullulate to jet showers full of lower virtuality partons through initial- and final- state QCD radiations. The jet parton showers are converted into clusters of on-shell quarks and anti-quarks through the string meting mechanism of AMPT model. After the melting process, both a quark and anti-quark plasma and jet quark showers are built up. In the following, Zhang's parton cascade (ZPC) model automatically simulates all possible elastic partonic interactions among medium partons and jet shower partons, but without

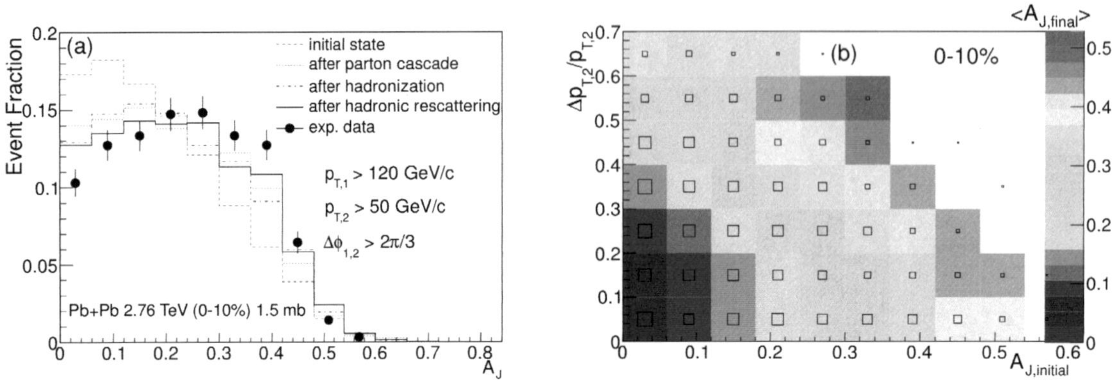

Figure 1. The AMPT results on the dijet asymmetry ratio A_J distributions at different evolution stages (a) and the dijet A_J evolution function $[\langle A_{J,final}\rangle(A_{J,initial}, \Delta p_{T,2}/p_{T,2})]$ (b) for most central Pb+Pb collisions.

including inelastic parton interactions or further radiations at present. When the partons freeze out, they are recombined into medium hadrons or jet shower hadrons via a simple coalescence model. The final-state hadronic interactions between jet shower hadrons and hadronic medium then are described by a relativistic transport (ART) model.

An anti-k_t algorithm from the standard Fastjet package is used to reconstruct jets [10]. The kinetic cuts are chosen as in the CMS experiments [3, 4]. The jet cone size (R) is set to be 0.5 or 0.3 for dijet or γ-jet study, respectively. A pseudorapidity strip of width $\Delta\eta$=1.0 centered on the jet position, with two highest-energy jets excluded, is used to estimate the background, which is subtracted from the reconstructed jet energy in Pb+Pb collisions.

3. Results and discussions

To describe dijet p_T asymmetry, an asymmetry ratio is defined as A_J=$(p_{T,1}-p_{T,2})/(p_{T,1}+p_{T,2})$ as LHC experiments did [2, 3]. Since heavy-ion collisions are dynamical evolutions including many important stages, Figure 1 (a) compares the A_J distributions at or after different evolution stages with experimental data in most central Pb+Pb collisions (0-10%) at $\sqrt{s_{\rm NN}}$ = 2.76 TeV. The dijet asymmetry ratio A_J increases from "initial state" to "after parton cascade" because jets, especially for subleading jets, lose much energy when they pass through the partonic matter. However, the following hadronization and hadronic rescattering processes do not seem to affect the formed A_J distribution any more.

To study how the initial A_J evolves into the final A_J, an asymmetry evolution function is defined as $\langle A_{J,final}\rangle$ $(A_{J,initial}, \Delta p_{T,2}/p_{T,2})$, which neglects the leading jet energy loss. Figure 1 (b) shows the dijet A_J evolution functions for most central Pb+Pb collisions (0-10%) at $\sqrt{s_{\rm NN}}$ = 2.76 TeV, where the color of cell denotes the $\langle A_{J,final}\rangle$ and the size of box in each cell represents the possibility of dijet events. It is found that final A_J of dijet events is driven by two sources. The first one is the remaining A_J part for which dijets keep their initial A_J with small jet energy loss fractions. The second one is the newly formed A_J part for which jets lose much more energy to increase their original A_J, which actually is the dominant source for the observed enhancement of large dijet asymmetry in the central collisions.

The γ-jet p_T imbalance is defined as the ratio of $x_{j\gamma}$=p_T^{jet}/p_T^{γ} to study γ-jet energy loss mechanism [4]. Figure 2 (a) gives the distributions of imbalance ratio $x_{j\gamma}$ at or after different evolution stages for most central Pb+Pb collisions (0-10%) at $\sqrt{s_{\rm NN}}$ = 2.76 TeV. Similarly as the formation of dijet asymmetry, γ-triggered jets mainly lose their energy in the parton

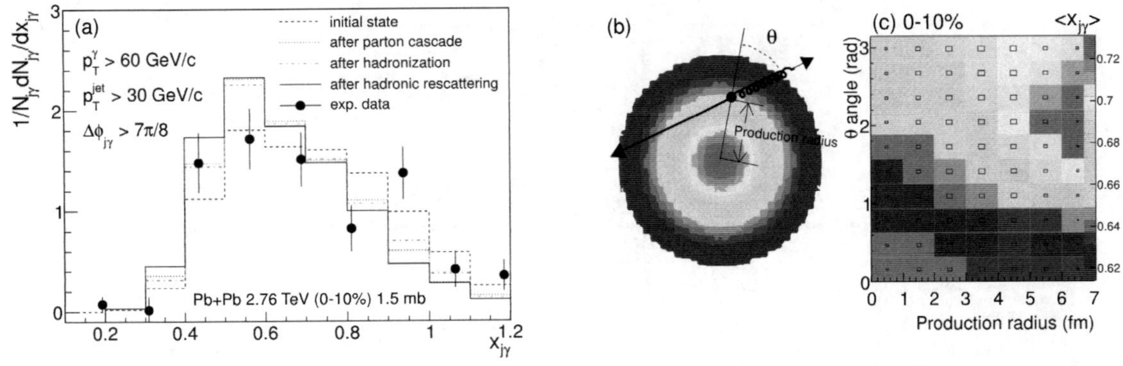

Figure 2. (a) The imbalance ratio $x_{j\gamma}$ distributions at different evolution stages for most central Pb+Pb collisions. (b) Illustration of the production of γ-jet which passes through the partonic medium in a central Pb+Pb collision. (c) The production radius and direction dependences of averaged imbalance ratio, $\langle x_{j\gamma}\rangle$(r, θ) for most central Pb+Pb collisions.

cascade process. Therefore, the γ-jet measurements can basically reflect the information about the interactions between jets and partonic matter.

However, the current measurements are the inclusive results from the average of all cases for γ-jets with any possible production position and direction. As Figure 2 (b) illustrates, a pair of photon and jet can be produced at a position (the black dot), where is at a distance of production radius r to the center, in a central Pb+Pb collision. The direction of photon can be represented by a θ angle which is the angle between the passing direction of photon and the vector from the center to production point. Figure 2 (c) shows the production radius and direction dependence of imbalance ratio of γ-jets, i.e. the averaged value $\langle x_{j\gamma}\rangle$ at (r, θ), for most central Pb+Pb collisions (0-10%) at $\sqrt{s_{NN}}$ = 2.76 TeV, where the color of cell denotes the $\langle x_{j\gamma}\rangle$ and the size of box in cell represents the possibility for γ-jet events with r and θ. The present measurements actually include all initial configurations of (r, θ) between γ-jets and the medium. They basically can be divided into three typical cases. (I) Punch-through jet cases with large radii and small θ angles, corresponding to small $x_{j\gamma}$ values. (II) Escaped-jet cases with large radii and large θ angles , corresponding to large $x_{j\gamma}$. (III) Tangential-jet cases with large radii and θ angles $\sim \pi/2$, corresponding to middle $x_{j\gamma}$. The two-dimensional dependence of $\langle x_{j\gamma}\rangle$(r, θ) in Figure 2 (c) indicates that final $x_{j\gamma}$ is sensitive to the initial γ-jet production information to a certain degree.

Based on the dependence of $\langle x_{j\gamma}\rangle$(r, θ), the different initial γ-jet production configurations could be classified by selecting $x_{j\gamma}$ range. For instance, Figure 3 (a)-(d) show the possibility distributions of measured γ-jet events in r-θ plane with different $x_{j\gamma}$ range selections. Figure 3 (e) shows the hadron azimuthal correlation associated with a triggered γ under the different $x_{j\gamma}$ selections, where the more enhanced away-side peak is observed with the decreasing of $x_{j\gamma}$. The inserted plot of Figure 3 (e) displays the corresponding ratios of γ-hadron correlations for away-side between central Pb+Pb and p+p collisions for different $x_{j\gamma}$ selections. It is pronounced to observe the large enhancement of two side peaks in small $x_{j\gamma}$ classes of Pb+Pb events, which indicates the deflection of both Mach-cone like medium excitation and jet shower by radial flow [11] are more likely formed for the punch-through configurations of γ-jet in central Pb+Pb collisions. Therefore, it provides an effective way to zoom in the medium responses by using γ-hadron correlations associated with different $x_{j\gamma}$ selections.

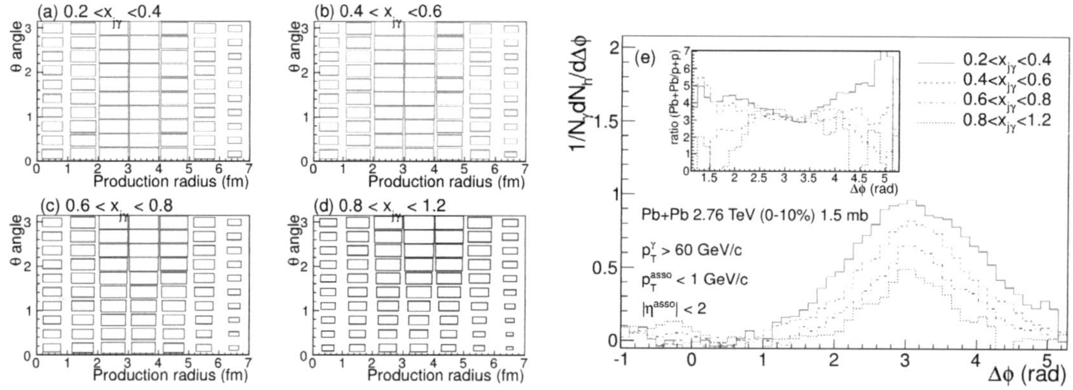

Figure 3. (a)-(d) The possibility distributions of measured γ-jet events in r-θ plane with different $x_{j\gamma}$ selections for most central Pb+Pb collisions. (e) The γ-hadron azimuthal correlations for the most central Pb+Pb collisions with different $x_{j\gamma}$ ranges, where the inserted panel shows the corresponding away-side γ-hadron correlation ratios of most central Pb+Pb to p+p collisions.

4. Summary

In summary, the p_T asymmetries for dijets and γ-jets are investigated with a multi-phase transport model for Pb+Pb collisions at $\sqrt{s_{\rm NN}} = 2.76$ TeV. A large asymmetry is produced by strong interactions between jets and partonic matter, and hardly affected by hadronization and hadronic rescatterings. The final asymmetry is basically driven by both initial asymmetry and partonic jet energy loss. The imbalance ratio, $x_{j\gamma}$, is sensitive to both production position and passing direction of γ-jet, which could enable a detail tomographic study on the formed partonic matter.

Acknowledgments

This work was supported by the Major State Basic Research Development Program in China under Contract No. 2014CB845404; the NSFC of China under Projects No. 11175232, No. 11035009, and No. 11375251; the Knowledge Innovation Program of CAS under Grant No. KJCX2-EW-N01; the Youth Innovation Promotion Association of CAS; and the project sponsored by SRF for ROCS, SEM; CCNU-QLPL Innovation Fund (Grant No. QLPL2011P01); the "Shanghai Pujiang Program" under Grant No. 13PJ1410600.

References

[1] Wang X N and Gyulassy M, 1992 Phys. Rev. Lett. **68**, 1480
[2] Aad G *et al.* (ATLAS Collaboration), 2010 Phys. Rev. Lett. **105**, 252303
[3] Chatrchyan S *et al.* (CMS Collaboration), 2011 Phys. Rev. C **84**, 024906
[4] Chatrchyan S *et al.* (CMS Collaboration), 2013 Phys. Lett. B **718**, 773
[5] Zhang H, Owens J F, Wang E and Wang X N, 2017 Phys. Rev. Lett. **98**, 212301
[6] Zhang H, Owens J F, Wang E and Wang X N, 2009 Phys. Rev. Lett. **103**, 032302
[7] Ma G L, 2013 Phys. Rev. C **87**, 064901
[8] Ma G L, 2013 Phys. Lett. B **724**, 278
[9] Lin Z W, Ko C M, Li B A, Zhang B and Pal S, 2005 Phys. Rev. C **72**, 064901
[10] Cacciari M, Salam G P and Soyez G, 2012 Eur. Phys. J. C **72**, 1896
[11] Li H, Liu F, Ma G L, Wang X N and Zhu Y, 2011 Phys. Rev. Lett. **106**, 012301

Journal of Physics: Conference Series **509** (2014) 012075 doi:10.1088/1742-6596/509/1/012075

Appearance of a quark matter phase in hybrid stars

Tomoki Endo

Division of Physics, Department of General Education, Kagawa National Colledge of Technology, 355 Chokushi-cho, Takamatsu, Kagawa 761-8058, Japan

E-mail: endo@t.kagawa-nct.ac.jp

Abstract. The appearance of quark matter in the core of hybrid stars is a fundamental issue in such compact stars. The central density of these stars is sufficiently high such that nuclear matter undergoes a further change into other exotic phases that consist of hyperons and quarks. However, the equation of state (EOS) for the high-density matter is still not clear and several recent observations have indicated the limitations of the EOSs; theoretical studies should try to elucidate the EOSs. It is believed that the inner regions of the stars should consist of a mixed hadron-quark phase. We study the mixed hadron-quark phase, taking into account finite-size effects, and find that that the mixed phase should be restricted to a narrower region. Therefore, a quark matter phase should appear in the central region.

1. Introduction

Currently accepted theories and many experimental results suggest that hadronic matter changes to quark matter in high-density and/or high-temperature regimes by way of the deconfinement transition. The properties of quark matter have been actively studied theoretically in terms of the quark–gluon plasma, color superconductivity [1, 2], magnetism [3, 4], and experimentally in terms of relativistic heavy-ion collisions [5], and early-universe studies and compact stars [6, 7]. Such studies are continuing to provide exciting results [8]. Presently, we consider that compact stars consist of not only nuclear matter but also other matter such as hyperons and quarks. We call such stars *hybrid stars*.

Because many theoretical calculations have suggested that the deconfinement transition is of the first order at low temperature and high density [9, 10], we assume that it is a first-order phase transition here. The Gibbs condition [11] then gives rise to various structured mixed phases (SMPs). The SMPs proposed by Heiselberg et al. [12] and Glenndening and Pei [13] suggest a crystalline structure for the mixed phase in the cores of hybrid stars. Such structures are called "droplets", "rods", "slabs", "tubes", and "bubbles". We present the equation of state (EOS) for the mixed phase taking into account the charge screening effect [14] without relying on any approximations. We investigate the inner structures of these stars [15]. In this paper, we review the inner structures of hybrid stars and apply our EOS to a stationary rotating star.

2. Formalism and Numerical Results

We use the EOS given in our paper [14], which is presented in our framework. Therefore, our approach is only briefly explained here. Thermal equilibrium is implicitly achieved at $T = 0$. We consider that the hadronic and quark matter and the mixed phase are β stable. We employ density functional theory (DFT) under the local density approximation [16, 17]. To account for the confinement, we introduce a sharp boundary between the two phases employing the bag model [12, 14] with a surface tension parameter σ. The determination of the surface tension between hadronic and quark matter is a difficult problem. Thus, many authors have treated the surface tension as a free parameter and have observed its effect [12, 13, 18]; we take the same approach in this study. To determine the charge screening effect, we also conduct the calculations without the screening effect. We then apply the EOS derived in our paper [14] to the Tolman–Oppenheimer–Volkoff (TOV) equation [15]. After that, we apply our EOS to a stationary rotating star.

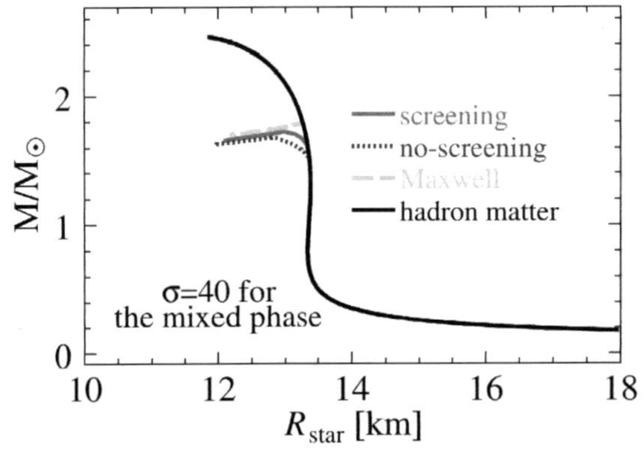

Figure 1. (Color online) Mass–radius relation of stars. The difference between screened and not screened mixed phases is clearly small.

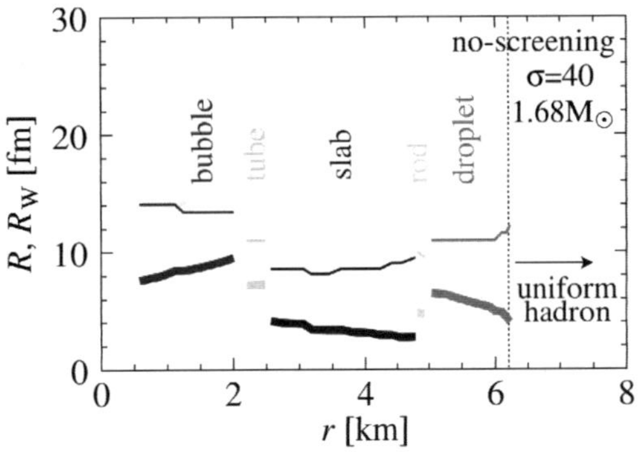

Figure 2. (Color online) Structure size and cell size in the core region of hybrid stars without screening. The radius is 12.6 km. Thick lines denote R and thin lines denote R_W.

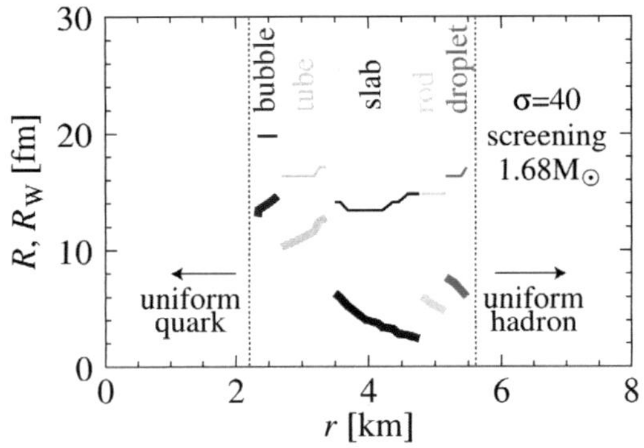

Figure 3. (Color online) Structure size and cell size in the core region of hybrid stars with screening. The mixed phase is restricted by charge screening. The radius is 12.9 km.

Figure 1 shows the mass–radius relations of stars with and without screening, using the Maxwell constructrion, and in the case of pure hadronic matter. There is a slight difference in the region around the maximum mass. However, as this difference is only about $0.05M_{\odot}$, it is small compared with the total mass of the star. Thus, charge screening does not strongly affect the bulk properties of the star. On the contrary, we find that the inner structure is

Journal of Physics: Conference Series **509** (2014) 012075 doi:10.1088/1742-6596/509/1/012075

greatly affected by charge screening. Figure 2 shows the inner core region of a hybrid star without screening. The mixed phase appears in a large part of the star and we hardly see the quark matter phase. In the case of screening shown in Fig. 3, on the other hand, the mixed phase region clearly becomes narrow and there is a quark matter phase in the central region because of charge screening. Our results suggest that quark matter could exist in the inner regions of compact stars. It is thus possible to attribute the magnetism of compact stars to spin polarization [19]. Many theoretical studies have used other models for the glitch phenomenon [20] and gravitational waves [21]. While we cannot simply apply our EOS to studying these phenomena, it is interesting to compare our results with those of other studies.

Recently, many theoretical studies tried to take into account the effect of rotation [22, 23, 24, 25]. We accordingly also try to take into account the effects of rotation in our study. However, rotation in general relativity is very difficult. Therefore, we assume: 1. Stationary rigid rotation ("uniform rotation"); 2. Axial symmetry with respect to the spin axis; and 3. The matter is a perfect fluid. There is a review of stationary rotation in general relativity [26] and in other papers [22]; we follow their calculation. Then, we apply our EOS to the stationary rotating star. Figure 4 shows the result for a rotating star with screening using our EOS. The red curve shows the maximum mass of the star and the blue curve shows the mass-shedding curve, which corresponds to the "Kepler frequency". The Kepler frequency occurs when the centrifugal force is equal to gravity. Therefore, the right-hand side of the blue curve is not valid. If the red curve is lower than the observations, the EOS should be ruled out. Fortunately, our EOS is consistent with these observations. We see an important relation between radius and rotation. Ordinarily,

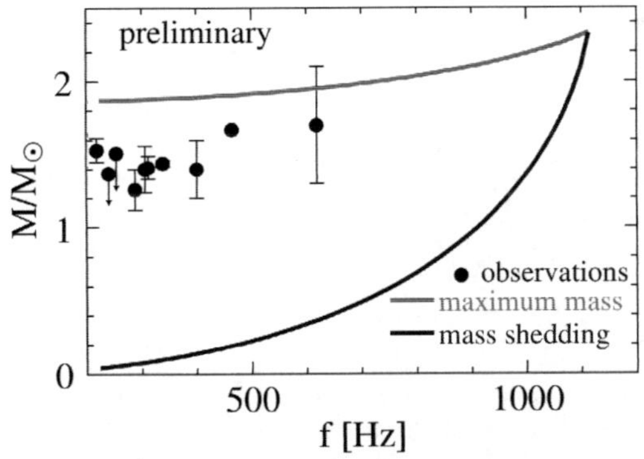

Figure 4. (Color online) Mass–frequency relation with our model (charge screening) plotted against the observational data listed in [22].

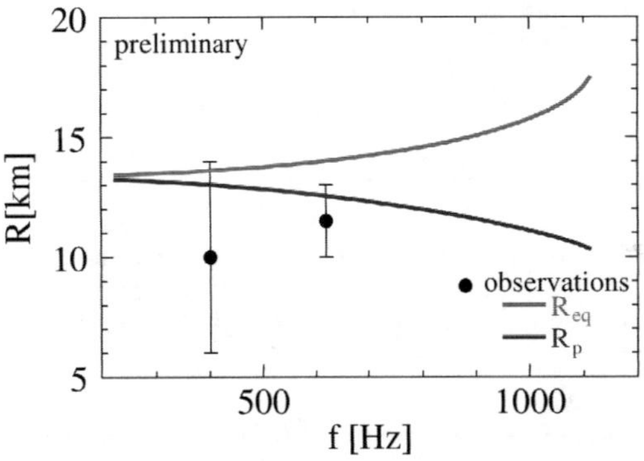

Figure 5. (Color online) Radius–frequency relation of our model plotted against observational data (SAXJ1808.4-3658 and 4U1608-52).

the "radius" of the star is single-valued because we consider that the star is spherical. However, if the star is rapidly rotating, we have to pay attention to the different "radii". Because of the effect of the rotation, a star deforms from a sphere to an ellipse. Therefore, we introduce two values, R_{eq} and R_{p}, which are the "equatorial radius" and the "polar radius", respectively. Figure 5 shows R_{eq} and R_{p} with respect to rotation. If the rotation rate is 400 Hz or faster, the two radii are different. Therefore, we have to note the effects of rotation on rapidly rotating stars.

3. Summary and Concluding Remarks

In this study, we demonstrated how charge screening affects the hadron–quark mixed phase in the cores of hybrid stars, taking into account rotation effects. We found that the inner structures are strongly affected. In particular, a core consisting of quark matter could appear due to the charge screening effect. Another case, kaon condensation, has been studied [27] and the results are similar to those of our papers [14, 28]. We used a simple model for quark matter and nuclear matter. To obtain a more realistic picture of the hadron–quark phase transition, we need to take into account color superconductivity [1, 18] and the relativistic mean field theory [29]. We will then be able to provide more realistic results. Neutron stars have other important physics – magnetic fields. However, the origin of these magnetic fields is still unknown. There are ways to explain magnetic fields based on the spin-polarization of the quark matter [4, 19]. However, whether the quark matter exists or not strongly depends on the EOS. In this calculation, we did not take into account magnetic fields. If we include a magnetic field, the resluts are very interesting with respect to the rotation of the star.

Acknowledgments

This work was supported in part by the Principal Grant of the Kagawa National College of Technology.

References

[1] As resent reviews, Alford M, Schmitt A, Rajagopal K and Schäfer T 2008 *Rev. Mord. Phys.* **80** 1455
[2] Alford M and Reddy S 2003 *Phys. Rev.* D **67** 074024
[3] Tatsumi T, Maruyama T and Nakano E 2004 *Prog. Theor. Phys. Suppl.* **153** 190
[4] Tatsumi T 2000 *Phys. Lett.* B **489** 280
[5] Adcox K, et al. (PHENIX collaboration) 2002 *Phys. Rev. Lett.* **88** 022301; Adler C, et al. (STAR collaboration) 2003 *Phys. Rev. Lett.* **90** 082302
[6] Madsen J 1999 *Lect. Notes Phys.* **516** 162
[7] Cheng K S, Dai Z G and Lu T 1998 *Int. Mod. Phys.* D **7** 139
[8] Rischke D H 2004 *Prog. Part. Nucl. Phys.* **52** 197
[9] Pisalski R D and Wilczek F 1984 *Phys. Rev. Lett.* **29** 338
[10] Gavai R V, Potvin J and Sanielevici S 1987 *Phys. Rev. Lett.* **58** 2519.
[11] Glendenning N K 1992 *Phys. Rev.* D **46** 1274; 2001 *Phys. Rep.* **342** 393
[12] Heiselberg H, Pethick C J and Staubo E F 1993 *Phys. Rev. Lett.* **70** 1355
[13] Glendenning N K and Pei S 1995 *Phys. Rev.* C **52** 2250
[14] Endo T, Maruyama T, Chiba S and Tatsumi T 2006 *Prog. Theor. Phys.* **115** 337
[15] Endo T 2011 *Phys. Rev.* C **83** 068801
[16] Parr R G and Yang W 1989 *Density-Functional Theory of atoms and molecules*, Oxford Univ. Press
[17] Gross E K U and Dreizler R M 1995 *Density functional theory*, Plenum Press
[18] Alford M, Rajagopal K, Reddy S, and Wilczek F 2001 *Phys. Rev.* D **64** 074017
[19] Tatsumi T arXiv:1107.0807[hep-ph]
[20] Bejger M, Haensel P and Zdunik J L 2005 *Mon. Not. Roy. Astron. Soc.* **359** 699
[21] Minuitti G, Pons J A, Berti E, Gualtieri L and Ferrari V 2003 *Mon. Not. Roy. Astron. Soc.* **338** 389
[22] Kurkela A, Romatschke P, Vuorinen A and Wu B arXiv:1006.4062[astro-ph.HE]
[23] Orsaria M, Rodrigues H, Weber F and Contrera G A 2013 *Phys. Rev.* D **87** 023001
[24] Weber F, Orsaria M, Negreiros R arXiv:1307.1103[astro-ph.SR]
[25] Belvedere R, Boshkayev K, Rueda J A, and Ruffini R arXiv:1307.2836[astro-ph.SR]
[26] Stergioulas N, Friedman J L 1995 *Astrophys. J* **444** 306
[27] Maruyama T, Tatsumi T, Voskresensky D N, Tanigawa T, Endo T and Chiba S 2006 *Phys. Rev.* C **73** 035802
[28] Maruyama T, Tatsumi T, Endo T and Chiba S 2006 *Recent Res. Devel. Phys.* **7** 1; nucl-th/0605075
[29] Shen H, Toki H, Oyamatsu K and Sumiyoshi K 1998 *Nucl. Phys.* A **637** 435

Journal of Physics: Conference Series **509** (2014) 012076 doi:10.1088/1742-6596/509/1/012076

Towards the dynamical study of heavy-flavor quarks in the Quark-Gluon-Plasma

H. Berrehrah[1], E. Bratkovskaya[1], W. Cassing[2], P.B. Gossiaux[3] and J. Aichelin[3]

[1] Frankfurt Institute for Advanced Studies and Institute for Theoretical Physics, Johann Wolfgang Goethe Universität, Ruth-Moufang-Strasse 1, 60438 Frankfurt am Main, Germany

[2] Institut für Theoretische Physik, Universität Giessen, 35392 Giessen, Germany

[3] Subatech, UMR 6457, IN2P3/CNRS, Université de Nantes, École des Mines de Nantes, 4 rue Alfred Kastler, 44307 Nantes cedex 3, France

E-mail: berrehrah@fias.uni-frankfurt.de

Abstract. Within the aim of a dynamical study of on- and off-shell heavy quarks Q in the quark gluon plasma (QGP) - as produced in relativistic nucleus-nucleus collisions - we study the heavy quark collisional scattering on partons of the QGP. The elastic cross sections $\sigma_{q,g-Q}$ are evaluated for perturbative partons (massless on-shell particles) and for dynamical quasi-particles (massive off-shell particles as described by the dynamical quasi-particles model "DQPM") using the leading order Born diagrams. We demonstrate that the finite width of the quasi-particles in the DQPM has little influence on the cross sections $\sigma_{q,g-Q}$ except close to thresholds. We, furthermore, calculate the heavy quark relaxation time as a function of temperature T within the different approaches using these cross sections.

1. Introduction

The heavy quarks Q are created in hard processes in relativistic heavy-ion collisions (HIC) and are initially isotropically distributed in the transverse momentum space. To explain the final azimuthal anisotropy observed in experiments, one has to study in detail the dynamics and the transport properties of the heavy quarks in the medium, i.e. in the quark gluon plasma (QGP). Since the experimental discovery of the strongly interacting QGP (sQGP), many models and theoretical schemes have been set up to describe the large coupling of heavy quarks to the hot QGP. Indeed, the early studies of Q scattering in the QGP based on pQCD cross sections failed to achieve a quantitative understanding of the experimental data [1]. The purpose of this contribution is to explore new descriptions for the interaction between heavy quarks and the QGP partons (q, g) at finite temperature. Therefore, we will present the corresponding elastic cross sections following different models and we will end by evaluating the heavy quark relaxation time as a function of the temperature T of the medium.

2. Cross sections at finite temperature

The central question about the heavy quark (Q) dynamics is the magnitude and detailed form of the interactions between the quark Q and the partons (q, g) of the finite temperature medium. Indeed, the evaluation of the corresponding cross sections depends on the (running) coupling in the interaction vertices and the quark and gluon propagators at finite T. Although the scattering of heavy quarks in vacuum and in the QGP for massless partons has extensively been studied in the literature [2] using pQCD, a full treatment of finite temperature effects is still needed to consider the case of non-perturbatively interacting massive light and heavy quarks and gluons. In this contribution, we present an effective approach for the derivation of the on- and off-shell cross sections for the interaction of massive dynamical quasi-particles as constituents of the finite temperature strongly interacting medium (sQGP). In order to consider all the effects of non-perturbative nature of the sQGP constituents, we refrain from a fixed-order thermal loop calculation and pursue instead a more phenomenological approach. The multiple strong interactions of quarks and gluons in the sQGP here are encoded in their effective propagators-extracted from lattice QCD data within the DQPM [3]- with dynamical spectral functions and an effective running coupling $g^2(T)$ which is enhanced in the infrared.

3. qQ and gQ elastic scattering

The matrix elements for the elastic scattering of a heavy quark with a light quark $qQ \to qQ$ (resp. gluon $gQ \to gQ$) have been calculated for the case of massless partons in Ref. [2]. These pQCD cross sections have to be supplemented by two parameters to allow for a quantitative evaluation: the coupling constant α_s and the infrared (IR) regulator μ which renders the cross section infrared finite. The value of the cut-off or of the gluon mass has been fixed according to phenomenological considerations. Therefore, the uncertainties in these scales is large which also leads to corresponding uncertainties in the cross sections. In our study we will present explicit values for these two parameters which are based on different theoretical considerations. Moreover, we will take into account the quasi-particle nature of the incoming and outgoing particles by incorporating spectral functions. In this way we will be able to test to what extent the quasi-particle nature of quarks and gluons will influence the heavy quark scattering. We briefly recall the ingredients of the different approaches studied in this paper:

- **HTL-GA**: from Gossiaux and Aichelin following a Hard-Thermal-Loop inspired approach [1, 4] for massless light quarks and gluons. As compared to former pQCD approaches this model differs in the description of $q, g-Q$ interactions in two respects:
  - an effective running non-perturbative coupling, which remains finite for vanishing 4-momentum transfer squared $t \to 0$, is employed [5],
  - an infrared regulator in the t-channel, which is determined by hard thermal loop calculations. The t-channel is adjusted to give the same energy loss as calculated in a HTL approach [1].

  In the HTL-GA approach we will also consider the case with fixed coupling α_s and the Debye mass ($m_D \approx \xi g_S T$) as an infrared regulator.
- **IEHTL** (Infrared Enhanced Hard Thermal Loop): the main characteristics of this approach are:
  - off-shell kinematics, in particular light quarks and gluons virtualities,
  - internal and external lines of light and heavy quarks as well as gluons are dressed with non-perturbative spectral functions,
  - parametrizations of the quark and gluon propagators from the DQPM that are matched to reproduce lQCD data in equilibrium,
  - qgq and QgQ vertices are modified compared to pQCD vertices by replacing the perturbative coupling with the full running coupling depending on temperature according to the DQPM.
- **DpQCD** (Dressed pQCD): in the same spirit as the IEHTL model with the difference that for DpQCD we only consider the DQPM pole masses for the QGP partons thus treating the partons on-shell.

Fig. 1 (a) (resp. (b)) present the qQ (resp. gQ) elastic scattering cross sections calculated within the models presented above as a function of $\sqrt{s}$ for different temperatures ($2T_c, 3T_c$). Apart from threshold effects $\sigma_{q,g-Q}$ are approximately independent on $\sqrt{s}$, however, differ substantially in magnitude. These figures show clearly the different values for $\sigma_{q,g-Q}$ for the different choices of couplings and infrared regulators μ. Since the values of μ increase with increasing temperature, all the cross sections decrease for the temperature of $3T_c$ (dashed lines) in comparison to the $2T_c$ case (solid lines).

The comparison of the DpQCD/IEHTL approaches demonstrates that the off-shell mass distributions only have a sizeable impact close to the threshold given by the pole masses. This is due to the moderate parton widths considered in the DQPM model. At energies below the on-shell threshold the off-shell cross section increases with $\sqrt{s}$ because more and more masses can contribute. Whereas the on-shell cross section diverges at the threshold the off-shell cross section shows a maximum at the on-shell threshold and decreases then due to the decrease of the on-shell cross section.

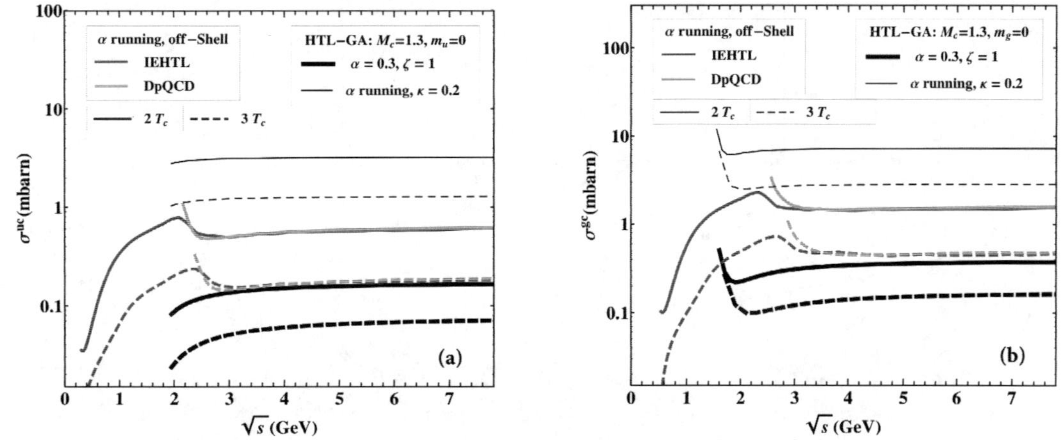

Figure 1. *Comparison of $\sigma_{u,g-c}$ calculated within the HTL-GA (black lines), DpQCD (yellow lines) and IEHTL (red lines) approaches as a function of $\sqrt{s}$ for different temperatures T (see legend). $m_{u,g}$ and M_c are given in GeV. Left (a): $uc \to uc$; Right (b): $gc \to gc$.*

4. Transition rate and relaxation time

To compute transport coefficients one needs first to evaluate the elastic transition rate $\omega(T)$ which is given for the case of on-shell ($\omega^{\rm on}_{jQ}$) and off-shell ($\omega^{\rm off}_{jQ}$) partons, with $j = q$, by [6]:

$$\omega^{\rm on}_{qQ}(m^i_q, M^i_Q, m^f_q, M^f_Q, T) = \int_{\rm Th}^{\infty} ds\ \sigma^{\rm on}_{qQ}(m^i_q, M^i_Q, m^f_q, M^f_Q, T, s)\ \mathcal{P}^{\rm on}(m^i_q, M^i_Q, T, s),$$
$$\omega^{\rm off}_{qQ}(T) = \int_{\rm Th}^{\infty} ds\ \sigma^{\rm off}_{qQ}(T,s)\ \mathcal{P}^{\rm off}(T,s), \qquad (1)$$

with the threshold $\mathrm{Th} = \max\{(m^i_q + M^i_Q)^2, (m^f_q + M^f_Q)^2\}$ and $\mathcal{P}$ denoting the probability for a qQ pair with the energy $\sqrt{s}$ in the medium at finite T:

$$\mathcal{P}^{\rm on}(m^i_q, M^i_Q, T, s) = C^{\rm on}\frac{E_1E_2}{\sqrt{s}} p_{cm}(s) v_{rel}(s) f(E_1) f(E_2)$$
$$\mathcal{P}^{\rm off}(T,s) = \int dm^i_q \int dM^i_Q\ \frac{E_1E_2}{\sqrt{s}}\ C^{\rm off}(T)\ p_{\rm cm}(s) v_{rel}(s)\ f_q(E_1) f_Q(E_2) A_{q^i}(m^i_q) A_{Q^i}(M^i_Q), \qquad (2)$$

with the center-of-mass momentum $p_{\rm cm}$, relative velocity v_{rel} and the normalization factors $C^{\rm on}$ and $C^{\rm off}$. In (2) $f_{q,Q}(E)$ is the Fermi-Dirac distribution for the light and heavy quark. Similar calculations can be performed for massive gluons by including the Bose-Einstein distribution in Eq. (2).

Figure 2-(a) (resp. (b)) shows the qQ (resp. gQ) elastic transition rate as a function of the medium temperature T/T_c for the different approaches presented above. We notice different power laws in T, i.e. $\sim T^{-\beta}$ for the HTL-GA and DpQCD/IEHTL models. In fact, one can find that ($\beta^{T<1.2T_c} \sim 2, \beta^{T>1.2T_c} \sim 1.7$) for the HTL-GA versions, whereas ($\beta^{T<1.2T_c} \sim 4, \beta^{T>1.2T_c} \sim 2$) for the DpQCD/IEHTL approaches that are practically identical. The higher power coefficients in the DpQCD/IEHTL approaches can be traced back to the infrared enhancement of the effective coupling. These different power laws in T will have a sizeable effect on the transport coefficients. We stress again that the effect of the DQPM spectral function on $\omega(T)$ is negligible by comparing $\omega(T)$ using the DpQCD/IEHTL approaches.

By means of the transition rates we can compute the relaxation time τ for particles within the different models. In the dilute gas approximation the relaxation time τ is obtained for on-shell particles $(\tau_c^{-1})_{DpQCD}, (\tau_c^{-1})_{HTL-GA}$ and for off-shell quasi-particles $(\tau_c^{-1})_{IEHTL}, (\tau_c^{-1})_{DQPM}$ by (3) [6], where n_i is

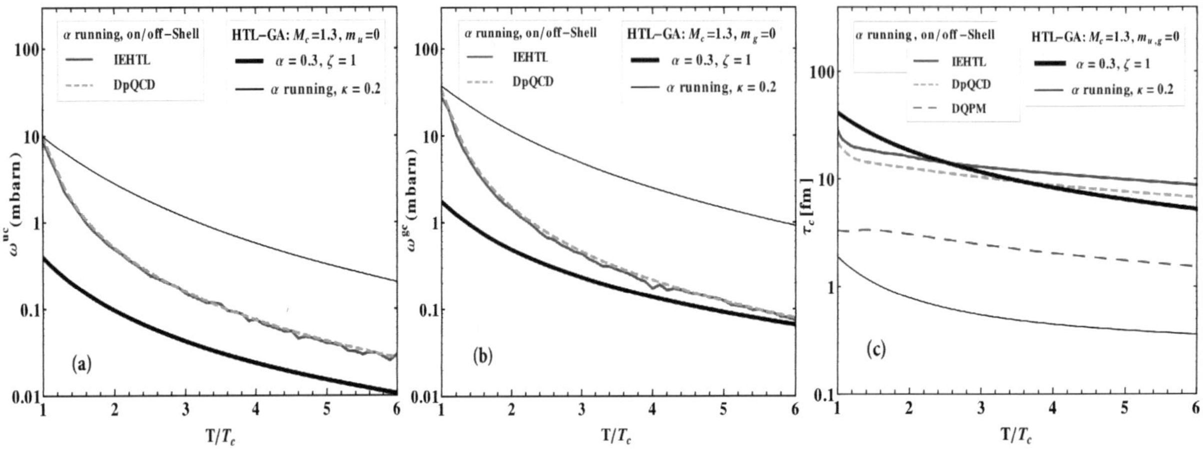

Figure 2. *uc (a) and gc (b) elastic transition rate, (c) heavy quark relaxation time as a function of T/T_c with $T_c = 158$ MeV. Presented here are different models for off-shell scattering with the DQPM spectral functions (IEHTL (DQPM), red silde (dashed) line) and on-shell partons with the DQPM pole masses (DpQCD, yellow dashed lines). Also shown is a comparison to results of the HTL-GA approaches with constant coupling (thick black line) and running coupling (thin black line), where M_c is given in GeV.*

the quark/antiquark or gluon density. For the DQPM we do not need the explicit cross sections since the inherent quasi-particle width $\gamma_c(T)$ directly provides the total interaction rate [3].

$$(\tau_c^{-1})_{DpQCD} = \sum_{i\in q,\bar{q},g} n_i^{\rm on}(T)\ \omega_{\rm i\,c}^{DpQCD}(T), \quad (\tau_c^{-1})_{HTL-GA} = \sum_{i\in q,\bar{q},g} n_i^{\rm on}(T)\ \omega_{\rm i\,c}^{HTL-GA}(T), \quad (\tau_c^{-1})_{DQPM} = \frac{\hbar c}{\gamma_c(T)}$$

$$(\tau_c^{-1})_{IEHTL} = \sum_{i\in q,\bar{q},g} n_i^{\rm off}(T)\ \omega_{\rm i\,c}^{IEHTL}(T), \quad n_i^{\rm off}(T) = \iint \frac{d^3p}{(2\pi)^3} A_i^{BW}(m_i) dm_i\ f_i/g_i(p,T,m_i,\mu_i). \tag{3}$$

Fig. 2-(c) shows the heavy quark relaxation time for the case of on-shell and off-shell partons at finite T following the DpQCD/IEHTL and HTL-GA approaches. We deduce that the heavy quark needs more interactions to relax when its interactions are described by σ^{HTL-GA} as compared to $\sigma^{DpQCD/IEHTL}$.

5. Summary

We have presented the elastic scattering of heavy quarks with quarks and gluons in a quark-gluon-plasma using two approaches based on different regularization schemes. *i*) The dynamical quasi-particle model (DQPM) [3, 4] in which quarks and gluons have a finite mass and width that vary with temperature T. *ii*) The Peshier-Gossiaux-Aichelin approach [1] which uses massless partons, a running coupling $\alpha_s(Q^2)$ and an infrared regulator which have been adjusted to reproduce the heavy quark energy loss in heavy-ion collisions at RHIC energies. The size of the elastic cross sections is dominated by the infrared regulator which in the finite temperature medium is determined by a dynamical gluon mass.

We have demonstrated that the finite width of the partons in the DQPM has little influence on the cross sections except close to thresholds. The total cross sections show a smooth dependence on the invariant energy. Even if the heavy quark relaxation time shows large differences between the different models, explicit transport calculations in comparison to experimental data are needed to figure out the appropriate scenario.

References

[1] Gossiaux P and Aichelin J 2008 *Phys.Rev.* **C78** 014904 (*Preprint* 0802.2525)
[2] Combridge B 1979 *Nuclear Physics B* **151** 429 – 456 ISSN 0550-3213
[3] Cassing W 2009 *The European Physical Journal Special Topics* **168**(1) 3–87 ISSN 1951-6355
[4] Berrehrah H, Bratkovskaya E, Cassing W, Gossiaux P, Aichelin J *et al.* 2013 (*Preprint* 1308.5148)
[5] Dokshitzer Y L, Marchesini G and Webber B 1996 *Nucl.Phys.* **B469** 93–142 (*Preprint* hep-ph/9512336)
[6] Marty R, Bratkovskaya E, Cassing W, Aichelin J and Berrehrah H 2013 (*Preprint* 1305.7180)

Journal of Physics: Conference Series **509** (2014) 012077 doi:10.1088/1742-6596/509/1/012077

Heavy vs. light flavor energy loss within a partonic transport model

Jan Uphoff,[1] Oliver Fochler,[1] Zhe Xu,[2] and Carsten Greiner[1]

[1] Institut für Theoretische Physik, Johann Wolfgang Goethe-Universität Frankfurt, Max-von-Laue-Str. 1, D-60438 Frankfurt am Main, Germany

[2] Department of Physics, Tsinghua University, Beijing 100084, China

E-mail: uphoff@th.physik.uni-frankfurt.de

Abstract. The full space-time evolution of gluons, light and heavy quarks in ultra-relativistic heavy-ion collisions is studied within the partonic transport model *Boltzmann Approach to MultiParton Scatterings* (BAMPS). We discuss for light and heavy quarks the elastic and radiative energy loss with a running coupling. Radiative processes, in particular, are implemented through an improved version of the Gunion-Bertsch matrix element, which is derived from comparisons to the exact result, explicitly taking finite heavy quark masses into account. Subsequently, we present results with and without radiative processes and compare them to experimental data at LHC. A focus is put on the nuclear modification factor of charged hadrons and D mesons.

1. Introduction

In ultra-relativistic heavy-ion collisions at the Relativistic Heavy Ion Collider (RHIC) at BNL [1, 2] and the Large Hadron Collider (LHC) at CERN [3] a hot and dense medium is produced. Since quarks and gluons form the relevant degrees of freedom, it is commonly named quark gluon plasma (QGP). Interesting properties are, for instance, a collective behavior like a nearly perfect fluid or the quenching of highly energetic particles.

In particular charm and bottom quarks provide a unique way to gain insight in the properties of this matter. Since they are heavy, their production time is at a very early stage of the heavy-ion collision when enough energy is available [4]. Therefore, they traverse the medium right from the beginning for a rather long time, interact with other medium particles, participate in the flow, and lose energy. The mechanism responsible for their strong interaction with the medium is actively debated, most recently, for instance, in [5, 6, 7, 8, 9, 10, 11, 12, 13, 14, 15].

In this paper we study the nuclear modification factor and elliptic flow of heavy flavor particles at LHC within the partonic transport model BAMPS. Furthermore, we calculate the elastic and radiative energy loss of light and heavy quarks in a static thermal medium. To this end, we also compare the D meson nuclear modification factor to that of charged hadrons obtained with elastic and radiative processes.

Journal of Physics: Conference Series **509** (2014) 012077 doi:10.1088/1742-6596/509/1/012077

2. Partonic transport model BAMPS

BAMPS (*Boltzmann Approach to MultiParton Scatterings*) [16, 17] is a partonic transport model that describes the full space-time evolution of the QGP by solving the Boltzmann equation,

$$\left(\frac{\partial}{\partial t} + \frac{\mathbf{p}_i}{E_i}\frac{\partial}{\partial \mathbf{r}}\right) f_i(\mathbf{r}, \mathbf{p}_i, t) = \mathcal{C}_i^{2\to 2} + \mathcal{C}_i^{2\leftrightarrow 3} + \dots \,, \tag{1}$$

for on-shell partons and pQCD interactions. For light partons, all possible $2 \to 2$ and $2 \leftrightarrow 3$ processes are included. On the heavy flavor sector, all relevant elastic and radiative collisions of heavy quarks with other medium constituents have been implemented in BAMPS, while the running of the coupling is explicitly taken into account for elastic as well as radiative light and heavy flavor processes. The divergent t channel of heavy flavor scatterings is regularized with a screening mass μ that is determined by matching elastic energy loss calculations with leading order pQCD cross sections to results from hard thermal loop (HTL) calculations. The comparison of both results shows that the screening mass μ is smaller than the usually employed Debye mass m_D, more precisely, $\mu^2 = \kappa m_D^2$ with $\kappa = 1/(2e) \approx 0.2$ [18, 19, 20].

For radiative processes we generalize the recently calculated improvement of the Gunion and Bertsch (GB) cross section [21] to finite masses [22]. The result is consistent with the calculation of Ref. [23]. The dead cone suppression [24, 25] of small angle radiation due to the finite heavy quark mass is explicitly present in our result.

3. Results and comparison with data

Quantitative BAMPS comparisons [26, 20, 27, 7] show that elastic processes with a running coupling and the improved screening procedure contribute significantly to the energy loss of heavy quarks. However, they alone cannot reproduce the data of the nuclear modification factor or the elliptic flow of any heavy flavor particle species. Before radiative heavy quark processes have been implemented in BAMPS, we mimicked their influence by effectively increasing the elastic cross section by a factor $K = 3.5$, which is tuned to the elliptic flow data of heavy flavor electrons at RHIC [7]. Simultaneously, the nuclear modification factor of heavy flavor electrons at RHIC can be described with the same parameter. Having fixed this parameter to the RHIC data, we find a good agreement with the experimentally measured nuclear modification factor and elliptic flow of all heavy flavor particles at LHC (see Fig. 1). Most of these calculations were predictions [7]. As a note, only (Peterson) fragmentation but no coalescence is considered, which might influence the exact value of K.

However, the need of the phenomenological K factor is rather unsatisfying from the theory perspective. Therefore, the question arises whether radiative processes can account for the missing contribution parameterized by the K factor. To this end, we present in the following BAMPS calculations including radiative processes for both light and heavy particles. Since the improved screening prescription is only derived for heavy quarks, we employ in the following the standard Debye screening ($\kappa = 1$) to treat light and heavy partons consistently.

In the left panel of Fig. 2 the elastic and radiative energy loss of light and heavy quarks in a static thermal medium is depicted. The radiative energy loss is on the same order or only slightly larger than the the elastic energy loss for all quark masses. Although the mass hierarchy is visible for the elastic energy loss, the radiative energy loss of all flavors is of the same size. The reason lies in the implementation of the Landau-Pomeranchuk-Migdal (LPM) effect producing a second dead cone at small emission angles that overshadows the dead cone due to the heavy quark mass. This similar energy loss of light and charm quarks is indeed part of the explanation why the measured nuclear modification factors of charged hadrons and D mesons in heavy-ion collision have the same values. Furthermore, mass effects in the fragmentation of gluons and light quarks to charged hadrons and charm quarks to D mesons lead to a very similar suppression of charged hadrons and D mesons in BAMPS, as is depicted in the right panel of Fig. 2.

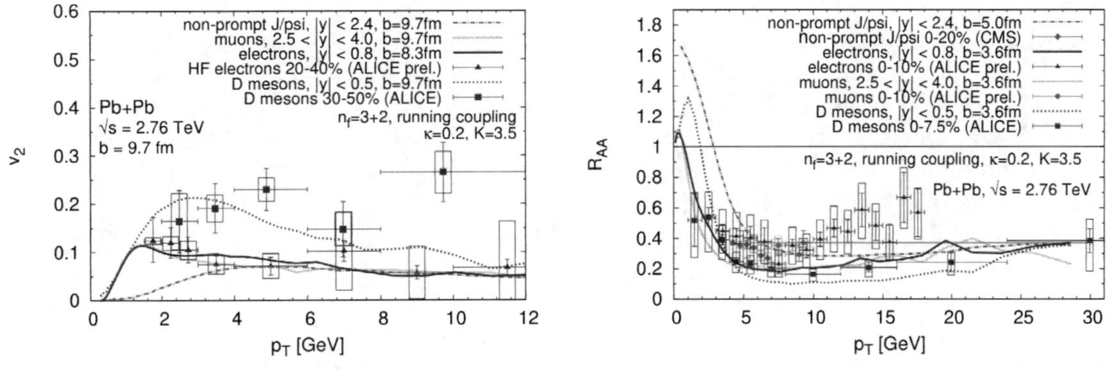

Figure 1. Elliptic flow v_2 (left) and nuclear modification factor R_{AA} (right) of various heavy flavor particles at LHC together with data [28, 29, 30, 31, 32, 33]. Only binary heavy flavor processes are considered and multiplied with $K = 3.5$.

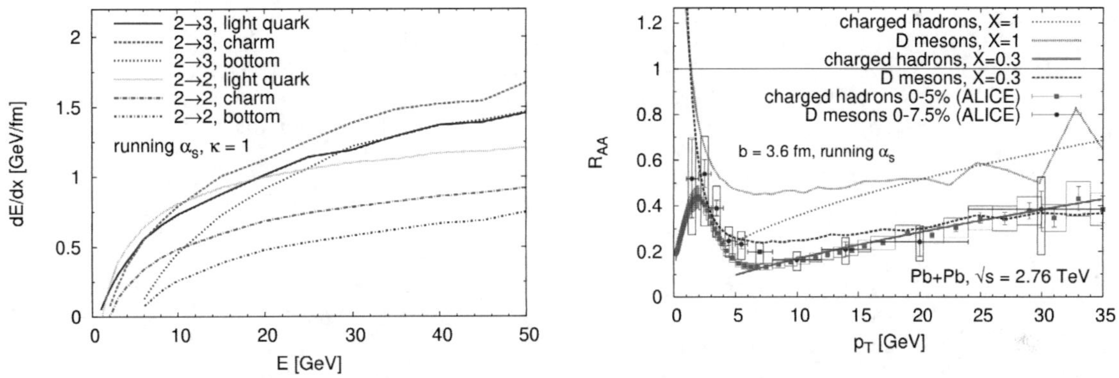

Figure 2. Left: Elastic and radiative energy loss per unit length of a light quark ($M = 0\,\mathrm{GeV}$), a charm quark ($M = 1.3\,\mathrm{GeV}$), and a bottom quark ($M = 4.6\,\mathrm{GeV}$) traversing a static thermal medium with temperature $T = 0.4\,\mathrm{GeV}$. The curves are calculated with running coupling and Debye mass prefactor $\kappa = 1$. Right: Nuclear modification factor R_{AA} of charged hadrons and D mesons at LHC in comparison to data [34, 32]. The LPM parameter X is set to 1 and 0.3.

Although the shape of the nuclear modification factor as a function of the transverse momentum is nicely reproduced by BAMPS (see $X = 1$ curve in the right panel of Fig. 2), the overall suppression is underestimated. The reason for this discrepancy is probably the effective implementation of the LPM effect in BAMPS, which discards all possible interference effects and only allows independent scatterings. If we introduce a factor $X < 1$ that modifies the LPM cut-off and effectively allows more radiative interactions, a good agreement with the experimental data of charged hadrons and D mesons is found for $X = 0.3$. Although the exact value of X is a free parameter, we expect that a more sophisticated implementation of the LPM effect would effectively correspond to an $X < 1$ and might make the need of the X parameter obsolete. We will study this in more detail and also compare to other theoretical calculations in a forthcoming study.

4. Summary

We presented heavy flavor calculations with the parton cascade BAMPS. Allowing only elastic interactions with a running coupling and an improved Debye screening, the experimental data of all heavy flavor particles at LHC can be described if the binary cross sections are multiplied with $K = 3.5$, which is tuned to the elliptic flow data at RHIC. Furthermore, we show first results including also radiative processes. To this end, the elastic and radiative energy loss of light and heavy quarks in a static medium is shown. Furthermore, we compare BAMPS results to the experimentally measured nuclear modification factor of charged hadrons and D mesons at LHC. It would be interesting for a future project to study the impact of radiative processes on heavy flavor correlations [35] or heavy flavor tagged jets [36].

Acknowledgements

This work was supported by the Bundesministerium für Bildung und Forschung (BMBF), the NSFC under grant No. 11275103, HGS-HIRe, and the Helmholtz International Center for FAIR within the framework of the LOEWE program launched by the State of Hesse. Numerical computations have been performed at the Center for Scientific Computing (CSC).

References

[1] Adams J *et al.* (STAR) 2005 *Nucl.Phys.* **A757** 102–183 (*Preprint* nucl-ex/0501009)
[2] Adcox K *et al.* (PHENIX) 2005 *Nucl.Phys.* **A757** 184–283 (*Preprint* nucl-ex/0410003)
[3] Müller B, Schukraft J and Wyslouch B 2012 *Ann.Rev.Nucl.Part.Sci.* **62** 361–386 (*Preprint* 1202.3233)
[4] Uphoff J, Fochler O, Xu Z and Greiner C 2010 *Phys.Rev.* **C82** 044906 (*Preprint* 1003.4200)
[5] Abir R, Jamil U, Mustafa M G and Srivastava D K 2012 *Phys.Lett.* **B715** 183–189 (*Preprint* 1203.5221)
[6] Meistrenko A, Peshier A, Uphoff J and Greiner C 2013 *Nucl.Phys.* **A901** 51–64 (*Preprint* 1204.2397)
[7] Uphoff J, Fochler O, Xu Z and Greiner C 2012 *Phys.Lett.* **B717** 430–435 (*Preprint* 1205.4945)
[8] He M, Fries R J and Rapp R 2013 *Phys.Rev.Lett.* **110** 112301 (*Preprint* 1204.4442)
[9] Horowitz W 2013 *Nucl.Phys.* **A904-905** 186c–193c (*Preprint* 1210.8330)
[10] Buzzatti A and Gyulassy M 2012 (*Preprint* 1207.6020)
[11] Cao S, Qin G Y, Bass S A and Müller B 2013 *Nucl.Phys.* **A904-905** 653c–656c (*Preprint* 1209.5410)
[12] Nahrgang M, Aichelin J, Gossiaux P B and Werner K 2013 (*Preprint* 1305.3823)
[13] Alberico W, Beraudo A, De Pace A, Molinari A, Monteno M *et al.* 2013 (*Preprint* 1305.7421)
[14] Mazumder S, Bhattacharyya T and Alam J e 2013 (*Preprint* 1305.6445)
[15] Lang T, van Hees H, Steinheimer J and Bleicher M 2013 (*Preprint* 1305.1797)
[16] Xu Z and Greiner C 2005 *Phys.Rev.* **C71** 064901 (*Preprint* hep-ph/0406278)
[17] Xu Z and Greiner C 2007 *Phys.Rev.* **C76** 024911 (*Preprint* hep-ph/0703233)
[18] Gossiaux P B and Aichelin J 2008 *Phys.Rev.* **C78** 014904 (*Preprint* 0802.2525)
[19] Peshier A 2012 *Nucl.Phys.* **A888** 7–22 (*Preprint* 0801.0595)
[20] Uphoff J, Fochler O, Xu Z and Greiner C 2011 *Phys.Rev.* **C84** 024908 (*Preprint* 1104.2295)
[21] Fochler O, Uphoff J, Xu Z and Greiner C 2013 *Phys.Rev.* **D88** 014018 (*Preprint* 1302.5250)
[22] Uphoff J, Fochler O, Xu Z and Greiner C 2013 In preparation
[23] Aichelin J, Gossiaux P B and Gousset T 2013 (*Preprint* 1307.5270)
[24] Dokshitzer Y L and Kharzeev D E 2001 *Phys.Lett.* **B519** 199–206 (*Preprint* hep-ph/0106202)
[25] Abir R, Greiner C, Martinez M, Mustafa M G and Uphoff J 2012 *Phys.Rev.* **D85** 054012 (*Preprint* 1109.5539)
[26] Uphoff J, Fochler O, Xu Z and Greiner C 2011 *Nucl.Phys.* **A855** 444–447 (*Preprint* 1011.6183)
[27] Fochler O, Uphoff J, Xu Z and Greiner C 2011 *J.Phys.* **G38** 124152 (*Preprint* 1107.0130)
[28] Abelev B *et al.* (ALICE Collaboration) 2013 (*Preprint* 1305.2707)
[29] Sakai S (ALICE Collaboration) 2013 *Nucl.Phys.* **A904-905** 661c–664c
[30] Conesa del Valle Z (ALICE Collaboration) 2013 *Nucl.Phys.* **A904-905** 178c–185c (*Preprint* 1212.0385)
[31] Abelev B *et al.* (ALICE Collaboration) 2012 *Phys.Rev.Lett.* **109** 112301 (*Preprint* 1205.6443)
[32] Grelli A (ALICE Collaboration) 2013 *Nucl.Phys.* **A904-905** 635c–638c (*Preprint* 1210.7332)
[33] Chatrchyan S *et al.* (CMS Collaboration) 2012 *JHEP* **1205** 063 (*Preprint* 1201.5069)
[34] Abelev B *et al.* (ALICE Collaboration) 2013 *Phys.Lett.* **B720** 52–62 (*Preprint* 1208.2711)
[35] Uphoff J, Senzel F, Xu Z and Greiner C 2013 (*Preprint* 1310.1340)
[36] Senzel F, Fochler O, Uphoff J, Xu Z and Greiner C 2013 (*Preprint* 1309.1657)

Open charm measurements in $p+p$ collisions at STAR

David Tlustý for the STAR collaboration
Nuclear Physics Institute, Academy of Sciences Czech Republic, Na Truhlářce 39/64, 180 86 Praha 8, Czech Republic
Czech Technical University in Prague, Faculty of Nuclear Sciences and Physical Engineering, Břehová 7, 11519, Prague 1, Czech Republic

E-mail: tlustdav@fjfi.cvut.cz

Abstract.
In this article, we will present the STAR results of open charm hadron and non-photonic productions at mid-rapidity in $p+p$ collisions at $\sqrt{s} = 200$ and 500 GeV. Open charm mesons are reconstructed directly via hadronic decay channels with daughter particles identified by STAR Time Projection Chamber (TPC) and Time Of Flight (TOF) detectors. Non-photonic electron yields are calculated by subtracting photonic electrons from inclusive electrons identified using TPC and Electromagnetic Calorimeter. These measurements are compared to theoretical model calculations and physics implications will be discussed.

1. Introduction

Heavy quark production at RHIC energies is dominated by initial gluon fusion and can be described by perturbative QCD (pQCD) due to their large mass [1]. The measurement of the charm quark production in $p+p$ collisions provides both pQCD test and baseline for any measurement in Heavy-Ion collisions. The study of heavy quark production in relativistic nuclear collisions follows two different approaches: (i) the direct reconstruction of heavy flavor mesons and (ii) the identification of electrons from semi-leptonic decays of such mesons, the so-called non-photonic electrons.

2. Data Analysis

2.1. The Direct Reconstruction of Open Charm Mesons

Invariant yield of charm quark Y is obtained from fitting the reconstructed invariant mass spectrum of open charm mesons through hadronic decays: $D^0(\overline{D^0}) \rightarrow K^{\mp}\pi^{\pm}$ (BR = 3.89%) and $D^{*\pm} \rightarrow D^0(\overline{D^0})\pi^{\pm}$ (BR = 67.7%) $\rightarrow K^-\pi^+\pi^{\pm}$ (total BR = 2.63%). The Y is calculated then as

$$Y \equiv \frac{\mathrm{d}^2 N_{c\bar{c}}}{2\pi p_T \mathrm{d}p_T \mathrm{d}y} = \frac{1}{N_{\mathrm{trig}}} \frac{N(p_T, y)}{2\pi p_T \Delta p_T \Delta y} \frac{f_{\mathrm{trg}}}{\mathrm{BR}\ f_{\mathrm{frag.}} \epsilon_{\mathrm{rec}}} \tag{1}$$

where N_{trig} is the total number of triggered events used for the analysis. $N(p_T, y)$ is the raw charm hadron signal in each p_T bin within a rapidity window $\Delta y = 2$. BR is the hadronic decay branching ratio for the channel of interest. ϵ_{rec} is the reconstruction efficiency including geometric acceptance, track selection efficiency, PID efficiency, and analysis cut efficiency. $f_{\mathrm{frag.}}$ represents the the ratio of charm quarks hadronized to open charm mesons. And f_{trg} is the correction factor to account for the bias between the minimum-bias sample used in this analysis and the total Non-single Diffractive (NSD) sample [2]. f_{trg} is found to be 0.65 for D^0 and 0.67 for D^* at $\sqrt{s} = 200$ GeV and 0.6 and 0.59 at $\sqrt{s} = 500$ GeV respectively.

The identification of daughter particles is done in the STAR experiment [3] at mid-rapidity $|y| < 1$ at $\sqrt{s} = 200$ and 500 GeV. The analysis presented herein is done using two datasets; the first one collected in year 2009 ($N_{\text{trig}} \sim 105$ million 200 GeV $p+p$ collisions), and the second one in 2011 ($N_{\text{MB}} \sim 50$ million 500 GeV $p+p$ collisions).

2.2. Identification of electrons from semi-leptonic decays

The invariant cross section for non-photonic electron production is calculated according to

$$E\frac{\mathrm{d}^3\sigma}{\mathrm{d}p^3} = \frac{1}{\mathcal{L}}\frac{1}{2\pi p_T \Delta p_T \delta y}\frac{N_{\text{npe}}}{\varepsilon_{\text{rec}}\varepsilon_{\text{trig}}\varepsilon_{\text{eid}}\varepsilon_{\text{BBC}}}, \tag{2}$$

where N_{npe} is the nonphotonic electron raw yield, ε_{rec} is the product of the single electron reconstruction efficiency and the correction factor for momentum resolution and finite spectrum bin width, $\varepsilon_{\text{trig}}$ is the high-tower trigger efficiency, ε_{eid} is the electron identification efficiency, $\mathcal{L}$ is the integrated luminosity with the z-position of vertex cuts, and $\varepsilon_{\text{BBC}} = 0.87 \pm 0.08$ is the BBC trigger efficiency.

The analysis of non-photonic electrons consists of three main steps: selection of a clean electron sample; subtraction of electron background arising from conversion and Dalitz decays; and residual corrections of the signal yield. The analysis details and a discussion of the sources of uncertainty can be found in [4]. The main background in this analysis is the substantial flux of photonic electrons from photon conversion in the detector material and Dalitz decay of π^0 and η mesons. These contributions need to be subtracted in order to extract the non-photonic electron yield according to formula

$$N_{\text{npe}} = N_{\text{inc}}\varepsilon_{\text{purity}} - \frac{N_{\text{pho}}}{\varepsilon_{\text{pho}}}, \tag{3}$$

where N_{npe} is the non-photonic electron yield, N_{inc} is the inclusive electron yield, N_{pho} is the photonic electron yield, ε_{pho} is the photonic electron reconstruction efficiency defined as the fraction of the photonic electrons identified through invariant mass reconstruction, and $\varepsilon_{\text{purity}}$ is the purity reflecting hadron contamination in the inclusive electron sample.

3. Results

3.1. The Direct Reconstruction of Heavy Flavor Mesons

Yields $N(p_T, y)$ are calculated in six p_T bins (first two for D^0, the next four for D^*) in $p+p$ 200 GeV and five p_T bins (first for D^0, the next four for D^*) in $p+p$ 500 GeV. The charm cross section at mid-rapidity $\mathrm{d}\sigma^{c\bar{c}}/\mathrm{d}y$ was obtained from the power-law function fit [2] to $\mathrm{d}^2\sigma^{c\bar{c}}/(2\pi p_T \mathrm{d}p_T \mathrm{d}y) = Y \cdot \sigma_{\text{NSD}}$, where Y is calculated according to (1). σ^{NSD} is the total NSD cross section, which is measured at STAR to be 30.0 ± 2.4 mb at $\sqrt{s} = 200$ GeV [5]. In the case of $\sqrt{s} = 500$ GeV, σ^{NSD} is extrapolated from 200 GeV measurement with the help of PYTHIA simulation to be 34 mb.

The charm production cross section at mid rapidity $\left.\frac{\mathrm{d}\sigma_{c\bar{c}}}{\mathrm{d}y}\right|_{y=0}$ is $170 \pm 45(\text{stat.})^{+37}_{-51}(\text{sys.})$ μb at $\sqrt{s} = 200$ GeV and is $217 \pm 86(\text{stat.}) \pm 73(\text{sys.})$ μb at $\sqrt{s} = 500$ GeV. FONLL predictions for p_T spectra [6] shown in Fig. 1. In order to compare STAR results with other experiments, we extrapolated $\left.\frac{\mathrm{d}\sigma_{c\bar{c}}}{\mathrm{d}y}\right|_{y=0}$ to $\sigma^{\text{NN}}_{c\bar{c}}$ using PYTHIA simulations with various parameter tunings giving extrapolation factors 4.7 ± 0.7 for 200 GeV and 5.6 ± 0.1 for 500 GeV collisions. The results are depicted in Figure 1 revealing agreement with with NLO prediction [7]. Note that the STAR result at 200 GeV was used in [7] to fit the μ_R and μ_F parameters, so only the STAR result at 500 GeV can be used as a test of this prediction.

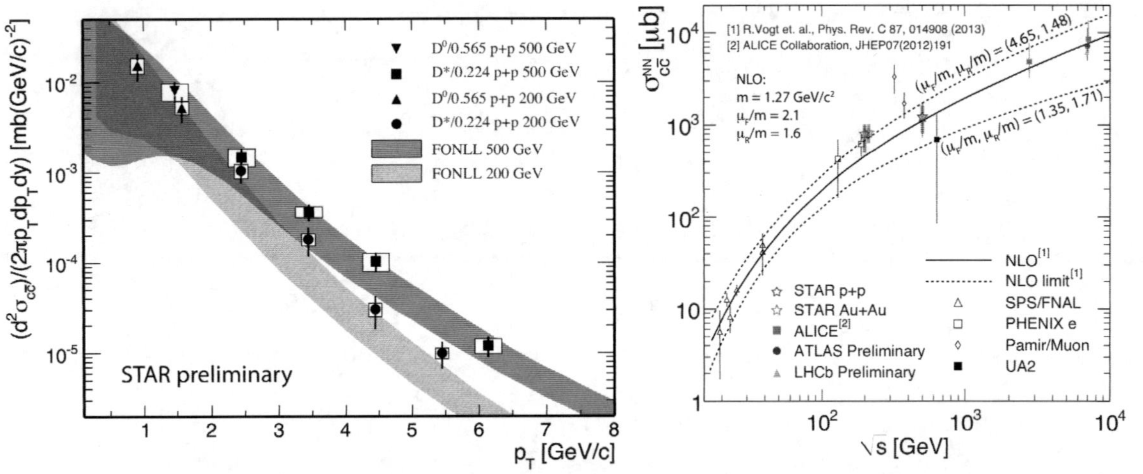

Figure 1. Left Panel: Charm quark production invariant cross section as a function of D meson p_T in 200 and 500 GeV $p+p$ collisions with two FONLL predictions [6] using normalization and factorization scale equal to charm quark mass m_c. Right Panel: Total charm cross section as a function of $\sqrt{s}$

3.2. Identification of electrons from semi-leptonic decays

Electrons from bottom and charm meson decays are the two dominant components of the non-photonic electrons. Mostly due to the decay kinematics, the azimuthal correlations between the daughter electron and daughter hadron are different for bottom meson decays and charm meson decays. A study of these azimuthal correlations has been carried out on STAR data and is compared with a PYTHIA simulation to obtain the ratio of the bottom electron yield to the heavy-flavor decay electron yield $\frac{e_b}{e_b+e_c}$ [8], where PYTHIA was tuned to reproduce STAR measurements of D mesons p_T spectra [9]. Using the measured $e_b/(e_b+e_c)$ together with the measured non-photonic electron cross section with the electrons from $J/\Psi, \Upsilon$ decay and Drell-Yan processes subtracted, we are able to disentangle these two components. Figure 2 shows the invariant cross section of electrons $\left(\frac{e^+ + e^-}{2}\right)$ from bottom (upper left) and charm (upper right) mesons as a function of p_T and the corresponding FONLL predictions, along with the ratio of each measurement to the FONLL calculations (lower panels).

From the measured spectrum, we determine the integrated cross section of electrons $\left(\frac{e^+ + e^-}{2}\right)$ at 3 GeV$/c < p_T <$ 10 GeV$/c$ from bottom and charm meson decays to be, respectively,

$$\left.\frac{d\sigma_{(B\to e)+(B\to D\to e)}}{dy}\right|_{y_e=0} = 4.0 \pm 0.5(\text{stat}) \pm 1.1(\text{syst})\text{nb}$$

$$\left.\frac{d\sigma_{D\to e}}{dy}\right|_{y_e=0} = 6.2 \pm 0.7(\text{stat}) \pm 1.5(\text{syst})\text{nb}$$

4. Conclusions

Open charm hadron (D^0, D^{*+}) cross section in $p+p$ minimum bias collisions at $\sqrt{s} = 200$ and 500 GeV at STAR is measured with results:
$d\sigma^{NN}_{c\bar{c}}/dy|_{y=0} = 170 \pm 45(\text{stat.})^{+37}_{-51}(\text{sys.})$ μb at 200 GeV, $217 \pm 86(\text{stat.}) \pm 73(\text{sys.})$ μb at 500 GeV. The total charm cross section at 500 GeV is within statistical and systematic uncertainties consistent with the latest NLO pQCD prediction.

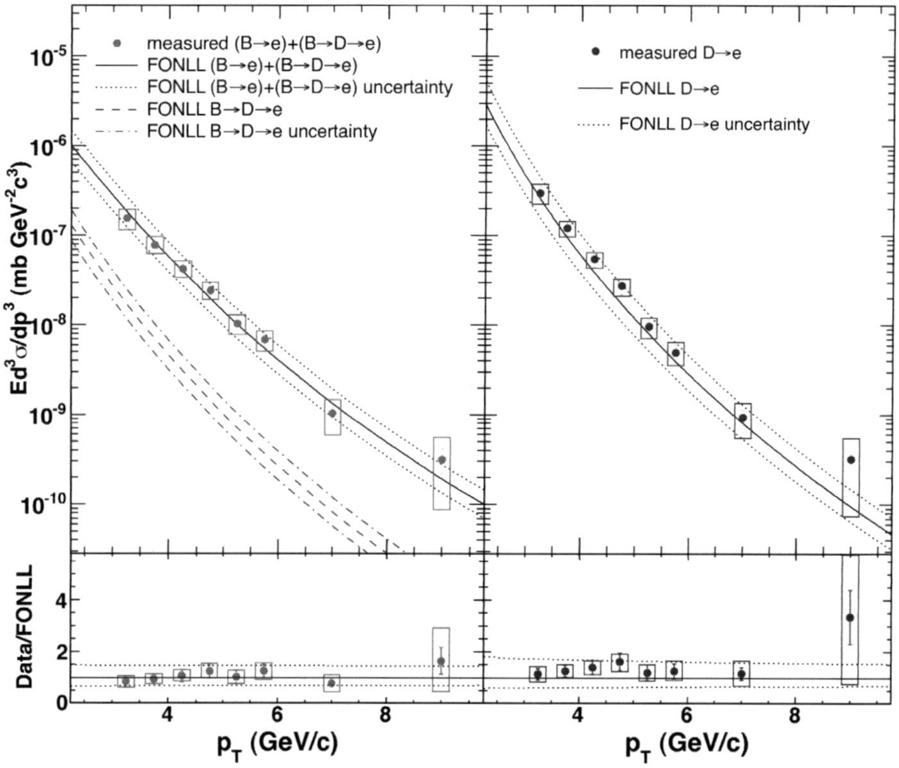

Figure 2. Invariant cross section of electrons from bottom (upper left) and charm meson (upper right) decay, together with the ratio of the corresponding measurements to the FONLL predictions for bottom (lower left) and charm electrons (lower right). The solid circles are experimental measurements. The error bars and the boxes are, respectively, the statistical and systematic uncertainties. The solid and dotted curves are the FONLL predictions and their uncertainties. The dashed and dot-dashed curves are the FONLL prediction for $B \to D \to e$.

Acknowledgments

This work has been supported by the grant 13-02841S of the Czech Science Foundation (GACR) and by the Grant Agency of the Czech Technical University in Prague, grant No. SGS13/215/OHK4/3T/14.

References

[1] Cacciari M, Nason P and Vogt R 2005 *Phys. Rev. Lett.* **95** 122001
[2] Adamczyk L et al. [STAR Collaboration] 2012,*Phys. Rev.* **D 86** 072013
[3] Shao M et al. 2003, *Nucl. Instrum. Methods* **A 499** 624
[4] Agakishiev B I et al. [STAR Collaboration] 2011, *Phys. Rev.* **D 83** 052006
[5] Adams J et al. [STAR Collaboration] 2003 *Phys. Rev. Lett.* **91** 172302
[6] Vogt R 2012 *private communication*
[7] Vogt R et al. 2013 *Phys. Rev.* **C 87** 014908
[8] Aggarwal M M et al. [STAR Collaboration] *Phys. Rev. Lett.* **105** 202301
[9] Lin X, `arXiv:hep-ph/0602067`

Journal of Physics: Conference Series **509** (2014) 012079 doi:10.1088/1742-6596/509/1/012079

Measurement of electrons from heavy-flavour decays in Pb-Pb collisions at $\sqrt{s_{\rm NN}} = 2.76$ TeV with ALICE

Deepa Thomas for the ALICE collaboration

ERC-Research Group QGP-ALICE, Utrecht University, Princetonplein 5, 3584 CC Utrecht, The Netherlands

E-mail: deepa.thomas@cern.ch

Abstract. The measurement of heavy-flavour (charm and beauty) production in ultra-relativistic heavy-ion collisions provides an important contribution to the study of the properties of the hot and dense medium created in such collisions. One approach to measure heavy-flavour production is via electrons from semi-leptonic decays of heavy-flavour hadrons. In this contribution we present the nuclear modification factor ($R_{\rm AA}$) and the azimuthal anisotropy (v_2) of heavy-flavour decay electrons in Pb-Pb collisions at $\sqrt{s_{\rm NN}} = 2.76$ TeV, which are sensitive to the interaction of c and b quarks with the medium. We also present, for the first time, the azimuthal angular correlations of heavy-flavour decay electrons and charged hadrons, $\Delta\phi(HFE, h)$, in Pb-Pb collisions and the corresponding near-side yield ratio ($I_{\rm AA}$).

1. Introduction

In collisions of lead ions at relativistic energies a high density, strongly interacting QCD medium is produced. Heavy-flavour particles produced in such collisions are a useful probe of the medium since they carry information about the full evolution of the system. One way to measure heavy-flavour production is through the measurement of electrons from heavy-flavour hadron decays (*HFE*). The observables discussed here reflect the interaction of heavy quarks with the medium. These are the nuclear modification factor ($R_{\rm AA}$), the elliptic flow of (v_2) of *HFE* and the azimuthal angular correlations between *HFE* and charged hadrons ($\Delta\phi(HFE, h)$). $R_{\rm AA}$, which can be interpreted in terms of parton energy loss in the medium, is defined as the ratio of $p_{\rm T}$-differential yields measured in nucleus-nucleus collisions to the corresponding cross-section in pp collisions scaled by the nuclear overlap function $< T_{\rm AA} >$.

Further insight into the transport properties of the medium can be obtained from the azimuthal anisotropy of particle production in non-central Pb-Pb collisions. The second harmonic coefficient of the Fourier expansion of particle azimuthal distributions, v_2, called elliptic flow, reflects the collective expansion of the medium at low $p_{\rm T}$ and the path-length dependence of parton energy loss. Further information on the parton fragmentation and energy loss can become accessible through azimuthal angular correlations $\Delta\phi(HFE, h)$.

2. Dataset and data analysis

Various detectors in the ALICE central barrel ($|\eta| < 0.9$) are used for the *HFE* measurements. The tracking system, which includes the Inner Tracking System (ITS) and the Time Projection Chamber (TPC) is used for the reconstruction of charged particle tracks and the measurement

Journal of Physics: Conference Series **509** (2014) 012079 doi:10.1088/1742-6596/509/1/012079

of their momenta. The particle identification system (PID) consists of a Time of Flight (TOF) detector, the TPC with which the specific ionization energy loss (dE/dx) of charged particles is measured and the Electromagnetic Calorimeter (EMCal), which registers electromagnetic showers in part of the central barrel acceptance ($|\eta| < 0.7$ and $80^\circ < \phi < 180^\circ$). At low p_T ($p_T < 6$ GeV/c for the present analysis) electrons are identified with the TPC and TOF detectors, while with the EMCal the measurement can be extended to higher p_T. In Pb-Pb collisions, the centrality was determined from the sum of the amplitudes of the signals in the VZERO detector [1] and defined in terms of percentiles of the total hadronic Pb-Pb cross section. VZERO detector is also used to measure the reaction plane for the v_2 measurement.

The results presented here are obtained from the Pb-Pb data at $\sqrt{s_{NN}} = 2.76$ TeV from 2010 and 2011. The reference measurements used are from pp collisions at 7 TeV.

The general analysis strategy used for all measurements presented here involves the identification of electrons and the statistical removal of the background from sources other than heavy-flavour hadron decays (*Non-HFE*). This background is dominated by electrons from photon conversions in the detector material and Dalitz decays of light neutral mesons (mostly π^0 and η). Two methods are available to remove the background, the invariant mass method and the cocktail method [2]. In the invariant mass method one makes use of the fact that electron-positron pairs from conversions and Dalitz decays have a small invariant mass and can thus be tagged. In the cocktail method, the *Non-HFE* background is calculated with a Monte-Carlo generator using the measured π^0 and η p_T-differential yields as input.

3. Results

The nuclear modification factor is obtained from the *HFE* yield measured in Pb-Pb collisions at $\sqrt{s_{NN}} = 2.76$ TeV and using the corresponding cross-section from pp collisions at $\sqrt{s} = 7$ TeV to determine the reference for $p_T < 8$ GeV/c [2]. The latter is achieved by scaling the cross-section measured at 7 TeV to 2.76 TeV using a FONLL pQCD [4] based procedure [3]. For $p_T > 8$ GeV/c the result from the FONLL pQCD calculation at $\sqrt{s} = 2.76$ TeV is used as reference.

The resulting R_{AA} of electrons from heavy-flavour hadron decays in central $(0-10\%)$ and semi-central $(40-50\%)$ Pb-Pb collisions is shown in Figure 1. The error bars represented by lines gives the statistical uncertainty and the box represents the systematic uncertainty. A strong suppression of the HFE yield is observed in central collisions in the measured p_T range from 3 to 18 GeV/c. In semi-central collisions the suppression is less pronounced but still significant. The *HFE* R_{AA} is consistent with that measured for heavy flavour decay muons at forward rapidity (2.5 <y< 4) in both centralities.

The elliptic flow of heavy-flavour decay electrons (v_2^{HFE}) is obtained using the v_2 measured for inclusive electrons (v_2^{Incle}) , the ratio R_{NP} of the *HFE* and *Non-HFE* yields ($N^{HFE}/N^{Non-HFE}$) and v_2 of *Non-HFE* [6].

$$v_2^{HFE} = \frac{(1+R_{NP})v_2^{Incle} - v_2^{Non-HFE}}{R_{NP}} \tag{1}$$

The electron v_2 is obtained using the event plane method in which the particle azimuthal angle (ϕ) is measured relative to an estimator of the reaction plane determined from the azimuthal distribution of signals in the VZERO detector [5]. The background v_2 ($v_2^{Non-HFE}$) is calculated with a Monte-Carlo generator where the measured $v_2(p_T)$ of $\pi^\pm$, π^0 and $K^\pm$ are used as input, assuming $v_2^{\pi^\pm} = v_2^{\pi^0}$, $v_2^{\eta} = v_2^{K^\pm}$. The v_2 of direct photons is assumed to be zero. The *HFE* v_2 is measured in $20-40\%$ central Pb-Pb collisions for $1.5 < p_T < 13$ GeV/c, as shown in Figure 2. A positive v_2 has been observed for *HFE*, which suggests the participation of heavy quarks in the collective flow at low p_T. At high p_T, it is more likely to be due to the path-length dependence of parton energy loss.

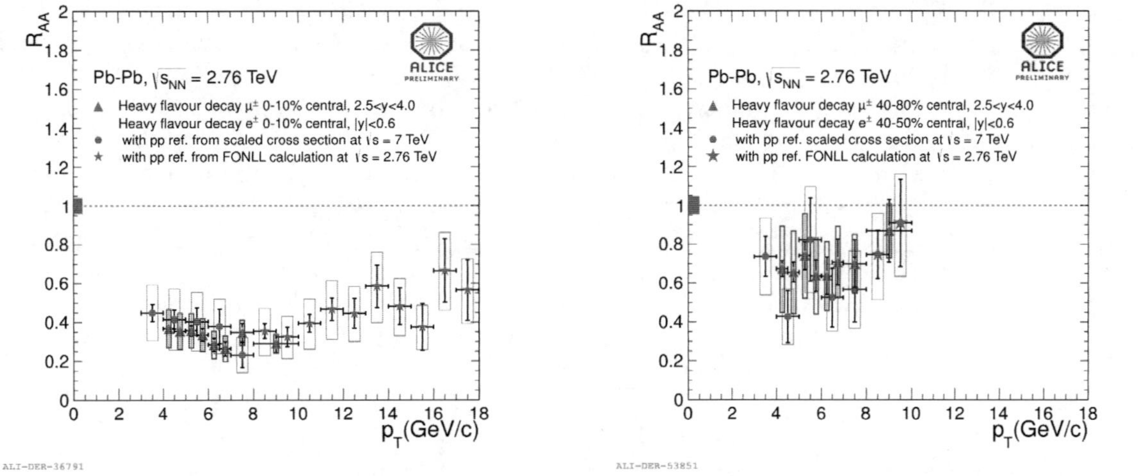

Figure 1. R_{AA} of heavy-flavour decay electrons measured in the two centrality ranges $0-10\%$ (left) and $40-50\%$ (right) in Pb-Pb collisions at $\sqrt{s_{NN}} = 2.76$ TeV, compared with the R_{AA} of heavy-flavour decay muons in the same centrality ranges.

Various theoretical models predict R_{AA} and v_2 for *HFE*, some of which are described here. In BAMPS approach [7] heavy quarks are transported through the medium while undergoing collisional and radiative energy loss, Rapp et al [8] include in-medium resonance scattering and coalescence of heavy quarks in the medium and in POWLANG [9] heavy quarks are transported following a Langevin approach and considering collisional energy loss.

The measured R_{AA} and v_2 values is compared with the theoretical predictions as shown in Figure 2. BAMPS gives a good description of heavy-flavour decay electron v_2 but predicts a larger in-medium suppression than measured. The prediction from Rapp et al is consistent with the measured R_{AA}, but it underestimates v_2^{HFE}. POWLANG describes the *HFE* R_{AA} in the range $3 < p_T^e < 12$ GeV/c but it underestimates v_2^{HFE}. The data provide a strong constraint on the theoretical description of heavy quark transport in the QGP.

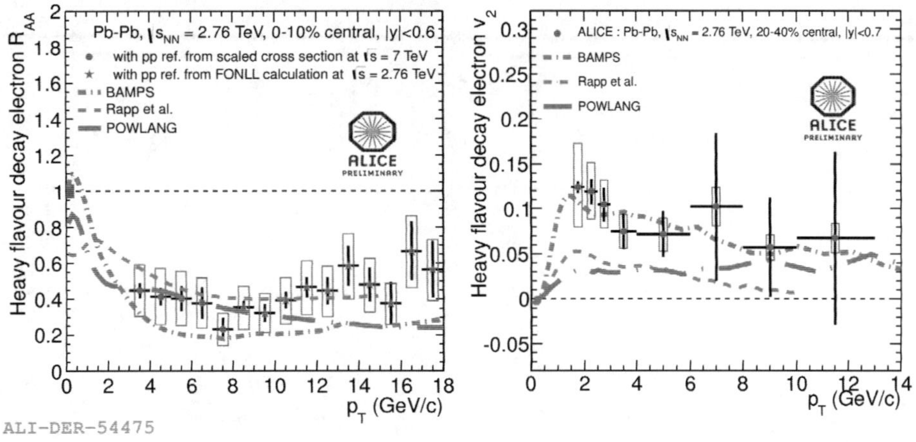

Figure 2. Comparison between measured R_{AA} and v_2 of heavy-flavour decay electrons and theoretical calculations [7, 8, 9].

The $\Delta\phi(HFE, h)$ correlation is measured for $4 < p_{\rm T}^{h} < 6$ GeV/c in two trigger electron $p_{\rm T}$ intervals, $6 < p_{\rm T}^{e} < 8$ GeV/c and $8 < p_{\rm T}^{e} < 10$ GeV/c for two centralities, $0-10\%$ and $20-50\%$, in Pb-Pb collisions at $\sqrt{s_{\rm NN}} = 2.76$ TeV and in pp collisions at 7 TeV. Since the vacuum fragmentation of heavy quarks is not expected to be different in pp collisions at $\sqrt{s} = 7$ and 2.76 TeV, the near-side yield in pp collisions at 7 TeV is used as reference. Figure 3 shows the $\Delta\phi(HFE, h)$ in Pb-Pb and pp collisions after removing the pedestal defined between $1.5 < \Delta\phi < 2$ rad. The correlation distribution in Pb-Pb collisions is consistent with pp collisions within statistical uncertainties. To address a possible modification of the parton fragmentation function in the QCD medium, the near-side ($-\pi/2 < \Delta\phi < \pi/2$) per-trigger yield is measured in Pb-Pb collisions and compared with the yield in pp collisions to obtain $I_{\rm AA}$, defined as ratio of the per-trigger yields in pp and Pb-Pb [10]. As depicted in Figure 4, the $I_{\rm AA}$ on the near-side is consistent with unity within the current uncertainties.

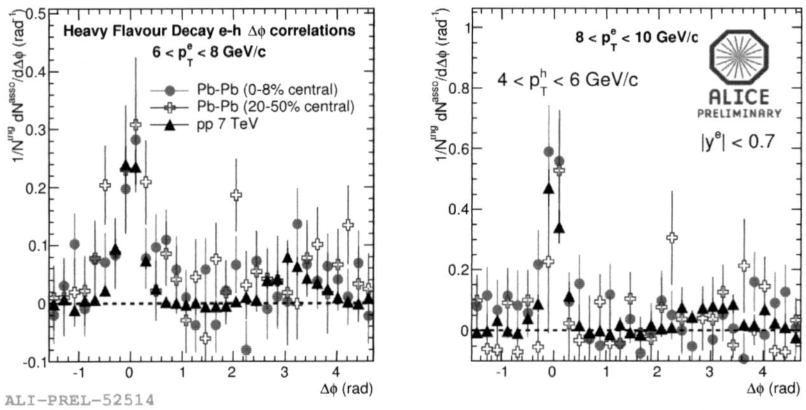

Figure 3. $\Delta\phi(HFE, h)$ distribution for $4 < p_{\rm T}^{h} < 6$ GeV/c in $0-8\%$ (red) and $20-50\%$ (blue) central Pb-Pb collisions at $\sqrt{s_{\rm NN}} = 2.76$ TeV compared to pp collisions at $\sqrt{s} = 7$ TeV (black).

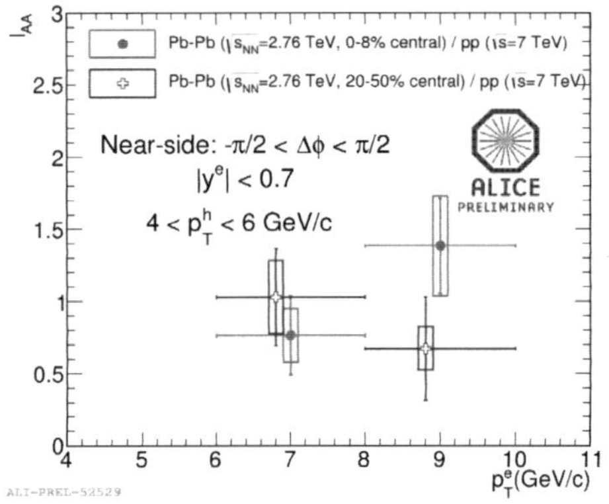

Figure 4. $I_{\rm AA}$ for near side ($-\pi/2 < \Delta\phi < \pi/2$) $\Delta\phi(HFE, h)$ correlations for $4 < p_{\rm T}^{h} < 6$ GeV/c in $0-8\%$ (red) and $20-50\%$ (blue) central Pb-Pb collisions at $\sqrt{s_{\rm NN}} = 2.76$ TeV.

Journal of Physics: Conference Series **509** (2014) 012079 doi:10.1088/1742-6596/509/1/012079

4. Conclusions

With ALICE, heavy-flavour production has been studied in Pb-Pb collisions via charm and beauty semi-leptonic decay channels to electrons. The nuclear modification factor, azimuthal anisotropy, and azimuthal angular correlations with hadrons have been measured for electrons from heavy-flavour hadron decays. R_{AA} in central Pb-Pb collisions demonstrates a significant energy loss of heavy quarks in the QCD medium. A non-zero v_2^{HFE} has been measured for non-central Pb-Pb collisions suggesting participation of heavy quarks in the collective motion of the medium. First measurements of $\Delta\phi(HFE, h)$ have been obtained for central and non-central Pb-Pb collisions. Higher statistics will help for more definite conclusions to be made in the future.

5. References

[1] Aamodt K et al. [ALICE Collaboration] 2008 *JINST* **3** S08002.
[2] Abelev B et al. [ALICE Collaboration] 2013 *Phys.Lett.* B **721** 1323.
[3] Averbeck R et al. 2011 *arXiv* **1107** 3243.
[4] Cacciari M et al. 2012 *JHEP* **1210** 137.
[5] Poskanzer A M, Voloshin S A 1998 *Phys.Rev.* C **58** 1671-1678.
[6] Adler S S et al. [PHENIX Collaboration] 2005 *Phys.Rev.* C **72** 024901.
[7] Uphoff J et al. 2012 *Phys.Lett.* B **71** 430-435.
[8] He M et al. 2012 *arXiv* **1208** 0256.
[9] Monteno M et al. 2011 *J.Phys.* G **38** 124144.
[10] Aamodt K et al. [ALICE Collaboration] 2012 *Phys.Rev.Lett* **108** 092301.

Journal of Physics: Conference Series **509** (2014) 012080 doi:10.1088/1742-6596/509/1/012080

D-meson nuclear modification factor and v_2 in Pb-Pb collisions at the LHC

Elena Bruna, for the ALICE Collaboration
Istituto Nazionale di Fisica Nucleare, Via P. Giuria 1, 10125 Torino
E-mail: bruna@to.infn.it

Abstract.
We present the ALICE results on open heavy flavour, focusing on the exclusive reconstruction of charmed mesons via displaced decay topologies. These measurements benefit from the large Pb-Pb statistics collected in 2011. The results on the nuclear modification factor R_{AA} for D mesons indicate a suppression of their yield in central collisions relative to binary-scaled pp collisions in a large momentum range. The comparison to the R_{AA} of non-prompt J/ψ (measured with CMS) indicates a difference in the suppression of charm and beauty, as expected according to the predicted mass hierarchy in energy loss models. The measurement of the azimuthal anisotropy of charmed mesons is also discussed. The observed positive second Fourier harmonic v_2 for transverse momentum $2 < p_T < 6$ GeV/c in semi-peripheral events is a hint for collective motion of charm quarks. The results discussed above are also compared to theoretical models.

1. Introduction
A quantitative picture of the medium created in high-energy Pb-Pb collisions can be obtained by measuring its properties with observables that give access to the energy-loss mechanism for different parton species and to the transport properties of the medium. Theoretical models of energy loss predict a hierarchical dependence on the colour charge and mass of the parton propagating through the medium. A larger energy loss is expected for gluons, together with a suppression of gluon radiation at small angles for partons with larger mass ("dead-cone" effect [1]). It is therefore interesting to compare medium effects (i.e. the path-length, colour-charge and mass dependence of the energy loss, as well as the collective motion) on heavy quarks versus light quarks and gluons. Such challenging goals require precise heavy-flavour measurements in Pb-Pb collisions and a clear understanding of the reference systems provided by pp and p-Pb collisions. The latter provides the control experiment needed to disentangle effects from the cold matter in absence of a hot and dense medium.

2. Heavy-flavour measurements in ALICE
The ALICE [2] detectors used in these analyses are the Inner Tracking System (ITS), the Time Projection Chamber (TPC) and the Time Of Flight (TOF) detector. They provide tracking, primary and secondary vertex reconstruction and particle identification, like K/π separation up to transverse momentum $p_T \sim 2.5$ GeV/c. These features are essential ingredients to perform open-heavy flavour analyses over a broad transverse momentum range. In addition,

the VZERO detector is used for triggering on collisions and for centrality measurements. Charmed mesons are reconstructed in ALICE through their hadronic decays: $D^0 \to K^-\pi^+$, $D^+ \to K^-\pi^+\pi^+$, $D^{*+} \to D^0\pi^+$ and $D_s^+ \to \phi(\to K^-K^+)\pi^+$. The analysis strategy exploits the particle identification and the detector capability to reconstruct secondary vertices displaced by a few hundred μm from the primary vertex. Topological cuts are applied on the reconstructed secondary vertices and the signals are extracted via fits to the invariant mass distributions.

In these proceedings we report on the prompt D-meson nuclear modification factor $R_{\rm AA}$, defined as the yield in Pb-Pb collisions divided by the cross-section in pp collisions scaled by the average nuclear overlap function in a given Pb-Pb centrality range, as a function of $p_{\rm T}$ and collision centrality. The prompt D-meson elliptic flow v_2, defined as the second harmonic coefficient of the Fourier azimuthal expansion, is also reported as a function of $p_{\rm T}$ in semi-peripheral collisions.

In the $R_{\rm AA}$ analysis, the prompt D-meson yield was obtained after subtracting the contribution of D mesons from beauty hadron decays based on perturbative QCD calculations (FONLL, [3]). The $R_{\rm AA}$ of non-prompt D mesons was assumed to be $R_{\rm AA}^{\rm feed-down} = 2R_{\rm AA}^{\rm prompt}$(on the basis of the CMS results of non-prompt J/ψ, instead of $R_{\rm AA}^{\rm feed-down} = R_{\rm AA}^{\rm prompt}$ as done in the previous analysis [4]) and the systematic uncertainty was assessed by varying its value between 1 and 3 times the prompt-D $\rm R_{\it AA}$. Due to the lack of statistics of the pp data at $\sqrt{s} = 2.76$ TeV, the pp reference needed for the $\rm R_{\it AA}$ analysis was determined by scaling the D-meson cross section measured at $\sqrt{s} = 7$ TeV to $\sqrt{s} = 2.76$ TeV using FONLL calculations [5]. The scaling uncertainty varies from $\sim 20\%$ to $\sim 5\%$ from low to high $p_{\rm T}$. The contribution of D mesons from beauty decays was also subtracted from the v_2 measurement with an assumption on the $R_{\rm AA}$ (as above) and v_2 of non-prompt D mesons ($0 \le v_2^{\rm feed-down} \le v_2^{\rm prompt}$ [6]). These analyses were performed on the data from the Pb-Pb runs at $\sqrt{s_{\rm NN}} = 2.76$ TeV collected in 2010 (minimum bias trigger, $\mathcal{L} = 2.12\ \mu{\rm b}^{-1}$) and 2011 ($\mathcal{L} = 28$ and $6\ \mu{\rm b}^{-1}$ for 0-7.5% and 10-50% respectively).

3. Results

The $R_{\rm AA}$ of prompt D mesons was measured with ALICE in the $0-7.5\%$ centrality class using the data sample collected in 2011, which extends the measurements to a wider transverse momentum range ($2 < p_{\rm T} < 36$ GeV/c) compared to the published results from the 2010 data [4]. As shown in Fig. 1 (left), the $R_{\rm AA}$ values for D^0, D^+ and D^{*+} agree within the uncertainties and show a strong suppression (factor of $4-5$ for $5 < p_{\rm T} < 16$ GeV/c) of the D-meson yields in Pb-Pb collisions relative to pp collisions. The first measurement of the D_s^+ meson in Pb-Pb collisions is also shown. A suppression of the D_s^+ is observed for $8 < p_{\rm T} < 12$ GeV/c, in agreement within the uncertainties with the non-strange charmed mesons $R_{\rm AA}$ in this $p_{\rm T}$ range. The D_s^+-meson yield could be less suppressed at lower $p_{\rm T}$ because of the predicted c-quark recombination with the enhanced strange quarks in the medium [8], however a more precise measurement is needed to draw a conclusion. The $R_{\rm AA}$ as a function of the number of participant nucleons weighted by the number of binary collisions is shown in Fig. 1 (right) for D^0 mesons with $2 < p_{\rm T} < 3$ GeV/c and compared to pions measured in the same centrality and $p_{\rm T}$ intervals. The main correlated systematic uncertainties for the D-meson yields come from the pp cross section and beauty feed down. The uncorrelated systematic uncertainties are mainly due to the D-meson yield extraction and cut variation. The suppression of D mesons with $2 < p_{\rm T} < 3$ GeV/c is almost independent of centrality, while the pion $R_{\rm AA}$ decreases with increasing centrality. The magnitude of the suppression is similar for D mesons and pions. A hint of a difference in central collisions is suggested but a larger dataset is needed to draw conclusion on the expected difference between D mesons and pion suppression due to the mass and colour charge dependence of the energy loss. In the higher $p_{\rm T}$ range the D-meson suppression is no longer flat, rather it increases from peripheral to central collisions (Fig. 2, left).

The $R_{\rm AA}$ of non-prompt J/ψ ($6.5 < p_{\rm T} < 30$ GeV/c and $|y| < 1.2$) measured with CMS is

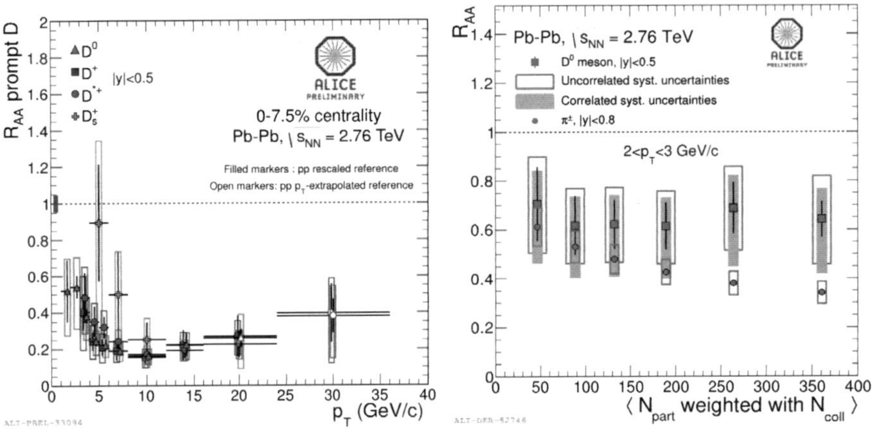

Figure 1. Left: R_{AA} as a function of p_T for prompt D^0, D^+, D^{*+} and D_s^+ for the $0-7.5\%$ most central Pb-Pb collisions at $\sqrt{s_{NN}} = 2.76$ TeV [7]. Right: R_{AA} as a function of centrality for prompt D^0 mesons in the transverse momentum range $2 < p_T < 3$ GeV/c, compared to pions.

reported in Fig. 2 (left) as a function of the number of participants weighted by the number of binary collisions [9]. The results are shown together with the D-meson R_{AA} from ALICE ($8 < p_T < 16$ GeV/c and $|y| < 0.5$). The selected p_T ranges correspond to similar kinematical ranges for the parent b and c quarks, but the measurements are performed in different rapidity ranges. An indication of a difference in the suppression of charm and beauty can be observed in central collisions, consistent with the mass hierarchy expected from various energy-loss models, such as those reported in the figure [10–12]. While the model [11] based on collisional and radiative energy loss in an anisotropic medium agrees with both prompt D mesons and non-prompt J/ψ, the others seem to underestimate both (i.e. [10], based on collisional energy loss in an expanding medium) or to underestimate the suppression of non-prompt J/ψ (i.e. [12], based on radiative energy loss and D-meson in-medium formation and dissociation).

The v_2 for prompt D mesons, reported in Fig. 2 (right), was measured as a function of p_T in the range $2 < p_T < 16$ GeV/c in the 30-50% centrality class [6]. We observe a positive v_2 for prompt D mesons in the transverse momentum range $2 < p_T < 6$ GeV/c, which is comparable to that of charged particles (not shown). This result suggests a collective motion also for heavy quarks at low p_T. At higher momentum a positive v_2 is expected due to the path-length dependence of the in-medium energy loss, given the different paths traveled by a parton propagating through the medium. The measurements of the prompt D-meson v_2 are also compared to various energy-loss models [10, 11, 13, 14, 16, 17]. While several theoretical models based on in-medium parton energy loss reproduce reasonably well the R_{AA} of prompt D mesons as a function of p_T (not shown, see [4]), they are challenged by simultaneously reproducing results from prompt D-meson R_{AA} and v_2.

4. Conclusions

The suppression of prompt D mesons observed in the momentum range $5 < p_T < 15$ GeV/c in central Pb-Pb collisions at $\sqrt{s_{NN}} = 2.76$ TeV is evidence for a strong modification of charm production in Pb-Pb compared to pp collisions due to energy loss. Results for the D_s^+ meson also show a suppression in $8 < p_T < 12$ GeV/c but suffer from statistics limitations that do not allow us to draw conclusions on a possible enhancement at low p_T due to a coalescence of charm quarks with strange quarks. The measurements of prompt D-meson R_{AA} as a function

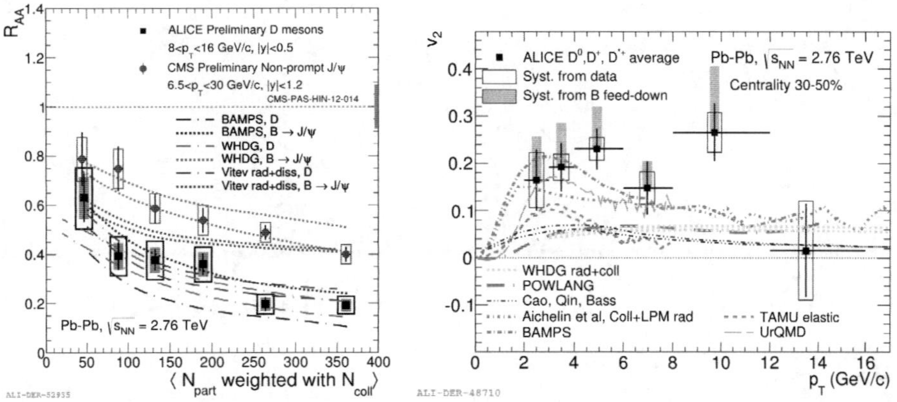

Figure 2. Left: R_{AA} as a function of centrality for prompt D mesons (average of D^0, D^+ and D^{*+}) in the transverse momentum range $8 < p_{\mathrm{T}} < 16$ GeV/c, compared to non-prompt J/ψ measured with CMS with $6.5 < p_{\mathrm{T}} < 30$ GeV/c. Results from theoretical calculations are superimposed (see text). Right: v_2 for prompt D mesons in the 30-50% centrality class as a function of p_{T} [6] superimposed to theoretical results.

of centrality indicate that the suppression tends to be constant with centrality in the lowest p_{T} range, $2 < p_{\mathrm{T}} < 3$ GeV/c, while it increases with centrality at intermediate/high p_{T}. The observed difference in the suppression of D mesons and non-prompt J/ψ from B-meson decays at high p_{T} in central collisions suggests a mass dependence in the energy loss of heavy quarks. The positive v_2 for prompt D mesons in the momentum range $2 < p_{\mathrm{T}} < 6$ GeV/c in semi-peripheral Pb-Pb collisions suggests that the interactions with the medium constituents transfer to charm quarks information on the azimuthal anisotropy of the system. Theoretical models of in-medium energy loss reproduce reasonably well the measured prompt D-meson R_{AA}. Nevertheless, a simultaneous description of both R_{AA} and v_2 remains a challenge for models, indicating that the energy-loss mechanisms and the participation of heavy quarks in the collective behavior of the Quark-Gluon Plasma are not yet fully understood today.

References

[1] Y. L. Dokshitzer and D. E. Kharzeev, Phys. Lett. B **519** (2001) 199 [hep-ph/0106202].
[2] K. Aamodt *et al.* [ALICE Collaboration], JINST **3** (2008) S08002.
[3] M. Cacciari *et al.*, JHEP **1210** (2012) 137 [arXiv:1205.6344].
[4] B. Abelev *et al.* [ALICE Collaboration], JHEP **1209** (2012) 112 [arXiv:1203.2160].
[5] R. Averbeck *et al.*, arXiv:1107.3243.
[6] B. Abelev *et al.* [ALICE Collaboration], Phys. Rev. Lett. **111** (2013) 102301 [arXiv:1305.2707].
[7] Z. Conesa del Valle [ALICE Collaboration], Nucl. Phys. A **904-905** (2013) 178c [arXiv:1212.0385].
[8] I. Kuznetsova *et al.*, Eur. Phys. J. C **51** (2007) 113. M. He *et al.*, Phys. Rev. Lett. **110** (2013) 112301. A. Andronic *et al.*, Phys. Lett. B **659** (2008) 149.
[9] S. Chatrchyan *et al.* [CMS Collaboration], JHEP **1205** (2012) 063 [arXiv:1201.5069], CMS-PAS-HIN-12-014.
[10] O. Fochler, J. Uphoff, Z. Xu and C. Greiner, J. Phys. G **38** (2011) 124152 [arXiv:1107.0130].
[11] W. A. Horowitz and M. Gyulassy, J. Phys. G **38** (2011) 124114 [arXiv:1107.2136].
[12] I. Vitev *et al.*, Phys. Rev. C **80** (2009) 054902,
[13] W. M. Alberico *et al.*, Eur. Phys. J. C **71** (2011) 1666 [arXiv:1101.6008].
[14] P. B. Gossiaux *et al.*, Phys. Rev. C **79** (2009) 044906, J. Phys. G **37** (2010) 094019.
[15] Rapp, He *et al.*, Phys. Rev. C 86 (2012) 014903.
[16] M. He, R. J. Fries and R. Rapp, Phys. Rev. C **86** (2012) 014903 [arXiv:1106.6006].
[17] T. Lang *et al.*, arXiv:1211.6912, arXiv:1212.0696.

Journal of Physics: Conference Series **509** (2014) 012081 doi:10.1088/1742-6596/509/1/012081

Open-charm production as a function of charged particle multiplicity in pp collisions at $\sqrt{s}$=7 TeV with ALICE

Renu Bala, for the ALICE Collaboration

Department of Physics, University of Jammu

E-mail: renu.bala@cern.ch

Abstract. Heavy quarks (charm and beauty) are an effective tool to investigate the properties of the Quark-Gluon Plasma created in heavy-ion collisions as they are produced in initial hard scattering processes and as they experience all the stages of the medium evolution. The measurement of heavy-flavour production cross sections in pp collisions at the LHC, besides providing a reference for heavy-ion studies, allows one to test perturbative QCD calculations. A brief review of ALICE results on the production of heavy-flavoured hadrons measured from fully reconstructed hadronic decay topologies in pp collisions at $\sqrt{s}$ = 7 TeV is presented. Furthermore, heavy-flavour production was also studied as a function of the particle multiplicity in pp collisions. This could provide insight into multi-parton scatterings. A measurement of the inclusive J/ψ yield as a function of the charged-particle pseudorapidity density was performed by the ALICE Collaboration at the LHC in pp collisions at $\sqrt{s}$ = 7 TeV. An increase of the J/ψ yield with increasing multiplicity was observed. In this context, the study of the yield of D mesons as a function of the charged-particle multiplicity could provide a deeper insight into charm-quark production in pp collisions. We will present the first results obtained for prompt D^0, D^+, and D^{*+} mesons using hadronic decay channels at midrapidity in pp collisions at $\sqrt{s}$=7 TeV as a function of the charged-particle multiplicity. The prompt D-meson yields as a function of multiplicity are measured in different p_T intervals. These yields will be compared to the results obtained for inclusive and non-prompt J/ψ.

1. Introduction

Heavy quarks are unique probes to study the Quark-Gluon Plasma produced in heavy ion collisions at the LHC. Due to their large masses, they are produced predominantly in hard parton scattering processes, during the initial stages of the collision. Therefore, they experience the entire evolution of the medium created in the collision and can act as probes of its properties. The measurement of heavy-quark production in pp collisions besides providing a necessary reference for the study of medium effects in Pb-Pb collisions, serves as a precision test for perturbative Quantum Chromodynamics (pQCD). In addition, the study of heavy-quark production as a function of charged particle multiplicity could provide insight into multiple hard parton scattering. This allows one to investigate the interplay between hard and soft QCD processes and to study the role of multi-parton interactions. It is interesting to remark that the highest multiplicities reached in pp collisions at 7 TeV are similar to those observed in Cu-Cu collisions at RHIC energies.

2. Open charm measurement with ALICE

ALICE (A Large Ion Collider Experiment) is the LHC experiment dedicated to heavy ion studies. ALICE [1] consists of two parts: a barrel at central rapidity and a muon spectrometer at forward rapidity. For the present analysis, we have used the information from a subset of the central barrel detectors, namely the Inner Tracking System (ITS), the Time Projection Chamber (TPC) and the Time Of Flight detector (TOF) for charged particle tracking and identification, the T0 for time zero measurement and the VZERO scintillator for triggering. The two tracking detectors, the ITS and the TPC, allow the reconstruction of charged-particle tracks in the pseudorapidity range -0.9 $< \eta <$ 0.9 with a momentum resolution better than 2% for $p_T <$ 20 GeV/c and they provide particle identification via a dE/dx measurement. The ITS, in particular, is a key detector for open heavy flavour studies because it allows us to measure the track impact parameter (i.e. the distance of closest approach of the track to the primary vertex) with a resolution better than 75 μm for $p_T >$ 1 GeV/c thus providing the capability to detect secondary vertices originating from heavy-flavour hadron decays. The TOF detector provides particle identification by time of flight measurement.

The results presented here are obtained from pp data recorded in 2010 at $\sqrt{s}$= 7 TeV. The analyzed sample was collected with a minimum bias trigger condition requiring at least one hit in the Silicon Pixel Detector (SPD) (equipping the two innermost layers of the ITS), or a signal in at least one of the two VZERO detectors, which cover the pseudorapidity ranges $-3.7 < \eta < -1.7$ and $2.8 < \eta < 5.1$.

D^0, D^+, D^{*+} and D_s^+ mesons are reconstructed in the central rapidity region from their hadronic decay channels $D^0 \to K^-\pi^+, D^+ \to K^-\pi^+\pi^+, D^{*+} \to D^0\pi^+ \to K^-\pi^+\pi^+$ and $D_s^+ \to \phi\pi^+ \to K^+K^-\pi^+$ (and charge conjugates). The D-meson yield is measured with an invariant mass analysis of fully reconstructed decay topologies displaced from the interaction vertex selected by requiring a large decay length and a good alignment between the reconstructed D-meson momentum vector and flight line. The identification of charged kaons in the TPC and TOF detectors helps to further reduce the background at low p_T.

3. D-meson production cross section measurement

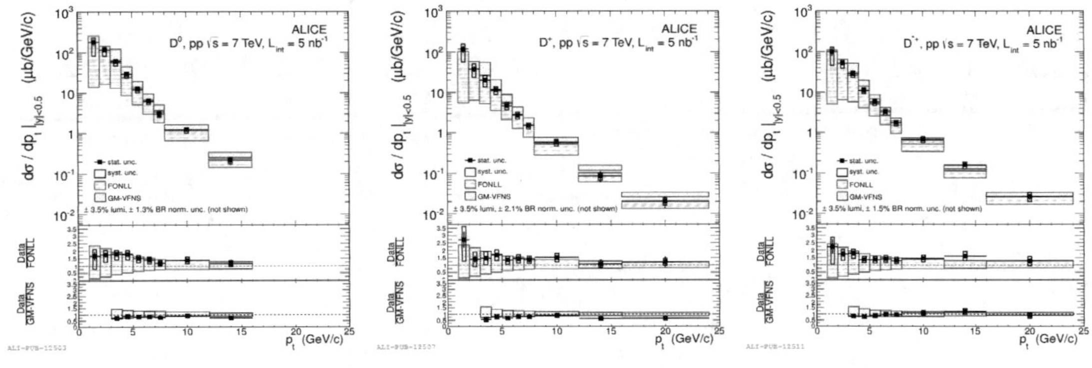

Figure 1. p_T differential cross section for D^0 (left panel), D^+ (middle panel) and D^{*+} (right panel) in pp collisions at $\sqrt{s}$=7 TeV, compared to FONLL [2] and GM-VFNS [3] theoretical predictions.

The differential cross section has been measured for prompt D^0, D^+, D^{*+} and D_s^+ mesons in pp collisions at $\sqrt{s} = 7$ TeV [4, 5]. To obtain the cross section for prompt D-mesons, the contribution due to beauty hadron decays, which is of about 10-15%, was evaluated based on Fixed Order with Next-to-Leading-Log resummation (FONLL) pQCD calculations [2] and subtracted from the inclusive D-meson yield. The resulting cross sections are compared to FONLL and General Mass Variable Flavour Number Scheme (GM-VFNS) [3] pQCD calculations. Figure 1 shows the p_T-differential cross section for prompt D^0, D^+ and D^{*+} mesons at $\sqrt{s} = 7$ TeV. The data are well described by pQCD calculation: at the upper limit of the uncertainty band when compared to FONLL and at the lower limit when compared to GM-VFNS predictions. Recently, it was reported that data are also well described by the k_T factorization approach [6].

4. D-meson yields as a function of charged-particle multiplicity

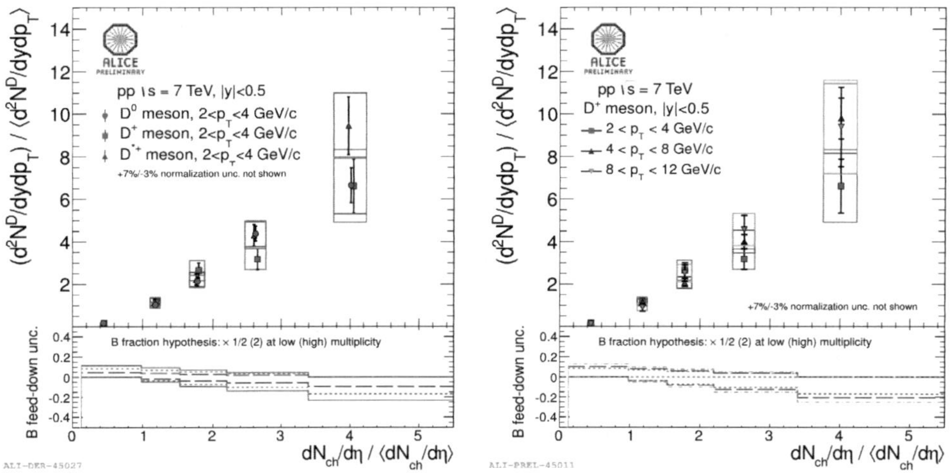

Figure 2. Left: Relative D^0 , D^+ and D^{*+} yields in $2 < p_T < 4$ GeV/c as a function of the relative multiplicity of charged particles produced in pp collision at $\sqrt{s} = 7$ TeV. Right: Relative D^+ yields in three p_T intervals as a function of the relative multiplicity of charged particles produced in pp collision at $\sqrt{s} = 7$ TeV

The production of D mesons is also studied in several p_T intervals as a function of the multiplicity of charged particle generated in the collision. The multiplicity estimator used in this analysis is the number of SPD tracklets (combination of two hits in the two layers of SPD, N_{trk}) in $|\eta| < 1.0$. Since the pseudorapidity coverage of the SPD changes with the z position of the interaction vertex, a correction to the measured N_{trk} is applied event-by-event. Using simulated events, it is verified that N_{trk} is proportional to $dN_{ch}/d\eta$ (pseudorapidity density of primary charged particles produced in the collision). The D-meson yields are extracted in five intervals of multiplicity. The left panel of figure 2 shows the yield of D^0, D^+ and D^{*+} in a given multiplicity interval divided by the value integrated over multiplicity for $2 < p_T < 4$ GeV/c as a function of charged-particle multiplicity, expressed as the ratio between the $dN_{ch}/d\eta$ in the considered multiplicity interval and the $< dN_{ch}/d\eta >$ for pp collisions at 7 TeV. The horizontal size of the error boxes represents the systematic uncertainty in the $dN_{ch}/d\eta/ < dN_{ch}/d\eta >$. The vertical size of the error boxes reflects all uncertainties on the relative yield except for the feed down contribution. The latter uncertainty is shown in the lower panel of each figure. To account for a possible difference in the multiplicity distribution of events with $c\bar{c}$ and $b\bar{b}$ production, the systematic uncertainty was estimated by allowing the B/D ratio predicted by FONLL to vary by a factor 2 up(down) at high(low) multiplicity. The results for D^0, D^+ and D^{*+} are compatible within the statistical and systematic uncertainties and show an increase of the D-meson yields with the charged-particle multiplicity of the event. The same trends are observed in all the measured p_T intervals as shown in the right panel of figure 2. Within the current uncertainties, no p_T dependence is observed.

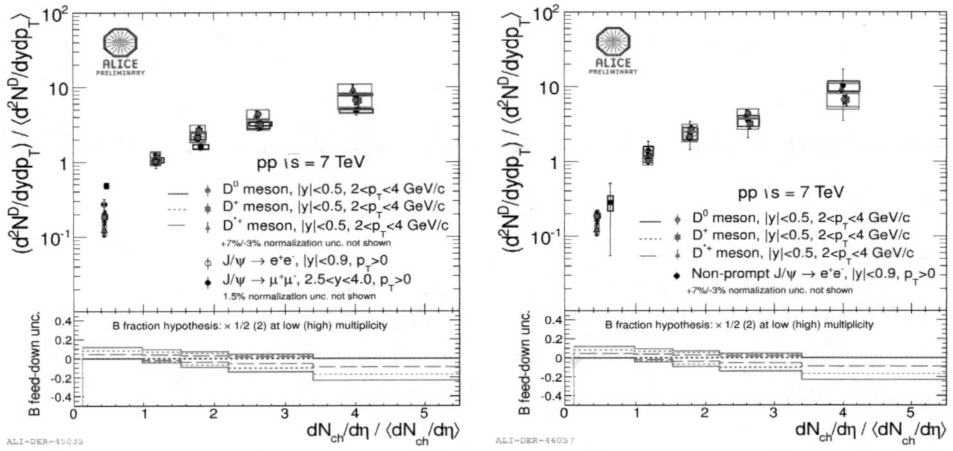

Figure 3. Left: Relative yield of D^0, D^+, D^{*+} for $2 < p_T < 4$ GeV/c and J/ψ for $p_T > 0$ as function of the event charged particle multiplicity. Right: Relative yield of D^0, D^+, D^{*+} for $2 < p_T < 4$ GeV/c and non prompt J/ψ as function of the event charged particle multiplicity.

The left panel of figure 3 shows the comparison of D meson and inclusive J/ψ yields as a function of multiplicity [7]. The J/ψ mesons were measured in two rapidity intervals, namely $|y| < 0.9$ and $2.5 < y < 4.0$, whereas the D-meson rapidity and p_T intervals are $|y| < 0.5$ and $2 < p_T < 4$ GeV/c respectively. Both open and hidden charm show similar behavior as a function of charged-particle multiplicity. Also the comparison with J/ψ from B-hadron decays in $|y| < 0.9$ (figure 3 right) shows a similar increase of yield.

5. Conclusions

ALICE detector provides excellent tracking, vertexing and particle identification to allow the measurement of charmed mesons via their hadronic decays, over a wide range of transverse momentum. The prompt D-meson production cross section is well described by NLO pQCD calculations. Furthermore, the D-meson yields was studied as a function of the charged particle multiplicity in pp collisions at $\sqrt{s}$=7 TeV. An increase of the D-meson yield as a function of multiplicity is observed. The comparison with the results for prompt and non-prompt J/ψ shows a similar increase as a function of charged-particle multiplicity. These results may indicate that either D-meson production in pp collisions is connected with a strong hadronic activity, or multi-parton interactions also affect the hard momentum scales relevant for charm quark production.

References

[1] Abelev B *et al.* 2008 *JINST* 0803:S08002
[2] Cacciari M, Frixione S, Houdeau N, Mangano M L, Nason P and GRidolfi 2012 (*Preprint* `hep-ph/1205.6344`)
[3] Kniehl B A *et al.* 2012 (*Preprint* `hep-ph/1202.0439`)
[4] Abelev B *et al.* 2012 *JHEP* **1201** 128
[5] Abelev B *et al.* 2012 *Phys. Lett. B* **718** 279
[6] Maciula R and Szczurek A 2013 *Phys. Rev.* D **87** 094022
[7] Abelev B *et al.* 2012 *Phys. Lett. B* **712** 165–175

Prospects for heavy flavour measurements with the ALICE inner tracker upgrade

Cristina Terrevoli (for the ALICE Collaboration)
Univerity and INFN Cagliari, I-09042 Monserrato (CA) ITALY
E-mail: cristina.terrevoli@cern.ch

Abstract.
ALICE is the general purpose heavy-ion detector at the CERN LHC. Its goal is to investigate the properties of the strongly interacting matter under the extreme conditions of density and temperature reached in Pb–Pb collisions, with the aim to characterize the Quark-Gluon Plasma (QGP). In this scenario, the upgrade of the ALICE inner tracker targets physics topics in which ALICE can bring a unique contribution to the QGP characterization via the heavy flavour probes. We present an overview of the inner tracker upgrade and the expected physics performance for heavy flavour measurements.

1. Introduction

Among the probes useful to investigate the properties of the QGP, heavy quarks play a special role because they are produced in the very initial stage of the collisions and they carry information about the properties of the traversed medium. An accurate measurement of heavy flavour provides information on fundamental properties of the medium, such as the transport coefficient, the thermalization and hadronization mechanisms. Interesting results have been obtained in the first three years of data-taking at the LHC, but there are still open questions, which would require measurements beyond the present capabilities of the ALICE apparatus. Among them, the most interesting are the measurements of the nuclear modification factor (R_{AA}) and anisotropy of the azimuthal distributions of charm and beauty mesons down to transverse momentum below 1 GeV/c. Another completely unexplored field is the measurement of the production of heavy flavour baryons, like the Λ_c, that can bring insight on the heavy quark hadronization mechanism in the presence of a partonic medium. At present such measurements are limited by the resolution of the Inner Tracking System (ITS), which, in particular, is not sufficient to measure in Pb–Pb collisions the production of Λ_c baryons, that have a mean proper decay length of only 60 μm. Another limitation of the present ALICE central barrel detectors to the measurement of heavy flavours at low momentum is the maximum achievable readout rate, which prevents the full exploitation of the luminosity expected to be delivered by the LHC after the Long Shutdown 2. An upgrade of the inner tracking system based on todays frontier technologies will improve the current performance in the pointing and momentum resolution, providing in addition a high tracking efficiency down to very low transverse momentum. Moreover, a faster readout, for all the central barrel detectors, will allow for a data collection rate, in Pb–Pb collisions, a factor 100 larger than at present, and this will contribute to enhance the ALICE physics performance very significantly.

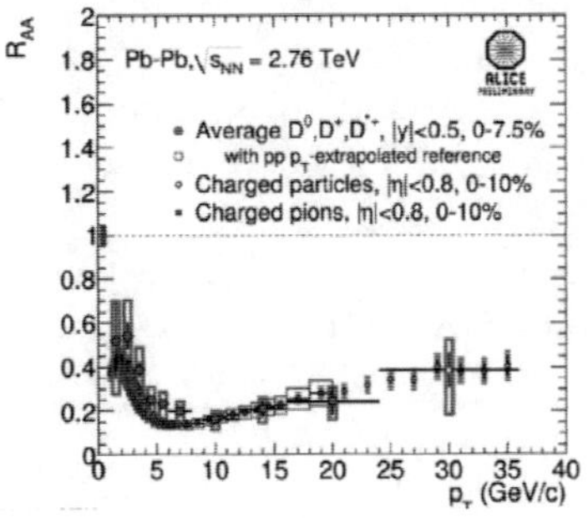

Figure 1. R_{AA} of D mesons compared with the R_{AA} of charged pions in Pb–Pb collisions at $\sqrt{s_{NN}}$= 2.76 TeV.

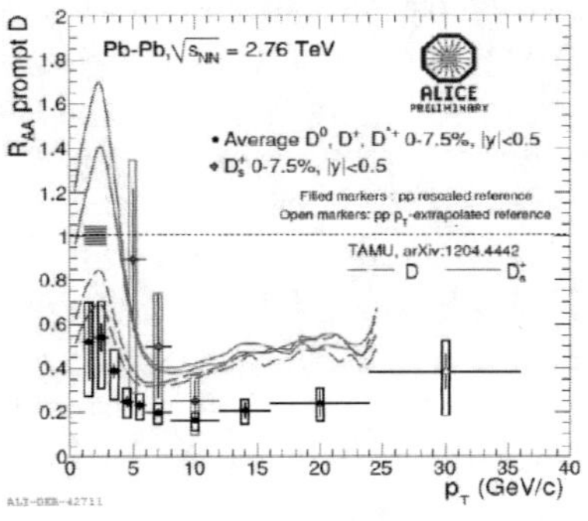

Figure 2. $\mathrm{D_s}$ R_{AA} compared with the theorethical models and D mesons R_{AA}.

2. ITS Upgrade concept

The current ITS has been designed to have an excellent capability to separate primary and secondary vertices of heavy flavour hadrons. It is composed of six layers of silicon detectors (pixels, drifts, and strips) with the innermost layer at a radius of 3.9 cm. The features of the ITS Upgrade [1] as compared with the present ITS are the following: (i) the innermost layer closer to the beam line at 2.2 cm, (ii) reduced material budget from the current 1.1% per pixel layer, to ~0.3%, (iii) smaller size pixel cells (from 50 × 425μm to (20-30)×(20-30)μm), (iv) a seventh detector layer. This will allow for a significant improvement of the tracking performance and the momentum resolution. In particular, the impact parameter resolution will be improved by a factor of three in the transverse direction and a factor of six in the longitudinal direction. Furthermore a continuous readout of Pb–Pb interactions at ~50 kHz will permit to exploit the upgrade LHC luminosity after 2018. The upgrade is targeted for the second long shutdown in 2017-2018.

3. Heavy Flavour results with the current Inner Tracking System

The energy-loss mechanism for different parton species is among the observables useful to investigate the properties of the QGP matter. Theoretical models predict a dependence on the colour charge and mass of partons propagating through the medium. In particular, a larger energy loss is expected for gluons, together with a suppression of gluon radiation at small angles for partons with larger mass (dead cone effect). This dependences can be investigated using the nuclear modification factor R_{AA}, the ratio of the p_{T} distribution in Pb–Pb and the p_{T} distribution in pp, scaled by the number of binary nucleon-nucleon collisions. The expected pattern is $R^{\pi}_{\mathrm{AA}} < R^{D}_{\mathrm{AA}} < R^{B}_{\mathrm{AA}}$ [2]. The R_{AA} of D mesons ($\mathrm{D^0}$, $\mathrm{D^+}$, $\mathrm{D^{*+}}$) was measured in central Pb–Pb collisions at $\sqrt{s_{NN}}$ =2.76 TeV and compared with that of charged pions [3], as shown in Fig. 1. The data suggest that the suppression of D mesons might be smaller than for pions in the low momentum region, but the large systematic uncertainties prevent a firm conclusion. The measurement of the B mesons R_{AA} is not accessible with the current setup of the ALICE detector. The upgrade of the ITS will open the possibility to study B mesons down to zero p_{T}. Moreover the improved tracking resolution and efficiency, together with the higher statistics, will improve the p_{T} coverage and the accuracy of D meson measurements. Another open question concerning the heavy-flavour interaction with the QGP medium, involves the hadronization of heavy quarks in the medium, which can be studied by measuring the baryon/meson ratio for charm and beauty (Λ_{c}/D and Λ_{b}/B): coalescence models predict an increase of baryon-to-meson

Journal of Physics: Conference Series **509** (2014) 012082 doi:10.1088/1742-6596/509/1/012082

ratio [5]. With the current set-up of ALICE, the analysis of Λ_c and Λ_b is not accessible in Pb–Pb due to the limited precision and statistics. Another important test for coalescence models is that, if coalescence contributes to the charm hadronization, the D_s is expected to be enhanced with respect to the other D mesons at low p_T. There is an hint of this enhancement [4](see Fig. 2), but it is necessary to improve tracking precision, statistics and to extend the measurement to low p_T to have a conclusive answer. The measurement of the elliptic flow v_2 is sensitive to the thermalitazion of charm and beauty in the QGP. Models predict a large D meson v_2 at low momentum and a mass dependence of v_2 ($v_2(B) < v_2(D)$) [6, 7]. The ALICE measurement of D meson v_2 down to 2 GeV/c shows a positive value of v_2 at intermediate p_T in semi-central collisions, but with a limited precision. The measurement is not accessible for B mesons. The new ITS will allow for a precise measurement of the D meson v_2 down to zero p_T and will open the possibility for the measurement of B meson v_2.

4. ITS Upgrade physics performance

Several simulation studies have been carried out to quantify the improved performance of the upgraded ITS and the higher statistics achievable thanks to the continuous read-out (up to 10 nb^{-1} of integrated luminosity). The D meson measurement will be extended at low and high p_T, with improved statistical precision. In Fig. 3 the expected R_{AA} of D^0 mesons is shown: the upgraded ITS allows to reach down to zero p_T. In Fig. 4 the statistical uncertainty of the D^{*+} measurement is shown. The analysis can be carried out in a wider range (currently 3-36 GeV/c) with high significance (20-100), taking advantage of the higher precision and statistics. A direct measurement of beauty will be possible via non-prompt D^0 and J/ψ, as shown in Fig. 5

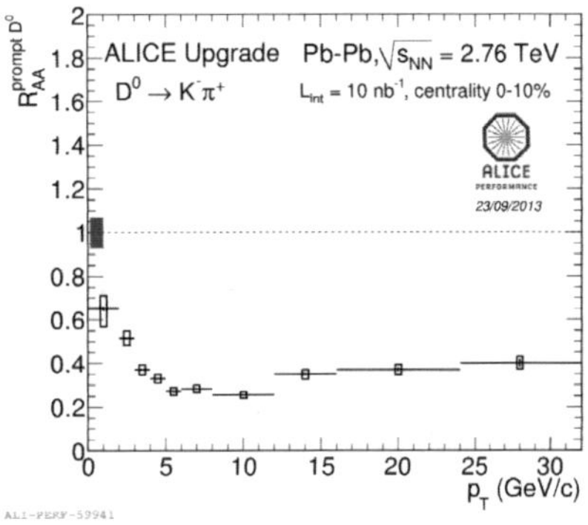

Figure 3. R_{AA} of D^0 in Pb–Pb at 5.5 TeV with the ALICE upgrade.

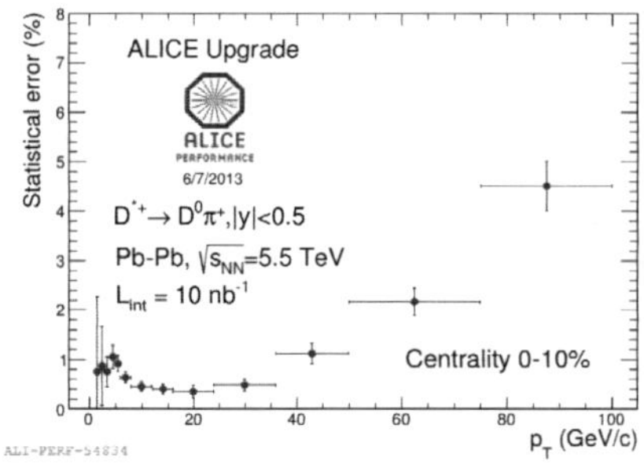

Figure 4. Statistical uncertainties of the D^* measurement in Pb–Pb 5.5 TeV with the ALICE upgrade.

(left), where the statistical uncertainty for non-prompt J/ψ is presented. This will allow for the measurement of the v_2 of B mesons for the first time. The Λ_c baryon will be accessible for $p_T > 2$ GeV/c in Pb–Pb thanks to the improved resolution, with a higher precision. In Fig. 5 (right) the double ratio Λ_c/D in Pb–Pb over pp is shown together with two theoretical predictions. Studies on the Λ_b reconstruction are also ongoing: the baryon will be accessible for the first time via its decay in Λ_c.

Also the D_s analysis will benefit for the much larger available statistics, with expected precise measurements of R_{AA} and v_2 down to low p_T (see Fig. 6).

5. Conclusions

ALICE has a strong upgrade physics programme for precision QGP studies, in particular for a deeper understanding of energy loss mechanisms, azimuthal anisotropy and in-medium

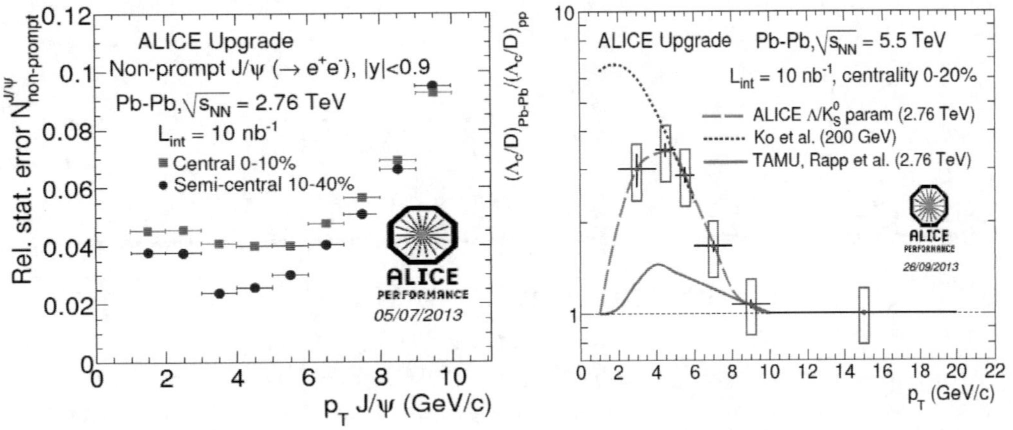

Figure 5. (Left) Non-prompt J/ψ yield: statistical uncertainty, for central and semi-central collisions, with the ALICE upgrade. (Right) Λ_c/D^0 ratio in Pb–Pb vs pp collisions, measured with the ALICE upgrade.

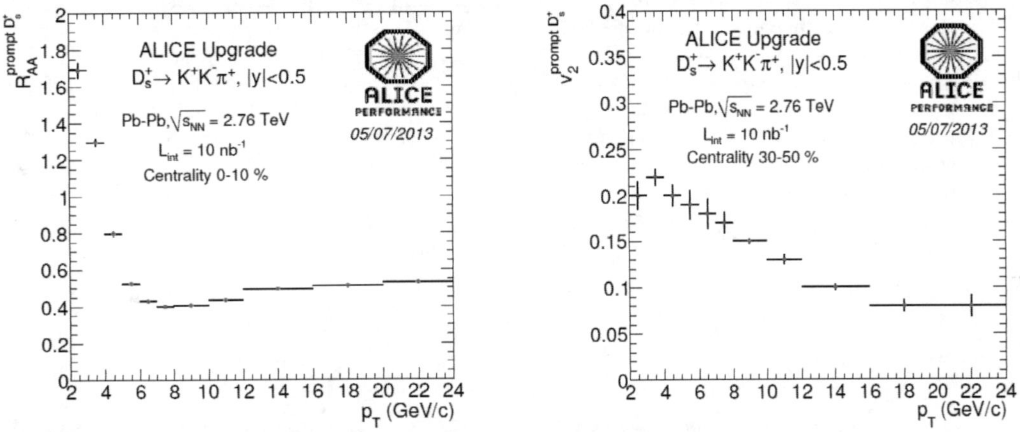

Figure 6. (Left) Statistical uncertainty on R_{AA} for D_s with the ALICE upgrade. (Right) v_2 for D_s with the ALICE upgrade.

hadronization, where heavy flavour measurements play a central role. The main requirements for the upgrade are enhanced rate capabilities and a new Inner Tracking System. These features will provide a strong increase of the statistical precision in the measurements of yields and spectra of heavy flavour mesons and baryons.

References

[1] ALICE Collaboration, 2012 ITS Upgrade Conceptual Design Report *CERN-LHCC-2012-013*
[2] Dokshitzer Y L and Kharzeev D, et al 2001 *Phys. Lett.* B **519** 199.
[3] Abelev B et al 2012 (ALICE Collaboration) arXiv:1203.2160.
[4] Innocenti G et al (ALICE Collaboration) 2012 *Quark Matter 2012* http://qm2012.bnl.gov
[5] Lee S H, Ohnishi K, Yasui S, Yoo I and Ko C, 2008 *Phys. Rev. Lett.* **100** 222301
[6] Aichelin J, Gossiaux P and Gousset T 2012 arXiv:1201.4192
[7] Uphoff J, Fochler O, Xu Z and Greiner C 2012 arXiv:1205.4945

14th International Conference on Strangeness in Quark Matter (SQM2013) IOP Publishing
Journal of Physics: Conference Series **509** (2014) 012083 doi:10.1088/1742-6596/509/1/012083

NA61/SHINE experiment upgrade with vertex detector for open charm measurements

Yasir Ali and Paweł Staszel for the NA61/SHINE collaboration
M. Smoluchowski Institute of Physics, Jagiellonian University, Reymonta 4, 30-059 Krakow Poland

E-mail: yasir.ali@uj.edu.pl

Abstract.
The feasibility study of direct open charm measurements by its decay into two daughter particles, pion and kaon, in central Pb-Pb collision at SPS energies of 158 AGeV and 40 AGeV, shows that such measurements are viable at NA61/SHINE experiment if it is supplemented with a dedicated Vertex Detector (VD). We use AMPT (A Multi-Phase Transport model) event generator to generate the physical input for simulation and employed GEANT4 application to describe particle transport through the experimental setup. The VD will allow for precise track reconstruction at the target proximity. The direct open charm measurements will be a challenging due to the low production yield and short life time of D^0 mesons. The obtained results are based on the predicted yields of D^0 meson by Hadron String Dynamic (HSD) model. This study also addresses the issue of VD optimization and the emphasis is put on the prospect of the development of a VD based on CMOS technology.

1. Introduction

Open heavy flavor is an important probe of the dense matter formed in heavy ion collision. The measurement of D^0 mesons provide a unique opportunity to test the validity of perturbative QCD-based and statistical models of nucleus-nucleus collisions at high energies. So far there are no direct open charm measurements at SPS energies while extensive studies have been done in the past and new experimental initiatives are proposed to measure charmonia states at SPS energies [1]. Recently, it was pointed out by Satz [2] that only precise measurement of both open and hidden charm will allow to construct observables that can provide information on the in-medium behaviour of quarkonia in a model independent way. Thus we developed a feasibility study dedicated to the measurement of the D^0 mesons (open charm) in Pb+Pb collision at SPS energies. This study is done for top SPS energy of 158 AGeV and for lower energy of 40 AGeV. This measurement is challenging due to the short life time and low production yield of D^0 mesons. D^0 meson can be reconstructed by its decay into hadronic and semi-leptonic decay channels. We have considered only the bench mark channel of two body hadronic decay in which D^0 meson decays into pion and kaon with the branching ratio of 3.9% [3]. The other channels will be addressed in the future consideration. For the physical input we generated 200k 0-10% central events with AMPT (A Multi-Phase Transport Model) event generator. It provides good background description which in case of this study are distributions of charged hadrons, mostly pions and kaons [4]. The AMPT model predicts the average production yield of 0.01 for $D^0+\overline{D}^0$ per central Pb+Pb event. This value seems to be underpredicted with respect

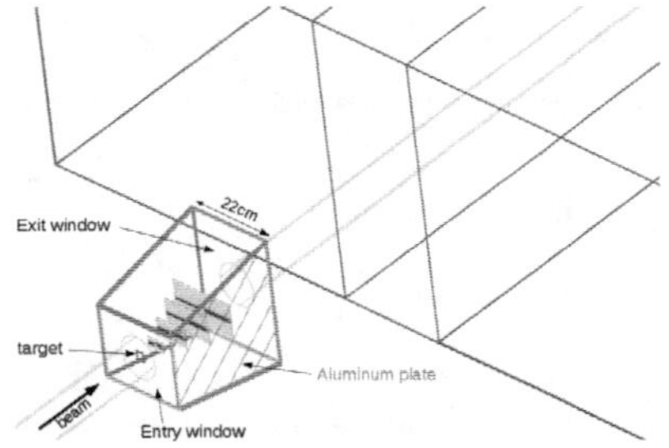

Figure 1. The helium vessel containing four vertex detector stations, carbon fiber support and water cooling tubes.

to PYTHIA results scaled to the number of binary nucleon-nucleon collisions in central Pb+Pb equal to 0.21 [5] as well as with respect to the prediction of the HSD model. Because HSD model was tuned to properly describe available p+A and π+A charm production data at SPS energies, we scaled the AMPT production yield to the prediction of the HSD model. The average yield of D^0 mesons predicted by HSD model is 0.1 [6].

GEANT4 is employed to describe the particle transport through experimental setup. In order to reconstruct D^0 mesons a dedicated vertex detector providing high position resolution was conceived.

2. Vertex Detector for open charm measurements

The NA61/SHINE (SHINE = SPS Heavy Ion and Neutrino Experiment) [7] at the CERN SPS is a fixed target experiment whose description is given by [8]. This experiment will be upgraded by a dedicated VD which allows for precise tracking at the target proximity. The VD modelled in GEANT4 consists of four detection stations VDS1, VDS2, VDS3 and VDS4 located respectively at distances of 5 cm, 10 cm, 15 cm, and 20 cm downstream of the target. The external transverse dimensions of the stations are 2×4 cm^2 (VDS1), 4×8 cm^2 (VDS2), 6×12 cm^2 (VDS3), 8×16 cm^2 (VDS4). These dimensions allow to cover more than 99% of the acceptance of pions and kaons originating from the decay of D^0 mesons. The range of kinematical acceptance is described by [9].

Each station has also a square hole (3×3 mm^2) in the central part which allows the beam ions to go through without an interaction. The stations and the target are closed in the aluminium vessel as shown in Figure 1. The vessel is equipped with thin front and back kapton windows and filled with helium gas at atmospheric pressure to minimize multiple scattering of the produced particles. The advantage of helium is that it reduces the primary beam, and spectator interactions.

The track reconstruction is based on the hits generated during particle transport through vertex detector stations, Vertex Time Projection Chamber (VTPC1 and VTPC2). Only the tracks accepted in VTPC detectors are considered in the subsequent analysis. The acceptance is defined in the similar way as it is done in the standard NA61/SHINE simulation which is described in detail in [10]. In the invariant mass distribution for kaon-pion pairs at the collision energy of 158A GeV we found that the combinatorial background is few orders of magnitude higher than the $D^0 + \overline{D^0}$ signal shown in red colour. In order to suppress the large combinatorial background, four cuts were developed namely, cut on the track transverse momentum p_T, cut on the track impact parameter d cut on the longitudinal position V_z of the track pair vertex relative to primary interaction point and cut on the parent particle impact parameter D. The first two are single-particle cuts, while the two other are two-particle cuts. It should be stressed

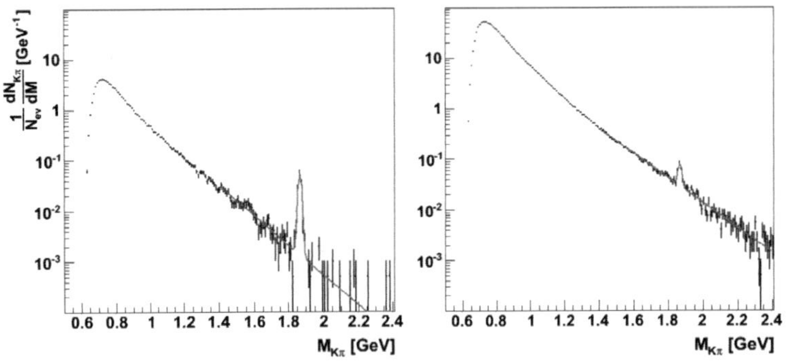

Figure 2. The invariant mass of pion-kaon pairs after full background reduction. The red curve represents fit with sum of exponential (background) and Gaussian function (D^0 signal). Left: analysis done using PID information, right: analysis that does not utilize the PID information

that last three cuts require the precise tracking information at the target proximity from the proposed VD.

The analysis assumes perfect particle identification (PID), i.e. for each track the related specie type was known. The invariant mass distribution with all the cuts applied, assuming perfect PID, at the collision energy of 158A GeV, is shown in the left panel of Figure 2. This plot is to be compared to the one in the right panel of Figure 2, which presents results of analysis that does not utilize the PID information. As can be seen from the comparison of the plots, analysis without PID suffers from a much higher background. Nevertheless, it is feasible to extract the signal yield even in such analysis. For the analysis assuming perfect PID we obtain signal to background (S/B) = 17 and with the case of no PID we obtain S/B = 1.0. When we extrapolate the results to 50 M events (anticipated event number collected during 1-2 months of the NA61/SHINE data taking) the total number of collected open charm mesons is $\sim 64000/(64000)$ with Signif = 246/(197) for perfect PID/(no PID) case. The reason to compare analysis with and without PID, is that in the NA61/SHINE experiment, exact identification on track by track basis is not possible in the whole phase space. Therefore, the real situation is between the two quoted analysis types.

We also developed simulation to determine the requirements for the detector. In the first vertex detector station for 0-10% central interactions, we observed the high hit occupancy in the inner region which is 5 hits/mm^2/event. The probability of empty frame is 95%, for single event is 4.7% and for pile-up is 2.5%. This high occupancy suggest to use pixel detector technology for the VD stations. The NA61/SHINE requirements in terms of resolution, material budget, radiation tolerance and time resolution are fullfilled by MIMOSA-26 device [11]. This chip is basically a sensor chip based on the Monolithic Active Pixel sensors with fast binary readout and is suitable for charge particles detection with the flux of 10^4 hits/mm^2/s and provides the resolution of 3.5 μm in both directions. The sensor is 50 μm thick and total thickness of detector is of the order of 0.3% X_0 [11].

To cover the active areas of the VD stations requires 126 MIMOSA-26 sensors. It was, however, realised that the protoype version of the VD has to build first to allow for testing the general concept of data reconstruction, the service systems responsible for mechanical support, sensor cooling, data readout and data synchronization. VDS1 and VDS2 will be equipped with 2 sensors each, and VDS3-VDS4 with 4 sensors. The VD prototype version will consist of 12 sensors. We developed simulations to find the most optimal locations of the sensors in order

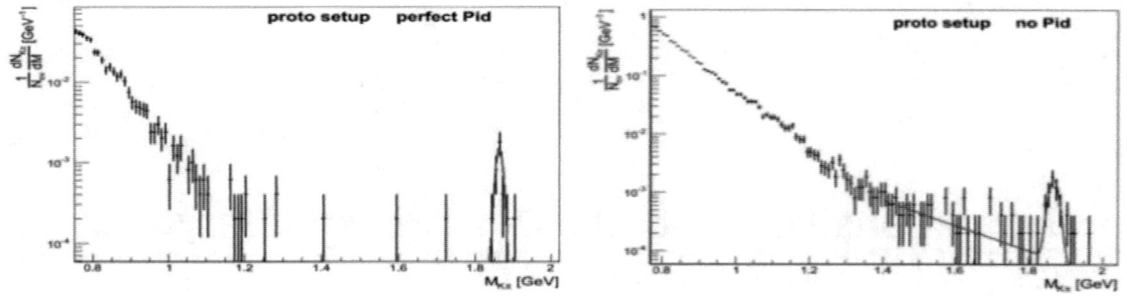

Figure 3. The combinatorial invariant mass spectrum for 0-10% central Ar+Ca collisions at 158 AGeV after background reduction for the prototype of VD.

to maximize the acceptance of the D^0 signal. We performed simulation of such measurement generating physical input by the AMPT event generator. We generated 500k 0-10% central Ar+Ca collisions at 158 AGeV. The results of the simulation with PID and no PID are shown in Figure 3 where we plotted invariant mass spectrum after all suppression cuts have been applied. From the plot one can see a well pronounced D^0 signal with the S/B = 11 and the Signif = 4.5. According to the schedule of the current NA61/SHINE scientific program we will be able to test the performance of prototype detector integrated with the rest of the experimental setup in February 2015 with the Ar beam colliding on Ca target at the top SPS energy of 158 AGeV. The results might be relevant not only for the test but also for physics output providing first measurement of open charm at the SPS energy.

3. Conclusion

Measurements of heavy flavor mesons gives better understanding of nucleus-nucleus reactions at relativistic energies. Open charm measurements are feasible with NA61/SHINE experiment at CERN-SPS energies if it is upgraded with high resolution VD. The MIMOSA-26 sensors fulfil all requirements imposed by NA61/SHINE application and will be used as a basic detection element in the VD stations. We will build prototype version of the VD which will be tested on the beam to prove the general concept of data reconstruction.

This contribution reports on the technical work motivated by the future experimental programme of the NA61/SHINE Collaboration at the CERN SPS. Authors gratefully acknowledge the contribution of NA61/SHINE to the results presented. This work is supported by the Foundation for Polish Science — MPD program, co-financed by the European Union within the European Regional Development Fund and the National Science Centre of Poland, grant UMO-2012/04/M/ST2/00816.

References

[1] Usai G, Scomparin E, 2012 Town meeting Relativistic Heavy-Ion Collisions CERN.
[2] Satz H. 2013, arXiv:1303.3493v2, BI-TP 2013/06.
[3] Ortona G, (ALICE collab.) 2011 *Intl. J. Mod. Phys. Conf Ser.* **2** 2530.
[4] Lin Z et al. 2005 *Phys.Rev.C* **72** 064901.
[5] Munzinger P, Stachel J, 2000 *PLB* **490** 196.
[6] Cassing W, Bratkovskaya E.L, Sibirtsev.A, 2001 *Nucl. Phys. A* **691**, 753.
[7] NA61/SHINE plans beyond the approved program, 2012 Addendum 6 to the proposal P330.
[8] Gazdzicki M, 2009 *J. Phys. G: Nucl. Part. Phys* **36** 064039.
[9] Abgrall N et al. (NA61/SHINE collab.), 2013 arXiv:1310.2417v1.
[10] Ali Y et al. 2013 *Acta Phy. Pol. B* **10**, 44.
[11] MIMOSA26 User Manual, Institut Pluridisciplinaire Hubert Curien IN2P3-CNRS.

Status and Challenges of the Compressed Baryonic Matter (CBM) Experiment at FAIR

Hans Rudolf Schmidt for the CBM Collaboration
University of Tuebingen, Auf der Morgenstelle14, Tuebingen, Germany

Hans-Rudolf.Schmidt@Uni-Tuebingen.de

Abstract. The CBM experiment is one of the four scientific pillars of the Facility for Antiprotons and Ion Research (FAIR) in Darmstadt, Germany. Its discovery potential – complementary to heavy-ion experiments at colliders – is based on high-luminosity ion beams. This enables access to extremely rare probes such as charmed particles, vector mesons or multi-strange hyperons with high statistics. However, 3rd generations readout systems and detectors are required to handle the large interaction rates (up to 10 MHz for Au+Au) with sufficient precision and bandwidth. In this contribution we will outline the unique CBM physics program focusing onto rare probes.

1. Introduction

The observation of a mass $(1.97 \pm 0.04)\, M_{\odot}$ neutron star [1] has spurred interest into models of nuclear matter capable to stabilize nuclear matter at high densities [2,3,4], i.e., to prevent the collapse of heavy neutron stars into black holes. The relevant degrees of freedom in nuclear matter at high densities are nucleons and hyperons; as well as quarks in the case of the onset of a deconfinement transition.

The future experiments at FAIR (Darmstadt) or at NICA (Dubna) will employ high intensity heavy-ion beams in the range 2-35 AGeV ($\sqrt{s_{NN}} = 2.7 - 8.3\, GeV$) and $\sqrt{s_{NN}} = 4 - 11\, GeV$), respectively. These experiments will allow to investigate rather exotic matter at baryon densities up to $7 \times \rho_0$ with good statistics for the first time. This will enable to explore yet uncharted territory with rare probes, such as charmed particles, di-leptons, double hypernuclei or multi-strange hyperons [5,6].

2. Rare Probes

Rare probes such as low mass di-leptons, charmonium and open charm will allow to characterize the medium at high net baryon densities much more detailed then previously possible. While bulk observables (flow, spectra of hadrons) as well as correlations and fluctuations are accessible to all experiments (STAR/PHENIX@RHIC, NA61@SPS, MPD@NICA) at the relevant low beam energies ($\sqrt{s_{NN}} \sim 8\, GeV$), the above mentioned rare probes can only be measured at CBM@FAIR.

In this paper, however, we will not further focus on leptons and charm. Instead, we will discuss more exotic probes of dense matter: strangeness in the form of multi-strange hyperons, double-strange hypernuclei and multi-strange di-baryon, the latter also called MEMOs (metastable exotic multi-hypernuclear objects).

2.1. Multi-strange Hyperons

The yields and phase-space distributions of multi-strange hyperons (Ξ, Ω) are promising tools to study the properties and the degrees of freedom of QCD matter at extreme nuclear densities. In particular, these particles may serve as unambiguous messengers of a high-density phase: for hyperon production binary reactions proceeds like $pp \rightarrow \Xi^- K^+ K^+ p$ or $pp \rightarrow \Omega^- K^+ K^+ K^0 p$ and have accordingly high thresholds of 3.7 and 7.0 GeV, respectively. In dense baryonic matter, however, Ξ^- and Ω^- can also be created via strangeness exchange reactions like $\Lambda\Lambda \rightarrow \Xi^- p$ and $\Lambda\Xi^- \rightarrow \Omega^- n$ or $\Lambda K^- \rightarrow \Xi^- \pi^0$, with the Λ and the K^- previously produced in independent reactions such as $pp \rightarrow K^+ \Lambda p$ and $pp \rightarrow K^+ K^- pp$, which require only 1.6.and 2.5 GeV, respectively. Alternatively, three-body collisions involving $\Lambda's$ or kaons open new production channels for Σ and Ω with respect to pp reactions. The production of multi-strange hyperons is thus expected to be enhanced at high densities, and their yield to be sensitive to the baryon density reached in the fireball. Moreover, the energy distributions of multi-strange hyperons provide information on the fireball temperature and the radial flow at the time, when they are emitted. Therefore, systematic measurements of Ξ^- and Ω^- production as function of beam energy and size of the colliding nuclei offer the possibility to study the nuclear matter equation of state, or baryon density fluctuations as they are expected to occur when the system undergoes a first-order phase transition or approaches the QCD critical endpoint.

Existing data on the production of multi-strange hyperons in nuclear collisions at SIS-100 beam energies are scarce [7]. In particular, no data on production below $40A\ GeV$ ($\sqrt{s_{NN}} = 8.8\ AGeV$) exist, except for 312 Ξ^- measured at 6 AGeV at the AGS [8]. A systematic measurement of multi-strange hyperons as diagnostic probes of dense nuclear matter at SIS-100 energies has thus a substantial discovery potential.

Energy [AGeV]	Σ^-	Ω^-	$\bar{\Lambda}^-$	$\bar{\Sigma}^-$	$\bar{\Omega}^+$
4.0	9.0×10^6	1.8×10^5	3.6×10^3	5.3×10^3	9.0×10^2
6.0	2.6×10^7	5.0×10^5	2.4×10^5	1.4×10^{34}	2.8×10^3
8.0	4.0×10^7	1.4×10^6	3.6×10^6	2.0×10^5	6.0×10^4
10.7	5.4×10^8	2.2×10^6	6.8×10^6	3.8×10^5	1.2×10^4

Table 1: Expected hyperon yields [9] measured per week at a minimum-bias interaction rate of $2 \times 10^4 s^{-1}$. The minimum bias multiplicity is assumed to be 25% of the one in central collisions.

A prerequisite for these measurements is sufficient statistics in finite time. **Table 1** shows that already at 2×10^4 minimum-bias interactions the accumulated data exceed previous measurements in terms of yield.

2.2. Search for Double Hypernuclei

Up to now, only a handful double hypernuclei have been seen, mostly in emulsion experiments. Hypernuclei, i.e. nuclei containing at least one hyperon in addition to nucleons, offer the fascinating perspective to explore the third, strange dimension of the chart of nuclei. Their investigation provides information on the hyperon-nucleon and even on the hyperon-hyperon interactions, which play an important role in neutron star models.

The conventional production mechanism is rather complicated, e.g., via the following chain reaction employing a kaon beam: $K^- + N \rightarrow K^+ + \Xi^-, \Xi^- + ^{12}C \rightarrow {}^{6}_{\Lambda\Lambda}He + ^4He + t$. Hence the production cross sections are exceedingly small. On the other hand, heavy-ion collisions offer the possibility to produce (double) hypernuclei via coalescence of $\Lambda's$ with nucleons or light fragments in the final state of the reaction. In high-energy nucleus-nucleus collisions, Λ hyperons are produced abundantly. Their maximum yield is observed at beam energies between 30 and $40AGeV$ (about 50 in central Pb + Pb collisions). At $10\ AGeV$ still 15 $\Lambda's$ are produced. The

Journal of Physics: Conference Series **509** (2014) 012084 doi:10.1088/1742-6596/509/1/012084

coalescence probability of Λ with light nuclei is highest at low energies yielding, according to hadron gas model calculations [10, see also 14], $2 \times 10^{-2}\ {}^{3}_{\Lambda}He$, $10^{-6}\ {}^{5}_{\Lambda\Lambda}He$ and $3 \times 10^{-8}\ {}^{6}_{\Lambda\Lambda}He$ per central collision at $8 - 10\,AGeV$. This would give, with the foreseen CBM interactions rates about 120 ${}^{5}_{\Lambda\Lambda}He/week$ and still 3.6 ${}^{6}_{\Lambda\Lambda}He/week$. For the detection of the double hypernuclei, the decay chain has to be reconstruction in the tracking detectors, e.g., ${}^{5}_{\Lambda\Lambda}He \rightarrow {}^{5}_{\Lambda}He + \pi^- \rightarrow {}^{4}He + p + \pi^-$.

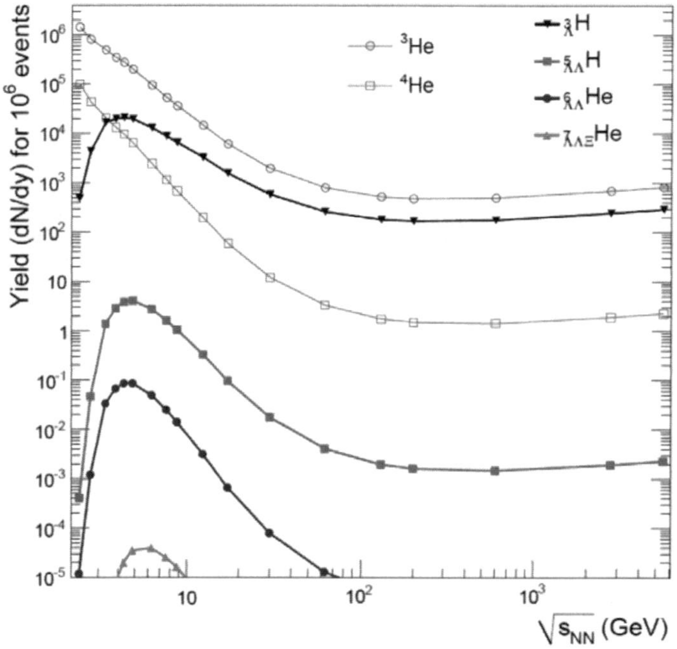

Figure 1: Energy dependence of hypernuclei yields at mid-rapidity for 10^6 central $Au + Au$ events as calculated with the statistical model [10]. The predicted yields of ${}^{3}He$ and ${}^{4}He$ are included for comparison.

2.3. Search for Meta-Stable Multi-Strange Objects

Metastable objects with strangeness, e.g. strangelets and (strange) di-baryons, were proposed long ago as collapsed states of matter, consisting of either baryons or quarks [11,12]. Up to date, none of these objects have been observed. Their existence or absence is an open issue in high-energy physics (cf. [13]).

High-energy nuclear collisions, with kaons and $\Lambda's$ being abundantly produced in a single event, could provide a tool to create such composites objects with multiple units of strangeness. Recent calculations within a hybrid (microscopic transport + hydrodynamics) model predict multiplicities/event of about $3 \cdot 10^{-2}$ for the $(\Xi^0\Lambda)_b$ in central collisions at 35A GeV [14]. Owing to the finite lifetime of such an object its decay, e.g., $(\Xi^0\Lambda)_b \rightarrow \Lambda\Lambda$, should be detectable via displaced vertices ($c\tau \approx 1 - 5\ cm$) [15]. Indeed, simulations show that $(\Xi^0\Lambda)_b's$ can be reconstructed with a signal-to-background ratio above 200 for 10^{12} central events [16,17].

Most of the strange di-baryon searches were done via the (K^+, K^-) reaction (cf. [18], and Refs. therein), which is not directly comparable with the production of di-baryons in heavy-ion collisions. The nearest equivalent to heavy-ion reactions is a result reported by the KTeV collaboration [19] at Fermilab, were lightly bound H^0-dibaryons produced in pN-collisions were searched for. However, the acceptance of the experimental apparatus was optimal for a decay length $c\tau = 160\ cm$, and thus the reported limit are not applicable for the search range of $c\tau \approx 1 - 5\ cm$ accessible in the CBM experiment.

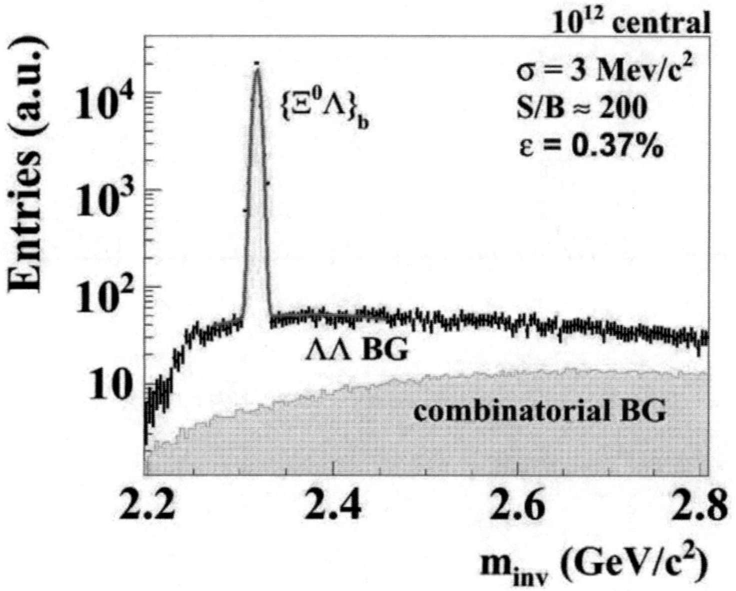

Figure 2: Reconstructed invariant-mass distribution of *ΛΛ* candidates

3. Summary

Heavy-ion collisions in the SIS-100/300 energy range are an ideal tool for the production of hadronic matter at neutron star core densities, and hence offer the unique opportunity to investigate fundamental properties of strongly interacting systems and its constituents: the nuclear matter equation of state, exotic new phases such as quarkyonic matter, in-medium modifications of hadrons as a signature for chiral symmetry restoration, hypernuclei and multi-strange objects, charm production at threshold beam energies, and charm propagation in nuclear matter.

References

[1] P. B. Demorest, T. Pennucci, S. M. Ransom, M. S. E. Roberts, and J. W. T. Hessels, *Nature* **467**, 1081 (2010).
[2] I. Sagert, L. Tolos, D. Chatterjee, J. Schaffner-Bielich and C. Sturm, *Phys. Rev.* C **86** , 045802 (2012)
[3] M. Orsaria, H. Rodrigues, F. Weber, and G. A. Contrera, *Phys. Rev.* D **87**, 023001 (2013)
[4] G. Colucci and A. Sedrakian, *arXiv*:1302.6925 (2013)
[5] B. Frimann et al. (eds.), *"The CBM Physics Book"*, Springer Lecture Notes in Physics 814 (2011)
[6] P. Senger and V. Friese, https://www.gsi.de/documents/DOC-2011-Aug-29.html
[7] C. Blume, *J. Phys.* G **31** (2005) S57
[8] P. Chung et al., *Phys. Rev. Lett.* **91**, 202301 (2003)
[9] A. Andronic, *private communication*
[10] A. Andronic et al., *Phys. Lett.* B **697** (2011) 203
[11] A. R. Bodmer, *Phys. Rev.* D **4** (1971) 1601
[12] R. L. Jaffe, *Phys. Rev. Lett.* **38** (1977) 195
[13] H. Clement, *Prog. Part. Nucl. Phys.* **67**, 486 (2012)
[14] J. Steinheimer et al, *Phys. Lett.* B**714** (2012) 85
[15] J. Schaffner-Bielich, R. Mattiello and H. Sorge, *Phys. Rev. Lett.* **84** (2000) 4305
[16] I.Vassiliev, *private communication*
[17] I. Vassiliev, I. Kisel, and D. Malakhov, *CBM Progress Report* 2009
[18] R.W. Stotzer et al, *Phys. Rev. Lett.* **79**, 3646 (1997)
[19] A.Alavi-Harati et al., *Phys. Rev. Lett.* **84**, 2593 (2000)

Journal of Physics: Conference Series **509** (2014) 012085 doi:10.1088/1742-6596/509/1/012085

Beam Energy Dependence of Strange Hadron Production from STAR at RHIC

Feng Zhao (for the STAR Collaboration)
University of California, Los Angeles
475 Portola Plaza, Knudsen Hall 5-117
Los Angeles, CA, 90095

E-mail: fengzhao@mail.physics.ucla.edu

Abstract. We report the STAR measurements of K_s, Λ, Ξ, and Ω at mid-rapidity from Au+Au collisions at the beam energies of 7.7, 11.5, 19.6, 27 and 39 GeV per nucleon from the RHIC Beam Energy Scan (BES) program. Nuclear modification factors and ratios of baryon to meson yields are discussed. We also investigate the strangeness enhancement and ratios of anti-baryon to baryon yields as a function of beam energy at RHIC, and compare the results with the statistical thermal model.

1. Introduction

Strangeness production is sensitive to the collision dynamics in heavy ion collisions, and it is used as a probe to study the nuclear medium created in the collisions [1]. The nuclear medium varies with the collision energy. In RHIC Au+Au 200 GeV collisions, the modification factors (R_{CP}) of baryons and mesons at intermediate p_T follow different trends, and the baryon-to-meson ratios have a strong centrality dependence and are enhanced in the intermediate p_T range in central collisions [2, 3]. Those phenomena may be explained by the recombination/coalescence model [4, 5, 6, 7, 8], which is an indication of the formation of the strongly coupled quark-gluon plasma (sQGP). At high p_T, the R_{CP} is much less than unity, presumably due to the partonic energy loss in the medium. Our measurement of strange particle R_{CP} and baryon-to-meson ratios may reveal how the collision dynamics changes with the beam energy.

The statistical thermal model has been used to describe the particle yields in nuclear collisions. However, one of the problematic uncertainties in the statistical thermal model comes from complicated feed-down contributions to light hadrons (π, K, p) from multiple strong decay resonances whose masses, widths, and decay properties are not well determined. It will be important to only use the yields of strange particles reconstructed from weak decay channels to test the statistical thermal model. We compare the anti-baryon to baryon ratios with the statistical thermal model.

STAR has collected high statistics data in year 2010 and 2011 in the RHIC Beam Energy Scan Program at 7.7, 11.5, 19.6, 27, and 39 GeV Au+Au collisions. In STAR, we reconstruct the strange hadrons (K^0_S, Λ, Ξ, Ω) using the topology of their weak decay channels [9, 10], $K^0_S \rightarrow \pi^+\pi^-$ (69.2% branching ratio), $\Lambda \rightarrow p\pi$ (63.9% branching ratio), $\Xi \rightarrow \Lambda\pi$ (99.9% branching ratio), and $\Omega \rightarrow \Lambda K$ (67.8% branching ratio). The decay daughters π, K, and p are identified by the Time Projection Chamber (TPC) [11] of the STAR detector system.

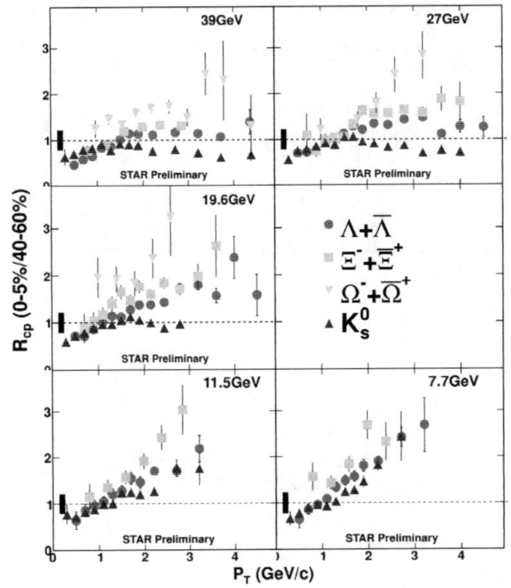

Figure 1. The nuclear modification factor (R_{CP}), as a function of p_T for K^0_S, $\Lambda(\bar{\Lambda})$, $\Xi^-(\bar{\Xi}^+)$, and $\Omega^-(\bar{\Omega}^+)$ at $|y| < 0.5$ from Au+Au collisions at $\sqrt{s_{NN}}$ from 39 to 7.7 GeV. Errors are statistical only.

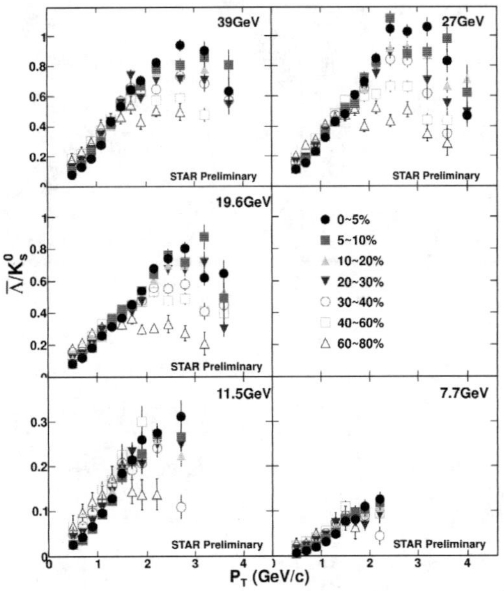

Figure 2. The $\bar{\Lambda}/K^0_S$ ratio as a function of p_T at $|y| < 0.5$ from Au+Au collisions at $\sqrt{s_{NN}}$ from 39 to 7.7 GeV. Errors are statistical only.

2. Results

The strange particle p_T spectra and yields (dN/dy) at mid-rapidity are obtained after correcting for geometrical acceptance and the reconstruction efficiency [12]. The $\Lambda(\bar{\Lambda})$ spectra have been corrected for the feed-down contribution from Ξ and Ξ^0 weak decays.

2.1. Strange Hadron Nuclear Modification Factor (R_{CP}) and Baryon-to-Meson Ratios

Figure 1 shows the R_{CP} for K^0_S, $\Lambda(\bar{\Lambda})$, $\Xi^-(\bar{\Xi}^+)$, $\Omega^-(\bar{\Omega}^+)$ as a function of p_T at beam energies from 7.7 to 39 GeV. The R_{CP} is calculated as the ratio of particle yields at most central collisions (0-5%) to yields at peripheral collisions (40-60%) scaled by the number of inelastic binary collisions (N_{bin}). The R_{CP} data are from Refs. [12, 13, 14]. The strange baryon R_{CP} has a similar behavior at 19.6, 27, and 39 GeV. There seems to be a difference between K^0_S R_{CP} and strange baryon R_{CP}. At lower energies (7.7 and 11.5 GeV), the R_{CP} of K^0_S, Λ, Ξ^- all increase with p_T. The R_{CP} of K^0_S with a p_T above 2 GeV/c is below or consistent with unity at 19.6, 27, and 39 GeV, however it is above unity at 7.7 and 11.5 GeV. Figure 2 shows the $\bar{\Lambda}/K^0_S$ ratios for different centralities from 7.7 to 39 GeV. There is an enhancement of $\bar{\Lambda}/K^0_S$ ratios at p_T around 3 GeV/c at 19.6, 27, and 39 GeV. And there is a strong centrality dependence of $\bar{\Lambda}/K^0_S$ ratios at the three energies. At 7.7 and 11.5 GeV, $\bar{\Lambda}/K^0_S$ ratios increase monotonically with increasing p_T, and the $\bar{\Lambda}/K^0_S$ ratios have a weaker centrality dependency than at higher collision energies. Figure 3 shows the $N(\Omega^- + \bar{\Omega}^+)/(2N(\phi))$ ratios scaled by the number of constituent quarks, n_q, as a function of p_T/n_q from central collisions. At high p_T/n_q, the ratios at 11.5 GeV are systematically lower than those at higher collision energies.

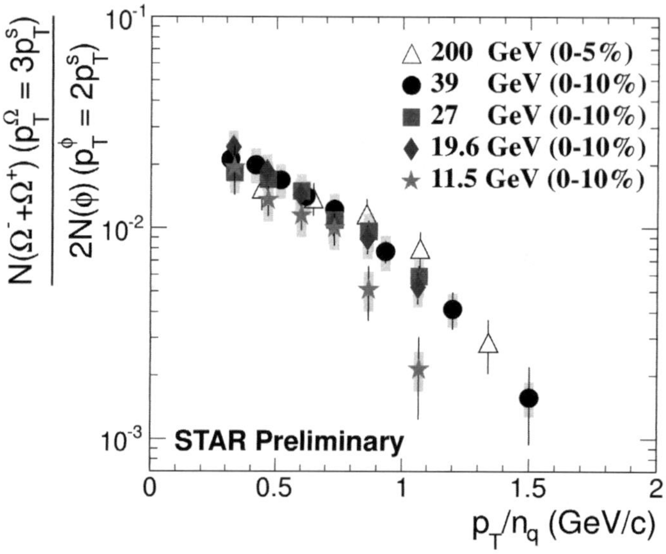

Figure 3. $N(\Omega^- + \bar{\Omega}^+)/(2N(\phi))$ ratios scaled by the number of constituent quarks (n_q), as a function of p_T/n_q from central Au+Au collisions from 11.5 to 200 GeV. The black error bars are the statistical errors, and the green bands are systematic errors.

2.2. Anti-baryon to Baryon Ratios and Statistical Thermal Model

In the statistical thermal model, particle yields can be described by invariant mass, degeneracy factor and chemical potential (μ) of each particle type, strangeness-suppression factor (γ_S), and chemical freeze-out temperature (T) [15]. When we take the anti-baryon to baryon ratios, the parameters are cancelled except μ_B/T and μ_S/T: $ln(ratio) = -2\frac{\mu_B}{T} + \frac{\mu_S}{T}\Delta S$, where μ_B is the baryon chemical potential, and μ_S is the strangeness chemical potential. The ΔS in the formula is the strangeness-number difference between baryon and anti-baryon. The above formula describes, for each particle type, a straight line described by the parameters $\frac{\mu_B}{T}$ and $\frac{\mu_S}{T}$. The $\frac{\mu_B}{T}$ and $\frac{\mu_S}{T}$ are the properties of the collision system, independent of the particle type. When we draw the $ln(ratio)$ of $\frac{\bar{\Lambda}}{\Lambda}$, $\frac{\bar{\Xi}^+}{\Xi^-}$, and $\frac{\bar{\Omega}^+}{\Omega^-}$ in one plot, with $\frac{\mu_B}{T}$ on the x-axis and $\frac{\mu_S}{T}$ on the y-axis, the three lines should cross at one point, if the statistical thermal model assumptions are valid [16]. With this method, we can test the statistical thermal model. The 7.7, 11.5, 19.6, 27, and 39 GeV results are shown in Fig. 4, and the statistical thermal model seems to describe our measurements well. For each collision energy, we use linear regression to determine the crossing point ($\frac{\mu_B}{T}$, $\frac{\mu_S}{T}$). The $\frac{\mu_B}{T}$ and $\frac{\mu_S}{T}$ from 7.7 to 39 GeV are shown in Fig. 5, and the $\frac{\mu_B}{T}$ values fall on the curve of parameterization with μ_B and T fitting of AGS, SPS, and RHIC 130 GeV central collision data [17].

3. Summary

In summary, we presented the strange particle R_{CP} and baryon-to-meson ratios at 7.7, 11.5, 19.6, 27 and 39 GeV. The behaviors of R_{CP} and baryon-to-meson ratios at 11.5 GeV are different from those at 19.6 GeV, which may indicate a change of collision dynamics from 19.6 GeV to 11.5 GeV. The statistical thermal model has been tested with the measured strange anti-baryon to baryon ratios, and describes the data well.

References

[1] Rafelski J and Muller B 1982 *Phys. Rev. Lett.* **48** 1066
[2] STAR Collaboration, Aggarwal M M *et al* *arXiv:nucl-ex*/1007.2613v1
[3] STAR Collaboration, Adams J *et al* 2007 *Phys. Rev. Lett.* **98** 062301
[4] Lamont M A C (for the STAR Collaboration) 2006 *J. Phys. Conf. Ser.* **50** 192

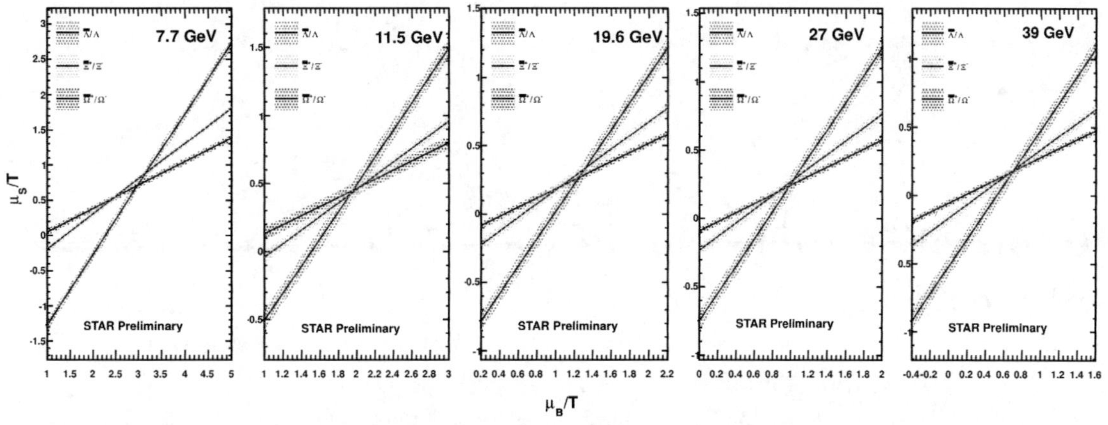

Figure 4. Testing of statistical thermal model in $\frac{\mu_B}{T}$ and $\frac{\mu_S}{T}$ parameter space with strange anti-baryon to baryon ratios in 7.7, 11.5, 19.6, 27, and 39 GeV Au+Au central collisions. Errors are statistical only.

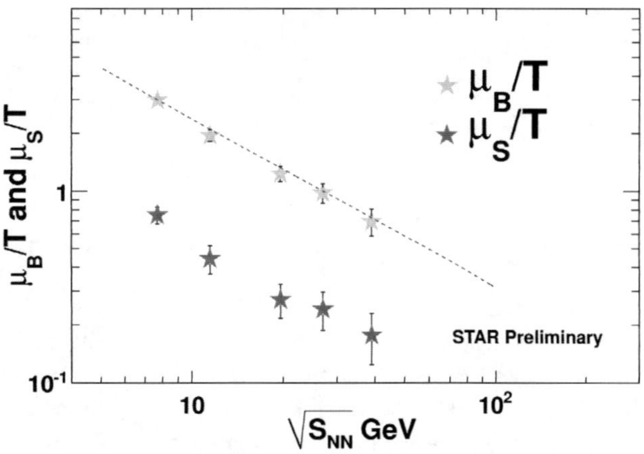

Figure 5. The $\frac{\mu_B}{T}$ and $\frac{\mu_S}{T}$ parameters as a function of beam energy. The curve is the parameterization with μ_B and T fitting of AGS, SPS, and RHIC 130 GeV data [17]. Errors are statistical only.

[5] Hwa R C and Yang C B 2002 *Phys. Rev. C* **66** 025205
[6] Greco V, Ko C M and Levai P 2003 *Phys. Rev. Lett.* **90** 202302
[7] Fries R J, Muller B, Nonaka C and Bass S A 2003 *Phys. Rev. Lett.* **90** 202303
[8] Molnar D and Voloshin S A 2003 *Phys. Rev. Lett.* **90** 092301
[9] STAR Collaboration, Adler C *et al* 2002 *Phys. Rev. Lett.* **89** 092301
[10] STAR Collaboration, Adams J *et al* 2004 *Phys. Rev. Lett.* **92** 182301
[11] Anderson M *et al* 2003 *Nucl. Instrum. Meth. A* **499** 659
[12] Zhu X (for the STAR Collaboration) 2012 *Acta Phys. Polon. Supp.* **5** 213
[13] Zhang X (for the STAR Collaboration) 2013 *Nucl. Phys. A* **904-905** 543
[14] Zhao F (for the STAR Collaboration) 2013 *PoS* **CPOD 2013** 036
[15] Cleymans J *arXiv:nucl-th*/9704046v1
[16] Zhao F (for the STAR Collaboration) 2012 *AIP Conf. Proc.* **1441** 797
[17] Becattini F *et al* 2006 *Phys. Rev. C* **73** 044905

Resonance Production in $\sqrt{s} = 7$ TeV pp collisions with ALICE.

Graham Richard Lee

School of Physics and Astronomy, University of Birmingham, Edgbaston, Birmingham, UK, B15 2TT

E-mail: graham.richard.lee@cern.ch

Abstract. The study of pp collisions at LHC energies is important not only as a baseline for future analysis in heavy-ion events at ALICE but aiding also in the calibration of QCD inspired models at LHC energies. ALICE has measured the $\Sigma^*(1385)$, $\phi(1020)$ and $K^*(892)$ resonances in pp collisions at $\sqrt{s} = 7$ TeV, measurements include the mass and width of the Σ^* resonance, normalised spectra and particle ratios. The Σ^*, ϕ and K^*, measurements are compared to QCD inspired models with additional ϕ measurements at $\sqrt{s} = 2.76$ TeV giving a greater range in transverse momentum than in earlier analysis.

1. Introduction

The production of resonances with proton-proton collisions at ALICE can be studied as a useful probe to help understand the underlying event in the collisions, due to the majority of the final state particles originating from these resonances [1]. These measurements can be used to tune QCD inspired models such as PYTHIA [2] and PHOJET [3], which are tuned to LHC energy values. In heavy-ion collisions, some resonances are sensitive probes of the environment within the created fireball. Due to the short lifetime of the order of 1-10 fm/c, resonances such as the Σ^* and ϕ decay within the medium before freeze-out. For heavy-ion collision analysis a baseline in proton-proton collisions is needed for comparison.

2. Analysis Method

Resonance studies at $\sqrt{s} = 7$ TeV exploit the Inner Tracking System (ITS) and Time Projection Chamber (TPC) for vertex reconstruction and tracking. The position of the primary vertex along the beam axis is required to be within ± 10 cm from the centre of the ALICE Barrel. Resonances are reconstructed at mid-rapidity through their dominant hadronic decay channel, $K^* \to \pi^{\pm}+K^{\mp}$, $\phi \to K^{+}+K^{-}$, $\Sigma^{*+} \to \Lambda+\pi^{+}$, $\Sigma^{*-} \to \Lambda+\pi^{-}$, $\bar{\Sigma}^{*+} \to \bar{\Lambda}+\pi^{+}$ and $\bar{\Sigma}^{*-} \to \bar{\Lambda}+\pi^{-}$. The ϕ and K^* analyses combined the information from the TPC and Time Of Flight (TOF) detectors for the particle identification of the decay product candidates. These candidates where identified by requiring a TPC energy loss or Time of Flight measurement within at least 3σ of the expected value for a given candidate. This requirement was relaxed in the $\sqrt{s} = 2.76$ TeV pp analysis of the ϕ resonance allowing the maximum p_T for ϕ reconstruction to be increased from 7 GeV/c to 20 GeV/c. The Σ^* analysis required the Λ candidate to be identified via its weak decay, with the daughters having a Distance of Closest Approach (DCA) of lower than 0.05 cm. Resonance peaks are reconstructed from the invariant-mass distribution of the identified daughters. The combinatorial background is estimated and removed using the event-mixing technique. The resonance peaks are fitted with a Breit-Wigner (for the K^{*0}) or a convolution of a Breit-Wigner with a gaussian (for the ϕ) function plus a polynomial (to describe the shape of the residual background). The Σ^* signal was also fitted with a Breit-Wigner function plus a polynomial (for the residual background), but in addition a Gaussian function was used to fit the $\Lambda-\pi$ pairs resulting from the $\Lambda(1520)$ contamination. The inclusion of this contamination within the fitting process was critical in producing a stable fit throughout the p_T range measured. The raw yields are corrected for acceptance and efficiency ($A \times \epsilon$) from Monte Carlo simulations with the PYTHIA 6.4 event generator (tune Perugia 0 for the mesonic analysis and tune Perugia 0 [4] for the baryonic analysis) and a GEANT3-based simulation of the ALICE detector response [5].

3. Results

The masses and widths of the extracted Σ^* as a function of transverse momentum p_T are shown in figure 1. Both the mass and width values show no overall p_T dependence and no significant deviation from the PDG values. The ratios of the yields of Σ^* over Ξ^-, π^- and K^- as a function of collision energy in figure 2 are in an agreement with the ratio obtained from the Σ^* STAR analysis [6] and the π, K and Ξ STAR analysis [7] [8]. The measured ratios are also in agreement with the predictions of a statistical hadronization model with $T = 170$ MeV and a strangeness suppression factor $\gamma_s = 0.6$ by Becattini [9]. The value for Σ^* has been provided by the author in private communication and was obtained with the same model.

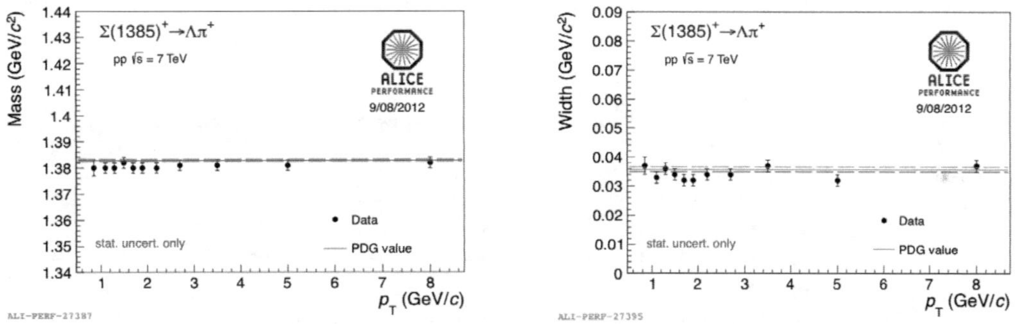

Figure 1. Mass and width values for $\Sigma^+(1385)$.

Figure 3 shows the extracted p_T spectra of ϕ, K^* and Σ^* at 7 TeV compared to theoretical models. There is little agreement for both ϕ and K^* with the best agreement found with PYTHIA D6T tune. PYTHIA ATLAS-CSC and PHOJET over-predict the ϕ and K spectra at low p_T while the PYTHIA and Perugia 0 tune underestimates both by nearly a factor two. The Σ^* spectrum is underestimated for all tunes with the largest deviation in line shape resulting from the D6T tune. The analysis of the ϕ spectra at the lower collision energy of $\sqrt{s} = 2.76$ TeV allows further investigation of the spectra at a higher p_T ($> 7GeV/c$) than the 7 TeV analysis. Figure 4 shows the ratio of ϕ/π as a function of p_T for different collision energies, including results from PHENIX at $\sqrt{s} = 200$ GeV [10] and preliminary plots from PHENIX at $\sqrt{s} = 62.4$ GeV. With the $\sqrt{s} = 2.76$ TeV analysis this ratio can be extended up to higher p_T where a saturation is observed in the ratio.

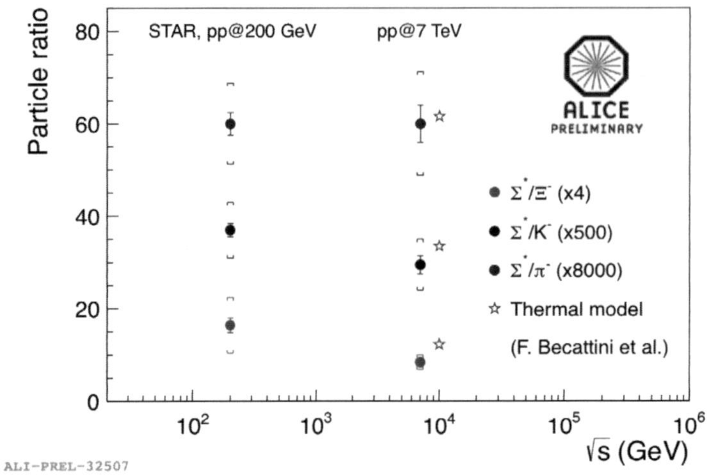

Figure 2. Σ^* ratios to stable particles as a function of collision energy with comparison to Becattini thermal model [9].

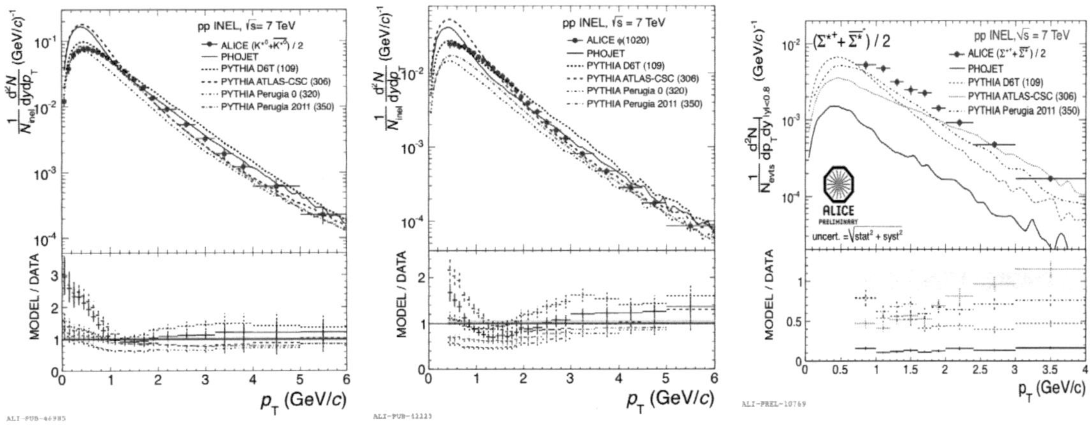

Figure 3. K^* (a), ϕ (b) and Σ (c) spectra as compared to theoretical predictions in pp collisions with PHOJET and PYTHIA tunes D6T (109), ATLAS-CSC (306) and Perugia 0 (320).

4. Conclusions

Resonances such as K^*, ϕ and Σ^* have been measured by ALICE in pp collisions. In particular, masses and widths show no p_T dependence and are within reasonable agreement with the PDG values. The particle ratios for the Σ^* resonance show agreement with previous STAR analysis and Becattini model for π^-, K^- and Ξ^-. There is little agreement between the extracted spectra of ϕ, K^* and Σ^* at 7 TeV compared to PYTHIA and PHOJET models, with the Σ^* predictions especially showing a large under-prediction for all model tunes. Analysis of the ϕ resonance at 2.76 TeV extends the spectra to a larger p_T range and shows a saturation in the ϕ/π ratio at the higher p_T.

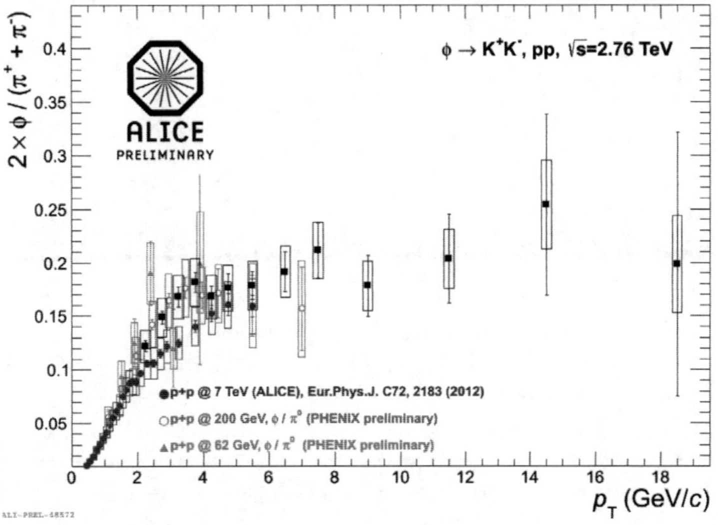

Figure 4. ϕ/π ratio as a function p_T for a number of collision energies including ALICE $\sqrt{s} =$ 2.76 TeV.

References

[1] Aguilar-Benitez M et al. 1991. *Z.Phys.*, C **50** 40.
[2] Mrenna S Torbjorn S and Peter S. 2006. *JHEP*, **0605** 026.
[3] Engel R and Ranft J. 1996. *Phys.Rev.*, D **54** 4244.
[4] Skands and Zeiler P. 2010. *Phys.Rev.*, D **82** 074018.
[5] Federico C Rene B and Simone G. 1994. *CERN* ***W5013***.
[6] Abelev B et al. 2006. *Phys.Rev.Lett.*, **97** 132301.
[7] Adams J et al. 2005. *Phys.Rev.*, C **71** 064902.
[8] Abelev B et al. 2007. *Phys.Rev.*, C **75** 064901.
[9] Milov A Becattini F, Castorina P and Satz H. 2011. *J.Phys.*, G **38** 025002.
[10] Adare A et al. 2011. *Phys.Rev.*, C **83** 024909.

14th International Conference on Strangeness in Quark Matter (SQM2013) IOP Publishing
Journal of Physics: Conference Series **509** (2014) 012087 doi:10.1088/1742-6596/509/1/012087

Hadronic resonances in heavy-ion collisions at ALICE

A G Knospe (for the ALICE Collaboration)

Department of Physics, The University of Texas at Austin, 1 University Station C1600, Austin, TX 78712-0264 USA

E-mail: anders.knospe@cern.ch

Abstract. Properties of the hadronic phase of high-energy heavy-ion collisions can be studied by measuring the ratios of hadronic resonance yields to the yields of longer-lived particles. These ratios can be used to study the strength of re-scattering effects, the chemical freeze-out temperature, and the lifetime between chemical and kinetic freeze-out. The restoration of chiral symmetry during the early hadronic phase or near the phase transition may lead to shifts in the masses and increases in the widths of hadronic resonances. The ALICE collaboration has measured the spectra, masses, and widths of the $\mathrm{K}^*(892)^0$ and $\phi(1020)$ resonances in Pb–Pb collisions at $\sqrt{s_{\mathrm{NN}}} = 2.76$ TeV. These results, including R_{AA} and the ratios of the integrated resonance yields to stable hadron yields, are presented and compared to results from other collision systems and to theoretical predictions.

Strongly interacting matter undergoes the transition from a partonic phase to the hadronic phase around a critical temperature of about 160 MeV [1, 2]. The yields of various particles at chemical freeze-out can be predicted using thermal models [3, 4], but the measurable resonance yields may differ from these predictions due to re-scattering and/or regeneration in the hadronic phase. Resonances with lifetimes of a few fm/c may decay during the hadronic phase and their decay hadronic daughters may be re-scattered in the medium, reducing the number of resonances that can be reconstructed [5]. Pseudo-elastic scattering of particles (*e.g.*, $\pi^-\mathrm{K}^+ \to \mathrm{K}^*(892)^0 \to \pi^-\mathrm{K}^+$) before kinetic freeze-out may increase the resonance yield (regeneration) [6–8]. The measurable resonance yields are therefore expected to depend upon the chemical freeze-out temperature and the elapsed time between chemical and kinetic freeze-out. The model described in [7, 9, 10] makes predictions for the ratios of the yields of resonance to stable-particle yields as functions of these parameters, and in these proceedings the measured $\mathrm{K}^*(892)^0/\mathrm{K}$ ratio is compared to that model to extract an estimate of the lower limit of the lifetime of the hadronic phase.

Near the phase transition, chiral symmetry is expected to be partially restored [11]. It is predicted that resonances which decay when chiral symmetry is at least partially restored will exhibit mass shifts or width broadening with respect to the vacuum values of those parameters [12, 13]. The fraction of resonances with vacuum properties will be increased by regeneration in the later hadronic phase. Re-scattering and regeneration are predicted [7, 14] to have the greatest effect on resonance yields at low p_{T} ($\lesssim 2$ GeV/c).

In these proceedings, measurements of the $\mathrm{K}^*(892)^0$ and $\overline{\mathrm{K}}^*(892)^0$ are averaged and these particles are collectively called K^{*0}. The mass is omitted from the symbols of the K^{*0} and ϕ.

This analysis uses the ALICE [15] Inner Tracking System (ITS, for tracking and vertex finding), Time Projection Chamber (TPC, for tracking and particle identification), and VZERO detector (for triggering and centrality estimation). The K^{*0} (ϕ) resonance signals are extracted in multiple centrality bins from about 9 M events. The z position of the primary vertex is required to be within 10 cm of the center of the ALICE detector. The resonances are reconstructed through their hadronic decay channels: $K^{*0}\rightarrow\pi^{\pm}K^{\mp}$ (branching ratio 66%) and $\phi\rightarrow K^{-}K^{+}$ (branching ratio 48.9%) [16]. Each decay pion or kaon candidate is required to have TPC energy loss dE/dx within $2\sigma_{TPC}$ of the expected value for the given particle species, where σ_{TPC} is the TPC dE/dx resolution.

The resonances are identified via invariant-mass reconstruction, with a combinatorial background estimated using event mixing. The background-subtracted distribution is fit with a peak function plus a polynomial (which parameterizes the residual background from correlated track pairs that are not described by the combinatorial background). The K^{*0} is fit with a Breit-Wigner peak, while the ϕ is fit with a Voigtian peak (convolution of a Breit-Wigner function with a Gaussian) to account for detector effects. The raw yields are corrected for the efficiency $\times$ acceptance. This is extracted from simulated Pb–Pb collisions, with particle production simulated using HIJING [17] and interactions with the ALICE detector simulated using GEANT3 [18]. A separate correction factor is applied to account for the energy-loss cuts (91% when a $2\sigma_{TPC}$ PID cut is applied to both decay daughters). The K^{*0} (ϕ) spectra are fit with a Lévy-Tsallis [19] (Boltzmann-Gibbs blast-wave [20]) function so that the yield can be extrapolated for values of p_T outside the measured region.

The mass and width of the K^{*0} and ϕ are shown in Fig. 1 as functions of p_T. The measured K^{*0} mass and width are consistent with the values extracted from simulations (the same simulations used to extract the efficiency $\times$ acceptance). Similarly, the measured ϕ width is consistent the simulated values. The measured ϕ mass has been corrected to account for the mass shift observed in simulations (*i.e.*, corrected for detector effects). Therefore, the ϕ mass reported in Fig. 1b should be compared to the vacuum value directly, rather than the simulation. The corrected ϕ mass is consistent with the vacuum value [16]. No centrality dependence is observed in the mass or width for either particle.

The ratios K^{*0}/K^{-} and ϕ/K^{-} (the ratios of dN/dy for $|y| < 0.5$ integrated over all p_T) are shown for different centrality bins in Fig. 2a (K^{-} yields from [21]). The ϕ/K^{-} ratio is independent of centrality, while the K^{*0}/K^{-} ratio appears to decrease for more central collisions. This suggests that the K^{*0} is affected by re-scattering, while the ϕ (with its longer lifetime) is not. The K^{*0}/K^{-} ratio as a function of energy for pp and central A–A collisions [22, 23] is shown in Fig. 2b. The K^{*0}/K^{-} ratio is independent of energy from RHIC to LHC energies, but a suppression is observed in A–A collisions with respect to pp collisions. A thermal model fit of ALICE particle yields [24] (the K^{*0} is excluded from the fit) gives a temperature of 156 MeV and a K^{*0}/K ratio of 0.30 (Fig. 2b, red line), which is 50% larger than the measured ratio 0.19 ± 0.05, however this model does not include re-scattering. The measured K^{*0}/K ratio can also be compared to the model described in [7, 9, 10] to estimate properties of the hadronic phase. This model uses statistical calculations plus re-scattering during the hadronic phase to predict the K^{*0}/K ratio (and others) as a function of the chemical freeze-out temperature and the time between chemical and kinetic freeze-out. If the lifetime of the hadronic phase is assumed to be 0 (*i.e.*, no re-scattering), the measured K^{*0}/K ratio would correspond to a chemical freeze-out temperature of $\sim$ 120 MeV. On the other hand, if a chemical freeze-out temperature of 156 MeV is assumed (based on [24]), the model predicts a $K^{*0}/K \approx 0.35$ in the case of no re-scattering (Fig. 2b, magenta line). However, if a non-zero hadronic lifetime is permitted and a temperature of 156 MeV is again assumed, the measured K^{*0}/K ratio gives an estimate of $\approx$ 1.5 fm/c as a lower limit on the lifetime of the hadronic phase, according to this model (this is only a lower limit because the model does not include regeneration).

In order to study the p_T dependence of resonance suppression, the measured K^{*0} and ϕ spectra for central Pb–Pb collisions are compared to predicted spectra based on the blast-wave model [20]. The blast-wave parameters used to generate the predicted spectra are taken from

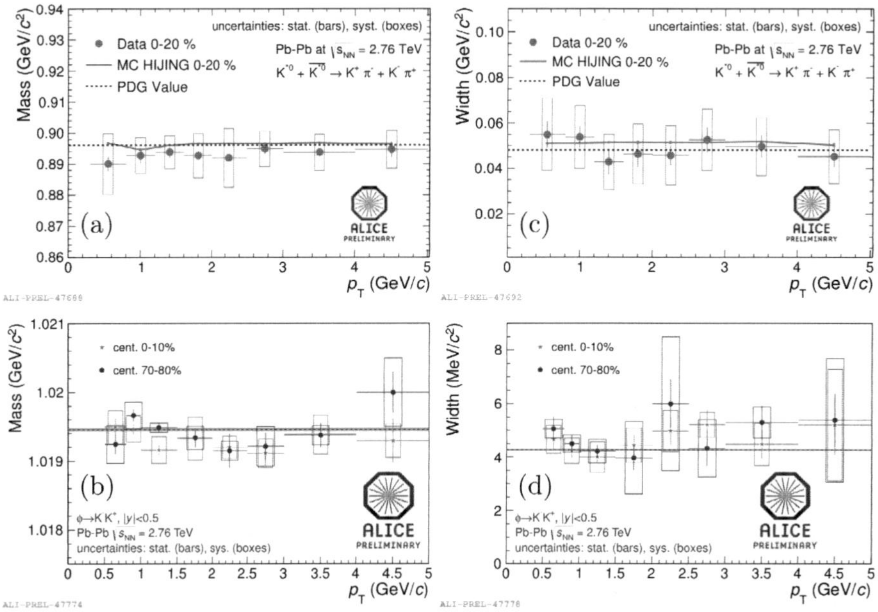

Figure 1. Mass and width of K^{*0} and ϕ.

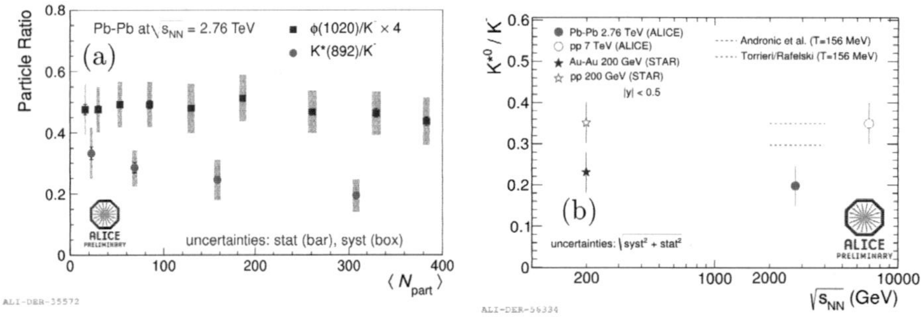

Figure 2. **(a)** K^{*0}/K^- and ϕ/K^- (scaled by 4) ratios as functions of centrality. **(b)** K^{*0}/K^- ratio as a function of $\sqrt{s_{NN}}$ for different collision systems and theoretical predictions.

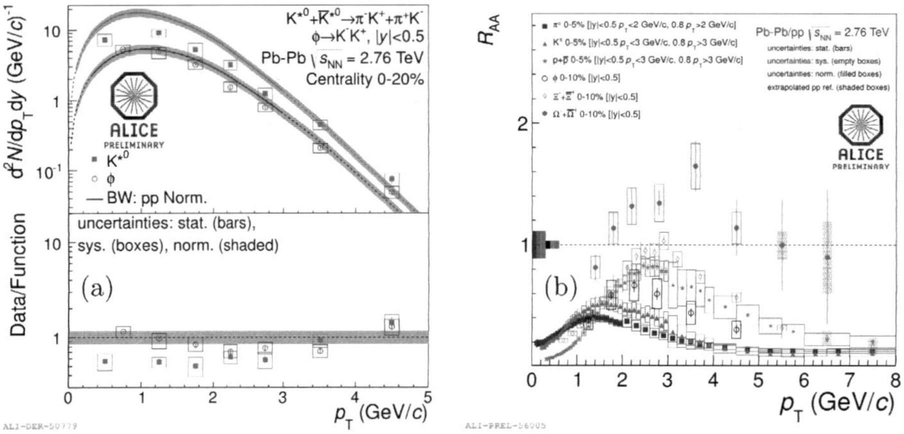

Figure 3. **(a)** Spectra of K^{*0} and ϕ (centrality 0-20%) with the blast-wave predictions. The lower panel shows the ratio of the data to the predictions. **(b)** R_{AA} of various particles.

global fits of the spectra of $\pi^{\pm}$, $K^{\pm}$, and (anti)protons [21]. It should be noted that these fits are not valid for $p_T \gtrsim 3$ GeV/c. The predicted K^{*0} (ϕ) spectrum is normalized so that its integral is equal to the K yield in Pb–Pb collisions at $\sqrt{s_{NN}} = 2.76$ TeV [21] times the K^{*0}/K (ϕ/K) ratio in pp collisions at $\sqrt{s} = 7$ TeV [23] (*i.e.*, assuming no re-scattering). The measured and predicted spectra are shown in Fig. 3a, with the ratios of the measured and predicted spectra shown in the lower panel. The measured ϕ spectrum is not suppressed but the measured K^{*0} spectrum is suppressed with respect to the predicted spectrum (normalized as described above). Furthermore, the K^{*0} suppression is flat (≈ 0.6) for $p_T < 3$ GeV/c. In peripheral collisions, neither resonance is observed to be suppressed with respect to the predicted spectra.

The nuclear modification factor R_{AA} is shown in Fig. 3b for six different particle species. R_{AA} is defined as the ratio of a particle yield in A–A collisions to the yield of the same particle in pp collisions, scaled by the mean number of nucleon-nucleon collisions in the A–A events. For $p_T \lesssim 2.5$ GeV/c, the ϕ R_{AA} appears to follow R_{AA} of p and Ξ, while for mid-to-high p_T, the ϕ R_{AA} tends to be between R_{AA} of the mesons (π and K) and baryons (p and Ξ). For peripheral collisions, the ϕ R_{AA} is consistent with the p and Ξ R_{AA} throughout the measured p_T range and consistent with one for $p_T > 1$ GeV/c.

In conclusion, the masses and widths of the K^{*0} and ϕ resonances (reconstructed via hadronic decays) are consistent with the vacuum values. The K^{*0}/K ratio decreases for central collisions (which suggests the importance of re-scattering), while the ϕ/K ratio is independent of centrality. An approximate lower limit of 1.5 fm/c for the lifetime of the hadronic phase has been estimated by comparing the measured K^{*0}/K^{-} ratio to a theoretical model prediction [7, 9, 10]. Comparisons of the resonance spectra to predicted spectra based on the blast-wave model indicate that the suppression of the K^{*0} yield is flat for $p_T < 3$ GeV/c. The nuclear modification factor R_{AA} of ϕ appears to follow R_{AA} of Ξ and Ω for $p_T \lesssim 2.5$ GeV/c, and lies between the R_{AA} of mesons (π and K) and R_{AA} of baryons (p and Ξ) at high p_T.

References

[1] Borsányi S *et al* 2010 *J. High Energy Phys.* **2010** No. 9 1–31
[2] Aoki Y *et al* 2009 *J. High Energy Phys.* **2009** No. 6 1–17
[3] Andronic A, Braun-Munzinger P and Stachel J 2011 *J. Phys.* G **38** 124081
[4] Braun-Munzinger P and Stachel J 2011 *From Nuclei To Stars: Festschrift in Honor of Gerald E Brown* ed Lee S (Singapore: World Scientific) (*Preprint* `1101.3167v1`)
[5] Bleicher M and Aichelin J 2002 *Phys. Lett.* B **530** 81–7
[6] Bleicher M and Stöcker H 2004 *J. Phys.* G **30** S111–8
[7] Markert C, Torrieri G and Rafelski J 2002 *Proc. of PASI 2002* (*Preprint* `hep-ph/0206260`)
[8] Vogel S and Bleicher M 2005 *Proc. of Nucl. Phys. Winter Meeting, Bormio* (*Preprint* `nucl-th/0505027v1`)
[9] Rafelski J, Letessier J and Torrieri G 2001 *Phys. Rev.* C **64** 054907
[10] Rafelski J, Letessier J and Torrieri G 2002 *Phys. Rev.* C **65** 069902(E)
[11] Petreczky P 2007 *Nucl. Phys.* A **785** 10–7
[12] Eletsky V L, Belkacem M, Ellis P J and Kapusta J I 2001 *Phys. Rev.* C **64** 035202
[13] Brown G E and Rho M 2002 *Rhys. Rep.* **363** 85–171
[14] Bleicher M *et al* 1999 *J. Phys.* G **25** 1859–96
[15] Aamodt K *et al* (ALICE Collaboration) 2008 *J. Inst.* **3** No. S08002 i–245
[16] Beringer J *et al* (Particle Data Group) 2012 *Phys. Rev.* D **86** 010001
[17] Wang X–N and Gyulassy M 1991 *Phys. Rev.* D **44** 3501
[18] Brun R, Carminati F and Giani S 1994 *CERN-W5013*
[19] Tsallis C 1988 *J. Stat. Phys.* **52** 479
[20] Schnedermann E, Sollfrank J and Heinz U 1993 *Phys. Rev.* C **48** 2462–75
[21] Abelev B *et al* (ALICE Collaboration) 2013 *Phys. Rev.* C **88** 044910
[22] Aggarwal M M *et al* (STAR Collaboration) 2011 *Phys. Rev. C* **84** 034909
[23] Abelev B *et al* (ALICE Collaboration) 2012 *Eur. Phys. J.* C **72** 2183
[24] Stachel J (for the ALICE Collaboration) *these proceedings*

Journal of Physics: Conference Series **509** (2014) 012088 doi:10.1088/1742-6596/509/1/012088

Low mass vector meson production in pp and Pb-Pb collisions measured with ALICE at the LHC

Alessandro De Falco for the ALICE Collaboration

E-mail: alessandro.de.falco@ca.infn.it

Abstract. Vector mesons are key probes of the hot and dense state of strongly interacting matter produced in heavy ion collisions. Their dileptonic decay channel is particularly suitable for these studies, since dileptons have negligible final state interactions in hadronic matter. In particular, strangeness production can be studied through the measurement of muon pairs coming from the decay of ϕ mesons. The ALICE apparatus at the LHC can access vector mesons produced at forward rapidity ($2.5 < y < 4$) through their decays in muon pairs. We present results on vector meson production in pp and Pb-Pb collisions at $\sqrt{s_{NN}} = 2.76$ TeV. The ϕ meson production cross section is measured in pp collisions. In Pb-Pb collisions, a measurement of the $\phi/(\rho+\omega)$ ratio and of the ϕ nuclear modification factor as a function of the collision centrality is obtained.

Low mass vector meson (ρ, ω, ϕ) production provides key information on the hot and dense state of strongly interacting matter produced in high-energy heavy ion collisions. In particular, strangeness production can be accessed through the measurement of ϕ meson production. The nuclear modification factor provides information about hadron production [1]. At low transverse momenta, $p_T < 2$ GeV/c, hadron production is governed by soft processes which can be described in terms of hydrodynamical and thermal models [2, 3], while at high $p_{\rm T}$ ($p_{\rm T} > 5$ GeV/c) hard processes dominate. In the latter region, yields are suppressed due to the parton energy loss caused by gluon bremsstrahlung [4]. At intermediate $p_{\rm T}$ ($2 < p_{\rm T} < 5$ GeV/c), light mesons are suppressed with respect to binary scaling, differently from protons and antiprotons. In this context, it is interesting to compare their suppression pattern with the one of the ϕ, which is a meson having a mass similar to the proton one.

The detection of the ϕ meson through its decay in lepton pairs has the advantage that, differently from the K^+K^- channel, dileptons are not influenced by final state interactions, like rescattering and absorption [5]. Previous results at the SPS [6, 7] showed a discrepancy between results in the two channels, while most recent measurements [8, 9] showed an agreement between the measurements in dileptons and kaons in Pb-Pb and In-In.

Vector meson production in pp collisions provides a reference for the determination of the nuclear modification factor. Moreover, it is interesting by itself, since it can be used to tune particle production models in the LHC energy range.

The ALICE experiment at the LHC can access vector mesons produced at forward rapidity through their decays in muon pairs. The detector is described in [10]. The measurement in the dimuon channel was performed using the forward muon spectrometer, that covers the pseudorapidity range $-4 < \eta < -2.5$ (although in the ALICE reference frame the muon spectrometer covers a negative η range, we chose to present our results with a positive y

notation). It consists of an hadron absorber, a set of cathode pad chambers (five stations, each one composed of two chambers) for the track reconstruction in a dipole magnetic field, two stations of two resistive plate chambers for the muon trigger and an iron wall acting as a muon filter. The centrality is determined with the VZERO detector, that consists of two arrays of plastic scintillators placed at 3.4 m and -0.9 m from the IP.

Data were taken in both pp and Pb-Pb collisions at $\sqrt{s_{NN}} = 2.76$ TeV. The data sample in pp collisions was collected in 2013 and amounts to an integrated luminosity $L_{\rm int} = 81.1$ nb^{-1}. Data relevant for this analysis were acquired with a dimuon trigger, that is provided by the coincidence of two single-muon trigger signals. The trigger selects tracks with a low $p_{\rm T}$ threshold of roughly 0.5 GeV/c. Offline selections were applied in order to remove beam-gas background. Muon tracks were selected asking that the tracks reconstructed in the tracking stations matched the ones in the trigger chambers and that their pseudorapidity was in the range $-4 < \eta_\mu < -2.5$. Muon pairs were selected requiring that the dimuon rapidity was within the interval $2.5 < y_{\mu\mu} < 4$.

The combinatorial background in the opposite sign dimuon mass spectrum was subtracted using the event mixing technique. The S/B ratio at the ϕ peak is ~2.

The mass spectrum after background subtraction was described as a superposition of light meson decays into muon pairs, with an additional contribution coming from charm and beauty semi-muonic decays. Low-mass resonance shapes, as well as the acceptance and efficiency for each process, were obtained through a Monte Carlo simulation with a parametric generator [11], while open charm and beauty were generated using a parametrization of PYTHIA. Alternative fits to the mass spectrum were performed replacing this hadronic cocktail with empirical functions. In order to extract the $p_{\rm T}$-differential cross section, the measured number of ϕ mesons $N_\phi^{\rm meas}(\Delta p_{\rm T})$ was measured in several $p_{\rm T}$ intervals. The differential cross section in $\Delta p_{\rm T}$ was calculated as $d\sigma_\phi/dydp_{\rm T} = N_\phi^{\rm meas}(\Delta p_{\rm T})/(A\varepsilon(\Delta p_{\rm T})BR_{\phi\to l^+l^-}L_{\rm int})$, where $BR_{\phi\to l^+l^-}$ is the branching ratio for the ϕ decay into lepton pairs (assuming lepton universality, the more precise measurement in e^+e^- is used), A is the acceptance and ε is the efficiency for the $\phi \to \mu^+\mu^-$ process. The sources of systematic uncertainty are those on the evaluation of the branching ratio (1%), on the integrated luminosity (1.8%), on the number of ϕ determined by the fit (ranging from 8.7% at low $p_{\rm T}$ to 4.6% at high $p_{\rm T}$), on the tracking and trigger efficiency (4.5% and 3%, respectively). The differential cross section is reported in Fig. 1. In the same figure, the predictions based on the PYTHIA tunes ATLAS-CSC, D6T, Perugia-0 and Perugia-11 and on PHOJET are shown [12, 13, 14, 15, 16]. It can be observed that, while the Perugia-0 and Perugia-11 tunes underestimate the cross section by about a factor of two, the other calculations agree with the measurement within 15%. Consistent results have been obtained in a previous measurement at $\sqrt{s} = 7$ TeV [11].

Data in Pb-Pb collisions were collected in 2011 at $\sqrt{s_{NN}} = 2.76$ TeV. For this data set, a higher $p_{\rm T}$ selection, with a threshold at about 1 GeV/c, was applied on the single muons at the trigger level. The selections applied were the same as in pp analysis, with an additional cut on the single muon $p_{\rm T}$ at 0.85 GeV/c that removes muons that, due to the fact that the trigger threshold is not sharp, have been accepted at the trigger level. Due to the limited acceptance at low $p_{\rm T}$ caused by the high $p_{\rm T}$ threshold imposed by the trigger, an additional cut on the dimuon $p_{\rm T}$ at 2 GeV/c was applied. The number of opposite sign muon pairs was $\sim 2 \cdot 10^6$.

The combinatorial background was evaluated with the event mixing technique also in Pb-Pb analysis. S/B ratio at the ϕ peak is ~ 0.1 for most central collisions, increasing up to ~ 3 for peripheral collisions.

The detector acceptance and efficiency was estimated by means of an embedding Monte Carlo technique, that consists in simulating signal particles and embedding the generated hits on the detector to those of real events. This technique allows one to account for the variation of the reconstruction efficiency with the detector occupancy and, thus, the collision centrality.

In Fig. 2, the ratio $BR_\phi\sigma_\phi/(BR_\rho\sigma_\rho + BR_\omega\sigma_\omega)$, is shown as a function of the number

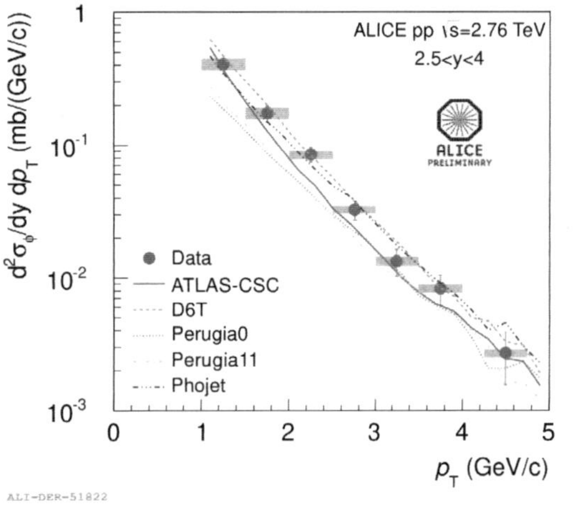

Figure 1. ϕ differential cross section vs p_T in pp collisions at $\sqrt{s}$ = 2.76 TeV compared with the predictions based on several PYTHIA tunes and PHOJET.

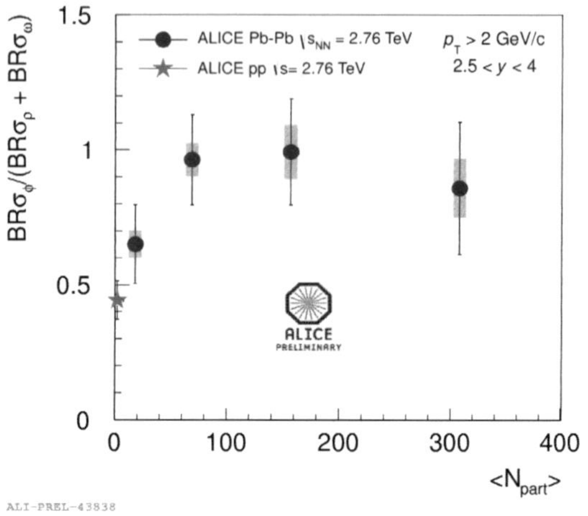

Figure 2. Ratio $BR_\phi\sigma_\phi/(BR_\rho\sigma_\rho + BR_\omega\sigma_\omega)$ as a function of the number of participating nucleons N_{part}.

of participating nucleons N_{part}. The pp result is also reported for comparison. Systematic uncertainties are mainly due to the variations related to p_T cut on single muon and to the description of the correlated background due to non-resonant components. Other components, like the uncertainty on the tracking and trigger efficiency and on the acceptance, mainly cancel out in the ratio and are thus neglected. The ratio increases from pp to Pb-Pb and tends to saturate when moving from semiperipheral to central collisions. The value for most central Pb-Pb collisions is about two times the one measured in pp collisions. A similar behaviour is observed in the ϕ multiplicity per participant, displayed in Fig. 3: the multiplicity increases faster than N_{part} from peripheral to semiperipheral collisions, and tends to saturate for higher N_{part} values.

The nuclear modification factor for a given centrality, p_T and y range is obtained as $R_{AA} = \langle\phi\rangle/(\sigma_{pp}\langle T_{AA}\rangle)$, where $\langle\phi\rangle$ is the ϕ multiplicity, σ_{pp} is the cross section in pp collisions in the same p_T and rapidity range and $\langle T_{AA}\rangle$ is the average nuclear overlap function of the considered centrality class [17, 18]. Deviations of R_{AA} from unity quantify the departure of the multiplicity in Pb-Pb from a superposition of incoherent nucleon-nucleon collisions. R_{AA} measured as a function of the number of participants is shown in Fig. 4. In addition to the sources of systematic uncertainties already discussed for the $BR_\phi\sigma_\phi/(BR_\rho\sigma_\rho + BR_\omega\sigma_\omega)$ ratio, also the uncertainty on tracking and trigger efficiency, on the acceptance, on σ_{pp} and T_{AA} give a sizeable contribution to the systematic uncertainty on R_{AA}. In peripheral collisions, the nuclear modification factor is compatible, within uncertainty, with unity, indicating that these collisions behave as a superposition of incoherent pp collisions. In most central collisions, R_{AA} is reduced to about 0.5, showing a clear suppression of the ϕ multiplicity with respect to the pp reference in the intermediate p_T region. In the same figure, a comparison with the ALICE measurement in the KK channel at midrapidity [19] is shown. Within uncertainties, each point measured in the dimuon channel is compatible to the corresponding one in the hadronic channel. On the other side, the decrease of R_{AA} vs N_{part} appears steeper than the one in the kaon channel, such that differences in the two channels, or rapidity ranges, cannot be excluded within the precision of the measurement.

In 2013, data were taken also in p-Pb collisions. Results are currently under analysis. Thanks

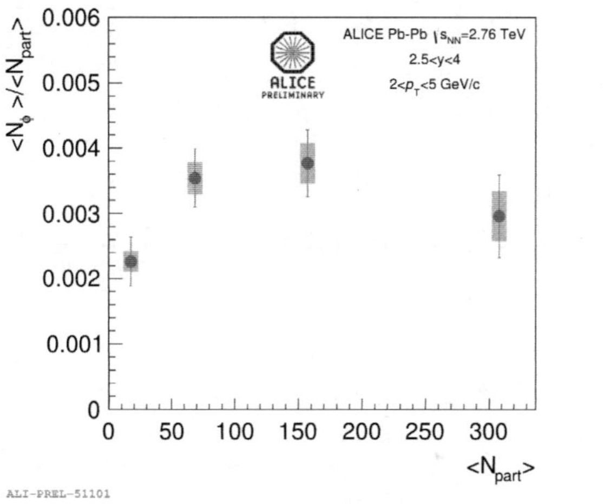

Figure 3. ϕ multiplicity per participant as a function of the number of participating nucleons.

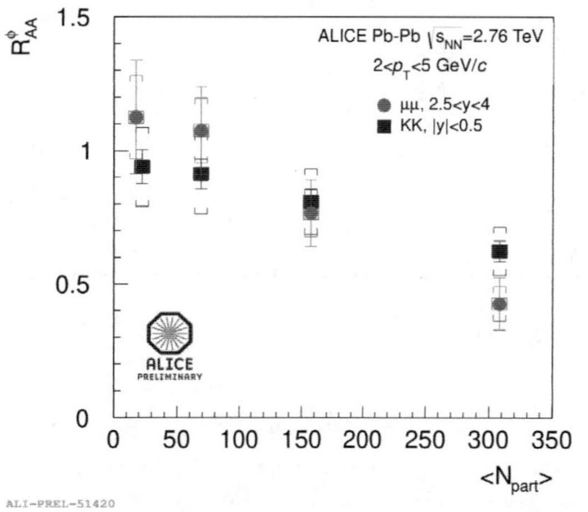

Figure 4. ϕ $R_{\rm AA}$ as a function of $N_{\rm part}$ measured in the dimuon channel for $2.5 < y < 4$, compared to the ALICE measurement in the K^+K^- channel for $|y| < 0.5$.

to the large statistics, cold nuclear matter effects will be investigated with good accuracy.

In conclusion, the ϕ production cross section was measured in pp collisions at $\sqrt{s} = 2.76$ TeV. The comparison with some commonly used PYTHIA tunes and PHOJET shows the ATLAS-CSC and D6T tunes, as well as PHOJET, reproduce the measured cross section within $\sim 15\%$, while Perugia-0 and Perugia-11 underestimate the cross section by about a factor of two. In Pb-Pb collisions, the ϕ multiplicity per participant increases from peripheral to semiperipheral collisions and tends to saturate in semicentral and central collisions. The nuclear modification factor is compatible with unity for peripheral collisions and decreases down to ~ 0.5 for most central collisions. Each point measured in the dimuon channel is compatible to the corresponding one in the hadronic channel, although the decrease of $R_{\rm AA}$ vs $N_{\rm part}$ in the dimuon channel at forward rapidity appears steeper than the one in the kaon channel at midrapidity.

References

[1] Adler S S *et al.* (PHENIX Collaboration) 2005 *Phys. Rev.* C **72** 014903
[2] Braun-Munzinger P *et al.* 2003 Invited review for *Quark-gluon Plasma* Vol. 3 (Singapore: World Scientific)
[3] Kolb P F and Heinz U W 2003 Invited review for *Quark-gluon Plasma* Vol. 3 (Singapore: World Scientific)
[4] d'Enterria D 2009 Jet Quenching *Preprint* arXiv:0902.2011 [nucl-ex]
[5] Pal S *et al.* 2002 *Nucl. Phys.* A **707** 525
[6] Alt C *et al.* (NA49 Collaboration) 2005 *Phys. Rev. Lett.* **94** 052301
[7] Alessandro B *et al.* (NA50 Collaboration) 2003 *Phys. Lett.* B **555** 147
[8] Adamova D *et al.* (CERES Collaboration) 2006 *Phys. Rev. Lett.* **96** 152301
[9] Arnaldi R *et al.* (NA60 Collaboration) 2011 *Phys. Lett.* B **699** 325
[10] Aamodt K *et al.* (ALICE Collaboration) 2008 *J. Instrum.* **3** S08002
[11] Abelev B *et al.* (ALICE Collaboration) 2012 *Phys. Lett.* B **710** 557
[12] Sjöstrand T *et al.* 2006 *J. High Energy Phys.* **05** 026
[13] Buttar C *et al.* 2004 *Acta Phys. Pol.* B **35** 433
[14] Field R 2008 *Acta Phys. Pol.* B **39** 2611
[15] Skands P Z 2010 *Phys.Rev.* D **82** 074018
[16] Engel R 1995 *Z. Phys.* C **66** 203; Engel R and Ranft J 1996 *Phys. Rev* D **54** 4244
[17] Abelev B *et al.* (ALICE Collaboration) 2013 *Phys. Rev.* C **88** 044909
[18] Toia A 2011 *J. Phys.* G **38** 124007
[19] Knospe A G *Hadronic resonances in heavy-ion collisions at ALICE* (these proceedings)

Journal of Physics: Conference Series **509** (2014) 012089 doi:10.1088/1742-6596/509/1/012089

Elliptic flow of Λ, Ξ and Ω in 2.76 A TeV Pb+Pb collisions

Huichao Song, Fanli Meng, Xianyin Xin, Yu-Xin Liu

Department of Physics and State Key Laboratory of Nuclear Physics and Technology, Peking University, Beijing 100871, China

E-mail: huichaosong@pku.edu.cn

Abstract. The elliptic flow v_2 for strange and multi-strange bayons in 2.76 A TeV Pb+Pb collisions is investigated with VISHNU hybrid model that connects 2+1-d viscous hydrodynamics with a hadron cascade model. It is found that VISHNU nicely describes $v_2(p_T)$ data for Λ, Ξ and Ω at various centralities. Comparing with the ALICE data, it roughly reproduces the mass-ordering of v_2 among π, K, p, Ξ and Ω within current statistics, but gives an inverse v_2 mass-ordering between p and Λ.

1. Introduction

The QGP viscosity is a hot research topic. Viscous hydrodynamic simulations showed that elliptic flow v_2 is very sensitive to the QGP shear viscosity to entropy density ratio η/s. Even the minimum value $1/(4\pi)$ from the AdS/CFT correspondence could lead to 20-25% suppression of v_2 [1, 2]. Generally, the QGP viscosity is extracted from the integrated v_2 for all charged hadrons, since it is directly related to the fluid momentum anisotropy and varies monotonously with η/s [3, 4]. The anisotropy flow for common hadrons (pions, kaons and protons) are developed in both QGP and hadronic stages. A precise extraction of the QGP viscosity from these soft hadron data requires a sophisticated hybrid model that nicely describes the kinetics and decoupling of the hadronic matter. In contrast, hadrons contain multiple strange quarks (Ξ, Ω and etc.) are predicted to decouple earlier from the system due to their smaller hadronic cross sections [5, 6]. They may directly probe the QGP phase and can be used to test the QGP viscosity extracted from the elliptic flow of common hadrons [7].

Recently, the ALICE collaboration measured the multiplicity, p_T spectra and elliptic flow for strange and multi-strange hadrons [8, 9], so this is the right time to perform a careful theoretical investigation of these data. In this article, we will briefly report VISHNU calculations of the elliptic flow for Λ, Ξ and Ω hyperons in 2.76 A TeV Pb+Pb collisions. Detailed investigation of these strange and multi-strange hadrons can be found in the incoming article [10]. The VISHNU predictions for the elliptic flow of ϕ meson are documented in Ref [11].

2. Set-ups

The theoretical tool used here is VISHNU hybrid model [12]. It connects 2+1-d hydrodynamics for the viscous QGP fluid expansion to a hadron cascade model for the kinetic evolution of the hadronic matter. For simplicity, we neglect the net baryon density, heat flow and bulk viscosity. The equation of state (EOS) s95p-PCE used for the hydrodynamic evolution is constructed from

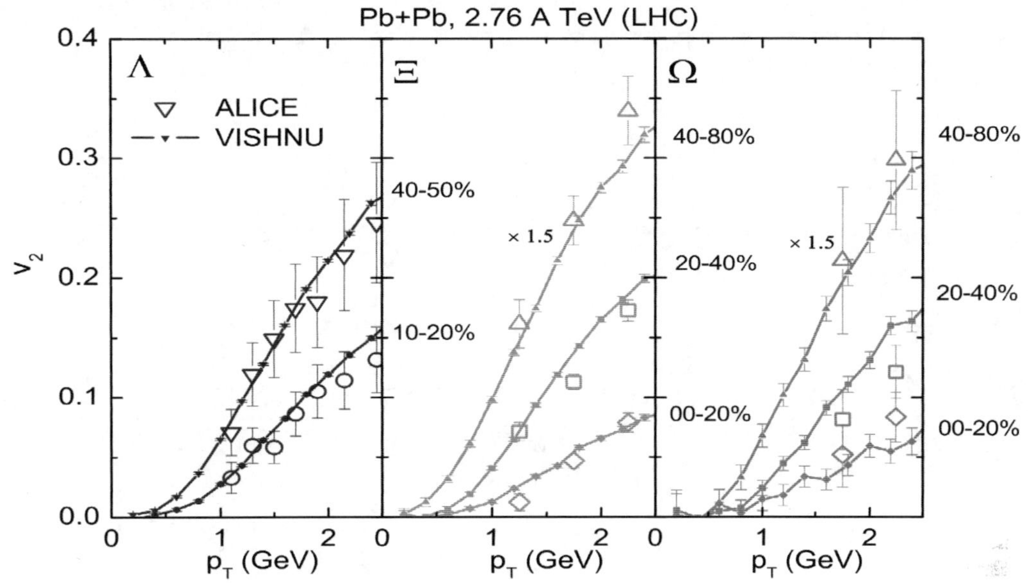

Figure 1. (Color online) Differential elliptic flow for Λ, Ξ and Ω in 2.76 A TeV Pb+Pb collisions. Experimental data are from ALICE [9], theoretical lines are from `VISHNU` simulations with $(\eta/s)_{QGP} = 0.16$ and `MC-KLN` initial conditions.

recent lattice data [13]. The hyper-surface switching between hydrodynamics and the hadron cascade is constructed by a switching temperature at 165 `MeV`, which is close to the QCD phase transition and chemical freeze-out temperature at top RHIC energy. Following Ref [14], we use MC-KLN initial conditions. The initial entropy density profiles for selected centralities are generated from MC-KLN model through averaging over a large number of fluctuating entropy density distributions with recentering and aligning the reaction plane. The QGP specific shear viscosity $(\eta/s)_{QGP}$ is set to 0.16, which gives a nice description of $v_2(p_T)$ for pion, kaon and protons at the LHC [11]. For $(\eta/s)_{QGP} = 0.16$, the starting time τ_0 is 0.9 fm/c, which is obtained from fitting the slope of the p_T spectra for all charged hadrons below 2 GeV [14]. With these inputs and parameters fixed, we predict the elliptic flow for strange and multi-strange bayons in 2.76 A TeV Pb+Pb collisions.

3. Results

`VISHNU` hybrid model is a successful tool to describe and predict the soft hadron productions at RHIC and the LHC. Using `VISHNU` [12], we extracted the QGP specific shear viscosity from the integrated elliptic flow for all charged hadrons in 200 A GeV Au+Au collisions and found $1/(4\pi) < (\eta/s)_{QGP} < 2.5/(4\pi)$ $(0.08 < (\eta/s)_{QGP} < 0.20)$, where the uncertainties are dominated by the undetermined initial conditions from `MC-Glauber` and `MC-KLN` models [3]. With that extracted QGP viscosity, `VISHNU` gives a nice description of the p_T spectra and elliptic flow for all charged and identified hadrons at RHIC [4].

Ref. [14] extrapolates the `VISHNU` calculations to the LHC and shows that `VISHNU` could nicely describe the p_T spectra and elliptic flow for all charged hadrons with MC-KLN initial conditions and $(\eta/s)_{QGP} = 0.20 - 0.24$, a slightly higher value than the one extracted at RHIC with MC-KLN initial conditions. With the identified soft hadron data becoming available, we

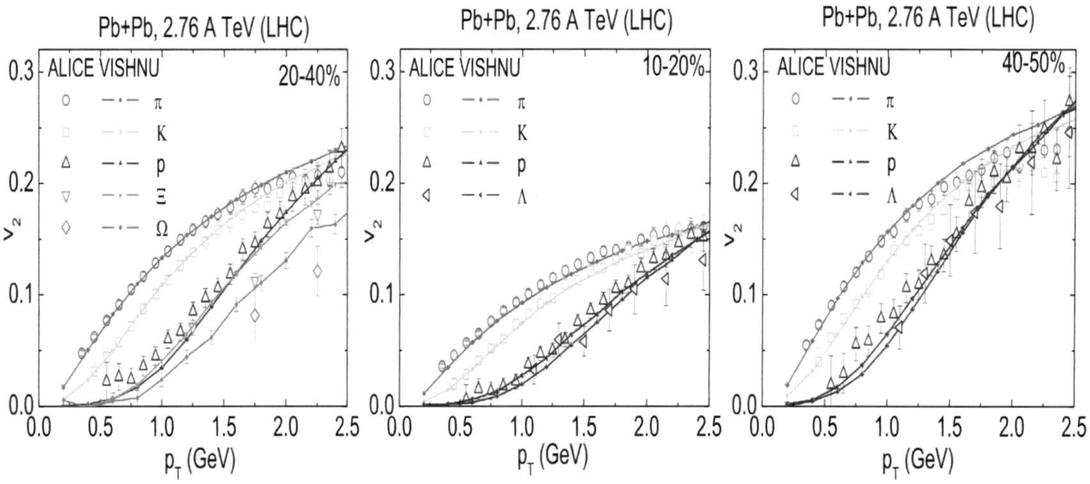

Figure 2. (Color online) Mass ordering of elliptic flow for various hadrons in 2.76 A TeV Pb+Pb collisions. The theoretical lines for pions, kaons and protons are from earlier VISHNU calculations in Ref. [11].

further calculated the multiplicity, p_T spectra and differential elliptic flow for pions kaons and protons and found these $v_2(p_T)$ prefer $(\eta/s)_{QGP} = 0.16$ for MC-KLN [11, 15]. It is noticed that the integrated and differential elliptic flow for all charged hadrons are measured by 4 particle cumulants method ($v_2\{4\}$), while the elliptic flow for identified hadrons are measured by scalar product method ($v_2\{sp\}$) based on 2 particle correlations. Comparing with $v_2\{4\}$, non-flow and fluctuations raise the flow signal of $v_2\{sp\}$, leading to a slightly lower value of $(\eta/s)_{QGP}$ to fit $v_2\{sp\}$ data for identified hadrons.

Recently, the ALICE collaboration measured the elliptic flow for Λ, Ξ and Ω in 2.76 A TeV Pb+Pb collisions using scalar product method [9]. Using VISHNU hybrid model, we calculate the differential elliptic flow $v_2(p_T)$ for these strange and multi-strange hadrons. Here we choose $(\eta/s)_{QGP} = 0.16$, MC-KLN initial conditions and other parameter sets as used in early calculations [11] that nicely describe the soft hadron data for pions kaons and protons at the LHC. Fig. 1 shows that VISHNU nicely describes $v_2(p_T)$ data for Ξ and Ω from central to peripheral collisions and nicely describes Λ data at 10% and 40% centralities as measured by ALICE. Due to the very low particle yields, the elliptic flow for Ω has large error bars in ALICE data and the VISHNU results. For better comparisons, the statistics for Ω should be further increased in both experimental and theoretical sides.

Fig. 2 further explores the mass-ordering of elliptic flow among various hadrons in 2.76 A TeV Pb+Pb collisions. Radial flow tends to push low p_T hadrons to higher transverse momenta. Such effects are more efficient for heavier particles, leading to the mass-ordering of v_2 as predicted in hydrodynamics [16]. Comparing with $v_2(p_T)$ for all charge hadrons, the mass-splitting of v_2 among various hadrons reflects the interplay of the radial and elliptic flow during hadronic evolution, which is more sensitive to details of theoretical calculations. Fig. 2 shows that, within current statistics, VISHNU roughly reproduces the v_2 mass ordering among π, K, p, Ξ and Ω in semi-central collisions, but gives an inverse mass-ordering between p and Λ for the two selected (10-20% and 40-50%) centralities. Early OSU calculations [17], using pure viscous hydrodynamics VISH2+1, nicely predicts the mass ordering among π, K, p, Ξ, Ω and Λ as measured in experiments later on, but over-predicts the elliptic flow for proton and Λ at 10-20% centrality bin. With the microscopic description of hadronic rescatterings and evolution, VISHNU hybrid model improves the descriptions of the elliptic flow for proton and Λ, while the slightly under-predicts of the proton $v_2(p_T)$ below 2 GeV lead to an inverse mass ordering of v_2 between p and Λ. To improve the description of $v_2(p_T)$ for proton as well as for Ξ and Ω, the related hadronic cross sections in UrQMD need to be improved. On the other hand, to further understand these rare multi-strange hadrons and to test theoretical models, the statistics for Ξ and Ω need to be increased in both experimental and theoretical sides.

4. Summary

In this proceeding, we present VISHNU hybrid model calculations of the elliptic flow for Λ, Ξ and Ω in 2.76 A TeV Pb+Pb collisions. VISHNU nicely describes $v_2(p_T)$ for these strange and multi-strange hadrons at various centralities measured by ALICE. It roughly reproduce the mass-ordering of v_2 among π, K, p, Ξ and Ω within current statistics, but gives an inverse mass-ordering between p and Λ. For a better comparison and understanding of the elliptic flow data for the rare multi-strange hadrons like Ξ and Ω, high statistic runs are needed in both experimental and theoretical sides.

Acknowledgments: This work was supported by the new faculty startup funding from Peking University. We gratefully acknowledge extensive computing resources provided by Tianhe-1A at National Supercomputing Center in Tianjin, China.

References

[1] Song H and Heinz U 2008 Phys. Lett. **B658** 279; 2008 Phys. Rev. C **77** 064901; 2008 Phys. Rev. C **78** 024902; Song H 2009 Ph.D Thesis, The Ohio State University (August 2009), arXiv:0908.3656 [nucl-th].
[2] Romatschke P and Romatschke U 2007 Phys. Rev. Lett. **99** 172301; Luzum M and Romatschke P 2008 Phys. Rev. C **78** 034915; Dusling K and Teaney D 2008 Phys. Rev. C **77** 034905; Molnar D and Huovinen P 2008 J. Phys. G **35** 104125.
[3] Song H, Bass S A, Heinz U, Hirano T and Shen C 2011 Phys. Rev. Lett. **106** 192301 [2012 Erratum-ibid. **109** 139904].
[4] Song H, Bass S A, Heinz U, Hirano T and Shen C 2011 Phys. Rev. C **83** 054910 [2012 Erratum-ibid. C **86** 059903].
[5] Adams J *et al.* [STAR Collaboration] 2004 Phys. Rev. Lett. **92** 182301.
[6] Estienne M [STAR Collaboration] 2005 J. Phys. G G **31** S873.
[7] Song H 2013 Nucl. Phys. A **904-905** 114c, Song H 2014 arXiv:1401.0079 [nucl-th].
[8] Abelev B B *et al.* [ALICE Collaboration] 2013 arXiv:1307.5543 [nucl-ex]; Abelev B B *et al.* [ALICE Collaboration] 2013 arXiv:1307.5530 [nucl-ex].
[9] Zhou Y [for the ALICE Collaboration] 2013 arXiv:1309.3237 [nucl-ex]; Noferini F [ALICE Collaboration] 2013 Nucl. Phys. A904-905 483c.
[10] Meng F, Song H, *et al.*, in preparation.
[11] Song H, Bass S A and Heinz U 2013 arXiv:1311.0157 [nucl-th].
[12] Song H, Bass S A and Heinz U 2011 Phys. Rev. C **83** 024912; Song H 2012 Eur. Phys. J. A **48** 163.
[13] Huovinen P and Petreczky P 2010 Nucl. Phys. **A837** 26.
[14] Song H, Bass S A and Heinz U 2011 Phys. Rev. C **83** 054912 [2013 Erratum-ibid. C **87** 019902].
[15] Heinz U, Shen C and Song H 2012 AIP Conf. Proc. **1441** 766.
[16] Huovinen P, Kolb P F, Heinz U W, Ruuskanen P V and Voloshin S A 2001 Phys. Lett. B **503** 58.
[17] Shen C, Heinz U, Huovinen P and Song H 2011 Phys. Rev. C **84** 044903.

Journal of Physics: Conference Series **509** (2014) 012090 doi:10.1088/1742-6596/509/1/012090

Multi-strange baryon production in Pb–Pb and pp collisions at $\sqrt{s_{\mathrm{NN}}} = 2.76$ TeV with the ALICE experiment at the LHC

Domenico Colella (for the ALICE Collaboration)
Dipartimento Interateneo di Fisica "M. Merlin" and Sezione INFN, Via Orabona 4, 70126 Bari, Italy
E-mail: domenico.colella@cern.ch

Abstract. The production of Ξ^- and Ω^- baryons and their anti-particles in Pb–Pb and pp collisions at $\sqrt{s_{\mathrm{NN}}} = 2.76$ TeV has been measured by the ALICE collaboration. The transverse momentum spectra at mid-rapidity ($|y| < 0.5$) in pp and in Pb–Pb collisions for five centrality intervals have been compared with model predictions. Hyperon yields and spectra in Pb–Pb collisions, normalized to the corresponding measurements in pp at the same centre-of-mass energy, allow the study of the strangeness enhancement and the nuclear modification factor as a function of the transverse momentum (p_{T}) and collision centrality.

1. Introduction
The study of strange and multi-strange particle production in relativistic heavy-ion (A–A) interactions is an important tool to investigate the properties of the strongly interacting system created in the collision, as there is no net strangeness content in the colliding nuclei. In particular, baryons with more than one unit of strangeness are very useful probes of the early partonic stages of the collision due to their small hadronic cross-section.

The enhancement of strangeness in heavy-ion collisions was one of the earliest proposed signals for the Quark-Gluon Plasma [1]. It rests on the expectation that in a deconfined state the abundances of parton species should quickly reach their equilibrium values, resulting in a higher abundance of strangeness per participant than that seen in pp interactions. This enhancement, first observed at SPS [2] and confirmed afterwards by data at RHIC [3], is more pronounced for particles with larger strangeness content and decreases as the centre-of-mass energy of the collision increases. Over the past 15 years, it has been found that the canonical suppression effect, for pp collisions, is important [4] and contributes to the overall hyperon enhancement in A–A. This interpretation can now be re-examined at the much higher LHC energy.

A suppression of high-momentum unidentified particle production in Pb–Pb compared to pp collisions has been observed by the ALICE collaboration and used as evidence for strong parton energy loss and large medium density at the LHC [5]. Measurements of the nuclear modification factors for identified particles could provide insight into particle production and energy loss mechanisms.

The ALICE experiment was specifically designed to study heavy-ion collisions at the LHC, namely the properties of strongly interacting matter at high energy density. The first LHC heavy-ion run took place at the end of 2010 with Pb–Pb ions accelerated at a centre-of-mass energy $\sqrt{s_{NN}} = 2.76$ TeV. Almost 30×10^6 minimum bias nuclear interaction triggers were recorded. The pp reference data were collected during March 2011. Almost 80×10^6 minimum bias nucleon interaction triggers were recorded. The tracking and vertexing are performed with the full tracking system: the Inner Tracking System (ITS, six layers of silicon detectors) and the Time Projection Chamber (TPC), which is also used for particle identification via specific ionization. The Silicon Pixel Detector (SPD, the two innermost layers of the ITS) and the VZERO detector (scintillation hodoscopes placed on either side of the interaction point) were used for triggering. The VZERO was also crucial for the collision centrality determination. A complete description of the ALICE sub-detectors can be found in [6].

2. Analysis method

Multi-strange baryons are reconstructed through their weak decay topologies, namely $\Xi^- \to \pi^- + \Lambda$, $\Omega^- \to \mathrm{K}^- + \Lambda$ (with $\Lambda \to \pi^- + \mathrm{p}$), and the corresponding charge conjugate decays for the anti-particles. The resulting branching ratios are 63.9% and 43.3%, for the Ξ and the Ω respectively. The Ξ and the Ω candidates are found by combining reconstructed charged tracks: cuts on geometry and kinematics are applied to select first the Λ candidate and then to match it with all the remaining secondary tracks (bachelor candidates). In addition, cuts on particle identification via specific energy loss in the TPC for the three daughter tracks are used.

To extract the signal in p_{T} intervals, a symmetric region around the invariant mass peak ($\pm 3\sigma$) is defined by fitting the distribution with the sum of a Gaussian and a polynomial. The background is sampled in two regions on both sides of the peak and fitted with a polynomial of first or second degree (depending on the colliding system and the p_{T} interval). In each p_{T} interval the signal in the peak region is obtained by subtracting the integral of the background fit function from the peak population. More details can be found in [7].

3. Results and conclusions

Transverse momentum spectra, corrected for acceptance and efficiency, in pp and Pb–Pb collisions are shown in Figure 1 and 2 respectively. In pp collisions, the comparison with predictions obtained using PYTHIA 6 (Perugia-2011 tune [8][9]) shows that this model underestimates the yields, both for Ξ and Ω, and does not reproduce the shape of the spectra. Spectra in Pb–Pb collisions, reconstructed in five centrality classes (0-10%, 10-20%, 20-40%, 40-60% and 60-80%), are compared with predictions obtained using four hydrodynamic models: VISH2+1 [10] , HKM [11] , Kraków [12] and EPOS [13] . Kraków provides a good description for both yields and shapes in the p_{T} range up to 3 GeV/c, and EPOS gives the most successful description of spectral shape in a wider p_{T} range. The overall description for the Ω is less successful while for both particles types it degrades going from central to more peripheral collisions.

The strangeness enhancements, calculated as the ratio between the yields in Pb–Pb collisions and those in pp interactions at the same energy, both normalized to the number of participants (N_{part}), are shown in Figure 3(a-b). The pp reference values were obtained by interpolating ALICE and STAR data at different energies [7] and checked with the preliminary measurements of the yields in pp collisions at the reference energy. The enhancements as a function of the mean number of participants increase with centrality and with the strangeness content of the particle as already observed at lower energies, and decrease as the centre-of-mass energy increases, continuing the trend established at lower energies (between SPS and RHIC energies).

The hyperon-to-pion ratios $\Xi/\pi \equiv (\Xi^- + \overline{\Xi}^+)/(\pi^- + \pi^+)$ and $\Omega/\pi \equiv (\Omega^- + \overline{\Omega}^+)/(\pi^- + \pi^+)$, for A–A and pp collisions, both at LHC and RHIC energies, are shown in Figure 3(c) as a function of $\langle N_{\mathrm{part}} \rangle$. The relative production of strangeness in pp collisions at the LHC is larger than at RHIC energy while it is almost the same at the two energies for A–A collisions, giving an

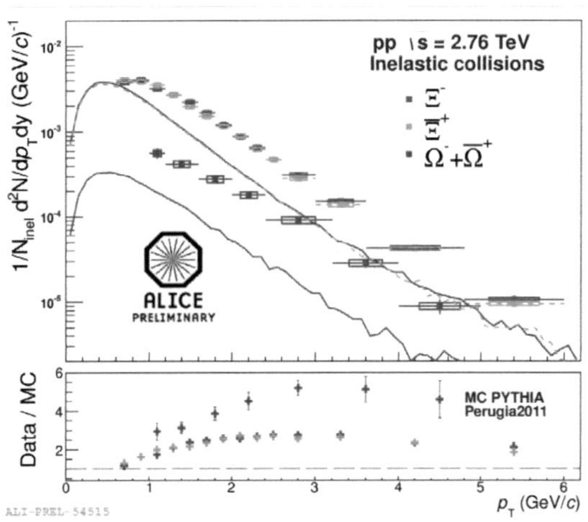

Figure 1. Transverse momentum spectra normalized to the number of inelastic collisions for Ξ^-, $\overline{\Xi}^+$ and Ω in pp collisions, compared with predictions obtained with PYTHIA 6 (Perugia-2011 tune [8][9]). Ratios of data to models are also shown. The error bars represent the statistical uncertainties and the boxes the systematic ones.

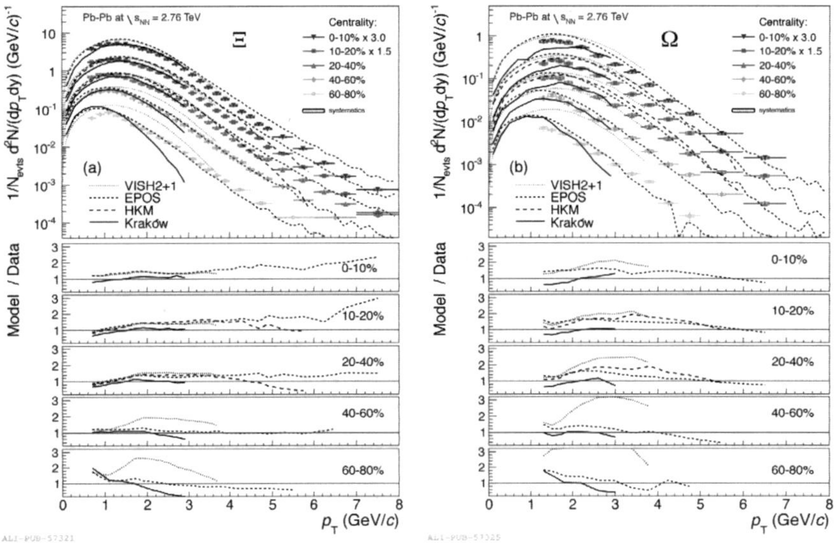

Figure 2. Transverse momentum spectra for Ξ (a) and Ω (b) hyperons (average of particle and anti-particle) in five different centrality classes, compared to hydrodynamic models. Ratios of models to data are also shown.

explanation of the decrease of enhancement with increasing energy. The measurements of these ratios in Pb–Pb collisions are in agreement with predictions from the thermal model, based on a grand canonical approach [14][15]. The hyperon-to-pion ratio increases when going from pp to A–A, showing a relative enhancement of strangeness production in A–A collisions (normalized to the pion yield) which is about half of that seen when normalizing to the mean number of participants. The enhancement rises with centrality up to about $\langle N_{\rm part}\rangle \sim 150$ and apparently saturates thereafter.

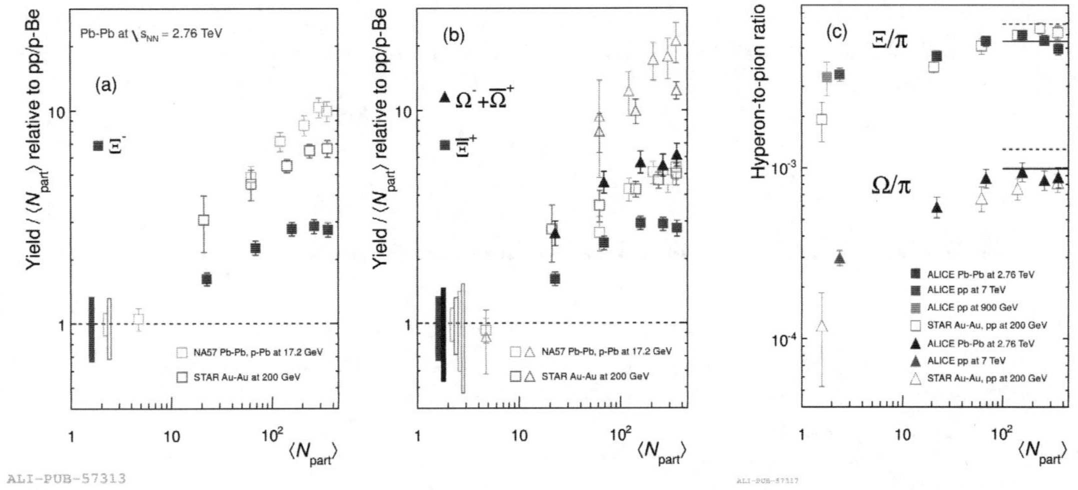

Figure 3. (a,b) Enhancements in the rapidity range $|y| < 0.5$ as a function of the mean number of participants $\langle N_{\text{part}} \rangle$. Boxes on the dashed line at unity indicate statistical and systematic uncertainties on the pp or p–Be reference. (c) Hyperon-to-pion ratios as a function of $\langle N_{\text{part}} \rangle$, for A–A and pp collisions at LHC and RHIC energies. The lines mark the thermal model predictions from [14] (full line) and [15] (dashed line).

The nuclear modification factor (R_{AA}) is defined as the ratio between the differential yield in A–A collisions ($\text{d}^2N^{\text{AA}}/\text{d}p_{\text{T}}dy$) and the differential cross-section of particle production in pp collisions ($\text{d}^2\sigma^{\text{NN}}/\text{d}p_{\text{T}}dy$), normalized by the geometric nuclear overlap function T_{AA} [16]. It has been measured by ALICE in different centrality intervals and both for multi-strange baryons and lighter particles [17] [18]. The R_{AA} for Ξ follows the same trend as the proton at high p_{T} (> 5 GeV/c), where the suppression does not depend on the particle mass. At intermediate p_{T} there are indications of mass-ordering among the baryons. The R_{AA} for the Ω seems to be strongly affected by the strangeness enhancement and the effect becomes weaker when going from the most central to the most peripheral class.

References

[1] Rafelski J and Müller B 1982 *Physics Review Letters* **48** 1066
[2] Antinori F *et al.* (NA57 Collaboration) 2010 *Journal of Physics* G **37** 045105
[3] Abelev B I *et al.* (STAR Collaboration) 2008 *Physical Review* C **77** 044908
[4] Hamieh J S, Redlich K and Tounsi A 2000 *Physics Letters* B **486** 61
[5] Aamodt K *et al.* (ALICE Collaboration) 2011 *Physics Letters* B **696** 30
[6] Aamodt K *et al.* (ALICE Collaboration) 2008 *JINST* **3** S08002
[7] Aamodt K *et al.* (ALICE Collaboration) 2013 *arXiv:1307.5543* [nucl-ex]
[8] Skands P Z 2010 *Physical Review* D **82** 074018
[9] Skands P Z 2011 *arXiv:1005.3457* [hep-ph]
[10] Shen C, Heinz U, Huovinen P and Song H 2011 *Physical Review* C **84** 044903
[11] Karpenko Y, Senyukov Y and Werner K 2012 *arXiv:1204.5351* [nucl-th]
[12] Bozek P 2012 *Acta Physica Polonica* B **43** 4
[13] Werner K *et al.* 2012 *Physical Review* C **85** 064907
[14] Andronic A, Braun-Munzinger P and Stachel J 2009 *Physics Letters* B **673** 142; Erratum 2009 *Physics Letters* B **678** 561
[15] Cleymans J, Kraus I, Oeschler H, Redlich K and Wheaton S 2006 *Physical Review* C **74** 034903
[16] Aamodt K *et al.* (ALICE Collaboration) 2013 *arXiv:1301.4361* [nucl-ex]
[17] Barnby L for the ALICE Collaboration, these Proceedings.
[18] Knospe A G for the ALICE Collaboration, these Proceedings.

Journal of Physics: Conference Series **509** (2014) 012091 doi:10.1088/1742-6596/509/1/012091

K^0_S & Λ Production in ALICE

Luke Hanratty for the ALICE Collaboration
School of Physics and Astronomy, The University of Birmingham, Edgbaston, Birmingham, B15 2TT, United Kingdom
E-mail: hanratty@cern.ch

Abstract. The production of Λ and K^0_S hadrons at the LHC can be measured through the reconstruction of their weak decay topologies via decay channels with only charged particles in the final state. The tracking and particle identification capabilities of the ALICE detector allow us to measure the spectra of these particles over a wide transverse momentum range ($0.4 < p_T < 12$ GeV/c), and to precisely determine the behaviour of the baryon-to-meson ratio Λ/K^0_S.

Transverse momentum spectra and production yields at mid-rapidity ($|y| < 0.5$) for Λ and K^0_S are presented for $\sqrt{s_{NN}} = 2.76$ TeV Pb–Pb collisions as a function of centrality. The evolution of the Λ/K^0_S ratio is discussed, in comparison with corresponding results in pp collisions at the LHC and in $\sqrt{s_{NN}} = 200$ GeV Au–Au collisions at RHIC. Comparisons to theoretical models will also be outlined.

1. Introduction

Relativistic collisions of heavy ions can be used to study the behaviour of strongly-interacting matter as it undergoes a phase transition from a quark-gluon plasma to a hadronic phase. By studying the transverse momentum (p_T) of hadrons emitted from the collisions, information about the production mechanisms of these hadrons can be inferred.

It was observed at the Super Proton Synchotron (SPS) [1] and at the Relativistic Heavy Ion Collider (RHIC) [2, 3] that ratios of p_T spectra for baryons and mesons, such as p/π and Λ/K^0_S are enhanced at intermediate p_T, when compared to the corresponding ratios in pp collisions. At low p_T, the hydrodynamical flow of the medium could contribute to this effect, as the heavier baryons are pushed to higher momentum than the lighter mesons. However, such models are typically only applicable to $p_T \sim 2$ GeV/c, while the enhancement is still evident at ~ 4 GeV/c. Another potentially contributing effect is coalescence, in which multiple soft quarks from the medium combine to form hadrons. The momentum of a baryon will then be given by the sum of the momenta of three quarks, and thus be typically higher than mesons formed from two quarks. This mechanism could extend up to higher p_T values (10 - 20 GeV/c) [4] at LHC energies, if the quarks from jet fragmentation also undergo recombination [5]. Above ~ 6 GeV/c the ratio is compatible with that measured in pp collisions, and so it is expected that vacuum-like fragmentation will dominate hadron production. A precise measurement of these ratios over a broad transverse momentum range for a wide selection of energies and centrality intervals is thus important for examining the interplay between these different mechanisms.

In these proceedings, the measurement of the Λ/K^0_S ratio by ALICE will be discussed. These singly-strange particles can be identified with a single technique over a wide momentum range, which allows for a precise measurement of the ratio.

2. Analysis Technique

ALICE is designed for excellent particle identification in high-multiplicity collisions. For this analysis, a two-station scintillator detector at forward rapidity (VZERO) was used for event and centrality selection, the Inner Tracking System (ITS) for vertex identification and particle tracking, and the Time Projection Chamber (TPC) for further particle tracking and particle identification (PID) through energy loss measurements. About 1.5×10^7 events from the 2010 Pb–Pb run were analysed in 7 different centrality bins spanning from 0-90% of the most central nuclear collisions.

Λ and K^0_S were identified by the topology of their decays into $p\pi^-$ and $\pi^+\pi^-$ respectively, with branching ratios of 63.9% and 69.2% [6]. Pairs of oppositely charged tracks which could emerge from a common secondary vertex were considered 'V0 candidates', as described in [7]. More stringent cuts were applied to reduce the large level of combinatorial background present in Pb–Pb collisions [8]. Cuts on the distance of closest approach between the two daughter tracks and also to the primary vertex were tightened, as were cuts on the decay radius and the cosine of the pointing angle (the angle between the V0s position and momentum vectors.) In addition, the rate of energy loss in the TPC for positive daughters of Λ candidates with p_T below 1.2 GeV/c was required to be within 3σ of that expected for a proton. A cut was further applied on the Armenteros-Podolanski diagram [9] in order to exclude Λ and $\overline{\Lambda}$ from being reconstructed as background to the K^0_S.

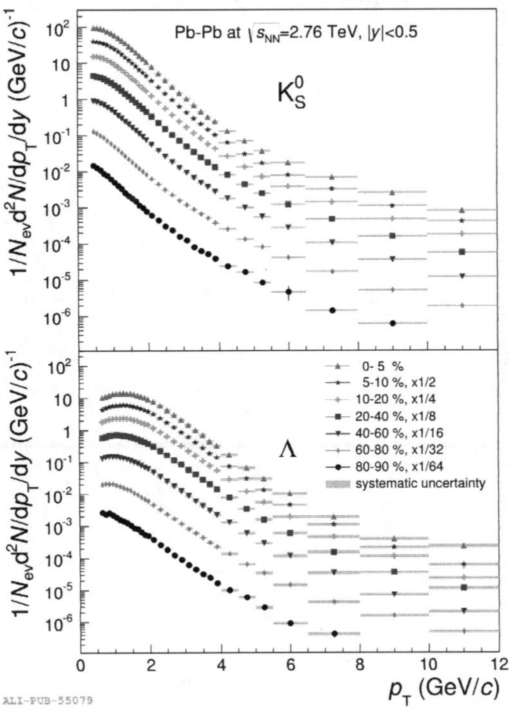

Figure 1. Λ and K^0_S spectra in log scale.

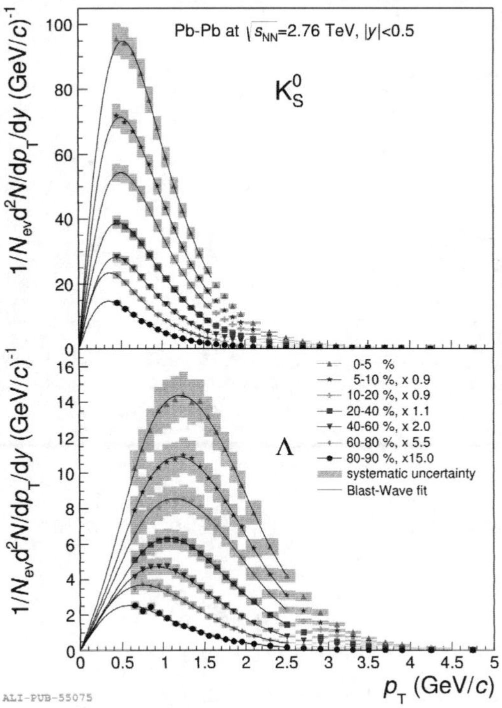

Figure 2. Λ and K^0_S spectra in linear scale.

An invariant mass distribution was created for Λ and K^0_S by assuming that the decay daughters for each V0 candidate were $p\pi^-$ and $\pi^+\pi^-$ respectively. The yield of each particle could then be extracted by subtracting the background from the revealed peak. The background was estimated by first using a Gaussian plus 2nd degree polynomial fit to identify the peak region, and then fitting the area to either side of the peak and extrapolating into the 'peak region' to estimate

Journal of Physics: Conference Series **509** (2014) 012091 doi:10.1088/1742-6596/509/1/012091

the background under the peak. The signal is then defined as the difference between the total number of counts in the 'peak region' and the background fit in the same region.

A Monte-Carlo study was used to correct these signals for efficiency and acceptance. Further, the yield of Λ was corrected to remove Λ's coming from weak decays of $\Xi^-$and Ξ^0, using the Ξ^-yields measured by ALICE [10]. The final spectra are shown in Figures 1 and 2.

3. Results

The $\Lambda/\mathrm{K}^0_\mathrm{S}$ ratio as measured by ALICE is shown in Figure 3 for different centralities [8].The figure also includes the ratio measured in pp collisions at energies which bracket the $\sqrt{s_\mathrm{NN}}$ = 2.76 TeV energy of the Pb–Pb collisions. It can be seen that the ratio in pp collisions does not appear to depend upon the collision energy, and also that the ratio in the most peripheral Pb–Pb collisions (80-90% centrality) is consistent with that measured in pp collisions. All centralities give a comparable result for p_T ~1.5 GeV/c, and for p_T > 7 GeV/c, while at the peak of the enhancement, the most central (0-5%) ratio is approximately 3 times that measured in pp collisions. This factor of 3, as well as the p_T range of the enhancement, is consistent with what is observed for the p/π ratio as shown in Figure 4 [11].

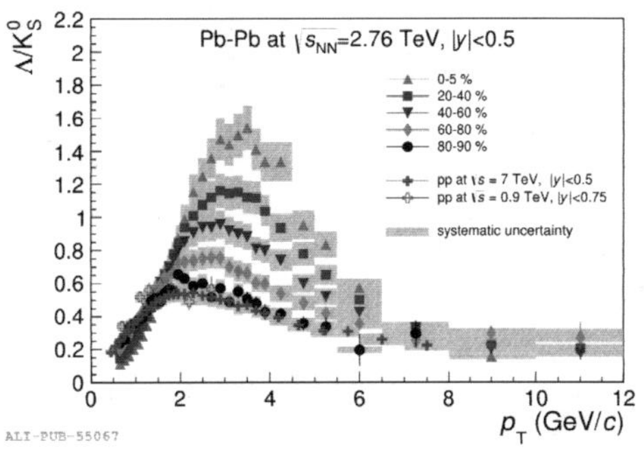

Figure 3. $\Lambda/\mathrm{K}^0_\mathrm{S}$ ratio for different centrality selections in Pb–Pb, and for pp.

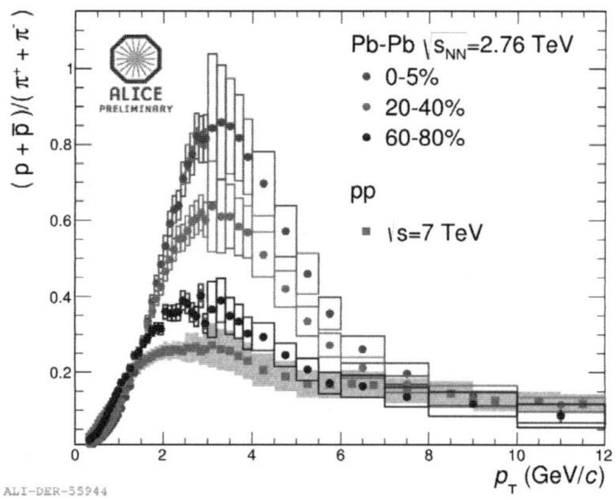

Figure 4. p/π ratio for different centrality selections in Pb–Pb, and for pp.

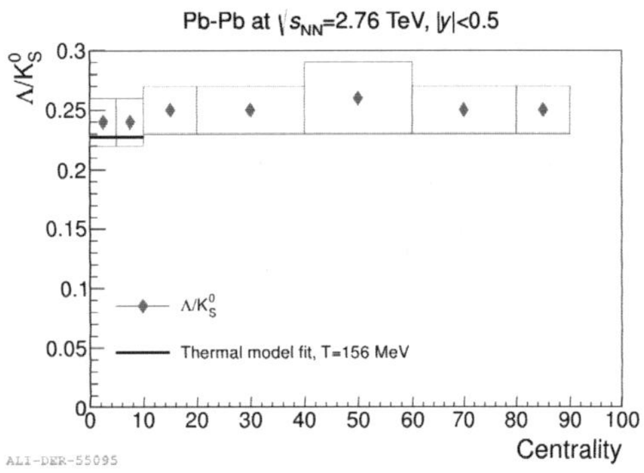

Figure 5. p_T-integrated $\Lambda/\mathrm{K}^0_\mathrm{S}$ ratio.

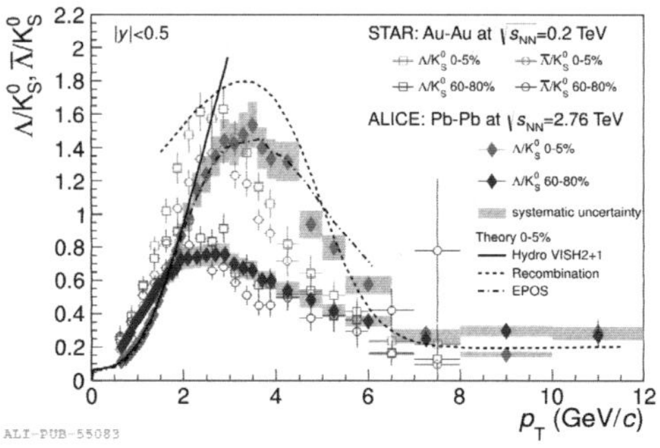

Figure 6. Comparison of $\Lambda/\mathrm{K}^0_\mathrm{S}$ ratio for central and peripheral collisions in ALICE, STAR, and theoretical predictions.

Studies of the p_T-integrated yields [8] of Λ and K^0_S suggest that a thermal model (with a temperature of 156 MeV) predicts accurately the total yields, and further that the ratio of integrated yields is constant within errors across all centrality intervals, as shown in Figure 5. This would suggest that the enhancement of the ratio of p_T spectra is due to a redistribution of particles in p_T, rather than the introduction of additional production channels. This picture is supported by the low p_T region of the Λ/K^0_S ratio in Figure 3, where there is a change in the ratio with centrality consistent with a depletion of Λ with respect to K^0_S, which is strongest for the most central collisions.

Figure 6 shows the measured Λ/K^0_S for central (0-5%) and peripheral (60-80%) centrality classes, for $\sqrt{s_{NN}} = 2.76$ TeV Pb–Pb collisions with ALICE and $\sqrt{s_{NN}} = 200$ GeV Au–Au collisions with STAR [12]. While the antibaryon/baryon ratio approaches unity at the LHC, it is $\sim$0.8 at RHIC, so both Λ/K^0_S and $\overline{\Lambda}/K^0_S$ ratios have been plotted separately for STAR. From this, it can be seen that the magnitude of the enhancement is very similar at both RHIC and LHC energies. The enhancement extends to slightly higher p_T at ALICE, but is still restricted to fairly modest p_T.

Also shown in Figure 6 are predictions by various theoretical models for the 0-5% Λ/K^0_S ratio. The viscous hydrodynamical model [13, 14, 15] describes the behaviour of the ratio accurately below 2 GeV/c, but then strongly deviates from the data as the ratio begins to turn over. A recombination model is also shown [16], which describes the general behaviour of the ratio well, but overestimates the magnitude of the data by approximately 15%. Finally, a result from the EPOS2.17v3 model is shown [17]. This model includes the interaction between jets and the hydrodynamical medium, and appears to describe well the behaviour of the Λ/K^0_S ratio over the full p_T range. It would be interesting to see how these and other models describe the variation of the ratio with energy and centrality.

4. Conclusions

The Λ and K^0_S spectra and their ratios have been measured in $\sqrt{s_{NN}} = 2.76$ TeV Pb–Pb collisions at ALICE over a wide range of centrality and p_T and compared to those measured in pp collisions at $\sqrt{s} = 900$ GeV and 7 TeV. At high p_T, there is no discernible difference between the Λ/K^0_S ratio in pp and Pb–Pb collisions, suggesting that the relative production rates are dominated by vacuum-like fragmentation. In the intermediate p_T region however, Λ production is strongly enhanced in Pb–Pb relative to K^0_S for central collisions, while the p_T-integrated ratio appears to remain constant with respect to centrality. This would indicate that particles are redistributed in p_T in Pb–Pb as compared to pp collisions. The enhancement is similar at STAR and ALICE, suggesting that any collision energy dependence is weak.

References

[1] Schuster T *et al.* (NA49 Collaboration) 2006 *J. Phys. G* **32** S479–S482 (*Preprint* nucl-ex/0606005)
[2] Adler S S *et al.* (PHENIX Collaboration) 2003 *Phys. Rev. Lett.* **91** 172301 (*Preprint* nucl-ex/0305036)
[3] Adams J *et al.* (STAR Collaboration) 2006 (*Preprint* nucl-ex/0601042)
[4] Hwa R C 2008 *J. Phys. G.* **35** 104017 (*Preprint* nucl-th/0804.3763)
[5] Hwa R C and Yang C 2004 *Phys. Rev. C* **70** 024905 (*Preprint* nucl-th/0401001)
[6] Beringer J *et al.* (Particle Data Group) 2012 *Phys. Rev. D* **86**(1) 010001
[7] Aamodt K *et al.* (ALICE Collaboration) 2011 *Eur. Phys. J. C* **71** 1594 (*Preprint* hep-ex/1012.3257)
[8] Abelev B *et al.* (ALICE Collaboration) 2013 *Phys. Rev. Lett.* **111**(22) 222301 (*Preprint* nucl-ex/1307.5530)
[9] Podolanski J and Armenteros R 1954 *Phil. Mag.* **45** 13
[10] Abelev B *et al.* (ALICE Collaboration) 2013 (*Preprint* arXiv:nucl-ex/1307.5543)
[11] Velasquez A O 2013 *Nuclear Physics A* **904905** 763c – 766c ISSN 0375-9474
[12] Agakishiev G *et al.* (STAR Collaboration) 2012 *Phys. Rev. Lett.* **108** 072301 (*Preprint* nucl-ex/1107.2955)
[13] Song H and Heinz U W 2008 *Phys. Lett. B* **658** 279–283 (*Preprint* nucl-th/0709.0742)
[14] Song H and Heinz U W 2008 *Phys. Rev. C* **77** 064901 (*Preprint* nucl-th/0712.3715)
[15] Song H and Heinz U W 2008 *Phys. Rev. C* **78** 024902 (*Preprint* nucl-th/0805.1756)
[16] Fries R J, Greco V and Sorensen P 2008 *Ann. Rev. Nucl. Part. Sci.* **58** 177–205 (*Preprint* nucl-th/0807.4939)
[17] Werner K 2012 *Phys. Rev. Lett.* **109**(10) 102301 (*Preprint* nucl-th/1204.1394)

14th International Conference on Strangeness in Quark Matter (SQM2013) IOP Publishing
Journal of Physics: Conference Series **509** (2014) 012092 doi:10.1088/1742-6596/509/1/012092

Strange Baryon to Meson Ratio

Eleazar Cuautle and Alejandro Ayala
Instituto de Ciencias Nucleares, Universidad Nacional Autónoma de México, Apartado Postal 70-543, México Distrito Federal 04510, México

E-mail: ecuautle@nucleares.unam.mx

Abstract. We present a model to compute baryon and meson transverse momentum distributions, and their ratios, in relativistic heavy-ion collisions. The model allows to compute the probability to form colorless bound states of either two or three quarks as functions of the evolving density during the collision. The qualitative differences of the baryon to meson ratio for different collision energies and for different particle species can be associated to the different density dependent probabilities and to the combinatorial factors which in turn depend on whether the quarks forming the bound states are heavy or light. We compare to experimental data and show that we obtain a good description up to intermediate values of *pt*.

1. Introduction
Long time ago it has been proposed [1] that strangeness is easier produced in the QGP than in a hadronic environment. At the Large Hadron Collider (LHC), new results on strangeness production [2] are being obtained, among them are the transverse momentum distributions for pions, kaons, protons and Λ as well as baryon to meson ratios. The distributions show disagreement with perturbative QCD results, especially on the strangeness sector. Consequently, the baryon to meson ratio presents an unexplained behavior at intermediate transverse momenta. The baryon to meson ratios measured at LHC energies could be different from the ratios measured at lower energies, since the former are driven predominantly by hard collisions. Many ideas have been developed to understand these ratios. Some studies include the percolation model [3], recombination [4], the statistical model [5], and also flow [6] in proton-proton collisions. The slight discrepancies of the statistical model with data provide the motivation to get further insight into the hadronization mechanism [7].

This work reports on an alternative model to study the production of hadrons in relativistic heavy ion collisions, where the evolving density during the collision and the strangeness abundance are key ingredients, the so called *dynamical quark recombination model* [8].

2. Transverse momentum distribution
The transverse momentum distributions are computed within the dynamical recombination model are given by

$$\begin{aligned}\frac{dN}{p_T dp_T dy} &= g\frac{m_T}{4\pi}\frac{\rho_{\rm nucl}^2}{\Delta\tau}\int_{\tau_0}^{\tau_f}\tau d\tau\mathcal{P}(\tau)I_0(p_T\sinh\eta_T/T)\\ &\times\int d\eta\cosh(y-\eta)e^{-[m_T\cosh(y-\eta)\cosh\eta_T]/T}\end{aligned}\tag{1}$$

Journal of Physics: Conference Series **509** (2014) 012092 doi:10.1088/1742-6596/509/1/012092

where $\Delta\tau = \tau_f - \tau_0$ is the proper-time interval spanned during the hadron emission, η_T is related to the transverse expansion velocity β_T by $\beta_T = \tanh(\eta_T)$, the temperature T is given by the cooling law $T = T_0(\tau_0/\tau)^{v_s^2}$ and $\mathcal{P}$ is the probability of forming a given hadron as a function of proper-time τ and accounts for the fact that hadronization is not instantaneous. The profile of this probability can be obtained by Monte Carlo simulation (see Ref [8] for details). The spectrum given by Eq. (1), has T, β_T and the overall normalization as the free parameters.

2.1. *Combinatorial probabilities*

The combinatorial analysis allows to estimate in a simple way the number of possible colorless baryons and mesons, when we take a certain number of heavy (Q) and light (q) quarks. Considering the case where one starts out with a set of n q-quarks ($q = u, d$) and m Q-quarks ($Q = s$), it is possible to form sets of two (mesons) or tree (baryons) quarks. The more simple case is when we have one q and one Q each comming in three colors. In that case the possible colorless combinations are show on the table 1. From this it is easy to compute the relative abundance of the mesons containing a Q quark with respect to the total combinations, so the number of relative mesons is $2mn/(n+m)^2$. In a similar way, the total number of possible baryons plus antibaryons is $2(n+m)^3$, and the relative abundances of the baryons containing one Q quark is $3n^2m/(n+m)^3$. In this way, one can compute the baryon to meson ratio as a function of the number of light (n) and heavy (m) quarks.

Table 1. Number of possible combination of 2 or 3 quarks

Kind	# of Mesons	kind	# of Baryons
$q\bar{q}$	$3n^2$	qqq	n^3
$q\bar{Q}$	$3nm$	qqQ	$3n^2m$
$Q\bar{q}$	$3nm$	qQQ	$3nm^2$
$Q\bar{Q}$	$3m^2$	QQQ	m^3

The previous exercise may be generalized for more light flavors, for instance, for the case of $q = u, d$ and $Q = s$. In this case, some of the combinations are $\Omega(sss)$ and $\Phi(s\bar{s})$. Considering an equal number (n) of u and d and m as the number of s-quarks, we can estimate the ratio Ω/Φ as,

$$\frac{\Omega}{\phi} = \frac{4m^2n + m^2}{8n^3 + 3n^2m + 6nm^2}. \tag{2}$$

Since the strange quarks are produced in lesser quantities than up and down, we introduce the variable $x > 1$ as $u = xs$ to write Eq. (2) as

$$\frac{\Omega}{\phi} = \frac{4x + 1}{8x^3 + 3x^2 + 6x}. \tag{3}$$

This equation could be simplified even more, when ones takes only one light and one heavy flavor. Let us take for instance only u and s quarks, then Eq. (3) becomes

$$\frac{\Omega}{\phi} = \frac{1}{x+1}. \tag{4}$$

The behavior of the Ω/Φ ratio as a function of the number of light and heavy flavors, described by Eq. (2), is shown in Fig. 1. For the case when we take $u = xs$ and the number of u's is equal to the number of d's, the baryon to meson ratio has the behavior shown with the dashed curve in Fig. 2. The case where there is only one heavy and one light quark ($d = 0$), described by Eq. (4), is shown in the same figure with the solid curve. Notice that low values of x correspond to higher collisions energies whereas high values correspond to less energetic collisions.

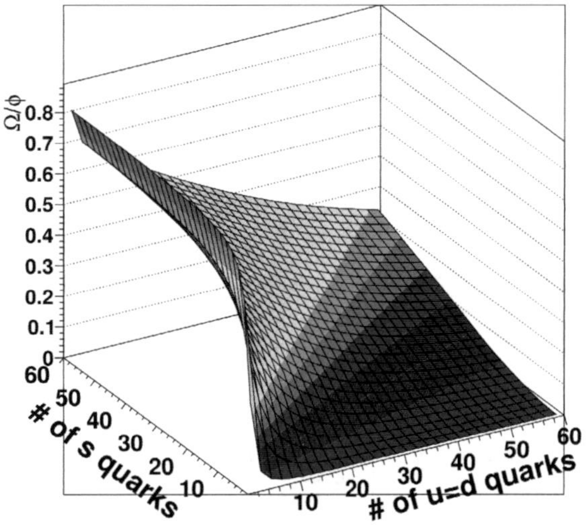

Figure 1. Ratio Ω/Φ as function of the number of s and light ($u = d$) quarks.

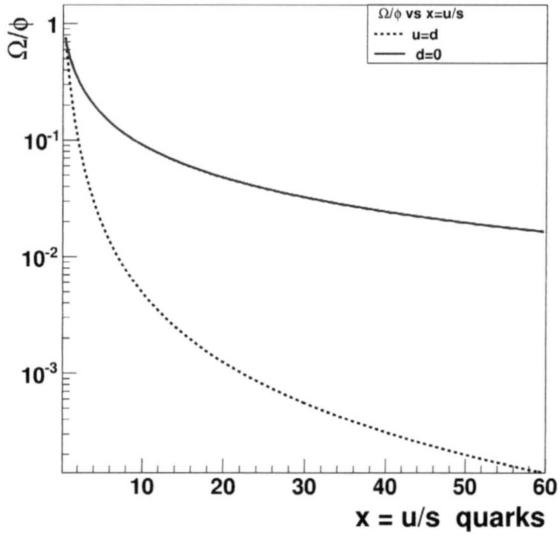

Figure 2. Ratio Ω/ϕ as function of proportionality factor $x = u/s$ between light u, d and s quarks.

3. Results

In order to estimate the momentum distribution, we take an initial hadronization time, $\tau_0 = 1$ fm, and initial and final temperatures as 200 and 100 MeV respectively, which means a final hadronization time $\tau_f = 8$ fm. Taking the Ω spectrum produced in Au+Au from STAR data [9] at 62.4 GeV, we can fit it and extract the transverse velocity β_T, as well as the normalization constant. The last parameters and the data on the ratio Ω/Φ are used to extract the distribution of the Φ, since this last one is not reported from experiment. The results are plotted in Fig. 3. Regarding the combinatorial probability and p_t spectra, let us take into account the constrains on the probability ratios to form baryons and mesons, and the the fact that there is a proportionality factor x between light and heavy quarks. Furthermore, taking into account the parameters of the fit and the constrains of the Eq. (4), it is possible to extract the proportionality factor x between the number of light and heavy quarks. The top of Fig. 4 shows, the fitted spectra to Ω data and the prediction of the Φ spectrum. The bottom part shows the ratio fitted with three different values of the x. One can see that after ≈ 3 GeV our curves grow above data. This behavior could be explained as a consequence of the absence of the fragmentation and energy loss in our approach.

In summary, the dynamical recombination model together with constrains on the Ω/ϕ ratio from combinatorial probabilities can describe RHIC data for Au+Au at two energies. We observe that the fits are better at low energy and lighter ions in the collisions. The constrains on the probabilities could provide a reasonable value for the proportionality between the number of light and heavy flavors, x.

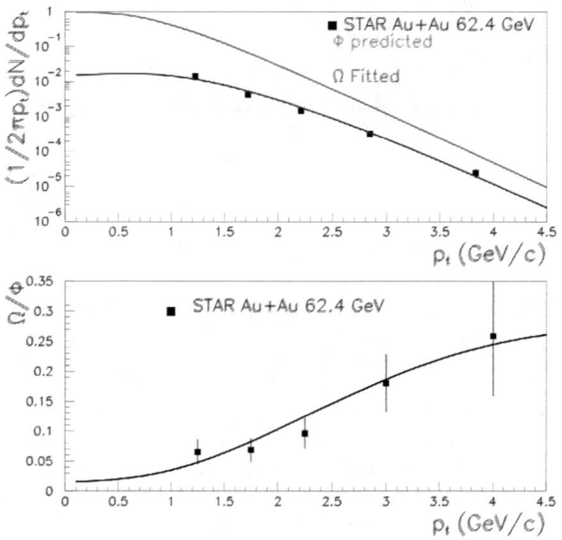

Figure 3. Experimental results of the spectra and the ratio Ω/Φ fitted to the model and extracted the $\beta = 0.40$

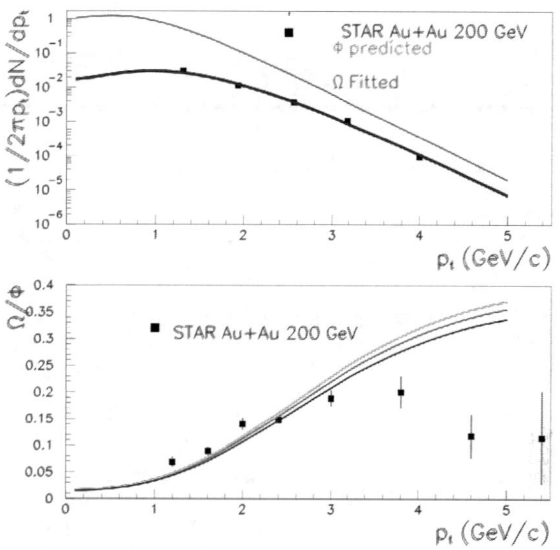

Figure 4. Spectra and ratio Ω/ϕ fitted to the model and extracted the $\beta = 0.45$, and the ratio for three values of x.

3.1. Acknowledgments
Support for this work has been received by CONACyT under grant numbers 101597 and 128534 and PAPIIT-UNAM under grant numbers IN107911,IN108414 and IN103811.

References

[1] Rafelski J and Müller B 1982 *Phys. Rev. Lett.* **48** 1066
[2] Abelev B *et al* (ALICE Collaboration) 2012 *Phys. Lett.* B **712** 309; Aamodt A *et al* (ALICE Collaboration) 2011 *Eur. Phys. J.* C **71** 1594
[3] Bautista I and Pajares C 2010 *Phys. Rev.* C **82** 34912
[4] Hwa R C and Zhu L 2011 *Phys. Rev.* C **84**, 64914; Fries R J, Müller B, Nonaka C and Bass S A 2003 *Phys.Rev. Lett.* **90** 202303
[5] Wheatron S, Cleymans J and Hauer M 2009 *Comp. Phys. Comm.* **180** 84; Braun-Munzinger P, Magestro D, Redlich K and Stachel J 2001 *Phys. Lett.* B **518** 41
[6] Cuautle E and Paic G 2008 *J. Phys. G* **35** 075103; Ortiz Velasquez A, Christiansen P, Cuautle Flores E, Maldonado Cervantes A I and Paic G 2013 *Phys. Rev. Lett.* **111** 042001
[7] Cleymans J, Kraus, L, Oeschler H, Redlich K and Wheaton S 2006 *Phys. Rev.* C **74** 34903
[8] Ayala A, Matínez M, Paic G and Toledo-Sanchez G 2008 *Phys. Rev.* C **77** 044901 Ayala A, Magnin J, Montaño L M and Toledo-Sanchez G 2009*Phys. Rev.* C **80** 064905
[9] Aggarwal M M *et al* (STAR Collaboration), Phys. Rev. C **83** 024901 (2011); Vasconcelos G M S (for the STAR Collaboration) 2010 *J. Phys.* G **37** 094034

Journal of Physics: Conference Series **509** (2014) 012093 doi:10.1088/1742-6596/509/1/012093

Quarkonia production and suppression in heavy ion collisions in STAR

Daniel Kikoła[1] **for the STAR Collaboration**

Purdue University, West Lafayette, IN, US

E-mail: kikola@if.pw.edu.pl

Abstract.
We report on recent STAR results on J/ψ and Υ production and suppression at midrapidity in Au+Au collisions at RHIC. We present the J/ψ nuclear modification factor as a function of centrality and transverse momentum at $\sqrt{s_{NN}}$ = 200, 62.4 and 39 GeV and discuss the energy dependence of J/ψ production. We also present the Υ nuclear modification factor in Au+Au collisions at 200 GeV calculated using high statistics p+p data taken in 2009.

1. Introduction

Measurements of production of various quarkonia states can provide insight into thermodynamic properties of the Quark-Gluon Plasma (QGP). Quarkonia production is anticipated to be suppressed in the presence of a QGP, compared to yields in $p + p$ collisions scaled by the number of binary nucleon-nucleon collisions, due to the Debye screening of the quark-antiquark potential in the deconfined, partonic medium. The Debye screening length depends on the temperature of the QGP. Different quarkonia states have different binding energies and thus different radii. Therefore, they are expected to disassociate at different temperatures. However, there are other effects that may affect the observed suppression. Cold nuclear matter effects (for instance shadowing and nuclear absorption), feed down from exited states, and secondary production via coalescence of uncorrelated heavy quarks can alter the observed yields. Each of these mechanisms has a different energy dependence: color screening increases with energy density thus increases with colliding energy, secondary production is bigger at higher energies due to larger cross section for heavy quark production, and nuclear absorption decreases with the energy [1]. Study of quarkonia production for different colliding systems, collision energies, and centralities can help to disentangle the interplay of these mechanisms and to understand the partonic medium properties.

2. Data analysis

STAR [2] is a large acceptance ($|\eta|$ <1, full coverage in ϕ) multipurpose detector composed of several subsystems making it well suited to measure quarkonia production. We reconstruct quarkonia (J/ψ, Υ) via the di-electron decay channel with a branching ratio B_{ee} = 5.9% for J/ψ and B_{ee} = 2.4% for Υ. In the analyses reported here, the STAR Time Projection Chamber (TPC), Barrel Electromagnetic Calorimeter (BEMC), and Time-Of-Flight (TOF) detectors are

[1] Present address: Faculty of Physics, Warsaw University of Technology, Warsaw, Poland

used. The TPC provides charged track reconstruction and particle identification via specific ionization energy loss (dE/dx). The BEMC is used for electron selection at high p_T and also for a high-p_T electron trigger for Υ and high-p_T J/ψ measurements. At low p_T, the TOF is used to improves electron identification. In this article, we present results for J/ψ in $|y| < 1$ and Υ in $|y| < 0.5$ obtained with $p+p$ data taken in 2009, and Au+Au data collected in 2010 at $\sqrt{s_{\rm NN}} = 200$, 62 and 39 GeV.

3. Results

To study the modification of quarkonium production in Au+Au collisions relative to $p+p$ interactions, we calculate the nuclear modification factor R_{AA}: $R_{AA} = \frac{\mathrm{d}^2N_{Au+Au}/dydp_T}{N_{\rm bin}\times \mathrm{d}^2N_{p+p}/dydp_T}$, where $\mathrm{d}^2N_{Au+Au}/dydp_T$ and $\mathrm{d}^2N_{p+p}/dydp_T$ are invariant yields in Au+Au and $p+p$ collisions, respectively, and $N_{\rm bin}$ is number of binary collisions in a given Au+Au centrality class.

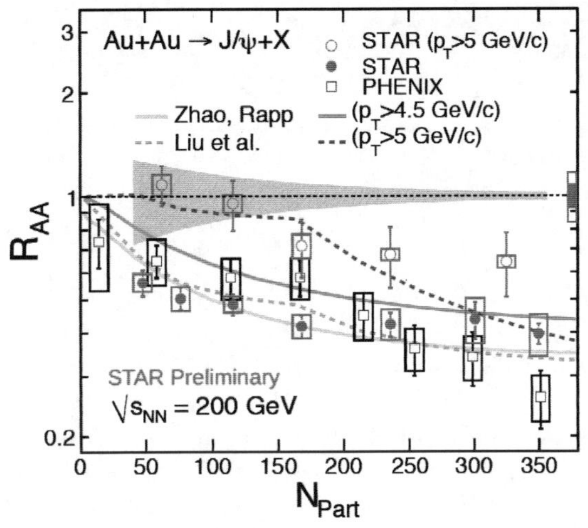

Figure 1. J/ψ R_{AA} as a function of centrality compared with data from PHENIX [3]. The statistical and systematic uncertainties are shown as vertical bars and boxes, respectively. The shaded band about unity shows the systematic uncertainties from estimates of the number of binary collisions and the box about unity shows the normalization uncertainty from the statistical and global systematic uncertainties of the $p+p$ reference data.

Figure 1 shows the R_{AA} for J/ψ as a function of centrality (represented by number of participants, $N_{\rm part}$) in Au+Au collisions at $\sqrt{s_{\rm NN}} = 200$ GeV. We compare p_T-integrated results (which are dominated by low-p_T part) from STAR and PHENIX to high-p_T ($p_T > 5$ GeV) data from STAR. In general, the R_{AA} decreases with the centrality and a strong suppression is observed in the most central collisions for p_T-integrated data. High-p_T J/ψ results are systematically higher than the low-p_T measurements. There is no suppression in peripheral events for high-p_T J/ψ while a significant suppression in central collisions is observed, although smaller than minimum bias results. Moreover, the J/ψ is not suppressed in d+Au collisions for $p_T > 5$ GeV/c [4] which indicates that cold-nuclear-matter effects are small at high p_T. Therefore, the observed suppression at high p_T is a manifestation of the QGP effects. Note that feed-down from $B \to J/\psi$ (which is $\sim 10-25\%$ for $4 < p_T < 12$ GeV/c in $p+p$ collisions) is not removed and part of the suppression could come from $b-$quark energy loss.

STAR data are also compared with two model predictions by Zhao and Rapp [5] and Liu et al. [6]. Both models assume suppression of J/ψ due to color screening and secondary production via recombination and both describe the low-p_T data reasonably well. In the case of high-p_T J/ψ, calculations by Liu et al. agree with the data but the model by Zhao and Rapp underpredicts the R_{AA}.

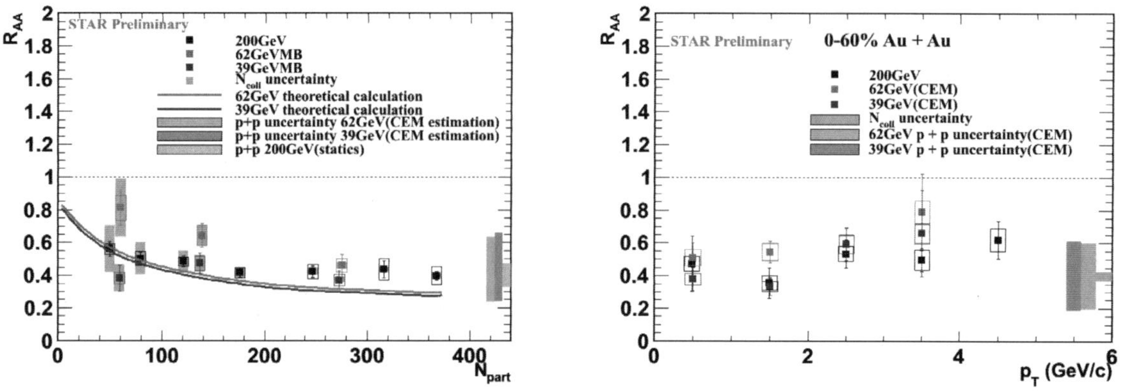

Figure 2. Nuclear modification factor R_{AA} for low-p_T ($p_T < 5$ GeV/c) J/ψ at $\sqrt{s_{\mathrm{NN}}} = 200$, 62.4 and 39 GeV as a function of N_{part} (left panel) and p_T (right panel).

Figure 2 shows the nuclear modification factor as a function of centrality and p_T for low$-p_T$ ($p_T < 5$ GeV/c) J/ψ in Au+Au collisions at $\sqrt{s_{\mathrm{NN}}} = 200$, 62.4 and 39 GeV. In the case of 39 and 62 GeV, predictions of Color Evaporation Model [7] are used as a baseline, with uncertainties indicated by the boxes on the right hand side of the plot. Point-to-point statistical and systematic errors are indicated by bars and boxes on each data point, respectively. We observe a significant suppression at all three energies. The STAR data are compared to predictions from model calculations by Zhao and Rapp which assumes two main components: suppression of direct J/ψ due to color screening and secondary production via recombination of c and $\bar{c}$ quarks. The model predicts very similar suppression for both 39 and 62 GeV, despite a significant difference in energy and thus different initial energy densities. In this model, larger suppression of direct J/ψ production at higher energy is offset by an enhanced regeneration rate due to a larger cross section for charm quarks. These calculations agree reasonably well with our data within errors.

In 2009, a new, high-statistics $p+p$ dataset was collected by STAR with an integrated luminosity of $L = 19.7$ pb^{-1}. This enabled more precise measurement of Υ production in $p+p$ compared to published results and, in turn, a better R_{AA} determination. STAR measured the $\Upsilon(1S+2S+3S)$ cross section at $|y| < 0.5$ of $B\mathrm{d}\sigma/\mathrm{dy} = 92 \pm 17 \pm 19$ pb. The left panel of Fig. 3 shows the energy dependence of Υ cross section; the STAR measurement is consistent with world data trend and Color Evaporation Model [8] predictions. The right panel of Fig. 3 shows the Υ nuclear modification factor as a function of centrality for in Au+Au collisions at $\sqrt{s_{\mathrm{NN}}} =$ 200 GeV. The suppression increases with centrality and a significant suppression is observed in the most central collisions. Data are compared to a model by Strickland et al. [9], which assumes a thermal dissociation of Υ in a thermalized, expending partonic medium, simulated with relativistic hydrodynamics. These calculations predict complete suppression of $\Upsilon(2S)$ and $\Upsilon(3S)$ states in the most central collisions. The model reproduces the STAR results well and the data favors the scenario when an internal energy is used for heavy quark potential. However, cold-nuclear-matter effects are not included in that model.

In the near future, the Muon Telescope Detector (MTD) [10] will extend STAR capabilities for quarkonia measurements. It will allow the study of J/ψ elliptic flow with unprecedented

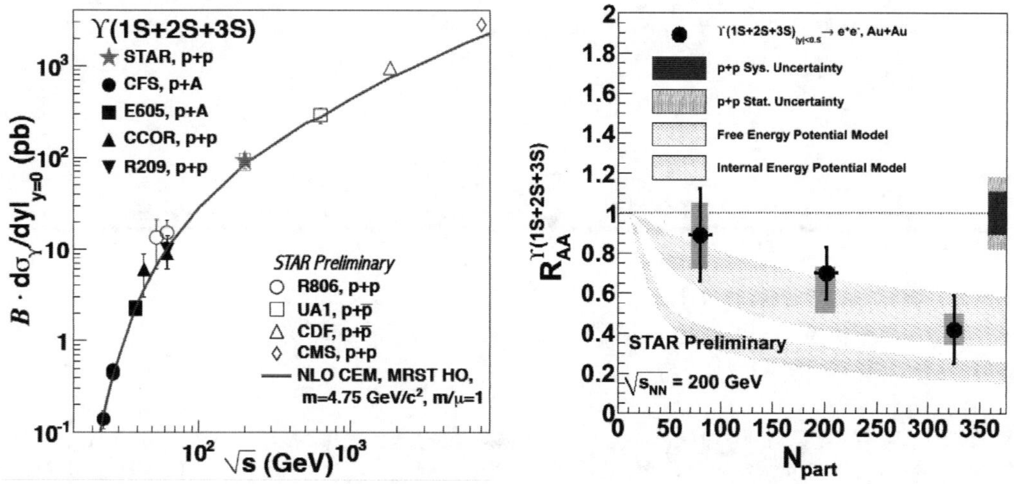

Figure 3. Left panel: The energy dependence of Υ cross section compared with predictions of Color Evaporation Model. Right panel: $R_{\rm AA}$ for Υ in Au+Au collisions at $\sqrt{s_{\rm NN}} = 200$ GeV.

precision and of suppression of the separate $\Upsilon(1S)$, $\Upsilon(2S)$ and $\Upsilon(3S)$ states. Measurement of suppression of different Υ states will advance the studies of the QGP thermodynamic properties. 63% of MTD was installed in 2013 and the construction will be completed for the RHIC run 2014.

4. Summary

J/ψ suppression at lower energies (39 and 62 GeV) is similar to at 200 GeV. The Υ suppression at $\sqrt{s_{\rm NN}} = 200$ GeV increases with centrality and it is consistent with a model that assumes complete suppression of $\Upsilon(2S)$ and $\Upsilon(3S)$ states. J/ψ suppression is systematically smaller at high p_T compared to low p_T. However, a significant J/ψ suppression at high p_T is observed in the most central Au+Au collisions.

References

[1] Scomparin E (NA60 Collaboration) 2009 *Nucl.Phys.* **A830** 239C–242C (*Preprint* 0907.3682)
[2] Anderson M *et al.* 2003 *Nucl. Instrum. Meth.* **A499** 659–678 (*Preprint* nucl-ex/0301015)
[3] Adare A *et al.* (PHENIX Collaboration) 2007 *Phys. Rev. Lett.* **98**(23) 232301
[4] Adare A *et al.* (PHENIX Collaboration) 2013 *Phys. Rev. C* **87**(3) 034904
[5] Zhao X and Rapp R 2010 *Phys.Rev.* **C82** 064905 (and private communication) (*Preprint* 1008.5328)
[6] Liu Y, Xu N and Zhuang P 2010 *Nuclear Physics A* **834** 317c – 319c ISSN 0375-9474
[7] Nelson R E, Vogt R and Frawley A D 2013 *Phys. Rev. C* **87**(1) 014908 (and private communication)
[8] Frawley A D, Ullrich T and Vogt R 2008 *Phys.Rept.* **462** 125–175 (*Preprint* 0806.1013)
[9] Strickland M 2011 *Phys.Rev.Lett.* **107** 132301 (*Preprint* 1106.2571)
[10] Ruan L, Lin G, Xu Z, Asselta K, Chen H *et al.* 2009 *J.Phys.* **G36** 095001 (*Preprint* 0904.3774)

14th International Conference on Strangeness in Quark Matter (SQM2013) IOP Publishing
Journal of Physics: Conference Series **509** (2014) 012094 doi:10.1088/1742-6596/509/1/012094

Jet suppression at LHC: theory vs. experiment

Magdalena Djordjevic

Institute of Physics Belgrade, University of Belgrade, Serbia

E-mail: magda@ipb.ac.rs

Abstract. Suppression of light and heavy flavor observables is one of the most important probes in studying the properties of QCD matter created at RHIC and LHC experiments. We will here summarize the most up-to-date light and heavy flavor suppression predictions for 2.76 TeV central Pb+Pb collisions at LHC. The predictions are based on our recent improvements in the energy loss calculations that take into account: i) theoretical formalism which includes finite size dynamical QCD medium with finite magnetic mass effects and running coupling, and ii) numerical procedure which includes path-length and multi-gluon fluctuations. Our theoretical predictions, obtained with no free parameters used in model testing, show a very good agreement with the experimental results for all available particle species. Our results show that the developed theoretical formalism is able to robustly explain suppression data at LHC, which strongly suggests that pQCD in Quark-Gluon Plasma can provide a reasonable description of the underlying jet physics in ultra relativistic heavy ion collisions.

Light and heavy flavor suppressions are considered to be excellent probes of QCD matter. Since suppression for a number of observables has been measured at LHC, their comparison with theoretical predictions allows testing our understanding of QCD matter. In this proceedings, we will summarize our theoretical predictions of jet suppression and compare them with recently available LHC measurements. Within this, we will then concentrate on some apparently puzzling results. We will also concentrate on measurements where a fine hierarchy between different probes can be observed.

To obtain suppression predictions, it is necessary to have reliable calculations of production, jet energy loss, fragmentation and decay. Regarding the energy loss, over the last several years we developed the energy loss formalism [1, 2, 3] in a finite size dynamical QCD medium of thermally distributed light quarks and gluons; we used the approach involving two hard thermal loops, which allows removing assumption of static scattering centers. We recently extended this formalism to the case of finite magnetic mass [4] and most recently we included running coupling [5]. To generate suppression predictions, we incorporated this formalism into a numerical procedure, which also includes i) light and heavy flavor productions [6, 7], ii) path-length [8, 9] and multi-gluon fluctuations [10], iii) up-to-date fragmentation functions for light [11] and heavy flavor [12, 13] and iv) in the case of heavy mesons, their decay to single electrons and J/ψ [6]. The temperature used in our predictions is 304 MeV, as extracted by ALICE, while the rest of the parameters are specified in [5].

We will now concentrate on the most recent LHC data for central Pb+Pb collisions, and our goal is to use the numerical procedure outlined above, in order to generate joint predictions for all available light and heavy flavor measurements. The predictions shown in Fig. 1 are generated by the same formalism, within the same numerical procedure, and with no free parameters used in model testing. As can be seen in Fig. 1, we compare our predictions with the experimental

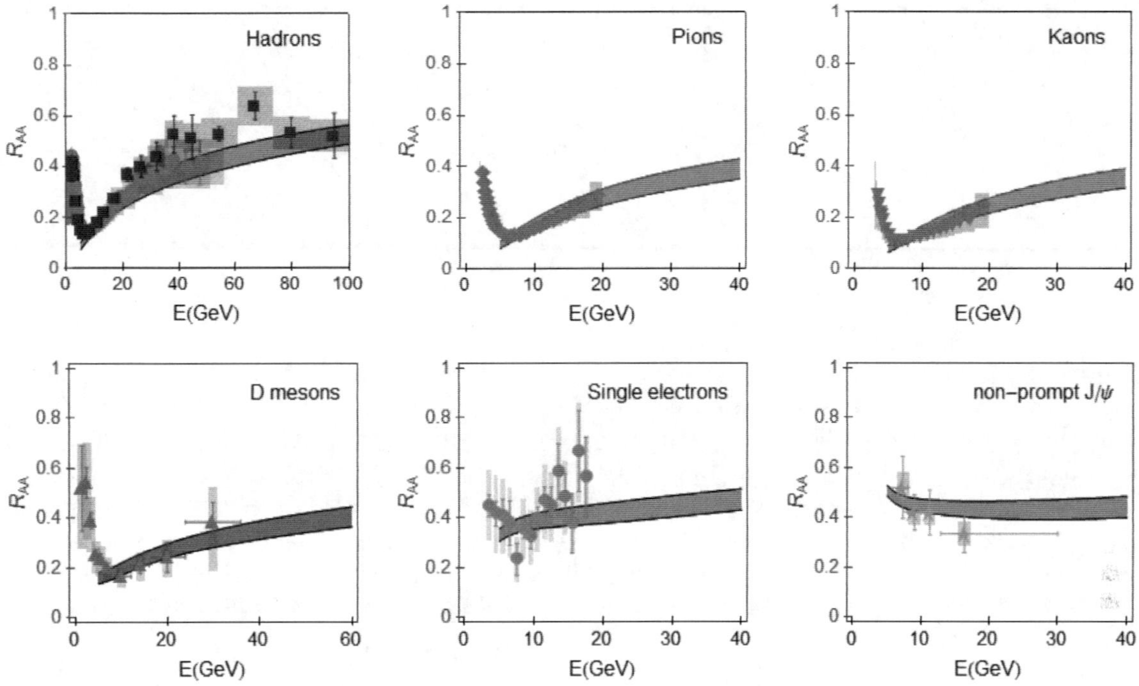

Figure 1. Our suppression predictions (shown as gray bands) are compared with the available experimental R_{AA} data for charged hadrons, pions, kaons, D mesons, non-photonic single electrons and J/ψ. The figure is adapted from [5]. The bands comes from the uncertainity in the magnetic mass value, see Ref. [5].

data for charged hadrons [15, 17], pions [16], kaons [16], D mesons [18], non-photonic single electrons [19] and non-prompt J/psi [20]. Note that the light flavor probes are shown in the upper panel of Figure 1, while the heavy flavor probes are shown in the lower panel. We see that we obtain excellent agreement for the light flavor and D mesons. We see that the single electron data are quite noisy, but we still obtain a good agreement with the predictions. There is also a good agreement for non-prompt J/psi, except for the last data point, which comes with large error bars.

If we now specifically concentrate on D mesons and charged hadron suppression data, we see that they reveal a puzzle which we call "heavy flavor puzzle at LHC", in a clear analogy to the well known "heavy flavor puzzle at RHIC" [21]. The following facts are relevant for this puzzle: i) while D meson suppression is a clear charm probe, charged hadrons are composed of both light quarks and gluons [22], ii) gluons have significant contribution to the charged hadron production [22], iii) charm and light quarks have about the same suppression, while the suppression of gluons is much higher [21]. Based on these facts, one can clearly infer that the suppression of charged hadrons should be smaller than the suppression of D mesons. However, contrary to this expectation, we see that experimental results actually show the same R_{AA} for charged hadrons and D mesons [18]. Having in mind the above, a naive conclusion would be that quarks and gluons loose the same amount of energy, which is however not in accordance with pQCD expectations. Even more surprisingly, Fig. 1 shows that that our pQCD based theoretical predictions are actually in an excellent agreement with the data. We therefore ask why both the experimental data and the theoretical predictions are not in an agreement with

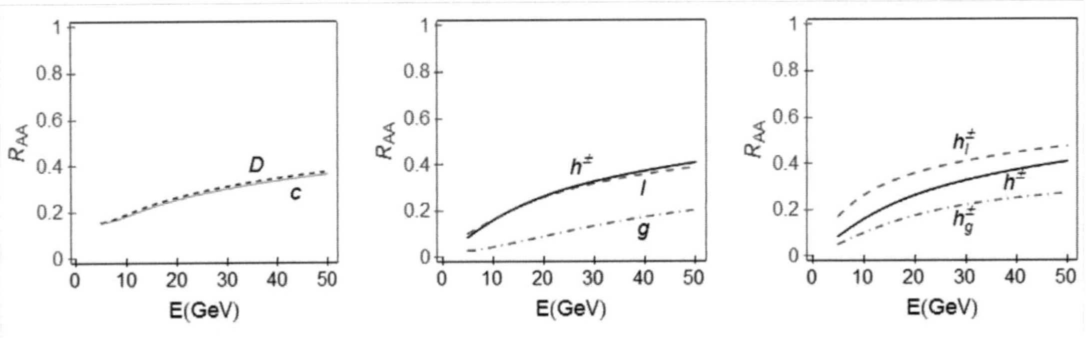

Figure 2. The right panel shows a comparison of the charm quark suppression predictions (full curve) with the D meson suppression predictions (dashed curve), as a function of momentum. The central panel shows the comparison of charged hadron suppression predictions (full curve) with light quark (dashed curve) and gluon (dot-dashed curve) suppression predictions. On the right panel, the dashed curve shows what would be the charged hadron suppression if only light quarks would contribute to charged hadrons. The dot-dashed curve shows what would be the charged hadron suppression if only gluons would contribute to charged hadrons, while the full curve shows the actual hadron suppression predictions. Figure adapted from [22].

the qualitative expectations outlined above.

To answer this question, in Fig. 2 we investigate how fragmentation functions modify the charged hadron and D meson suppressions, since these functions define the transfer from the parton to the hadron level. For D mesons (the left panel of Fig. 2), we see that the fragmentation does not modify bare charm quark suppression. Consequently, D meson suppression is indeed a genuine probe of the charm quark suppression in QCD medium. However, situation is much more complicated for the charged hadrons. In the central panel of Fig. 2, we compare the charged hadron suppression with the bare light quark and gluon suppression patterns. Surprisingly, we get that the charged hadron suppression is almost the same as the bare light quark suppression. This may suggest that gluon jets do not contribute to the charged hadron suppression, which is however inconsistent with the significant (even dominant) gluon contribution in charged hadrons. To investigate this, in the right panel of Fig. 2, we show what would be the charged hadron suppression if the hadrons were only composed of light quark jets, or only of gluon jets. We see that the charged hadron suppression is clearly in between these two suppression alternatives, so that both light quarks and gluons indeed significantly contribute to the charged hadron suppression. However, by comparing these two figures, we see that charged hadron fragmentation functions modify the bare light quark and gluon suppressions so that, coincidentally, their "resultant" charged hadron suppression almost identically reproduces the bare light quark suppression.

Therefore, the main result that we obtained is that the charged hadron suppression is almost identical to the light quark suppression, despite dominant gluon contribution in the charged hadron production. Since we also have that the D meson suppression is the same as the charm suppression, and that the light quarks have the same suppression as charm, we obtain the solution of the puzzle, i.e. that the charged hadron and D meson R_{AA}s should indeed be similar.

We next concentrate on whether, and under which conditions, the theory can explain fine hierarchy between the suppression measurements. For this, we come back to the Fig. 1, where we now concentrate on pion and kaon suppression data. That is, the experimental data for kaons and pions reveal an interesting fine resolution hierarchy, i.e. kaons are measured to consistently have somewhat higher suppression compared to pions [16]. From Fig. 1, we see that our predictions can very well reproduce this hierarchy; note that in obtaining these predictions, we used most up-to-date DSS fragmentation functions. On the other hand, the most widely used KKP fragmentation functions [23] would lead to a reversal of the observed hierarchy, and the reasons for this discrepancy are briefly discussed below.

As can be seen in [24], DSS fragmentation functions predict larger contribution of gluons in kaons compared to pions, while KKP predict the opposite. Since gluons have a larger suppression compared to light quarks, this evidently leads to the fact that DSS fragmentation functions will lead to a larger suppression for kaons compared to pions, which is in agreement with the experimental data. On the other hand, KKP fragmentation functions will lead to a larger suppression of pions compared to kaons, which is a reversed hierarchy and obviously in a disagreement with the experimental data. So, in distinction to the heavy flavor puzzle at LHC that we previously discussed, fine hierarchy between the pion and kaon suppressions represents an example where large gluon energy loss has a clear impact on the experimental observations.

In summary, in this proceedings, we showed that the same theoretical framework, with the same numerical procedure, and with no free parameters used in model testing, can simultaneously explain experimental data for a diverse set of probes at LHC. This also includes explanation of seemingly puzzling data such as the heavy flavor puzzle at LHC, and fine resolution hierarchy for the suppression of different probes. We also obtained an unintuitive, but important qualitative result that suppression of charged hadrons is a genuine probe of light quark suppression, which can considerably simplify interpretation of the relevant data.

Acknowledgments
This work is supported by Marie Curie International Reintegration Grant within the 7th European Community Framework Programme (PIRG08-GA-2010-276913) and by the Ministry of Science and Technological Development of the Republic of Serbia, under projects No. ON171004 and ON173052 and by L'OREAL-UNESCO National Fellowship in Serbia.

References

[1] Djordjevic M 2009 *Phys. Rev.* C **80**, 064909.
[2] Djordjevic M and Heinz U 2008 *Phys. Rev. Lett.* **101**, 022302.
[3] Djordjevic M 2006 *Phys. Rev.* C **74**, 064907.
[4] Djordjevic M and Djordjevic M 2012 *Phys. Lett.* B **709**, 229.
[5] Djordjevic M and Djordjevic M 2013 *Preprint* arXiv:1307.4098
[6] Cacciari M, Frixione S, Houdeau N, Mangano M L, Nason P and Ridolfi G 2012 *JHEP* **1210**, 137.
[7] Kang Z B, Vitev I and Xing H 2012 *Phys. Lett. B* **718**, 482; Sharma R, Vitev I and Zhang B W 2009 *Phys. Rev.* C **80**, 054902.
[8] Wicks S, Horowitz W, Djordjevic M and Gyulassy M 2007 *Nucl. Phys.* A **784**, 426.
[9] Dainese A 2004 *Eur. Phys. J.* C **33**, 495.
[10] Gyulassy M, Levai P and Vitev I 2002 *Phys. Lett.* B **538**, 282.
[11] de Florian D, Sassot R and Stratmann M 2007 *Phys. Rev. D* **75**, 114010.
[12] Cacciari M, Nason P 2003 *JHEP* **0309**, 006, Braaten E, Cheung K M, Fleming S and Yuan T C 1995 *Phys. Rev.* D **51**, 4819.
[13] Kartvelishvili V G, Likhoded A K, Petrov V A 1978 *Phys. Lett.* B **78**, 615.
[14] Wilde M (for the ALICE collaboration) 2013 *Nucl. Phys.* A **904-905** 573c.
[15] Aamodt K *et al*: (ALICE collaboration) 2013 *Phys. Lett.* B **720**, 52.
[16] Velasquez A O (for the ALICE collaboration) 2013 *Nucl. Phys.* A **904905** 763c
[17] Chatrchyan S *et al*: (CMS collaboration) 2012 *Eur. Phys. J.* C **72**, 1945.
[18] Grelli A (for the ALICE collaboration) 2013 *Nucl. Phys.* A **904905** 635c
[19] Sakai S (for the ALICE collaboration) 2013 *Nucl. Phys.* A **904-905**, 661c.
[20] Mihee J (for the CMS collaboration) 2013 *Nucl. Phys.* A **904-905**, 657c.
[21] Djordjevic M 2012 *Phys. Rev.* C **85**, 034904.
[22] Djordjevic M 2013 *Preprint* arXiv:1307.4702
[23] Kniehl B A, Kramer G and Potter B 2000 *Nucl. Phys.* B **582**, 514.
[24] Djordjevic M and Djordjevic M 2013 *Preprint* arXiv:1307.4714

Semi-classical approach to J/ψ suppression in high energy heavy-ion collisions

R Katz and P B Gossiaux
SUBATECH, CNRS/IN2P3, Université de Nantes, Ecole des Mines de Nantes, 4 rue Alfred Kastler, 44307 Nantes cedex 3, France
E-mail: katz@subatech.in2p3.fr

Abstract. We study the heavy quark/antiquark pair dynamics in strongly-coupled quark gluon plasma. A semi-classical approach, based on the Wigner distribution and Langevin dynamics, is applied to a color screened $c\bar{c}$ pair, in a hydrodynamically cooling fireball, to evaluate the total J/ψ suppression at both RHIC and LHC energies. Although its limitation is observed, this approach results to a J/ψ suppression of around 0.30 at RHIC and 0.25 at LHC.

1. Introduction

Quarkonia suppression was predicted by Matsui and Satz [1] as a sign of Quark-Gluon Plasma production in heavy-ion collisions. It has been experimentally observed but some aspects are still poorly understood, e.g. the unexpected "suppression" of the J/ψ suppression at low p_T as the collision energy increases. This paper aims to describe the J/ψ suppression from a dynamical point of view, as an alternative scheme to stationary sequential suppression, recombination...

Instead of studying the bound states survival in a stationary medium, we focus on the $Q\bar{Q}$ pair (Q being a heavy quark) dynamics in a non-stationary QGP until the freeze out. The QGP is considered here as a color screening medium at thermal equilibrium with a homogeneous time dependent temperature. The elliptic flow observations at RHIC and LHC suggest that the quarkonia thermalise partially with the medium. Our $Q\bar{Q}$ pair will therefore undergo a color screened, temperature dependent, self potential and some Langevin forces coming from its direct interaction with the thermal medium. We expect the thermalisation effect to be the answer to the observed suppression of the suppression as it should tend to hold the $Q\bar{Q}$ together.

Ideally a full quantum formalism would be desirable, however the thermalisation of a quantum state being still an unsolved problem, we base this study on a semi-classical formalism inspired by Young and Shuryak work [2]. As a test of reliability, a comparison to pure quantum results is carried out in the case without Langevin dynamics.

2. The Semi-classical approach without Langevin dynamics

2.1. The Wigner approach

In usual quantum mechanics, the $Q\bar{Q}$ pair probabilistic information is described by a wavefunction Ψ, which evolution is given by the Schrödinger equation. Equivalently, the $Q\bar{Q}$ pair can be described as a phase space distribution called the Wigner distribution $F(\vec{r},\vec{p},t)$, derived from the Wigner transformation of the wavefunction [3]. Its evolution is then given by the Wigner-Moyal equation, which classical limit (equivalent to the Liouville equation) is used within the semi-classical frame. To evaluate the probability (or "J/ψ weight") of the $Q\bar{Q}$ pair to bind as a J/ψ state, one projects its evolved Wigner distribution onto the J/ψ Wigner distribution, computed from the vacuum potential. This semi-classical approach allows us the access to numerical simulations through the practical use of the test particles method.

2.2. The color screened potentials and temperature scenarios

The binding $Q\bar{Q}$ potential in the QGP is a temperature dependent potential that takes into account the presence of color charges in their vicinity. We choose to lead this study with the potential $U(\vec{r}, T)$, evaluated by Kaczmarek et al. [4] from lQCD results and re-parameterised by us. It corresponds to the color singlet internal energy, i.e. assuming no energy exchange between the pair and the medium (a redundancy with the stochastic forces is therefore avoided).

In the common heavy-ion collisions simplified scenario, the $Q\bar{Q}$ pair is first produced within a short lapse of time ($\tau \lesssim 0.3$ fm/c), promptly followed by a warm cooling QGP phase ($T \gtrsim T_c$ and $0.3 \lesssim \tau \lesssim 5$ fm/c) in which the $Q\bar{Q}$ pair evolves. Finally, a cooler hadronisation phase ($T^{had} < T_c$) follows until the freeze out. The critical temperature T_c is taken equal to 165 MeV. We choose the space-time temperature distribution derived by [6] through the use of a hydrodynamic evolution of an initial thermalised state [7]. To simplify, the spatial distribution is reduced to the temperature at the centre of the fireball (see figure 1) and the $Q\bar{Q}$ pair is assumed to be produced at $t = 0.6$ fm/c, i.e. from which the QGP is at thermal equilibrium.

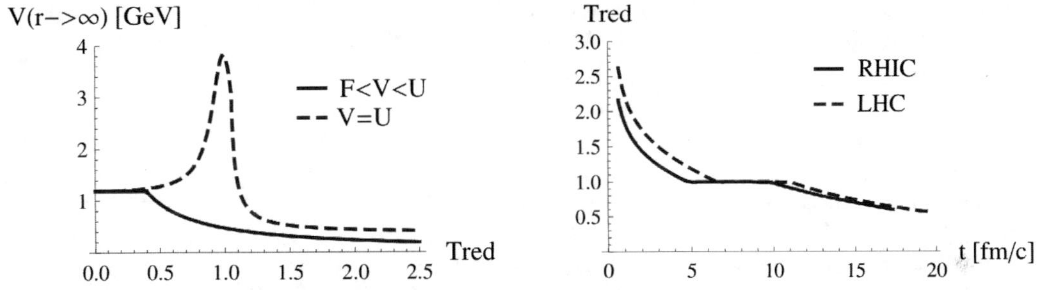

Figure 1. *Left*: "Intermediate" $F < V < U$ [5] and internal energy U potential asymptotic dependences on temperature. *Right*: Evolution of the reduced temperature $T_{red}(t) = T/T_c$ over time at the centre of the fireball at RHIC ($\sqrt{s_{NN}} = 200$ GeV) and LHC ($\sqrt{s_{NN}} = 2.76$ TeV) energies.

2.3. Initial distribution

As we are only interested in S states, the $\Psi_{Q\bar{Q}}(\mathbf{r}, t)$ wave-function is reduced to its radial part $R_{Q\bar{Q}}(r, t)$. The initial radial wave-function is a Gaussian wavepacket with a 0.165 fm r.m.s. for the $c\bar{c}$ pair. The latter can be estimated by applying the uncertainty principle to the intermediate quark of the $Q\bar{Q}$ production Feynman diagrams (LO u or t channel): $\Delta r \sim \hbar c/m_c \sim 0.16$ fm.

2.4. Semi-classical and quantum results without thermalisation processes

In order to check the limits of the semi-classical approximation, the results are here compared to pure quantum results (from the time-dependent Schrödinger equation). If V=0, the J/ψ weights are identical as expected from the Erhenfest theorem. However, in the case of our screened potential, the semi classical results exhibit strong discrepancies (Fig. 2): a "lump" for $t < 1$ fm/c and a difficulty to reach the continuum (the J/ψ normed weight remains close to 1). The observation of the test particle paths in phase space shows that the lump is due to their loss of momentum while climbing the potential barrier (making them enter the "J/ψ zone"). Consequently, if the Langevin dynamics does not lead the evolution, the validity of the semi-classical "thermalised" results (section 3 and [2]) is questionable. As a positive point, one observes for both formalism oscillations between eigenstates due to high asymptotic values of the potential when $5 \lesssim t \lesssim 12$ fm/c.

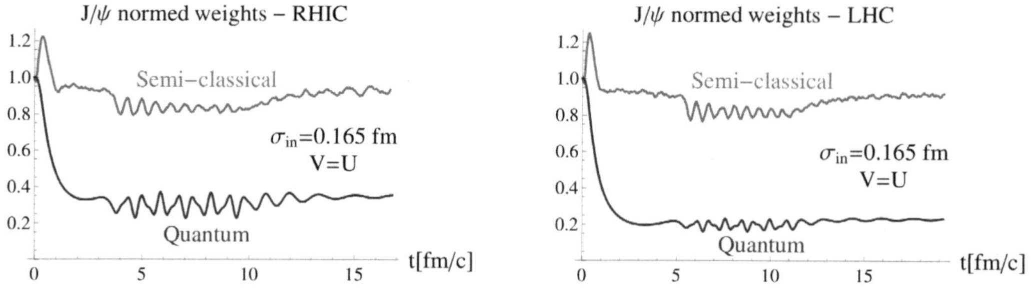

Figure 2. Semi-classical and quantum results for the J/ψ weights function of time with RHIC (*left*) and LHC (*right*) temperature scenarios.

3. The Semi-classical approach with Langevin dynamics

3.1. Semi-classical formalism with Langevin dynamics

A way to take into account the thermalisation of the $Q\bar{Q}$ pair in this dynamical model is to consider the random interactions between the $Q\bar{Q}$ pair and the QGP constituents. By analogy with the Fokker-Planck equation of motion in momentum space (equivalent to Langevin forces), we introduce additional stochastic terms in the Wigner Moyal equation.

In practice (particle test method), Langevin forces are introduced in Newton's equation of motion: a stochastic force $\vec{\xi}$, defined by $\langle \vec{\xi} \rangle$=0 and $\langle \xi_i(t)\xi_j(t') \rangle = \kappa\delta_{i,j}\delta$(t-t') (the fluctuations are uncorrelated over time), and a "friction" term $-A\vec{p}$ where A is the drag coefficient. The Einstein relation can then be deduced from quadratic and average momentum calculations: $\kappa/2 = mT/A$. A microscopic calculation from Gossiaux and Aichelin and a fit to experimental R_{AA} lead to a T-dependent drag: A[c/fm] $\cong 3T$[GeV] $+ 2.5T^2$ [8] for charm quarks. By mean of comparison, we will also use Young and Shuryak's drag $A = 4\pi T^2/(3\hbar cm)$ [2].

3.2. Results with RHIC and LHC temperature scenarios

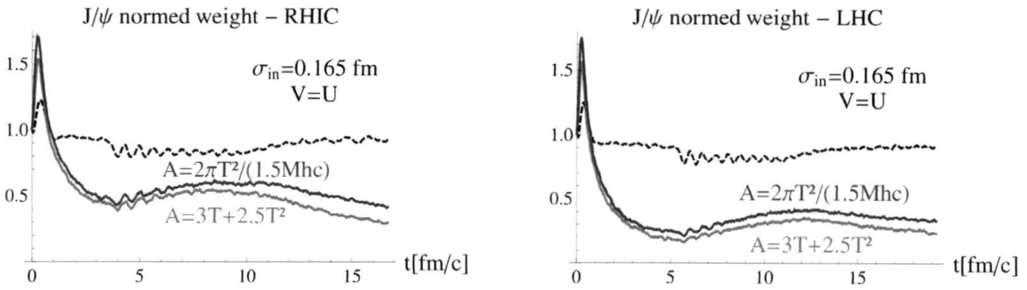

Figure 3. Semi-classical results with stochastic forces for the J/ψ weights function of time at RHIC (*left*) and LHC (*right*) and the two drags. Dashed lines: with no stochastic forces.

As shown in Fig.3, the additional Langevin dynamics leads to an actual evolution of the J/ψ weights that was missing in section 2.4: not only the $c\bar{c}$ pair thermalises with the QGP, but it also solves the previous test particles difficulty to reach the continuum. It also enhances the "lump" for $t < 1$ fm/c, thanks to rapid thermalisation of the distribution in momentum space [2] due to the drag term. The weight slow decrease, for $1 \lesssim t \lesssim 5$ fm/c, comes from the spatial diffusion of the $c\bar{c}$ pair distribution due to the stochastic force term. From $t \gtrsim 5$ fm/c, the light weight variations follow the important variations of the U potential: high asymptotic values lead to a narrowing of the spatial distribution and low asymptotic values to its spatial diffusion.

The two drag coefficients give similar evolutions with a difference of ~ 0.1 at the freeze out. As a robustness test, an initial wavepacket r.m.s. (see section 2.3) variation of ± 0.01 fm/c (12%) leads to a maximum normed J/ψ weight deviation of 0.03 (8%) at the freeze out.

4. Conclusion

In order to better understand the J/ψ suppression at RHIC and LHC, the dynamics of a $c\bar{c}$ pair has been studied through a semi-classical formalism proposed by Young and Shuryak [2]. This formalism, i.e. the classical evolution of a $Q\bar{Q}$ quantum Wigner distribution, has been chosen as a convenient way to introduce a classical thermalisation process. The normed J/ψ weights obtained at the freeze out with this formalism are summed up in table 1, and compared to some extent to pure quantum results and experimental data. Of course the comparison to data may not be taken too seriously as we have not considered cold nuclear matter effects, feed downs from other quarkonia... Including them would mostly have a positive effect as our results (especially at RHIC) underestimate the data at high p_T, where color-screening effects are expected to be relatively more important [9].

	Semi-classical		Quantum		Experimental data	
Thermalisation ? →	No	Yes	No	Yes	high p_T	low p_T
RHIC	0.9	0.29	0.31	?	0.64 $\pm$0.14	0.26 $\pm$0.05
LHC	0.9	0.23	0.23	?	0.20 $\pm$0.03	0.83 $\pm$0.14

Table 1. J/ψ normed weights for RHIC and LHC at the freeze out. — RHIC AuAu $\sqrt{S_{NN}} = 200$ GeV collisions: 1) high p_T STAR data [9] (inclusive (prompt and non prompt) J/ψ, $5 < p_T < 14$ GeV/c, $|y| < 1$ and 0-10% centrality) and 2) low p_T PHENIX data [10] (inclusive J/ψ, $p_T < 5$ GeV/c, $|y| < 0.35$ and 0-5% centrality). The non prompt contribution is estimated to 10-25% of the inclusive production. — LHC PbPb $\sqrt{S_{NN}} = 2.76$ TeV collisions: 1) high p_T CMS data [11] (prompt J/ψ, $6.5 < p_T < 30$ GeV/c, $|y| < 2.4$ and 0-10% centrality) (inclusive J/ψ: $R_{AA} = 0.24 \pm 0.03$) and 2) low p_T ALICE data [12] (inclusive J/ψ, $0 < p_T < 8$ GeV/c, $|y| < 0.9$ and 0-10% centrality).

The comparison between quantum and semi-classical results has revealed the limitation of the latter, and makes questionable the accuracy of the results obtained with the additional Langevin dynamics (Table 1 and [2]). They could however still be relevant if the stochastic forces appear to be the leading ingredient of the evolution. It seems to be indeed the case within the semi-classical frame, but remains to be checked in a full quantum study. In a near future work, a full dynamical quantum approach including some stochastic forces will be developed.

Acknowledgments

We are grateful for support from TOGETHER project Région Pays de la Loire.

References

[1] Matsui T and Satz H 1986 J/ψ suppression by quark-gluon plasma formation *Phys. Lett.* B **178**, 416
[2] C. Young and Shuryak E 2009 *Phys. Rev.* C **79**: 034907 (*Preprint* arXiv: 0803.2866)
[3] Liboff R L 1990 *Kinetic Theory: Classical, Quantum, and Relativistic Descriptions* (Wiley-interscience)
[4] Kaczmarek O, Karsch F and Zantow F 2004 (*Preprint* arXiv:hep-lat/0406036v1) Antonov D, Debbio L and Giacomo A 2003 JHEP 0308 011 (*Preprint* arXiv:hep-lat/0302015v2)
[5] Mocsy A and Petreczky P 2007 *Phys. Rev.* D **77**, 014501 (*Preprint* arXiv: 0705.2559v2 [hep-ph])
[6] Kolb P F and Heinz U 2003 Hwa, R.C. (ed.) et al.: Quark gluon plasma 634-714
[7] Glauber R J 1959 Lecture on Theoretical Physics Vol. 1 (New York: W.E. Brittin, Interscience)
[8] Gossiaux P B and Aichelin J 2008 *Phys. Rev.* C **78** 014904 (*Preprint* arXiv:0802.2525 [hep-ph]).
[9] Adamczyk L et al. [STAR Collaboration] 2013 *Phys. Lett.* B **722** 55 (*Preprint* arXiv:1208.2736 [nucl-ex])
[10] Adare A et al. [PHENIX Collaboration] 2011 *Phys. Rev.* C **84** 054912 (*Preprint* arXiv:1103.6269 [nucl-ex]).
[11] Chatrchyan S et al. [CMS Collaboration] 2012 *JHEP* **1205** 063 (*Preprint* arXiv:1201.5069 [nucl-ex])
[12] Maire A et al. [ALICE Collaboration] 2012 PoS ConfinementX 211 (*Preprint* arXiv:1301.4058 [hep-ex]).

Journal of Physics: Conference Series **509** (2014) 012096 doi:10.1088/1742-6596/509/1/012096

Studies of Onia and Strangeness Production in Proton-proton Collisions at LHCb

N H Brook
on behalf of the LHCb collaboration
HH Wills Physics Laboratory, University of Bristol, Bristol. BS8 1TL. UK.
E-mail: n.brook@bristol.ac.uk

Abstract. Studies of strangeness and quarkonia production in the forward region provide important input to the understanding of QCD models in a kinematic range where they have large uncertainties. The LHCb experiment has collected a dataset corresponding to an integrated luminosity of about 3 fb^{-1} in proton-proton collisions at $\sqrt{s} = 7$ and 8 TeV, as well as smaller samples recorded with other LHC operating modes. The latest studies of the production of kaons, Λ baryons, and quarkonia states are presented.

1. Introduction

The LHCb experiment is a single-arm forward spectrometer covering the pseudorapidity range $2 < \eta < 5$, designed for the study of hadrons containing b or c quarks [1].The studies of the production of quarkonia are a stringent test of QCD. Quarkonia production is usually described in the framework of NRQCD, where short-distance calculations are used for the partonic calculations of the production of a heavy quark-antiquark pair and a large distance matrix elements for the evolution of the pair to quarkonia. The NRQCD calculations depend on the colour-singlet (CS) [2] and colour-octet (CO) matrix elements [3] that account for the probability of a heavy quark-antiquark pair, in a particular colour state, to evolve into heavy quarkonium. In proton-proton (pp) collisions quarkonia can be produced directly, through feed-down from higher quarkonia states or in the case charmonium via the decay of b hadrons.

The study of strangeness production provides sensitive tests of soft hadronic interactions in the non-perturbative regime. Both onia and strangeness production in pp-collision act as an important benchmark for the study of the quark-gluon plasma in heavy-ion collisions.

2. Onia production

The differential production cross-sections of prompt J/ψ and Υ mesons at $\sqrt{s} = 8$ TeV has been measured by LHCb [4], in the range of rapidity $2.0 < y < 4.5$ and $p_T < 14$ GeV/c(J/ψ) or $p_T < 15$ GeV/c(Υ) with an integrated luminosity of 51 pb^{-1}. Similar measurements have also been made by LHCb at $\sqrt{s} = 7$ TeV [5] and $\sqrt{s} = 2.76$ TeV [?] The J/ψ and Υ mesons are detected through their decays to $\mu^+\mu^-$ pairs. The fraction of J/ψ mesons from b-hadron decays is measured in the same fiducial region. The yields of both prompt J/ψ mesons and J/ψ from b are determined from a two-dimensional fit in (p_T, y) bins to the distributions of invariant mass and pseudo decay time of the signal candidates, following the approach outlined in ref. [7]. The integrated cross section for prompt J/ψ meson production in the fiducial region is measured to be $\sigma(\text{prompt}J/\Psi) = 10.94 \pm 0.02 \pm 0.79$ μb, where the first error is statistical and the second is systematic, which is experimentally dominated by the timing fit, the trigger and luminosity measurement. The integrated cross section for the production of J/ψ from b in the same fiducial region is $\sigma(J/\Psi$ from $b) = 1.28 \pm 0.01 \pm 0.11$ μb.

The integrated cross section for Υ production (times the dimuon braching factors $B^{iS} = B(\Upsilon(iS) \to \mu\mu)$) in the range $p_T < 15$ GeV/c and $2.0 < y < 4.5$ are measured to be

$$\begin{aligned}
\sigma(pp \to \Upsilon(1S)X) \times B^{1S} &= 3.241 \pm 0.018 \pm 0.231 \text{ nb} \\
\sigma(pp \to \Upsilon(2S)X) \times B^{2S} &= 0.761 \pm 0.008 \pm 0.055 \text{ nb} \\
\sigma(pp \to \Upsilon(3S)X) \times B^{3S} &= 0.369 \pm 0.005 \pm 0.027 \text{ nb.}
\end{aligned}$$

In Figure 1 the measured differential cross-section for the production of prompt J/ψ mesons as a function of p_T are compared to three theoretical models (assuming no polarisation.) Both the NNLO* CSM and the NLO NRQCD models provide reasonable descriptions of the experimental data, whilst the CSM at NLO underestimates the cross-section by an order of magnitude. In Figure 2 shows the Υ(1S) meson differential cross-section as a function of p_T and again the NNLO* CSM model is in agreement. The prompt production of higher states of charmonium

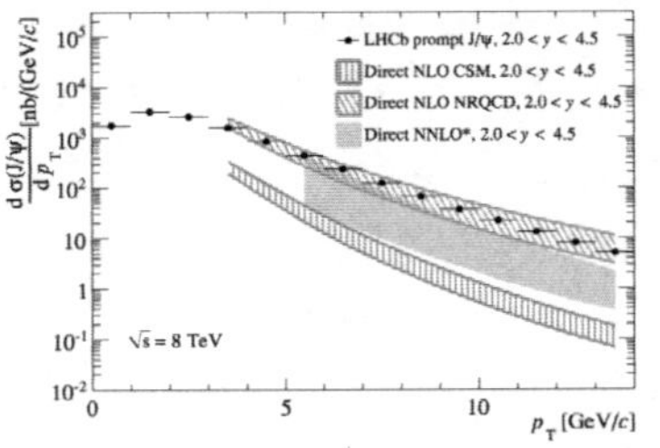

Figure 1. Comparison of the differential cross-section for the production of prompt J/ψ meson as a function of p_T with direct production in an NLO NRQCD [8] (orange diagonal shading), an NNLO* CSM [9] (solid yellow) and an NLO CSM [10] (blue vertical shading).

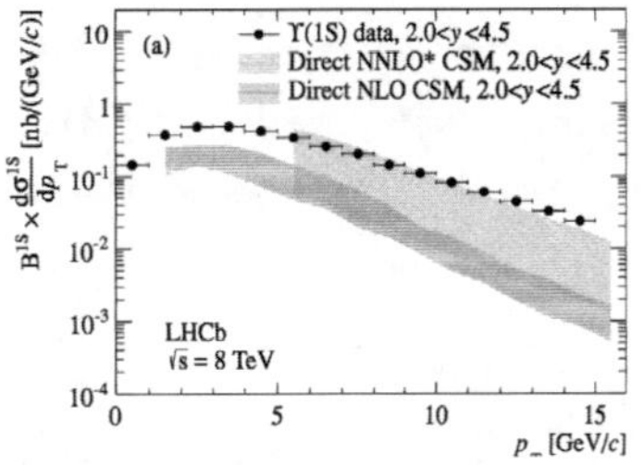

Figure 2. Comparison of the differential production cross-sections times dimuon branching fractions for Υ(1S) mesons as a function of p_T with direct production in an NNLO* CSM [11] (solid yellow) and an NLO CSM [10] (blue vertical shading) model. The points show the measurements reported in this analysis.

(χ_{c0}, χ_{c1} and χ_{c2}) have also been studied at LHCb at $\sqrt{s} = 7$ TeV [12] . In Ref. [12] the χ_c mesons are identified through their decay to $J/\psi(\to \mu\mu)\gamma$, using converted photons. The signal yields are extracted to a fit to the $\Delta M = M(J/\psi\gamma) - M(J/\psi)$ spectrum. The χ_{c0} meson is observed for the first time at high energy hadron collisions. The signal is observed with a statistical significance of 4.3σ with a yield of 705 ± 163, for $4 < p_T^{J/\psi} < 20$ GeV/c. Figure 3 shows the ratio of χ_{c2} to χ_{c1} as a function of p_T with a comparison with a next to leading order [13] and leading order [14] calculations. The major uncertainties in extracting the yields in these measurements are the fitting procedure, the photon efficiency and $p_T^{\chi_c}$ spectrum.

The production of J/ψ meson accompanied by open charm (and pairs of open charm) has been measured by LHCb [16]. The production cross section of J/ψ and open charm mesons in the fiducial volume $2 < y_{J/\psi}, y_C < 4, p^T_{J/\psi} < 12$ GeV/c$, 3 < p^T_C < 12$ GeV/c are summarised in Table 1. The major sources of the experimental systematics come from the knowledge of the tracking, luminosity and trigger. In addition the lack of knowledge of the J/ψ polarisation and the charm mesons' branching ratios add a major uncertainty. The predictions from gluon-gluon

Journal of Physics: Conference Series **509** (2014) 012096 doi:10.1088/1742-6596/509/1/012096

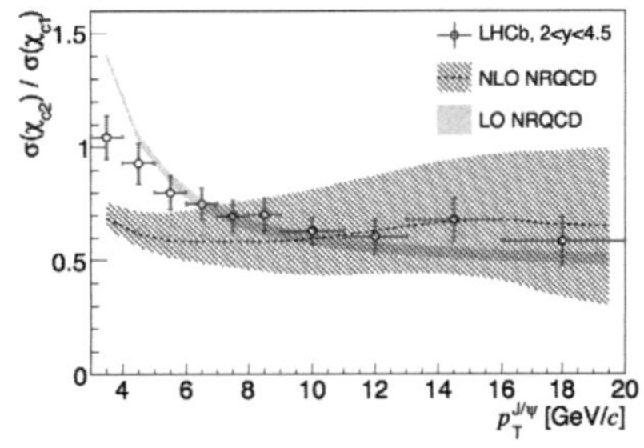

Figure 3. Ratio of χ_{c2} to χ_{c1} cross-sections in the rapidity range $2.0 < y < 4.5$. The results results are compared with the NLO NRQCD calculation from Ref. [13] (blue shading) and the LO NRQCD calculation of Ref. [14] (solid green). The LHCb results are obtained assuming the χ_c mesons are produced unpolarized.

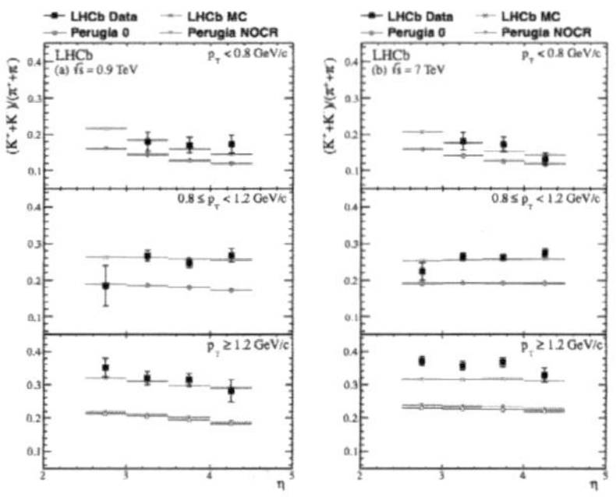

Figure 4. Result for the k/π ratios at $\sqrt{s}$ = 0.9 and 7.0 TeV. The data are compared to different tune on the PYTHIA event generator [19].

fusion [17] are significantly smaller than the observed cross section, which are better described with double parton scattering models [18].

Table 1. Production cross-sections for $J/\psi C(= D^0, ...)$. The first uncertainty is statistical, and the second is systematic.

Mode	σ(nb)
$J/\psi D^0$	$161.0 \pm 3.7 \pm 12.2$
$J/\psi D^+$	$56.6 \pm 1.7 \pm 5.9$
$J/\psi D_s^+$	$30.5 \pm 2.6 \pm 3.4$
$J/\psi \Lambda_c^+$	$43.2 \pm 7.0 \pm 12.0$

3. Strangeness and baryon production

The charged particle species production ratio have been measured at LHCb in three separate p_T regions: $p_T < 0.8$ GeV/c, $0.8 \leq p_T < 1.2$ GeV/c and $p_T \geq 1.2$ GeV/c [15]. For the lowest p_T bin the pseudorapidity range of the measurements is $3.0 < \eta < 4.5$ and $2.5 < \eta < 4.5$ for the higher p_T ranges. The particle identification (PID) algorithm uses information from the RICH and tracking detectors to construct a negative log likelihood for the particle hypothesis. The dominant uncertainty is associated with the understanding of the PID performance. Figure 4 shows the kaon and pion ratio as a function of η in the three p_T ranges. There is a clear rise in the ratio as p_T rises. The data are compared to the predictions of several tunes of the PYTHIA 6.4 [19] generator. The data are higher than the Perugia 0 and NOCR [20] tunes but in agreement with the LHCb tune [21]. LHCb recently published measurements of baryons containing both strange and bottom quarks [22]. In particular LHCb presented the mass measurements of Ξ_b^- and Ω_b^- using the decays $\Xi_b^- \to J/\psi\ \Xi^-$ and $\Omega_b^- \to J/\psi\ \Omega^-$. The mass measurements of the Ω_b^- is of interest given the discrepancy between the CDF [23] and D0 [24] collaborations. The invariant mass distributions are shown in Figure 5 with a signal yield of 111 ± 12 Ξ_b^- and

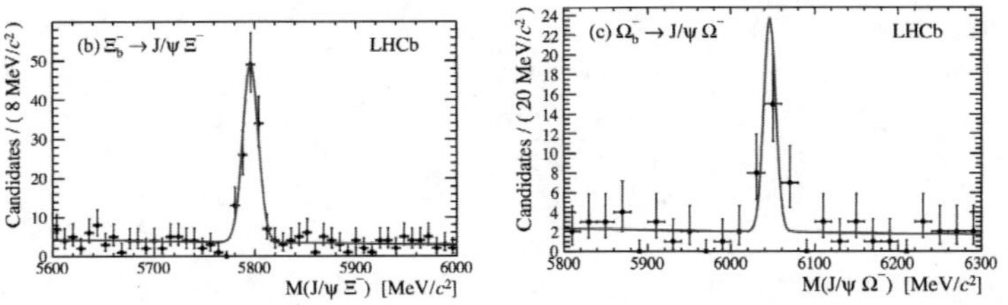

Figure 5. The invariant mass distribution for Ξ_b^- and Ω_b^- candidates. The solid lines are the results of an unbinned likelihood fit.

19 ± 5 Ω_b baryons. The mass measurements are $M(\Xi_b^-) = 5795.8 \pm 0.9 \pm 0.4$ MeV/c^2 and $M(\Omega_b^-) = 6046.0 \pm 2.2 \pm 0.5$ MeV/c^2 and are the most precise to date. The Ω_b^- result is in agreement with the CDF measurement.

4. Summary

LHCb has made many important productions results in the forward rapidity region. The results include evidence for double parton scattering and investigations of excited quarkonia. Since the conference LHCb have published their results on J/ψ polarisation [25]. There are further studies on quarkonia ongoing making using of the full 2011 ($\mathcal{L} = 1.1$ fb^{-1}) and 2012 ($\mathcal{L} = 2.08$ fb^{-1}) datasets as well as data from pA collisions.

References

[1] LHCb collaboration, Alves Jr A A i*et al* 2008 *JINST* **3** S08005
[2] Baier R and Ruckl R 1981 *Phys. Lett.* B **102** 364
[3] Braaten E and Fleming S 1995 *Phys. Rev. Lett.* **74** 3327
[4] LHCb collaboration, Aaij R *et al* 2013 *JHEP* **06** 064
[5] LHCb collaboration, Aaij R *et al* 2011 *Eur. Phys. J.* C**71** 1645, LHCb collaboration, Aaij R *et al* 2012 *Eur. Phys. J.* C**72** 2025
[6] LHCb collaboration, Aaij R *et al* 2013 *JHEP* **02** 041
[7] LHCb collaboration, Aaij R *et al* 2011 *Eur. Phys. J.* C **71** 1645
[8] Butsenschoen M and Kniehl B A 2011 *Phys. Rev.* D **84** 051501, Butsenschoen M and Kniehl B A 2011 *Phys. Rev. Lett.* **106** 022003
[9] Lansberg J 2009 *Eur. Phys. J.* C **61** 693
[10] Campbell J M,Maltoni F and Tramontano F 2007 *Phys. Rev. Lett.* **98** 252002
[11] Artoisenet P *et al* 2008 *Phy. Rev. Lett.* **101** 152001
[12] LHCb collaboration, Aaij R *et al* 2013 LHCb collaboration, Aaij R *et al* 2013 Measurement of the relative rate of prompt χ_{c0}, χ_{c1} and χ_{c2} production at $\sqrt{s} = 7$ TeV. hep-ex/1307.4285v2
[13] Ma Y Q *et al* 2011 *Phys. Rev.* D **83** 111503
[14] Likhoded A *et al* 2013 Hadronic Properties of χ_c mesons at LHC hep-ph/1305.2389
[15] LHCb collaboration, Aaij R *et al* 2012 *Eur. Phys. J.* C **72** 2168
[16] LHCb collaboration, Aaij R *et al* 2012 *JHEP* **06** 141
[17] Baranov S P 2006 *Phys. Rev.* D **73** 074021
[18] Baranov S P *et al* 2011 *Phys. Lett.* B **705** 116
[19] Sjöstrand T *et al* 2006 *JHEP* **05** 026
[20] Skands P Z 2010 *Phys. Rev.* D**82** 074018
[21] Belyaev I *et al* 2010 *Nuclear Science Symposium Conf. Record (NSS/MIC)* 1155
[22] LHCb collaboration, Aaij R *et al* 2013 *Phys. Rev. Lett.* **110** 182001
[23] CDF collaboration, Aaltonen T *et al* 2009 *Phys. Rev.* D **80** 072003
[24] D0 collaboration, Abrzov V *et al* 2008 *Phys. Rev. Lett.* **101** 232002
[25] LHCb collaboration, Aaij R *et al* 2013 Measurement of J/ψ polarization in pp-collsiions at $\sqrt{s} = 7$ TeV. hep-ex/1307.6379

Production of Strange, Non-strange particles and Hypernuclei in an Excluded-Volume Model

S. K. Tiwari, and C. P. Singh

Department of Physics, Banaras Hindu University, Varanasi-221005, India

E-mail: sktiwari4bhu@gmail.com

Abstract. We present a systematic study of production of strange and non-strange hadron yields and their ratios obtained in various experiments using our thermodynamically consistent excluded-volume model. We also analyze the production of light nuclei, hypernuclei and their antinuclei in terms of our excluded-volume model over a broad energy range starting from Alternating Gradient Synchrotron (AGS) to Large Hadron Collider (LHC) energies. Further, we extend our model for studying rapidity spectra of hadrons produced in heavy-ion collisions.

1. Introduction

The production of strange particles in heavy-ion collisions is considered as one of the most important signals for the phase transition from hot, dense hadron gas (HG) to quark-gluon plasma (QGP) [1]. Recently we have analyzed the production of strange hadrons over a broad energy range from AGS to LHC energies using our new thermodynamically consistent excluded-volume model [2]. We have suitably incorporated the attractive interactions between hadrons through the addition of hadron resonances in our model. We assign an equal hard-core radius to each baryon in order to include repulsive interactions between them while mesons in the model are treated as pointlike particles. Using chemical freeze-out criteria as proposed by us [2], we calculate strange and non-strange hadron ratios and compare our results with the experimental data. We also study the production of light nuclei, hypernuclei and their antinuclei from AGS to LHC energies using our model and confront our model predictions with the experimental results. Moreover, we analyze the experimental data on rapidity distributions of various hadrons in terms of our model and elucidate the role of the longitudinal flow present in the medium formed in heavy-ion collisions. In the last, we present the conclusions and summary.

2. Formulation of Model

The grand canonical partition function for baryons in our thermodynamically consistent excluded-volume model can be written as follows [2, 3] :

$$lnZ_i^{ex} = \frac{g_i}{6\pi^2 T}\int_{V_i^0}^{V-\sum_j N_j V_j^0} dV \int_0^\infty \frac{k^4 dk}{\sqrt{k^2+m_i^2}}\frac{1}{\left[exp\left(\frac{E_i-\mu_i}{T}\right)+1\right]}, \qquad (1)$$

where g_i is the degeneracy factor of ith species of baryons, E_i is the energy of the ith particle ($E_i = \sqrt{k^2+m_i^2}$), V_i^0 is the eigen volume assigned to each baryon of ith species. We have assigned an equal volume $V^0 = 4\pi r^3/3$ for each baryon with a hard-core radius $r = 0.8\ fm$. Mesons in our model are considered as pointlike as they can penetrate each others volume. We have included all the baryons, mesons and their resonances having masses upto 2 GeV/c^2 in our model. To ensure strangeness conservation, we have imposed the condition of strangeness neutrality by considering $\sum_i S_i(n_i^s - \bar{n}_i^s) = 0$ where S_i is the strangeness of ith hadron.

Recently, Gorenstein [4] has criticized our model on the basis of thermodynamical and statistical consistency and claimed that the thermodynamical average of number of baryons is not equal to its statistical average in our model. We calculate these two averages in our model and find a relationship between them as follows [5] :

$$\overline{n} = \langle n \rangle \ - T\langle \frac{n^0}{(1-R)} V^0 \frac{\partial \overline{n}}{\partial \mu} \rangle. \tag{2}$$

We see that the thermal average of number of baryons is not exactly equal to its statistical average in our model and an extra term appears because $\overline{N}$, present in the available volume $(V \ - \ V^0\overline{N})$, is a function of μ. We call this extra term as "correction term" and plotted this term and $\overline{n}$ with respect to T at constant $\mu_B = 400 MeV$ in Fig. 1. We find that "correction term" gives an almost negligible contribution and hence can be ignored. Thus, our model can safely be considered as thermodynamically consistent [5].

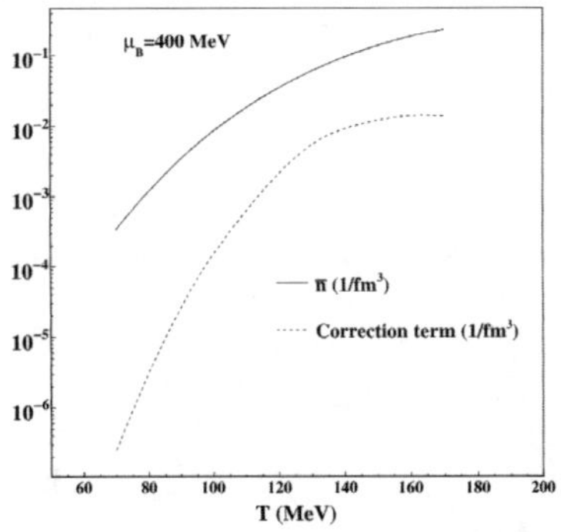

Figure 1. Variation of thermodynamical average of the number density of baryons and the "correction term" with respect to T at constant $\mu_B = 400\ MeV$ [5].

The rapidity distributions of baryons can be calculated by using following formula [6] :

$$\left(\frac{dN}{dy}\right)_{th} = \frac{g_i V \lambda_i}{2\pi^2}\left[(1-R) - \lambda_i \frac{\partial R}{\partial \lambda_i}\right] exp\left(\frac{-m_i\ coshy}{T}\right)\left[m_i^2 T + \frac{2 m_i T^2}{coshy} + \frac{2T^3}{cosh^2 y}\right], \tag{3}$$

where m_i is the mass of the ith species and V is the freeze-out volume of the system. After incorporating the longitudinal flow in our thermal model the formula for the rapidity distributions of baryons takes the following form [6] :

$$\frac{dN_i}{dy} = \int_{-\eta_{max}}^{\eta_{max}} \left(\frac{dN_i}{dy}\right)_{th} (y-\eta)\ d\eta, \tag{4}$$

where η_{max} is a free parameter related with the longitudinal flow velocity (β_L) [6].

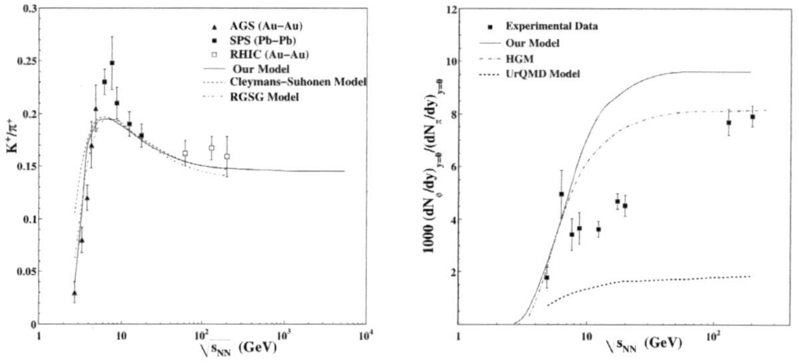

Figure 2. Left panel : Variation of K^+/π^+ ratio with respect to $\sqrt{s_{NN}}$. Solid line is the results of our model [2]. Dashed and dash-dotted lines are the results calculated in Cleymans-Suhonen model [7] and RGSG model [8], respectively. Symbols are the experimental data [2]. Right panel : Energy dependence of the ϕ/π, ($\pi = 1.5(\pi^+ + \pi^-)$) ratio at midrapidity. Dash-dotted line is the result of statistical hadron gas model (HGM) [9]. Symbols are the experimental data points [10]. Solid line represents our model calculation [2] and dashed line is the result of UrQMD model [11].

3. Results and Discussions

Figure 2 represents the variation of K^+/π^+ ratio with respect to center-of-mass energies ($\sqrt{s_{NN}}$). We have compared our model results with that of other models [7, 8] and find that our model result is closer to experimental data in comparison to other models. We have shown the variations of the ϕ/π, ($\pi = 1.5(\pi^+ + \pi^-)$) ratio at midrapidity ($y = 0$) with $\sqrt{s_{NN}}$ in Fig. 3. We compare our results with the statistical hadron gas model (HGM) [9] and the experimental data [10]. We find that our model provides a good description of the experimental data at lower energies but fails to describe the data at higher energies. HGM model describes the data well at the lower and upper energies while it completely disagrees with the experimental data at the intermediate energies. We also show the results of the hadronic transport model Ultra-relativistic Quantum Molecular Dynamical (UrQMD) [11] in which kaon coalescence mechanism is used for the production of ϕ. The predictions of UrQMD lie well below the experimental data at all energies which suggests that this mechanism fails badly in explaining the production mechanism of ϕ.

In Fig. 4, we present the rapidity distribution of anti-proton ($\bar{p}$) for central $Au-Au$ collisions at $\sqrt{s_{NN}} = 200\ GeV$ over full rapidity range. Dotted line shows the distribution of $\bar{p}$ due to stationary thermal source which describes only the midrapidity data while it fails to describe the experimental data at other rapidities. Solid line shows the rapidity distributions of $\bar{p}$ after incorporation of the longitudinal flow in our thermal model and it gives a good agreement with the experimental data [12]. After fitting the experimental data, we get the value of $\eta_{max} = 3.20$ and hence the longitudinal flow velocity $\beta_L = 0.922$ at $\sqrt{s_{NN}} = 200\ GeV$. In Fig. 5, we show the variations of $\Lambda/p, d/p$, H^3_Λ/He^3 and $\bar{H}^3_{\bar{\Lambda}}/\bar{H}e^3$ with $\sqrt{s_{NN}}$. Comparisons with the experimental data yield a good agreement between them [6].

4. Conclusions

In conclusion, we find that our model provides a good fit to the experimental data on ratios of hadron yields as well as mid-rapidity densities of various particles. We present an analysis of rapidity distributions of hadrons at various center-of-mass energies using our model. We see that

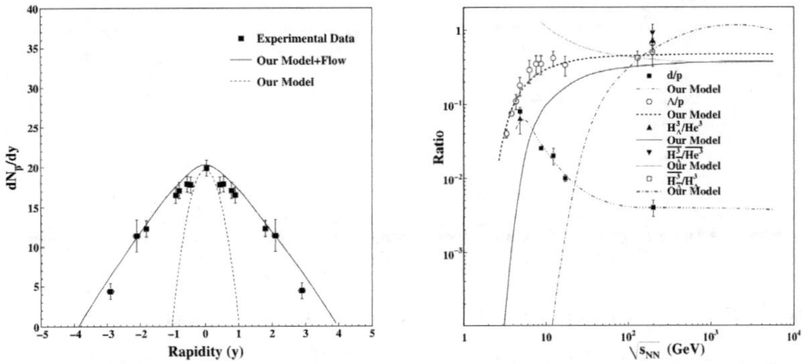

Figure 3. Left panel : Rapidity distributions for $\bar{p}$ in central $Au - Au$ collisions at $\sqrt{s_{NN}} = 200\ GeV$. Lines are our model calculations and symbols represent the experimental data. Right panel : The energy dependence of ratios of various baryons, nuclei, and their antinuclei. Lines are our model calculations and symbols represent the experimental data [6].

our thermal model alone cannot describe the experimental data fully unless we incorporate flow velocity in the longitudinal direction and as a result our modified model predictions show a good agreement with the experimental data. Although, thermal model fails in case of multistrange particles which suggests that a somewhat different type of mechanism e.g., coalescence model is required. Further, we have analyzed the production of nuclei and hypernuclei in terms of our model which describe the experimental data very well.

Acknowledgments
SKT is grateful to Council of Scientific and Industrial Research (CSIR), New Delhi for providing a research grant.

References

[1] Singh C P 1993 *Phys. Rep.* **236** 147-224; 1992 *Int. J. Mod. Phys.* A **7** 7185.
[2] Tiwari S K, Srivastava P K and Singh C P 2012 *Phys. Rev.* C **85** 014908.
[3] Singh C P, Srivastava P K and Tiwari S K 2009 *Phys. Rev.* D **80** 114508 ; 2011 *Phys. Rev.* D **83** 039904(E); Srivastava P K, Tiwari S K and Singh C P 2010 *Phys. Rev.* D **82** 014023; 2011 *Nucl. Phys.* A **862-863 C** 424-426.
[4] Gorenstein M I 2012 *Phys. Rev.* C **86** 044907.
[5] Tiwari S K and Singh C P 2013 *Advances in High Energy Physics* **2013** 805413.
[6] Tiwari S K, Srivastava P K and Singh C P 2013 *J. Phys.* G **40** 045102.
[7] Cleymans J and Suhonen E 1987 *Z. Phys.* C **37** 51-56.
[8] Rischke D H, Gorenstein M I, Stöcker H and Greiner W 1991 *Z. Phys.* C **51** 485-489.
[9] Andronic A, Braun-Munzinger P and Stachel J 2006 *Nucl. Phys.* A **772** 167-199.
[10] Blume C and Markert C 2011 *Prog. Part. Nucl. Phys.* **66** 834-879.
[11] Bleicher M *et al.* 1999 *J. Phys.* G **25** 1859.
[12] Bearden I G *et al.* 2004 *Phys. Rev. Lett.* **93** 102301.

Journal of Physics: Conference Series **509** (2014) 012098 doi:10.1088/1742-6596/509/1/012098

Strangeness at high temperatures

Christian Schmidt (for BNL-Bielefeld Collaboration)
Universität Bielefeld, Fakultät für Physik, Postfach 100131, D-33501 Bielefeld, Germany
E-mail: schmidt@physik.uni-bielefeld.de

Abstract. We use up to fourth order cumulants of net strangeness fluctuations and their correlations with net baryon number fluctuations to extract information on the strange meson and baryon contribution to the low temperature hadron resonance gas, the dissolution of strange hadronic states in the crossover region of the QCD transition and the quasi-particle nature of strange quark contributions to the high temperature quark-gluon plasma phase.

1. Introduction

At RHIC and the LHC, quark gluon plasma is created by means of heavy ion collisions. The initial conditions of such collisions are free of net strangeness. It is theoretically well understood [1] and experimentally verified [2] that many strange, anti-strange quarks are created during these collisions. This will ultimately lead to event-by-event fluctuations of strangeness, as long as only a fixed subsystem (specific kinetic window) is observed. To what extent the created fireball reaches chemical equilibrium before its hadronic freeze out — depending on collision energy and system size — is, however, still under debate. Furthermore, it is questionable whether strange hadrons also freeze-out in the critical crossover region ($T \approx T_c$) that is controlled by the light quark sector. Based on the very smooth behavior of strangeness fluctuations [3, 4] it has been suggested [5, 6] that deconfinement for strange quarks takes place at a temperature significantly larger than T_c, determined by chiral symmetry breaking. This would imply the existence of strange bound sates inside the QGP for some temperatures $T > T_c$. A detailed analysis of various cumulants of net strangeness fluctuations and baryon number strangeness correlations obtained by the BNL-Bielefeld Collaboration [7] supports, however, the statement that the strangeness liberation temperature is consistent with T_c. We will briefly repeat this study here. We obtain these cumulants as derivatives of the logarithm of the QCD partition function, or equivalently of the pressure (P) in units of T^4, with respect to the chemical potentials. In general we define

$$\chi_{mn}^{XY} = \left. \frac{\partial^{(m+n)}[P(T,\hat{\mu}_X,\hat{\mu}_Y)/T^4]}{\partial\hat{\mu}_X^m \partial\hat{\mu}_Y^n} \right|_{\vec{\mu}=0} , \quad (1)$$

where $X, Y = B, S, Q$ are the quantum numbers of net baryon (B), net strangeness (S) and net electric charge (Q) and $\hat{\mu}_X = \mu_X/T$ are the corresponding chemical potentials with $\vec{\mu} = (\mu_B, \mu_S, \mu_Q)$. We will also use the notations $\chi_{0n}^{XY} \equiv \chi_n^Y$ and $\chi_{m0}^{XY} \equiv \chi_m^X$. The various χ_{mn}^{XY} cary information on the quantum numbers of the relevant degrees of freedom in the system [8, 9]. Note, however, that the quantities χ_{mn}^{XY} also define the Taylor expansion coefficients of the pressure with respect to $\hat{\mu}_X$ and $\hat{\mu}_Y$. Their study is thus important also for other reasons such as a critical point search [10, 11] or the analysis of the freeze out properties of heavy ion collisions [12]. In the following, we will compare our non-perturbative lattice QCD results to an uncorrelated gas of hadrons at low temperatures and to an uncorrelated gas of quarks at high temperatures, with special emphasis on the strangeness fluctuations. Here the lattice calculations have been performed with highly improved staggered quarks (HISQ) and almost physical quark masses (the ratio of light to strange quark mass is 1/20, with a strange quark mass that was fixed to its physical value). The lattice sizes have been chosen as $N_\sigma^3 \times N_\tau$, with aspect ratio $N_\sigma/N_\tau = 4$ and $N_\sigma = 24$ and 32. A continuum limit is still to be taken and will require at least one even finer lattice spacing. However, from the stability of our current results we estimate very small effects of the nonzero lattice spacing. For the comparison with the HRG we take all particles and resonances up to a mass cutoff of 2.5 GeV into account.

2. Strangeness in the hadronic phase

Strange hadrons are relatively heavy. Even the lightest strange meson ($K^\pm$) is with 493 MeV heavy enough to be treated in Boltzmann approximation, in the relevant temperature interval just around the chiral crossover ($T_c = 154 \pm 9$ MeV [13]). The pressure that is build up within an uncorrelated gas of all known strange hadrons and resonances (P_S^{HRG}) receives contributions from 4 different strangeness sectors, namely from strange mesons ($|S| = 1, M$) and strange baryons ($|S| = i, B$) with $i = 1, 2, 3$, respectively. Assuming Boltzmann statistics, the μ_B and μ_S dependence is simple and solely determined by the baryonic and strange charges of the hadrons, *i.e.* grouped into the 4 different sectors we find for the partial pressure of all strange hadrons:

$$\begin{aligned} P_S^{\mathrm{HRG}}(T,\hat\mu_B,\hat\mu_S) &= P_{|S|=1,M}^{\mathrm{HRG}}(T)\cosh(\hat\mu_S) + P_{|S|=1,B}^{HRG}(T)\cosh(\hat\mu_B-\hat\mu_S) \\ &+ P_{|S|=2,B}^{HRG}(T)\cosh(\hat\mu_B-2\hat\mu_S) + P_{|S|=3,B}^{HRG}(T)\cosh(\hat\mu_B-3\hat\mu_S)\,. \end{aligned} \quad (2)$$

As long as strange hadrons, with baryonic and strange charges that fall into the before mentioned strangeness sectors, are a valid description of the strangeness carrying degrees of freedom in the system, Eq. (2) will imply various relations between correlations of net baryon number and net strangeness χ_{mn}^{BS}, that can be checked in (lattice) QCD. In particular we can use these relations to calculate the partial pressures of the different strangeness sectors from all fluctuations and correlations χ_{mn}^{BS} up to fourth order, *i.e.* with $0 \le m \le 4$, $1 \le n \le 4$, $(m+n) \le 4$ and $m+n$ even, we can use 6 different BS-correlations. It is easy to see that according to Eq. (2), all 6 χ_{mn}^{BS} that we use, can be written as linear combinations of the 4 partial pressures. To obtain the partial pressures we have to invert this linear mapping. We find that the linear mapping has a kernel of dimension 2, consequently our solution is not uniquely defined but will depend on two parameters c_1 and c_2. As a result, we find the operators that project onto the 4 different partial pressures, given by

$$M(c_1,c_2) = \chi_2^S - \chi_{22}^{BS} + c_1v_1 + c_2v_2\,, \quad (3)$$

$$B_1(c_1,c_2) = \frac{1}{2}\left(\chi_4^S - \chi_2^S + 5\chi_{13}^{BS} + 7\chi_{22}^{BS}\right) + c_1v_1 + c_2v_2\,, \quad (4)$$

$$B_2(c_1,c_2) = -\frac{1}{4}\left(\chi_4^S - \chi_2^S + 4\chi_{13}^{BS} + 4\chi_{22}^{BS}\right) + c_1v_1 + c_2v_2\,, \quad (5)$$

$$B_3(c_1,c_2) = \frac{1}{18}\left(\chi_4^S - \chi_2^S + 3\chi_{13}^{BS} + 3\chi_{22}^{BS}\right) + c_1v_1 + c_2v_2\,, \quad (6)$$

where $v_1 = \chi_{31}^{BS} - \chi_{11}^{BS}$ and $v_2 = \frac{1}{3}(\chi_2^S - \chi_4^S) - 2\chi_{13}^{BS} - 4\chi_{22}^{BS} - 2\chi_{31}^{BS}$ are 2 linear independent basis vectors of the kernel. Therefore, as long as the uncorrelated gas of hadrons is a valid description of QCD, the combinations v_1 and v_2 have to vanish[1] and do not contribute to

[1] Of course, the uncorrelated gas model will be an approximation at all temperatures and v_1 and v_2 are thus not be expected to vanish exactly , *i.e.* they are no order parameters.

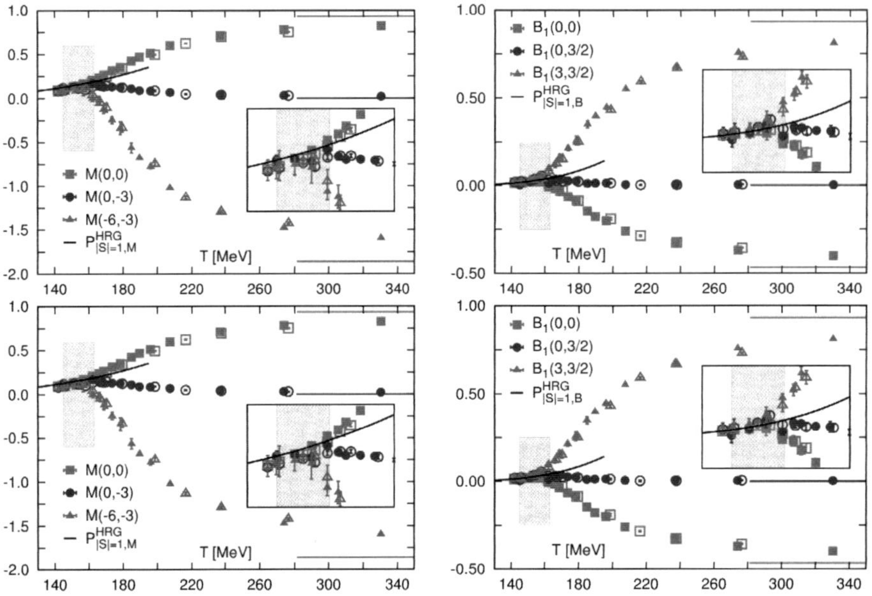

Figure 1. Operators that project for low temperatures onto the partial pressure of strange mesons (top-left), baryons with strangeness $|S| = 1$ (top-right), $|S| = 2$ (bottom-left) and $|S| = 3$ (bottom-right). Results from $N_\tau = 6$ and 8 lattices are shown by open and full symbols, respectively. Also shown are results from the HRG model (black solid lines). Vertical bands indicate the chiral corossover temperature [13].

Eqs. (3-6). In Fig. 1, we show our results for the partial pressure, obtained in the 4 different strangeness sectors, *i.e.* we plot our data for the strangeness fluctuations and correlations in the combinations given by Eqs. (3-6) for 3 different choices of the parameter c_1 and c_2, respectively (color coded). We find that for temperatures $T \lesssim 160$ MeV, all three different combinations agree, in each of the four sectors. In addition we find agreement with the HRG results shown as solid black lines, in the same temperature range. One can thus conclude that for $T \lesssim 160$ MeV, an uncorrelated gas of hadrons is compatible with the relevant strange degrees of freedom in QCD. The same conclusion can be drawn from Fig. 2 (left), where we show the combinations v_1 and v_2, that serve as an indicator for the validity of the HRG, *i.e.* they vanish as long as the strange degrees of freedom can be described by hadrons. Here we also show the difference of second and fourth order net baryon number fluctuations ($\chi_2^B - \chi_4^B$). This quantity also vanishes in the HRG but, in contrast to v_1, v_2 receives contributions from all baryons, including the light sector. All quantities shown in Fig. 2 (left) behave rather similar and in particular break away from the HRG in the crossover region $T \sim (154 \pm 9)$ MeV. We thus have no indication that strange hadrons survive the chiral crossover.

3. The plasma phase

Finally we probe the strange degrees of freedom in the high temperature phase, were we expect to find strangeness to be carried by quasi free quarks. As strange quarks carry baryon number $1/3$ and electric charge $-1/3$, an interchange of derivatives with respect to $\hat{\mu}_B$ and $\hat{\mu}_Q$ in a gas of free strange quarks will only lead to a sign change. The ratio $-\chi_{211}^{BQS}/\chi_{121}^{BQS}$ thus reaches unity for large temperatures were quarks behave as free particles. In Fig. 2 (right), where we plot this ratio, we find deviations from the free quark gas result, which are of the order of $(10-15)\%$ for

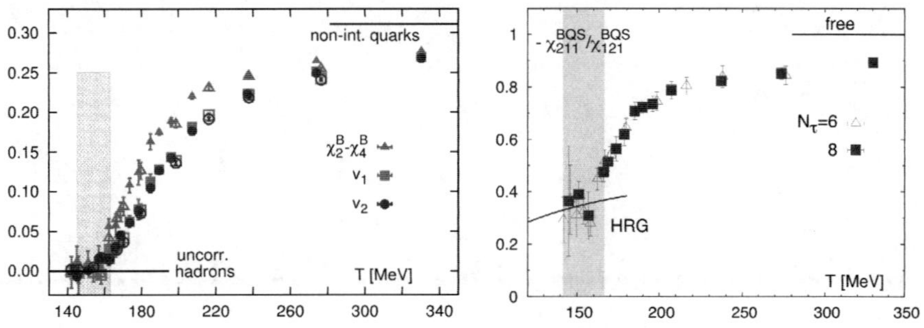

Figure 2. Left: quantities that vanish within and uncorrelated gas of hadrons (Boltzmann approximation). Right: quantity that reaches unity in an uncorrelated gas of quark and gluons. In both cases we show results from $N_\tau = 6$ and 8 lattices by open and full symbols, respectively

$T \gtrsim 300$ MeV. Similar deviations have also been found in other bulk thermodynamic quantities like pressure and energy density. In the hadronic sector, this ratio projects onto strange charged baryons. Here the charge one and two sectors are weighted differently by the cumulants χ_{211}^{BQS} and χ_{121}^{BQS} and the ratio will be less than unity. We have also plotted the HRG result in Fig. 2 (right), and find agreement with the HRG for $T \lesssim 160$ MeV.

Acknowledgement
CS acknowledges support by the BMBF under grant 05P12PBCTA, the numerical calculations have been performed using the USQCD GPU-clusters at JLab, the Bielefeld GPU cluster and the NYBlue at the NYCCS.

References
[1] Jacak B V and Muller B 2012 *Science* **337** 310
Jacak B and Steinberg P 2010 *Phys. Today* **63N5** 39
[2] Blume C and Markert C 2011 *Prog. Part. Nucl. Phys.* **66** 834
Agakishiev G *et al.* [STAR Collaboration] 2012 *Phys. Rev. Lett.* **108** 072301
Abelev B I *et al.* [STAR Collaboration] 2010 *Phys. Rev.* C **81** 044902
Nicassio M [ALICE Collaboration] 2012 *Acta Phys. Polon. Supp.* **5** 237
Lara C E P [ALICE Collaboration] arXiv:1303.6496 [nucl-ex]
[3] Borsanyi S *et. al* 2012 *JHEP* **1201** 138
[4] Bazavov A *et al.* [HotQCD Collaboration] 2012 *Phys. Rev.* D **86** 034509
[5] Aoki Y, Fodor Z, Katz S D and Szabo K K 2006 *Phys. Lett.* B **643** 46
Aoki Y *et al.* 2009 *JHEP* **0906** 088
Borsanyi S *et al.* [Wuppertal-Budapest Collaboration] 2010 *JHEP* **1009** 073
[6] Ratti C, Bellwied R, Cristoforetti M and Barbaro M 2012 *Phys. Rev.* D **85** 014004
Bellwied R, Borsanyi S, Fodor Z, Katz S D and Ratti C arXiv:1305.6297 [hep-lat]
[7] Bazavov A, Ding H T, Hegde P, Kaczmarek O, Karsch F, Laermann E, Maezawa Y and Mukherjee S *et al.* 2013 *Phys. Rev. Lett.* 111 **082301**
[8] Koch V, Majumder A and Randrup J 2005 *Phys. Rev. Lett.* **95** 182301
[9] Ejiri S, Karsch F and Redlich K 2006 *Phys. Lett.* B **633** 275
[10] Datta S, Gavai R V and Gupta S 2013 *Nucl. Phys.* A904-905 **2013** 883c
[11] Schmidt C 2010 *Prog. Theor. Phys. Suppl.* **186** 563
[12] Bazavov A, Ding H T, Hegde P, Kaczmarek O, Karsch F, Laermann E, Mukherjee S and Petreczky P *et al.* 2012 *Phys. Rev. Lett.* **109** 192302
[13] Bazavov A *et al.* 2012 *Phys. Rev.* D **85** 054503

Journal of Physics: Conference Series **509** (2014) 012099 doi:10.1088/1742-6596/509/1/012099

Systematic Properties of the Tsallis Distribution: Energy Dependence of Parameters.

J. Cleymans

UCT-CERN Research Centre and Department of Physics
University of Cape Town, Rondebosch, South Africa

E-mail: `jean.cleymans@uct.ac.za`

Abstract. Changes in the transverse momentum distributions with beam energy are studied using the Tsallis distribution as a parameterization. The dependence of the Tsallis parameters q, T and the volume on beam energy is determined. The Tsallis parameter q shows a weak but clear increase with beam energy with the highest value being approximately 1.15. The Tsallis temperature and volume are consistent with being independent of beam energy within experimental uncertainties.

In the analysis of new data, a Tsallis-like distribution gives excellent fits to the transverse momentum distributions as shown by the STAR [1] and PHENIX [2] collaborations at RHIC and by the ALICE [3], ATLAS [4] and CMS [5] collaborations at the LHC. In this talk we review the parameterization used by these groups and propose a slightly different one which leads to a more consistent interpretation and has the bonus of being thermodynamically consistent [7].

In the framework of Tsallis statistics [6, 8, 9, 10, 11] the entropy, S, the particle number, N, the energy E and the pressure P are given by corresponding integrals over the Tsallis distribution:

$$f \equiv \left[1+(q-1)\frac{E-\mu}{T}\right]^{-\frac{1}{q-1}} . \tag{1}$$

where T and μ are the temperature and the chemical potential, q is a new variable often refereed to as Tsallis parameter. The relevant thermodynamic quantities are given by (see e.g. [11]

$$S = -gV\int\frac{d^3p}{(2\pi)^3}\left[f^q\ln_q f - f\right], \tag{2}$$

$$N = gV\int\frac{d^3p}{(2\pi)^3}f^q, \tag{3}$$

$$E = gV\int\frac{d^3p}{(2\pi)^3}\sqrt{p^2+m^2}f^q, \tag{4}$$

$$P = g\int\frac{d^3p}{(2\pi)^3}\frac{p^2}{3E}f^q. \tag{5}$$

V is the volume, g is the degeneracy factor. The short-hand notation

$$\ln_q(x) \equiv \frac{x^{1-q}-1}{1-q}, \tag{6}$$

often referred to as q-logarithm has been used. These expressions are thermodynamically consistent, e.g. it can be shown [11] that consistency relations of the type

$$N = V \left.\frac{\partial P}{\partial \mu}\right|_T , \tag{7}$$

and

$$\epsilon + P = Ts + \mu n \tag{8}$$

(where n, s, ϵ refer to the densities of the corresponding quantities) are indeed satisfied.
Following from Eq. (3), the momentum distribution is given by,

$$\frac{d^3N}{d^3p} = \frac{gV}{(2\pi)^3}\left[1 + (q-1)\frac{E-\mu}{T}\right]^{-q/(q-1)} , \tag{9}$$

or, expressed in terms of transverse momentum, p_T, transverse mass, $m_T \equiv \sqrt{p_T^2 + m^2}$, and rapidity y

$$\frac{d^2N}{dp_T\, dy} = gV\frac{p_T m_T \cosh y}{(2\pi)^2}\left[1 + (q-1)\frac{m_T\cosh y - \mu}{T}\right]^{-q/(q-1)} . \tag{10}$$

At mid-rapidity, $y = 0$, and for zero chemical potential, as is relevant at the LHC, this reduces to

$$\left.\frac{d^2N}{dp_T\, dy}\right|_{y=0} = gV\frac{p_T m_T}{(2\pi)^2}\left[1 + (q-1)\frac{m_T}{T}\right]^{-q/(q-1)} . \tag{11}$$

In the limit where the parameter q goes to 1 it is well-known that this reduces the standard Boltzmann distribution. The parameterization given in Eq. (10) is close to the one used in [1, 2, 3, 4, 5]:

$$\frac{d^2N}{dp_T\, dy} = p_T\frac{dN}{dy}\frac{(n-1)(n-2)}{nC(nC + m_0(n-2))}\left[1 + \frac{m_T - m_0}{nC}\right]^{-n} , \tag{12}$$

where n (not to be confused with the particle density) and C are fit parameters and has been discussed in detail in Ref. [7]. At mid-rapidity $y = 0$ and zero chemical potential, they have a similar dependence on the transverse momentum as Eq. (11) apart from an additional factor m_T on the right-hand side. However, the inclusion of the rest massis not in agreement with the Tsallis distribution as it breaks m_T scaling which is present in Eq. (10) but not in Eq. (12). The inclusion of the factor m_T leads to a more consistent interpretation of the variables q and T.

A very good description of transverse momenta distributions at RHIC has also been obtained in Refs [12, 13] on the basis of a coalescence model where the Tsallis distribution is used for quarks. Tsallis fits have also been considered in Ref. [14, 15, 16] but with a different power law leading to smaller values of the Tsallis parameter q.

Interesting results were obtained in Refs. [17, 18] where spectra for identified particles were analyzed and the resulting values for the parameters q and T were considered.
It has been shown that, at lower energies it is possible to describe transverse momentum spectra in $p-p$ collisions with Boltzmann distributions provided one takes into account the decays of resonances [19]. This is not in contradiction with the results presented here as the transverse momentum distributions become more and more Boltzmann like at lower energies as can be seen in the values of q which are closer to one as the energy is decreased.

The energy dependence in $p-p$ collisions can be determined by studying data for charged particles at beam energies of 0.54 [20], 0.9, 2.36 and 7 TeV [3, 4, 5]. These involve distributions summed over charged particles. The fits were performed using a sum of three Tsallis distributions, the first one for π^+, the second one for K^+ and the third one for protons p. The relative weights between these were simply determined by the corresponding degeneracy factors, i.e. 1 for for π^+ and K^+ and 2 for protons. The fit was taken at mid-rapidity and for $\mu = 0$ using the following expression:

$$\left.\frac{1}{2\pi p_T}\frac{d^2N(\text{charged particles})}{dp_T dy}\right|_{y=0} = \frac{2V}{(2\pi)^3}\sum_{i=1}^{3} g_i m_{T,i}\left[1+(q-1)\frac{m_{T,i}}{T}\right]^{-\frac{q}{q-1}}, \tag{13}$$

where $i = (\pi^+, K^+, p)$ and $g_{\pi^+} = 1$, $g_{K^+} = 1$ and $g_p = 2$. The factor 2 in front of the right hand side of this equation takes into account the contribution of the antiparticles $(\pi^-, K^-, \bar{p})$. A comparison with charged particle distributions is shown in Fig. 1. It is to be noted that the Tsallis distribution presented above also gives an excellent description of transverse momentum distributions in $p - Pb$ collisions at all pseudorapidity intervals obtained by the ALICE collaboration [21]. The Tsallis parameters q and T needed to describe the transverse momentum distributions of charged particles are shown in Fig. 2. The value of qt has a tendency to increase slowly with increasing energy [7] while no clear energy dependence for T can be discerned.

In conclusion, the Tsallis distribution leads to excellent fits to the transverse momentum

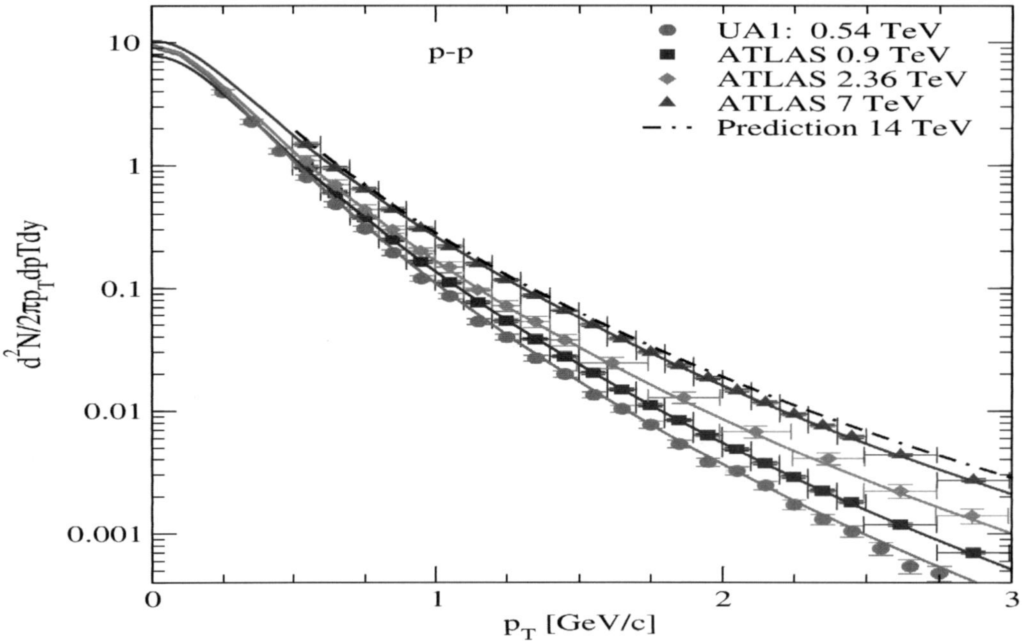

Figure 1. Fits to transverse momentum distributions of charged particles [4, 20] using the Tsallis distribution.

distributions in high energy $p - p$ and $p - Pb$ collisions. By comparing results from UA1 [20] to results obtained at the LHC [4, 5] it has been possible to extract the parameters q, T and V for a wide range of energies [7]. A consistent picture emerges from a comparison of fits using the Tsallis distribution in high energy $p - p$ collisions.

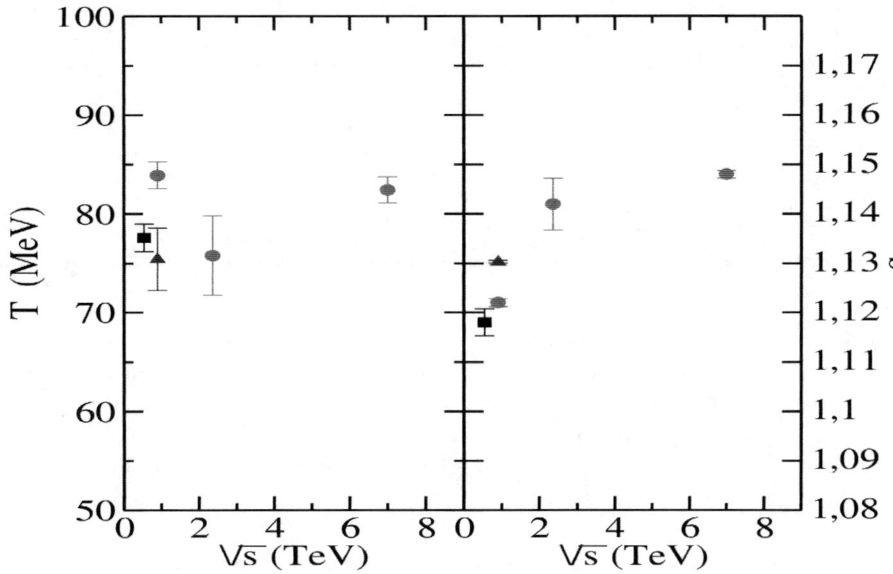

Figure 2. Values of the Tsallis parameters T (left) and q (right), as a function of beam energy, obtained from fits to the transverse momentum distributions of charged particles as described in the text. The square point is from [20], the round points are from [4] while the triangle point is from [3].

References

[1] Abelev B I *et al* 2007 *Phys. Rev. C* **75** 064901
[2] Adare A *et al* 2010 *Phys. Rev. C* **83** 052004; 2011 *Phys. Rev. C* **83** 064903
[3] Aamodt K *et al* 2011 *Eur. Phys. J. C* **71** 1655; 2010 *Phys. Lett. B* **693** 53
[4] Aad G *et al* 2011 *New J. Phys.* **13** 053033
[5] Khachatryan V *et al* 2010 *Phys. Rev. Lett.* **105** 022002
[6] Tsallis C 1988 *J. Statist. Phys.* **52** 479
[7] Cleymans J, Lykasov G I, Parvan A S, Sorin A S, Teryaev O V, Worku D 2013 *Phys. Lett. B* **723** 351
[8] Tsallis C, Mendes R S, Plastino A R 1998 *Physica A* **261** 534
[9] Conroy J M, Miller H G and A. R. Plastino A R 2010 *Phys. Lett. A* **374** 4581
[10] Cleymans J and Worku D 2012 *J. Phys. G* **39** 025006
[11] Cleymans J and Worku D 2012 *Eur. Phys. J. A* **48** 160
[12] Ürmössy K, Biró T S 2010 *Phys. Lett. B* **689** 14
[13] Ürmössy K, Biró T S 2009 *J. Phys. G* **36** 064044
[14] Wong C-Y, Wilk G 2012 *Acta Physica Polonica* **43** 2047
[15] Wibig T 2010 *J. Phys. G: Nucl. Part. Phys.* **37** 115009
[16] Wibig T and Kurp I 2003 *JHEP* **0312** 039
[17] Marques L, Andrade-II E, Deppman A arXiv:1210.1725[hep-ph]
[18] Sena I, Deppman A 2013 *Eur. Phys. J. A* **49** 17
[19] Becattini F and Passaleva G 2002 *Eur. Phys. J. C* **23** 551
[20] Albajar C *et al* 1990 *Nucl. Phys. B* **335** 261
[21] Abelev B I *et al* 2013 *Phys. Rev. Lett.* **110** 082302

Journal of Physics: Conference Series **509** (2014) 012100 doi:10.1088/1742-6596/509/1/012100

Elliptic Flow from fKLN Initial Conditions

F. Scardina , M. Ruggieri, S. Plumari, V.Greco

Department of Physics and Astronomy, University of Catania, Via S. Sofia 64, I-95125 Catania
INFN-Laboratori Nazionali del Sud, Via S. Sofia 62, I-95123 Catania, Italy

E-mail: scardinaf@lns.infn.it

Abstract. A current goal of ultra-relativistic heavy ion collisions experiments is to verify a picture in which the two colliding nuclei can be described as two tiny disks of a Color Glass Condensate (CGC) that is the limiting state of QCD matter at very high density and energy. The searching of CGC in uRHIC experiments is strongly related to the evaluation of the η/s. In fact in viscous hydrodynamics simulations a standard Glauber initial condition leads to estimate $\eta/s \sim 1/4\pi$, while the fKLN initialization, which model the melting of Color Glass Condensate, leads to at least a factor of 2 larger η/s. We have found in the framework of kinetic theory that the out-of-equilibrium initial distribution described by the fKLN reduces the efficiency in building-up the elliptic flow. In particular from our work emerges that the available data on v_2 are in agreement with a $\eta/s \sim 1/4\pi$ also for fKLN initial conditions.

1. Introduction

The experiments of Ultra-relativistic heavy-ion collisions (uRHICs) performed at the Relativistic Heavy-Ion Collider (RHIC) and at the Large Hadron Collider (LHC) offer the possibility to create a strongly interacting quark-gluon plasma (QGP) [1, 2]. For non central collisions the overlap region between the two colliding nuclei is strongly deformed in the transverse plane and a measure of such a deformation is given by the spatial eccentricity. As a consequence of this spatial deformation the pressure gradient in the x direction of the transverse plane is larger than the pressure gradient in the y direction and this leads to a deformation into the momentum space. Such a deformation can be evaluated in terms of a fourier expansion of the transverse momentum spectra. The second coefficient, who accounts for the largest contribution, is the elliptic flow v_2. The elliptic flow is both sensitive to the initial condition in the overlap region and to the ratio between the shear viscosity and the entropy density (η/s) of the expanding fireball [3, 4]. The smaller is η/s the larger is the elliptic flow. Therefore measuring the v_2 gives the possibility to know the viscosity of the plasma [3, 4]. However there is a source of uncertainty in the determination of η/s due to the lack of knowledge regarding the initial spatial eccentricity, for which one does not have a direct experimentally access and that can only be estimated using theoretical models.

A widely used model to describe the initial condition is the Glauber model in which the initial profile of the fireball is given by the geometrical superposition of the profiles of the two colliding nuclei. However the heavy ion collisions at very high energies offer the possibility to verify the picture in which the two colliding nuclei can be described as two tiny disks of a Color Glass Condensate [5, 6]. In this proceedings we discuss the results obtained for the elliptic flow using these two different initial conditions [7].

2. The factorized KLN model

One model widely used to describe the distribution of particle $(dN_g/dyd^2\boldsymbol{x}_\perp)$ created after the collisions of the two nuclei consisting of color glass condensate is the factorized KLN model (fKLN) [5, 8, 6] in which:

$$\frac{dN_g}{dyd^2\boldsymbol{x}_\perp} = \int d^2\boldsymbol{p}_T\, p_A(\boldsymbol{x}_\perp)p_B(\boldsymbol{x}_\perp)\Phi(\boldsymbol{p}_T, \boldsymbol{x}_\perp, y) \,, \tag{1}$$

where Φ rapresent the momentum space distribution

$$\Phi(\boldsymbol{p}_T, \boldsymbol{x}_\perp, y) = \frac{4\pi^2 N_c}{N_c^2-1}\frac{1}{p_T^2}\int^{p_T} d^2\boldsymbol{k}_T \alpha_S(Q^2) \times \phi_A(x_1, k_T^2; \boldsymbol{x}_\perp) \times \phi_B(x_2, (\boldsymbol{p}_T - \boldsymbol{k}_T)^2; \boldsymbol{x}_\perp) \tag{2}$$

here $x_{1,2} = p_T \exp(\pm y)/\sqrt{s}$ and the ultraviolet cutoff p_T is assumed to be equal to 3 GeV/c; α_S is the strong coupling constant and $p_{A,B}$ appearing in Eq.1 denotes the probability to find one nucleon at a given transverse coordinate, $p_A(\boldsymbol{x}_\perp) = 1-(1-\sigma_{in}T_A(\boldsymbol{x}_\perp)/A)^A$, in which σ_{in} is the inelastic cross section and T_A corresponds to the usual thickness function of the Glauber model. The saturation behaviour is built in the unintegrated gluon distribution functions (uGDF) $\phi_A(x_1, k_T^2; \boldsymbol{x}_\perp)$ and $\phi_B(x_1, k_T^2; \boldsymbol{x}_\perp)$ which refer respectively to nucleus A and B that can be written as follows

$$\phi_A(x_1, k_T^2; \boldsymbol{x}_\perp) = \frac{\kappa\, Q_s^2}{\alpha_s(Q_s^2)}\left[\frac{\theta(Q_s - k_T)}{Q_s^2+\Lambda^2} + \frac{\theta(k_T - Q_s)}{k_T^2+\Lambda^2}\right] \tag{3}$$

a similar equation holding for partons belonging to nucleus B. Following [8] we take the saturation scale for the nucleus A as

$$Q_{s,A}^2(x, \boldsymbol{x}_\perp) = Q_0^2\left(\frac{T_A(\boldsymbol{x}_\perp)}{1.53p_A(\boldsymbol{x}_\perp)}\right)\left(\frac{0.01}{x}\right)^\lambda \,, \tag{4}$$

and similarly for nucleus B, with $\lambda = 0.28$ and $Q_0^2 = 2\ GeV^2$. This choice is the one adopted in fKLN or MC-KLN and in hydro simulations [8, 3] to study the dependence of $v_2(p_T)$ on η/s.

3. Dynamical evolution of the QGP

We employ transport theory to describe the evolution of the QGP created in relativistic heavy-ion collisions. [9, 10, 11], The time evolution of the gluons distribution function $f(\boldsymbol{x}, \boldsymbol{p}, t)$ is governed by the Boltzmann equation:

$$\begin{aligned} p_\mu \partial^\mu f_1 &= \int d\Gamma_2 d\Gamma_{1'} d\Gamma_{2'}(f_{1'}f_{2'} - f_1 f_2) \\ &\quad \times|\mathcal{M}|^2\delta^4(p_1+p_2-p_{1'}-p_{2'}) \,, \end{aligned} \tag{5}$$

where $d^3\boldsymbol{p}_k = 2E_k(2\pi)^3 d\Gamma_k$ and $\mathcal{M}$ corresponds to the transition amplitude. Usually a key input ingredient of a transport approach is the knowledge of the cross section; here we reverse it and, with the aim of creating a more direct link with viscous hydrodynamics, we determine $|\mathcal{M}|^2$ in order to have a fixed η/s, for more details see [12, 13, 7]. We have carried out simulations of the QGP created in $Au+Au$ collision at $\sqrt{s} = 200\,AGeV$ with $b = 7.5\,$fm, belonging to the $20-30\%$ centrality class. The standard initial condition for simulations of the plasma fireball created at RHIC is a $\boldsymbol{x}$-space distribution given by the Glauber model and a $\boldsymbol{p}$-space thermalized spectrum in the transverse plane at a time $\tau = 0.6\,$fm/c with a maximum initial temperature $T_0 = 340$MeV. In this case, for a standard mixture of N_{part} and N_{coll} we find

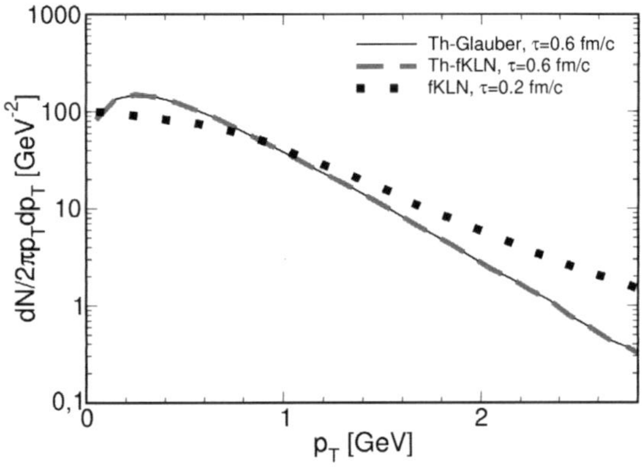

Figure 1. Transverse momentum distributions within $|y| \leq 0.5$ for the different initial conditions as in the legend for Au+Au at $\sqrt{s} = 200GeV$ and $b = 7.5\ fm$

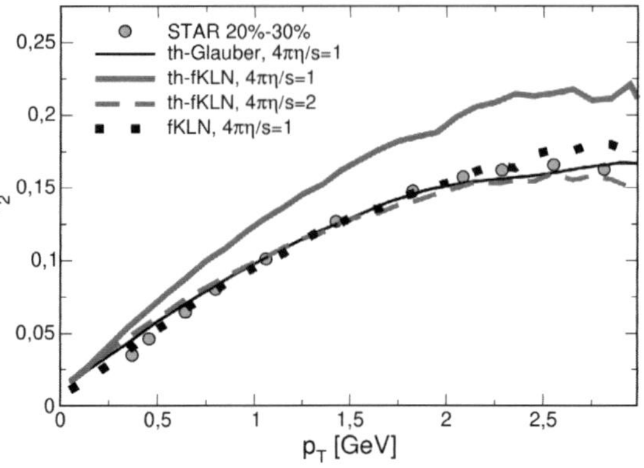

Figure 2. Elliptic flow for the different initial conditions at RHIC (Au+Au collisions at $\sqrt{s} = 200GeV$), with $b = 7.5$

$\epsilon_x = 0.284$. We will refer to this case as *Th-Glauber* and it is indicated by the thin line in Fig. 1. Instead the study of the impact of an initial CGC state has been performed considering an $\boldsymbol{x}$-space distribution given by the fKLN and a thermalized p_T distribution at $\tau_0 \sim 0.6 fm/c$ and $\mathrm{T}_0 = 340$MeV, henceforth we will refer to it as *Th-fKLN* and it is shown by the dashed line in Fig.1. In this case we have an initial spatial eccentricity equal to 0.357. It represents the case implemented in hydrodynamics, that has lead to the conclusion that the CGC suggests a $\eta/s \sim 2/4\pi$ [8, 3, 14]. The third initial conditions is the full fKLN initial conditions where both the coordinate space and the momenta space are implemented according to fKLN model and it is indicated by the dotted line in Fig. 1. We notice that initially this last spectrum is quite far from thermalization reflecting the saturation behaviour typical of the melted CGC; however it thermalizes in approximatelly $1fm/c$ [7]. As initial time we take $\tau_0 = 0.2\ fm/c$ because in this case there is no pre-assumption of thermalization. This is not usually considered in hydrodynamics because there it is implicitly assumed a distribution function in $\boldsymbol{p}$-space in local equilibrium, at least in the transverse plane. For all the previous cases, as usually done, a Bjorken scaling at the initial time is assumed and the particle multiplicity, dN/dy, is fixed to correctly reproduce the experimental data.
In figure 2 the v_2 for the cases *Th-Glauber* and *Th-fKLN* is shown for a fixed $\eta/s \sim 1/4\pi$ respectively by the thin line and by the thick line. Using the *Th-Glauber* initial conditions we get results very close to the data, indicated by the circle, instead if we use the *Th-fKLN* case we get an higher v_2. For such initial conditions the agreement with the data is achieved only if the η/s is increased by a factor of two as indicated by the dashed line in the same figure. Our result are in agreement with hydrodynamics calculations [14, 3, 15, 4] and this indicates the validity of our approach in the limit where hydrodynamics is applicable. Instead the dotted line in Fig. 2 shows the full KLN initialization. In this case the data are reproduced with a viscosity $4\pi\eta/s \sim 1$ because the larger eccentricity is compensated by the non equilibrium distribution that is less efficient in converting the initial spatial anisotropy in the momentum space. This is the novelty of our work which shows that the build-up of the elliptic flow can be affected also by off-equilibrium features of the distribution function in $\boldsymbol{p}$-space.
We have found a similar effect also at LHC energies ($\sqrt{s} = 2.7\ ATeV$) as shown in Fig.3. In this case the starting time for the simulation, when the termalized initial conditions are considered, is $\tau = 0.3\ fm/c$ with maximum initial temperature $\mathrm{T}_0 = 510\ MeV$, while the Q_0^2 is equal to 4.5

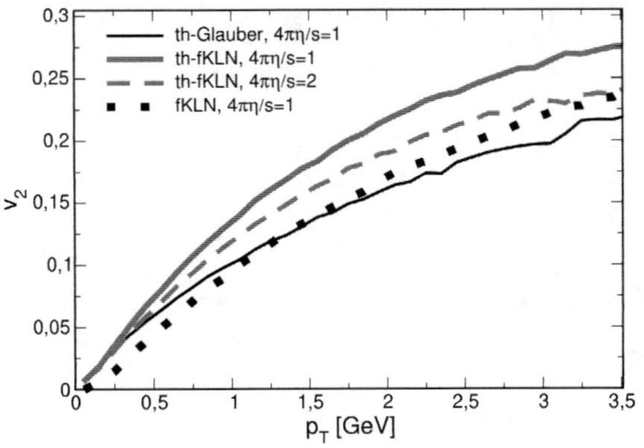

Figure 3. Elliptic flow for the different initial conditions as described in the legend. All the calculations refer to Pb+Pb collisions at $\sqrt{s} = 2.76TeV$, with $b = 7.5$

GeV^2. We observe that when the *Th-fKLN* initial conditions are considered we get an elliptic flow quite higher compared to the v_2 that we get with the *Th-Glauber* initial conditions even using η/s a factor two larger η/s, nevertheless the full KLN initialization gives a v_2 quite similar to that obtained with *Th-Glauber* with $\eta/s \sim 1/4\pi$.

4. Conclusions

In the framework of kinetic theory at fixed η/s we have found that the amount of elliptic flow produced in heavy ion collisions depends not only on the pressure gradients and on the ratio η/s of the system, but also on the initial distribution in momentum space. In particular a momentum distributions with a saturated behavior generates smaller v_2 respect to the thermal one. Using the fKLN distribution we have found that the effect of the initial non equilibrium distribution affects the estimate of η/s of about a factor two. Our results open the way to study the effect of initial momentum distribution on higher harmonics $v_n = \langle cos(n\varphi_p)\rangle$ offering a promising tool to further narrow the evaluation of the η/s of the plasma. Moreover we are setting our code in such way to include also the initial state fluctuations in order to study the impact of non-equilibrium initial conditions on the v_3. This should be very interesting since the hydrodinamics results show that the Th-fKLN with $\eta/s = 2/4\pi$ would generate a low v_3 respect to the available data leading to the conclusion that Th-fKLN is not able to describe for the experimental observations [14]. On the light of our results this last conclusion may be revised.
Acknowledgements. V.G. and F.S. acknowledges the QGPDyn grant.

References

[1] Adams J *et al.* (STAR Collaboration) 2005 *Nucl.Phys.* **A757** 102–183 (*Preprint* nucl-ex/0501009)
[2] Aamodt K *et al.* (ALICE Collaboration) 2010 *Phys.Rev.Lett.* **105** 252302 (*Preprint* 1011.3914)
[3] Song H, Bass S A, Heinz U, Hirano T and Shen C 2011 *Phys.Rev.* **C83** 054910 (*Preprint* 1101.4638)
[4] Luzum M and Romatschke P 2008 *Phys.Rev.* **C78** 034915 (*Preprint* 0804.4015)
[5] Kharzeev D, Levin E and Nardi M 2005 *Nucl.Phys.* **A747** 609–629 (*Preprint* hep-ph/0408050)
[6] Hirano T, Heinz U W, Kharzeev D, Lacey R and Nara Y 2006 *Phys.Lett.* **B636** 299–304 (*Preprint* nucl-th/0511046)
[7] Ruggieri M, Scardina F, Plumari S and Greco V 2013 *Phys.Lett.* **B727** 177–181 (*Preprint* 1303.3178)
[8] Drescher H J and Nara Y 2007 *Phys.Rev.* **C75** 034905 (*Preprint* nucl-th/0611017)
[9] Ferini G, Colonna M, Di Toro M and Greco V 2009 *Phys.Lett.* **B670** 325–329 (*Preprint* 0805.4814)
[10] Plumari S, Baran V, Di Toro M, Ferini G and Greco V 2010 *Phys.Lett.* **B689** 18–22 (*Preprint* 1001.2736)
[11] Scardina F, Colonna M, Plumari S and Greco V 2013 *Physics Letters B* **724** 296 – 300 ISSN 0370-2693
[12] Plumari S, Puglisi A, Scardina F and Greco V 2012 *Phys.Rev.* **C86** 054902 (*Preprint* 1208.0481)
[13] Plumari S, Puglisi A, Colonna M, Scardina F and Greco V 2013 *J.Phys.Conf.Ser.* **420** 012029 (*Preprint* 1209.0601)
[14] Adare A *et al.* (PHENIX Collaboration) 2011 *Phys.Rev.Lett.* **107** 252301 (*Preprint* 1105.3928)
[15] Alver B H, Gombeaud C, Luzum M and Ollitrault J Y 2010 *Phys.Rev.* **C82** 034913 (*Preprint* 1007.5469)

Measurement of D-meson production in p-Pb collisions with the ALICE detector

Grazia Luparello (for the ALICE Collaboration)
Nikhef, Science Park 105, 1098 XG Amsterdam, The Netherlands
E-mail: E-mail: Grazia.Luparello@cern.ch

Abstract. The ALICE Collaboration has measured the production of prompt D mesons in p–Pb collisions at $\sqrt{s_{\rm NN}} = 5.02$ TeV in the rapidity range $-0.04 < y_{\rm cms} < 0.96$ via the exclusive reconstruction of their hadronic decays: $\rm D^0 \rightarrow K^-\pi^+$, $\rm D^+ \rightarrow K^-\pi^+\pi^+$, $\rm D^{*+} \rightarrow D^0\pi^+$ and $\rm D_s^+ \rightarrow \phi\pi^+$. The $p_{\rm T}$-differential production cross sections and the $p_{\rm T}$-dependent nuclear modification factors with respect to a proton-proton reference, $R_{\rm pPb}$, are presented.

1. Introduction

Proton-nucleus interactions at the LHC allow for the study of fundamental properties of Quantum Chromodynamics (QCD) at low parton fractional momentum x and high gluon densities [1] and they provide a reference for the studies of deconfined matter created in nucleus-nucleus collisions. Measurements of charm particle production in central Pb-Pb collisions at $\sqrt{s_{\rm NN}} = 2.76$ TeV showed a suppression of the D-meson yield with respect to the binary-scaled pp yield in a broad $p_{\rm T}$ range [2]. This effect is consistent with the energy loss of charm quarks while they traverse the hot and dense medium formed in such collisions. To come to a quantitative understanding of the energy loss results, it is important to disentangle hot medium effects from initial-state effects due to cold nuclear matter, such as a modification of the parton distribution functions in the nucleus [3] and gluon saturation effects [4]. Initial-state effects can be isolated by studying the D meson production in p-Pb collisions, where a hot medium is not expected to form. A sensitive observable is the nuclear modification factor, $R_{\rm pPb}$. In case of minimum bias p–Pb interactions $R_{\rm pPb}$ is defined as $R_{\rm pPb} = \frac{1}{\rm A}\frac{{\rm d}\sigma_{\rm pPb}/{\rm d}p_{\rm T}}{{\rm d}\sigma_{\rm pp}/{\rm d}p_{\rm T}}$, where A is the number of nucleons in the lead nucleus, $\sigma_{\rm pp}$ and $\sigma_{\rm pPb}$ are the production cross sections of D mesons in pp and p–Pb collisions respectively. In these proceedings, we present the results obtained from a data sample of $\sim$100M p–Pb collisions at $\sqrt{s_{\rm NN}} = 5.02$ TeV collected with a Minimum Bias trigger defined in ALICE by the coincidence of signals in the two VZERO detectors, consisting of two arrays of scintillator counters covering the pseudorapidity regions $2.8 < \eta < 5.1$ and $-3.7 < \eta < -1.7$.

2. Measurement of D-meson production in p–Pb collisions

The analysis strategy is based on the invariant mass analysis of fully reconstructed decay topologies. $\rm D^0$, $\rm D^+$, $\rm D^{*+}$ and $\rm D_s^+$ mesons and their antiparticles are reconstructed via their hadronic decay channels $\rm D^0 \rightarrow K^-\pi^+$ (with branching ratio, BR, of 3.88 ± 0.05 %), $\rm D^+ \rightarrow K^-\pi^+\pi^+$ (BR of 9.13 ± 0.19 %), $\rm D^{*+} \rightarrow D^0\pi^+$ (BR of 67.7 ± 0.5 %) and $\rm D_s^+ \rightarrow \phi\pi^+$

with the subsequent decay $\phi \rightarrow K^+K^-$ (BR of 2.28±0.12 %) [5]. The extraction of the D-meson signals out of the large combinatorial background from uncorrelated tracks is based on the reconstruction and selection of secondary vertex topologies with a separation of a few hundred micrometers from the primary vertex. The Time Projection Chamber (TPC) and the Inner Tracking System (ITS) detectors, in particular its two innermost layers equipped with silicon pixel detectors, provide a resolution of the track impact parameter of about 70 μm for tracks with $p_T = 1$ GeV/c. To further suppress the combinatorial background, particle identification (PID) on the D-meson decay products is employed. PID is performed using the information on specific energy deposit in the TPC and on the time of flight measured with the Time Of Flight (TOF) detector. The signal yield is extracted by fitting the invariant mass distribution using a Gaussian function for the signal peak. The background is fitted using an exponential shape in the case of D^0, D^+ and D_s^+. For the D^{*+}, the distribution of $\Delta M = M(K^-\pi^+\pi^+) - M(K^-\pi^+)$ is fitted with a threshold function convoluted with an exponential. The raw yields extracted by the fit are divided by 2 to give the average (particle and antiparticle) cross section. The correction for acceptance and efficiency is performed using Monte Carlo simulations based on PYTHIA with the Perugia-0 tuning to generate the D-meson signals on top of an underlying event generated with HIJING. The efficiencies are weighted to take into account the dependence of the D-meson production on the number of charged particles produced in the collisions.

In order to extract the cross section of prompt D mesons, the contribution of D mesons from B decays is subtracted. This B feed-down contribution is evaluated using the B-meson cross-section from FONLL [6], the $B \rightarrow D + X$ decay kinematics from EvtGen [7], and the efficiencies for prompt and feed-down D mesons from simulations. FONLL calculations are used because they describe well the beauty hadron cross sections measured at the Tevatron [8] and at the LHC [9], [10]. To account for possible initial-state effects on the beauty-quark production, the resulting yield of D mesons from B feed-down is multiplied by a nuclear modification factor $R_{\rm pPb}^{\rm feed-down}$ of D mesons from B decays. The central value for the calculation is chosen as $R_{\rm pPb}^{\rm feed-down} = R_{\rm pPb}^{\rm prompt}$ while the systematic uncertainty is evaluated by varying the hypothesis on the $R_{\rm pPb}$ of feed-down D mesons in the range $0.9 < R_{\rm pPb}^{\rm feed-down}/R_{\rm pPb}^{\rm prompt} < 1.3$. This hypothesis is tuned on the basis of calculations obtained combining NLO pQCD cross sections from MNR [11], EPS09 [3] parameterizations of the nuclear parton distribution functions and B-decay kinematics from the EvtGen package [7]. The large uncertainty on the predictions at low p_T is taken into account.

The D-meson production cross section in pp collisions at $\sqrt{s_{\rm NN}} = 5.02$ TeV needed for the $R_{\rm pPb}$ calculation is obtained by scaling the measured D meson production cross section at $\sqrt{s} = 7$ TeV using the ratio of FONLL predictions at $\sqrt{s} = 5.02$ and 7 TeV with the procedure described in [12]. The uncertainty on the scaling decreases with increasing p_T from 17% to 3% as estimated by varying the parameters entering the FONLL calculations, namely the factorization and renormalization scales and the charm quark mass.

Figure 1 shows the D^{*+} and the D_s^+ cross sections in the momentum ranges $1 < p_T < 24$ GeV/c and $2 < p_T < 12$ GeV/c respectively, compared with the pp reference scaled by the number of nucleons in the lead nucleus A. Several sources of uncertainties are considered. The systematic uncertainty on the yield extraction is determined in each p_T interval by repeating the fit in a different mass range, varying the background fit function (a parabola instead of an exponential for D^0, D^+, D_s^+ and a power law multiplied by an exponential or a polynomial for the D^{*+}) and by counting the histogram entries in the invariant mass region of the signal after subtracting the background estimated from the side bands. The effect of the imperfect implementation of the detector description in the Monte Carlo simulations is estimated by repeating the analysis with different selections on the track quality and on the decay topology. The uncertainty associated to the particle identification is studied by comparing the cross sections obtained with and without PID. The effect on the selection efficiency due to the shape of the simulated D meson spectrum is estimated from the relative difference between the Monte

Journal of Physics: Conference Series **509** (2014) 012101 doi:10.1088/1742-6596/509/1/012101

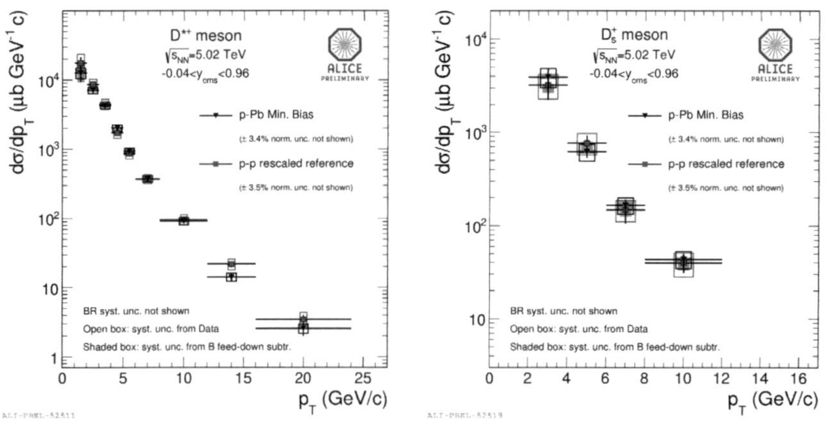

Figure 1. D^{*+} (left) and D_s^+ (right) p_T-differential cross section for p–Pb collisions at $\sqrt{s_{NN}} = 5.02$ TeV compared to the pp scaled cross sections. The shaded area represents the total systematic uncertainty.

Carlo efficiencies obtained using different p_T shapes (Pythia, FONLL, flat p_T distribution). The estimated systematic uncertainties vary from ~15% to ~6% depending on the p_T interval and the D-meson species. The systematic uncertainty on the feed-down subtraction was discussed above.

3. D-meson nuclear modification factor

The p–Pb spectra are quantitatively compared to the pp reference by computing the nuclear modification factor, R_{pPb}. Figure 2 shows the R_{pPb} of the D^0, D^+, D^{*+} and D_s^+. The nuclear modification factors of the four meson species are in agreement with each other, and they are compatible with unity within the uncertainties in the full p_T-range considered. This measurement suggests that the large suppression of the D-meson yield observed in Pb–Pb collisions is mainly due to final-state effects. The weighted average of the R_{pPb} of D^0, D^+, D^{*+} in the p_T range $1 < p_T < 24$ GeV/c is calculated using the statistical uncertainties as a weight. The systematic error on the average is calculated by propagating the uncertainties through the weighted average, where the contributions from tracking efficiency, B feed-down correction and scaling of the pp reference are taken as fully correlated among the three species. The bottom plot of figure 2 shows R_{pPb} compared with pQCD calculations based on the MNR [11] code combined with nuclear parton distributions functions from EPS09 [3] parameterizations and with a model based on the Color Glass Condensate [13]. Both models describe the data within uncertainties.

4. Conclusions

The measurement of the nuclear modification factor of prompt D^0, D^+, D^{*+} and D_s^+ in p–Pb collisions at $\sqrt{s_{NN}} = 5.02$ TeV is presented. The measured R_{pPb} is compatible with unity within systematic uncertainties over the full p_T range exploited confirming that the strong suppression observed in central Pb–Pb interactions can be interpreted as a final state effect due to in-medium parton energy loss. Theoretical models including nuclear shadowing or saturation effects can describe the data within uncertainties.

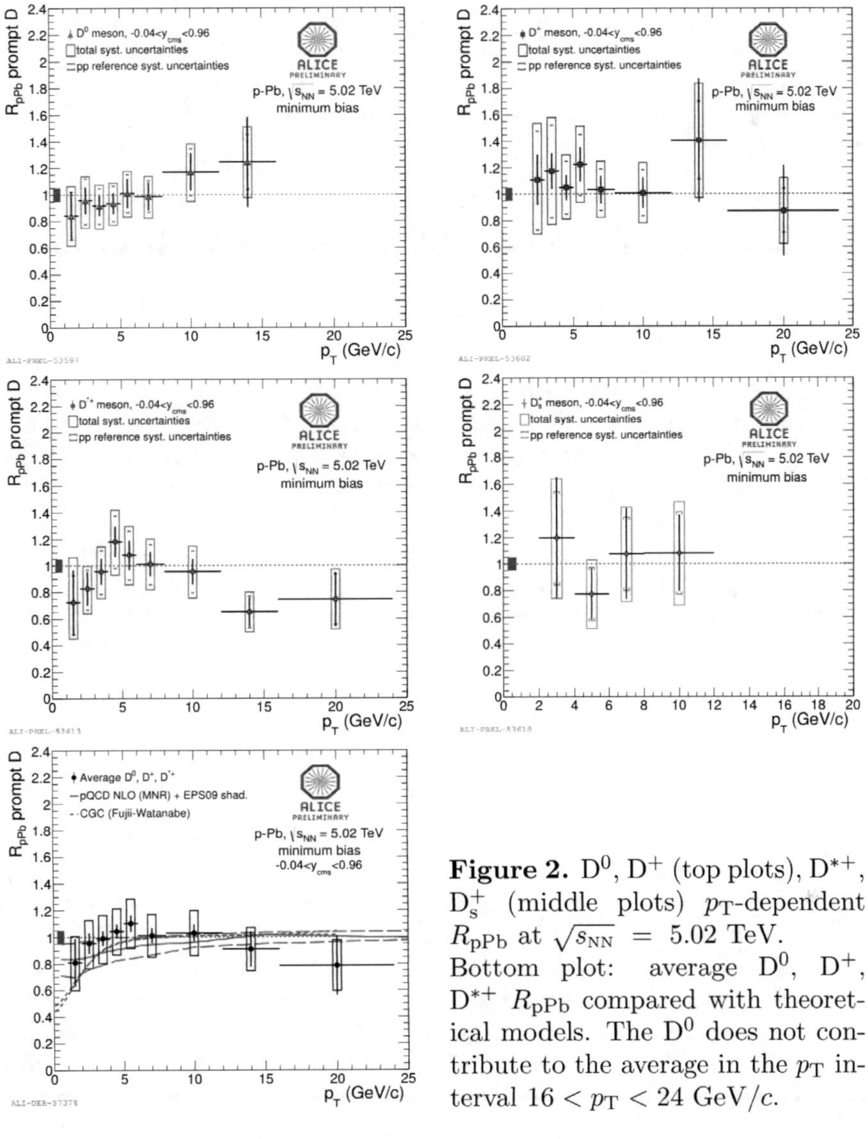

Figure 2. D^0, D^+ (top plots), D^{*+}, D_s^+ (middle plots) p_T-dependent R_{pPb} at $\sqrt{s_{NN}} = 5.02$ TeV. Bottom plot: average D^0, D^+, D^{*+} R_{pPb} compared with theoretical models. The D^0 does not contribute to the average in the p_T interval $16 < p_T < 24$ GeV/c.

References

[1] Salgado C A *et al* 2012 *J. Phys. G* **39** 015010
[2] Abelev B *et al* 2012 *JHEP* **1209** 112
[3] Eskola K, Paukkunen H and Salgado C A 2009 *JHEP* **0904** 065
[4] Jujii H, Gelis F and Venugopalan R 2006 *Nucl. Phys. A* **780** 146
[5] Beringer J *et al* (Particle Data Group) 2012 *Phys. Rev. D* **86** 010001
[6] Cacciari M, Greco M, Nason P 1998 *JHEP* **9805** 007; Cacciari M *et al* 2001 *JHEP* **0103** 006
[7] Lange D 2001 *Nucl. Instrum. Meth. A* **462** 152
[8] Cacciari M *et al* 2004 *JHEP* **0407** 033
[9] Aaij R *et al* [LHCb Coll.] 2010 Phys. Lett. B **694** 209-216
[10] Khachatryan V *et al* [CMS Coll.], 2011 *Eur. Phys. J. C* **71** 1575
[11] Mangano M, Nason P and Ridolfi G 1992 *Nucl. Phys. B* **373** 295
[12] Averbeck R, Bastid N, Conesa del Valle Z, Crochet P, Dainese A, Zhang X 2011 *arXiv:1107.3243*
[13] Fujii H and Watanabe K 2013 *Nucl. Phys. A* **920** 78

Journal of Physics: Conference Series **509** (2014) 012102 doi:10.1088/1742-6596/509/1/012102

Measurement of electrons from heavy-flavour hadron decays in pp and p-Pb collisions with ALICE

M Heide (for the ALICE collaboration)
Institut für Kernphysik, Wilhelm-Klemm-Str. 9, 48149 Münster, Germany
E-mail: mheide@wwu.de

Abstract. ALICE measured electrons from inclusive heavy-flavour hadron decays and beauty-hadron decays at mid-rapidity in pp collisions at $\sqrt{s} = 7$ TeV and 2.76 TeV and p-Pb collisions at $\sqrt{s_{\rm NN}} = 5.02$ TeV. For both pp collision energies, $p_{\rm T}$-differential electron production cross sections are presented and compared to pQCD predictions. For p-Pb collisions, the $p_{\rm T}$-dependent nuclear modification factor of electrons from inclusive heavy-flavour hadron decays is presented.

(Some figures in this article are in colour only in the electronic version)

1. Introduction

Measurements of electrons from heavy-flavour hadron decays in pp collisions test perturbative QCD (pQCD) predictions and are references for Pb-Pb collisions, where heavy quarks are probes for in-medium energy loss [1]. Cold nuclear matter effects can be disentangled from hot-medium effects via measurements of heavy-flavour hadron decay electrons in p-Pb collisions [2].

The large branching ratio of semi-electronic charm and beauty-hadron decays ($\approx 10\%$) makes electron measurements an alternative to analyses of heavy-flavour production via reconstruction of exclusive hadronic decays. With the ALICE [3] experiment inner barrel detectors, electrons can be identified down to low transverse momenta $p_{\rm T}$. Moreover, electrons from beauty-hadron decays can be measured separately, exploiting the spatial resolution provided by the ALICE Inner Tracking System (ITS) to select single electrons or secondary vertices with large displacements from the primary vertex. In addition, the ratio of electrons from beauty to those from inclusive heavy-flavour decays can be determined via azimuthal angular electron-hadron correlations.

2. Electron identification

All analyses of electrons from heavy-flavour hadron decays select a pure electron sample at mid-rapidity, using signals from the Time Projection Chamber (TPC) and from at least one of the devices Time of Flight (TOF), Electromagnetic Calorimeter (EMCal), and Transition Radiation Detector (TRD), with a track quality selection using ITS and TPC tracking information. TPC particle identification is based on energy deposition per unit of length, $\mathrm{d}E/\mathrm{d}x$, while TOF relates particle mass hypotheses to the time of flight and momentum p. The EMCal identifies electrons by their ratio $E/p \approx 1$, with the energy E deposited by an absorbed particle. Electron-hadron separation in the TRD uses a signal composed of the energy deposited by a charged particle in the TRD gas volume and the transition radiation emitted by electrons with $p \gtrsim 1$ GeV/c.

3. Additional requirements of beauty-hadron decay analyses

Kinematical differences between decays of B and D mesons can be used to identify electrons from beauty-hadron decays. Owing to the large decay length of B mesons ($c\tau \approx 500$ μm), their decay electron tracks and vertices are on average further displaced from the primary vertex than others. The ALICE ITS is well suited to detect this difference. Moreover, the reconstruction of simulated events shows that the near-side peak of azimuthal angular correlations between electrons from beauty decays and charged hadrons is about twice as wide as for charm decays. In ALICE, electrons from beauty-hadron decays were measured via three approaches:

- selection of electrons with a large transverse impact parameter to the primary vertex [4].
- reconstruction of electron-hadron vertices displaced from the primary vertex, with requirements on the invariant mass of the electron-hadron pair and on the hadron p_{T}.
- measurement of azimuthal angular correlations between electrons and charged hadrons. The near-side peak in the angular distribution around the electron track is fitted with PYTHIA templates for beauty and charm decays to determine their relative contribution.

4. Electron background subtraction and corrections

All selected electron samples contain electron background from non-heavy-flavour decays, which is subtracted by two alternative methods. In the first, p_{T}-differential non-heavy-flavour electron spectra are calculated based on hadron cross sections measured at the LHC. For the analyses of beauty decays via impact parameter and displaced secondary vertices, simulations determine the fraction of electron background remaining after all selection criteria. All measured spectra are corrected for detector acceptances and selection efficiencies and undergo an unfolding procedure to correct for bremsstrahlung effects and momentum smearing in the reconstruction.

The second background subtraction method is based on the identification of photonic electrons by reconstruction of $e^+ - e^-$ pairs with small invariant mass. The combinatorial background to this measurement is estimated via reconstruction of like-sign electron pairs and subtracted from the photonic electron yield. The resulting background is corrected for the pair-finding efficiency. This method is used in analyses of the relative beauty contributions to inclusive electron spectra from heavy-flavour hadron decays via electron-hadron correlations.

5. Results from proton-proton collisions at $\sqrt{s} = 7$ TeV and $\sqrt{s} = 2.76$ TeV

Fig. 1 shows measurements of heavy-flavour decay electrons in pp collisions at $\sqrt{s} = 7$ TeV. The left panel presents the p_{T}-differential electron cross section from heavy-flavour hadron decays for 0.5 GeV/$c < p_{\mathrm{T}} <$ 8 GeV/c in the rapidity range $|y| < 0.5$ [5]. In the middle and right panel, results from all three analyses of beauty decay electrons are summarized: the p_{T}-differential cross section determined by the impact parameter method is shown for 1 GeV/$c < p_{\mathrm{T}} <$ 8 GeV/c and $|y| < 0.8$ [4]. In comparison to the electron cross section from charm decays, calculated from ALICE measurements of charm-hadron cross sections, beauty decays become dominant for $p_{\mathrm{T}} \gtrsim 4$ GeV/c. The measurements of the p_{T}-differential cross section via displaced electron-hadron vertices and of the ratio $e(b)/e(b+c)$ of electrons from beauty to inclusive heavy-flavour decays via electron-hadron correlations, both for $|y| < 0.7$, are complementary to the first method. Their results are comparable and extend the p_{T} range of beauty decay measurements to 13 and 18 GeV/c, respectively. All spectra are described by pQCD predictions at fixed-order plus next-to-leading logarithms (FONLL [6]) within uncertainties. Predictions from the k_{T}-factorization [7] and GM-VFN scheme [8] achieve similarly good descriptions of the data.

At $\sqrt{s} = 2.76$ TeV, the reference energy for Pb-Pb collisions, the p_{T}-differential electron cross section from inclusive heavy-flavour hadron decays was measured for 2 GeV/$c < p_{\mathrm{T}} <$ 12 GeV/c and $|y| < 0.7$ (Fig. 2, left). For the same rapidity range, the ratio $e(b)/e(b+c)$ was determined via azimuthal angular electron-hadron correlations for 2.5 GeV/$c < p_{\mathrm{T}} <$ 10 GeV/c (Fig. 2,

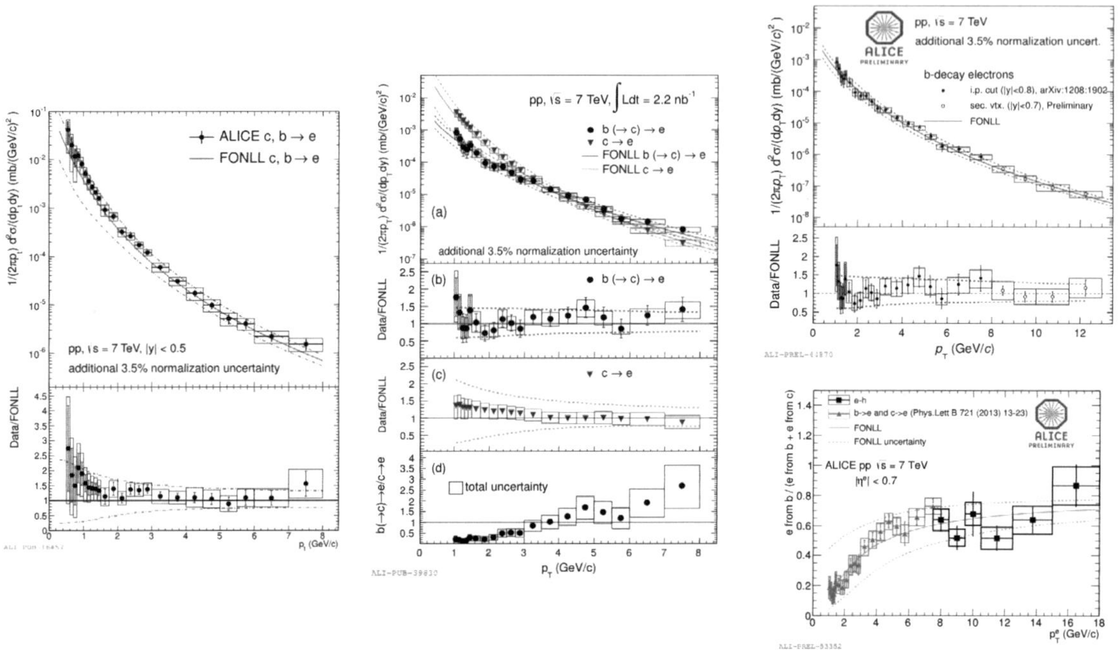

Figure 1. Electron cross section from inclusive heavy-flavour hadron decays [5] (left), beauty-hadron decays measured via impact parameter [4], and charm-hadron decays (middle) in pp collisions at $\sqrt{s} = 7$ TeV. Upper right: electron cross section from beauty-hadron decays from [4], with preliminary result of a measurement via displaced secondary vertices for $p_{\mathrm{T}} > 8$ GeV/c. Lower right: ratio of electron yield from beauty-hadron decays to inclusive heavy-flavour hadron decays, measured via azimuthal angular electron-hadron correlations (black) and as ratio of results from middle and left panel (red). All results are compared to FONLL calculations [6].

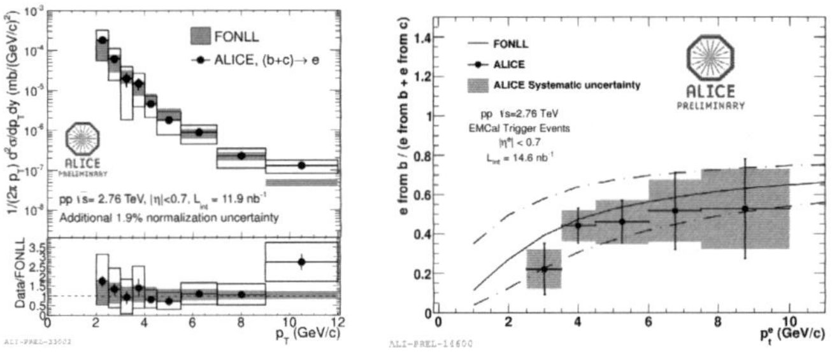

Figure 2. Electron cross section from heavy-flavour hadron decays (left) and ratio of electrons from beauty to inclusive heavy-flavour hadron decays (right) in pp collisions at $\sqrt{s} = 2.76$ TeV, compared to FONLL predictions [6].

right). All measurements are described by FONLL predictions [6], calculations from the k_T-factorization framework [7], and the GM-VFN scheme [8] within uncertainties.

6. Results from p-Pb collisions at $\sqrt{s_{\mathrm{NN}}} = 5.02$ TeV

Inclusive heavy-flavour decay electrons were measured in p-Pb collisions for $-0.14 < y < 1.06$ via two electron identification methods: in the p_{T} range from 0.5 GeV/c to 6 GeV/c, TPC and TOF were utilized, while TPC and EMCal were used between 2 GeV/c and 12 GeV/c. The p_{T}-differential cross sections measured by both methods agree in the common range, where the TPC-TOF result has smaller uncertainties and is therefore considered for the preliminary result.

Fig. 3 shows the nuclear modification factor $R_{\mathrm{pPb}}(p_{\mathrm{T}}) = \frac{1}{\langle T_{\mathrm{pPb}} \rangle} \frac{\mathrm{d}N_{\mathrm{pPb}}/\mathrm{d}p_{\mathrm{T}}}{\mathrm{d}\sigma_{\mathrm{pp}}/\mathrm{d}p_{\mathrm{T}}}$ for electrons from

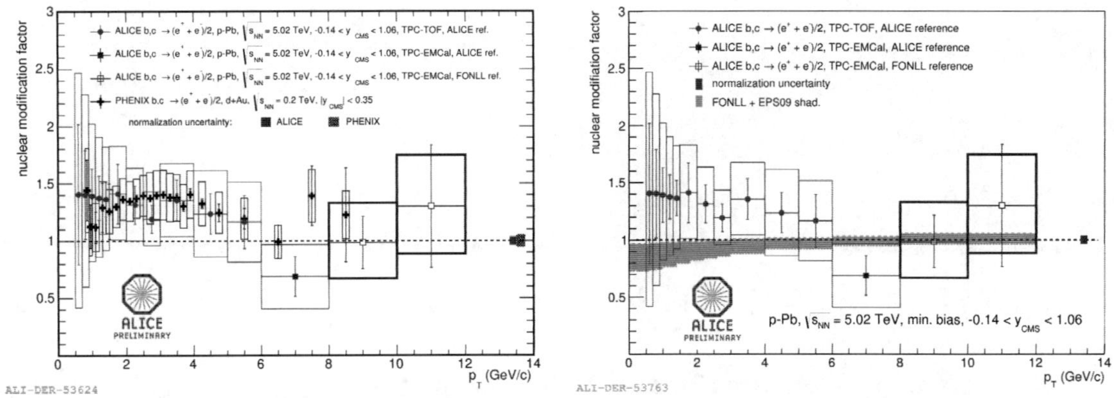

Figure 3. R_{pPb} of electrons from inclusive heavy-flavour decays, compared to PHENIX results from d-Au collisions at 0.2 TeV [9] (left) and EPS09 nuclear shadowing predictions [10] (right).

heavy-flavour hadron decays. $\langle T_{\mathrm{pPb}} \rangle$ is the average nuclear overlap function, $\mathrm{d}N_{\mathrm{pPb}}/\mathrm{d}p_{\mathrm{T}}$ the p_{T}-differential electron yield in p-Pb collisions, and $\mathrm{d}\sigma_{\mathrm{pp}}/\mathrm{d}p_{\mathrm{T}}$ the p_{T}-differential cross section in pp collisions. The electron spectrum in p-Pb collisions is compared to a pp reference spectrum that for $p_{\mathrm{T}} < 8$ GeV/c is based on the result from pp collisions at $\sqrt{s} = 7$ TeV (see Fig. 1), scaled to $\sqrt{s} = 5.02$ TeV using FONLL calculations. At higher p_{T}, FONLL predictions were used as a reference. The p_{T} dependence of R_{pPb} is similar to the behaviour observed for non-photonic electrons at RHIC in d-Au collisions at $\sqrt{s_{\mathrm{NN}}} = 0.2$ TeV [9] (Fig. 3, left), which showed an enhancement at intermediate p_{T}. However, the uncertainties of the ALICE measurements also allow for an agreement with predictions for the nuclear modification of the parton distribution functions from EPS09 parameterizations [10] that do not imply such an enhancement (Fig. 3, right). Future reductions of these uncertainties are expected from an improved pp reference, interpolating with measurements at $\sqrt{s} = 2.76$ TeV, and the usage of a TRD electron trigger.

7. Summary

ALICE measured the p_{T}-differential production cross sections of electrons from inclusive heavy-flavour and beauty-hadron decays at mid-rapidity for pp collisions at $\sqrt{s} = 7$ TeV and $\sqrt{s} = 2.76$ TeV. Results from three independent analyses of beauty-hadron decays are in mutual agreement. All measured cross sections from pp collisions are described by pQCD calculations within uncertainties. In p-Pb collisions at $\sqrt{s_{\mathrm{NN}}} = 5.02$ TeV, the p_{T} dependence of the nuclear modification factor of electrons from inclusive heavy-flavour hadron decays was determined. Within uncertainties, it agrees with the results from d-Au collisions at $\sqrt{s_{\mathrm{NN}}} = 0.2$ TeV by the PHENIX collaboration and with EPS09 nuclear shadowing predictions.

References

[1] Armesto N, Dainese A, Salgado C A and Wiedemann U A 2005 *Phys. Rev. D* **71**(5) 054027
[2] Salgado C A *et al.* 2012 *Journal of Physics G: Nuclear and Particle Physics* **39** 015010
[3] Aamodt K *et al.* (ALICE Collaboration) 2008 *JINST* **3** S08002
[4] Abelev B *et al.* (ALICE Collaboration) 2013 *Phys.Lett.* **B721** 13–23
[5] Abelev B *et al.* (ALICE Collaboration) 2012 *Phys. Rev. D* **86**(11) 112007
[6] Cacciari M, Greco M and Nason P 1998 *JHEP* **9805** 007; private communication
[7] Maciula R and Szczurek A 2013 *Proc. XXI. Int. Workshop on Deep-Inelastic Scattering and Related Subjects DIS 2013 (Marseille)* (*Preprint* `arXiv:1306.6808`); private communication
[8] Bolzoni P and Kramer G 2013 *Nucl.Phys.* **B872** 253–264; private communication
[9] Adare A *et al.* (PHENIX Collaboration) 2012 *Phys. Rev. Lett.* **109**(24) 242301
[10] Eskola K, Paukkunen H and Salgado C 2009 *JHEP* **0904** 065

D meson-hadron angular correlations in pp and p-Pb collisions with ALICE at the LHC

Fabio Colamaria, for the ALICE collaboration

Dipartimento Interateneo di Fisica M. Merlin and Sezione INFN, Bari, Italy

E-mail: fabio.colamaria@ba.infn.it

Abstract. The comparison of angular correlations between charmed mesons and charged hadrons produced in pp, p-Pb and Pb-Pb collisions can give insight into charm quark energy loss mechanisms in the hot nuclear medium formed in heavy-ion collisions and can help to spot possible modifications of charm quark hadronization induced by the presence of the medium. The analysis of pp and p-Pb data and the comparison with predictions from pQCD calculations, besides constituting the necessary baseline for the interpretation of Pb-Pb results, can provide relevant information on charm production and fragmentation processes.

We present a study of azimuthal correlations between D^0 and D^{*+} mesons and charged hadrons measured by the ALICE experiment in pp collisions at $\sqrt{s} = 7$ TeV and p-Pb collisions at $\sqrt{s_{NN}} = 5.02$ TeV. D mesons were reconstructed from their hadronic decays at central rapidity and in the transverse momentum range $2 < p_T < 16$ GeV/c, and they were correlated to charged hadrons reconstructed in the pseudorapidity range $|\eta| < 0.8$. The current status of the analysis in pp and p-Pb collisions and the prospects for Pb-Pb collisions are discussed.

1. Physics motivations

ALICE [1] measured p_T-differential cross sections for D-meson production at central rapidity in pp [2–4] and Pb-Pb [5] collisions. A significant suppression of D-meson yields was found in central Pb-Pb collisions for $p_T > 4$-5 GeV/c, with respect to cross sections in pp collisions, scaled by the nuclear overlap function. This can be attributed, at least partially, to the energy loss of charm quarks in the Quark Gluon Plasma formed in such collisions. The study of angular correlations between D mesons and unidentified charged hadrons provides more insight into this topic, giving the possibility to:

- study the charm quark fragmentation in the various collision systems;
- address the production processes of $c\bar{c}$ pairs and their interplay with the underlying event;
- characterize from a different point of view the charm quark energy loss in heavy-ion collisions compared to cross section measurements [5] (i.e., a suppression of the away side peak at $\Delta\phi \sim \pi$ is expected in the azimuthal correlation distributions).

To this end, the pp and p-Pb measurements provide a reference for the comparison of the studied observables, i.e. the yields in the near side/away side peaks, the integrated yield, and the height of the correlation baseline.

Journal of Physics: Conference Series **509** (2014) 012103 doi:10.1088/1742-6596/509/1/012103

2. D-meson reconstruction with ALICE

ALICE is a general purpose detector, optimized to operate in a high track density environment as that of Pb-Pb collisions. It comprises a central barrel in which tracks with $|\eta| < 0.9$ and $p_{\mathrm{T}} > 80$ MeV/c are reconstructed and a forward muon spectrometer covering the rapidity range $2.5 < y < 4$.

The main detectors used in this analysis are:

- Inner Tracking System (ITS), a silicon detector composed of two layers of pixel (SPD), two of drift (SDD) and two of strip (SSD) detectors, with tracking and vertexing purposes;
- Time Projection Chamber (TPC), a gaseous detector for 3-dimensional tracking and particle identification (PID) via measurement of the specific ionization energy loss;
- Time Of Flight (TOF), which provides PID, exploited in this analysis for separating pions from kaons up to 2.5 GeV/c of momentum.

D mesons are reconstructed at central rapidity via their hadronic decays by performing an invariant mass analysis of displaced secondary vertices, exploiting the topology of the decay and the secondary vertex displacement together with particle (K and π) identification.

In particular, D^0, D^+, D^{*+} and D_s^+ mesons have been reconstructed in the $\mathrm{D}^0 \to \mathrm{K}^-\pi^+$, $\mathrm{D}^+ \to \mathrm{K}^-\pi^+\pi^+$, $\mathrm{D}^{*+} \to \mathrm{D}^0\pi^+$, $\mathrm{D}_s^+ \to \phi\pi^+ \to \mathrm{K}^-\mathrm{K}^+\pi^+$ decay channels.

As an example, the left panel of Fig. 1 shows the cross section for prompt D^0 production measured by ALICE in pp collisions at $\sqrt{s} = 7$ TeV [2], compared with predictions from perturbative QCD calculations (FONLL and GM-VFNS). The right panel of Fig. 1 shows the ratios of p_{T}-integrated cross sections for D^0, D^+, D^{*+} and D_s^+ production in pp collisions at $\sqrt{s} = 7$ TeV [3], compared with results from other experiments, PYTHIA simulations and Statistical Hadronization Model predictions. Cross sections for all the D mesons show a good compatibility with FONLL and GM-VFNS theoretical predictions within the uncertainties; their ratios are also in good agreement with the other measurements and with predictions within the uncertainties.

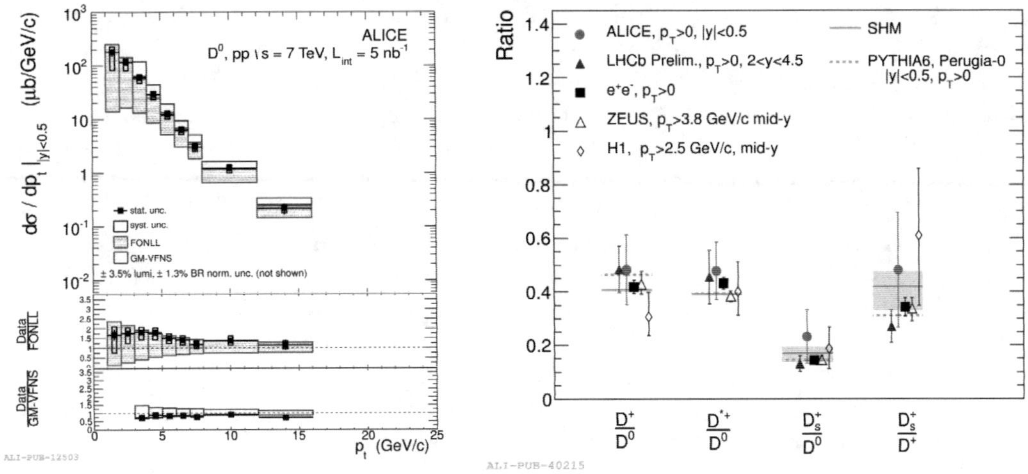

Figure 1. Left: cross section for prompt D^0 production at central rapidity in pp collisions at $\sqrt{s} = 7$ TeV, compared to FONLL [6] (red boxes) and GM-VFNS [7] (blue boxes) calculations. Ratios between data and calculations are shown in the bottom panels. Right: ratio of cross sections for D^0, D^+, D^{*+} and D_s^+ production in pp collisions at $\sqrt{s} = 7$ TeV (red circles). Results from other experiments [8–11] and from PYTHIA 6.4.21 simulations (green dotted line) and Statistical Hadronization Model [12] (gray line) are shown.

3. D-hadron correlation: analysis strategy and results

The analysis has been performed on a minimum bias sample of $3.14 \cdot 10^8$ pp events at $\sqrt{s} = 7$ TeV and on a minimum bias sample of $1.3 \cdot 10^8$ p-Pb events at $\sqrt{s_{\rm NN}} = 5.02$ TeV. D-meson candidates (*trigger* particles) are correlated with charged hadrons (*associated tracks*) with $|\eta| < 0.8$ and the difference in pseudorapidity $\Delta\eta$ and azimuthal angle $\Delta\phi$ is evaluated. The contribution of background candidates is subtracted using the correlation distribution of candidates in the sidebands of the invariant mass distribution, normalized by the ratio of background candidates in signal and sideband regions. Corrections are applied to account for:

- Associated track reconstruction efficiency. Each D-hadron correlation is weighted by the inverse of the track reconstruction efficiency of the associated track, computed from a MC simulation as a function of the z coordinate of the primary vertex and of the track $p_{\rm T}$ and η (while the ϕ distribution of the associated track sample was found to be isotropic);
- D-meson selection efficiency. Each correlation entry is weighted by the D-meson reconstruction and selection efficiency, evaluated as a function of $p_{\rm T}$ and the event multiplicity from a MC simulation; other minor dependencies are neglected to avoid introducing fluctuations in the efficiency, due to the limited statistics of the MC sample;
- Limited detector acceptance and detector spatial inhomogeneities. A correction factor is obtained by correlating D mesons from an event with tracks from other events (*Event mixing*) with similar z of the primary vertex and multiplicity and normalizing the correlation distribution obtained to its value in $(\Delta\phi,\Delta\eta) = (0,0)$;
- Beauty feed-down. The fraction of D mesons coming from B meson decays is evaluated using the reconstruction efficiencies of prompt and secondary D mesons and the $p_{\rm T}$-differential cross-section of the latter, obtained using FONLL calculations and the EvtGen package, as described in [2]. MC simulations based on PYTHIA are used to obtain a template of angular correlations between D mesons from B meson decays and charged hadrons, which is then subtracted from the inclusive D-hadron correlation distribution obtained from data.

In the current analysis, the azimuthal correlations are not normalized to the number of D mesons and are not subtracted of the feed-down component. After performing the steps listed above, 2-dimensional correlation plots $(\Delta\eta,\Delta\phi)$ are obtained. Due to the limited statistics available, the 2D correlations are projected on the $\Delta\phi$ axis, producing azimuthal correlation plots. Figure 2 shows D^0-hadron correlations for $3 < p_{\rm T}(D^0) < 5$ GeV/c (left) and $8 < p_{\rm T}(D^0) < 16$ GeV/c (right) in pp collisions at $\sqrt{s} = 7$ TeV, with associated tracks with $p_{\rm T}^h > 0.3$ GeV/c. Near $(\Delta\phi \sim 0)$ and away side $(\Delta\phi \sim \pi)$ peaks are visible, though with significant fluctuations induced by the limited statistics available. In Fig. 3 D^{*+}-hadron correlations are shown, for $5 < p_{\rm T}(D^*) < 8$ GeV/c and a $p_{\rm T}^h$ threshold of 0.3 GeV/c. In all correlations a fit is superimposed to guide the eye; it consists of two Gaussians (with mean fixed at 0 and π) plus a constant baseline evaluated from points around $\pm\pi/2$.

The feasibility of the analysis in p-Pb collisions at $\sqrt{s_{\rm NN}} = 5.02$ TeV and Pb-Pb collisions at $\sqrt{s_{\rm NN}} = 2.76$ TeV with the currently available statistics is being studied. Figure 4 shows D^0-hadron correlations for $8 < p_{\rm T}(D^0) < 16$ and associated tracks with $p_{\rm T}^h > 1$ GeV/c in p-Pb collisions at $\sqrt{s_{\rm NN}} = 5.02$ TeV, with the same fit function described above superimposed.

4. Conclusions and outlook

We have presented measurements of D meson-hadron angular correlations in pp collisions at $\sqrt{s} = 7$ TeV and in p-Pb collisions at $\sqrt{s_{\rm NN}} = 5.02$ TeV.

The extension of the analysis to Pb-Pb collisions is challenged by the very low D-meson signal/background ratio and the large amount of tracks uncorrelated to D mesons. Improvement in the performance is expected with the ALICE upgrade (~2018-2019) [13], where a huge increment of statistics (×100) and a striking increase of the S/B ratio for D-meson reconstruction,

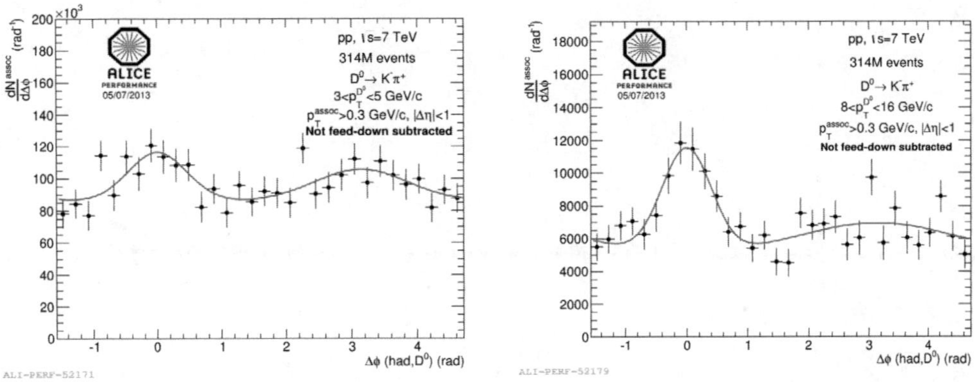

Figure 2. D^0-hadron azimuthal correlations for $3 < p_T(D^0) < 5$ GeV/c (left) and $8 < p_T(D^0) <$ 16 (right) for associated tracks with $p_T^h > 0.3$ GeV/c in pp collisions at $\sqrt{s} = 7$ TeV.

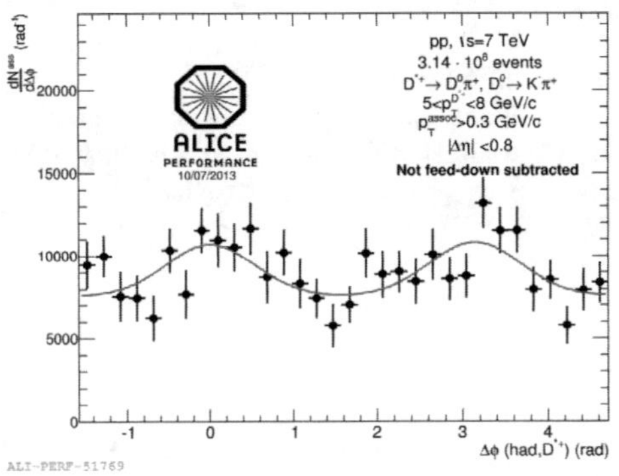

Figure 3. D^{*+}-hadron azimuthal correlations for $5 < p_T(D^{*+}) < 8$ GeV/c and associated tracks with $p_T^h > 0.3$ GeV/c, in pp collisions at $\sqrt{s} = 7$ TeV.

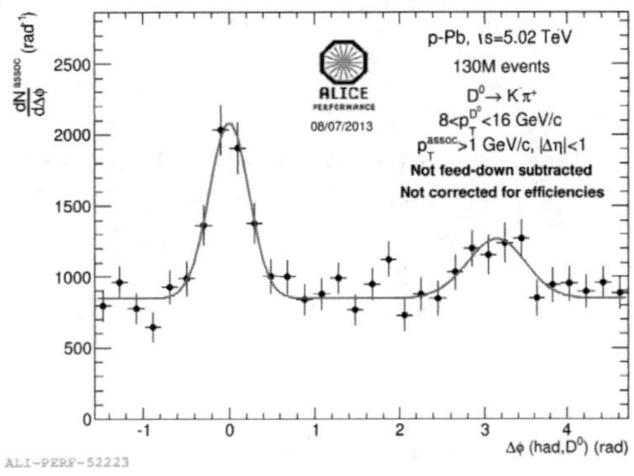

Figure 4. D^0-hadron azimuthal correlations for $8 < p_T(D^0) < 16$ GeV/c and associated tracks with $p_T^h > 1$ GeV/c, in p-Pb collisions at $\sqrt{s_{NN}} = 5.02$ TeV.

thanks to the improved resolution on the track impact parameter provided by the new ITS, are expected in Pb-Pb collisions. This will reduce the uncertainty on the background subtraction, one of the main limiting factor in the Pb-Pb analysis, allowing thus to study nuclear matter effects on D-hadron correlations by comparison with pp and p-Pb results.

References

[1] ALICE Collaboration 2012, *JINST* **3** S08002
[2] ALICE Collaboration 2012, *JHEP* **01** 128
[3] ALICE Collaboration 2012, *Phys. Lett.* B **718** 27994
[4] ALICE Collaboration 2012, *JHEP* **1207** 191
[5] ALICE Collaboration 2012, *JHEP* **09** 112
[6] M. Cacciari et al. 2011, CERN-PH-TH-2011-227
[7] B.A. Kniehl, G. Kramer, I. Schienbein, H. Spiesberger 2012, *Eur. Phys. J.* C **72** 2082
[8] LHCb Collaboration 2010, LHCb-CONF-2010-013.
[9] ZEUS Collaboration 2005, *Eur. Phys. J.* C **44** 351
[10] H1 Collaboration 2005, *Eur. Phys. J.* C **38** 447
[11] L. Gladilin 1999, arXiv:hep-ex/9912064
[12] A. Andronic, F. Beutler, P. Braun-Munzinger, K. Redlich, J. Stachel 2009, *Phys. Lett.* B **678** 350
[13] ALICE Collaboration 2012, CERN-LHCC-2012-012

Journal of Physics: Conference Series **509** (2014) 012104 doi:10.1088/1742-6596/509/1/012104

Inclusive J/ψ production in p-Pb collisions with ALICE at the LHC

Igor Lakomov for the ALICE Collaboration

Institut de Physique Nucléaire d'Orsay, Université Paris-Sud, 91406 Orsay, France

E-mail: ilakomov@cern.ch

Abstract. In 2013, inclusive J/ψ measurements in p-Pb collisions were performed at the LHC at $\sqrt{s_{\rm NN}}$ = 5.02 TeV, in order to measure the effects related to the cold nuclear matter on the J/ψ production. Their evaluation is important in order to be able to disentangle hot and cold nuclear matter effects in Pb-Pb collisions. In this contribution, results on the inclusive J/ψ production in p-Pb collisions, in the dimuon decay channel at forward and backward rapidity, down to zero transverse momentum, are presented and compared to theoretical models.

1. Introduction

Production of charmonia, bound state of c and $\bar{c}$ quarks, is an intense research activity, both experimentally and theoretically. The peculiar properties of some of the charmonium states, like their small size (< 1 fm) and strong binding energy (several hundred MeV), make them ideal probes of the strongly interacting deconfined matter, the so-called Quark-Gluon Plasma (QGP), formed in ultra-relativistic heavy-ion collisions. ALICE (A Large Ion Collider Experiment) [1] was designed to study heavy-ion collisions and the properties of the QGP. One of the signatures of the QGP formation is the so-called J/ψ "melting" [2] that results in a suppression of the J/ψ in heavy-ion collisions. The J/ψ suppression can be quantified in terms of the nuclear modification factor:

$$R_{\rm AA} = \frac{Y_{\rm AA}}{< T_{\rm AA} > \times \sigma_{\rm pp}^{\rm J/\psi}}, \tag{1}$$

where $Y_{\rm AA}$ is the invariant yield of the J/ψ measured in nucleus-nucleus (A-A) collisions, $T_{\rm AA}$ is the nuclear thickness function, $\sigma_{\rm pp}^{\rm J/\psi}$ is the J/ψ cross-section in pp collisions at the same energy. If $R_{\rm AA}$ differs from unity nucleus-nucleus collisions cannot be considered as a simple superposition of elementary nucleon-nucleon collisions. ALICE measured $R_{\rm AA}^{0-80\%} = 0.545 \pm 0.032(stat.) \pm 0.083(syst.)$ in Pb-Pb collisions at $\sqrt{s_{\rm NN}}$ = 2.76 TeV [3]. The J/ψ suppression observed in this measurement was well described in terms of hot nuclear matter effects (related to QGP formation). However other effects, related to cold nuclear matter (CNM) are expected to play a role. They can be divided into three main groups:

- **Initial state effects.** Gluon shadowing [4] (or gluon saturation [5]) is an effect of depletion of low-momentum gluons in nuclei compared to protons. It is currently poorly constrained by data at the LHC energy.
- **Nuclear absorption.** A J/ψ , or a pre-resonant $c\bar{c}$ pair can be dissociated by interacting with cold nuclear matter [6]. At the LHC, due to the large Lorentz boost of the colliding

nuclei at mid and forward rapidity in p-Pb collisions the $c\bar{c}$ pair spends a very short time within cold nuclear matter. Therefore such effect (also known as nuclear absorption) is expected to be negligible.

- **Coherent parton energy loss.** An effect related to the medium-induced radiative loss by the initial (partons) or final ($c\bar{c}$ pair) state [7].

To investigate the role of CNM effects and disentangle them from the hot nuclear matter effects in Pb-Pb collisions, p-Pb collisions were studied at the LHC at $\sqrt{s_{\rm NN}} = 5.02$ TeV in 2013. First results are already submitted for publication [8].

2. Analysis

The ALICE detector allows to study the J/ψ production down to zero transverse momentum in two decay channels: dimuon at forward and dielectron at mid-rapidity. In this work, we will concentrate on the dimuon channel, studied in the muon spectrometer [1] with an acceptance $-4 < y_{lab} < -2.5$. Due to the LHC beam energy asymmetry, the nucleon-nucleon center-of-mass system is shifted by $\Delta y = 0.465$ in the direction of the proton beam with respect to the laboratory system. Two configurations were adopted, obtained by inverting the direction of the two beams in the machine. The measurements of the J/ψ production in ALICE were then performed in two different rapidity ranges: backward ($-4.46 < y_{cms} < -2.96$) and forward ($2.03 < y_{cms} < 3.53$) with integrated luminosities of 5.81 ± 0.19 nb^{-1} and 5.03 ± 0.18 nb^{-1}, respectively. This analysis is based on the dimuon trigger selection, requiring a coincidence of minimum bias (MB) interaction with two opposite sign muon tracks detected in the trigger chambers of the muon spectrometer. The MB trigger efficiency for non-single-diffractive events is $> 99\%$ [9]. The dimuon trigger efficiency reaches 96% for $p_{\rm T,\mu} > 1.5$ GeV/c. To reject single muons at the edge of the acceptance of the spectrometer, the condition of $-4 < \eta_\mu < -2.5$ was required.

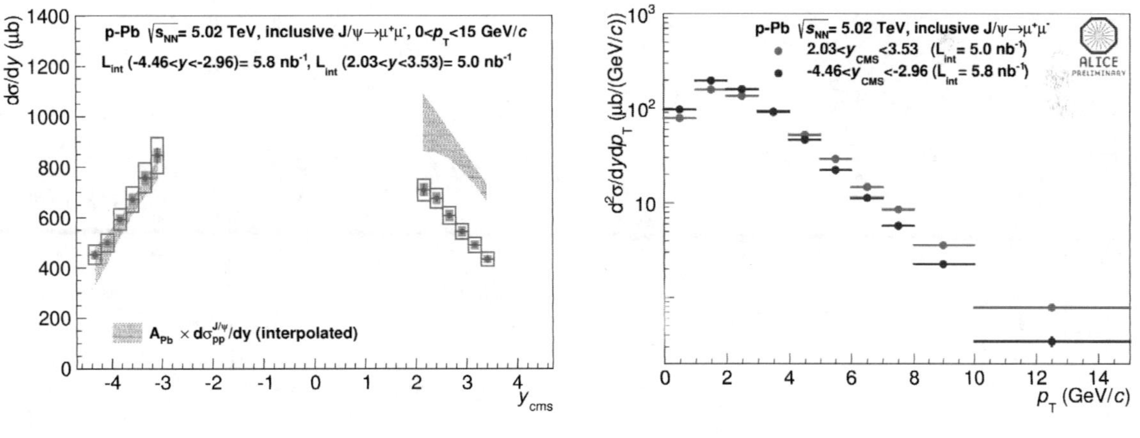

Figure 1. The inclusive J/ψ production cross section, as a function of rapidity and $p_{\rm T}$. Error bars represent statistical uncertainties, open boxes uncorrelated systematic uncertainties, shaded boxes show partially correlated systematic uncertainties. Bands in the left panel correspond to the result of the interpolation of the inclusive $\sigma_{\rm pp}^{{\rm J}/\psi}$ scaled by the Pb-nucleus mass number $A_{\rm Pb}$. Left panel is from [8].

The raw J/ψ yield was extracted by fitting the opposite-sign dimuon invariant mass spectrum by the sum of a background and a signal function. The numbers of J/ψ and its systematic uncertainty were evaluated from different fits varying background, signal ([10, 11]) functions and the fitting range. The acceptance times efficiency ($A \times \epsilon$) was estimated using a pure J/ψ signal

Monte-Carlo simulation tuned on the real data kinematic distributions. The integrated values of $A \times \epsilon$ are $(17.1 \pm 1.2)\%$ and $(25.4 \pm 1.3)\%$ for the backward and forward rapidity regions, respectively. The difference in the values results from different tracking efficiency in the two periods of data taking. In order to study the nuclear effects on the J/ψ production in p-Pb collisions, the nuclear modification factor $R_{\rm pPb}$ defined as (1) was used. Given the fact that there are no measurements of $\sigma_{\rm pp}^{{\rm J}/\psi}$ at $\sqrt{s} = 5.02$ TeV, an interpolation procedure was followed, based on the following steps:

- **Energy interpolation:** existing ALICE pp results at $\sqrt{s}$ = 2.76 and 7 TeV (covering $2.5 < y < 4.0$ in six bins) were interpolated with empirical shapes (linear, power law and exponential).
- **Rapidity interpolation/extrapolation:** due to the y-shift in p-Pb, the results from the previous step were fitted with empirical shapes (gaussian, polynomial) to reach the desired y-coverage.

The results of the procedure were validated by comparison with theoretical models: CEM [12] and actual FONLL calculation for $c\bar{c}$ production [13]. The values for $\sigma_{\rm pp}^{{\rm J}/\psi}$ scaled by the Pb-nucleus mass number are shown in the left panel of Fig. 1 by a blue band representing the uncertainties estimated by this procedure.

In addition to $R_{\rm pPb}$ the forward-to-backward ratio $R_{\rm FB}$, defined as:

$$R_{\rm FB}(|y_{cms}|) = \frac{R_{\rm pA}(y_{cms})}{R_{\rm pA}(-y_{cms})} = \frac{Y_{\rm pA}(y_{cms})}{Y_{\rm pA}(-y_{cms})}. \tag{2}$$

and independent of the pp reference cross section, was also studied.

$R_{\rm FB}$ was measured in a narrower rapidity range $(2.96 < |y_{cms}| < 3.53)$ than $R_{\rm pPb}$. The advantages of $R_{\rm FB}$ are that the nuclear thickness function and $\sigma_{\rm pp}^{{\rm J}/\psi}$ cancel out in the ratio and therefore, a pp interpolation is not needed. However the overall statistics were reduced by 30%.

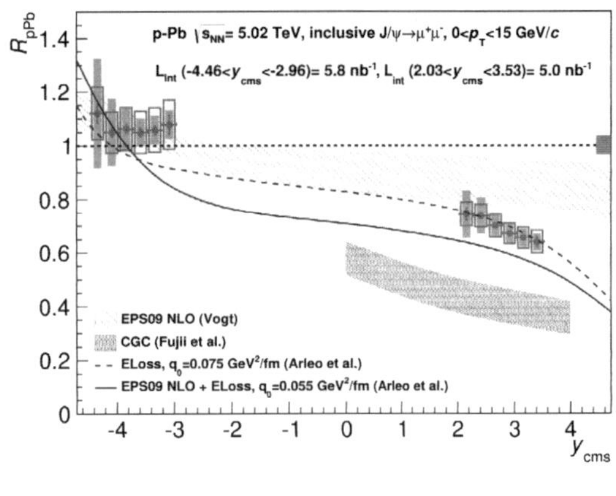

Figure 2. The y-dependence of nuclear modification factor for inclusive J/ψ production at $\sqrt{s_{\rm NN}} = 5.02$ TeV, compared to theorethical predictions [14, 15, 16]. Error bars correspond to statistical uncertainties, open boxes represent uncorrelated systematic uncertainties, shaded boxes show partially correlated systematic uncertainties. Box around $R_{\rm pPb} = 1$ shows correlated uncertainties. From [8].

3. Results

The measured J/ψ cross-sections in p-Pb collisions at $\sqrt{s_{\rm NN}} = 5.02$ TeV for the full statistics are shown in Fig. 1 differentially in y (left panel) and in $p_{\rm T}$ (right panel). In the backward region, the J/ψ production is larger, and softer in $p_{\rm T}$. The nuclear modification factors $R_{\rm pPb}$ integrated over $p_{\rm T}$ for the forward and backward rapidity ranges are:

$R_{\rm pPb}$ $(2.03 < y_{cms} < 3.53) = 0.70 \pm 0.01^{stat.} \pm 0.04^{syst.uncorr.} \pm 0.03^{syst.part.corr.} \pm 0.03^{syst.corr.}$

$R_{\rm pPb}$ $(-4.46 < y_{cms} < -2.96) = 1.08 \pm 0.01^{stat.} \pm 0.08^{syst.uncorr.} \pm 0.07^{syst.part.corr.} \pm 0.04^{syst.corr.}$

The uncertainty is dominated by the pp interpolation and the tracking efficiency. In Fig. 2 the rapidity dependence of R_{pPb} is presented. In the forward rapidity region the Color Glass Condensate (CGC)-based model underestimates the data while predictions of models including parton energy loss and/or shadowing are in fair agreement with the data.

An integrated value of $R_{\mathrm{FB}} = 0.60 \pm 0.01(stat.) \pm 0.06(syst.)$ was measured. In Fig. 3 the results on R_{FB} are shown as a function of rapidity (left) and p_{T} (right). The uncertainty on R_{FB} is dominated by the tracking efficiency. A model which includes shadowing as the only CNM effect slightly overestimates the data integrated over p_{T}, while the inclusion of an energy loss component leads to a better agreement. A sizeable p_{T} dependence of R_{FB} with a stronger suppression at low p_{T} is measured. Theoretical models including energy loss and shadowing show strong nuclear matter effects at low p_{T} in fair agreement with the data.

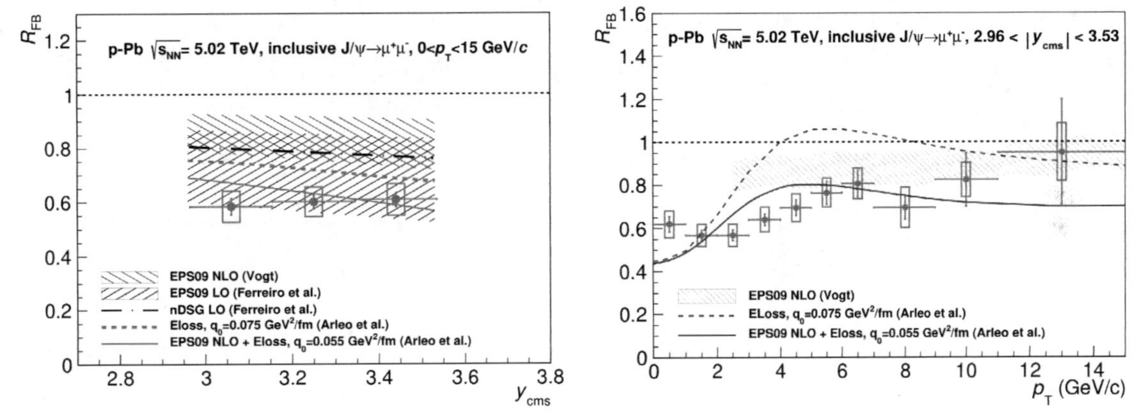

Figure 3. The forward to backward ratio R_{FB} for inclusive J/ψ production, as a function of y and p_{T}, compared to theoretical models [14, 15, 16, 17, 18]. Statistical uncertainties are shown as bars, uncorrelated systematic uncertainties correspond to open boxes. From [8].

4. Conclusions

Inclusive J/ψ suppression was measured by ALICE in p-Pb collisions at $\sqrt{s_{\mathrm{NN}}} = 5.02$ TeV. The R_{FB} (R_{pPb}) were measured vs y and p_{T} (vs y) and compared with theoretical models. Within uncertainties, both the shadowing and the coherent energy loss approaches give a fair description of the data, while the CGC-inspired model predicts too much suppression.

References

[1] Aamodt K et al. ALICE collaboration 2008 *JINST* **3** S08002.
[2] Matsui T and Satz H 1986 *Phys. Lett. B* **178** 416.
[3] Abelev B et al. ALICE collaboration 2012 *Phys. Rev. Lett.* **109** 072301.
[4] Eskola K et al. 2009 *JHEP* **0904** 065.
[5] Kharzeev D E et al. 2012 arXiv:1205.1554; Dominguez F et al. 2012 *Phys. Lett. B* **710** 182.
[6] Lourenco C et al. 2009 *JHEP* **0902** 014.
[7] Arleo F, Peigne S 2012 *Phys. Rev. Lett.* **109** 122301.
[8] Abelev B et al. ALICE Collaboration 2013 arXiv:1308.6726.
[9] Abelev B et al. ALICE Collaboration 2013 *Phys. Rev. Lett.* **110** 032301.
[10] Gaiser J 1982 SLAC - SLAC-255 194p, http://www.slac.stanford.edu/cgiwrap/getdoc/slac-r-255.pdf.
[11] Shahoyan R 2001 Ph.D.Thesis IST, Lisbon, Portugal, http://www.cern.ch/NA50/theses/ruben.ps.gz
[12] Gluck M, Owens J F and Reya E 1978 *Phys. Rev. D* **17** 2324.
[13] Cacciari M, Greco M and Nason P 1998 JHEP **9805** 007.
[14] Vogt R 2013 arXiv:1301.3395 and priv. comm.

[15] Arleo F, Peigne S 2012 arXiv:1212.0434.
[16] Fujii H, Watanabe K 2013 arXiv:1304.2221.
[17] Ferreiro E et al. 2013 arXiv:1305.4569.
[18] Arleo F et al. 2013 arXiv:1304.0901.

Journal of Physics: Conference Series **509** (2014) 012105 doi:10.1088/1742-6596/509/1/012105

Two-particle correlations in p-Pb collisions at the LHC with ALICE

Leonardo Milano, on behalf of the ALICE Collaboration

CERN - Organisation européenne pour la recherche nucléaire

E-mail: Leonardo.Milano@cern.ch

Abstract. The double ridge structure previously observed in Pb-Pb collisions has also been recently observed in high-multiplicity p-Pb collisions at $\sqrt{s_{\rm NN}} = 5.02$ TeV. These systems show a long-range structure (large separation in $\Delta\eta$) at the near- ($\Delta\varphi \simeq 0$) and away-side ($\Delta\varphi \simeq \pi$) of the trigger particle. In order to understand the nature of this effect the two-particle correlation analysis has been extended to identified particles. Particles are identified up to transverse momentum $p_{\rm T}$ values of 4 GeV/c using the energy loss signal in the Time Projection Chamber detector, complemented with the information from the Time of Flight detector. This measurement casts a new light on the potential collective (i.e. hydrodynamic) behaviour of particle production in p-Pb collisions.

1. Introduction

The study of particle correlations is a powerful tool to probe the mechanism of particle production in collisions of hadrons and nuclei at high beam energy. This is achieved by measuring the distributions of relative angles $\Delta\varphi$ and $\Delta\eta$, where $\Delta\varphi$ and $\Delta\eta$ are the differences in azimuthal angle φ and pseudorapidity η between two particles. In small systems, such as minimum-bias proton–proton (pp) collisions, the correlation at ($\Delta\varphi \approx 0$, $\Delta\eta \approx 0$) is dominated by the "near-side" jet peak, and at $\Delta\varphi \approx \pi$ by the recoil or "away-side" structure due to particles originating from jet fragmentation [1]. Additional ridge-like structures, which persist over a long range in $\Delta\eta$, emerge in nucleus–nucleus (A–A) collisions in addition to the jet-related correlations [2–4]. A similar long-range ($2 < |\Delta\eta| < 4$) near-side ($\Delta\varphi \approx 0$) structure has been observed in pp collisions at a centre-of-mass energy $\sqrt{s} = 7$ TeV in events with significantly higher-than-average particle multiplicity [5] and in high-multiplicity proton–lead (p–Pb) collisions at a nucleon–nucleon centre-of-mass energy $\sqrt{s_{\rm NN}} = 5.02$ TeV [6]. Recent measurements in p–Pb collisions employed a procedure for removing the jet contribution by subtracting the correlations extracted from low-multiplicity events, revealing essentially the same long-range structures on the away side in high-multiplicity events [7–9]. These ridge structures have been attributed to mechanisms that involve initial-state effects, such as gluon saturation [10] and colour connections forming along the longitudinal direction [11], and final-state effects, such as parton-induced interactions [12], and collective effects developing in a high-density system possibly formed in these collisions [13]. To further characterize this effect in p–Pb collisions at $\sqrt{s_{\rm NN}} = 5.02$ TeV, these ridge structures are studied via a Fourier decomposition [14] and the v_2 of pions, kaons and protons[1] has been measured.

[1] Pions, kaons and protons, as well as the symbols π, K and p, refer to the sum of particles and antiparticles.

2. Analysis

A detailed description of the ALICE detector and the event and track selection can be found in Ref. 7, 15. Events are classified in four classes defined as fractions of the analyzed event sample, based on the charge deposition in the VZERO-A detector, and denoted "0–20%", "20–40%", "40–60%", "60–100%" from the highest to the lowest multiplicity. Particle identification is based on the difference (expressed in units of the resolution σ - N_σ) between the measured and the expected signal for π, K, or p in the Time Projection Chamber (TPC) and the Time Of Flight detector (TOF) detectors. The $N_{\sigma,\mathrm{TPC}}$ versus $N_{\sigma,\mathrm{TOF}}$ correlation is reported in Fig. 1 for tracks with momentum (p) $1.5 < p < 1.6\,\mathrm{GeV}/c$, when the kaon mass is assumed. For a given species, particles are selected with a circular cut defined as

$$\sqrt{N^2_{\sigma,\mathrm{TPC}} + N^2_{\sigma,\mathrm{TOF}}} < 3.$$

In the region where the areas of two species overlap, the identity corresponding to the smaller distance is assigned. Contamination from misidentified particles is significant only for K above 1.5 GeV/c and is less than 15%.

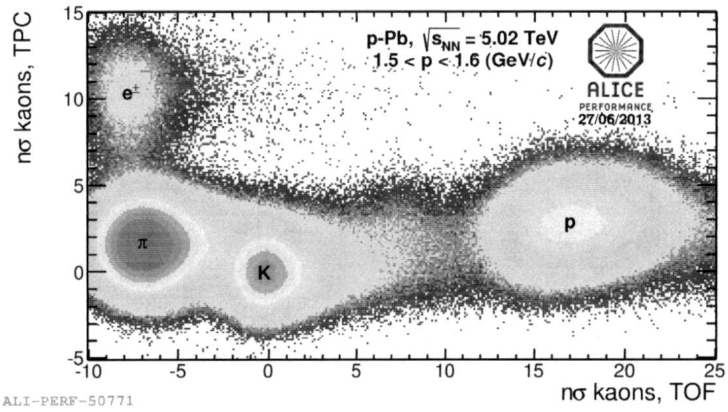

Figure 1. $N_{\sigma,\mathrm{TPC}}$ versus $N_{\sigma,\mathrm{TOF}}$ for tracks with $1.5 < p < 1.6\,\mathrm{GeV}/c$ in the kaon mass hypothesis.

This analysis uses unidentified charged tracks as trigger particles and combines them with either unidentified charged hadrons or with π, K and p as associated particles (denoted $h-h$, $h-\pi$, $h-\mathrm{K}$ and $h-\mathrm{p}$, respectively). The correlation is expressed in terms of the associated yield per trigger particle where both particles are from the same transverse momentum p_T interval in a fiducial region of $|\eta| < 0.8$:

$$\frac{1}{N_\mathrm{trig}}\frac{\mathrm{d}^2N_\mathrm{assoc}}{\mathrm{d}\Delta\eta\mathrm{d}\Delta\varphi} = \frac{S(\Delta\eta,\Delta\varphi)}{B(\Delta\eta,\Delta\varphi)} \tag{1}$$

where N_trig is the total number of trigger particles in the event class and p_T interval. The signal distribution $S(\Delta\eta,\Delta\varphi) = 1/N_\mathrm{trig}\ \mathrm{d}^2N_\mathrm{same}/\mathrm{d}\Delta\eta\mathrm{d}\Delta\varphi$ is the associated yield per trigger particle for particle pairs from the same event. The background distribution $B(\Delta\eta,\Delta\varphi) = \alpha\ \mathrm{d}^2N_\mathrm{mixed}/\mathrm{d}\Delta\eta\mathrm{d}\Delta\varphi$ is constructed by correlating the trigger particles in one event with the associated particles from other events of the same event class and within the same 2 cm-wide z_vtx interval and corrects for pair acceptance and pair efficiency.

3. Results

The per-trigger yield of the 60–100% event class is subtracted from that in the 0–20% event class in order to reduce the jet contribution as in Ref. 7. In the left panel of Fig. 2 the resulting $h-\mathrm{p}$ correlation for $1.5 < p_\mathrm{T} < 2\,\mathrm{GeV}/c$ is shown. Fourier coefficients can be extracted from the $\Delta\varphi$ projection of the per-trigger yield by a fit with:

$$\frac{1}{N_\mathrm{trig}}\frac{\mathrm{d}N_\mathrm{assoc}}{\mathrm{d}\Delta\varphi} = a_0 + 2\,a_1\cos\Delta\varphi + 2\,a_2\cos 2\Delta\varphi + 2\,a_3\cos 3\Delta\varphi. \tag{2}$$

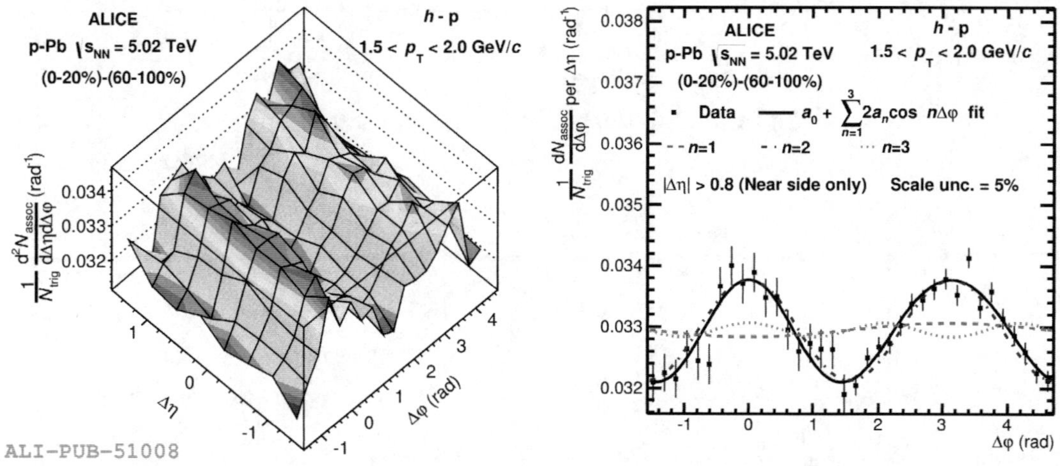

Figure 2. Left panel: associated yield per trigger particle as a function of $\Delta\varphi$ and $\Delta\eta$ for $h-\mathrm{p}$ correlations for $1.5 < p_\mathrm{T} < 2\,\mathrm{GeV}/c$ for the 0–20% event class where the corresponding correlation from the 60–100% event class has been subtracted. Right panel: projection of the left panel to $\Delta\varphi$ averaged over $0.8 < |\Delta\eta| < 1.6$ on the near side and $|\Delta\eta| < 1.6$ on the away side. The figure contains only statistical uncertainty. Systematic uncertainties are mostly correlated and are less than 5%.

The projection is averaged over $0.8 < |\Delta\eta| < 1.6$ on the near side and $|\Delta\eta| < 1.6$ on the away side. From the relative modulations $V_{n\Delta}^{h-i}\{\mathrm{2PC,sub}\} = a_n^{h-i}/(a_0^{h-i}+b)$, where a_n^{h-i} is the a_n extracted from $h-i$ correlations and b is the combinatorial baseline of the lower-multiplicity class which has been subtracted (b is determined on the near side within $1.2 < |\Delta\eta| < 1.6$), the $v_n\{\mathrm{2PC,sub}\}$ coefficient of order n for a particle species i (out of h, π, K, p) are then defined as:

$$v_n\{\mathrm{2PC,sub}\} = \sqrt{V_{n\Delta}^{h-h}} \qquad\qquad v_n\{\mathrm{2PC,sub}\} = V_{n\Delta}^{h-i}/\sqrt{V_{n\Delta}^{h-h}}. \tag{3}$$

Figure 3 shows the extracted $v_2\{\mathrm{2PC,sub}\}$ coefficients for h, π, K and p as a function of p_T. The coefficient v_2^p is significantly lower than v_2^π for $0.5 < p_\mathrm{T} < 1.5\,\mathrm{GeV}/c$, and larger than v_2^π for $p_\mathrm{T} > 2.5\,\mathrm{GeV}/c$. The crossing occurs at $p_\mathrm{T} \approx 2\,\mathrm{GeV}/c$. The coefficient v_2^K is consistent with v_2^π above $1\,\mathrm{GeV}/c$; below $1\,\mathrm{GeV}/c$ there is a hint that v_2^K is lower than v_2^π. The mass ordering and crossing is qualitatively similar to observations in nucleus–nucleus collisions [16].

4. Summary

The Fourier coefficient v_2 of the double-ridge structure in p–Pb collisions, obtained using a procedure for removing the jet contribution, exhibits a dependence on p_T that is reminiscent of the one observed in collectivity-dominated Pb–Pb collisions at the LHC. These observations and their qualitative similarity to measurements in A–A collisions [16] are rather intriguing. Furthermore, a mass ordering at low transverse momenta [17] can be described by hydrodynamic model calculations [18, 19]. Their theoretical interpretation is promising to give further insight into the unexpected phenomena observed in p–Pb collisions at the LHC.

References

[1] Wang X N 1993 *Phys.Rev.* **D47** 2754–2760 (*Preprint* `hep-ph/9306215`)

[2] Aamodt K *et al.* (ALICE) 2012 *Phys.Lett.* **B708** 249–264 (*Preprint* `1109.2501`)

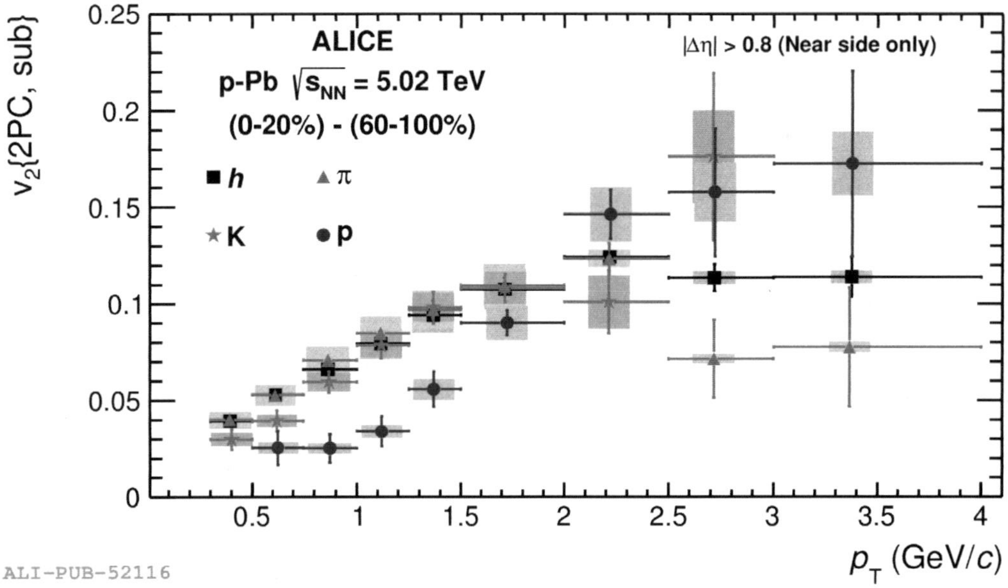

Figure 3. The Fourier coefficient $v_2\{2\mathrm{PC}, \mathrm{sub}\}$ for hadrons (black squares), pions (red triangles), kaons (green stars) and protons (blue circles) as a function of p_T from the correlation in the 0–20% multiplicity class after subtraction of the correlation from the 60–100% multiplicity class. The data are plotted at the average-p_T for each considered p_T interval and particle species under study. Error bars show statistical uncertainties while shaded areas denote systematic uncertainties.

[3] Chatrchyan S *et al.* (CMS) 2012 *Eur.Phys.J.* **C72** 2012 (*Preprint* 1201.3158)
[4] Aad G *et al.* (ATLAS) 2012 *Phys.Rev.* **C86** 014907 (*Preprint* 1203.3087)
[5] Khachatryan V *et al.* (CMS) 2010 *JHEP* **09** 091 (*Preprint* 1009.4122)
[6] Chatrchyan S *et al.* (CMS) 2013 *Phys. Lett.* **B718** 795 – 814 (*Preprint* 1210.5482)
[7] Abelev B *et al.* (ALICE) 2013 *Phys.Lett.* **B719** 29–41 (*Preprint* 1212.2001)
[8] Aad G *et al.* (ATLAS) 2013 *Phys.Rev.Lett.* **110** 182302 (*Preprint* 1212.5198)
[9] Chatrchyan S *et al.* (CMS) 2013 *Phys.Lett.* **B724** 213–240 (*Preprint* 1305.0609)
[10] Dusling K and Venugopalan R 2013 *Phys.Rev.* **D87** 094034 (*Preprint* 1302.7018)
[11] Arbuzov B, Boos E and Savrin V 2011 *Eur.Phys.J.* **C71** 1730 (*Preprint* 1104.1283)
[12] Alderweireldt S and Van Mechelen P 2012 (*Preprint* 1203.2048)
[13] Bozek P and Broniowski W 2013 *Phys. Rev. C 88,* **014903** (*Preprint* 1304.3044)
[14] Voloshin S and Zhang Y 1996 *Z.Phys.* **C70** 665–672 (*Preprint* hep-ph/9407282)
[15] Aamodt K *et al.* (ALICE) 2008 *JINST* **3** S08002
[16] Abelev B *et al.* (ALICE) 2013 *Phys.Lett.B* **719** 18–28 (*Preprint* 1205.5761)
[17] Abelev B *et al.* (ALICE) 2013 *Phys.Lett.* **B726** 164–177 (*Preprint* 1307.3237)
[18] Werner K, Bleicher M, Guiot B, Karpenko I and Pierog T 2013 (*Preprint* 1307.4379)
[19] Bozek P, Broniowski W and Torrieri G 2013 *Phys.Rev.Lett.* **111** 172303 (*Preprint* 1307.5060)

Transverse Momentum Distributions of Identified Particles in p–Pb Collisions at $\sqrt{s_{\mathrm{NN}}} = 5.02$ TeV

J. Anielski

Institut für Kernphysik, Universität Münster, Wilhelm-Klemm-Str. 9, 48149 Münster, Germany

E-mail: j.anielski@wwu.de

Abstract. Transverse momentum (p_{T}) distributions of identified hadrons produced in p–Pb collisions at $\sqrt{s_{\mathrm{NN}}} = 5.02$ TeV have been measured at mid-rapidity ($0 < y_{\mathrm{CMS}} < 0.5$) by ALICE. Particle tracks are reconstructed using the central barrel detectors. Particle identification is performed via specific energy loss, time-of-flight or their characteristic decay topology over a wide transverse momentum range (0 GeV/c up to 8 GeV/c).

Spectral shapes and particle ratios are measured in six multiplicity classes. They are compared with several model calculations and results from Pb–Pb collisions at $\sqrt{s_{\mathrm{NN}}} = 2.76$ TeV and pp collisions at $\sqrt{s_{\mathrm{NN}}} = 7$ TeV at the LHC. The results are discussed with respect to possible collective effects in p–Pb collisions.

1. Introduction

Recent measurements of di-hadron correlations in p–Pb collisions at $\sqrt{s_{\mathrm{NN}}} = 5.02$ TeV revealed a double-ridge pattern, reminiscent of the one observed in Pb–Pb collisions [1, 2, 3, 4]. This raises the question of the possible existence of collective effects in high multiplicity p–Pb collisions. Further insight into the observed phenomena can be gained by studying the evolution of spectral shapes with the particle mass and particle ratios as a function of charged-particle density.

Primary charged particles ($\pi^{\pm}, \mathrm{K}^{\pm}, \mathrm{p}$ and $\bar{\mathrm{p}}$) are identified by their specific energy loss ($\mathrm{d}E/\mathrm{d}x$) in the Inner Tracking System (ITS) and the Time Projection Chamber (TPC) [5] and by their time-of-flight with the TOF detector [6]. Weakly decaying particles ($\mathrm{K}^0_{\mathrm{s}}, \Lambda$ and $\bar{\Lambda}$) are identified by their characteristic decay topology. In this work, we present results for primary particles, defined as all particles from the collision, excluding products of weak decays of strange particles. The transverse momentum distributions have been reported in [7]. The particle identification techniques are explained in detail in [8, 9, 10].

2. Particle ratios as a function of p_{T} and $\mathrm{d}N_{\mathrm{ch}}/\mathrm{d}\eta$

The ratios p/π and $\Lambda/\mathrm{K}^0_{\mathrm{s}}$ as a function of transverse momentum are shown in Fig. 1 for a high and a low multiplicity class in p–Pb and in Pb–Pb collisions. For high multiplicity events, an increase of protons at intermediate p_{T} and a corresponding depletion at low p_{T} are evident. The change of shape with multiplicity can be observed for protons (and Λ) in both collision systems, but the relative increase in multiplicity and the absolute magnitude of the effect, however, are much smaller in p–Pb collisions. In Pb–Pb collisions this effect is generally attributed to collective flow or quark recombination [11, 12, 13].

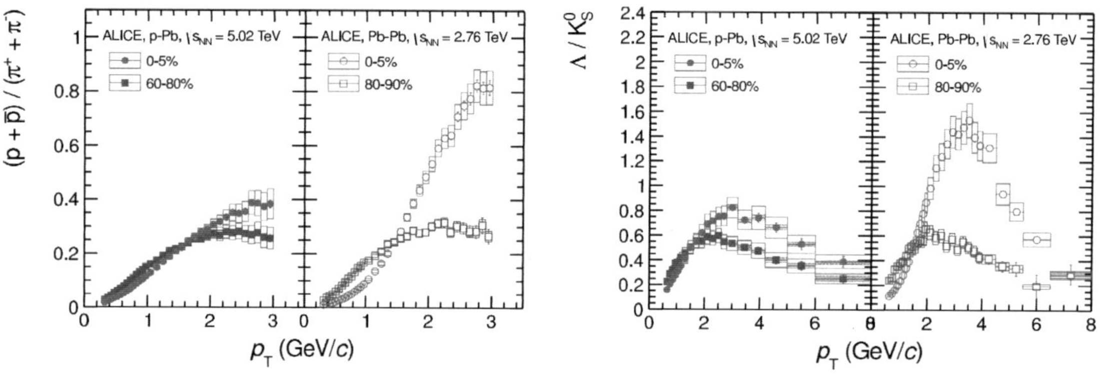

Figure 1: p/π and Λ/K^0_s ratio as a function of p_T for p–Pb and Pb–Pb collisions. Low percentiles correspond to the highest multiplicity events. Systematic errors are largely correlated across multiplicity. Multiplicity uncorrelated errors are drawn as a band for p–Pb. From [7].

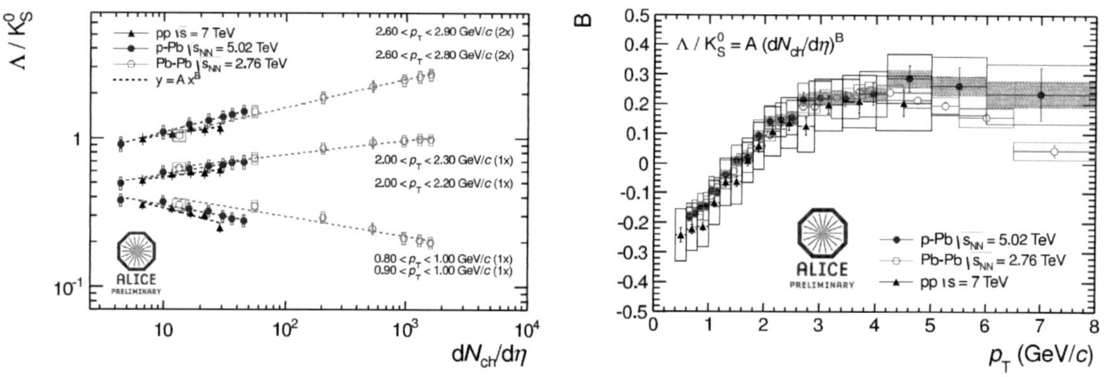

Figure 2: Fit of the Λ/K^0_s ratio as a function charged multiplicity with power law (dashed lines) for three different p_T slices and for pp, p-Pb and Pb-Pb (left). Extracted exponents B as a function of p_T (right). Based on [7].

To further quantify this observation, we plot the Λ/K^0_s ratio as a function of charged-particle multiplicity for three different momentum bins and for pp, p–Pb and Pb–Pb collisions in Fig. 2 (left). For each momentum slice, the particle ratios follow a similar trend for the three collision systems and are fitted individually with a power-law. The exponents of the fits, plotted as a function of p_T in Fig. 2 (right), show similar values in all collision systems and follow the same trend. This observation indicates a universal behavior of the Λ/K^0_S ratio as a function of multiplicity, regardless of the collision system. A similar scaling behavior applies to the p/π ratios.

3. Global Blast-Wave Fit

To compare the evolution of the spectral shape with multiplicity between different collision systems a simultaneous blast-wave fit to the π, K, p and Λ spectra was performed for each multiplicity bin. The blast-wave framework assumes a locally thermalized and collectively expanding medium and a common freeze-out for all particles [14]. It is known, as discussed e.g. in [8], that the fit parameters depend significantly on the fitting range. Nevertheless, the blast-wave framework is a useful tool to describe spectra with a small set of parameters. The extracted freeze-out temperatures (T_{kin}) and the transverse flow (β_T) show a similar behavior

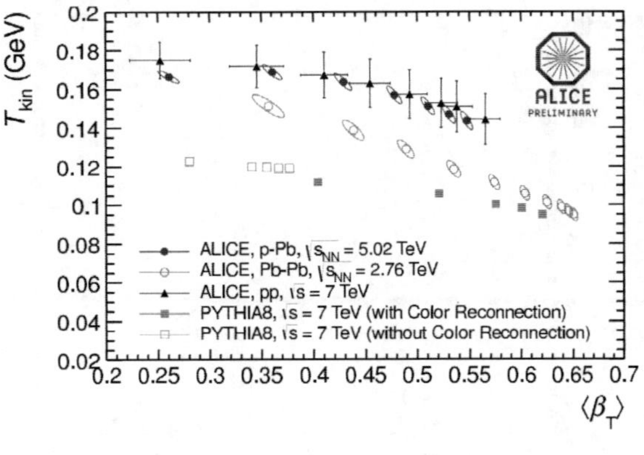

Figure 3: Fit parameters from combined Blast-Wave fit for several multiplicity/centrality classes in pp, p–Pb and Pb–Pb. Charged-particle multiplicity increases from left to right. Based on [7].

in p–Pb and Pb–Pb collisions (Fig. 3). This trend is generally understood in Pb-Pb collisions as due to an increasingly stronger radial flow as a function of centrality in heavy-ion collisions. To further test this picture, a simultaneous blast-wave fit to transverse momentum distributions in pp collisions as a function of multiplicity has been performed. Due to the limited available range, the fit has been performed for π, K and p for 0.5-1.0, 0.3-1.5 and 0.5-2.5 GeV/c, respectively. These are slightly different from the ranges used for p–Pb [7]. It should also be noted that the pp data have been divided in multiplicity intervals based on charged tracks at mid-rapidity, which could introduce a bias on the shape of the spectra. The extracted parameters show a very similar trend to the p–Pb points, within the caveats mentioned above.

Despite this suggestive observation, it has to be noted that effects other than radial flow can produce the observed trend. To illustrate this, pp spectra generated with PYTHIA 8 (tune 4C) have been fitted for different multiplicity slices. The open symbols show the results for PYTHIA without the color reconnection (CR) mechanism [15, 16]. For the spectra with CR (closed symbols) a trend qualitatively similar to the one observed in pp, p–Pb and Pb–Pb collisions can be seen.

4. Comparison with Models

The measured p_T distributions in the 5-10% bin are compared with calculations from DPMJET [17], Kraków [18], and EPOS LHC 1.99 v3400 [19] in Fig 4. DPMJET is a QCD-inspired generator, which treats soft and hard scattering processes in an unified way. It can reproduce the pseudorapidity distribution of p–Pb collisions at the LHC [20], but it fails to describe the transverse momentum spectra of identified particles. The hydrodynamical Kraków model uses a Glauber model based on Monte Carlo to create fluctuating initial conditions. It can reproduce the shape of the measured pion and kaon spectra reasonably well up to 1 GeV/c, but fails for higher momenta. This could, in the hydrodynamical framework, indicate the onset of a non-thermal component. The EPOS model is founded on parton-based Gribov Regge theory and implements an hydrodynamical evolution for bulk particles. It can describe the pion and proton distributions within 20% over the full measured range, but shows larger deviations for kaons and lambdas.

5. Discussion

The measurement of identified particle spectra adds valuable information to address the presence of possible collective effects in p–Pb collisions. The p/π and Λ/K^0_S ratios show a similar behavior in p–Pb and Pb–Pb collisions at the LHC and the increase of the ratio scales with

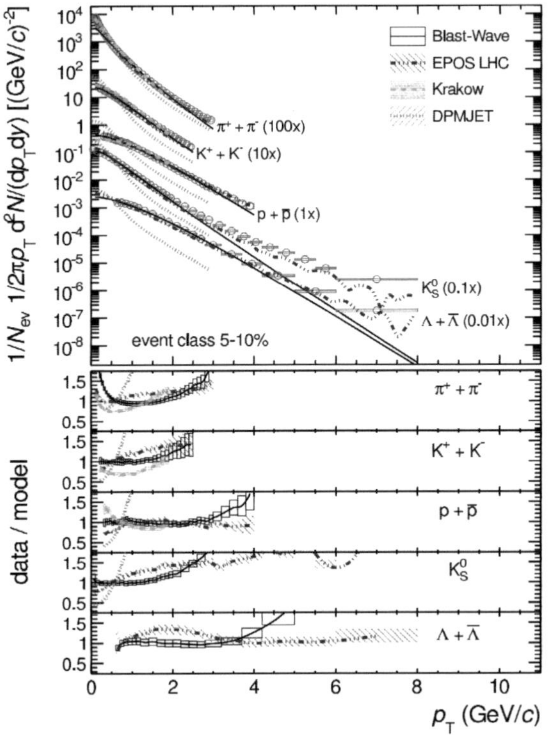

Figure 4: Comparisons of the spectra (5-10% multiplicity bin) with several models. From [7].

the charged-particle multiplicity for pp, p–Pb and Pb-Pb collisions. The blast-wave analysis and the comparison with models indicate that final state effects seem to be needed to reproduce the measured spectra. The blast-wave analysis of Monte Carlo spectra, generated with PYTHIA, shows that other effects like color-reconnection can produce a pattern similar to radial flow.

References

[1] Abelev B *et al.* (ALICE Collaboration) 2012 *Phys. Lett.* **B719** 29–41
[2] Chatrchyan S *et al.* (CMS Collaboration) 2013 *Phys. Lett.* **B718** 795–814
[3] Aad G *et al.* (ATLAS Collaboration) 2012 (*Preprint* hep-ex/1212.5198)
[4] Khachatryan V *et al.* (CMS Collaboration) 2010 *JHEP* **1009** 091
[5] Alme J, Andres Y, Appelshauser H, Bablok S, Bialas N *et al.* 2010 *Nucl. Instrum. Meth.* **A622** 316–367
[6] Akindinov A, Alici A, Agostinelli A, Antonioli P, Arcelli S *et al.* 2013 *Eur. Phys. J. Plus* **128** 44
[7] Abelev B *et al.* (ALICE Collaboration) 2013 (*Preprint* nucl-ex/1307.6796)
[8] Abelev B *et al.* (ALICE Collaboration) 2012 *Phys. Rev. Lett.* **109** 252301
[9] Abelev B *et al.* (ALICE Collaboration) 2013 (*Preprint* hep-ex/1303.0737)
[10] Abelev B *et al.* (ALICE Collaboration) 2013 (*Preprint* nucl-ex/1307.5530)
[11] Fries R, Muller B, Nonaka C and Bass S 2003 *Phys. Rev. Lett.* **90** 202303
[12] Bozek P 2011 (*Preprint* nucl-th/1111.4398)
[13] Muller B, Schukraft J and Wyslouch B 2012 *Ann. Rev. Nucl. Part. Sci.* **62** 361–386
[14] Schnedermann E, Sollfrank J and Heinz U W 1993 *Phys. Rev.* **C48** 2462–2475
[15] Skands P Z and Wicke D 2007 *Eur. Phys. J.* **C52** 133–140
[16] Schulz H and Skands P 2011 *Eur. Phys. J.* **C71** 1644
[17] Roesler S, Engel R and Ranft J 2000 1033–1038 (*Preprint* hep-ph/0012252)
[18] Bozek P 2012 *Phys. Rev.* **C85** 014911
[19] Pierog T, Karpenko I, Katzy J, Yatsenko E and Werner K 2013 (*Preprint* hep-ph/1306.0121)
[20] Abelev B *et al.* (ALICE Collaboration) 2013 *Phys. Rev. Lett.* **110** 032301

Recent results on ultra-peripheral Pb-Pb and p-Pb collisions at the ALICE experiment

J.D. Tapia Takaki
IPN Orsay, Université Paris-sud, CNRS-IN2P3, France
The University of Kansas, Lawrence, United States
E-mail: Daniel.Tapia.Takaki@cern.ch

Abstract. In this paper, we will discuss results on coherent and incoherent J/ψ photoproduction in ultra-peripheral Pb-Pb collisions at $\sqrt{s_{\rm NN}} = 2.76$, as well as preliminary results on exclusive J/ψ and two-photon production in ultra-peripheral p-Pb collisions at $\sqrt{s_{\rm NN}} = 5.02$ TeV. Results on four-pion production in ultra-peripheral Pb-Pb collisions will also be given. Finally, the prospects for other UPC measurements in ALICE will be briefly mentioned.

Introduction

The LHC is not only the most powerful collider for proton-proton and heavy-ion collisions, but also for photon-photon and photon-hadron (γp and γPb) interactions, offering a unique opportunity to study fundamental aspects of QED and QCD via photon-induced processes. In ultra-peripheral collisions (UPC), two ions (or protons) pass by each other with an impact parameter at least two times larger than the nuclear radii, and here the hadronic interactions are strongly suppressed. The number of photons scales like Z^2 for a single source, so that in heavy-ion collisions exclusive particle production is dominated by electromagnetic processes. Typical examples are $\gamma\gamma \to l^+l^-$ (where $l^+l^- = e^+e^-$, $\mu^+\mu^-$, $\tau^+\tau^-$) or exclusive $\gamma + p \to$ J/ψ + p, which is modeled in pQCD by the exchange of two-gluons with no net-color transfer.

The experimental challenge for such measurements consists in having dedicated UPC triggers that are often orthogonal to the general trigger strategy of the experiments. Moreover, validating an exclusive analysis requires a good understanding of the trigger efficiency for the exclusivity conditions imposed at the online and offline levels, for which control triggers are usually required.

1. The ALICE experiment

The ALICE experiment [1] is well suited to study UPC vector meson production in a systematic way. The UPC triggers used the VZERO forward detector for vetoing, and the muon spectrometer, Time-of-Flight and Silicon Pixel Detector as triggers. The muon spectrometer covers the range $-4.0 < y < -2.5$, while a central barrel placed inside a large solenoid magnet (0.5 T) covers the pseudorapidity range $|\eta| < 0.9$. The pseudorapidity range $2.8 < \eta < 5.1$ is covered by VZERO-A and $-3.7 < \eta < -1.7$ by VZERO-C, which are scintillator tile arrays with a time resolution better than 1 ns allowing us to distinguish between beam-beam and beam-gas collisions. For the J/ψ analysis at mid-rapidity the TPC provides particle ID information. This is complemented with information provided by other detectors at the offline level, including the

ZDCs, which are located at about 116 m at both sides from the IP, and allow us to study both neutrons and protons emitted in the very forward region. Additional details on the way in which these detectors were used for these analyses are given in [2, 3]

2. Coherent and incoherent J/ψ production in UPC Pb-Pb

The first LHC measurements on exclusive photoproduction of J/ψ vector mesons produced in Pb-Pb collisions at $\sqrt{s_{\mathrm{NN}}} = 2.76$ TeV were carried out by ALICE [2, 3]. This measurement is particularly interesting as the UPC J/ψ cross section depends on the nuclear gluon distribution squared [5], making it a very good probe to study nuclear shadowing at low Bjorken-x values. At mid-rapidity J/ψs have a Bjorken-x value of about 10^{-3}. One of ALICE's advantage is that J/ψs can be measured down to zero p_T, which is important here as UPC J/ψs are characterized by their very low p_T. Coherent J/ψ production, where the photon couples coherently to all nucleons, is characterized by an average J/ψ p_T of about 60 MeV/c. For these events the target nucleus usually stays intact. For the incoherent case, the photon couples to a single nucleon, representing the quasi-elastic scattering off a single nucleon and the J/ψ has a somewhat larger average p_T of about 500 MeV/c. The p_T distributions for both coherent-enhanced and incoherent-enhanced data samples are well described by fitting together the expected signals and backgrounds using templates built from MC and data. This was used to extract the measured yield of coherent and incoherent J/ψ candidates. The measured cross sections are $d\sigma^{J/\psi}_{Coh}/dy = 1.00 \pm 0.18$ (stat) $^{+0.24}_{-0.26}$ (syst) mb at $-3.6 < y < -2.6$ and $d\sigma^{J/\psi}_{Coh}/dy = 2.38$ $^{+0.34}_{-0.24}$ (stat+syst) mb at $-0.9 < y < 0.9$. These results were compared to the available model calculations as described in [2, 3]. Best agreement is found with models that include nuclear gluon shadowing, consistent with EPS09 (see Figure 1). Recent model calculations provide ways on how these measurements can be used to extract information on the nuclear gluon density [4]. Furthermore, the incoherent J/ψ cross section was also recently published [3] and provides additional constraints to model calculations.

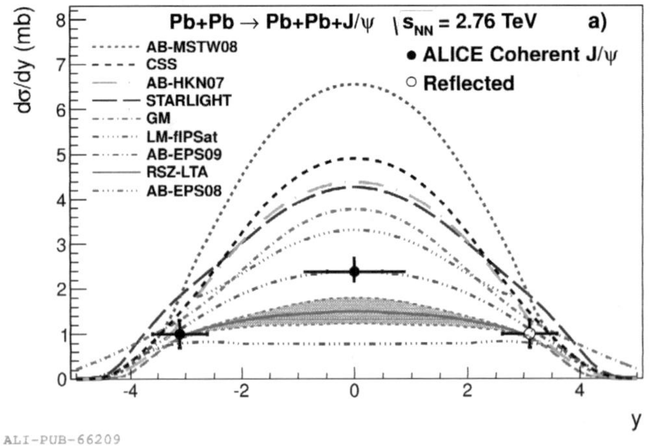

Figure 1. Coherent J/ψ cross section in ultra-peripheral Pb-Pb collisions at $\sqrt{s_{\mathrm{NN}}} = 2.76$ TeV measured by ALICE. Results are compared to theoretical calculations. See text for details.

3. Exclusive J/ψ photoproduction off protons in UPC p-Pb at $\sqrt{s_{\mathrm{NN}}} = 5.02$ TeV

The lead nucleus acts as a photon emitter (enhanced flux by factor Z^2 ~7000 compared with the proton photon flux). At LHC the Bjorken-x goes down to 10^{-5} reaching for the first time the 1 TeV scale for the γp centre-of-mass energy ($W_{\gamma p}$). Once the photon emitter is known, $W_{\gamma p}$ is determined by the J/ψ rapidity: $W_{\gamma p}^2 = 2E_p M_{J\psi} \exp(-y)$, where $M_{J/\psi}$ is the J/ψ mass and $E_p = 4$ TeV is the energy of the proton beam. For the 2013 p-Pb run, apart from the central and forward J/ψ measurements, we introduced a new trigger that allowed us pairing a single muon low p_T (with a trigger threshold of 0.5 GeV/c) using the forward muon spectrometer, together with a muon that can be reconstructed by the central barrel detector (particle ID is provided by the TPC). Because the muon spectrometer is a single-arm detector and we had both p-Pb and Pb-p colliding systems, the following measurements were possible: a) muon-arm p-Pb ($21\leq W_{\gamma p} \leq 45$ GeV); b) muon+barrel p-Pb ($45\leq W_{\gamma p} \leq 82$ GeV); c) central barrel p-Pb or Pb-p ($100\leq W_{\gamma p} \leq 250$ GeV); d) muon+barrel Pb-p ($300\leq W_{\gamma p} \leq 550$ GeV), and e) muon-arm Pb-p ($500\leq W_{\gamma p} \leq 1160$ GeV). In what follows, we will only present results for the muon-arm p-Pb. Figure 2 shows the J/ψ p_T distribution, after fitting together signals and backgrounds to data as described above. Contributions from events where the proton dissociates are shown in red. This non-exclusive J/ψ p_T shape was estimated from data by requiring events with more than two hits in VZERO-C. To validate this procedure several cross checks were carried out using control samples. Our results shown in Figure 3 are compatible with HERA measurements. Although not shown in this figure, LHCb recently extracted $\sigma(W_{\gamma\mathrm{p}})$ from their measurement of exclusive J/ψ production in pp collisions [6]. Owing to the ambiguity over which proton emitted the photon, for each measured $d\sigma/dy$ there are two coupled solutions for $\sigma(W_{\gamma\mathrm{p}})$, one at low energy and one at high energy. In addition, a functional form with a built-in power law dependence was assumed for $\sigma(W_{\gamma\mathrm{p}})$. For these reasons, pp measurements cannot provide an unambiguous test of the power-law dependence of the cross section as a function of the γp centre-of-mass energy, which is needed for observing new phenomena at high energies such as saturation.

4. Exclusive two-photon production in γp from UPC p-Pb

Preliminary cross sections for the exclusive two-photon process were also presented for the first time at this conference. They correspond to the first measurement in p-Pb collisions. Indeed, one of the main challenges of measuring J/ψ photoproduction is to have under control background from inelastic processes such as proton dissociation. The measured cross section for this process in the invariant mass range $1.5 < M_{\mathrm{inv}} < 2.5$ GeV/c^2, using the same technique as for the J/ψ, is $\sigma(\gamma\gamma \to \mu^+\mu^-) = 1.76 \pm 0.12$ (stat) $^{+0.16}_{-0.15}$ (syst) μb at forward rapidity. The STARLIGHT prediction is 1.8 μb which is in good agreement with our measurements. This measurement could be used to constraint the inelastic background associated mainly to proton dissociation in hadronic colliders [7].

5. Four-pion production in UPC Pb-Pb

It is interesting to look for excited states of photo-produced ρ^0 mesons. It is not clear how many excited states exist or their quantum numbers (see special PDG review [11]). STAR reported on four-pion production in UPC [12], albeit at much lower centre-of-mass energies. No HERA publications on photo-production of a ρ^0 excited states exist. Figure 4 shows the four-pion p_T distribution, where a clear coherent peak can be seen at low transverse momenta. This corresponds to data collected during the 2011 Pb-Pb run, where we have 10 times more statistics than those published by STAR. One of the current research interesest is understanding the possible production mechanics [13].

6. Summary and outlook

ALICE is the first LHC experiment to provide results on ultra-peripheral collisions. Thanks to our collected UPC data we are exploring novel kinematic regimes with the potential for novel and high-profile physics. In UPC J/ψ analyses in γPb we have found a good agreement with models including moderate nuclear gluon shadowing. Two publications are available [2, 3], which include

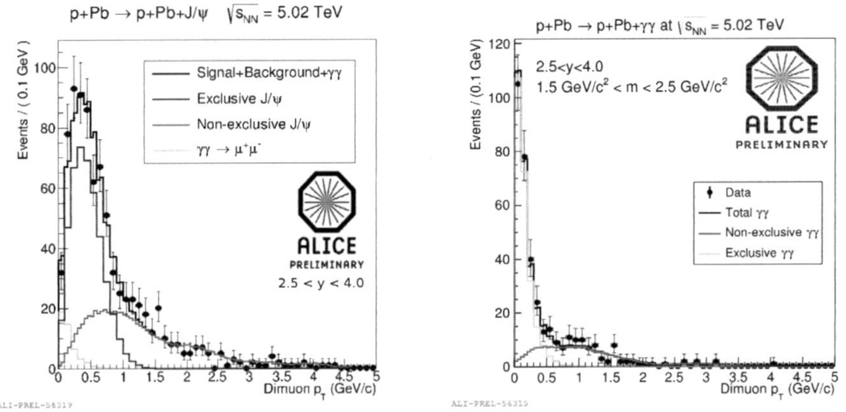

Figure 2. Transverse momentum distribution for events with exactly two oppositely charged muons satisfying the event selection for the forward dimuon samples, in the invariant mass range of the exclusive J/ψ ($2.8 < M_{\rm inv} < 3.3$ GeV/c^2) and ($1.5 < M_{\rm inv} < 2.5$ GeV/c^2) in the right and left plots, respectively. The templates for exclusive J/ψ and $\gamma\gamma \to \mu^+\mu^-$ were obtained from STARLIGHT. The non-exclusive templates were obtained from data, as described in the text.

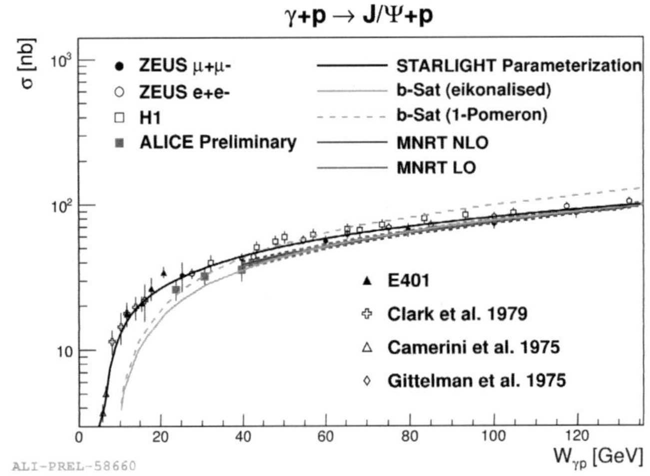

Figure 3. Exclusive J/ψ photoproduction cross section off protons measured by ALICE and compared to existing data from colliders and fixed-target experiments. Comparisons to STARLIGHT, MNRT [8] and the b-Sat models [9] are also shown.

results for coherent and incoherent J/ψ as well as results from exclusive two-photon production. Other ongoing analyses include ρ^0 and four-pion photoproduction in UPC Pb-Pb and p-Pb. In UPC p-Pb, we have studied J/ψ photoproduction from γp at $W_{\gamma\rm p} \sim 30$ GeV and two-photon process with their exclusive and inelastic components. Results at higher γp energies from Pb-p collisions reaching the 1 TeV energy scale will be available soon. We also collected UPC triggers in pp running, during the years 2011 and 2012, where similar studies can be carried out.

Journal of Physics: Conference Series **509** (2014) 012107 doi:10.1088/1742-6596/509/1/012107

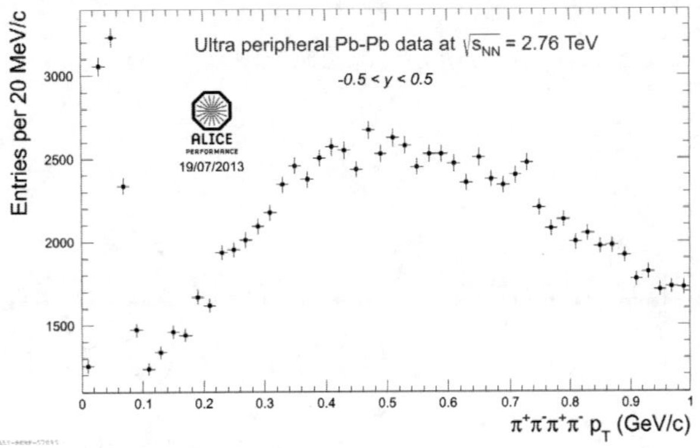

Figure 4. Transverse momentum distribution for the UPC four-pion production. A coherent four-pion peak is clearly visible.

Acknowledgments

I would like to thank the organisers of SQM 2013 held at the University of Birmingham for supporting my participation to this conference.

References

[1] Aamodt K *et al* [ALICE Collaboration] 2008 JINST **3** S08002
[2] Abelev B *et al* [ALICE Collaboration] 2013 Phys. Lett. B **718** 1273
[3] Abbas E *et al* [ALICE Collaboration] 2013 Eur. J. Phys. **C73** 2617
[4] Guzey V *et al* 2013 Phys. Lett. B **726** 290
[5] Ryskin M G 1993 Z. Phys. C **57** 89
[6] RAaij *et al* [LHCb Collaboration] 2013 J. Phys. G **40** 045001
[7] Harland-Lang L A *et al* 2012 Eur. Phys. J. C **72** 2110
[8] Martin A D *et al* 2008 Phys. Lett. B **662** 252
[9] Motyka L and Watt G 2008 Phys. Rev. D **78** 014023
[10] Abelev B I *et al* [STAR Collaboration] 2010 Phys. Rev. C **81** 044901
[11] Beringer J *et al* [Particle Data Group Collaboration] 2012 Phys. Rev. D **86** 010001
[12] Abelev B I *et al* [STAR Collaboration] 2010 Phys. Rev. C **81** 044901
[13] Klusek-Gawenda M and Szczurek A 2013 arXiv:1309.2463 [nucl-th]

Journal of Physics: Conference Series **509** (2014) 012108 doi:10.1088/1742-6596/509/1/012108

Production of hypernuclei in Pb-Pb collisions at $\sqrt{s_{\rm NN}} = 2.76$ TeV with ALICE at the LHC

Ramona Lea (for the ALICE Collaboration)

Physics Department, University and INFN of Trieste, Via A. Valerio 2, 34127 Trieste, Italy

E-mail: ramona.lea@cern.ch

Abstract. Results on (anti-)hypertriton production in Pb–Pb collisions at $\sqrt{s_{\rm NN}} = 2.76$ TeV, obtained by ALICE at the LHC, are reported. The (anti-)$^{3}_{\Lambda}$H signal is extracted exploiting its mesonic decay ($^{3}_{\Lambda}$H $\rightarrow$ ^{3}He $+\pi$) , via the topological identification of secondary vertices and the analysis of the invariant mass distributions of (^{3}He, π) pairs. The (^{3}He,π) invariant mass distributions in different transverse momentum and in proper time intervals are shown.
Search for $\Lambda\Lambda$ state (H-Dibaryon) via its decay H-Dibaryon $\rightarrow \Lambda p\pi^{-}$ is discussed. As no H-Dibaryon signal is observed, an upper limit on its production yield has been determined.

1. Introduction

Ultra-relativistic heavy-ion collisions provide a unique opportunity for understanding the strong interaction: many strange hadrons are produced in the collision and the study of their production yield helps to understand the nature of the space-time evolution and chemical equilibrium of the hadronic fireball. In high-energy heavy-ion collisions hyperon-baryon bound systems, called hypernuclei, can emerge from the hot and dense fireball region of the reaction. Their production yield can be estimated by employing two distinct models: they can be formed via the coalescence of nucleons and hyperons produced in the collision or they can be produced directly in the hadronization process. The study of the production yield of hypernuclei at the LHC energy can help to distinguish between the two models, as the production yield is expected to be different in the two approaches.
Thanks to its unique performance for particle identification, the ALICE detector [1, 2] allows for the measurement of hypernuclei and gives the opportunity to search for predicted states such as H-Dibaryon, a bound state formed by six quarks (*uuddss*) already predicted by Jaffe in 1977 [3] using a bag model calculation. The energy reached at the LHC leads to large production probabilities of such particles, as described for example by thermal models [4, 5, 6].

2. (Anti-) Hypertriton

Hypernuclear physics was born in 1952 when the Polish scientists M. Danysz and J. Pniewski observed the first hypernuclear decay event in a photographic emulsion exposed to cosmic rays at around 26 km above the ground [7]. The hypertriton $^{3}_{\Lambda}$H is the lightest known hypernucleus and is formed by a proton, a neutron and a Λ.
The study of the production of ($^{3}_{\Lambda}\overline{\rm H}$)$^{3}_{\Lambda}$H detected via its decay $^{3}_{\Lambda}$H $\rightarrow$ ^{3}He + π^{-} ($^{3}_{\Lambda}\overline{\rm H}\rightarrow^{3}\overline{\rm He}+\pi^{+}$) with ALICE will be presented in this section.

2.1. Analysis

For the present study, nearly 20 million of central (0-10%) and nearly 18 million of semi-central (10-50%) events from Pb–Pb collisions at $\sqrt{s_{\rm NN}} = 2.76$ TeV collected by ALICE during the 2011 run are analysed. The main detector used in the analysis is the Time Projection Chamber (TPC) [8] which has a full azimuthal acceptance for tracks in the pseudo-rapidity region $|\eta| < 0.9$. Both daughter particles of the ${}^3_\Lambda$H can be identified using the TPC over a wide range of momentum. Once both daughter tracks are identified, it is possible to obtain the hypertriton signal candidates by reconstructing their decay vertices. A set of topological cuts has been implemented in order to reduce the combinatorial background. These cuts include: distance of closest approach (DCA) between the two tracks ($< 0.7\ \sigma$), DCA of the "pion" track from the primary vertex (> 0.4 cm) and cosine of the pointing angle between the line joining the primary and secondary vertex and the the total momentum vector direction (cosine of pointing angle > 0.99). Using the collected statistics it has been possible to obtain a (${}^3_\Lambda\overline{\rm H}$) ${}^3_\Lambda$H signal, both in central and in semi-central events.
Fig. 1 shows the invariant mass distribution of [(^{3}He+π^-)+(${}^3\overline{\rm He}$+π^+)] for semi-central events (10-50%) in the transverse momentum range $2 \leq p_{\rm T} < 10$ GeV/c. Black empty points are the data points while the red histogram is the "like-sign" LS invariant mass spectrum [(^{3}He,π^+)+(${}^3\overline{\rm He}$,π^-)]. The light curve is the sum of a third-degree polynomial function, used to evaluate the combinatorial background, and a Gaussian function for the signal. The full circles at the bottom of the plot are data after the polynomial background subtraction, and the superimposed black line is the Gaussian fit. A signal with a significance of 3.94 is clearly visible.

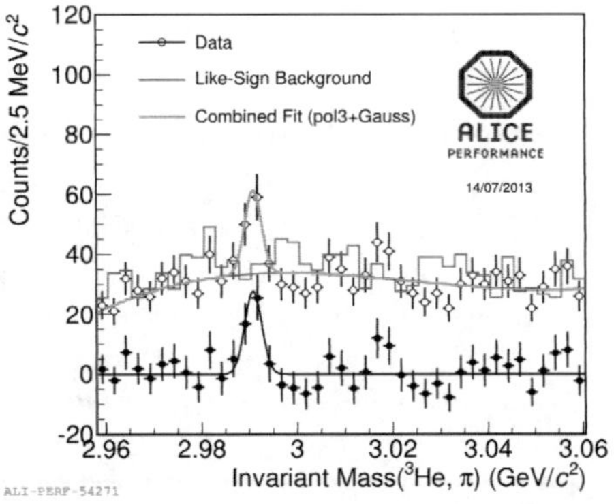

Figure 1. Invariant mass of [(^{3}He,π^-)+(${}^3\overline{\rm He}$,π^+)] for semi-central (10-50%) events. See text for more details.

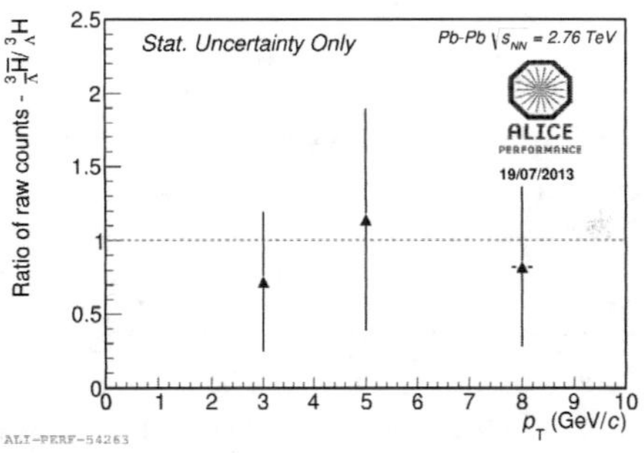

Figure 2. ${}^3_\Lambda\overline{\rm H}/{}^3_\Lambda$H ratio of raw yields versus $p_{\rm T}$: this ratio is consistent with unity (reference line) in the whole $p_{\rm T}$ range. Uncertainties are statistical only.

The collected data allow for the extraction of ${}^3_\Lambda$H(${}^3_\Lambda\overline{\rm H}$) signals in 3 $p_{\rm T}$ bins: $2 \leq p_{\rm T} < 4$ GeV/c; $4 \leq p_{\rm T} < 6$ GeV/c and $6 \leq p_{\rm T} < 10$ GeV/c. The same procedure to extract the signal described in the previous paragraph has been applied for each $p_{\rm T}$ bin, and quality checks on the stability of the mean and the width of the signal have been performed. The peak width versus $p_{\rm T}$ is stable within 2σs, the mean is stable and consistent with the value of ${}^3_\Lambda$H mass from literature [9].
Fig. 2 shows the ${}^3_\Lambda\overline{\rm H}/{}^3_\Lambda$H ratio of the raw yields versus $p_{\rm T}$: this ratio is consistent with the unity (reference line) in the whole considered $p_{\rm T}$ range.
Using the statistics recorded in 2011 it has also been possible to extract the (${}^3_\Lambda$H+${}^3_\Lambda\overline{\rm H}$) signal in four ct bins, where ct has been defined as $ct = \frac{MLc}{p}$, M being the nominal value of the ${}^3_\Lambda$H mass=2.991 GeV/c^2, L the decay length and p the total momentum of the ${}^3_\Lambda$H candidate.

Fig.3 shows the invariant mass spectrum of the sum of ($^3_\Lambda$H+$^3_\Lambda\overline{\text{H}}$) in two ct bins : $1 \leq ct < 4$ cm (left panel) and $10 \leq ct < 28$ cm (right panel). The empty circle points are the data while full line is the "like-sign" invariant mass spectrum. The light curve is a function which is the sum of a third degree polynomial, used to evaluate the"like-sign" combinatorial background, and a Gaussian for the signal. The full line in the lower part of the plot is the signal extracted after the polynomial background subtraction. A signal is visible in each panel. As soon as the efficiency correction is available the $^3_\Lambda$H lifetime can be determined.

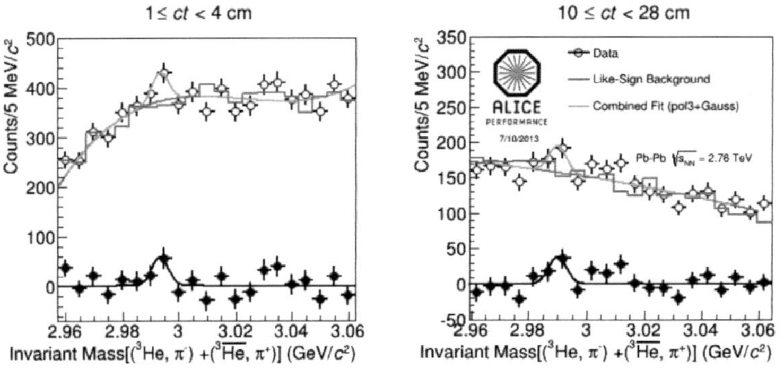

Figure 3. $[(^3\text{He}, \pi^-) + (^3\overline{\text{He}}, \pi^+)]$ invariant mass spectra for 2 ct bins. a) [1-4] cm b) [10-28] cm. See text for more details.

3. H-Dibaryon

Recent lattice calculations [10, 11] suggest that H-Dibaryon should be a bound state, even if those calculations have been performed at an unphysical pion mass ($m_\pi \approx 390$ GeV/c^2). When the results are extrapolated chirally to the pion physical mass [12, 13] the H-Dibaryon is likely to be unbound by 13 ± 14 MeV/c^2 above the $\Lambda\Lambda$ (2.231 GeV/c^2) threshold or even close to the Ξp threshold (2.26 GeV/c^2). A binding energy of around 1 MeV/c^2 is also favored from the observed double-Λ hypernuclei, which gives the current constraints on the $\Lambda\Lambda$ interaction (for a recent review [14] and the references therein). In this analysis decay of the H-Dibaryon into $\Lambda p\pi$ has been investigated. Other possible decay channels contain a neutron which is difficult to detect with the ALICE setup. The expected branching ratios depend on the binding energy, as shown in [15]. Since a low binding energy is favored by the current theoretical discussions, the region of mass between 2.2-2.3 GeV/c^2 has been investigated. Considering the H $\rightarrow \Lambda p\pi$ decay channel, a signal for a bound state would result in a peak in the invariant mass or in a broad structure above the $\Lambda\Lambda$ threshold in case of a resonance. The results shown here for the H-Dibaryon are based on the analysis of about 13.8 million Pb–Pb events in the centrality class of 0-80% taken with the ALICE apparatus in 2010. The reconstructed invariant mass distribution is shown in Fig.4. No evidence of a signal for the H-Dibaryon was found. Fig. 4 also shows the expected signal for the H-Dibaryon for two assumed binding energies: 21 MeV/c^2, which leads to an expected mass of 2.21 GeV/c^2, and 1 MeV/c^2, which corresponds to an expected mass of 2.23 GeV/c^2.
The expected signal was computed estimating the acceptance $\times$ efficiency (from a Monte-Carlo simulation), the production rates as predicted by the thermal-model prediction [16] and the expected branching ratios [17]. In the Monte-Carlo simulation the lifetime of the free Λ hyperon was assumed for the H-Dibaryon. The upper limits for the production yield in the two cases described above, and assuming the thermal model predictions, are reported in Tab. 1. Upper limits have been computed by using a frequentist approach which assumes Poissonian likelihood distributions for background and signal [18, 19].

Particle (Mass /GeV/c^2)	Upper limit of dN/dy (99% CL)
H-Dibaryon (2.23 GeV/c^2)	$\leq 2.0 \times 10^{-4}$
H-Dibaryon (2.21 GeV/c^2)	$\leq 8.4 \times 10^{-4}$

Table 1. Upper limits of dN/dy for the H-Dibaryon at 99% confidence level.

The extracted limits are a factor of 10 lower than the thermal-model predictions used to estimate the expected signal while these successfully describe the yields measured by STAR for the hypertriton [20] within uncertainties. The comparison with other theoretical models can be found in [21].

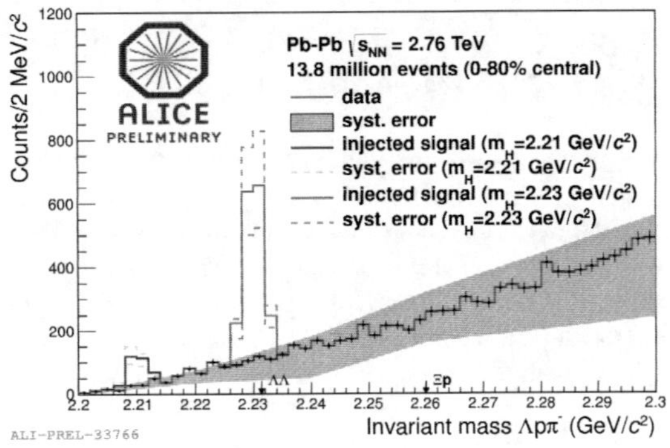

Figure 4. Invariant mass of $\Lambda + p + \pi$. The expected signal using thermal-model predictions and the estimated acceptance $\times$ efficiency for two different possible bound states of the H-Dibaryon are also shown.

References

[1] Aamodt K *et al.* (ALICE collaboration) 2006 *J. Phys. G: Nucl. Part. Phys.* **32** 1295
[2] Aamodt K *et al.* (ALICE collaboration) 2008 *JINST* **3** S08002
[3] Jaffe R 1977 *Phys. Rev. Lett.* **38** 195 and erratum ibid 1977 **38** 617
[4] Andronic A *et al.* 2011 *Phys. Lett. B* **697** 203
[5] Cleymans J *et al.* 2011 *Phys. Rev. C* **84** 054916
[6] Petráň M, Letessier J, Petráček J and Rafelski J 2013 CERN-PH-TH/2012-262, arXiv:1303.2098v2 [hep-ph]
[7] Danysz M, Pniewski J 1953 *Phil. Mag.* *44* **350** 348
[8] Alme J *et al.* 2010 *Nucl. Instrum. Meth. A* **622** 316-367
[9] Juric M *et al.* 1973 **Nucl. Phys. B 52** 1
[10] Beane S R *et al.* (NPLQCD Collaboration) 2011 *Phys. Rev. Lett.* **106** 162001
[11] Inoue T *et al.* (HAL QCD Collaboration) 2011 *Phys. Rev. Lett.* **106** 162002
[12] Shanahan P E, Thomas A W, Young R D, 2011 *Phys. Rev. Lett.* **107** 092004
[13] Heidenbauer J, Meiner U G 2011 *Phys. Lett. B* **706** 100
[14] Botta E, Bressani T, Garbarino G 2012 *Eur. Phys. J. A* **48** 41
[15] Schaffner-Bielich J, Mattiello R, Sorge H 2000 *Phys. Rev. Lett.* **84** 4305
[16] Andronic A, private communication
[17] Schaffner-Bielich J private communication
[18] Rolke W, Lopez A 2001 *Nucl. Inst. Meth. A* **458** 745
[19] Rolke W, Lopez A, Conrad J and James F 2005 *Nucl. Instr. Meth. A* **551** 493
[20] Abelev B I *et al.* (STAR Collaboration) 2010 *Science* **328** 58
[21] Dönigus B (for the ALICE Collaboration) these proceedings

Journal of Physics: Conference Series **509** (2014) 012109 doi:10.1088/1742-6596/509/1/012109

Search for exotic hyper-matter and measurement of nuclei with ALICE at the LHC

Benjamin Dönigus (for the ALICE Collaboration)
Institut für Kernphysik, Goethe-Univeristät Frankfurt, Max-von-Laue-Str. 1, 60438 Frankfurt am Main

E-mail: benjamin.doenigus@cern.ch

Abstract. The high collision energies reached at the LHC lead to significant production yields of light (anti)-nuclei and hyper-matter in proton-proton and, in particular, Pb–Pb collisions. The excellent particle identification capabilities of the ALICE apparatus, in particular the specific energy loss in the Time Projection Chamber (TPC) and Time-of-Flight (TOF) measurement, allow for the detection of such rarely produced particles. Further, the Inner Tracking System (ITS) gives the possibility to separate primary nuclei from those coming from the weak decay of heavier systems. This offers the unique opportunity to search for exotica like the bound state of a Λ and a neutron which would decay into deuteron and pion, or the bound states of two Λs. We show here the performance of the ALICE apparatus in the sector on (anti-)nuclei and in particular the results on deuteron production in Pb–Pb collisions at $\sqrt{s_{\rm NN}}$ = 2.76 TeV. These are compared with thermal model predictions. We further present the current state of the searches, by their upper limits on the production yields, and compare the results to thermal and coalescence model expectations.

1. Introduction

The measurement of light (anti-)nuclei and search for hyper-matter, created in heavy-ion collisions, is challenging since the production probability decreases with the increasing mass of the investigated particles. The production yields of these particles can be predicted by thermal models based on an analysis of lighter particles [1, 2, 3]. The chemical freeze-out temperature T, the volume V and the baryo-chemical potential μ_B are the only three free parameters in this approach as long as one assumes an equilibrated system. To account for possible non-equilibrium effects the factors γ_s and γ_q are very often introduced (e.g. [3]). An open question in the field of heavy-ion collisions is if the production of nuclei and other multi-baryon states, such as hyper-nuclei or also hyper-matter, is entirely described by a thermal model or if those objects are formed by coalescence of hadrons or even directly from quarks.

Hyper-matter is used as a synonym of matter composed either partly or entirely by strange matter. Here we focus on exotic states like a possible Λn bound state or the H-Dibaryon. The H-Dibaryon is a bound state of *uuddss* ($\Lambda\Lambda$) and was predicted by Jaffe in a bag model calculation [4]. Recent lattice QCD calculations [5, 6] also suggest a bound state, with binding energies in the range 13-50 MeV/c. A chiral extrapolation of these lattice calculations to a physical pion mass resulted in a H-Dibaryon which is unbound by either 13 ± 14 MeV/c [7] or it lies close to the Ξp threshold [8]. This renewed the interest in the experimental searches for the H-Dibaryon which were ongoing since its prediction in 1977 [4]. Details of the H-Dibaryon analysis can be found in [9]. The results of these searches are compared to model expectations at the end of these proceedings.

2. Deuteron spectra and yields in Pb–Pb

The excellent performance of the Time-Projection Chamber (TPC) [10] and the Time-Of-Flight detector (TOF) [11] allows for the clear identification of charged stable particles over a range of 0.15 to 5 GeV/c in rigidity $R = p/z$, where p is the track momentum and z is the charge number. Combining the specific energy loss ($\mathrm{d}E/\mathrm{d}x$) in the TPC and the TOF information, anti-alpha nuclei were identified, as discussed in some more detail in [12]. The same technique is used to measure the yields of deuterons.

This measurement is affected by a huge background, as for all nuclei, coming from knockout from material. This is not relevant for the measurement of anti-nuclei, like the anti-alpha. Rejection of this background is possible by restricting the distance-of-closest-approach (dca) of the track to the primary vertex in z-direction dca_z and then fitting the dca_{xy} distribution to extract the signal in the p_{T} window between 0.6 and 1.9 GeV/c (technique is described in more detail in [13]). The deuterons are identified using only the dE/dx measurement in the TPC

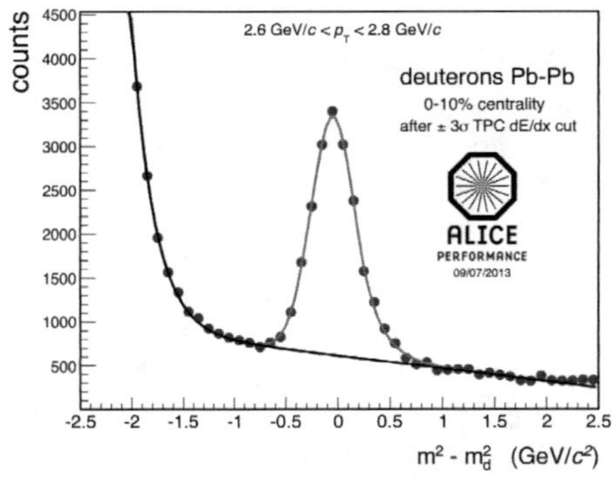

Figure 1. $m^2 - m_{\mathrm{d}}^2$ distribution measured by TOF for one example p_{T} bin ($2.6\,\mathrm{GeV}/c < p_{\mathrm{T}} < 2.8\,\mathrm{GeV}/c$), after preselection done by the specific energy loss measurement in the TPC of $\pm 3\sigma$ for deuterons. Fitted by a Gaussian function plus exponential tail to extract the yield.

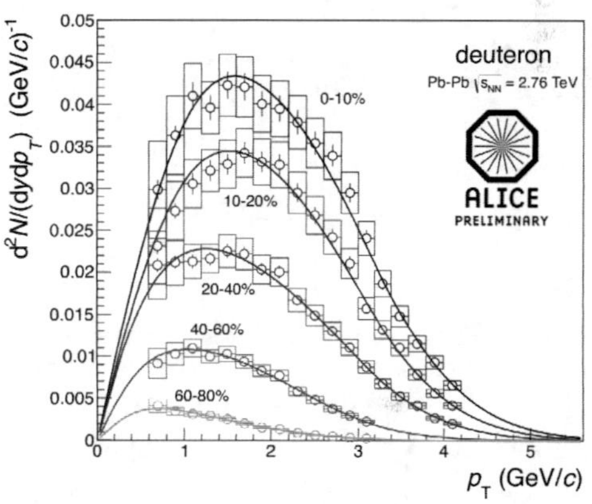

Figure 2. Measured deuteron spectra at $\sqrt{s_{\mathrm{NN}}} = 2.76\,\mathrm{TeV}$ for five different centralities. The lines through the measured points are individual blast-wave fits to extrapolate towards low p_{T}.

for pt ¡ 1.5 GeV and combining the TPC and TOF information for higher pt. A sample of deuteron selected with a 3 sigma cut around the expected dE/dx value is initially used to build the $m^2 - m_{\mathrm{d}}^2$-distribution, where m is the mass measured with TOF and $m^2 = p^2/(\gamma^2 - 1)$. Then, this distribution is fitted with a Gaussian function + exponential tail in p_{T} intervalls (shown in Figure 1). The extracted yields are then efficiency and acceptance corrected and the final spectra are shown in Figure 2 for five different collision centralities. The spectra show a characteristic hardening with increasing centrality, qualitatively similar to proton spectra. To extract p_{T} integrated yields the spectra in the different centrality bins are fitted by individual blast-wave distributions. The ratios of these yields with those of protons are shown in Figure 3 for five different centralities. The measured ratio agrees well with the mean of the d/p and the $\bar{\mathrm{d}}/\bar{\mathrm{p}}$ from PHENIX[14], which is a fair comparison since the μ_B at the LHC is getting close to 0 MeV [2].

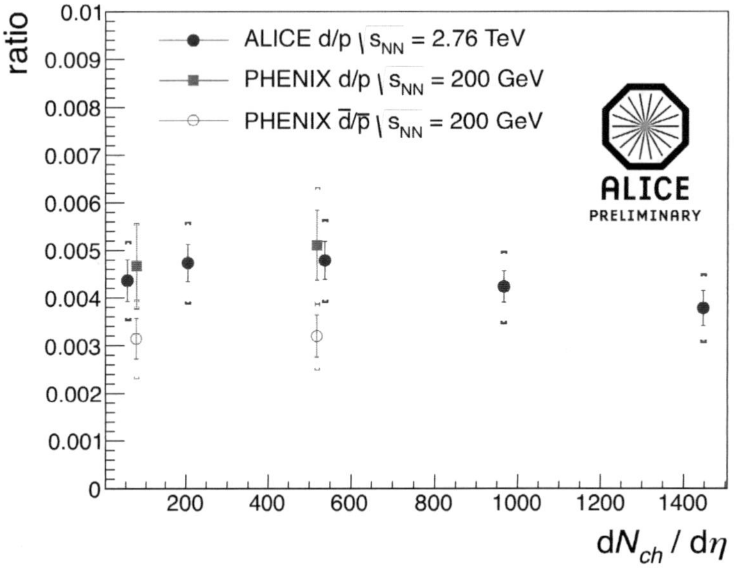

Figure 3. Deuteron-to-proton ratio over centrality compared with the PHENIX measurement [14] of deuteron-to-proton ratio and the anti-deuteron-to-anti-proton ratio at $\sqrt{s_{\rm NN}} = 200\,{\rm GeV}$.

3. Search for exotica: Λn bound state

The HypHI collaboration at GSI found evidence for a possible Λn bound state, decaying into a deuteron and a pion. The signal was detected on top of a background which can be described by the mixed events technique [15]. We investigate here only the $\overline{\Lambda \rm n}$ bound state, because of the much lower background compared to the corresponding particle (Λn) state. Monte Carlo simulations have been used to estimate the efficiency for the detection, the $\overline{\Lambda \rm n}$ was generated flat in rapidity y and in transverse momentum $p_{\rm T}$. We assume the lifetime to be that of the Λ particle, as suggested by the HypHI experiment. The $p_{\rm T}$ distribution of the $\overline{\Lambda \rm n}$ was extrapolated from the blast-wave fits for π, K, p done at the same energy. The estimated efficiency×acceptance versus $p_{\rm T}$ is convoluted with the normalised blast-wave function to get a weighted efficiency. The unknown $p_{\rm T}$ shape is the main source of uncertainty, therefore, different functions are used to study systematics (the boundary cases: blast-wave of deuteron and ^{3}He). The analysis strategy is to search for displaced decay vertices of anti-deuterons and pions. If one of the daughter tracks is identified as an anti-deuteron and the second daughter as a pion (both via specific energy loss in the TPC), the invariant mass of these candidates is calculated. The result is shown in Figure 4 together with the systematic error as a band and the peak which we expect from the Monte Carlo estimation and its corresponding error. Since no signal is observed, an upper limit on the production of this particle was set: ${\rm d}N/{\rm d}y \leq 1.5 \cdot 10^{-3}$ (at 99 % CL).

4. Summary and comparision with different models

The measured yield of deuterons presented here compares well with the current best equilibrium thermal model fit [16] which gives a temperature of about 156 MeV. One can now contrast the upper limits set for the exotica currently under investigation by ALICE, i.e. H-Dibaryon and Λn bound state, with different models. Figure 5 shows a comparison of the upper limits to the equlibrium thermal model [17] (164 MeV, which was the prediction for LHC and 156 MeV, the current best fit), the non-equilibrium thermal model prediction [18] from their current best fit [3]

Journal of Physics: Conference Series **509** (2014) 012109 doi:10.1088/1742-6596/509/1/012109

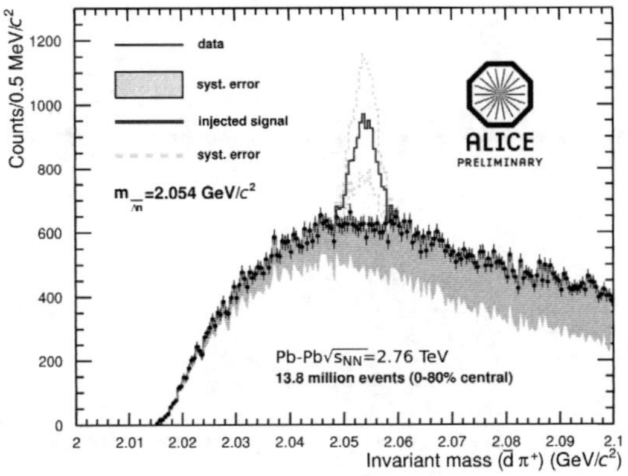

Figure 4. Invariant mass of anti-deuterons and pions. The expected signal is also shown.

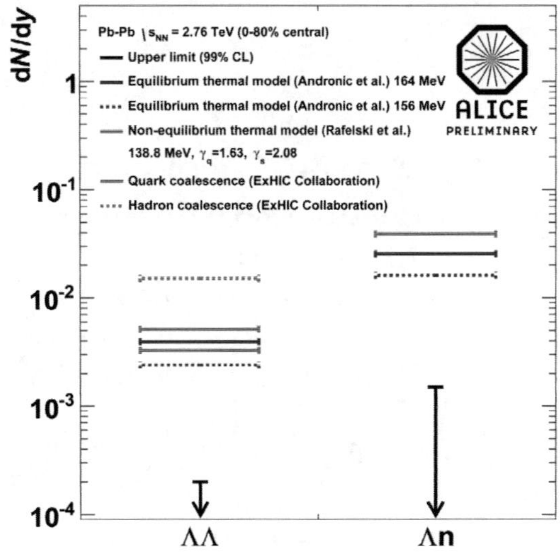

Figure 5. Upper limits of production yields for H-Dibaryon (ΛΛ) and Λn bound state compared with different models.

and coalescence predictions (quark and hadron coalescence) from the ExHIC collaboration [19]. All models are at least a factor 10 above the upper limits. Currently an effort continues to further reduce these upper limits by utilising more statistics than was previously used. Around 20 million central events will enhance the statistics. This will allow to make an even stricter decision about the existence of these particles.

References

[1] Andronic A, Braun-Munzinger P, Stachel J and Stöcker H 2011 *Phys. Lett.* B **697** 203
[2] Cleymans J, Kabana S, Kraus I, Oeschler H, Redlich K and Sharma N 2011*Phys. Rev.* C **84** 054916
[3] Petráň M, Letessier J, Petráček V and Rafelski J 2013 CERN-PH-TH/2012-262, arXiv:1303.2098v2 [hep-ph]
[4] Jaffe R 1977 *Phys. Rev. Lett.* **38** 195 and erratum ibid 1977 **38** 617
[5] Inoue T et al. (HAL QCD Collaboration) 2011 *Phys. Rev. Lett.* **106** 162001
[6] Beane S R *et al.* (NPLQCD Collaboration) 2011 *Phys. Rev. Lett.* **106** 162002
[7] Shanahan P E, Thomas A W and Young R D 2011 *Phys. Rev. Lett.* **107** 092004
[8] Haidenbauer J and Meißner U-G 2011 *Phys. Lett.* B **706** 100
[9] Lea R (for the ALICE Collaboration) these proceedings
[10] Alme J *et al.* (ALICE TPC Collaboration) 2010 *Nucl. Instr. Meas.* A **622** 316
[11] Aamodt K *et al.* (ALICE Collaboration) 2008 *JINST* **3** S08002
[12] Dönigus B (for the ALICE Collaboration) 2013 *Nucl. Phys.* A **904-905** 457c
[13] Abelev B *et al.* (ALICE Collaboration) 2013 CERN-PH-EP-2013-019, arXiv:1303.0737v2 [hep-ex]
[14] Adler S S 2005 *Phys. Rev. Lett.* **94** 122302
[15] http://www.bnl.gov/hhi/files/talks/TakehikoSaito.pdf, as shown 1.3.2012 at the Riken-BNL workshop on "Hyperon-Hyperon Interactions and Searches for Exotic Di-Hyperons in Nuclear Collisions"
[16] Stachel J these proceedings
[17] Andronic A private communication
[18] Petráň M private communication
[19] Cho S *et al.* (ExHIC Collaboration) 2011 *Phys. Rev. C* **84** 064910

Journal of Physics: Conference Series **509** (2014) 012110 doi:10.1088/1742-6596/509/1/012110

Measurements of $J/\psi \rightarrow e^+e^-$ with ALICE at the LHC

Fiorella Fionda on behalf of the ALICE Collaboration
Dipartimento Interateneo di Fisica "M. Merlin" and Sezione INFN, Bari, Italy
E-mail: fiorella.fionda@ba.infn.it

Abstract. The ALICE detector provides excellent capabilities to study quarkonium production at the Large Hadron Collider (LHC). Heavy quarkonia, bound states of charm or beauty quark anti-quark pairs such as the J/ψ, are expected to be produced by initial hard processes. Thus they will provide insight into the earliest and hottest stages of AA collisions where the formation of a Quark-Gluon Plasma (QGP) is expected. Furthermore, high-precision data from pp collisions represent an essential baseline for the measurement of nuclear modifications in heavy-ion collisions and serve also as a crucial test for several models of quarkonium hadroproduction. In addition, the study of pA collisions allows to investigate nuclear modifications due to Cold Nuclear Matter (CNM) effects. In ALICE, J/ψ were measured in pp and Pb–Pb collisions down to $p_T = 0$ via their di-electron decay channel in the central barrel ($|y| < 0.8$). Results on the nuclear modification factor (R_{AA}) at central rapidities in Pb–Pb collisions at $\sqrt{s_{NN}} = 2.76$ TeV will be shown and their implications discussed. A separation of the prompt and non-prompt components is also possible down to p_T of the J/ψ of 2 GeV/c.

The production of heavy quarkonia involves both perturbative and non-perturbative mechanisms of Quantum-Chromo-Dynamics (QCD). In proton-proton collisions, several models [1, 2] attempted to describe the quarkonia production, but failed in reproducing simultaneously cross-sections, polarization, transverse momentum and rapidity dependence as measured at the Tevatron [3, 4] and RHIC [5] colliders. Results in proton-proton collision at the new LHC energies have provided additional constraints to those models as well as the baseline reference for AA analyses. At the high temperatures and large energy densities reached in relativistic heavy-ion collisions, the matter consists of deconfined quarks and gluons, in the state referred to as "Quark-Gluon-Plasma" (QGP) [6]. According to the prediction by Matsui and Satz [7], in the deconfined medium formed in nucleus-nucleus collisions quarkonium production is suppressed relative to that in proton-proton collisions due to the color analogue of the Debye screening mechanism. The observable to quantify the nuclear medium effects is the so-called "nuclear modification factor" R_{AA}, defined as

$$R_{AA} = \frac{1}{\langle T_{AA}\rangle} \frac{d^2N^{AA}_{J/\psi}/dp_T dy}{d^2\sigma^{pp}_{J/\psi}/dp_T dy}$$

where $d^2N^{AA}_{J/\psi}/dp_T dy$ represents the J/ψ production yield in AA collisions, $d^2\sigma^{pp}_{J/\psi}/dp_T dy$ is the cross section in pp collisions at the same energy and $\langle T_{AA}\rangle$ is the nuclear overlap function determined by Glauber model calculations. The J/ψ suppression observed at SPS and RHIC [8, 9, 10] is not completely understood yet. Open points are in particular the observation of a similar J/ψ suppression at the two different centre-of-mass energies and a stronger suppression at forward-rapidity ($\sim$40%) compared to mid-rapidity at RHIC. Two theoretical models were proposed in order to reproduce RHIC and SPS data and provide predictions for the LHC: *i)* the "regeneration" mechanism from deconfined quarks in the medium to compete the J/ψ suppression in the QGP [11, 12]; *ii)* the statistical hadronization of charm quarks at phase boundary [13, 14]. The interpretation of the experimental results is still under debate due to the large experimental uncertainty on the total $c\bar{c}$ production cross section, which prevents more precise model calculations, as well as the lack of an exhaustive understanding of Cold Nuclear Matter (CNM) effects, strongly dependent on the centre-of-mass energy of the system and determined from pA collisions. Furthermore, other important contributions are J/ψ from the decays of higher mass charmonium states (e.g. χ_c and ψ') and beauty hadrons (non-prompt J/ψ). In particular, measuring the fraction of non-prompt J/ψ, f_B, gives access to the R_{AA} of both prompt and non-prompt produced J/ψ mesons and the latter reflects directly the nuclear modification factor of beauty hadrons. According to the QCD predictions [15] the parton energy loss in the QGP implies the following hierarchy in the measured R_{AA}: $R^{\pi}_{AA} < R^{D}_{AA} < R^{B}_{AA}$. Therefore the comparison between the R_{AA} of non-prompt J/ψ with the R_{AA} of other hadrons could offer an important insight onto the parton energy loss mechanisms in the QGP.

In ALICE the J/ψ production is measured at central rapidity ($|y| < 0.8$) in the dielectron channel J/$\psi \rightarrow e^+e^-$ and at forward rapidity ($2.5 < y < 4$) in the dimuon channel J/$\psi \rightarrow \mu^+\mu^-$ reaching in both cases $p_T = 0$. The focus of this paper is on the results obtained at mid-rapidity. The main tracking detectors used in this analysis are the Inner Tracking System (ITS), which allows for the measurement of the J/ψ fraction from beauty hadron decays, and the Time Projection Chamber (TPC), which is used for tracking and electron identification via specific energy deposition measurement.

In the left-hand panel of Fig. 1 the inclusive J/ψ cross sections in pp collisions as a function of rapidity are shown at both $\sqrt{s} = 7$ TeV ($L^{e^+e^-}_{int} = 5.6$ nb^{-1} and $L^{\mu^+\mu^-}_{int} = 15.6$ nb^{-1}) and $\sqrt{s} = 2.76$ TeV ($L^{e^+e^-}_{int} = 1.1$ nb^{-1} and $L^{\mu^+\mu^-}_{int} = 19.9$ nb^{-1}) [16]. The measurement at $\sqrt{s} = 2.76$ TeV represents the reference used for the R_{AA} analysis in Pb–Pb. The currently large uncertainties on this reference limit the accuracy on the R_{AA} determination.

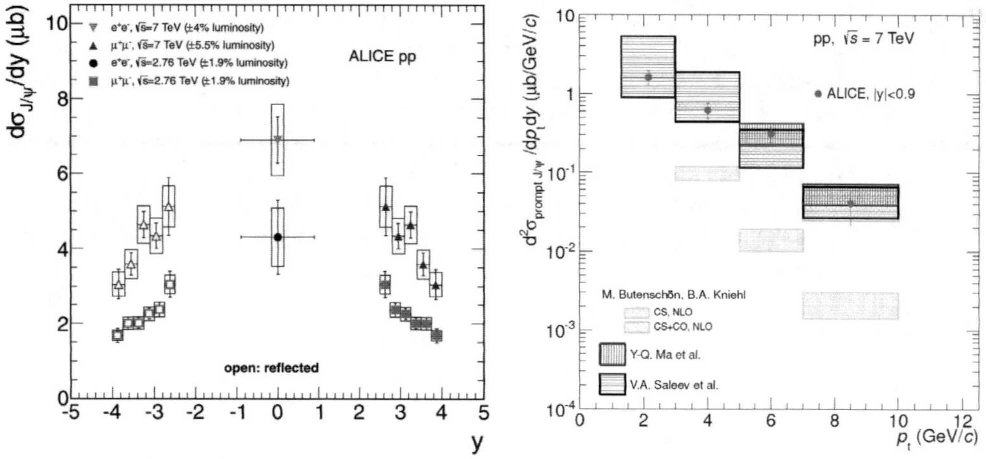

Figure 1. Left: Inclusive J/ψ cross section as a function of rapidity measured in pp collisions at $\sqrt{s} = 7$ TeV and 2.76 TeV [16]. Right: Prompt J/ψ cross section as a function of p_T compared to several theoretical predictions [17].

The fraction of J/ψ coming from beauty hadron decays was measured in proton-proton collisions at $\sqrt{s} = 7$ TeV down to $p_T = 1.3$ GeV/c. In the right-hand panel of Fig. 1 the prompt

Journal of Physics: Conference Series **509** (2014) 012110 doi:10.1088/1742-6596/509/1/012110

J/ψ cross section as a function of transverse momentum is compared to next-to-leading order (NLO) non-relativistic QCD (NRQCD) theoretical calculations (see [17] and references therein), which include color-singlet (CS) and color-octet (CO) contributions; heavier charmonium feed-down is also included. The comparison suggests that the CO processes are important to describe the data.

In Pb–Pb collisions at $\sqrt{s_{\rm NN}} = 2.76$ TeV the nuclear modification factor $R_{\rm AA}$ of inclusive J/ψ was measured as a function of centrality for $p_{\rm T} > 0$ ($L_{\rm int} = 15\ \mu{\rm b}^{-1}$). This is shown in the left-hand panel of Fig. 2 as a function of the mean number of participant nucleons $\langle N_{\rm part}\rangle$ (estimated from Glauber model) and it is compared with the inclusive J/ψ $R_{\rm AA}$ measured at mid rapidity by PHENIX in Au-Au collisions at $\sqrt{s_{\rm NN}} = 200$ GeV [18]. The comparison indicates a reduced suppression for most central collisions in ALICE w.r.t. PHENIX, and this behaviour is in qualitative agreement with a regeneration scenario at LHC energies.

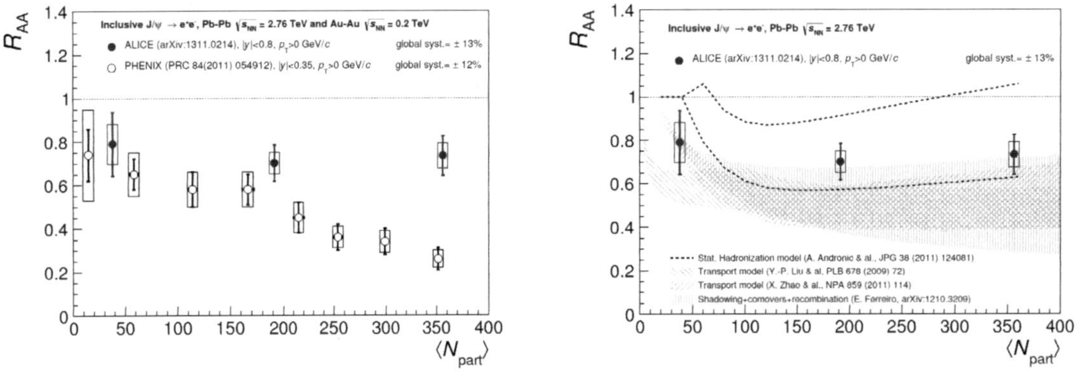

Figure 2. Nuclear modification factor $R_{\rm AA}$ measured in Pb–Pb collisions at $\sqrt{s_{\rm NN}} = 2.76$ TeV as a function of the mean number of participants $\langle N_{\rm part}\rangle$ compared with PHENIX [18] results at lower energy (left-hand panel) and with theoretical models [11, 12, 19, 20] (right-hand panel).

In the right-hand panel of Fig. 2 the inclusive J/ψ $R_{\rm AA}$ is compared with theoretical models that include the (re)combination of deconfined charm quark pairs from the QGP. In particular, the hashed bands represent the results from two transport models [11, 12] and from the comover interaction model [19] where up to 50% of the J/ψ were produced from deconfined $c\bar{c}$ pairs recombination. The prediction from the statistical hadronization model [20] is also shown by solid lines. All models exhibit a good agreement with data albeit with large uncertainties, due to the large uncertainty on the inclusive $c\bar{c}$ production cross section and CNM effects (e.g. nuclear shadowing). The latter are currently being addressed by measuring J/ψ production in p–Pb collisions.

The fraction of non-prompt J/ψ were measured in Pb–Pb collisions at $\sqrt{s_{\rm NN}} = 2.76$ TeV as a function of centrality down to $p_{\rm T} = 2$ GeV/c, as shown in the left-hand side of Fig. 3. No significant dependence of the non-prompt J/ψ fraction on centrality can be observed. In the right-hand panel of Fig. 3 the fraction of non-prompt J/ψ as a function of transverse momentum measured by ALICE and CMS [21] in Pb–Pb collisions and integrated over centrality is shown, along with the results in pp collisions at $\sqrt{s} = 7$ TeV by ALICE [17], ATLAS [22] and CMS [23]. The CDF data in p$\bar{\rm p}$ collisions at $\sqrt{s} = 1.96$ TeV [3] are also reported. Considering the ALICE and CMS results together, an indication for a similar trend of $f_{\rm B}$ as a function of $p_{\rm T}$ in proton-proton and Pb–Pb collisions is observed.

In summary, transverse momentum spectra and rapidity distributions of inclusive J/ψ were measured down to $p_{\rm T} = 0$ for pp collisions at $\sqrt{s} = 2.76$ TeV and $\sqrt{s} = 7$ TeV. NRQCD cal-

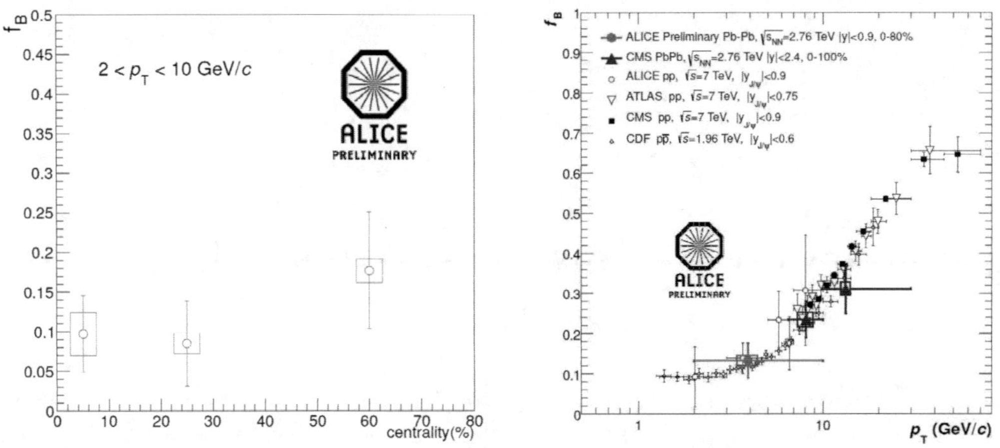

Figure 3. Left: Non-prompt J/ψ fraction measured in Pb–Pb collisions at $\sqrt{s_{NN}} = 2.76$ TeV as a function of centrality (statistical and systematic uncertainties shown by bars and boxes respectively). Right: Fraction of non-prompt J/ψ in Pb–Pb collisions at $\sqrt{s_{NN}} = 2.76$ TeV measured by ALICE and CMS [21] at central rapidity as a function of p_T. Results in pp [17, 22, 23] and p$\bar{\text{p}}$ [3] collisions are also shown.

culations are consistent with the measured prompt J/ψ production cross section. The nuclear modification factor R_{AA} was measured in Pb–Pb collisions at $\sqrt{s_{NN}} = 2.76$ TeV down to $p_T = 0$, as a function of centrality. The comparison with PHENIX results and theoretical predictions provides an indication for (re)generation of J/ψ from deconfined charm quarks. The non-prompt J/ψ fraction were also measured in Pb–Pb collision as a function of centrality. The combination of ALICE and CMS results in Pb–Pb collisions suggests a trend of f_B as a function of p_T that is similar to that in pp.

References

[1] Brambilla N *et al.* 2011 *Eur. Phys. J.* C **71** 1534
[2] Lansberg J P 2009 *Eur. Phys. J.* C **60** 693
[3] Acosta D *et al.* (CDF Collaboration) 2005 *Phys. Rev.* D **71** 032001
[4] Abulencia A *et al.* (CDF Collaboration) 2007 *Phys. Rev. Lett.* **99** 132001
[5] Adare A *et al.* (PHENIX Collaboration) 2007 *Phys. Rev. Lett.* **98** 232002
[6] Shuryak E V 1978 *Sov. Phys. JETP* **47** 212
[7] Matsui T and Satz H 1986 *Phys. Lett.* B **178** 416
[8] Alessandro B *et al.* (NA50 Collaboration) 2005 *Eur. Phys. J.* C **39** 355
[9] Adare A *et al.* (PHENIX Collaboration) 2007 *Phys. Rev. Lett.* **98** 232201
[10] Adare A *et al.* (PHENIX Collaboration) 2011 *Phys. Rev.* C **84** 054912
[11] Zhao X and Rapp R 2008 *Phys. Lett. B* **664** 253
[12] Liu Y, Zhen Q, Xu N and Zhuang P 2009 *Phys. Lett.* B **678** 72
[13] Braun-Munzinger P and Stachel J 2000 *Phys. Lett.* B **490** 196
[14] Andronic A, Braun-Munzinger P, Redlich K and Stachel J 2007 *Phys. Lett.* B **652** 259
[15] Armesto N, Salgado C A and Wiedemann U A 2004 *Phys. Rev.* D **69** 114003
[16] Abelev B *et al.* (ALICE Collaboration) 2012 *Phys. Lett.* B **718** 295
[17] Abelev B *et al.* (ALICE Collaboration) 2012 *J. High Energy Phys.* JHEP11(2012)065
[18] Adare A *et al.* (PHENIX Collaboration) 2007 *Phys. Rev. Lett.* **98** 232301
[19] Ferreiro E arXiv:1210.3209 and private communication
[20] Andronic A, Braun-Munzinger P, Redlich K and Stachel J 2011 *J. Phys.* G **38** 124081
[21] Chatrchyan S *et al.* (CMS Collaboration) 2012 *J. High Energy Phys.* JHEP05(2012)063
[22] Aad G *et al.* (ATLAS Collaboration) 2011 *Nucl. Phys.* B **850** 387
[23] Chatrchyan S *et al.* (CMS Collaboration) 2012 *J. High Energy Phys.* JHEP02(2012)011

14th International Conference on Strangeness in Quark Matter (SQM2013) IOP Publishing
Journal of Physics: Conference Series **509** (2014) 012111 doi:10.1088/1742-6596/509/1/012111

Charmonium production measurements in Pb-Pb collisions with ALICE at the LHC

Lizardo Valencia Palomo for the ALICE Collaboration
Université de Paris-Sud and Institut de Physique Nucléaire d'Orsay, France
E-mail: lvalenci@cern.ch

Abstract. ALICE, one of the four main experiments at the LHC, aims to study the Quark-Gluon-Plasma, a state of matter created in high-energy heavy-ion collisions where quarks and gluons behave as free particles, rather than confined into hadrons. In this proceedings I discuss the measurements of charmonium production in Pb-Pb collisions at $\sqrt{s_{\rm NN}} = 2.76$ TeV performed at forward rapidity in the dimuon decay channel.

1. Introduction

Ultra-relativistic heavy-ion collisions are used to recreate in the laboratory the high energy densities which are thought to have existed during the first microseconds after the Big Bang. It is expected that, at extreme conditions of temperature and baryon density, ordinary nuclear matter undergoes a phase transition to a deconfined state of quarks and gluons called Quark-Gluon-Plasma (QGP).

Among the possible probes of the QGP, heavy quarks are of particular interest because they are produced in the initial stage of the collision and experience the whole evolution of the system. In particular, in 1980's the charmonium family was proposed as a tool to characterize the QGP: color screening of the heavy quarks potential, in the deconfined state, leads to a suppression of charmonium production [1]. Different charmonium states with different binding energies will have different melting temperatures, since the screening radius depends on the temperature of the medium. The J/ψ suppression is predicted to occur above a critical temperature of the medium ($T_{\rm C}$) and sequentially, in order of binding energy [2].

Charmonium production in high energy heavy-ion collisions at the SPS was extensively studied by three different experiments: NA38 [3], NA50 [4, 5] and NA60 [6], all of which performed the measurements in the dimuon decay channel. In central Pb-Pb collisions at 158 GeV/nucleon, the NA50 results show a clear J/ψ suppression (factor of 2), suggesting the creation of a deconfined medium. At RHIC, the PHENIX experiment has studied the J/ψ production both at mid rapidity ($|y| < 0.35$) via $e^+ e^-$ decay and at forward rapidity ($|y| \in$ [1.2,2.2]) via $\mu^+ \mu^-$ decay in Au-Au collisions at $\sqrt{s_{NN}} = 200$ GeV. These measurements were performed as a function of the centrality, rapidity and transverse momentum. An important aspect is that PHENIX found a larger suppression at forward rapidity with respect to mid rapidity. Another unexpected feature is that the suppression was found to be similar to that at the SPS energy [8].

The modification of the J/ψ production is quantified by means of the nuclear modification factor:

Journal of Physics: Conference Series **509** (2014) 012111 doi:10.1088/1742-6596/509/1/012111

$$R_{\mathrm{AA}} = \frac{Y}{\langle T_{\mathrm{AA}} \rangle \times \sigma^{\mathrm{pp}}},$$

Y is the invariant yield of the measured particle in nucleus-nucleus collisions, $\langle T_{\mathrm{AA}} \rangle$ the nuclear thickness function, σ^{pp} the cross section measured in pp at the same energy in the center of mass.

2. The ALICE experiment

ALICE (A Large Ion Collider Experiment) [9] is the LHC experiment designed to study heavy-ion collisions. It consists of a central barrel covering $|\eta| < 0.9$ embedded in a solenoidal magnet and a forward Muon Spectrometer with a $-4 < \eta < -2.5$ acceptance. Given the variety and complexity of all the subsystems located in the central barrel, only the detectors employed in this analysis are mentioned. The VZERO detector consists of two hodoscope scintillators located at both sides of the interaction point. The VZERO triggers the data taking and determines the centrality of the collisions by using a Glauber based Monte Carlo fit to the VZERO amplitude [10].

The Muon Spectrometer [11] is composed of absorbers, dipole magnet, tracking and trigger chambers. The front absorber filters out hadrons created in the collisions, the beam shield protects the chambers from particles produced at large rapidities, the iron wall absorbs hadrons that punch-through the front absorber and the rear absorber protects the trigger chambers from the background generated by the accelerator. The dipole magnet bends charged particles, allowing to extract the sign of their electric charge and their momenta. Ten tracking chambers, grouped by pairs in five stations, are used to reconstruct the trajectory of the particles. A dedicated trigger system is provided by four trigger chambers with a time resolution of 2 ns.

3. Analysis

The $\mathrm{J}/\psi \rightarrow \mu^{+}\mu^{-}$ analysis with the 2011 Pb-Pb collisions was performed over an integrated luminosity of 70 $\mu\mathrm{b}^{-1}$, corresponding to approximately 17 million dimuon events collected with the opposite sign dimuon trigger.

The raw yield is extracted with two different methods. In the first one the opposite-sign dimuon invariant mass spectrum is fitted with the sum of a modified Crystal Ball function (CB2) to model the signal plus a Variable Width Gaussian (VWG) for the background. The CB2 consists of a Gaussian core portion plus two power-law tails, one in the low mass range and another in the high mass one with independent parameters. The VWG contains four free parameters, one of which indicates the relation between the Gaussian width and the mass value. In the second approach, the combinatorial background is subtracted using the event mixing technique. The remaining uncorrelated background is fitted with an exponential function, while the CB2 is once again used to describe the signal. The results obtained from the different methods are then combined to extract a weighted mean and the systematic uncertainties on the signal extraction. The raw J/ψ yield in the 0-90% centrality range with $0 < p_{\mathrm{T}} < 8$ GeV/c and $2.5 < y < 4$ amounts to approximately 40 000 counts.

The acceptance times efficiency value, $A \times \epsilon$, is obtained by embedding Monte Carlo J/ψ into real events. The integrated $A \times \epsilon$ is 14%, with a 8% relative decrease from peripheral to central collisions. The inclusive J/ψ cross section in pp used for normalization was measured by the ALICE experiment at $\sqrt{s} = 2.76$ TeV [12].

The most important source of systematic uncertainties (9%) was found to be the pp reference. Other contributions to the systematic uncertainties are the choice of the MC input, the nuclear thickness function, the trigger and tracking efficiencies.

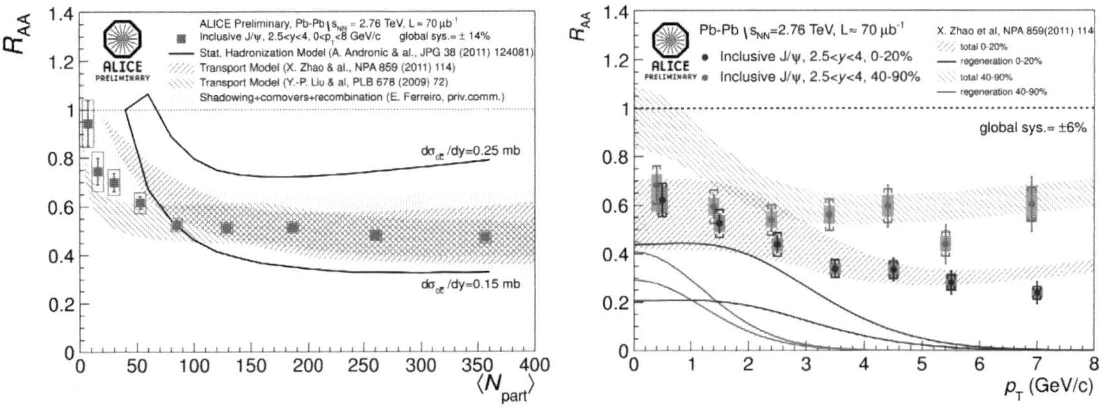

Figure 1. *Left:* R_{AA} vs N_{part} and the predictions from the Statistical Hadronization Model, Transport Models and shadowing plus comovers plus recombination model. *Right:* R_{AA} vs p_{T} for central and peripheral collisions compared to a transport model.

4. Results

The inclusive J/ψ R_{AA} measured by ALICE at $\sqrt{s_{\mathrm{NN}}} = 2.76$ TeV in the range $2.5 < y < 4$ and $0 < p_{\mathrm{T}} < 8$ GeV/c is shown on the left panel of figure 1 as a function of N_{part}, indicating no significant centrality dependence for $N_{\mathrm{part}} > 70$. The centrality integrated value is $R_{\mathrm{AA}}^{0-90\%} = 0.497 \pm 0.006$ (stat.) $\pm$ 0.078 (syst.), indicating a clear J/ψ suppression. This result is in agreement with the measurement performed using the 2010 Pb-Pb collisions data [13]. When compared to RHIC results, the ALICE R_{AA} is approximately three times larger than the R_{AA} measured by PHENIX for $N_{\mathrm{part}} > 180$ at forward rapidity [14]. The centrality dependence of the data is compared to four theoretical predictions. In the Statistical Hadronization Model the charmonium production occurs at the phase boundary by statistical hadronization of charm quarks [15]. The two Transport Models use different rate equations of the J/ψ dissociation and regeneration. In both models, 50% of the measured yield in central collisions is due to regeneration and the other 50% from the initial production [16, 17]. In the model including comovers plus recombination, the J/ψ are suppressed by partonic comovers [18].

The right panel of figure 1 presents the R_{AA} vs p_{T} for central (0-20%) and peripheral collisions (40-90%) and the results are compared with the predictions of the transport model [16]. The R_{AA} presents a notable p_{T} dependence only in the most central collision. In the 40-90% centrality collisions the model underestimates the suppression for $p_{\mathrm{T}} < 3$ GeV/c. In both centrality ranges, according to the model, J/ψ from regeneration have notable dominant contribution in the low-p_{T} regime. In particular, the regenerated J/ψ are the dominant contribution for the most central collisions.

The left panel of figure 2 presents the R_{AA} as a function of the centrality for the low-p_{T} ($0 < p_{\mathrm{T}} < 2$ GeV/c) and high-p_{T} ($5 < p_{\mathrm{T}} < 8$ GeV/c) J/ψ. On one side, the suppression for the low-p_{T} J/ψ shows to be centrality independent when $N_{\mathrm{part}} > 100$. On the other side, for high-p_{T} J/ψ, the nuclear modification factor presents a clear centrality dependence.

On the right panel of figure 2 the J/ψ $\langle p_{\mathrm{T}} \rangle$ measured by ALICE is shown as a function of the centrality together with measurements at RHIC by PHENIX. The $\langle p_{\mathrm{T}} \rangle$ presents a clear decreasing trend, a striking difference with respect to the one obtained by the PHENIX experiment in Au-Au and Cu-Cu collisions at $\sqrt{s_{\mathrm{NN}}} = 200$ GeV/c in the rapidity interval $1.2 < |y| < 2.2$

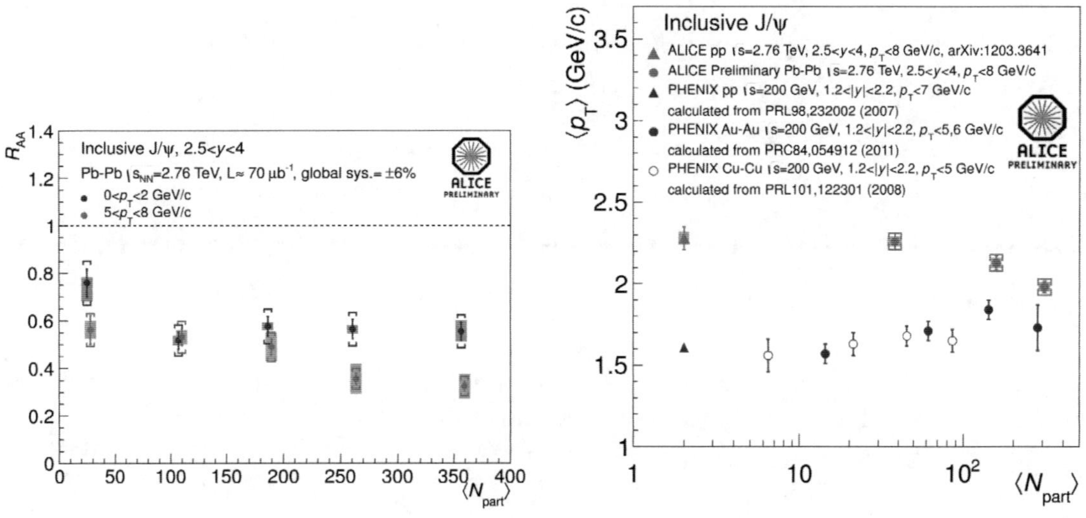

Figure 2. *Left:* R_{AA} vs N_{part} for low and high p_{T} J/ψ. *Right:* J/ψ $\langle p_{\mathrm{T}} \rangle$ compared to PHENIX measurements.

5. Conclusions

The inclusive J/ψ R_{AA} measured with the ALICE detector in Pb-Pb collisions at $\sqrt{s_{\mathrm{NN}}} = 2.76$ TeV has no significant centrality dependence for $N_{\mathrm{part}} > 70$ and, for the most central collisions, it is approximately three times larger than the one measured at forward rapidity at RHIC. The clear decrease of the $\langle p_{\mathrm{T}} \rangle$ with increasing N_{part} is a completely different behaviour with respect to lower energy results. Theoretical models which include a full and partial J/ψ production from charm quarks in a dense partonic phase successfully describe the data. In all the theoretical predictions, the (re)generation component is dominant in the low-p_{T} regime.

References

[1] T. Matsui and H. Satz, Phys. Lett. B 178 (1986) 416.
[2] S. Digal *et al.*, Phys. Rev. D 64 (2001) 094015.
[3] NA38 Collaboration, Phys. Lett. B 255 (1991) 459.
[4] NA50 Collaboration, Phys. Lett. B 477 (2000) 28.
[5] NA50 Collaboration, Eur. Phys. J. C 39 (2005) 335.
[6] NA60 Collaboration, Phys. Rev. Lett. 99 (2007) 132302.
[7] PHENIX Collaboration, Phys. Rev. Lett. 98 (2007) 232301.
[8] ALICE Collaboration, JINST 3 (2008) S08002.
[9] ALICE Collaboration, arXiv:1301.4361.
[10] ALICE Collaboration, CERN-LHCC-1999-022.
[11] ALICE Collaboration, Phys. Lett. B 718 (2012) 295.
[12] ALICE Collaboration, Phys. Rev. Lett. 109 (2012) 072301.
[13] PHENIX Collaboration, Phys. Rev. C 84 (2011) 054912.
[14] A. Andronic et al, J.Phys. G 38 (2011) 124081.
[15] X. Zhao and R. Rapp, Nucl. Phys. A 859 (2011) 114.
[16] Y.-p. Liu *et al.*, Phys. Lett. B 678 (2009) 72.
[17] E. Ferreiro, arXiv:1210.3209.

Upsilon Production in Pb-Pb and p-Pb Collisions at Forward Rapidity with ALICE at the LHC

Palash Khan for the ALICE Collaboration

Saha Institute of Nuclear Physics, Kolkata, India

E-mail: palash.khan@cern.ch

Abstract. The ALICE apparatus at the LHC was designed and built to perform dedicated studies of the Quark-Gluon Plasma (QGP), a strongly interacting phase of QCD matter, expected to be created in heavy-ion collisions, where quarks and gluons are deconfined. In such collisions heavy flavors are produced at the very early stage of the interaction by the initial hard scattering processes and hence can be used to characterize the hot and dense medium. In particular the sequential suppression of quarkonia (charmonia and bottomonia) was proposed as a thermometer of the deconfined medium. The inclusive $\Upsilon(1S)$ production has been measured down to zero transverse momentum in its dimuon decay channel at forward rapidity ($2.5 < y_{\rm lab} < 4.0$) using the muon spectrometer. Here results on the $\Upsilon(1S)$ nuclear modification factor ($R_{\rm AA}$) in Pb-Pb collisions at $\sqrt{s_{\rm NN}} = 2.76$ TeV are discussed and compared to the measurement at mid-rapidity by the CMS Collaboration and to theoretical predictions. Also recent results on $R_{\rm pPb}$ and forward-to-backward yield ratio ($R_{\rm FB}$) in p-Pb collisions at $\sqrt{s_{\rm NN}} = 5.02$ TeV are discussed.

1. Introduction

At very high temperature and energy density the hadronic matter can turn into a state of deconfined quarks and gluons known as Quark-Gluon Plasma (QGP) [1]. This state of matter can be created by colliding two large nuclei at ultra-relativistic energies. In such collisions, heavy flavors, specially charm and bottom, are produced at the very early stage by the hard scattering and hence can be used to characterize the hot and dense medium. It is predicted that the production of quarkonia (charmonia and bottomonia) in nucleus-nucleus collisions will be suppressed relative to that in proton-proton due to the Color-Debye Screening mechanism [2]. In particular, a sequential suppression of quarkonia, depending on their binding energy has been proposed as a thermometer of the deconfined medium [3].

The suppression of quarkonia can be quantified by measuring the nuclear modification factor $R_{\rm AA}$, which is the ratio of yield in nucleus-nucleus collisions to that in proton-proton collisions scaled by the number of binary collisions.

The suppression of quarkonium states depends on several factors, as the feed-down contributions from higher-mass resonances into the observed quarkonium yield, the b-hadrons decay into charmonium and the quarkonium regeneration due to the $q\bar{q}$ recombination (expected to be important for $c\bar{c}$ pairs at LHC energies) [4]. Other important competing mechanisms are the Cold Nuclear Matter (CNM) effects, such as nuclear shadowing or gluon saturation, which can break the binary scaling even in absence of the QGP. Hence, for the interpretation of results in heavy-ion collisions, data from pA are crucial since they allow to disentangle the CNM effects

Journal of Physics: Conference Series **509** (2014) 012112 doi:10.1088/1742-6596/509/1/012112

from those related to the formation of a hot medium. With this goal in mind, the ALICE [5] collaboration has measured the $\Upsilon(1S)$ production in Pb-Pb collisions at $\sqrt{s_{\rm NN}} = 2.76$ TeV and p-Pb collisions at $\sqrt{s_{\rm NN}} = 5.02$ TeV. For p-Pb, data were taken with two beam configurations corresponding to the proton going towards or opposite to the direction of the muon spectrometer (labeled p-Pb and Pb-p in the following).

2. Results

A data sample corresponding to an integrated luminosity of $L_{\rm int} = 69.2\ \mu{\rm b}^{-1}$ (for Pb-Pb), 5.3 $\rm nb^{-1}$ (for p-Pb) and 6.1 $\rm nb^{-1}$ (for Pb-p) have been collected. In Pb-Pb collisions, the measurement of the nuclear modification factor of the inclusive $\Upsilon(1S)$ production has been performed in the $2.5 < y_{\rm cms} < 4$ rapidity range, down to $p_{\rm T} = 0$ and in the $0\% - 90\%$ centrality class at $\sqrt{s_{\rm NN}} = 2.76$ TeV.

As the $\Upsilon(1S)$ production in pp collisions at $\sqrt{s} = 2.76$ TeV and 5.02 TeV has not been measured at forward rapidity, $\sigma^{pp}_{\Upsilon(1S)}$ was obtained with a data driven method. The y-differential cross-section at mid-rapidity was interpolated from the available measurements at different energies [6, 7, 8, 9]. The mid-rapidity $\sqrt{s}$-interpolation is either based on empirical shapes or on FONLL pQCD predictions assuming that the $\Upsilon(1S)$ production is proportional to the $\rm b\bar{b}$ cross-section. The extrapolation from mid to forward rapidity relies on the predictions of the $\Upsilon(1S)$ production performed with different settings of Pythia 6.4. Rapidity-differential distributions were generated at $\sqrt{s} = 2.76$ TeV and 5.02 TeV on a large rapidity range ($0 < y < 4.5$) with different Pythia settings selected according to their ability in reproducing mid-rapidity [9] and forward rapidity [10] data available at $\sqrt{s} = 7$ TeV. Distributions were normalized to the y-differential cross-section resulting from the $\sqrt{s}$-interpolation at mid-rapidity. In ($2.5 < y < 4.0$), $\sigma^{pp}_{\Upsilon(1S)}$ cross-section is $18.75^{+2.85}_{-1.80}$(extrap.)±1.65(norm.) nb. The first uncertainty component (extrap.) is associated to the extrapolation from mid to forward rapidity and comes from the spread of calculations obtained with the different Pythia settings. The second component (norm.) includes the uncertainties on the functional fit and on the FONLL calculations obtained by varying the QCD parameters.

The integrated $R_{\rm AA}$ value is 0.439 ± 0.065(stat.)±0.028(uncorrelated syst.)$^{+0.092}_{-0.076}$(correlated syst.) and implies a significant suppression of inclusive $\Upsilon(1S)$. Two centrality ranges were studied and a stronger suppression in more central collisions has been observed (left plot of Fig. 1). In addition, two rapidity ranges were studied and no significant variation of the suppression with rapidity was observed (right plot of Fig. 1). The $\Upsilon(1S)$ $R_{\rm AA}$ as a function of both rapidity and centrality was found to be comparable with that observed for the J/ψ measured by ALICE in the same kinematic range (not shown here). The $\Upsilon(1S)$ $R_{\rm AA}$ is compared to CMS results for the kinematic range $|y| < 2.4$ [11] (left plot of Fig. 1). A similar suppression as a function of centrality has been found by both ALICE and CMS. Small variation with rapidity of the inclusive $\Upsilon(1S)$ suppression has been observed over the present range accessed by the two experiments (right plot of Fig. 1).

Our results are also compared with models. The Anisotropic Hydro Model [12] includes feed-down of $\Upsilon(1S)$ by higher mass states, but does not include recombination effects nor cold nuclear matter effects. Data is described with the hypothesis of a boost invariant plateau temperature profile with minimum shear viscosity at forward-rapidity (left plot of Fig. 2). The Transport Model [13] includes small $\rm b\bar{b}$ regeneration, feed-down from higher mass (around 50 %) and CNM effects by an overall absorption cross-section of 0-2 mb. The model is in fair agreement with data within uncertainties (right plot of Fig. 2).

In case of p-Pb and Pb-p the inclusive $\Upsilon(1S)$ $R_{\rm pPb}$ was measured at forward ($2.04 < y_{cms} < 3.54$) and backward rapidity ($-4.46 < y_{cms} < -2.96$). A stronger suppression is observed for inclusive $\Upsilon(1S)$ at the forward rapidity than backward rapidity (Fig. 3). The results are compared to that of the J/ψ in the same kinematic range. The observed suppression for the

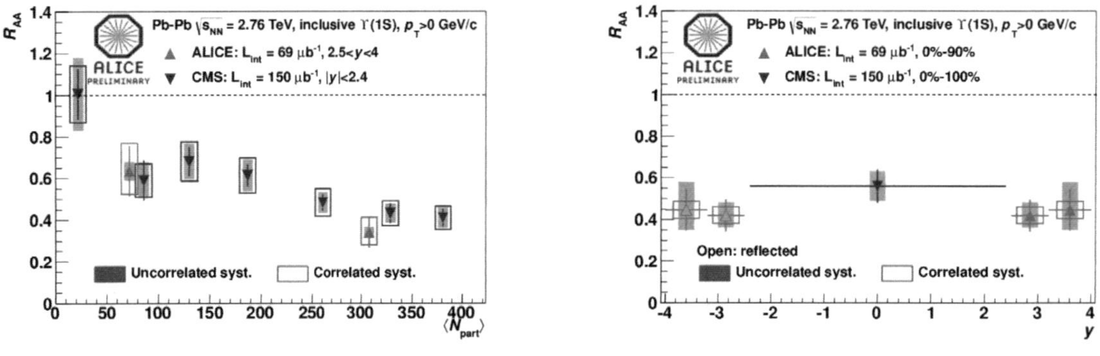

Figure 1. $R_{\rm AA}$ as a function of $\langle N_{\rm part}\rangle$ and y for $\Upsilon(1S)$ as measured by ALICE and CMS.

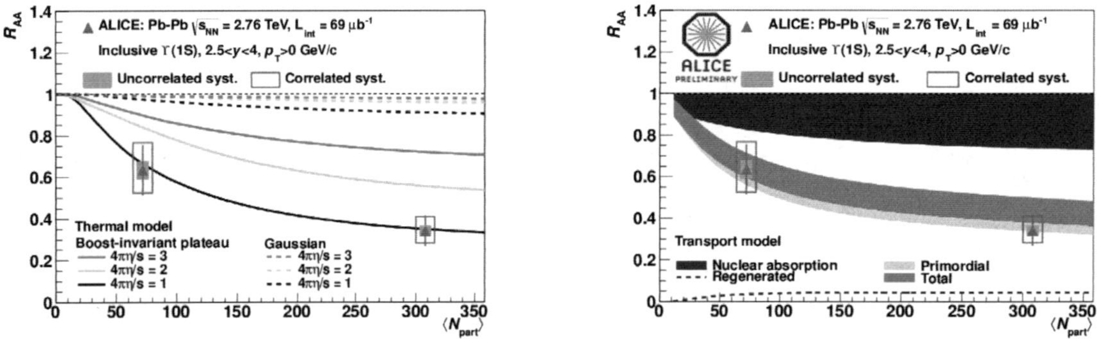

Figure 2. ALICE $R_{\rm AA}$ as a function of $\langle N_{\rm part}\rangle$ for $\Upsilon(1S)$ compared with models from Strickland [12] and Emerick et al. [13].

$\Upsilon(1S)$ is comparable with the J/ψ one at forward rapidity, while at backward rapidity the suppression of $\Upsilon(1S)$ is stronger.

The experimental results are compared with two models. The EPS09 Shadowing calculation at LO [16] and NLO [14] agrees reasonably well with the data at forward rapidity but tend to underestimate the $\Upsilon(1S)$ suppression at backward rapidity (left and right plot of Fig. 3). Within the present uncertainties, neither of the calculations are favored by the data.

The ratio $R_{\rm FB}$ between forward and backward $\Upsilon(1S)$ production was measured in the common rapidity region ($2.96 < |y_{cms}| < 3.54$) of p-Pb and Pb-p. It is found to be larger than that of the J/ψ measured in the same kinematic window (left plot of Fig. 4). The results are compared with an EPS09 shadowing calculation at both LO [16] and NLO [14] and with the Coherent Parton Energy Loss model [15] (right plot of Fig. 4). The agreement is reasonably good, except for the LO case with larger shadowing.

3. Conclusions

The nuclear modification factor for inclusive $\Upsilon(1S)$ has been measured at forward rapidity $2.5 < y < 4.0$. In Pb-Pb collisions, the suppression of inclusive $\Upsilon(1S)$ is stronger in central than semi-peripheral collisions. The $\Upsilon(1S)$ suppression pattern is comparable with forward-rapidity J/ψ result from ALICE within uncertainties. No strong variation with rapidity has been observed within the large range probed by ALICE and CMS.

In p-Pb data, we observe a small suppression of $\Upsilon(1S)$, which tends to increase from

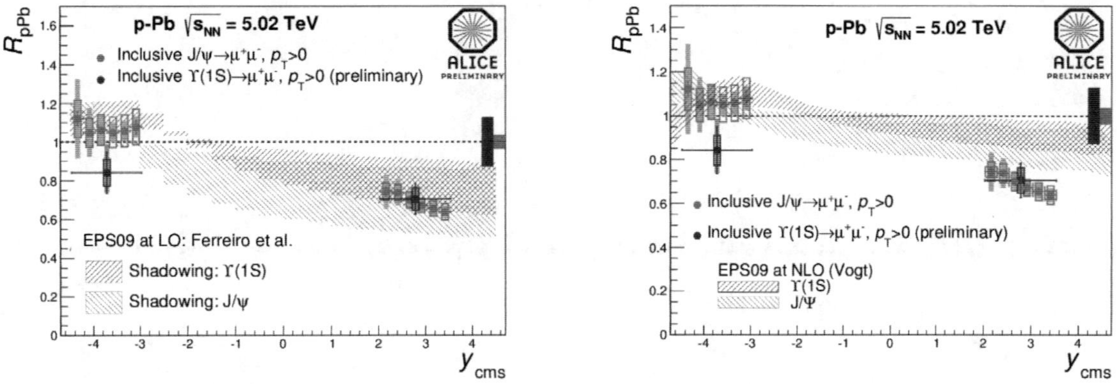

Figure 3. Comparison of the inclusive $\Upsilon(1S)$ $R_{\rm pPb}$ with inclusive J/ψ data and models from Ferreiro et al. [16] and Vogt et al. [14].

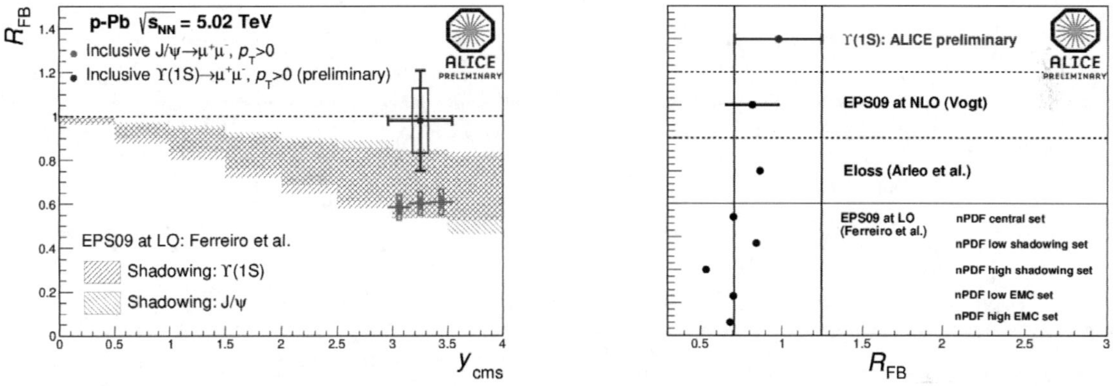

Figure 4. Comparison of the inclusive $\Upsilon(1S)$ $R_{\rm FB}$ with various models [14], [15], [16].

backward to forward rapidity. At forward rapidity the J/ψ suppression is comparable with the one of $\Upsilon(1S)$ within uncertainties. The J/ψ $R_{\rm FB}$ is significantly lower than the one measured for $\Upsilon(1S)$.

References

[1] E.V. Shuryak, Theory of Hadronic Plasma, Sov. Phys. JETP **47** 212 (1978).
[2] T. Matsui and H. Satz, Phys. Lett. B **178**, 416 (1986).
[3] S. Digal, P. Petreczky, and H. Satz, Phys. Rev. D **64** (2001) 094015 [arXiv:hep-ph/0106017].
[4] A. Andronic, P. Braun-Munzinger, K. Redlich and J. Stachel, Phys. Lett. B **571** (2003) 36 [nucl-th/0303036].
[5] K. Aamodt et al., JINST 3 (2008) S08002.
[6] D. Acosta et al. [CDF Collaboration], Phys. Rev. Lett. **88** (2002) 161802.
[7] V. M. Abazov et al. [D0 Collaboration], Phys. Rev. Lett. **94** (2005) 232001 [arXiv:hep-ex/0502030].
[8] S. Chatrchyan et al. [CMS Collaboration], JHEP **1205**, 063 (2012) [arXiv:1201.5069 [nucl-ex]].
[9] V. Khachatryan et al. [CMS Collaboration], Phys. Rev. D **83** (2011) 112004 [arXiv:1012.5545 [hep-ex]].
[10] R. Aaij et al. [LHCb Collaboration], Eur. Phys. J. C **72** (2012) 2025 [arXiv:1202.6579 [hep-ex]].
[11] S. Chatrchyan et al. [CMS Collaboration], Phys. Rev. Lett. **109** (2012) 222301 [arXiv:1208.2826 [nucl-ex]].
[12] M. Strickland, arXiv:1207.5327 [hep-ph].
[13] A. Emerick, X. Zhao and R. Rapp, Eur. Phys. J. A **48** (2012) 72 [arXiv:1111.6537 [hep-ph]].
[14] J. L. Albacete et al., Int. J. Mod. Phys. E **22** (2013) 1330007 [arXiv:1301.3395 [hep-ph]].
[15] F. Arleo, S. Peigné, arXiv:1212.0434 [hep-ph].
[16] E. G. Ferreiro et al., Eur. Phys. J. C **73**, 2427 [arXiv:1110.5047 [hep-ph]].

Open Heavy Flavor Production in Heavy-ion Collisions from STAR

Yifeii Zhang, for the STAR Collaboration
State Key Laboratory of Particle Detection and Electronics, University of Science and Technology of China, Hefei, China
E-mail: ephy@ustc.edu.cn

Abstract. We report on the STAR measurements of open charm hadron and heavy flavor decay electron (NPE) production in Au+Au collisions at $\sqrt{s_{NN}}$ = 200 GeV. The centrality dependences of D-meson and NPE p_T spectra, nuclear modification factors as well as the NPE elliptic flow v_2 are presented. The model comparisons and interpreted physics are discussed.

1. Introduction

Charm quarks are a unique tool to probe the strongly interacting matter created in relativistic heavy-ion collisions at RHIC energies, which consists of the carriers (quarks) and the mediators (gluons) of the strong force [1]. Their large masses ($\simeq$1.3 GeV/c^2) are not easily affected by the strong interaction with QCD medium. They are believed to be created mainly via initial hard parton scatterings. Therefore they carry clean information of the system at early stage in heavy-ion collisions [2]. Energetic charm quarks are predicted to lose less energy than light quarks, due to the suppression of gluon radiation by the large mass, when they traverse a Quark-Gluon Plasma [3, 4]. In contrast, recent measurements of non-photonic electrons (NPE) from heavy quark decays at high transverse momentum (p_T) show a jet quenching level similar to that of the light hadrons [5, 6]. This observation renews the interest in charm production and the interactions of heavy charm quarks with the hot and dense matter produced in nuclear collisions at RHIC. Open charm production at low p_T , in particular in the kinematic region sensitive to the radial flow in hot QCD medium or cold nuclear matter effect, is crucial for understanding the charm and light quark hadronization mechanisms and the bulk/cold medium propterties.

In this proceedings, we present the measurements of open charm via two independent decay channels: hadronic $D^0 \to K\pi$ and semi-leptonic ($c\&b \to e + X$). The centrality depdence of the transverse momentum distributions as well as the nuclear modification factor respect to p + p result show less suppression in perpheral collisions but strong suppression in most central collisions above p_T = 3 GeV/c for both D^0 and NPE measurements, which indicates significant energy loss of charm quark in the hot dense medium. However, this remain us a puzzle whether bottom lose similar amount of energy as charm quark, since in the NPE measurement, especially at high p_T, charm and bottom is hardly to be separated. Benifit from our D^0 measurement, the first preliminary separation of charm and bottom in NPE measurement without model constraints will be presented. The first observation of the low p_T bump structure is consistent with model calculations assuming strong charm-medium interactions.

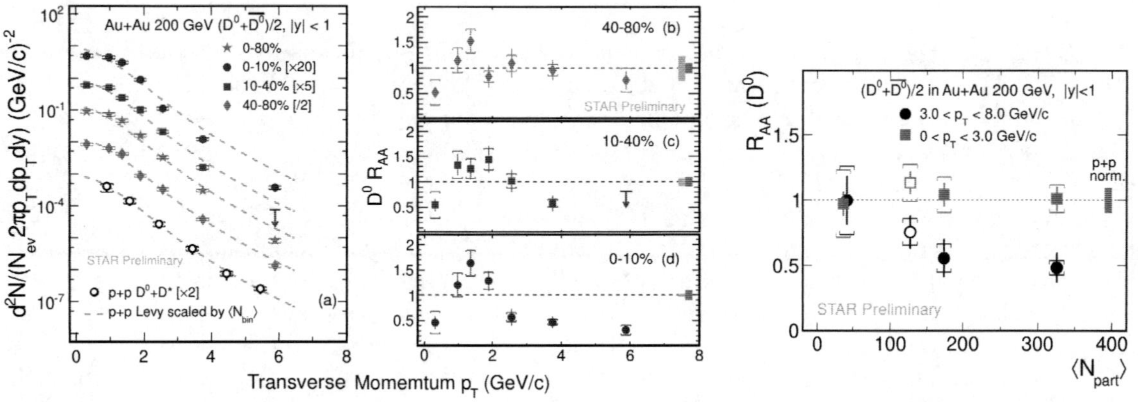

Figure 1. Left panel: (a) Centrality dependence of D^0 p_T differential invariant cross section. Data from peripheral (40-80%) to central (0-10%) are shown from bottom to top. The curves are number of binary collisions scaled Levy function from fitting to the $p+p$ result (open circles). D^0 R_{AA} are shown in 40-80% (b), 10-40% (c) and 0-10% (d) centrality. Right panel: Integrated D^0 R_{AA} as a function of number of participants.

2. Open charm measurement with STAR

The Solenoidal Tracker at RHIC (STAR) covers large acceptance with mid-rapidity within ± 1 and full azimuth. The main subsystems used in this analysis are the Time Projection Chamber (TPC) [7], the Time-Of-Flight (TOF) detector [8], the Barrel Electromagnetic Calorimeter (BEMC) and the Barrel Shower Maximum Detector (BSMD) [9]. The TPC measures the track momentum and ionization energy loss ($\mathrm{d}E/\mathrm{d}x$) for particle identification. The TOF detector greatly extends the particle identification kinematics via measuring time of flight, especially for low p_T pion and kaon separation up to 1.7 GeV/c, which is crucial for current D^0 reconstruction statistically without any vertex detector. The BEMC and BSMD are effective for high p_T electron trigger and hadron rejection, which is used for high p_T NPE analysis. The other detector used is the Vertex Position Detectors (VPD) for the online minimum bias trigger. The central trigger was defined with a combination cut using the spectator signals in the Zero Degree Calorimeter (ZDC) and the multiplicity signal in the Time-Of-Flight (TOF) at midrapidity.

The pion-kaon invariant mass was reconstructed by combinning all pion and kaon candidates with large combinatorial background. A mix-event technique was used to reproduce the combinatorial background. The D^0 signals are then obtained after the mix-event background subtraction in each centrality and transverse momentum bin. The D^0 reconstruction efficiency and detector acceptance were studied via Monte Carlo simulation with full STAR detector geometry. The other detailed techniques, such as signal extraction, cut efficiency, tof matching efficiency, double-counting effect, etc, are similar as the D^0 measurement in $p+p$ collisions [10].

The electron was identified via $\mathrm{d}E/\mathrm{d}x$ and momentum over energy ratio from combination of TPC and BEMC. The further hadron rejection was done according to the different shower size between electrons and hadrons in the BSMD. The photonic electron background was reconstructed by pairing the electron and positron and select small mass (less than 150 MeV) region, which covers almost all the photon conversions and large fraction of π^0 and η Dalitz decays. The photonic electron reconstruction efficiency was also studied via standard STAR simulation. The NPE signals are obainted by subtracting the photonic background from the inclusive electron candidates corrected by the purity. The details are similar as the NPE analysis in $p+p$ collisions [11].

3. Results

The D^0 p_T spectra in each centrality bin in Au+Au collisions are shown as solid symbols in Fig. 1(a). From bottom to top are 40-80% (diamonds), 0-80% (stars), 10-40% (squares) and 0-10% (circles). The open circles represent the D-meson p_T spectrum in p+p collisions [10]. The dashed curves are number of binary collisions ($N_{\rm bin}$) scaled Levy function from fitting to the p+p result. Peripheral 40-80% is consistent with th e curve, while central 0-10% deviates from the curve at high p_T. Fig. 1(b),(c),(d) show the D^0 R_{AA} for 40-80%, 10-40% and 0-10%, respectively. Within uncertainties there is no obvious suppression in peripheral collisions but strong suppression in central collisions observed. Vertical lines denote statistic uncertainties while brackets are for bin-by-bin systematic uncertainties. The vertical bars around unity from left to right represent the overall uncertainty for Au+Au and p+p collisions, respectively. In 0-10% central collisions, the suppression level is around 0.4, which is consistent with NPE measurement [5, 6] and light hadron R_{AA} [12]. This indicates strong interactions between charm quark and the surrounding medium in Au+Au central collisions at RHIC. Figure 1 right panel shows the integrated R_{AA} as a function of number of participants. The low p_T data points (0-3 GeV/c, squares) agree with unity and show a number-of-binary-collisions scaling behavior. The high p_T data points (3-8 GeV/c, circles) show strongly centrality dependent: no suppression in peripheral collisions and strong suppression (~0.5) in central collisions. Open symbols are for minimum bias 0-80%.

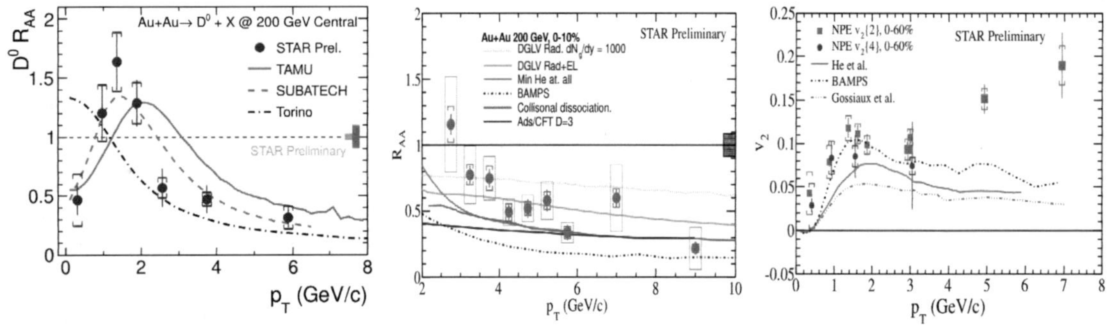

Figure 2. Left panel: D^0 R_{AA} for Au+Au central 0-10% (blue circles) compare with model calculations. The vertical bars around unity denote the uncertainties for N_{coll} in central 0-10% and p+p normalization from left to right, respectively. Right panel: NPE elliptic flow v_2 in 200 GeV Au+Au collisions. Curves are theoretical calculations.

Recent model calculations present interesting results compared with our data, especially for most central collisions, see Fig. 2 left panel. For example, the calculations from M. He and R. Rapp, performed a T-Matrix dynamic evolution of the charm quark in a Langevin + Hydro simulation plus a coalescence at low and intermediate p_T and fragmentation at high p_T for the D-meson hadronization [13, 14]. Their calculations can also reproduce the high p_T suppression with elastic scatterings. In the model of P. B. Gossiaux et. al from SUBATECH group [15, 16], the low p_T bump structure is mainly caused by the charm quark recombined with light quark in the medium with strong radial flow. In their calculation they used pure collisional energy loss and collisional + radiative energy loss (LPM) can both describe the D^0 data with different rescaling factor K. The calculations from Torino group [17, 18] with tuned transport coefficients can not reproduce the low p_T bump structure due to the pure fragmentation without any quark coalescence mechanism. Such a "bump" structure could also be explained by cold nuclear matter effect, such as Cronin effect with p_T broadening.

The NPE p_T spectra with improved statistics was reported in [19]. The NPE R_{AA} is strong suppressed in central 0-10% collisions, similar as D^0 and light hadrons, shown in 2 middle panel. The detailed comparison with models is illustrated in [19]. Figure 2 right panel shows the NPE elliptic flow v_2 in 200 GeV Au+Au collisions. Each of the models predicts a non-zero v_2 of charm quarks. The finite NPE v_2 at low p_T suggest strong charm-medium interactions. At $p_T > 3$ GeV/c, v_2 increases with p_T which is likely due to jet-like correlations.

4. Conclusions

In summary, we report the STAR preliminary results on the charmed meson (D^0) and NPE production at mid-rapidity in Au+Au collisions at $\sqrt{s} = 200$ GeV. The charm production cross sections at low p_T (< 3 GeV/c) show number of binary collisions scaling, which indicates that charm quark is mainly produced via initial hard scatterings. The centrality dependence of the transverse momentum distributions as well as the nuclear modification factor respect to p+p result show less suppression in peripheral collisions but strong suppression in most central collisions for both D^0 and NPE, which indicates significant energy loss of charm quark in the hot dense medium. The first observation of the low p_T bump structure of D^0 R_{AA} is consistent with models with strong charm-medium interactions. However, current precision of the data does not allow us to distinguish the coalescence approach and the cold nuclear matter effect. The strong charm-medium interactions result in the sizeable NPE v_2 at low p_T.

5. Acknowledgment

This research is supported partially by the Major State Basic Research Development Program in China with grant No. 2014CB845402, the National Natural Science Foundation of China with grant No. 11375184, the 985 Project II of Chinese Ministry of Education with grant No. ZC9850290172, and the Fundamental Research Funds for the Central Universities of China with grant No. WK2030040028.

References

[1] Adams J *et al* [STAR Collaboration] 2005 *Nucl. Phys.* A **757** 102
[2] Muller B 2004 *Nucl. Phys.* A **12** 067
[3] Dokshitzer Y and Kharzeev D 2001 *Phys. Lett.* B **519** 199
[4] Djordjevic M, Gyulassy M and Wicks S 2005 *Phys. Rev. Lett.* **94** 112301
[5] Abelev B *et al* [STAR Collaboration] 2007 *Phys. Rev. Lett.* **98** 192301
[6] Adler S *et al* [PHENIX Collaboration] 2006 *Phys. Rev. Lett.* **96** 032301
[7] Anderson M *et al* 2003 *Nucl. Instr. Meth.* A **499** 659
[8] STAR TOF proposal http://drupal.star.bnl.gov/STAR/files/future/proposals/tof-5-24-2004.pdf
[9] Beddo M *et al* 2003 *Nucl. Instr. Meth.* A **499** 725
[10] Adamczyk L *et al* [STAR Collaboration] 2012 *Phys. Rev.* D **86** 072013
[11] Agakishiev H *et al* 2011 *Phys. Rev.* D **83** 052006
[12] Abelev B *et al* [STAR Collaboration] 2006 *Phys. Rev. Lett.* **97** 152301
[13] He M *et al* 2012 *Phys. Rev.* C **86** 014903
[14] He M *et al* 2013 *Phys. Rev. Lett.* **110** 112301
[15] Gossiaux P *et al* 2010 *J. Phys.* G **37** 094019
[16] Gossiaux P *et al* 2013 *J. Phys.* G **904-905** 992c-995c
[17] Alberico W *et al* 2011 *Eur. Phys. J.* C **71** 1666
[18] Alberico W *et al* 2013 *Preprint* hep-ph/1305.7421
[19] Kikola D [STAR Collaboration] the proceedings for SQM2013

Journal of Physics: Conference Series **509** (2014) 012114 doi:10.1088/1742-6596/509/1/012114

Dynamical freeze-out criterion in event-by-event hydrodynamics

Hannu Holopainen[1] **and Pasi Huovinen**[1]

[1] Frankfurt Institute for Advanced Studies, Ruth-Moufang-Straße 1, D-60438 Frankfurt am Main, Germany

E-mail: huovinen@fias.uni-frankfurt.de

Abstract. In hydrodynamical modelling of ultrarelativistic heavy-ion collisions the freeze-out is typically assumed to take place on a surface of constant temperature or energy density. In this work we apply a dynamical freeze-out criterion, which compares the hydrodynamical expansion rate with the pion scattering rate, to an event-by-event ideal hydrodynamics at the full RHIC collision energy. We present hadron spectra and elliptic flow calculated using (2+1)-dimensional ideal hydrodynamics, and show the differences between constant temperature and dynamical freeze-out criteria. We find that when the freeze-out ratio is fixed to one, the different freeze-out criteria lead to slightly different spectra and $v_2(p_T)$ in the event-by-event calculations.

1. Introduction

In hydrodynamical modelling of heavy-ion collisions the fluid dynamical description is usually assumed to break down, and the momentum distributions to freeze-out, on a thin surface of constant temperature or energy density. However, it has been argued for a while ago that a more physical criterion would be when the expansion rate of the fluid exceeds the scattering rate of the constituents of the fluid [1]. Such a dynamical criterion has been applied to hydrodynamical modelling before [2, 3], but it has not been widely used. In this contribution we study how a dynamical freeze-out criterion affects the final pion and proton distributions and their elliptic anisotropy in event-by-event hydrodynamics. We extend our previous results shown in Ref. [4] by using an equation of state (EoS) for a chemically frozen system.

2. Fluid dynamics

Since we want to concentrate on the effects of the freeze-out criterion, we ignore all the complications of viscosity, and use a modified version of the ideal fluid model described in Ref. [5]. The model is boost invariant, and includes finite net baryon density, which, however, does not contribute to pressure. As an EoS we use the $s95p$-PCE-$v1$ parametrisation [6], which contains a chemical freeze-out at temperature $T_{\text{chem}} = 150$ MeV to reproduce the observed π/p ratio at RHIC [7].

We initialise the system using a Monte Carlo Glauber approach. The entropy density is given as a mixture of binary (75%) and wounded nucleon (25%) profiles. A Monte Carlo Glauber model gives only the positions of the wounded nucleons and binary collisions and we need to distribute entropy around these positions before we can initialise hydrodynamical evolution. Our choice is

Journal of Physics: Conference Series **509** (2014) 012114 doi:10.1088/1742-6596/509/1/012114

a 2-dimensional Gaussian:

$$s(x,y) = \text{const.} \sum_{\text{wn,bc}} \frac{1}{2\pi\Sigma^2} \exp\left[-\frac{(x-x_i)^2+(y-y_i)^2}{2\Sigma^2}\right], \tag{1}$$

where Σ is a free parameter controlling the width of the Gaussian. Typical values for this fluctuation size parameter is order of 0.5 fm, and our choice here is 0.8 fm (see Ref. [4]). For net-baryon density a pure wounded nucleon profile is used. To obtain averaged initial state, we evaluate an average of 1000 event-by-event fluctuating initial states, which include the above mentioned smearing by Gaussians. As initial time of the evolution we use $\tau_0 = 0.6$ fm.

3. Dynamical freeze-out criterion

A usual requirement for the validity of fluid dynamics is that the Knudsen number must be much larger than one. Since the ratio of the expansion rate θ to the scattering rate Γ is the ratio of an inverse of a macroscopic scale to an inverse of a microscopic scale, it can be identified as a Knudsen number. Thus the requirement that the scattering rate is larger than the expansion rate is equivalent to requiring that the Knudsen number is larger than one. We assume that fluid dynamics is valid until $K = \theta/\Gamma = 1$, and use $K_\text{f} = 1$ as the freeze-out criterion in the dynamical case. As a baseline for comparison we use freeze-out at constant temperature of $T_\text{f} = 120$ MeV.

The expansion rate of the system, $\theta = \partial_\mu u^\mu$, can be evaluated once the flow velocity is known, but the scattering rate is not given by hydrodynamics, but requires microscopic calculations. Several evaluations of the scattering rate of pions in equilibrated pion gas can be found in the literature [2, 3, 8], but since we are interested in pion scatterings in chemically frozen hadron gas, we calculate the rate ourselves. In semiclassical approximation the scattering rate of pions in kinetically (but not necessarily chemically) equilibrated hadron-resonance gas is given by

$$\Gamma = \frac{1}{n_\pi(T,\mu_\pi)} \sum_i \int \mathrm{d}^3p_\pi \mathrm{d}^3p_i\, f_\pi(T,\mu_\pi) f_i(T,\mu_i)\, v_{\pi i}(s)\, \sigma_{\pi i}(s), \tag{2}$$

where n_π is the density of pions, $f_i(T,\mu_i)$ is the thermal distribution function, $v_{\pi i}$ is the relative velocity, $\sigma_{\pi i}$ is the cross section for scattering of pion with particle i, and the summation over i runs over all particle species included in the EoS. Cross sections are evaluated as in the UrQMD hadron cascade [9], *i.e.* the main contribution comes from resonance formation which is evaluated using the relativistic Breit-Wigner formula. The scattering rates are different for different hadrons, and therefore different hadrons should freeze-out at different times. Implementation of such a sequential freeze-out in fluid dynamics is very difficult, and it has also been argued that a proton requires more than one scattering with pions to change its momentum significantly [10]. To simplify the problem we thus assume that all hadrons freeze-out simultaneously.

Once the constant temperature or constant Knudsen number surface is found, the particle and resonance distributions are calculated using the conventional Cooper-Frye prescription, and the contributions from resonance decays are added to particle distributions.

4. Results

We consider Au+Au collisions at Relativistic Heavy Ion Collider (RHIC) with $\sqrt{s_{NN}} = 200$ GeV. In Fig. 1 we show the average temperature and flow velocity on the freeze-out surface as function of centrality. The initial state is the average one, and we compare our results to those of the blast-wave fit by the STAR collaboration [11]. Trivially constant temperature freeze-out leads to a centrality independent temperature, whereas constant Knudsen number leads to the hotter surface the more peripheral the collision. As well known, the more central the collision, the larger the flow velocity, but the dependence on centrality is stronger for the dynamical freeze-out criterion. For both criteria the temperature and velocity differs from the blast-wave fits,

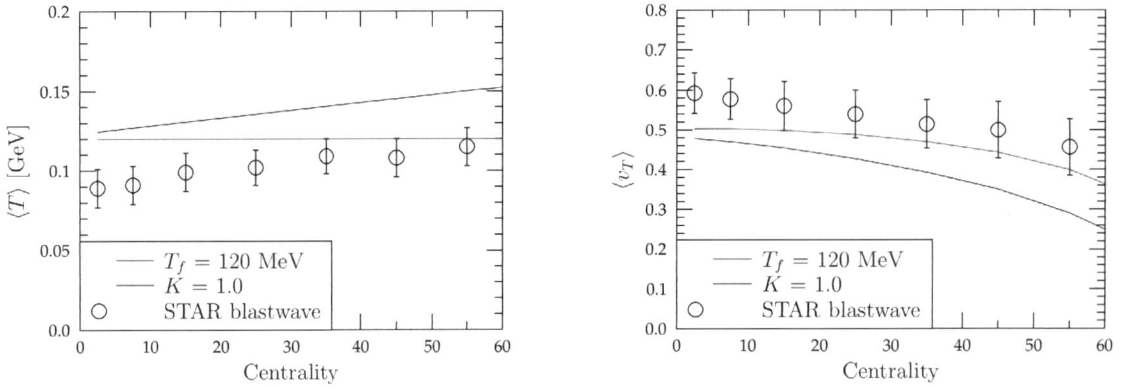

Figure 1. Average temperature (left) and flow velocity (right) on the freeze-out surface as function of centrality in $Au + Au$ collisions at $\sqrt{s_{\mathrm{NN}}} = 200$ GeV using freeze-out at constant temperature (red) or at constant Knudsen number (blue). Ideal fluid calculations are compared with the STAR blast-wave fit [11].

but this is understandable: The shape of the freeze-out surface affects the favoured values of temperature and flow velocity, and fluid dynamical calculation leads to more complicated shape of the freeze-out surface than a simple blast-wave fit. What is interesting although, is that the slopes of the centrality dependence of temperature and flow velocity are similar in the blast-wave fit and in the calculation with dynamical freeze-out criterion.

Fig. 2 depicts the p_T-spectra of positive pions and protons at $20 - 30\%$ centrality. As well known [5], when all the other parameters are kept unchanged, event-by-event calculation leads to stiffer spectra than the averaged initial state calculation. The constant Knudsen number freeze-out leads to slightly steeper spectra than the constant temperature freeze-out, except for pions in the event-by-event case. Thus, to obtain similar p_T-distributions slightly larger Knudsen number $K_{\mathrm{f}} > 1$ might be required for the averaged initial state. On the other hand, since the pion spectrum is independent of the freeze-out criterion in the event-by-event case, a slightly different initial time, or initial shape might be required to reproduce the $T_{\mathrm{f}} = 120$ MeV result in that case.

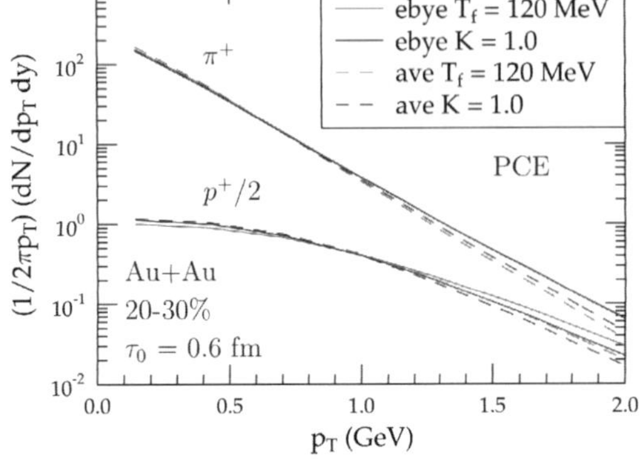

Figure 2. The p_T-distributions of positive pions and protons in $Au + Au$ collisions at $\sqrt{s_{\mathrm{NN}}} = 200$ GeV using freeze-out at constant temperature (red) or at constant Knudsen number (blue). Calculations are done for both averaged (dashed) and fluctuating (solid) initial states.

In Fig. 3 the pion and proton p_T-differential v_2 are shown in collisions with $20-30\%$ centrality. Since our p_T-distributions were not tuned to be the same in all the cases, one cannot draw

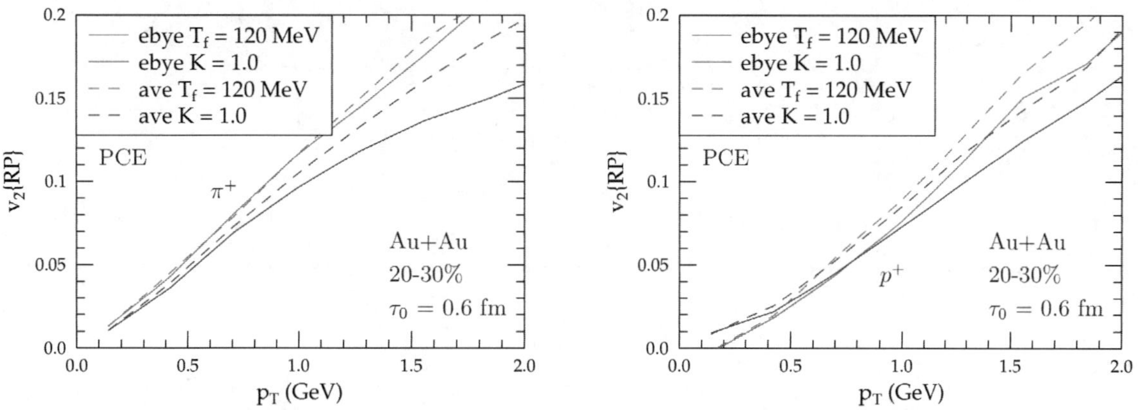

Figure 3. The p_T-differential elliptic flow of pions (left) and protons (right) in $Au+Au$ collisions at $\sqrt{s_{NN}} = 200$ GeV using averaged (dashed) or fluctuating (solid) initial states, and freeze-out at constant $T = 120$ MeV temperature (red) or at constant Knudsen number $K = 1$ (blue).

any strong conclusions based on the $v_2(p_T)$ results shown here. Nevertheless, when constant temperature freeze-out is used, the pion $v_2(p_T)$ is not sensitive to the event-by-event fluctuations of the initial state [5], but the dynamical criterion creates some sensitivity to fluctuations at $p_T \gtrsim 1$ GeV. More strikingly, the dynamical criterion reduces pion $v_2(p_T)$ by $\sim 10\%$ both in the average initial state and in the event-by-event calculation. For protons the situation is reversed: The fluctuations have now a clear reducing effect independent of the freeze-out criterion, whereas the freeze-out criterion has a weaker and p_T-dependent effect. At very low p_T the dynamical criterion increases and at large p_T decreases the proton $v_2(p_T)$.

5. Conclusions

We argued that the constant temperature/density freeze-out is an oversimplification, but the effects of the freeze-out criterion on spectra and $v_2(p_T)$ turned out to be small. The $\sim 10\%$ reduction in pion $v_2(p_T)$ is interesting, but it is unknown whether it survives if K_f is taken as a free parameter, and chosen to lead to the same p_T-distributions than T_f.

Acknowledgments

The work of H.H. was supported by the Extreme Matter Institute (EMMI) and the work of P.H. by BMBF under contract no. 06FY9092.

References

[1] Bondorf J P, Garpman S I A and Zimanyi J 1978 *Nucl. Phys.* A **296** 320
[2] Hung C M and Shuryak E V 1998 *Phys. Rev.* C **57** 1891
[3] Eskola K J, Niemi H and Ruuskanen P V 2008 *Phys. Rev.* C **77** 044907
[4] Holopainen H and Huovinen P 2012 *J. Phys. Conf. Ser.* **389** 012018
[5] Holopainen H, Niemi H and Eskola K J 2011 *Phys. Rev.* C **83** 034901
[6] Huovinen P and Petreczky P 2010 *Nucl. Phys.* A **837** 26
[7] Huovinen P 2008 *Eur. Phys. J.* A **37** 121
[8] Daghigh R G and Kapusta J I 2002 *Phys. Rev.* D **65** 064028
[9] Bass S A *et al.* 1998 *Prog. Part. Nucl. Phys.* **41** 255
[10] Prakash M, Prakash M, Venugopalan R and Welke G 1993 *Phys. Rept.* **227** 321
[11] Adams J *et al.* (STAR Collaboration) 2004 *Phys. Rev. Lett.* **92** 112301

AUTHOR INDEX

AUTHOR INDEX